00 1479505 00101
71

S0-AIH-290

Springer

Tokyo
Berlin
Heidelberg
New York
Barcelona
Hong Kong
London
Milan
Paris

Handbook of Glycosyltransferases and Related Genes

Edited by:
Naoyuki Taniguchi, Koichi Honke, Minoru Fukuda

Co-editors:

Henrik Clausen
Kiyoshi Furukawa
Gerald W. Hart
Reiji Kannagi
Toshisuke Kawasaki
Taroh Kinoshita
Takashi Muramatsu
Masaki Saito
Joel H. Shaper
Kazuyuki Sugahara
Lawrence A. Tabak
Dirk H. Van den Eijnden
Masaki Yanagishita

James W. Dennis
Koichi Furukawa
Yoshio Hirabayashi
Masao Kawakita
Koji Kimata
Ulf Lindahl
Hisashi Narimatsu
Harry Schachter
Pamela Stanley
Akemi Suzuki
Shuichi Tsuji
Katsuko Yamashita

With 95 Figures, Including 5 in Color

Springer

Naoyuki Taniguchi, M.D., Ph.D.
Professor, Department of Biochemistry
Osaka University Medical School
2-2 Yamadaoka, Suita, Osaka 565-0871, Japan

Koichi Honke, M.D., Ph.D.
Associate Professor, Department of Biochemistry
Osaka University Medical School
2-2 Yamadaoka, Suita, Osaka 565-0871, Japan

Minoru Fukuda, Ph.D.
Professor, Glycobiology Program
Cancer Research Institute
The Burnham Institute
10901 North Torrey Pines Road, La Jolla, CA 92037, USA

QP
606
.G6
H36
2002

Cover Illustration by Koichi Honke

ISBN 4-431-70311-X Springer-Verlag Tokyo Berlin Heidelberg New York

Library of Congress Cataloging-in-Publication Data
Handbook of glycosyltransferases and related genes / N. Taniguchi, K. Honke, M. Fukuda, eds.
    p. ; cm.
    Includes bibliographical references.
    ISBN 443170311X (hardcover : alk. paper)
    1. Glycosyltransferases—Handbooks, manuals, etc.   2. Glycosyltransferase
genes—Handbooks, manuals, etc.   I. Taniguchi, Naoyuki, 1942–   II. Honke, K. (Koichi),
1957–   III. Fukuda, Minoru, 1945–
    [DNLM: 1. Glycosyltransferases—Handbooks.   2. Glycosyltransferases—genetics—
Handbooks. QU 39 H2355 2002]
    QP606.G6 H36 2002
    572'.792—dc21

                                                                                    2001049605

Printed on acid-free paper

© Springer-Verlag Tokyo 2002
Printed in Japan
This work is subject to copyright. All rights are reserved whether the whole or part of the material is concerned, specifically the rights of translation, reprinting, reuse of illustrations, recitation, broadcasting, reproduction on microfilms or in other ways, and storage in data banks.
The use of registered names, trademarks, etc. in this publication does not imply, even in the absence of a specific statement, that such names are exempt from the relevant protective laws and regulations and therefore free for general use.
Product liability: The publisher can give no guarantee for information about drug dosage and application thereof contained in this book. In every individual case the respective user must check its accuracy by consulting other pharmaceutical literature.

Typesetting: Best-set Typesetter Ltd., Hong Kong
Printing and binding: Hicom, Japan
SPIN: 10757455

Preface

The so-called postgenomic research era has now been launched, and the field of glycobiology and glycotechnology has become one of the most important areas in life science because glycosylation is the most common post-translational modification reaction of proteins in vivo. On the basis of Swiss-Prot data, over 50 proteins are known to undergo glycosylation, but in fact the actual functions of most of the sugar chains in the glycoconjugates remain unknown.

The complex carbohydrate chains of glycoproteins, glycolipids, and proteoglycans represent the secondary gene products formed through the reactions of glycosyltransferases. The regulation of the biosynthesis of sugar chains is under the control of the expression of glycosyltransferases, their substrate specificity, and their localization in specific tissue sites. There is a growing body of evidence to suggest that these enzymes play pivotal roles in a variety of important cellular differentiation and developmental events, as well as in disease processes. Over 300 glycosyltransferases appear to exist in mammalian tissues. If the genes that have been purified and cloned from various species such as humans, cattle, pigs, rats and mice are counted as one, approximately 110 glycogenes that encode glycosyltransferases and related genes have been cloned at present, and this number continues to grow each day. However, most of the functions of the glycosyltransferase genes and related genes are unknown. This fact has stimulated numerous new and interesting approaches in molecular biological investigations.

The removal of a specific glycogene in knockout mice indicates that some of the glycosyltransferases are essential for survival, development, and oncogenesis. Experiments using cells or animals that had been transfected with a specific glycosyltransferase gene show that these enzymes relate to cancer metastases, cellular invasion, and the suppression of xenoantigen expression. Although a number of excellent investigations have been conducted, the issue of the real functions of glycoconjugates of interest remains to be elucidated. The reason for the difficulty in exploring the functional significance of glycosyltransferase genes, or glycogenes, is that even if we are able to disrupt the gene and obtain phenotypic changes in mice, we cannot explain this as a consequence or a cause of disease. Moreover, and more importantly, we cannot identify the actual target protein(s) in vivo that may be affected by the lack of sugar chains. The phenotypic changes observed are actually not due to a disrupted glycogene but, rather, to the secondary effect of the glycogene which may affect the aberrant glycosylation of other glycoproteins. The same would be true for transgenic mice that overexpress specified glycogenes or for Congenital Disorders of Glycosylation (CGD) patients with various symptoms that cannot be explained via the mechanism by which the symptoms appeared. In the future, therefore, the identification of target proteins in patients or in glycogene-disrupted animals is a prerequisite for understanding the real cause of the pathophysiology of diseases. The field of functional glycomics focuses on carrying out this type of research. This handbook is useful for experimental design for this approach, which may identify the target molecules of such glycosyltransferases and related genes.

The aim of this book is to provide comprehensive coverage of all glycosyltransferase genes and their related genes known at present. The nearly 100 chapters are designed to summarize the present knowledge of these enzymes. The presentations are brief and concise and are presented in a format that we hope will make comprehension of the data relatively easy for readers.

More than fifty experts who have worked on their "pet" enzymes that were largely purified and/or cloned by their group have contributed chapters on individual glycosyltransferase genes and their related genes. We express our sincere thanks to all of them for their enormous effort and detailed work. We are particularly grateful for the assistance of the staff at Springer-Verlag Tokyo for their patience during the preparation of the original draft of this book and its editing.

This publication was supported in part by the Naito Foundation, a Grant-in-Aid for Publication of Scientific Research Results No. 135313 from the Japan Society for the Promotion of Science, and a Grant-in-Aid for Scientific Research on Priority Area No. 10178105 from the Ministry of Education, Science, Sports and Culture of Japan.

July 2001

Naoyuki Taniguchi
Koichi Honke
Minoru Fukuda

EDITORS

Co-editors

Henrik Clausen (University of Copenhagen, Copenhagen, Denmark)
James W. Dennis (Mount Sinai Hospital, Toronto, Canada)
Kiyoshi Furukawa (Tokyo Metropolitan Institute of Gerontology, Tokyo, Japan)
Koichi Furukawa (Nagoya University, Nagoya, Japan)
Gerald W. Hart (Johns Hopkins University, Baltimore, USA)
Yoshio Hirabayashi (Riken, Wako, Japan)
Reiji Kannagi (Aichi Cancer Center, Nagoya, Japan)
Masao Kawakita (Tokyo Metropolitan Institute of Medical Science, Tokyo, Japan)
Toshisuke Kawasaki (Kyoto University, Kyoto, Japan)
Koji Kimata (Aichi Medical University, Nagoya, Japan)
Taroh Kinoshita (Osaka University, Osaka, Japan)
Ulf Lindahl (Uppsala University, Uppsala, Sweden)
Takashi Muramatsu (Nagoya University, Nagoya, Japan)
Hisashi Narimatsu (National Institute of Advanced Industrial Science and
    Technology, Tsukuba, Japan)
Masaki Saito (National Cancer Center Research Institute, Tokyo, Japan)
Harry Schachter (The Hospital for Sick Children, Toronto, Canada)
Joel H. Shaper (Johns Hopkins University, Baltimore, USA)
Pamela Stanley (Albert Einstein College of Medicine, New York, USA)
Kazuyuki Sugahara (Kobe Pharmaceutical University, Kobe, Japan)
Akemi Suzuki (Riken, Wako, Japan)
Lawrence A. Tabak (University of Rochester, Rochester, USA)
Shuichi Tsuji (Ochanomizu University, Tokyo, Japan)
Dirk H. Van den Eijnden (Vrije Universiteit, Amsterdam, The Netherlands)
Katsuko Yamashita (Sasaki Institute, Tokyo, Japan)
Masaki Yanagishita (Tokyo Medical and Dental University, Tokyo, Japan)

Contents

N-Acetylgalactosaminyltransferases

Fucosyltransferases

Sialyltransferases

Glucuronyltransferases

GAG Synthesis

Sulfotransferases

Nucleotide Sugar Transporters

Dolichol Pathway/GPI-Anchor

N-Glycan Processing Enzymes

Appendix

Contributors

Glucosyltransferase

GlcCer Synthase (UDP-Glucose:Ceramide Glucosyltransferase, UGCG)

Introduction

Glycosphingolipids (GSLs) occur in vertebrates and lower animals, as well as in plants. They have been regarded as an enigmatic class of membrane lipids. Recent progress in biomembrane research indicates the existence of GSL microdomains (or "rafts") in cell surface membranes (Simon and Ikonen 1997) GSLs form microdomains by a clustering of cholesterol and glycosyl-phosphatidyl inositol (GPI)-anchored proteins. The microdomains are thought to involve a variety of biological events such as cell signaling and cell–cell interaction. Another characteristic feature of GSLs is their chemical diversity: over 400 GSLs with different sugar chain structures have been isolated and determined. The structural studies to date indicate that almost all GSLs are derived from the simplest GSL, glucosylceramide. This is synthesized by a ceramide-specific β-glucosyltransferase (GlcT-I). Recent success in the molecular cloning of the gene encoding GlcT-I has provided new insights into the previously unrecognized roles of GSLs (Ichikawa et al. 1996). When the GlcT-I function is knocked out in mice by gene-targeting technology, the mouse dies at embryonic day 8 (Yamashita et al. 1999). This study clearly shows that GlcT-I plays an essential role during embryonic development and differentiation.

Yoshio Hirabayashi and Shinichi Ichikawa*

Neuronal Circuit Mechanisms Research Group, RIKEN Brain Science Institute, 2-1 Hirosawa, Wako-shi, Saitama 351-0198, Japan
Tel. and Fax: +81-48-467-6372 or +070-5026-3783
e-mail: hirabaya@postman.riken.go.jp
* Present address: GENE CARE Research Institute, 200 Kajiwara, Kamakura 247-0063, Japan

Databanks

GlcCer synthase (UDP-glucose:ceramide glucosyltransferase, UGCG)

NC-IUBMB enzyme classification: E.C.2.4.1.80

Species	Gene	Protein	mRNA	Genomic
Homo sapiens	*UGCG*	BAA09451.1	D50840	–
		NP_003349.1	NM_003358	
Mus musculus	*ugcg*	BAA28782.1	D89866	AB012799–AB012807 (exons 1–9)
Rattus norvegicus	*ugcg*	CAA11853.1	AJ224156	–
		AAD02464.1	AF047707	
Drosophila melanogaster	–	–	AC004365	–
Caenorhabditis elegans	–	AAK31528	U53332	
		AAC48147	U58735	
Synechocystic sp.	–	BAA18121.1	D90911	–

Name and History

UDP-glucose ceramide glucosyltransferase (UGCG) (Ichikawa et al. 1998) is often abbreviated to GlcT-I (Ichikawa et al. 1994) or to GCS (glucosylceramide synthase) (Paul et al. 1996). Ceramide glucosyltransferase activity was first discovered in chick embryonic brain by Basu et al. (1968). Since then, this enzyme activity has been found in all animal tissues and also in plants. Since glucosylceramide is the precursor lipid for most GSLs, this enzyme is a key factor for their expression. Many groups have tried to purify the enzyme protein, but none of them were successful, mainly because the enzyme protein is tightly membrane-bound and an extremely minor component. Partial purification of the enzyme protein was achieved from rat liver Golgi membrane by Paul et al. (1996). A cDNA for human ceramide glucosyltransferase was isolated by expression cloning using the enzyme-deficient cells of mouse melanoma (Ichikawa et al. 1996).

Enzyme Activity Assay and Substrate Specificity

Ceramide glucosyltransferase catalyzes the transfer of glucose from UDP-glucose to ceramide (Cer) to form glucosylceramide (GlcCer):

$$Cer + UDP\text{-}Glc \rightarrow GlcCer + UDP$$

The reaction product is a β-linked glucosylceramide. The same enzyme protein can utilize UDP-galactose instead of UDP-Glc to synthesize galactosylceramide, but much less efficiently (about 10% of glucosylceramide synthetic activity) (Sprong et al. 1998; Wu et al. 1999).

The enzyme strictly recognizes the ceramide moiety of the acceptor: it is highly specific to D,L-*erythro*- but not *threo*-ceramide (Pagano and Martin 1988). As for the nucleotide specificity of the glucose donor, not only UDP-Glc but also CDP- and TDP-Glc are efficiently utilized, although the latter two donors do not occur naturally (Paul et al. 1996). UDP-Man, UDP-Xyl, UDP-GlcNAc, and GDP-Glc have no activity as hexose donors. A fatty acyl group in the ceramide is not critical, since even fluorescent labeled C6-NBD-ceramide is a good substrate as well as a truncated ceramide (Vunnam and Radin 1979; Jeckel et al. 1992; Paul et al. 1996).

The assay systems typically include 5–10 mM $MnCl^{2+}$ or Mg^{2+}. The pH optimum of its enzyme is about 7.4. The liposomal substrate is used for ceramide glucosyltrans-ferase because of its hydrophobic property (Basu et al. 1973). Phosphatidylcholine is a good liposomal lipid for the assay since the phospholipid stimulates glucosylce-ramide synthesis (Morell et al. 1970). For sensitive detection of the enzymic activity without using radioactive compounds, a fluorescent labeled C6-NBD-ceramide is commonly used. The addition of nicotinamide nucleotides such as NAD or NADP is an effective inhibitor or of UDP-Glc degradation by a pyrophosphatase. The purified preparation requires the absolutely exogenous phospholipid, dioleoyl phosphotidyl-choline (Paul et al. 1996).

D-*threo*-PDMP ([(R,R)-1-phenyl-2-decanoylamino-3-morpholino-1-propanol]) is most commonly used as a specific inhibitor for ceramide glucosyltransferase (Inokuchi and Radin 1987). This inhibitor is commercially available from Matreya (pleasant Gap, PA. USA). *N*-butyldeoxygalactonojirimycin (ND-DGJ) is also a potent inhibitor of mammalian ceramide glucosyltransferase (Platt et al. 1994).

Structural Chemistry

Mammalian ceramide glucosyltransferase is a hydrophobic, membrane-bound enzyme. The molecular mass of the rat enzyme is approximately 38 kDa on SDS-polyacrylamide gels, although the value calculated from the cDNA sequence is about 45 kDa (394 amino acid). Biochemical and immunochemical studies with Golgi membranes show that both the C terminus and a hydrophilic loop near the *N*-terminus of the enzyme are localized in the cytosolic side of the Golgi membrane (Marks et al. 1999). It is suggested that the rat enzyme forms a dimer or oligomer with another protein in the Golgi membrane (Marks et al. 1999). Histidine residue at 193 in the rat enzyme protein is shown to be involved in the binding of both UDP-Glc and the enzyme inhibitor PDMP (Wu et al. 1999). Since it is very difficult to prepare a large amount of the enzyme protein, progress in structural studies on this protein has been slow.

Preparation

The enzyme is distributed in all mammalian tissues, including brain, liver, kidney, skin, testis, spleen, lung, etc. The highest activity is found in the brain. The enzyme has been partially purified from rat liver. In the purification procedure, the enzyme is solubilized from the Golgi membrane by a detergent, CHAPSO or CHAPS. Purifi-cation is achieved using dye-agarose chromatography (Paul et al. 1996). The cloned enzyme has been expressed in mammalian cells in *E. coli* with the pET3a expres-sion system (Ichikawa et al. 1996), and also in insect cells with a baculovirus system. However, production of the recombinant protein in large quantities has been unsuc-cessful because a high level of expression is toxic to host cells.

Biological Aspects

Cell-surface GSLs are believed to play important roles in a variety of cellular processes such as cell–cell interaction, cell growth, development, and differentiation. Most GSLs are synthesized from glucosylceramide. Thus, ceramide glucosyltransferase is a key regulatory factor controlling levels of both cell-surface GSLs and intracellular ceramide. The enzymic activity is highly regulated during keratinocyte differentiation (Watanabe et al. 1998), and during the acute phase response in liver induced by LPS (Memon et al. 1999). Up-regulation of glucosylceramide synthase in association with the multidrug resistance of cancer cells has been demonstrated (Liu et al. 1999). However, little is known about the molecular mechanisms regulating ceramide glucosyltransferase activity in these processes. Excess generation of ceramide in B16 melanoma cells by bacterial sphingomyelinase treatment causes up-regulation of the synthetic activity (Komori et al. 1999). In this and other cases, the activity is believed to be regulated by both transcriptional and posttranslational levels (Komori et al. 1999; Boldin and Futerman 2000).

The gene for mouse ceramide glucosyltransferase, *Ugcg*, has been isolated and characterized (Ichikawa et al. 1998). The gene, of approximate size 32 kb, is composed of nine exons and eight introns. The promoter region lacks TATA and CAAT boxes, but contains Spl binding sites, indicating the housekeeping gene. The motifs for AhR, NF-kappaB, AP-2, and GATA-1 binding sites were also found. The data are useful for understanding the regulation of *Ugcg*. Homology searches using the mouse sequence reveal that *Ugcg* is phylogenetically highly conserved from mammals to cyanobacteria (see database).

A B16 mouse melanoma mutant, designated GM-95 and without ceramide glucosyltransferase activity, has been established (Ichikawa et al. 1994). Although GM-95 has no ability to synthesize any GSLs, it does not require any exogenous GSLs for growth and survival. This fact indicates that GSLs are dispensable for cell survival, at least in this cell line. GM-95 expresses sphingomyelin as a sole sphingolipid. When all sphingolipids are removed from the plasma membrane by exogenous sphingomyelinase, GM-95 cannot attach to the substratum and loses the ability to proliferate. This observation indicates that expression of either GSLs or sphingomyelin is necessary to maintain the cell adhesion and proliferation (Hidari et al. 1996).

Studies in vitro do not necessarily indicate that GSLs have no biological functions in vivo. Indeed, studies with a *Ugcg* knockout mouse (Yamashita et al. 1999) and our own study (unpublished results, 1999) demonstrate that GSL synthesis is essential in vivo. Until embryonic day (E) 6.5, null mutant embryos are morphologically indistinguishable from wild-type embryos, although the size of the mutant embryos is much smaller, but at E8.5, all null embryos die. A significant enhancement of apoptosis is observed in the ectoderm at E7.5 of the gastrulation stage. The knockout mice prove that GSLs are indispensable for embryogenesis and neuronal differentiation.

Future Perspectives

It is not yet clear why disruption of the ceramide glucosyltransferase gene in the mouse leads to embryonic lethality. A generation of conditional knockout and transgenic mice would be useful for understanding the in vivo functions of the enzyme and GSLs synthesis. Since the highly homologous genes are found in *C. elegans* and *Drosophila*, these animals are also powerful tools in attempts to solve this issue, and also to understand how the expression of ceramide glucosyltransferase is regulated genetically and epigenetically.

Since it is very difficult to produce and purify ceramide glucosyltransferase in large quantities, the enzyme protein is not available in a pure form. It is necessary to develop a method to produce and purify the protein with a hydrophobic property. To obtain structural information by X-ray crystallography, it is essential to crystallize the protein. Once obtained, the secondary and tertiary structural information will facilitate the development of specific inhibitors as therapeutic agents for human diseases, including GSLs storage diseases and cancer.

Further Reading

For reviews, see Ichikawa and Hirabayashi (1998, 2000), and Hirabayashi and Ichikawa (2000).

Details of the analysis of ceramide glucosyltransferase are described in Methods in Enzymology, Vol. 311, by Ichikawa and Hirabayashi (2000).

References

Basu S, Kaufman B, Roseman S (1968) Enzymatic synthesis of ceramide–glucose and ceramide–lactose by glycosyltransferases from embryonic chicken brain. J Biol Chem 243:5802–5804

Basu S, Kaufman B, Roseman S (1973) Enzymatic synthesis of glucocerebroside by a glucosyltransferase from embryonic chicken brain. J Biol Chem 248:1388–1394

Boldin SA, Futerman AH (2000) Up-regulation of glucosylceramide synthesis upon stimulation of axonal growth by basic fibroblast growth factor. Evidence for post-translational modification of glucosylceramide synthase. J Biol Chem 275:9905–9909

Hidari KIPJ, Ichikawa S, Fujita T, Sakiyama H, Hirabayashi Y (1996) Complete removal of sphingolipids from the plasma membrane disrupts cell to substratum adhesion of mouse melanoma cells. J Biol Chem 271:14636–14641

Hirabayashi Y, Ichikawa S (2000) Roles of sphingolipids and glycosphingolipids in biological membranes. In: Fukuda M, Hindsgaul O (eds) Frontiers in molecular biology, Series 30. Molecular and cellular glycobiology. Oxford University Press, New York, pp 220–248

Ichikawa S, Hirabayashi Y (1998) Glucosylceramide synthase and glycosphingolipid synthesis. Trends Cell Biol 8:198–202

Ichikawa S, Hirabayashi Y (2000) Genetic approaches for studies of glycolipid synthetic enzymes. Methods Enzymol 311:303–318

Ichikawa S, Nakajo N, Sakiyama H, Hirabayashi Y (1994) A mouse B16 melanoma mutant deficient in glycolipids. Proc Natl Acad Sci USA 91:2703–2707

Ichikawa S, Sakiyama H, Suzuki G, Hidari KI, Hirabayashi Y (1996) Expression cloning of a cDNA for human ceramide glucosyltransferase that catalyzes the first glycosylation step of glycosphingolipid synthesis. Proc Natl Acad Sci USA 93:4638–4643

Ichikawa S, Ozawa K, Hirabayashi Y (1998) Molecular cloning and characterization of the mouse ceramide glucosyltransferase gene. Biochem Biophys Res Commun 253:707–711

Inokuchi J, Radin NS (1987) Preparation of the active isomer of 1-phenyl-2-decanoylamino-3-morpholino-1-propanol, inhibitor of murine glucocerebroside synthetase. J Lipid Res 28:565–571

Jeckel D, Karrenbauer A, Burger KN, van Meer G, Wieland F (1992) Glucosylceramide is synthesized at the cytosolic surface of various Golgi subfractions. J Cell Biol 117:259–267

Komori H, Ichikawa S, Hirabayashi Y, Ito M (1999) Regulation of intracellular ceramide content in B16 melanoma cells. Biological implications of ceramide glycosylation. J Biol Chem 274:8981–8987

Liu YY, Han TY, Giuliano AE, Cabot MC (1999) Expression of glucosylceramide synthase, converting ceramide to glucosylceramide, confers adriamycin resistance in human breast cancer cells. J Biol Chem 274:1140–1146

Marks DL, Wu K, Paul P, Kamisaka Y, Watanabe R, Pagano RE (1999) Oligomerization and topology of the Golgi membrane protein glucosylceramide synthase. J Biol Chem 274:451–456

Memon RA, Holleran WM, Uchida Y, Moser AH, Ichikawa S, Hirabayashi Y, Grunfeld C, Feingold KR (1999) Regulation of glycosphingolipid metabolism in liver during the acute phase response. J Biol Chem 274:19707–19713

Morell P, Costantino-Ceccarini E, Radin NS (1970) The biosynthesis by brain microsomes of cerebrosides containing nonhydroxy fatty acids. Arch Biochem Biophys 141: 738–748

Pagano RE, Martin OC (1988) A series of fluorescent N-acylsphingosines: synthesis, physical properties, and studies in cultured cells. Biochemistry 27:4439–4445

Paul P, Kamisaka Y, Marks DL, Pagano RE (1996) Purification and characterization of UDP-glucose:ceramide glucosyltransferase from rat liver Golgi membranes. J Biol Chem 271:2287–2293

Platt FM, Neises GR, Karlsson GB, Dwek RA, Butters TD (1994) N-butyldeoxygalactonojirimycin inhibits glycolipid biosynthesis but does not affect N-linked oligosaccharide processing. J Biol Chem 269:27108–27114

Simon K, Ikonen E (1997) Functional rafts in cell membranes. Nature 387:569–572

Sprong H, Kruithof B, Leijendekker R, Slot JW, van Meer G, van der Sluijs P (1998) UDP-galactose:ceramide galactosyltransferase is a Class I integral membrane protein of the endoplasmic reticulum. J Biol Chem 273:25880–25888

Vunnam RR, Radin NS (1979) Short chain ceramides as substrates for glucocerebroside synthetase. Differences between liver and brain enzymes. Biochim Biophys Acta 573: 73–82

Watanabe R, Wu K, Paul P, Marks DL, Kobayashi T, Pittelkow MR, Pagano RE (1998) Up-regulation of glucosylceramide synthase expression and activity during human keratinocyte differentiation. J Biol Chem 273:9651–9655

Wu K, Marks DL, Watanabe R, Paul P, Rajan N, Pagano RE (1999) Histidine-193 of rat glucosylceramide synthase resides in a UDP-glucose- and inhibitor (D-threo-1-phenyl-2-decanoylamino-3-morpholinopropan-1-ol)-binding region: a biochemical and mutational study. Biochem J 341:395–400

Yamashita T, Wada R, Sasaki T, Deng C, Bierfreund U, Sandhoff K, Proia RL (1999) Proc Natl Acad Sci USA 96:9142–9147

Galactosyltransferases

β4-Galactosyltransferase-I

Introduction

The enzyme β4-galactosyltransferase-I (β4GalT-I; UDP-Gal:GlcNAc β4-galactosyltransferase; EC 2.4.1.38) is a constitutively expressed, *trans*-Golgi resident, type II membrane-bound glycoprotein that is widely distributed in vertebrates. The protein domain structure established for β4GalT-I consists of: (1) a short NH_2-terminal cytoplasmic domain of 11 or 24 amino acids depending on the protein isoform (Shaper et al. 1988; Russo et al. 1990); (2) a large COOH-terminal luminal domain containing the catalytic center (~270 amino acids) linked to a single transmembrane domain (19 amino acids) through a glycosylated peptide segment (~86 amino acids) termed the stem region. In essentially all vertebrate tissues, the primary function of β4GalT-I is to catalyze the transfer of Gal from UDP-Gal to GlcNAcβ-R, forming the *N*-acetyllactosamine (Galβ1-4GlcNAcβ1-R) or poly-*N*-acetyllactosamine structures assembled on glycoconjugates.

In mammals, β4GalT-I has been recruited for a second biosynthetic function, the tissue-specific production of the disaccharide lactose (Galβ1-4Glc), which takes place exclusively in the lactating mammary gland. The synthesis of lactose is carried out by a protein heterodimer assembled from β4GalT-I and α-lactalbumin that is termed lactose synthase (EC 2.4.1.22) (Brew et al. 1968). α-Lactalbumin is an abundant, non-catalytic milk protein that shares an ancestral gene in common with the hydrolytic enzyme lysozyme (Grobler et al. 1994). α-Lactalbumin is expressed de novo exclusively in the epithelial cells of the mammary gland during lactation. The net result of the binding of α-lactalbumin to β4GalT-I is to lower the K_m for Glc about three orders

Nancy L. Shaper[1] and Joel H. Shaper[1,2]

[1] Cell Structure and Function Laboratory, Johns Hopkins Oncology Center CRB-345, The Johns Hopkins University School of Medicine, 1650 Orleans Street, Baltimore, MD 21231-1000, USA
Tel. +1-410-955-8374; Fax +1-410-502-5499
e-mail: nshaper@jhmi.edu
[2] Department of Pharmacology and Molecular Sciences, The Johns Hopkins University School of Medicine, Baltimore, MD 21231-1000, USA

of magnitude. Consequently, Glc becomes an efficient acceptor substrate at physiological concentrations. Interestingly, β4GalT-I from nonmammalian vertebrates (e.g., chicken) can also functionally interact with α-lactalbumin in vitro to synthesize lactose. This observation indicates that the α-lactalbumin-binding domain on β4GalT-I predates the evolution of mammals (Shaper et al. 1997).

Databanks

β4-Galactosyltransferase-I

NC-IUBMB enzyme classification: EC 2.4.1.38 (which is indistinguishable from EC 2.4.1.90) and EC 2.4.1.22

Species	Gene	Protein	mRNA	Genomic	LocusLink (NCBI)
Homo sapiens	*B4GALTI*	P15291	X14085	AL161445	2683
			D29805		
			X55415		
Mus musculus	*B4galtI*	P15535	J03880	–	–
Bos taurus	–	P08037	X14558	–	–
Gallus gallus	–	AAB05218	U19890	–	–

A new resource, LocusLink (at http://www.ncbi.nlm.nih.gov), provides a single query interface to the curated sequence and descriptive information about genes. It presents information on official nomenclature, aliases, sequence accession numbers, phenotypes, EC numbers, MIM numbers, UniGene clusters, map information, and relevant web sites

Name and History

β4-Galactosyltransferase-I denotes the α-lactalbumin-responsive UDP-galactose:*N*-acetylglucosamine β4-galactosyltransferase that has been localized to human chromosome 9p13, mouse chromosome 4, and chicken chromosome Z. This enzyme is also referred to as UDP-galactose:*N*-acetylglucosamine β4-galactosyltransferase, glycoprotein 4β-galactosyltransferase, β1,4-galactosyltransferase, β4-*N*-acetyllactosamine synthase, or the A-protein of the lactose synthase heterodimer. β4GalT-I, β4GalT, β4GT, and GalTase are frequently used abbreviations.

Demonstration of the dual role of β4GalT-I in oligosaccharide biosynthesis and lactose biosynthesis is a major cornerstone of glycobiology. By the early 1960s the sequence of the nonreducing terminal trisaccharide Sia-Galβ1-4GlcNAc-R had been established for a number of secreted mammalian glycoproteins (Spiro 1962). Using a cell-free system from liver, McGuire et al. (1965) demonstrated the transfer of Gal from UDP-Gal to an asialoagalactoglycoprotein. Using mammary gland extracts, Watkins and Hassid (1962) also showed that the last step in lactose biosynthesis was the transfer of Gal (from UDP-Gal) to Glc. The puzzling observation was that the liver galactosyltranferase activity could not use Glc efficiently as an acceptor substrate. Conversely, the β4-galactosyltransferase in mammary glands (termed lactose synthase or lactose synthetase) could not use GlcNAc efficiently as a substrate. These observations suggested the presence of two separate β4-galactosyltransferase activities and provided the impetus to purify the different putative enzymes for comparative analysis.

The purification of lactose synthase proved to be an intractable problem due to the apparent lability of partially purified activity. Brodbeck and Ebner (1966) made the pivotal observation that the partially purified lactose synthase activity could be separated into two protein components (A-protein and B-protein) by gel filtration chromatography. Neither component had the ability to synthesize lactose by itself, but when the A- and B-proteins were combined lactose synthase activity was reconstituted. Subsequently, the authors showed that the B-protein was the abundant milk protein α-lactalbumin. The last piece of the puzzle was provided when Brew et al. (1968) demonstrated that the A-protein was a β4-galactosyltransferase activity (β4GalT-I) that transferred Gal to GlcNAc. In addition, when α-lactalbumin was added to the liver β4-galactosyltransferase, the acceptor sugar specificity was also changed from GlcNAc to Glc.

Enzyme Activity Assay and Substrate Specificity

Two reactions can be catalyzed by β4GalT-I.

$$\text{UDP-Gal} + \text{GlcNAc}\beta\text{1-R} \rightarrow \text{Gal}\beta\text{1-4GlcNAc}\beta\text{1-R} + \text{UDP} \tag{1}$$

in which the enzyme catalyzes the transfer of Gal, in the presence of a transition metal ion (e.g., Mn^{+2}), from the nucleotide sugar UDP-α-D-Gal to the 4-hydroxyl group of the acceptor sugar GlcNAc, which may be the free monosaccharide or the non-reducing terminal monosaccharide of a carbohydrate side-chain of a glycoprotein or glycolipid.

$$\text{UDP-Gal} + \text{Glc} \rightarrow \text{Gal}\beta\text{1-4Glc} + \text{UDP} \tag{2}$$

In the presence of the modifier protein α-lactalbumin, β4GalT-I catalyzes the transfer of Gal to Glc, forming lactose.

Assay Procedure

A number of methods to assay β4GalT-I enzyme activity, ranging from spectrophotometric techniques to sodium dodecyl sulfate (SDS)-acrylamide gels, have been developed (Ebner et al. 1972; Snow et al. 1999 and references therein). One of the most convenient assays measures the transfer of radiolabeled Gal ([^{3}H]Gal or [^{14}C]Gal) from the corresponding UDP-Gal to an appropriate acceptor sugar substrate (e.g., GlcNAc). The product ([^{3}H/^{14}C]Galβ4GlcNAc) is easily and rapidly separated from the radiolabeled reactant by ion-exchange chromatography on a Dowex-1-X8 column (chloride/formate form; 100–200 mesh) prepared in a plugged glass pasteur pipette (Brew et al. 1968). The assay system typically includes 0.3 mM UDP-Gal, 10 mM GlcNAc, and 10 mM Mn^{+2} in an appropriate buffer that does not precipitate the manganous ion (e.g., 100 mM sodium cacodylate or Tris-maleic acid). The pH optimum of the enzyme is about 6.8–7.2. When lactose synthase enzymatic activity is assayed, 10 mM Glc is substituted for GlcNAc, and α-lactalbumin (10–100 μg) is added. A number of sensitive methods have also been developed to assay β4GalT-I enzyme activity that avoid the use of radiolabeled UDP-Gal. These methods take advantage of analytical equipment that can identify the glycan product directly (ion chromatography) or use appropriate fluorescent derivatives of GlcNAc as acceptor substrates to enable product detection by high-performance liquid chromatography (HPLC) or capillary electrophoresis (CE) (Snow et al. 1999 and references therein).

Strategies based on the synthesis of donor nucleotide sugar and acceptor sugar analogs have been introduced to develop specific inhibitors for β4GalT-I (Hashimoto et al. 1997; Chung et al. 1998; Takayama et al. 1999). Based on in vitro assays these inhibitors are effective in the 1–200 μM range, but none is commercially available. Although these inhibitors can selectively inhibit β4GalT-I compared with α3-galactosyltransferase, it is an open question whether other members of the β4-galactosytransferase gene family (e.g., Lo et al. 1998) are also inhibited.

Preparation

Although β4GalT-I is widely distributed in somatic tissues from mammals (and non-mammalian vertebrates), the biological source of choice for large-scale preparation of the native enzyme remains colostrum or fresh, unpasteurized skim milk for two reasons. First, the level of enzyme activity is significant in this abundant, readily available source. Second, the β4GalT-I enzyme is present as an enzymatically active, soluble form due to proteolysis at multiple sites located in the stem region of the intact protein. The net result of proteolytic cleavage in the stem region is release of the catalytic domain that is enzymatically active and behaves in solution as a soluble globular protein, thereby faciltating subsequent purification. This soluble form of bovine β4GalT-I isolated from milk is available from several commercial sources including Sigma (St. Louis, MO, USA), Calbiochem (San Diego, CA, USA), and US Biological (Swampscott, MA, USA). Recombinant soluble bovine β4GalT-I expressed in the insect cell line *Spodoptera frugiperda*, is also available from Calbiochem. Recombinant soluble human β4GalT-I has been expressed in the yeast *Pichia pastoris* and as an enzymatically active protein A fusion protein in high yield (4.8–6.0 mg/L) in insect cells (Zhou et al. 2000). The full-length form of β4GalT-I has been isolated and characterized from Golgi membranes from the lactating sheep mammary gland (Smith and Brew 1977) and rat liver (Bendiak et al. 1993).

The specific interaction between β4GalT-I and α-lactalbumin was exploited more than 30 years ago for purification of this galactosyltransferase. Andrews (1970) introduced a two-step strategy for isolation of human β4GalT-I from milk. This strategy consisted of gel filtration followed by affinity chromatography on an absorbant prepared by covalently linking α-lactalbumin to cyanogen bromide-activated Sepharose 6B. Trayer and Hill (1971) reported purification of the bovine α-lactalbumin to constant specific activity by sequential chromatography on diethylaminoethyl (DEAE)-Sepharose, cellulose phosphate, and α-lactalbumin-Sepharose 4B. Subsequently Barker et al. (1972) introduced a purification strategy that consisted solely of sequential affinity chromatography using affinity ligands that mimic both the nucleotide donor substrate (UDP-hexanolamine) and the acceptor sugar substrate

(GlcNAc-hexanolamine). Using this strategy, a soluble form of β4GalT-I was purified from bovine milk to apparent homogeneity by a two-step procedure.

Biological Aspects

The β4GalT-I transcript is constitutively expressed in all human and murine tissues analyzed, although steady-state β4GalT-I mRNA levels in the brain are reduced about 10-fold compared to that in other somatic tissues (Lo et al. 1998). Human, mouse, and chicken β4GalT-I genes are each distributed in six exons. The human β4GalT-I gene spans ~54 kb of genomic DNA, whereas the chicken gene is distributed over ~20 kb (Shaper et al. 1997 and references therein).

At present, detailed analysis of transcriptional regulation of the β4GalT-I gene has been carried out only in the mouse. The organization of the 5'-end of the murine β4GalT-I gene is unusual in that three transcriptional start sites are contained within a ~725-bp contiguous piece of genomic DNA. The most distal start site is used exclusively during the later stages of spermatogenesis, primarily in haploid round spermatids (Harduin-Lepers et al. 1992). Recently it has been shown that a remarkably short 87-bp genomic fragment that flanks this β4GalT-I male germ cell specific start site is sufficient to drive correct in vivo male germ cell-specific expression of a reporter gene in transgenic mice (Charron et al. 1999).

Transcription of the β4GalT-I gene in somatic tissues takes place primarily from a start site positioned ~500 bp downstream of the germ cell start site. The resulting transcript is 4.1 kb in length. The third, most proximal start site is used predominantly in the mammary gland during lactation and results in production of a 3.9-kb mRNA (Shaper et al. 1988). Because the 4.1-kb start site is positioned upstream of the first in-frame ATG, and the 3.9-kb start site is positioned between the first two in-frame ATGs (which are 39 bp apart), translation of each mRNA results in synthesis of two catalytically identical, structurally related protein isoforms that differ only in the length of their respective, short, NH_2-terminal cytoplasmic domains (24 versus 11 amino acids). The position of the two start sites also dictates that the 4.1-kb mRNA has a rather long 5'-untranslated region of ~200 nt, whereas the 3.9-kb mRNA has a short 5'-untranslated region of ~25 nt (Shaper et al. 1988; Russo et al. 1990).

β4GalT-I and Lactose Biosynthesis

During lactation β4GalT-I enzyme levels in the mammary gland increase about 50-fold. With the recruitment of β4GalT-I for lactose biosynthesis, the regulatory problem arose about how to increase the levels of this enzyme specifically in the lactating mammary gland while maintaining the comparatively low level of constitutively expressed enzyme required for glycoconjugate biosynthesis in all somatic tissues. A comparative analysis of the structure of the mouse and chicken β4GalT-I gene in combination with an understanding of the regulation of the mouse gene has provided insight into how nature has solved this problem. The main insight from this analysis was that mammals have evolved a two-step mechanism to generate the levels of β4GalT-I required for lactose biosynthesis. In step 1, steady-state β4GalT-I mRNA levels are up-regulated as a result of the switch to the 3.9-kb transcriptional start site that is governed by a stronger mammary gland promoter (Rajput et al. 1996). In step 2, the 3.9-kb β4GalT-I mRNA was demonstrated to be translated more efficiently (three- to fivefold), relative to the constitutively expressed 4.1-kb transcript owing to deletion of the long GC-rich 5′-untranslated region characteristic of the 4.1-kb mRNA (Charron et al. 1998).

Generation of β4GalT-I Null Mice

Two groups have reported targeted deletion of the β4GalT-I gene. Asano et al. (1997) reported that about 60% of the homozygous mutants die within 4 weeks of birth and only 20% survive to 16 weeks. Female animals fail to lactate. Skin lesions develop within a few days of birth; and these lesions have thickened epidermis and a horny cell layer. Both male and female null mutant mice that survive to adulthood are fertile. The observation that adult males are fertile should be considered in the context of a model for mouse sperm–egg binding that is currently under investigation. This model posits that a sperm surface β4GalT-I binds to acceptor sugar substrate (GlcNAc) located on an extracellular matrix glycoprotein that forms in part the zona pellucida of the mouse egg (reviewed by Nixon et al. 2001).

Lu et al. (1997) reported that about 90% of the homozygous mutants die within 2–3 weeks of birth. Mutants are characterized by growth retardation, thin skin, sparse hair, poorly developed lungs, abnormal adrenal cortices, and dehydration. It was suggested that these defects are consistent with an endocrine insufficiency, the polyglandular nature of which is indicative of anterior pituitary failure. Consistent with this suggestion, the pituitary gland of null mutants is reduced in size, and β4GalT-I enzyme activity could not be detected. Homozygous mutant females fail to lactate. The 10% of the homozygous mutants that survive neonatal lethality have puffy skin.

Future Perspectives

As a result of the combined efforts of many investigators, β4GalT-I is the most completely characterized glycosyltransferase to date. We have a good understanding of the enzymology, gene regulation, and most recently the three-dimensional-structure (Gastinel et al. 1999). Despite these fundamental insights, a number of questions must still be answered. For example, at the protein level, β4GalT-I has been demonstrated to be phosphorylated on at least one serine residue located in a polypeptide fragment that includes the NH_2-terminal cytoplasmic domain, the transmembrane domain, and part of the stem region (Strous et al. 1987). What is the role of this posttranslational modification on the regulation of intrinsic enzymatic activity? At the cellular level, how is β4GalT-I organized in the Golgi membrane? X-ray inactivation data suggest that membrane-bound β4GalT-I is a dimer (Fleischer et al. 1993). Is this β4GalT-I dimer present in a supramolecular structure that includes other glycosyltransferases that participate in terminal glycosylation?

Lastly, with the recent success in obtaining high-resolution structural information on β4GalT-I, can these insights be exploited to develop specific inhibitors for β4GalT-

I that function in vivo? This question is particularly interesting in the context of the recent demonstration of a β4GalT-I gene family in which specific family members exhibit overlapping function based on in vitro enzymatic assay.

Further Reading

Shaper et al. (1997) compare the organization of a nonmammalian (chicken) vertebrate β4GalT-I gene with its mammalian counterpart. Based on this comparison, plausible changes introduced into the mammalian gene during the evolution of mammals and the recruitment of β4GalT-I for lactose biosynthesis are discussed.

Charron et al. (1998) noted that transcription of the β4GalT-I gene in somatic tissues yields a 4.1-kb mRNA (constitutively expressed) and a 3.9-kb mRNA (primarily expressed in the lactating mammary gland). The primary difference between the two transcripts is the length of their respective 5'-untranslated regions. The study demonstrates a biological rationale for the presence of the 3.9-kb transcript. This mRNA is translated more efficiently relative to its constitutive 4.1-kb counterpart because of deletion of most of the 5'-untranslated region. As a consequence, the increased efficiency of translation of the shorter mRNA contributes in part to the 50-fold increase in β4GalT-I enzyme activity, which is required for lactose biosynthesis.

Charron et al. (1999) defined the promoter that regulates in vivo expression of β4GalT-I in murine postmeiotic male germ cells. Using transgenic mice as the "test tube," the promoter was found to be completely contained within a remarkably short 87-bp genomic fragment.

References

Asano M, Furukawa K, Kido M, Matsumoto S, Umesaki Y (1997) Growth retardation and early death of β1,4-galactosyltransferase knockout mice with augmented proliferation and abnormal differentiation of epithelial cells. EMBO J 16:1850–1855

Andrews P (1970) Purification of lactose synthetase A protein from human milk and demonstration of its interaction with α-lactalbumin. FEBS Lett 9:297–300

Barker R, Olsen KW, Shaper JH, Hill RL (1972) Agarose derivatives of uridine diphosphate and N-acetylglucosamine for the purification of a galactosyltransferase. J Biol Chem 247:7135–7147

Bendiak B, Ward LD, Simpson RJ (1993) Proteins of the Golgi apparatus: purification to homogeneity, N-terminal sequence, and unusually large Stokes radius of the membrane-bound form of UDP-galactose:N-acetylglucosamine β1–4galactosyltransferase from rat liver. Eur J Biochem 216:405–417

Brew K, Vanaman TC, Hill RL (1968) The role of α-lactalbumin and the A protein in lactose synthetase: a unique mechanism for the control of a biological reaction. Proc Natl Acad Sci USA 59:491–497

Brodbeck U, Ebner KE (1966) Resolution of a soluble lactose synthetase into two protein components and solubilization of microsomal lactose synthetase. J Biol Chem 241:1391–1397

Charron M, Shaper JH, Shaper NL (1998) The increased level of β1,4-galactosyltransferase required for lactose biosynthesis is achieved in part by translational control. Proc Natl Acad Sci USA 95:14805–14810

Charron M, Shaper NL, Rajput B, Shaper JH (1999) A novel 14-base-pair regulatory element is essential for in vivo expression of murine β4-galactosyltransferase-1 in late pachytene spermatocytes and round spermatids. Mol Cell Biol 19:5823–5832

Chung SJ, Takayama S, Wong C-H (1998) Acceptor substrate-based selective inhibition of galactosyltransferases. Bioorg Med Chem Lett 8:3359–3364

Ebner KE, Mawal R, Fitzgerald DK, Colvin B (1972) Lactose synthetase (UDP-D-galactose:acceptor β-4-galactosyltransferase) from bovine milk. Methods Enzymol 28:500–510

Fleischer B, McIntyre JO, Kempner ES (1993) Target sizes of galactosyltransferase, sialyltransferase, and uridine diphosphatase in Golgi apparatus of rat liver. Biochemistry 32:2076–2081

Gastinel LN, Cambillau C, Bourne Y (1999) Crystal structures of the bovine β4-galactosyltransferase catalytic domain and its complex with uridine diphosphogalactose. EMBO J 18:3546–3557

Grobler JA, Rao KR, Pervaiz S, Brew K (1994) Protein sequences of two highly divergent canine type c lysozymes: implications for the evolutionary origins of the lysozyme α-lactalbumin superfamily. Arch Biochem Biophys 313:360–366

Harduin-Lepers A, Shaper NL, Mahoney JA, Shaper JH (1992) Murine β1,4 galactosyltransferase: round spermatid transcripts are characterized by an extended 5′-untranslated region. Glycobiology 2:361–368

Hashimoto H, Endo T, Kajihara Y (1997) Synthesis of the first tricomponent bisubstrate analog that exhibits potent inhibition against GlcNAc: β-1,4-galactosyltransferase. J Org Chem 62:1914–1915

Lo N-W, Shaper JH, Pevsner J, Shaper NL (1998) The expanding β4-galactosyltransferase gene family: messages from the databanks. Glycobiology 8:517–526

Lu Q-X, Hasty P, Shur BD (1997) Targeted mutation in β1,4-galactosyltransferase leads to pituitary insufficiency and neonatal lethality. Dev Biol 181:257–267

McGuire EJ, Jourdian GW, Carlson DM, Roseman S (1965) Incorporation of D-galactose into glycoproteins. J Biol Chem 240:PC4112–4115

Nixon B, Lu Q, Wassler MJ, Foote CI, Ensslin MA, Shur BD (2001) Galactosyltransferase function during mammalian fertilization. Cells Tissues Organs 168:46–57

Rajput B, Shaper NL, Shaper JH (1996) Transcriptional regulation of murine β1,4 galactosyltransferase in somatic cells: analysis of a gene that serves both a housekeeping and a mammary gland-specific function. J Biol Chem 271:5131–5142

Russo RN, Shaper NL, Shaper JH (1990) Bovine β1,4-galactosyltransferase: two sets of mRNA transcripts encode two forms of the protein with different amino-terminal domains. J Biol Chem 265:3324–3331

Shaper NL, Hollis GF, Douglas JG, Kirsch IR, Shaper JH (1988) Characterization of the full length cDNA for murine β-1,4-galactosyltransferase: novel features at the 5′-end predict two translational start sites at two in-frame AUG's. J Biol Chem 263:10420–10428

Shaper NL, Meurer JA, Joziasse DH, Chou TD, Smith EJ, Schnaar RL, Shaper JH (1997) The chicken genome contains two functional nonallelic β1,4-galactosyltransferase genes: chromosomal assignment to syntenic regions tracks fate of the two gene lineages in the human genome. J Biol Chem 272:31389–31399

Smith CA, Brew K (1977) Isolation and characteristics of galactosyltransferase from Golgi membranes of lactating sheep mammary glands. J Biol Chem 252:7294–7299

Snow DM, Shaper JH, Shaper NL, Hart GW (1999) Determination of β1,4-galactosyltransferase enzymatic activity by capillary electrophoresis and laser-induced fluorescence detection. Anal Biochem 271:36–42

Spiro RG (1962) Studies on the monosaccharide sequence of the serum glycoprotein fetuin. J Biol Chem 237:646–652

Strous GJ, van Kerkhof P, Fallon RJ, Schwartz AL (1987) Golgi galactosyltransferase contains serine-linked phosphate. Eur J Biochem 169:307–311

Takayama S, Chung SJ, Igarashi Y, Ichikawa Y, Sepp A, Lechler RI, Wu J, Hayashi T, Siuzdak G, Wong C-H (1999) Selective inhibition of β-1,4- and α-1,3-galactosyltransferases: donor sugar-nucleotide based approach. Bioorg Med Chem 7:401–409

Trayer IP, Hill RL (1971) The purification and properties of the A protein of lactose synthetase. J Biol Chem 246:6666–6675

Watkins W, Hassid WZ (1962) The synthesis of lactose by particulate enzyme preparations from guinea pig and bovine mammary glands. J Biol Chem 237:1432–1440

Zhou D, Malissard M, Berger EG, Hennet T (2000) Secretion and purification of recombinant galactosyltransferase from insect cells using pFmel-protA, a novel transponsition-based baculovirus transfer vector. Arch Biochem Biophys 373:3–7

3

β4-Galactosyltransferase-II, -III, -IV, -V, -VI, and -VII

Introduction

The Galβ1→4GlcNAc structure is commonly found in the outer chain moieties of N- and O-linked oligosaccharides, and in the oligosaccharides of lacto-series glycolipids. The terminal galactose itself is involved in galectin-mediated biological events, including apoptosis (Perillo et al. 1995), and many biologically active carbohydrate determinants involved in cell adhesion processes, such as polysialic acid, HNK-1 carbohydrate, poly-N-acetyllactosamine, Lewis X, and sialyl Lewis X, are expressed on the disaccharide groups. Therefore, β4-galactosyltransferase (β4GalT), which transfers galactose from UDP-Gal to N-acetylglucosamine, is one of the key enzymes in glycobiology. Targeted inactivation of the mouse β4GalT gene showed that the galactose and/or galactose-containing oligosaccharides are important for cell growth and differentiation, and that murine cells contain another β4GalT which produces the residual Galβ1→4GlcNAc groups found in the N-linked oligosaccharides of the mutant mouse (Asano et al. 1997; Kido et al. 1998; Lu et al. 1997). Since Lewis X-containing oligosaccharides inhibited the compaction of mouse embryos (Bird and Kimber 1984), and since a β4GalT-knockout mouse can survive beyond birth (Asano et al. 1997; Lu et al. 1997), the galactose-containing oligosaccharides synthesized by novel β4GalT could be important for the early development of mammalian embryos. To date, six novel human genes, five novel mouse genes, one novel rat gene, and one novel chicken gene which encode proteins with β4GalT activity have been cloned (see databank) (reviewed by Amado et al. 1999; Furukawa and Sato 1999). In addition to these, the mouse contains one more gene (accession No. D3779) which encodes a protein with β4GalT activity toward N-acetylglucosamine (Uehara and Muramatsu

Kiyoshi Furukawa[1] and Henrik Clausen[2]

[1] Department of Biosignal Research, Tokyo Metropolitan Institute of Gerontology, Itabashi-ku, Tokyo 173-0015, Japan
Tel. +81-3-3964-3241 ext. 3071; Fax +81-3-3579-4776
e-mail: furukawa@tmig.or.jp
[2] Faculty of Health Sciences, School of Dentistry, DK-2200 Copenhagen N, Denmark
Tel. +45-35326835; Fax +45-35326505
e-mail: henrik.clausen@odont.ku.dk

1997). However, this enzyme is totally different from β4GalT-I–VII based on the sequence similarities. There are also several genes homologous to mammalian β4GalTs in invertebrates such as the nematode and snail (reviewed by Lo et al. 1998; Amado et al. 1999). This review mainly summarizes the properties of six novel human β4GalTs, although their acceptor specificities have not yet been established.

Databanks

β4-Galactosyltransferase-II, -III, -IV, -V, -VI, and -VII

No EC number has been allocated to β4GalT-II, -III, -IV, -V, -VI, and -VII enzymes yet

Species	Gene	Protein	mRNA	Genomic
β4GalT-II				
Homo sapiens	–	BAA75819.1	AB024434	–
	–	AAC39733.1	AF038660	–
	β4Gal-T2	CAA73112.1	Y12510	–
Mus musculus	–	BAA34385.1	AB019541	–
	–	AAF22220.1	AF142670	–
Gallus gallus	*CKII*	AAB05217.1	U19889	–
β4GalT-III				
Homo sapiens	–	BBA75820.1	AB024435	–
	–	AAC39734.1	AF038661	–
Mus musculus	*β4Gal-T3*	CAA73111.1	Y12509	–
	–	AAF22221.1	AF142671	–
β4GalT-IV				
Homo sapiens	–	BAA75821.1	AB024436	–
	–	AAC72493.1	AF022367	–
Mus musculus	–	AAC39735.1	AF038662	–
	–	AAF22222.1	AF142672	–
β4GalT-V				
Homo sapiens	–	BAA25006.1	AB004550	–
	–	AAC39736.1	AF038663	–
Mus musculus	*mbgt*	BAA94791.1	AB004786	–
	–	AAF22223.1	AF142673	–
β4GalT-VI				
Homo sapiens	–	BAA76273.2	AB024742	–
	–	AAC39737.1	AF038664	–
Mus musculus	–	AAD41694.1	AF097158	–
	–	AAF22224.1	AF142674	–
Rattus norvegicus	–	AAC24515.1	AF048687	–
β4GalT-VII				
Homo sapiens	–	CAB56424.1	AJ005382	–

Name and History

As described in chapter 2, β4GalT was initially found in bovine mammary glands and milk, and was shown to be involved in the synthesis of lactose with the aid of α-lactalbumin in the middle of 1960. Later, β4GalT was shown to galactosylate *N*-

acetylglucosamine-terminating oligosaccharides. Although there were several papers describing the presence of another β4GalT (reviewed by Furukawa and Sato 1999), β4GalT was believed to be a single gene product which governs the galactosylation of all glycoconjugates. However, analysis of the β4GalT-knockout mouse revealed that the tissues examined contained residual β4GalT activity and Galβ1→4GlcNAc structure on the *N*-linked oligosaccharides, in addition to the marked changes in cell growth and differentiation, especially in the skin and small intestine, and the early motality of the animal (Asano et al. 1997; Kido et al. 1998; Lu et al. 1997). This encouraged us to clone another β4GalT gene from a human breast tumor cell cDNA library by a homology-based cloning method (Sato et al. 1998b). In the same time period, two other groups cloned several more human β4GalT genes (Almeida et al. 1997; Lo et al. 1998; Schwientek et al. 1998) by searching nucleotide sequences similar to mammalian β4GalT gene in EST sequences using the BLAST program (reviewed by Amado et al. 1999). These enzymes are named β4GalT-II, -III, -IV, -V, -VI, and -VII according to their homology distances from the previous enzyme, which is referred to as β4GalT-I, and show 55%, 50%, 41%, 37%, 33%, and 25% similarity to β4GalT-I, respectively, which is shown schematically in Fig. 1.

Multiple sequence alignment (ClustalW) and phylogenetic analysis of the predicted amino acid sequences indicate that the genes are classified into four groups: β4GalT-I and -II, β4GalT-III and -IV, β4GalT-V and -VI, and β4GalT-VII (Lo et al. 1998). Like other Golgi-residing glycosyltransferases, they are type II-membrane proteins, and contain four cystein residues in the putative catalytic domains of β4GalT-I–VI but not of β4GalT-VII. Each β4GalT appears to have a different length of stem region. There are some short sequence motifs common to all human β4GalTs (FNRA, NVG, DVD, and WGWG(G/R)EDD(D/E) in the conserved domain) and these could be involved in

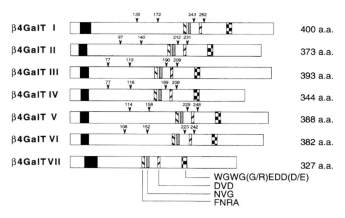

Fig. 1. Schematic illustration of human β4GalT proteins. For each transferase protein, the *solid box* indicates a putative transmembrane domain, and *other boxes* indicate short sequence motifs, FNRA, NVG, DVD, and WGWG(G/R)EDD(D/E), common to the seven enzymes. *Wedges* with numbers indicate the positions of Cys residues conserved among the six enzymes. The total number of composing amino acid residues in each enzyme is shown on the *right-hand* side

binding to UDP-Gal and/or *N*-acetylglucosamine-terminating oligosaccharides, and also in some enzymatic mechanism common to these β4GalTs.

Enzyme Activity Assay and Substrate Specificity

Like β4GalT I, β4GalT-II–VI catalyze the transfer of galactose from UDP-Gal to *N*-acetylglucosamine (Almeida et al. 1997; Sato et al. 1998b; Sato and Furukawa 1999; Schwientek et al. 1998), but β4GalT-VII transfers galactose to xylβ1→MU or xylβ1→pNP (Almeida et al. 1999; Okajima et al. 1999). Therefore, β4GalT-VII is considered to be an enzyme that is involved in the biosynthesis of the proteoglycan core region. The reaction proceeds at pH 6–7.5 and requires Mn^{2+} (2–20 mM) in the presence of Triton X-100 or CF-54 (0.1%–1%), especially for assays using glycolipid-based acceptors. β4GalT-I and -II synthesize *N*-acetyllactosamine, and can synthesize lactose in the presence of α-lactalbumin. β4GalT-III and -IV synthesize *N*-acetyllactosamine but not lactose in the presence of α-lactalbumin. β4GalT-III and -IV effectively utilize GlcNAcβ1→3Galβ1→4Glcα1→Cer as a substrate. Since β4GalT-IV failed to galactosylate glycoprotein acceptors, it may not function in glycoprotein biosynthesis (Schwientek et al. 1998). β4GalT-V and -VI galactosylate Glcα1→Cer (Nomura et al. 1998; Sato et al. 2000; Takizawa et al. 1999). In the case of β4GalT-V, the lactosylceramide synthase activity was only associated with the membrane-bound form and not with the soluble form (Sato et al. 2000).

Although β4GalT-IV and -V belong to different phylogenetic groups, they effectively galactosylate the GlcNAcβ1→6(Galβ1→3)GalNAcα1→pNP (core 2 *O*-linked oligosaccharide) (Ujita et al. 1998; Van Die et al. 1999) but not the GlcNAcβ1→6-(GlcNAcβ1→3)GalNAcα1→octyl (core 4 *O*-linked oligosaccharide), which is galactosylated effectively with β4GalT-I (Ujita et al. 2000). Poly-*N*-acetyllactosamine often extends on *N*- and *O*-linked oligosaccharides and those of glycolipids, and only β4GalT-I can effectively promote the extension of *N*-acetyllactosamine units together with β-1,3-*N*-acetylglucosaminyltransferase (Ujita et al. 1998). When we transfected individual full-length human β4GalT cDNAs into Sf-9 cells which contain little β4GalT activity, but do contain endogenous acceptor-oligosaccharides for β4GalT and UDP-Gal (Hollister et al. 1998; Yoshimi et al. 2000), and then examined whether galactosylated oligosaccharides are expressed on endogenous glycoproteins by lectin blot analysis, all the samples showed that several protein bands become positive to *Ricinus communis* agglutinin-I (RCA-I), which preferentially interacts with oligosaccharides terminated with β-1,4-linked galactose (Baenziger and Fiete 1979). Since RCA-I-positive bands disappeared upon treatment of blots with diplococcal β-galactosidase or *N*-glycanase, novel β4GalT-II–VI could be functioning in the galactosylation of *N*-linked oligosaccharides in vivo in addition to β4GalT-I (Guo et al. 2001). Further studies are necessary to establish the physiological roles of these novel β4GalTs.

Preparation

Since all β4GalTs are considered to be Golgi–resident, they could be prepared by the conventional methods applied for β4GalT-I preparation (Furukawa and Roth 1985). β4GalT-VI was purified from rat brain microsomal fraction by extracting with 1%

Triton X-100 followed by chromatography using WGA-agarose, UDP-hexanolamine-agarose and hydroxylapatite columns (Nomura et al. 1998). For other β4GalTs, enzymes in soluble form secreted into culture media from the gene-transfected Sf-9 cells were partly purified by chromatography using DEAE-Sepharose and S-Sepharose columns (Almeida et al. 1997; Schwientek et al. 1998), soluble-form enzymes fused with protein A secreted into culture media from the gene-transfected CHO cells were purified by IgG-Sepharose (Sato et al. 1998b; Okajima et al. 1999), and membrane-bound-form enzymes were solubilized with detergent and used directly as an enzyme source (Almeida et al. 1997; Sato and Furukawa 1999; Sato et al. 2000; Schwientek et al. 1998). Since the presence or absence of the transmembrane and cytoplasmic domains alters enzymatic properties (Sato et al. 2000), some of the enzymatic properties so far determined have to be reevaluated by a more careful examination using both soluble and membrane-bound proteins.

Biological Aspects

Previous research has not provided sufficient data to confirm any biological aspects of novel β4GalTs. However, Northern blot analysis revealed that β4GalT-I, -V, and -VI genes appear to be expressed in most human tissues (Sato et al. 1998b; Takizawa et al. 1999). In adult human brain, the β4GalT-III, -V, and -VI genes, but not the β4GalT-I, -II, and -IV genes, were expressed, while all β4GalT genes were expressed in human fetal brain (Lo et al. 1998). Since unique glycosylation is observed for mouse brain N-linked oligosaccharides (Nakakita et al. 1999), it is important to determine which of the β4GalTs are functioning in the galactosylation in the brain. In the case of β4GalT-I and -II, they both synthesize lactose in the presence of α-lactalbumin. However, only β4GalT-I gene is expressed in the mammary gland (Lo et al. 1998; Sato et al. 1998a), indicating that β4GalT-II is not a lactose synthase, and this is also supported by the fact that no lactose is produced in the β4GalT-I-knockout mouse (Asano et al. 1997).

During malignant transformation of cells, alteration of the cell surface glycosylation is amply documented. Although β4GalT activity showed no significant change after malignant transformation of cells, as observed in several cell lines, changes in the gene expression of novel β4GalTs were observed between NIH3T3 and its malignant transformant MTAg: the gene expression of β4GalT-V increased three-fold, that of β4GalT-II decreased to one-tenth, and those of other β4GalTs, including β4GalT-I, were relatively constant (Shirane et al. 1999). Since β4GalT-V preferentially galacto-sylates the GlcNAcβ1→6Man group of the GlcNAcβ1→6(GlcNAcβ-1→2)Manα1→6Man branch (Shirane et al. 1999), and its gene expression is correlated with that of N-acetylglucosaminyltransferase (GnT)-V (Sato et al. 1999), which is a key enzyme for the malignant transformation-associated alteration of N-linked oligosaccharides, the expression of tumor antigens which are often formed on the Galβ1→4-GlcNAcβ1→6Man outer chain could be regulated by β4GalT-V activity in addition to GlcNAcT-V activity. Since these β4GalTs appear to compete with each other for the galactosylation of N-linked oligosaccharides in vivo, it is of consequence to establish the fine acceptor specificities of individual β4GalTs in order to distinguish intrinsic functions of individual galactose-containing oligosaccharides.

References

Almeida R, Amado M, David L, Levery SB, Holmes EH, Merkx G, van Kessel AG, Rygaard E, Hassan H, Bennett E, Clausen H (1997) A family of human β4-galactosyltransferases: cloning and expression of two novel UDP-galactose:β-N-acetylglucosamine β1,4-galactosyltransferases, β4Gal-T2 and β4Gal-T3. J Biol Chem 272:31979–31991

Almeida R, Levery SB, Mandel U, Kresse H, Schwientek T, Bennett EP, Clausen H (1999) Cloning and expression of a proteoglycan UDP-galactose:β-xylose β1,4-galactosyltransferase I. A seventh member of the human β4-galactosyltransferase gene family. J Biol Chem 274:26165–26171

Amado M, Almeida R, Schwientek T, Clausen H (1999) Identification and characterization of large galactosyltransferase gene families: galactosyltransferases for all functions. Biochim Biophys Acta 1473:35–53

Asano M, Furukawa K, Kido M, Matsumoto S, Umesaki Y, Kochibe N, Iwakura Y (1997) Growth retardation and early death of β4-galactosyltransferase knockout mice with augmented proliferation and abnormal differentiation of epithelial cells. EMBO J 16:1850–1857

Baenziger JU, Fiete D (1979) Structural determination of *Ricinus communis* agglutinin and toxin specificity for oligosaccharides. J Biol Chem 254:9795–9804

Bird JM, Kimber SJ (1984) Oligosaccharides containing fucose-linked α(1–3) and α(1–4) to *N*-acetylglucosamine cause decompaction of mouse morula. Dev Biol 104:449–460

Furukawa K, Roth S (1985) Co-purification of galactosyltransferases from chick embryo liver. Biochem J 227:573–582

Furukawa K, Sato T (1999) β-1,4-Galactosylation of *N*-glycans is a complex process. Biochim Biophys Acta 1473:54–66

Guo S, Sato T, Shirane K, Furukawa K (2001) Galactosylation of N-linked oligosaccharides by human β-1,4-galactosyltransferases I, II, III, IV, V, and VI expressed in Sf-9 cells. Glycobiology, in press

Hollister JR, Shaper JH, Jarvis DL (1998) Stable expression of mammalian β-1,4-galactosyltransferase extends the *N*-glycosylation pathway in insect cells. Glycobiology 8:473–480

Kido M, Asano M, Iwakura Y, Ichinose M, Miki K, Furukawa K (1998) Presence of polysialic acid and HNK-1 carbohydrate on brain glycoproteins from β-1,4-galactosyltransferase-knockout mice. Biochem Biophys Res Commun 245:860–864

Lo NW, Shaper JH, Pevsner J, Shaper NL (1998) The expanding β4-galactosyltransferases. Glycobiology 8:517–526

Lu Q, Hasty P, Shur BD (1997) Targeted mutation in β1,4-galactosyltransferase leads to pituitary insufficiency and neonatal lethality. Dev Biol 181:257–267

Nakakita S, Menon KK, Natsuka S, Ikenaka K, Hase S (1999) β1,4-galactosyltransferase activity of mouse brain as revealed by analysis of brain-specific complex-type *N*-linked sugar chains. J Biochem 126:1161–1169

Nomura T, Takizawa M, Aoki J, Arai H, Inoue K, Wakisaka E, Yoshizuka N, Imokawa G, Dohmae N, Takio K, Hattori M, Matsuo N (1998) Purification, cDNA cloning, and expression of UDP-Gal:glucosylceramide β-1,4-galactosyltransferase from rat brain. J Biol Chem 273:13570–13577

Okajima T, Yoshida K, Kondo T, Furukawa K (1999) Human homolog of *Caenorhabditis elegans* sqv-3 gene is galactosyltransferase I involved in the biosynthesis of the glycosaminoglycan–protein linkage region of proteoglycans. J Biol Chem 274: 22915-22918

Perillo NL, Pace KE, Seilhamer JJ, Baum LG (1995) Apoptosis of T cells mediated by galectin-1. Nature 378:736–739

Sato T, Furukawa K (1999) Differences in *N*-acetyllactosamine synthesis between β-1,4-galactosyltransferases I and V. Glyconj J 16:73–76

Sato T, Aoki N, Matsuda T, Furukawa K (1998a) Differential effect of α-lactalbumin on β-1,4-galactosyltransferase IV activities. Biochem Biophys Res Commun 244:637–641

Sato T, Furukawa K, Bakker H, van den Eijnden DH, van Die I (1998b) Molecular cloning of a human cDNA encoding β-1,4-galactosyltransferase with 37% identity to mammalian UDP-Gal:GlcNAc β-1,4-galactosyltransferase. Proc Natl Acad Sci USA 95:472–477

Sato T, Shirane K, Furukawa K (1999) Changes in β-1,4-galactosylation of glycoproteins in malignantly transformed cells. Recent Res Dev Cancer 1:105–114

Sato T, Guo S, Furukawa K (2000) Involvement of recombinant human β-1,4-galactosyltransferase V in lactosylceramide biosynthesis. Res Commun Biochem Cell Mol Biol 4:3–10

Schwientek T, Almeida R, Levery SB, Holmes EH, Bennett E, Clausen H (1998) Cloning of a novel member of the UDP-galactose:β-*N*-acetylglucosamine β1,4-galactosyltransferase family, β4Gal-T4, involved in glycosphingolipid biosynthesis. J Biol Chem 273:29331–29340

Shirane K, Sato T, Segawa K, Furukawa K (1999) Involvement of β-1,4-galactosyltransferase V in malignant transformation associated changes in glycosylation. Biochem Biophys Res Commun 265:434–438

Takizawa M, Nomura T, Wakisaka E, Yoshizuka N, Aoki J, Arai H, Inoue K, Hattori M, Matsuo N (1999) cDNA cloning and expression of human lactosylceramide synthase. Biochim Biophys Acta 1438:301–304

Uehara K, Muramatsu M (1997) Molecular cloning and characterization of β-1,4-galactosyltransferase expressed in mouse testis. Eur J Biochem 244:706–712

Ujita M, McAuliffe J, Schwientek T, Almeida R, Hindsgaul O, Clausen H, Fukuda M (1998) Synthesis of poly-*N*-acetyllactosamine in core 2-branched *O*-glycans: the requirement of novel β-1,4-galactosyltransferase IV and β-1,3-*N*-acetylglucosaminyltransferase. J Biol Chem 273:34843–34849

Ujita M, Misra AK, McAuliffe J, Hindsgaul O, Fukuda M (2000) Poly-*N*-acetyllactosamine extention in *N*-glycans and core 2- and core 4-branched *O*-glycans is differentially controlled by I-extension enzyme and different members of the β1,4-galactosyltransferase gene family. J Biol Chem 275:15868–15875

Van Die I, van Tetering A, Schiphorst WECM, Sato T, Furukawa K, van den Eijnden DH (1999) The acceptor substrate specificity of human β4-galactosyltransferase V indicates its potential function in *O*-glycosylation. FEBS Lett 450:52–56

Yoshimi Y, Sato T, Ikekita M, Guo S, Furukawa K (2000) Presence of monoantennary complex-type and hybrid-type oligosaccharides terminated with β-*N*-acetylglucosamine in lepidopteran insect Sf-9 cells. Res Commun Biochem Cell Mol Biol 4:163–170

4

β3-Galactosyltransferase-I, -II, and -III

Introduction

In higher eukaryotes, galactose is commonly found in all classes of glycoconjugates, where it is bound as either α- or β-anomer through 1,3- or 1,4-linkage to various carbohydrate acceptor substrates. Families of galactosyltransferases are defined according to the type of linkage catalyzed. Purification studies have suggested the existence of several enzymes in each galactosyltransferase family, assumptions which have been confirmed by the recent cloning of genes encoding galactosyltransferases. However, the number of galactosyltransferase genes isolated has far surpassed these early predictions. The characterization of the members of each galactosyltransferase family has revealed differences in the patterns of tissue expression and in acceptor substrate specificity, although a certain degree of redundancy prevails between galactosyltransferases from a given family. For example, four β3-galactosyltransferase (β3GalT) genes have been described that direct the expression of enzymes linking Galβ1,3 to GlcNAc (Hennet et al. 1998; Kolbinger et al. 1998; Amado et al. 1998; Isshiki et al. 1999; Zhou et al. 1999a). A comparison between β3GalT proteins unraveled several conserved domains not found in other galactosyltransferases. Surprisingly, a β3-*N*-acetylglucosaminyltransferase enzyme as well as proteins homologous to the *Drosophila* signaling proteins Brainiac and Fringe were also identified among the β3GalT-related proteins. β3GalTs participate in the shaping of several oligosaccharide structures in *O*-glycans, *N*-glycans and glycolipids. This review summarizes the properties of three β3GalT enzymes that direct the formation of type-1 chains, the support of Le[a] and Le[b] antigens.

THIERRY HENNET and ERIC G. BERGER

Institute of Physiology, University of Zürich, Winterthurerstrasse 190, 8057 Zürich, Switzerland
Tel. +41-1-635-5080; Fax +41-1-635-6814
e-mail: Thennet@access.unizh.ch

Databanks

β3-Galactosyltransferase-I, -II, and -III

No EC number has been allocated to β3GalT-I, -II, and -III enzymes yet
However, the β3GalT enzymatic activity associated to EC 2.4.1.86 is compatible with that
of β3GalT-I, -II, and -III

Species	Gene	Protein	mRNA	Genomic
β3GalT-I				
Mus musculus	*β3galt1*	AAC53523	AF029790	–
Homo sapiens	*β3GALT1*	AAD23451	E07739	–
			AF117222	
β3GalT-II				
Mus musculus	*β3galt2*	AAC53524	AF029791	–
Homo sapiens	*β3GALT2*	CAA75245	Y15014	–
		CAA75344	Y15060	
β3GalT-III				
Mus musculus	*β3galt3*	AAC53525	AF029792	–
Homo sapiens	*β3GALT3*	CAA75346	Y15062	–

Name and History

A β3GalT purified from swine trachea mucosa was first described by Mendicino et al. (1982). This β3GalT activity, attributed to the *O*-glycan core1 β3GalT enzyme, was directed to GalNAc-*O*-Ser/Thr acceptors. The first β3GalT activity towards GlcNAc-based acceptors was reported the same year (Sheares et al. 1982) as constituent of GalT activity measured in cell extracts treated with GlcNAc levels inhibitory to β4GalT activity. Subsequent purification (Sheares and Carlson 1983; Holmes 1989) confirmed the differences between β4GalT and β3GalT enzymes with respect to their affinity for lactalbumin and to their sensitivity to elevated GlcNAc concentrations. The first molecular cloning of a type-1 β3GalT gene was achieved by expression cloning strategies based on the detection of the Le^a antigen (Sasaki et al. 1994) and the GM$_1$ ganglioside (Miyazaki et al. 1997). By providing gene sequences, these studies have opened the way to the isolation of further β3GalT genes from genome databases using similarity search algorithms. Such approaches led to the isolation of several β3GalT genes (Hennet et al. 1998; Kolbinger et al. 1998; Amado et al. 1998; Isshiki et al. 1999; Zhou et al. 1999a), which allowed the delination of conserved domains in the β3GalT protein family (Fig. 1).

The identification of a β3-*N*-acetylglucosaminyltransferase enzyme (Zhou et al. 1999b), which included the same conserved domains, suggested that this similarity may in fact reflect a structural constraint for the catalysis of a β1,3-linkage. In addition, similarity searches unraveled striking homologies between β3GalTs and the *Drosophila melanogaster* proteins Brainiac and Fringe (Yuan et al. 1997). Fringe (Panin et al. 1997) and Brainiac (Goode et al. 1992) have been described as signaling molecules, which interact with proteins of the Notch and the epidermal growth factor (EGF)-receptor pathways.

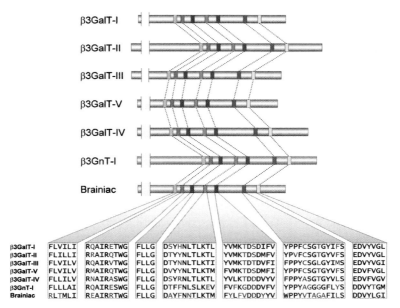

Fig. 1. Alignment of β3GalT, Brainiac, and β3GnT-I proteins. Proteins are schematically represented by *horizontal bars*. Type-1 β3GalT-I, -II, -III, and -V enzymes are shown at the top, followed by the β3GalT-IV GM$_1$ synthase, the β3GnT β1,3-*N*-acetylglucosaminyltransferase-I enzyme the *Drosophila* Brainiac protein. The positions of the conserved domains are indicated as *colored rectangles*. The *bottom panel* shows the amino acid sequences of the respective proteins in single-letter code, whereas conserved residues appear in *black* and nonconserved residues in *gray*

Enzyme Activity Assay and Substrate Specificity

β3GalT-I–III enzymes belong to the classical Leloir-type glycosyltransferases since they catalyze the transfer of Gal from the donor substrate UDP-Gal to GlcNAc-based acceptors. The reaction takes place over a broad pH range (5.5–7.5) and shows a strict requirement for Mn^{2+}, whereas minimal activity is detectable using Co^{2+}. As reported previously (Sheares et al. 1982), and in contrast to the β4GalTs, the β3GalT enzymes are not inhibited by elevated GlcNAc concentrations. Also, the β3GalT enzymes have no affinity to α-lactalbumin, which was conveniently used to separate β4GalT from β3GalT activities in purification procedures (Sheares and Carlson 1983; Holmes 1989). Of note is that the early studies with β3GalT proteins purified from pig trachea and adenocarcinoma cells represent the β3GalT-V enzyme (Isshiki et al. 1999; Zhou et al. 1999a), since β3GalT-I–III genes are either not, or hardly, expressed in these cell types. However, studies based on the heterologous expression of recombinant β3GalT enzymes have largely confirmed the findings obtained with the purified proteins. In vitro, β3GalT-I–III enzymes show a specificity toward GlcNAc-based acceptors. A low β3GalT activity was detected with β3GalT-I and β3GalT-II when using GalNAc (Hennet et al. 1998) or Gal (Kolbinger et al. 1998) acceptors. It remains unclear whether the β3GalT side-activity towards Gal accounts for the same activity reported earlier in human kidney (Bailly et al. 1988). When tested on complex acceptors, it appears that β3GalT-I–III enzymes are capable of transferring Gal to a broad range of oligosaccharides. In addition, some discrepancies were observed between the findings obtained in different laboratories. For example, on the one hand, β3GalT-II, but not β3GalT-I, was found to transfer Gal to ovalbumin (Amado et al. 1998), whereas on the other hand, a significant β3GalT activity toward ovalbumin was detected for the three β3GalT-I–III enzymes (Zhou et al. 1999a; D. Zhou 1999 unpublished data). A significant β3GalT activity was also noticed on the lactoceramide acceptors Lc3 and nLc5 for the β3GalT-I and β3GalT-II enzymes (Amado et al. 1998). In addition, for these three enzymes, no activity was detected on the Gal(β1,4)Xyl glycosaminoglycan core (D. Zhou 1999 unpublished data), indicating that β3GalT-I–III do not represent the β3GalT enzyme involved in the biosynthesis of the proteoglycan core. Finally, the ability to direct the formation of type-1 chains in vivo, as detected by sLe^a staining, has been confirmed in CHO cells stably expressing the β3GalT-II gene (Kolbinger et al. 1998).

Preparation

The first purification of a type-1 β3GalT enzyme was achieved from pig trachea (Sheares and Carlson 1983). The tissue was sonicated, and membrane proteins were solubilized in 1% Triton X-100. After the removal of small molecules by BioGel A filtration, the β3GalT activity was separated from the β4GalT and the mucin-type β3GalT activities by affinity chromatography on α-lactalbumin-Sepharose and asialo-ovine submaxillary mucin-DEAE, respectively. A similar purification protocol was applied by Holmes (1989) for the isolation of the β3GalT enzyme from the adenocarcinoma Colo205 and SW403 cell lines. As mentioned above, the purified β3GalT represents the β3GalT-V enzyme (Isshiki et al. 1999; Zhou et al. 1999a). The purification of the β3GalT-I–III enzymes, either in native form or as recombinant proteins, has not been reported to date.

Biological Aspects

The mRNAs encoding β3GalT-I–III are mainly expressed in brain tissue (Hennet et al. 1998; Kolbinger et al. 1998; Amado et al. 1998). Examination of the expression pattern during mouse development indicated that the three genes are progressively expressed starting from embryonic day 12 (D. Zhou 1999 unpublished data) up to the adult stage. A significant portion of N-glycan oligosaccharides isolated from rat brain contain β1,3-linked Gal branches (Zamze et al. 1998), which may be synthesized by the action of the β3GalT expressed in the brain. However, the functional relevance of this particular "brain-type" glycosylation is yet to be determined. Elevated β3GalT activity has been noted in several instances in adenocarcinoma cells, and was associated with an increase in Le^a and sLe^a reactivity found in gastric, colonic, and pancreatic tumors (Magnani et al. 1983). This increase in sLe^a antigens usually correlates with poor prognosis and increased metastatic potential (Nakamori et al. 1993). Note that sLe^a is a selectin ligand, and selectins have been involved in the control of metastasis (Kim et al. 1998).

The extensive similarity of the *Drosophila* signaling protein Brainiac to β3GalT enzymes (see Fig. 1) suggests that the functions of Brainiac are mediated by a glycosyltransferase activity. Notch and its ligands are large membrane proteins with several EGF-like repeats, and which bear glycosylation sites (Moloney et al. 2000). The potential involvement of glycosyltransferases in the regulation of developmental pathways raises interesting perspectives for the relevance of oligosaccharides in biological systems.

References

Amado M, Almeida R, Carneiro F, Levery SB, Holmes EH, Nomoto M, Hollingsworth MA, Hassan H, Schwientek T, Nielsen PA, Bennett EP, Clausen H (1998) A family of human β-3-galactosyltransferases—characterization of four members of a UDP-galactose-β-*N*-acetylglucosamine/β-*N*-acetylgalactosamine β-1,3-galactosyltransferase family. J Biol Chem 273:12770–12778

Bailly P, Piller F, Cartron JP (1988) Characterization and specific assay for a galactoside β-3-galactosyltransferase of human kidney. Eur J Biochem 173:417–422

Goode S, Wright D, Mahowald AP (1992) The neurogenic locus *brainiac* cooperates with the *Drosophila* EGF receptor to establish the ovarian follicle and to determine its dorsal–ventral polarity. Development 116:177–192

Hennet T, Dinter A, Kuhnert P, Mattu TS, Rudd PM, Berger EG (1998) Genomic cloning and expression of three murine UDP-galactose:β-*N*-acetylglucosamine β1,3-galactosyltransferase genes. J Biol Chem 273:58–65

Holmes EH (1989) Characterization and membrane organization of β1,3- and β1,4-galactosyltransferases from human colonic adenocarcinoma cell lines Colo 205 and SW403: basis for preferential synthesis of type-1 chain lacto-series carbohydrate structures. Arch Biochem Biophys 270:630–646

Isshiki S, Togayachi A, Kudo T, Nishihara S, Watanabe M, Kubota T, Kitajima M, Shiraishi N, Sasaki K, Andoh T, Narimatsu H (1999) Cloning, expression, and characterization of a novel UDP-galactose:β-*N*-acetylglucosamine β1,3-galactosyltransferase (β3Gal-T5) responsible for synthesis of type-1 chain in colorectal and pancreatic epithelia and tumor cells derived therefrom. J Biol Chem 274:12499–12507

Kim YJ, Borsig L, Varki NM, Varki A (1998) P-selectin deficiency attenuates tumor growth and metastasis. Proc Natl Acad Sci USA 95:9325–9330

Kolbinger F, Streiff MB, Katopodis AG (1998) Cloning of a human UDP-galactose:2-acetamido-2-deoxy-D-glucose 3β-galactosyltransferase catalyzing the formation of type-1 chains. J Biol Chem 273:433–440

Magnani JL, Steplewski Z, Koprowski H, Ginsburg V (1983) Identification of the gastrointestinal and pancreatic cancer-associated antigen detected by monoclonal antibody 19-9 in the sera of patients as a mucin. Cancer Res 43:5489–5492

Mendicino J, Sivakami S, Davila M, Chandrasekaran EV (1982) Purification and properties of UDP-Gal:*N*-acetylgalactosaminide mucin: β1,3-galactosyltransferase from swine trachea mucosa. J Biol Chem 257:3987–3994

Miyazaki H, Fukumoto S, Okada M, Hasegawa T, Furukawa K (1997) Expression cloning of rat cDNA encoding UDP-galactose:G$_{D2}$ β1,3-galactosyltransferase that determines the expression of G$_{D1b}$/G$_{M1}$/G$_{A1}$. J Biol Chem 272:24794–24799

Moloney DJ, Shair LH, Lu FM, Xia J, Locke R, Matta KL, Haltiwanger RS (2000) Mammalian Notch1 is modified with two unusual forms of *O*-linked glycosylation found on epidermal growth factor-like modules. J Biol Chem 275:9604–9611

Nakamori S, Kameyama M, Imaoka S, Furukawa H, Ishikawa O, Sasaki Y, Kabuto T, Iwanaga T, Matsushita Y, Irimura T (1993) Increased expression of sialyl Lewis X antigen correlates with poor survival in patients with colorectal carcinoma: clinicopathological and immunohistochemical study. Cancer Res 53:3632–3637

Panin VM, Papayannopoulos V, Wilson R, Irvine KD (1997) Fringe modulates Notch–ligand interactions. Nature 387:908–912

Sasaki K, Sasaki E, Kawashima K, Hanai N, Nishi T, Hasegawa M (1994) Beta-galactosyltransferase DNA and protein—useful for production of saccharide chains. Japanese Patent JP 6181759

Sheares BT, Carlson DM (1983) Characterization of UDP-galactose:2-acetamido-2-deoxy-D-glucose 3β-galactosyltransferase from pig trachea. J Biol Chem 258:9893–9898

Sheares BT, Lau JT, Carlson DM (1982) Biosynthesis of galactosyl-β1,3-N-acetylglucosamine. J Biol Chem 257:599–602

Yuan YP, Schultz J, Mlodzik M, Bork P (1997) Secreted Fringe-like signaling molecules may be glycosyltransferases. Cell 88:9–11

Zamze S, Harvey DJ, Chen YJ, Guile GR, Dwek RA, Wing DR (1998) Sialylated N-glycans in adult rat brain tissue—a widespread distribution of disialylated antennae in complex and hybrid structures. Eur J Biochem 258:243–270

Zhou D, Berger EG, Hennet T (1999a) Molecular cloning of a human UDP-galactose:GlcNAcβ1,3GalNAc β1,3 galactosyltransferase gene encoding an O-linked core3-elongation enzyme. Eur J Biochem 263:571–576

Zhou D, Dinter A, Gutierrez Gallego R, Kamerling JP, Vliegenthart JFG, Berger EG, Hennet T (1999b) A β-1,3-N-acetylglucosaminyltransferase with poly-N-acetyllactosamine synthase activity is structurally related to β-1,3-galactosyltransferases. Proc Natl Acad Sci USA 96:406–411

β3-Galactosyltransferase-IV (GM1 Synthase)

Introduction

β3-Galactosyltransferase-IV (β3GalT-IV) is an enzyme which catalyzes the conversion of GM2, GD2, and asialo-GM2 to GM1, GD1b, and asialo-GM1 (GA1), respectively (Miyazaki et al. 1997). This step is critical for the synthesis of major complex gangliosides such as GM1, GD1a, GD1b, GT1b, and GQ1b, which are enriched in the nervous system of vertebrates. Therefore, all major complex gangliosides are synthesized via the direct products of this enzyme. The cDNAs of β3GalT-IV were isolated by a eukaryocyte expression cloning system in 1997 (Miyazaki et al. 1997). This enzyme utilizes only glycolipid acceptors, not glycoproteins, and no other enzymes (genes) catalyzing similar functions have been detected to date. In the mouse genome (*Mus musculus* major histocompatibility locus class II region), a highly homologous gene as an orthologue to rat β3-galactosyltransferase-IV was reported by Rowen et al. (see AF100956, AF110520).

Databanks

β3-Galactosyltransferase-IV (GM1 synthase)

NC-IUBMB enzyme classification: E.C.2.4.1.62

Species	Gene	Protein	mRNA	Genomic
Homo sapiens	GalT4		Y15061	AL031228
	β3Galt4		AB026730	–
Mus musculus	–		AF082504	AF100956
				AF110520
Rattus norvegicus	–		AB003478	–

KOICHI FURUKAWA

Department of Biochemistry II, Nagoya University School of Medicine, 65 Tsurumai, Showa-ku, Nagoya 466-0065, Japan
Tel. +81-52-744-2070; Fax +81-52-744-2069
e-mail: koichi@med.nagoya-u.ac.jp

Name and History

GM1 synthase, β1,3-galactosyltransferase, β3GalT, β3GalT-IV, Gal-T2.

GM1/GD1b/GA1 Synthase

Since the cDNA cloning of the gene, the identity of GM1 synthase and GD1b synthase (and GA1 synthase) have been confirmed experimentally. Consequently, this enzyme is now called GM1/GD1b/GA1 synthase.

In PC cloning, Amado et al. (1998) defined this gene as β3GalT-IV with no evidence of the expected β3GalT activity onto GlcNAc-R.

Enzyme Activity Assay and Substrate Specificity

Enzyme activity is measured by the incorporation of UDP-[^{14}C]Gal onto acceptors (Miyazaki et al. 1997). The reaction (50 µl) mixture contains 150 mM sodium cacodylate-HCl (pH 7.0), 15 mM MnCl$_2$, 0.375% Triton CF-54, 325 µM GM2 (for GM1 synthesis), 400 µM UDP-Gal, UDP-[^{14}C]Gal (2.0×10^5 d.p.m.), and membranes containing 100 µg protein. The mixture was incubated at 37°C for 2 h. The products were isolated with a C18 Sep-Pak cartridge, and analyzed by thin-layer chromatography and fluorography.

The substrate structures used (measured with extracts from a transfectant line) are: GM2, 100.0%; GM3, 0.0%; GD2, 47.5%; GD1b, 0.0%; GlcCer, 0.0%; GT1b, 0.0%; GA2, 41.1%.

Preparation

Source (biological/commercial): Rat brain tissue

Expression systems: pMIKneo/M2T1-1 expression vector, DEAE-dextran or lipofectin (Miyazaki et al. 1997).

Isolation/purification: The enzyme was biochemically characterized by radioimmunoassay using cholera toxin, and showed the optimum pH 6.5–7.0, the Mn^{2+} requirement, and Km values for UDP-Gal and GM2 were 0.12 mM and 6 µM, respectively (Honke et al. 1986).

Biological Aspects

Gene Promoters

Not defined.

Trafficking

By transfection of GA1/GM1/GD1b synthase cDNA revealed that the enzyme is active with 43 kDa and is Golgi-located. Its *N*-glycan was metabolically labeled from [^{3}H]mannose and was *Endo*-H sensitive. Tsunicamycin treatment or point mutation of the *N*-glycosylation site resulted in the 40-kDa enzyme losing activity, and it was then concentrated in the endoplasmic reticulum. These results indicate that GM1 synthase depends on *N*-glycosylation for its activity and for proper trafficking (Martina et al. 2000).

Distribution in Tissues

In the expression analysis of mouse Gal-T2 (GM1 synthase), the enzyme is localized in Golgi. The highest mRNA expression level was found in the testes (Daniotti et al. 1999). In the postnatal neural retina, Gal-T2 mRNA increased after day 3, maintained high levels of expression during days 4–7, and then decreased to its initial value by day 10.

Normal Function/Substrates

Disease Involvement

Among human tumor cells, the β3GalT-IV gene is expressed at a high level in almost all the cell lines examined. GM1 is one of the major gangliosides in vertebrate nervous tissue. GM1 has been used in various experiments to examine the neurotrophic activity of gangliosides, i.e. in therapeutic trials of experimental Parkinson's disease (Schneider et al. 1992).

Future Perspectives

Complex gangliosides synthesized via the action of β3GalT-IV are important in the functions and maintenance of the nervous system. GM1 has been considered to be a marker of glycolipid-enriched microdomains or rafts. However, it is not known whether GM1 has its own biological role in the microdomains. Molecular mechanisms for the functions of complex gangliosides, including GM1, remain to be investigated.

As described above, the GM1 synthase gene is localized in the mouse "major histocompatibility locus class II region," and in the synthenic region of the human genome (ch. 6p) (Shiina et al. 2000), suggesting that this gene product may be implicated in the immune system.

References

Amado M, Almeida R, Carneiro F, Levery SB, Holmes EH, Nomoto M, Hollingsworth MA, Hassan H, Schwientek T, Nielsen PA, Bennett EP, Clausen H (1998) A family of human β3-galactosyltransferases. Characterization of four members of a UDP-galactose:β-*N*-acetyl-glucosamine/beta-nacetyl-galactosamine β-1,3-galactosyltransferase family. J Biol Chem 273:12770–12778

Daniotti JL, Martina JA, Zurita AR, Maccioni HJ (1999) Mouse β1,3-galactosyltransferase (GA1/GM1/GD1b synthase): protein characterization, tissue expression, and developmental regulation in neural retina. J Neurosci Res 58:318–327

Honke K, Taniguchi N, Makita A (1986) A radioimmune assay of ganglioside GM1 synthase using cholera toxin. Anal Biochem 155:395–399

Martina JA, Daniotti JL, Maccioni HJ (2000) GM1 synthase depends on N-glycosylation for enzyme activity and trafficking to the Golgi complex. Neurochem Res 25:725–731

Miyazaki H, Fukumoto S, Okada M, Hasegawa T, Furukawa K, Furukawa K (1997) Expression cloning of rat cDNA encoding UDP-galactose: GD2 β1,3-galactosyltransferase that determines the expression of GD1b/GM1/GA1. J Biol Chem 272:24794–24799

Schneider JS, Pope A, Simpson K, Taggart J, Smith MG, DiStefano L (1992) Recovery from experimental parkinsonism in primates with GM1 ganglioside treatment. Science 256:843–846

Shiina T, Kikkawa E, Iwasaki H, Kaneko M, Narimatsu H, Sasaki K, Bahram S, Inoko H (2000) The β-1,3-galactosyltransferase-4 (B3GALT4) gene is located in the centromeric segment of the human MHC class II region. Immunogenetics 51:75–78

β3-Galactosyltransferase-V

Introduction

β3-Galactosyltransferase (β3GalT) transfers a galactose from UDP-Gal to *N*-acetylglucosamine (GlcNAc) with a β1,3-linkage. To date, five members of the β3GalT family have been cloned and analyzed (Issiki et al. 1999). β3GalT-I was first cloned by an expression cloning method (Sasaki et al. 1994). The other three members, β3GalT-II–IV, which have homologous sequences to β3GalT-I were found in the EST database and cloned (Issiki et al. 1999; Kolbinger et al. 1998; Amado et al. 1998; Hennet et al. 1998).

We had been looking for the β3GalT responsible for the synthesis of sLe[a], which is a famous tumor marker known as CA19-9 (Magnani et al. 1983). The antigenic epitope of CA19-9 is defined as the sLe[a] structure, of which the biosynthetic pathways are shown in Fig. 1 of Chap. 28.

At least three glycosyltransferases are required for the synthesis of the sLe[a] epitope. First, β3GalT transfers Gal to GlcNAc with a β1,3-linkage, resulting in the synthesis of a type-1 chain, Galβ1-3GlcNAc, and then galactose-α3-sialyltransferase (ST3Gal) transfers a sialic acid (SA) to the Gal residue of the type-1 chain with an α2,3-linkage, resulting in a sialyl type-1 (sialyl Lewis c; sLe[c]) chain, SAα2-3Galβ1-3GlcNAc, synthesis. Finally, α3/4-fucosyltransferase (Fuc-TIII, FUT3, Lewis enzyme) transfers a fucose (Fuc) to the GlcNAc residue of the sialyl type-1 chain with an α1,4-linkage to complete the synthesis of the sLe[a] structure, SAα2-3Galβ1-3(Fucα1-4)GlcNAc. Fuc-TIII (FUT3) is the only enzyme determining the expression of sLe[a] antigen in colorectal cancer (Narimatsu et al. 1996, 1998).

The CA19-9 value in serum is frequently elevated in cancer patients, in particular in pancreatic, colorectal, and gastric cancer patients, and frequently used for the clinical diagnosis of those cancers. As well as its usefulness as a tumor marker, sLe[a] antigen is known to be a ligand for selectins (Takada et al. 1993). Clinical statistical

Hisashi Narimatsu

Laboratory of Gene Function Analysis, Institute of Molecular and Cell Biology (IMCB), National Institute of Advanced Industrial Science and Technology (AIST), Central-2, 1-1-1 Umezono, Tsukuba, Ibaraki 305-8568, Japan
Tel. +81-298-61-3200; Fax +81-298-61-3201
e-mail: h.narimatsu@aist.go.jp

analysis demonstrated that colorectal cancer patients who express abundant sLe^a antigens have a worse prognosis for liver metastasis than patients who do not express sLe^a antigens (Nakayama et al. 1995). Thus, it is of interest that sLe^a antigens may confer some metastatic capacity on cancer cells.

None of the four β3GalTs, β3GalT-I–IV were determined to be responsible for the expression of CA19-9 antigen in cancer cells (Issiki et al. 1999). Using a degenerate primer strategy based on the amino acid motifs conserved in the four β3GalTs, a new β3GalT, named β3GalT-V, was cloned from a Colo205 cDNA library (Issiki et al. 1999). The members of the β3GalT family share amino acid motifs in three positions, as seen in Fig. 1.

Databanks

β3-Galactosyltransferase-V

No EC number has been allocated

Species	Gene	Protein	mRNA	Genomic
Homo sapiens	*β3Gal-T5*	NP_149361	AB020337	AF064860
Pan troglodytes	*β3Gal-T5*	BAA94499	–	AB041414
Pan paniscus	*β3Gal-T5*	BAA94500	–	AB041415
Gorilla gorilla	*β3Gal-T5*	BAA94495	AB041410	–
Mus musculus	*β3Gal-TV*	AAF86241	AF254738	–

Name and History

β3-Galactosyltransferase V is abbreviated to β3GalT-V. Other synonyms for β3GalT-V have not been used. A cDNA encoding a human β3GalT-V was isolated from Colo205 cells (Issiki et al. 1999).

The CA19-9 epitope is usually carried on mucins produced by cancer cells. Holmes reported that partially purified β3GalT(s) from Colo205 cells exhibit preferential activity towards lactotriaosylceramide (Lc$_3$), GlcNAcβ1-3Gal β1-4Glcβ1-1Cer (Holmes 1989). The expression levels of type-1 Lewis antigens, Le^a, Le^b, and sLe^a antigens, were reported to correlate well with the β3GalT activity detected in the homogenates of human colorectal cancer cell lines (Valli et al. 1998). β3GalT activity was detected in some cancer cell lines, Capan-2, Colo201, Colo205, SW1116, etc., derived from pancreatic and colorectal cancers (Issiki et al. 1999).

The expression levels of the other four β3GalTs, β3GalT-I–IV were not found to be correlated with type-1 Lewis antigen expression in the cell lines derived from gastrointestinal and pancreatic tissues (Issiki et al. 1999). The expression level of β3GalT-V correlated well with Le^a, Le^b, and sLe^a antigen expression in a series of cultured cells, indicating that β3GalT-V is responsible for the synthesis of CA19-9 antigen (Fig. 2).

Enzyme Activity Assay and Substrate Specificity

The following reaction can be catalyzed by β3GalT-V:

$$\text{UDP-Gal} + \text{GlcNAc}\beta1\text{-R} \rightarrow \text{Gal}\beta1\text{-3GlcNAc}\beta\text{-R} + \text{UDP}$$

```
hβ3Gal-T1  ------------------------MASKVC̲L̲Y̲L̲V̲L̲T̲V̲V̲C̲W̲A̲S̲A̲L̲W̲Y̲L̲S TRPTSSYTGSK------P----FSHLTVARKN
hβ3Gal-T2  MLQWRRHCCFAKMTWNAKRSLFRHLIGVLSLVFLFAMFLFFNHHDWLPGRAGFKENPVTYTFPGFRSTKSETNHSSLRNIWKETVPQTLRPQTATNSNNTDLSPQGVTGLENTLSANG
hβ3Gal-T3  ------------MASALWTVLPSRMSLRSLKWSLLLLSLLSFFVWMYLSLPHYNVIERVNW------------------MYFEYE
hβ3Gal-T4  ---------------MQLRLFRRLLLAALLLVIVWTLFGFSGLGEELLSLS------------LASLLPAP
hβ3Gal-T5  ---------------MAFPKMRLKYICLLVLGALCLYFSMYSHNPFK------------------EQS

                                                                              *   *  *** **

                                    motif 1

hβ3Gal-T1  FTFGNIRTRPINPHSFEFLINEPNKCHK--NIPFLVILISTTHKEFDARQAIRETWGDENNFKGIKIATLFLLGKNADP----VLNQMVEQESQIFHDIIVEDFIDSYHNITLKTLMGM
hβ3Gal-T2  SIYNEKGTGHPNSYHFKYIINEPEKCQE--KSPFLILLIAAEPGQIEARRAIRQTWGNESLAPGIQTRIFLLGLSIKLN--G--YIQRAILEESRQYHDIIQQEYLDTYYNLTLKTLMGM
hβ3Gal-T3  PIYRQDFH----FTLREHSNCSH--QNPFLVILVTSHPSDVKARQAIRVTWGEKKSWWGYEVLTFFLLGQEAEKE---DKMLALSLEDEHLLYGDIIRQDFLDTYNNLTLKTIMAF
hβ3Gal-T4  ASPGPPLALP--R---LLIPNQPACSGPAPPLLLIVCTAPENLNQRNAIRASWGGLREARGLRVQTLFLLGEPNAQHFVWGSQGSDLASESAAQGDILQAAFQDSYRNLTLKTLSGL
hβ3Gal-T5  FVYKKDGN------FLKLPDTDCDQ--TPPFLVILVTSHKQLAERMAIRQTWGKERMVGKQLKTFFLLGTSSA---AETKEVDQESQRHGDIIQKDFLDVYYNLTLKTMMGI

              *        ***  ****  **          * ****  **          **      *    **  **  * **** **

                  motif 2

hβ3Gal-T1  RWVATFCSKAKYVMKTDSDIFVNMDNLLYKLLK---------PSTKPRR--------RYFTGVIING-GPIRDVRSKWYMPRDLYP--DSNYPPFCSGTGYIFSADVAELIY
hβ3Gal-T2  NWVATYCPHIPYVMKTDSDMFVNTEYLINKLLK---------PDLPPRH--------NYFTGYLMRGYAPNRNKDSKWYMPDLYP--SERYPVFCSGTGYVFSGDLAEKIF
hβ3Gal-T3  RWVTBFCPNAKYVMKTDTDVFINTGNLVKYLLN---------LNHSEK--------FFTGYPLIDNYSYRGFYQKTHISYQEYP--FKVFPPYCSGLGYIMSRDLVPRIY
hβ3Gal-T4  NWAEKHCPMARYVLKTDDDVYVNPELVSELVLRGGRWGQWERSTEPQREAEQEGGQVLHSEVPLLYLGRVHWRVNPSRTPGGRHRVSEEQWPHTWGPFPPYASGTGYVLSASAVQLIL
hβ3Gal-T5  EWVHRFCPQAAFVMKTDSDMFINVDYLTELLLK---------KNRLTR--------FFTGFLKLNEFPIRQPFSKWFVSKSEYP--WDRYPPFCSGTGYVFSGDVASQVY

              *   ** **  * **** **                           *               *      ** ** *

                   motif 3

hβ3Gal-T1  KTSLHTRLLHLEDVYVGLCLRKLGIHFPQNSG---FNHWKMAYSLCRYRRVITVHQISPEEMHRIWNDMSSKKH-----LRC--------
hβ3Gal-T2  KVSLGIRRLHLEDVYVGICLAKLRIDEVPPPNEFVFNHWRVSYSSCKYSHLITSHQFQPSELIKYWNHLQQNKH----NACANAAKEKAGRYRHRKLH---
hβ3Gal-T3  EMMGHVKPIKFEDVYVGICLNLLKVNIHIPEDTNLFFLYRIHLDVQLRRVIAAHGFSSKEIITFWQVMLRN-------TTCHY---
hβ3Gal-T4  KVASRAPLLPLEDVFVGVSARRGGLAPTQCVRKLAGATHYPLDR--CCYGKFLLTSHRLDFWKMQEAWKLVGGSDGERTAPFCSWFQGVLGILRCRAIAWLQS
hβ3Gal-T5  NVSKSVPYIKLEDVTFVGLCLERLNIRLEELHSQPTFPGGLRFSVCLFRRIVACHFIKPRTLLDYWQALENSRG---EDCPFV---

              *** **                                                         *
```

Fig. 1. Multiple sequence alignment (ClustalW) of the five human β3GalTs. Multiple amino acid sequences of the five human β3GalTs are shown. Introduced gaps are shown as *hyphens*. Three conserved motifs used for the design of degenerate primers are *squared*. Putative transmembrane domains are *shaded*, the four conserved cysteine residues are *boxed*, and the conserved possible *N*-glycosylation site is *double-underlined*. *Asterisks* indicate the amino acids conserved in the five β3GalTs

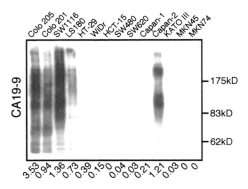

Fig. 2. Western blot analysis of various cancer cells with 19-9 (anti-sLe^a) and quantitative analysis of β3GalT-V transcripts by competitive RT-PCR. The results of Western blot analysis are shown in the actual gel, and the relative amounts of β3GalT-V transcripts (β3GalT-V/β-actin × 10^3) in various cancer cells are presented as numbers below the gel

or

$$\text{UDP-Gal} + \text{GalNAc}\beta\text{1-R} \rightarrow \text{Gal}\beta\text{1-3GalNAc}\beta\text{-R} + \text{UDP}$$

The enzyme assay and acceptor specificities of the five β3GalTs, β3GalT-I–V, were performed using cell homogenates of Namalwa cells transfected stably with each β3GalT cDNA. The full-length cDNA of human β3GalTs was inserted in a mammalian expression vector, pAMo, and stably expressed in Namalwa cells. The cells were solubilized in 20 mM HEPES buffer (pH 7.2) containing 2% Triton X-100. Lacto-*N*-neotetraose (LNnT) was pyridylaminated and digested with 20 milliunits/ml streptococcal β-galactosidase to remove the galactose residue at the nonreducing end. Thus, agalacto-LNnT-PA was prepared for the acceptor substrate for β3GalT. The β3GalT activity was assayed in 14 mM HEPES buffer (pH 7.4), 75 μM UDP-Gal, 11 μM MnCl$_2$, 0.01% Triton X-100, and 25 μM acceptor substrate. After incubation at 37°C for 2 h, the enzyme reactions were terminated by boiling for 3 min, followed by dilution with water. After centrifugation of the reaction mixtures at 15 000 r.p.m. for 5 min, 10 μl of each supernatant was subjected to HPLC analysis on a TSK-gel ODS-80Ts column (4.6 × 300 mm). The reaction products were eluted with 20 mM ammonium acetate buffer (pH 4.0) at a flow rate of 1.0 ml/min at 25°C and monitored with a Jasco FP-920 fluorescence spectrophotometer (Tokyo, Japan).

β3GalT-V exhibited the strongest β3GalT activity toward agalacto-LNnT-PA among the five β3GalTs (Issiki et al. 1999). Zhou et al. (1999) reported that β3GalT-V exhibits a marked preference for the O-linked core 3-GlcNAcβ1-3GalNAc substrate, and it was recently found that β3GalT-V efficiently transfers Gal to the GlcNAc of the glycolipid Lc3Cer and to the terminal GalNAc residue of the globoside Gb4, thereby synthesizing the glycolipid Gb5 that is known as the stage-specific embryonic antigen-3 (SSEA-3) (Zhou et al. 2000).

Preparation

Recombinant β3GalT-V is available. It is expressed with mammalian expression vectors (Isshiki et al. 1999) and with a baculovirus expression system (Zhou et al. 1999, 2000). Some human pancreatic and colorectal cell lines, Colo201, Colo205, SW1116, and Capan-2, express substantial amounts of β3GalT-V (Issiki et al. 1999) (Fig. 2).

Biological Aspects

In human tissue, the β3GalT-V expression is restricted in the colorectum, small intestine, stomach, and pancreas, which are known to express the CA19-9 antigen frequently when they become cancerous (Isshiki et al. 1999) (Table 1). Transfection experiments with the *β3Gal-T5* gene showed that β3GalT-V caused the cells to express type-1 Lewis antigens, such as Le^a, Le^b, and sLe^a, on the cell surface. Interestingly, it was observed that the expression of type-2 Lewis antigens, Le^x, L^y, and sLe^x, in those cells markedly decreased, in contrast to the increase of type-1 Lewis antigens (unpublished results, 1999). Although the biological function of the type-1 chain is unclear, the cells transfected with the *β3Gal-T5* gene will be a useful tool in analyzing the different functions of type-1 and type-2 chains.

As reported previously (Zhou et al. 1999, 2000; Isshiki et al. 1999), β3GalT-V exhibits the strongest activity and a wide specificity for acceptor substrates compared with the other β3GalTs, and is expressed in a tissue-specific manner. This indicates that β3GalT-V may be involved in some intestinal functions. The knockout mouse experiment will clarify this question.

The genomic structure of the human *β3Gal-T5* gene was easily determined by comparing the full-length cDNA sequence with its genome sequence, which has been registered in the Genome Project Database (registration no. AF064860). The *β3Gal-T5* gene is localized to human chromosome 21q22.3 and consists of four exons. The ORF of the *β3Gal-T5* gene was found to be encoded by a single exon (exon 4), as in the cases of the other four *β3Gal-T* genes (Isshiki et al. 1999). Three exons encoding the 5′ untranslated region are alternatively spliced to give rise to the transcript isoforms (Isshiki et al. 1999). The nucleotide sequence in the upstream region of the *β3Gal-T5* gene is available in the Genome Project Database (no. AF064860), but the promoter analysis of this gene has not been reported.

Table 1. Tissue distribution of β3GalT-V

Tissue	Relative amount of β3GalT-V transcript (β3GalT-V/β-actin × 10^3)
Brain	0.12
Lung	0.00
Esophagus	0.07
Stomach (body)	0.47
Stomach (antrum)	0.73
Jejunum	0.47
Colon	0.83
Liver	0.00
Pancreas	0.56
Spleen	0.00
Kidney	0.05
Adrenal gland	0.00
Uterus	0.06
Peripheral blood	0.00

Future Perspectives

The biological function of the type-1 chain is unclear. However, it is noteworthy that the type-1 chain is expressed in a tissue-specific manner, in contrast to the ubiquitous expression of the type-2 chain.

The following points will be the focus of future studies of β3GalTs.

1. The precise substrate specificity of the five or more β3GalTs should be determined using as many acceptor substrates as possible.
2. The mechanism of up-regulation of CA19-9 in cancer patients will be clarified on the molecular basis of β3GalT-V.
3. The transcriptional regulation of the *β3Gal-T5* gene is of interest in relation to the oncology and development of digestive tract tissues.
4. A structural analysis of β3GalTs by crystallography will differentiate the substrate specificity of β3GalT from that of β4GalT.
5. Knockout mouse analysis of β3GalT will tell us the biological function of β3GalT-V.

References

Amado K, Almeida R, Carneiro F, Leverly SB, Holmes EH, Nomoto M, Hollingsworth MA, Hassan H, Schwientek T, Nielsen PA, Bennett EP, Clausen H (1998) A family of human β3-galactosyltransferases. Characterization of four members of a UDP-galactose: β-N-acetylglucosamine, β-N-acetylgalactosamine, β-1,3-galactosyltransferase family. J Biol Chem 278:12770–12778

Hennet T, Dinter A, Kuhnert P, Mattu TS, Rudd MP, Berger EG (1998) Genomic cloning and expression of three murine UDP-galactose:β-N-acetylglucosamine β1,3-galactosyltransferase genes. J Biol Chem 273:58–65

Holmes EH (1989) Preparative in vitro generation of lacto-series type-1 chain glycolipids catalyzed by β-1-3-galactosyltransferase from human colonic adenocarcinoma Colo 205 cells. Arch Biochem Biophys 270:630–646

Issiki S, Togayachi A, Kudo T, Nishihara S, Watanabe M, Kubota T, Kitajima M, Shiraishi N, Sasaki K, Andoh T, Narimatsu H (1999) Cloning, expression, and characterization of a novel UDP-galactose:β-N-acetylglucosamine, β1,3-galactosyltransferase (β3GalT-V) responsible for synthesis of type-1 chain in colorectal and pancreatic epithelia and tumor cells derived therefrom. J Biol Chem 274:12499–12507

Kolbinger F, Streiff MB, Katopodis AG (1998) Cloning of a human UDP-galactose:2-acetamido-2-deoxy-D-glucose 3β-galactosyltransferase catalyzing the formation of type-1 chains. J Biol Chem 273:433–440

Magnani JL, Steplewski Z, Koprowski H, Ginsburg V (1983) Identification of the gastro-intestinal and pancreatic cancer-associated antigen detected by monoclonal antibody 19-9 in the sera of patients as a mucin. Cancer Res 43:5489–5492

Nakayama T, Watanabe M, Katsumata T, Teramoto T, Kitajima M (1995) Expression of sialyl Lewis(a) as a new prognostic factor for patients with advanced colorectal carcinoma. Cancer 75:2051–2056

Narimatsu H, Iwasaki H, Nishihara S, Kimura H, Kudo T, Yamauchi Y, Hirohashi S (1996) Genetic evidence for the Lewis enzyme, which synthesizes type-1 Lewis antigens in colon tissue, and intracellular localization of the enzyme. Cancer Res 56:330–338

Narimatsu H, Iwasaki H, Nakayama F, Ikehara Y, KudoT, Nishihara S, Sugano K, Okura H, Fujita S, Hirohashi S (1998) Lewis and secretor gene dosages affect CA19-9 and DU-

PAN-2 serum levels in normal individuals and colorectal cancer patients. Cancer Res 58:512–518

Sasaki K, Sasaki E, Kawashima K, Hanai N, Nishi T, Hasegawa M (1994) JP0618759 A 940705

Takada A, Ohmori K, Yoneda T, Tsuyuoka K, Hasegawa A, Kiso M, Kannagi R (1993) Contribution of carbohydrate antigens sialyl Lewis A and sialyl Lewis X to adhesion of human cancer cells to vascular endothelium. Cancer Res 53:354–361

Valli M, Gallanti A, Bozzaro S, Trinchera M (1998) β-1,3-galactosyltransferase and α-1,2-fucosyltransferase involved in the biosynthesis of type-1-chain carbohydrate antigens in human colon adenocarcinoma cell lines. Eur J Biochem 256:494–501

Zhou D, Berger EG, Hennet T (1999) Molecular cloning of a human UDP-galactose: GlcNAcβ1,3GalNAcβ1,3-galactosyltransferase gene encoding an O-linked core-3-elongation enzyme. Eur J Biochem 263:571–576

Zhou D, Henion T, Jungalwala FB, Berger EG, Hennet T (2000) The β1,3-galactosyltransferase β3GalT-V is a stage-specific embryonic antigen-3 (SSEA-3) synthase. J Biol Chem 275:22631–22634

7

α3-Galactosyltransferase

Introduction

UDP-Gal:Galβ1→4GlcNAcβ-R α3-galactosyltransferase (α3GalT) is a glycosyltransferase (GT) that acts late in the process of protein and lipid glycosylation. By the action of the enzyme, a Gal residue is attached to exposed Galβ1→4GlcNAc (lacNAc) termini yielding a Galα1→3Galβ1→4GlcNAc product structure that cannot be further elongated. The formation of this uncharged sequence provides an alternative for the common chain termination by sialic acid. Although no sugars can be added to the α-Gal residue, the chain underlying the α3-galactosylated terminus may be further modified by the addition of Fuc in α1,3-linkage to certain GlcNAc residues, and by the addition of GlcNAc in a β1,6-linkage to specific Gal residues (reviewed in Van den Eijnden 2000). Molecular cloning of α3GalT (Joziasse et al. 1989) has revealed that this enzyme is a type-II membrane protein showing a domain structure that is typical for Golgi-resident GTs. It shows extensive sequence similarity to three other GTs (blood group A and B transferase; Forssman glycolipid synthase), and can be grouped with these enzymes in one GT family. The enzyme is expressed in most mammals, with the notable exception of man, apes, and Old World monkeys.

Dirk H. Van den Eijnden and David H. Joziasse

Department of Medical Chemistry, Vrije Universiteit, Van der Boechorststraat 7, 1081 BT Amsterdam, The Netherlands
Tel. +31-20-444-8151; Fax +31-20-444-8144
e-mail: dh.van_den_eijnden.medchem@med.vu.nl

Databanks

α3-Galactosyltransferase

NC-IUBMB enzyme classification: E.C.2.4.1.124 and 2.4.1.151

Species	Gene	Protein	mRNA	Genomic
Homo sapiens	HGT2	–	–	M60263
				J05421
	GGTA1P	–	–	M60263
Pan paniscus	–	–	–	M72426
Gorilla gorilla	–	–	–	M73304
Pongo pygmaeus	–	–	–	M73305
Macaca mulatta	–	–	–	M73306
Cercopithecus aethiops	–	–	–	M73307
Erythorcebus patas	–	–	–	M73308
Ateles geoffroyi	–	–	–	M73309
Saimiri sciureus	–	–	–	M73310
Alouatta caraya	–	–	–	M73311
Mus musculus	Ggta-1	–	M26925	–
			M85153	–
Bos taurus	–	–	J04989	–
Sus scrofa	GGTA1	–	L36152	–
			L36535	–

Name and History

α3GalT activity was first detected in mouse Ehrlich ascites tumor cells (Blake and Goldstein 1981). Somewhat later, α3GalT was found in calf thymus (Van den Eijnden et al. 1983) in a search for the GT responsible for the synthesis of a terminal glycan structure reported to occur on calf thymus glycoproteins. This structure had first been believed to be Galβ1→3Galβ1→4GlcNAc, but this was later corrected in view of the enzyme's specificity. The enzyme was also described in rabbit stomach mucosa, where it could be distinguished from a human blood group B like α3GalT (Betteridge and Watkins 1983). The cDNA encoding α3GalT was first cloned from calf thymus, the enzyme being the third mammalian GT ever cloned (Joziasse et al. 1989). It has to be distinguished from the blood group B transferase, another α3-galactosyltransferase which, although related in the primary sequence, differs in acceptor substrate specificity.

Enzyme Activity Assay and Substrate Specificity

α3GalT catalyzes the reaction

$$\text{UDP-Gal} + \text{Gal}\beta\text{1-4GlcNAc}\beta\text{-R} \rightarrow \text{Gal}\alpha\text{1-3Gal}\beta\text{1-4GlcNAc}\beta\text{-R} + \text{UDP}$$

where R may be OH, oligosaccharide, oligosaccharide-(poly)peptide, oligosaccharide lipid, or an aromatic aglycon. The enzyme has an absolute requirement for Mn^{2+} ions, and shows a broad pH optimum (5.5–7.5) with a maximum at pH 6.0. UDP and UMP are inhibitory.

In a volume of 50 μl, a typical incubation mixture for assaying α3GalT activity contains: 5.0 μmol sodium cacodylate buffer, pH 6.0; 50 nmol Galβ1→4GlcNAc or 1.5 μmol Galβ1→4Glc (lactose) as acceptor substrate; 25 nmol UDP-[^{3}H] or [^{14}C]Gal (specific radioactivity ≈1 Ci/mol) as donor substrate; 2.5 μmol MnCl$_2$ (activator); 0.25 mmol ATP (to counteract enzymatic breakdown of UDP-Gal); 0.4 μl Triton X-100 (to solubilize membrane-bound enzyme); 50 μg bovine serum albumin (to stabilize the enzyme); an enzyme preparation (0.01–0.5 mU). The mixture is incubated for 10–120 min at 37°C. After incubation, the reaction is stopped by adding 0.5 ml ice-cold water and cooling the incubation vial on ice. The amount of radioactive Gal incorporated into the acceptor is estimated by passing the diluted mixture over small columns containing 0.5–1 ml Dowex 1-X8 (Cl⁻ form). The eluate and three washes of 0.5 ml water each are combined, and the radioactivity this contains is counted by liquid scintillation. A correction is made for incorporation into endogenous acceptors and for hydrolysis of UDP-Gal by conducting incubations lacking the exogenously added acceptor. The amounts of radioactivity obtained in these controls is subtracted from those obtained with the complete incubation mixtures, before the amount of Gal incorporated is calculated.

Alternatively, glycoproteins such as asialo-α$_1$-acid glycoprotein, glycolipids (lacto-series sphingoglycolipids) such as paragloboside, or oligosaccharides linked to an aromatic aglycon (e.g., lactose-paranitrophenol) may be used as the acceptor substrate. In these instances, the excess radioactivity of the donor substrate is removed from the incorporated radioactivity by precipitation of the acceptor with 0.5 M HCl containing 1% phosphotungstic acid (glycoproteins), or by passing the mixture over a Sep-Pac C-18 cartridge and eluting the product with methanol (glycolipids and acceptors with hydrophobic aglycons).

α3GalT preferentially acts on acceptor substrates carrying a nonreducing terminal Gal residue in β1,4-linkage to GlcNAc or Glc, and tolerates protein as well as lipid aglycons (Table 1). It shares its oligosaccharide specificity with α6NeuAcT, α2FucT, and α3FucT, which can also act on lacNAc. With these enzymes, α3GalT acts in a mutually exclusive way as it acts sluggishly on α6-sialylated, and α2- and α3-fucosylated structures. In addition to the OH group at C-3 of the Gal in Galβ1-4GlcNAc, which is the site to which α-Gal is introduced, the OH group at C-4 of the Gal is one that is required for activity. Deoxygenation of this OH yields a

Table 1. Acceptor specificity of bovine α3GalT

Acceptor	Relative activity (%)	K_m (mM)
Galβ1→4GlcNAc (lacNAc)	100	1.39
Galβ1→4[Fucα1→3]GlcNAcβ1→2Man (Lewis^x)	<0.3	nd[a]
Fucα1→2Galβ1→4GlcNAc (H-determinant)	2.8	nd
Galβ1→4Glc (lactose)	20.3	9.0
NeuAcα2→6Galβ1→4Glc	<0.3	nd
Galβ1→3GlcNAcβ1→3Galβ1→4Glc (LNT)	15.0	16.0
Galβ1→4GlcNAcβ1→2Manα1→3Manβ1→4GlcNAc	165	0.57
Asialo-α$_1$-acid glycoprotein	77.0	1.25
Galβ1→4Glcβ1-Cer (lactosylceramide)	<1.5	nd
Galβ1→4GlcNAcβ1→3Galβ1→4Glcβ1-Cer	88.8	nd

[a] nd, not determined

inactive acceptor. OH groups that may be deoxygenated but may not be substituted without loss of activity are the ones at C-2 and C-6 of Gal and C-3 of GlcNAc (Sujino et al. 1997). Although α6NeuAcT and α3GalT potentially compete for common lacNAc acceptor sites on protein-linked glycans, they may in fact cooperate in the case of branched glycans because they prefer different branches on such substrates (α6NeuAcT: Manα1-3Man branch; α3GalT: Manα1-6Man branch) (Van den Eijnden 2000).

Preparation

α3GalT can be obtained from calf thymus in a highly active form. After solubilization by Triton X-100, it is purified by repeated affinity chromatography on a column of UDP-hexanolamine-Sepharose (Blanken and Van den Eijnden 1985). Separation from β4GalT can be achieved by chromatography on a column of α-lactalbumin to which α3GalT does not bind. Recombinant bovine enzyme can be prepared by expression in insect cells using a recombinant baculovirus (Joziasse et al. 1990). The introduction of an insect signal peptide in the expression construct permits secretion of enzymatically active, soluble α3GalT into the culture medium, allowing a one-step purification to homogeneity. This recombinant enzyme form has been used in an efficient one-pot synthesis of the Galα1-3Galβ1-4GlcNAc sequence (Hokke et al. 1996). Unglycosylated, enzymatically active recombinant α3GalT may also be prepared by expression in *E. coli* (Fang et al. 1998). α3GalT obtained this way has been used succesfully to crystallize the enzyme (Gastinel et al. 2000).

Biological Aspects

Expression of α3GalT is species-specific, and inducible upon cellular differentiation and activation. Expression seems to be restricted mainly to mammals (with the exception of man, apes, and Old World monkeys) (Galili et al. 1987, 1988), although related enzymes may be active in certain species of fishes, reptiles, and amphibians. A cDNA was first isolated from bovine thymus (Joziasse et al. 1989). The human genome contains two α3GalT homologues, both pseudogenes (Joziasse et al. 1991; Larsen et al. 1990). One gene may be the once-functional orthologue of the α3GalT gene present in other mammals, while the second gene has the characteristics of a processed pseudogene.

In the mouse, α3GalT is expressed in most tissues and organs, with the exception of liver (Peters and Goldstein 1979; Ikematsu et al. 1993), and male germ cells (Johnston et al. 1995). Gene expression is up-regulated in thioglycollate-elicited, exudate peritoneal macrophages of a mouse or rat. As a result, these macrophages show an increased α3-galactosylation of cell surface glycoconjugates when compared with resident peritoneal macrophages (Mercurio and Robbins 1985; Sato and Hughes 1994; Sheares and Mercurio 1987), which may create binding sites for cell-surface lectin Mac-2 (galectin-3). Retinoic acid treatment of murine F9 embryonal carcinoma cells, which induces their differentiation into parietal endoderm, results in an induction of α3GalT expression via enhanced transcription, and in an increase in α-galactosylation (Cummings and Mattox 1988; Cho et al. 1996).

Murine α3GalT is encoded by a multiexon, single-copy gene which spans at least 80 kb (Joziasse et al. 1992). Exons 1–3 encode the 5′-untranslated sequence, whereas the protein coding sequence is distributed over exons 4–9. The start codon is located in exon 4, whereas exons 5–7 encode the "stem region." The largest exon, exon 9, contains almost the entire catalytic domain, and in addition encodes ≈1.9 kb of the 3′-untranslated sequence. Recently, the porcine gene was found to have a similar organization. The murine α3GalT mRNA is alternatively spliced in the region that encodes the stem region (Joziasse et al. 1992), which predicts the production of protein isoforms differing in the length of the stem. Similar splicing patterns have been detected in the porcine α3GalT (Vanhove et al. 1997). The stem region may play a role in determining the half-life of the enzyme within the cell, since it has been reported that α3GalT is continuously secreted from murine, canine, and bovine tissue culture cells into the growth medium as a result of proteolysis occurring at specific sites in the stem region (Cho et al. 1997). It is conceivable that alternative splicing may render α3GalT more or less sensitive to proteolysis, and affect targeting to the Golgi apparatus or to the secretory pathway.

Transgenic mice in which the α3-galactosyltransferase gene has been inactivated by homologous recombination appear to develop normally (Tearle et al. 1996; Thall et al. 1995), although they may develop eye cataracts. It has been suggested previously that terminal α1,3-linked galactose on the surface of the mouse egg zona pellucida is essential for initial sperm binding during fertilization (Bleil and Wassarman 1988), but Thall et al. (1995) showed that α3GalT knockout mice, which are unable to produce α3-galactosylated glycans, are still fertile. More recently, it was found that rather than a terminal α3-linked galactose, α3-linked fucose is essential to form a high-affinity sperm binding ligand (Johnston et al. 1998). Studies on xenotransplantation have shown that the Galα1→3Gal epitope, present on porcine tissues, is the major target for natural antibodies that cause the hyperacute rejection of pig organs transplanted to primates (reviewed in Joziasse and Oriol 1999). The same antibody was suggested to protect humans against pathogenic microorganisms and the interspecies transmission of retroviruses (reviewed by Rother and Squinto 1996).

Future Perspectives

Inactivation of the α3GalT gene in pigs by knockout or nuclear transfer technology will help to overcome the hyperacute rejection that occurs upon organ transplantation from pig to man, and in this way contribute to the success of xenotransplantation. The resolution of the crystal structure of α3GalT (Gastinel et al. 2000) may provide more insight into its catalytic mechanism, and forms the basis for the design of a transferase-specific enzyme inhibitor.

Further Reading

Galili et al. 1987. Report on the reciprocal relationship between natural anti-Galα3Gal antibody and the presence of the Galα1→3Gal epitope in mammals.

Galili et al. 1988. The authors show that the evolutionary pattern of expression of Galα1→3Gal epitopes on mammalian cells, observed using lectin or antibody staining, is consistent with enzyme assays showing that α3GalT is present in nonprimate mammals and prosimians, but not in Old World primates.

Joziasse et al. 1989. First report on the primary sequence of α3GalT. Homologous sequences are present in the human genome, but are not transcriptionally expressed.

Joziasse et al. 1991, and Larsen et al. 1990. Taken together, these papers show that the
human genome contains two α3GalT genes, but that both genes are pseudogenes.

Rother and Squinto 1996. Review of the importance of the natural anti-Galα1,3Gal antibody as a protection against parasites, and against the interspecies transmission of
retroviruses.

Thall et al. 1995. The authors describe the production of an α3GalT knockout mouse. The
report shows that the gene can be inactivated without affecting fertility. Moreover,
mice develop normally, and exhibit no gross phenotypic abnormalities.

References

Betteridge A, Watkins WM (1983) Two α-3-D-galactosyltransferases in rabbit stomach
mucosa with different acceptor substrate specificities. Eur J Biochem 132:29–35

Blake DA, Goldstein IJ (1981) An α-D-galactosyltransferase activity in Ehrlich ascites
tumor cells. Biosynthesis and characterization of a trisaccharide α-D-galactose1→
3-N-acetyllactosamine. J Biol Chem 256:5387–5393

Blanken WM, Van den Eijnden DH (1985) Biosynthesis of terminal Galα1→3Galβ1→
4GlcNAc-R oligosaccharide sequences on glycoconjugates. Purification and acceptor
specificity of a UDP-Gal:N-acetyllactosaminide α1→3-galactosyltransferase from calf
thymus. J Biol Chem 260:12927–12934

Bleil JD, Wassarman PM (1988) Galactose at the nonreducing terminus of O-linked
oligosaccharides of mouse egg zona pellucida glycoprotein ZP3 is essential for the
glycoprotein's sperm receptor activity. Proc Natl Acad Sci USA 85:6778–6782

Cho SK, Yeh J, Cho M, Cummings RD (1996) Transcriptional regulation of α1,3-galacto-
syltransferase in embryonal carcinoma cells by retinoic acid. Masking of Lewis[x] antigens by α-galactosylation. J Biol Chem 271:3238–3246

Cho SK, Yeh J, Cummings RD (1997) Secretion of α1,3-galactosyltransferase by cultured
cells and presence of enzyme in animal sera. Glycoconjugate J 14:809–819

Cummings RD, Mattox SA (1988) Retinoic acid-induced differentiation of the mouse
teratocarcinoma cell line F9 is accompanied by an increase in the activity of UDP-
galactose:β-D-galactosyl α1,3-galactosyltransferase. J Biol Chem 263:511–519

Fang J, Li J, Chen X, Zhang Y, Wang J, Guo Z, Zhang W, Yu L, Brew K, Wang PG (1998) Highly
efficient chemoenzymatic synthesis of α-galactosyl epitopes with a recombinant
α1→3-galactosyltransferase. J Am Chem Soc 120:6635–6638

Galili U, Clark MR, Shohet SB, Buehler J, Macher BA (1987) Evolutionary relationship
between the natural anti-Gal antibody and the Galα1→3Gal epitope in primates. Proc
Natl Acad Sci USA 84:1369–1373

Galili U, Shohet SB, Kobrin E, Stults CL, Macher BA (1988) Man, apes, and Old World
monkeys differ from other mammals in the expression of α-galactosyl epitopes on
nucleated cells. J Biol Chem 263:17755–17762

Gastinel LN, Bignon C, Misra AK, Hindsgaul O, Shaper JH, Joziasse DH (2001) Bovine
alpha1,3-galactosyltransferase catalytic domain structure and its relationship with AB
histo-blood group and glycosphingolipid glycosyltransferases. EMBO J 20(4):638–
649

Hokke CH, Zervosen A, Elling L, Joziasse DH, Van den Eijnden DH (1996) One-pot enzymatic synthesis of the Galα1→3Galβ1→4GlcNAc sequence with in situ UDP-Gal
regeneration. Glycoconjugate J 13:687–692

Ikematsu S, Kaname T, Ozawa M, Yonezawa S, Sato E, Uehara F, Obama H, Yamamura K,
Muramatsu T (1993) Transgenic mouse lines with ectopic expression of α1,3-
galactosyltransferase: production and characteristics. Glycobiology 3:575–580

Johnston DS, Shaper JH, Shaper NL, Joziasse DH, Wright WW (1995) The gene encoding
murine α1,3-galactosyltransferase is expressed in female germ cells but not in male
germ cells. Dev Biol 171:224–232

Johnston DS, Wright WW, Shaper JH, Hokke CH, Van den Eijnden DH, Joziasse DH (1998) Murine sperm-zona binding: a fucosyl residue is required for a high-affinity sperm-binding ligand. A second site on sperm binds a nonfucosylated, α-galactosyl-capped oligosaccharide. J Biol Chem 273:1888–1895

Joziasse DH, Oriol R (1999) Xenotransplantation: the importance of the Galα1,3Gal epitope in hyperacute vascular rejection. Biochim Biophys Acta 1455:403–418

Joziasse DH, Shaper JH, Van den Eijnden DH, Van Tunen AJ, Shaper NL (1989) Bovine α1,3-galactosyltransferase: isolation and characterization of a cDNA clone. Identification of homologous sequences in human genomic DNA. J Biol Chem 264:14290–14297

Joziasse DH, Shaper NL, Salyer LS, Van den Eijnden DH, van der Spoel AC, Shaper JH (1990) α1→3-Galactosyltransferase: the use of recombinant enzyme for the synthesis of α-galactosylated glycoconjugates. Eur J Biochem 191:75–83

Joziasse DH, Shaper JH, Jabs EW, Shaper NL (1991) Characterization of an α1,3-galactosyltransferase homologue on human chromosome 12 that is organized as a processed pseudogene. J Biol Chem 266:6991–6998

Joziasse DH, Shaper NL, Kim D, Van den Eijnden DH, Shaper JH (1992) Murine α1,3-galactosyltransferase. A single gene locus specifies four isoforms of the enzyme by alternative splicing. J Biol Chem 267:5534–5541

Larsen RD, Rivera-Marrero CA, Ernst LK, Cummings RD, Lowe JB (1990) Frameshift and nonsense mutations in a human genomic sequence homologous to a murine UDP-Gal:Galβ1,4GlcNAc α1,3-galactosyltransferase cDNA. J Biol Chem 265:7055–7061

Mercurio AM, Robbins PW (1985) Activation of mouse peritoneal macrophages alters the structure and surface expression of protein-bound lactosaminoglycans. J Immunol 135:1305–1312

Peters BP, Goldstein IJ (1979) The use of fluorescein-conjugated *Bandeiraea simplicifolia* B₄-isolectin as a histochemical reagent for the detection of α-D-galactopyranosyl groups. Exp Cell Res 120:321–334

Rother RP, Squinto SP (1996) The α-galactosyl epitope: a sugar coating that makes viruses and cells unpalatable. Cell 86:185–188

Sato S, Hughes RC (1994) Regulation of secretion and surface expression of Mac-2, a galactoside-binding protein of macrophages. J Biol Chem 269:4424–4430

Sheares BT, Mercurio AM (1987) Modulation of two distinct galactosyltransferase activities in populations of mouse peritoneal macrophages. J Immunol 139:3748–3752

Sujino K, Malet C, Hindsgaul O, Palcic MM (1997) Acceptor hydroxyl group mapping for calf thymus α-(1→3)-galactosyltransferase and enzymatic synthesis of α-D-Gal*p*-(1→3)-β-D-Gal*p*-(1→4)-β-D-Glc*p*NAc analogs. Carbohydr Res 305:483–489

Tearle RG, Tange MJ, Zannettino ZL, Katerelos M, Shinkel TA, Van Denderen BJW, Lonie AJ, Lyons I, Nottle MB, Cox T, Becker C, Peura AM, Wigley PL, Crawford RJ, Robins AJ, Pearse MJ, d'Apice AJF (1996) The α1,3-galactosyltransferase knockout mouse. Transplantation 61:13–19

Thall AD, Maly P, Lowe JB (1995) Oocyte Galα1,3Gal epitopes implicated in sperm adhesion to the zona pellucida glycoprotein ZP3 are not required for fertilization in the mouse. J Biol Chem 270:21437–21440

Van den Eijnden DH, Blanken WM, Winterwerp H, Schiphorst WECM (1983) Identification and characterization of a UDP-Gal:*N*-acetyllactosaminide α1,3-galactosyltransferase in calf thymus. Eur J Biochem 134:523–530

Van den Eijnden DH (2000) On the origin of oligosaccharide species. Glycosyltransferases in action. In: Ernst B, Hart G, Sinay P (eds) Oligosaccharides in chemistry and biology Part I, Vol. 2, pp. 589–624. Wiley/VCH, Weinheim

Vanhove B, Goret F, Soulillou J-P, Pourcel C (1997) Porcine α1,3-galactosyltransferase: tissue-specific and regulated expression of splicing isoforms. Biochim Biophys Acta 1356:1–11

GalCer Synthase (Ceramide Galactosyltransferase, CGT)

Introduction

The multilayer membrane system of myelin ensheathing axons of the central nervous system (CNS) and the peripheral nervous system (PNS) is synthesized and assembled by oligodendrocytes and Schwann cells, respectively. The lipid bilayer contributes approximately 70% to 85% of the total myelin mass. The lipid composition is characterized by a high content of cholesterol and glycolipids, of which galactosylceramides (cerebrosides, GalC) and sulfatides, the 3'-sulfoesters of galactosyl-ceramides (sGalC), have been recognized as the dominant components and also oligodendrocyte-specific myelin markers. Both are expressed in only a minute amount in kidney. The molar ratio of cholesterol:phospholipids (including plasmalogen–phospholipids):galactolipids is between $2:2:1$ and $4:3:2$ and reasonably stable in different species (human, bovine, and murine myelin).

Galactosylceramides and sulfatides confer special features upon the lipid bilayer of myelin: their unusually high content of long-chain fatty acyl residues (between C_{20} and C_{26}) largely present as D-α-hydroxy fatty acids, which are linked to the NH_2-group of the long-chain bases sphingosine and dihydrosphingosine (sphinganine), are the molecular basis for the unusual thickness (5 nm) of the lipid bilayer of myelin. The *all trans* configuration of the fatty acyl chains leads to a tight packing of the hydrophobic core of the bilayer, which guarantees the insulation by the internodes which is essential for rapid saltatory conduction. Galactosyl ceramide synthase (CGT) is the key enzyme of GalC synthesis. GalC and sGalC synthesis deficiency is incompatible with life, as demonstrated by the recently generated $cgt^{-/-}$ ("knockout") mouse. This mouse model also demonstrates that other postulated biosynthetic pathways are irrelevant for galactosylceramide synthesis.

WILHELM STOFFEL

Laboratory of Molecular Neuroscience, Institute of Biochemistry, Faculty of Medicine, University of Cologne, Joseph-Stelzmann-Straße 52, D-50931 Köln, Germany
Tel. +49 (0)221-478-6881; Fax +49 (0)221-478-6882
e-mail: wilhelm.stoffel@uni-koeln.de

Databanks

CalCer synthase (ceramide galactosyltransferase, CGT)

NC-IUBMB enzyme classification: E.C.2.4.1.45

Species	Gene	Protein	mRNA	Genomic
Homo sapiens	UGT8	–	NM_003360	–
			U30930	–
Mus musculus	cgt	–	X92122	X92123–X92126 (exons 1–4)
			AA245658	X92177 (exon 5)
Bos taurus	CGT	–	AF129810	–
Gallus gallus	CGT	–	AF129809	–

Name and History

Several studies in vivo using the intracerebral application of radiolabeled glucose, galactose, fatty acids, and serine led to the de novo synthesis of galactocerebrosides (GalC) in the central nervous system (Moser and Karnovsky 1959). The application in vivo of labeled stearoyl sphingosine, however, was not incorporated into GalC (Kopaczyk and Radin 1965).

Experiments in vitro with brain homogenates of different species as the enzyme source during the past three decades revealed three enzyme activities, which led to the suggestion of three biosynthetic pathways for GalC.

1. The psychosine pathway, according to which the galactosyl residue from UDP-galactose is transferred to sphingosine bases yielding psychosine (Clealand and Kennedy 1960) with subsequent acylation of psychosine.
2. The epimerization pathway, which is the synthesis of glucosylceramide from UDP-glucose and ceramide catalyzed by the UDP-glucosyl:ceramide glucosyl transferase (Basu et al. 1968; Brady 1962) followed by the epimerization of the glucose moiety to galactosyl ceramide.
3. The direct galactosylceramide pathway, in which UDP-galactose transfers galactose to *N*-2D-hydroxyacylsphingosine to yield galactocerebrosides (Morell and Radin 1969).

The transfer of galactose to naturally occurring *N*-2D-hydroxyacylsphingosine from UDP-galactose was first demonstrated in 1969 by Morell and Radin (Morell and Radin 1969) in experiments in vitro using rat brain homogenate as the enzyme source. The enzyme was named UDP-galactose ceramide galactosyltransferase or cerebroside synthase (CGT) (EC 2.4.1.45). *N*-2-hydroxy-fatty-acyl-containing ceramides were the preferred acceptor molecules of CGT. Regarding the pathways A and B, no enzymatic acylation of psychosine has been achieved and no glucosylceramide–galactosylceramide epimerase has been discovered to date.

Genetic evidence for the occurrence of only one biosynthetic pathway with galactosyl:ceramide galactosyl transferase (synthase) as the key enzyme came from the purification, cloning (Schulte and Stoffel 1993), elucidation of the gene organization in mouse and human (Bosio et al. 1996b, 1998a; Coetzee et al. 1996b), and the *cgt* "knockout" mouse model (Bosio et al. 1996a; Coetzee et al. 1996a), described under Biological Aspects. The *cgt* gene ablation leads to a complete deficiency of galactosyl cerebrosides and sulfatides, and excludes pathways 1 and 2 deduced from in vitro experiments.

Enzyme Activity Assay and Substrate Specificity

On the basis of the early observation that short-chain 2D-hydroxy fatty acylsubstituted ceramides are more than ten times better acceptor substrates for the galactosyl transfer, N-2D-hydroxy-butanoyl, -hexanoyl, -octanoylsphingosine and N-2D-hydroxyoctanoylsphinganine were synthesized from commercially available racemic 2-hydroxy fatty acids and the nonderivatized long-chain bases via their respective N-hydroxysuccinimide esters (Ong and Brady 1972). The diastereomers can be separated by thin-layer chromatography (TLC) on silica gel G plates (solvent system chloroform:methanol, 93:7 v/v).

The enzyme assay of UDP-galactose:ceramide galactosyl transferase has been elaborated by Neskovic et al. (1974, 1981, 1986). In the assay, the delipidated and therefore inactive CGT is reactivated by phosphatidyl choline or -ethanolamine. The sensitivity of the assay is enhanced by synthetic short-chain N-2D-hydroxy-acylceramides dissolved in Triton X-100. Long-chain fatty-acyl sphingosines show low enzymatic activity due to their low solubility. The transfer of (U-^{14}C) galactose from UDP-(U-^{14}C) galactose to the acceptor ceramide yields the chloroform–methanol-soluble ^{14}C-labeled GalC reaction product, which is measured by scintillation counting of aliquots of the organic phase. Another aliquot is used for radio TLC (solvent system chloroform:methanol:water, 65/25/4 v/v/v) using a TLC scanner or preferably a phospho-imager (Nescovic et al. 1974).

Preparation

CGT resides as a membrane-bound enzyme in the endoplasmic reticulum. As with all glycosyltransferases known to date, CGT is a glycoprotein. In the last 15–20 years, a partial purification and characterization of this labile enzyme has been achieved. Its activity is stabilized by phospholipids (Nescovic et al. 1974, 1981, 1986). Starting with the Triton X-100 extract of rat brain, the enzyme was enriched 300-fold in relation to the brain homogenate by DE52 ion exchange chromatography, ammonium sulfate precipitation, and DE Sephadex and Cibachrom Blue Sepharose chromatography. SDS-polyacrylamide electrophoresis (PAGE) revealed a prominent 53-kDa protein band among a few contaminations in the Coomassie R250-stained gel. The 53-kDa protein was regarded as the CGT, although antibodies raised against the protein were unable to inhibit the enzyme activity.

Enzyme purification was achieved by Schulte and Stoffel (1993, 1995). Following the first purification steps of the Neskovic procedure, CGT was isolated as a homogeneous band of 64 kDa by lectin affinity chromatography. CGT, recognized as a glycoprotein (Schulte and Stoffel 1993, 1995), has been purified to homogeneity by adsorption to immobilized concanavalin A and the *Lens culinaris* lectin, and desorption by methyl-mannoside or methylglucoside.

Starting with microsomes from myelinating rats (p18–p28), the hydrogenated Triton X-100 extract was processed via DE52 cellulose-, Blue Sepharose-, and lentil lectin Sepharose affinity chromatography, which resulted in a CGT with a specific activity of 13 000 nmol/h/mg protein in a yield of 7.6%, homogenous in SDS–PAGE.

The glycoprotein structure of CGT was verified by endoglycosidase treatment, which reduced the mol mass from 64 to 58 kDa, a difference which accounts for two of the three *N*-glycosylation antennae. CGT nitrocellulose electroblots were reacted with specific lectin–digoxygenin conjugates. Only concanavalin A, which has a broad specificity, and *Galanthus nivalis* lectin (GNA), which specifically recognizes terminal mannose residues, were reactive. Therefore, CGT belongs to the mannose-rich or hybrid glycoproteins. This is in agreement with the observation that most lipid synthesizing enzymes reside in the ER.

Tryptic digests of the purified CGT were separated by HPLC and yielded 10 different peptides, some of which showed significant homology to UDP-glucuronyl transferases (Schulte and Stoffel 1993). These peptides served as templates for the design of degenerated oligonucleotides, which were used for PCR reactions of the λgt library of myelinating rat brain. Using the PCR fragments as probes, complete cDNA clones of CGT were obtained by screening the λgt brain library, the open reading frame (ORF) of 1623 bp code for a 541 amino acid residue polypeptide. The hydrophobicity plot suggests one transmembrane domain in the terminal region of CGT. A N-terminal signal sequence of 20 residues is released during cotranslational translocation and yields the 521 amino acid residue mature form of CGT. This was verified by Edman degradation of the isolated purified rat brain CGT. Three potential *N*-glycosylation sites are at positions 78, 333, and 442, the first two being posttranslationally modified. Figure 1 schematically summarizes these structural features of CGT, including the ER-residing signal Lys-Lys-X-Lys at the C-terminus oriented toward the cytosol.

Biological Aspects

The pivotal question about the biological (in vivo) function of CGT in oligodendrocytes and Schwann cells can be answered most comprehensively and unambiguously with genetic tools.

The structural and functional aspects of the enzyme were analyzed in cell lines (HEK 293) transfected with and expressing mutated CGT cDNA constructs. Glycosylation mutants of the CGT protein showed that *N*-glycosylation is required for enzyme activity. The CGT protein truncated by the cystosolic, C-terminal domain or the transmembrane or both is inactive (C. Cytosolic Niemand and W. Stoffel unpublished results 1995).

Ablation of the *cgt* in the mouse by gene targeting by homologous recombination in embryonic stem (ES) cells was achieved by the isolation of the mouse CGT gene, the elucidation of the gene organization, and the construction of a mutated *cgt* vector for the replacement of the wild-type *cgt* locus in ES cells. After mouse blastocyst injection of homologously recombined ES cells and transfer into pseudopregnant mice, a homozygous *cgt* "knockout" mouse line was generated (Bosio et al. 1996a, Coetzee et al. 1996a). The CGT-deficient mouse line is completely devoid of cerebrosides, sulfatides, and 3-galactosyl-1,2-diacylglycerol in the myelin of CNS, PNS, and kidney.

Fig. 1. Schematic presentation of the proposed topology of cerebroside synthase (CGT) in the membrane of the endoplasmic reticulum. Note the luminal localization of the catalytic N-terminus with the three putative *N*-glycosylation sites of CGT

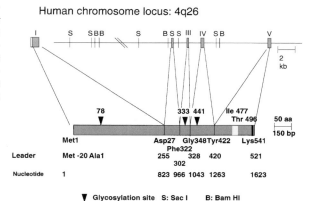

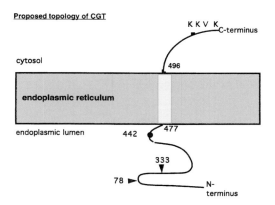

A compensatory transfer of glucosyl residues to ceramides substituted by normal and 2D-hydroxylated fatty acids normally used for GalC synthesis was observed in CNS and PNS myelin. Also, the ceramide part of brain sphingomyelins corresponded largely to these moieties. The loss of GalC and sGalC in the myelin membranes causes abnormal CNS but an ultrastructurally normal appearance of the PNS myelin membrane structure (Bosio et al. 1998a, 1998b). CNS myelin shows vacuolation, and abundant myelin with a loss of the axonal contact wrapping axons, which are disoriented, often repeatedly. Peripheral nerves have lost their saltatory conduction, and the velocity of conduction has decreased to that of unmyelinated axons (Bosio et al. 1998b). Tremor and seizures lead to premature death around p30, the end of the myelination period. Biophysical studies (Langmuir balance isotherms) revealed that the tightly packed lipid hydrophobic zone is lost owing to the lack of galactocerebrosides, with their long-chain saturated fatty acid residues and their stabilizing contribution by the hydrogen bond belt between cholesterol and the GalC and sGalC molecules in the outer leaflet of the asymmetric lipid bilayer.

The complete lack of GalC and sGalC in the $cgt^{-/-}$ mouse also proves that the UDP-galactose ceramide galactosyltransferase (GalC synthase) reaction is the only biosynthetic pathway relevant for cerebroside and sulfatide synthesis in vivo.

Ceramides substituted with normal and 2D-hydroxy fatty acids are both acceptor molecules in the synthase reaction, and this suggests that specificities for different ceramide species are artefacts of the in vitro systems used in earlier studies, presumably caused by the different solubilization of the acceptor substrates.

Future Perspectives

The $cgt^{-/-}$ mouse generated as a conventional "knockout" model has a lifespan reduced to that of the myelination period (p30). We need to know more about the function of CGT in the normal maintenance of the CNS and PNS myelin membrane in the adult, and the function of CGT in myelin regeneration of nerve lesions.

Therefore, one future goal will be the generation of a conditional, time-inducible cgt-knockout using the doxycyclin inducible $rtetTA$ $Cre/loxP$ system (Gossen and Bujard 1992; Sauer and Henderson 1988). Cell-specific cgt-ablation (oligodendrocytes, Schwann cells, and kidney cells) can be achieved with the appropriate promoter controlling the relevant cell-specific expression of the reversed tetracyclin-inducible transactivator of the Cre recombinase.

Further Reading

Stoffel W, Bosio A (1997) Myelin glycolipids and their functions. Curr Opin Neurobiol 7:654–660

References

Basu S, Kaufman B, Roseman SJ (1968) Enzymatic synthesis of ceramide glucose and ceramide lactose by glycosyltransferases from embryonic chicken brain. J Biol Chem 243:5802–5804

Bosio A, Binczek E, Stoffel W (1996a) Functional breakdown of the lipid bilayer of the myelin membrane in central and peripheral nervous system by disrupted galacto-cerebroside synthesis. Proc Natl Acad Sci USA 93:13280–13285

Bosio A, Binczek E, Stoffel W (1996b) Molecular cloning and characterization of the mouse CGT gene encoding UDP-galactose ceramide-galactosyltransferase (cerebroside synthetase). Genomics 35:223–226

Bosio A, Binczek E, Haupt WF, Stoffel W (1998a) Composition and biophysical properties of myelin lipid define the neurological defects in galactocerebroside–sulfatide-deficient mice. J Neurochem 70:308–315

Bosio A, Binczek E, Haupt WF, Stoffel W (1998b) Biophysical properties of myelin lipid define the neurological defects in galactocerebroside- and sulfatide-deficient mice. J Neurochem 70:308–315

Brady RO (1962) Studies on the total enzymatic synthesis of cerebrosides. J Biol Chem 237:2416–2417

Clealand WW, Kennedy EP (1960) The enzymatic synthesis of psychosine. J Biol Chem 235:45–51

Coetzee T, Fujita N, Dupree J, Shi R, Blight A, Suzuki K, Popko B (1996a) Myelination in the absence of galactocerebroside and sulfatide: normal structure with abnormal function and regional instability. Cell 86:209–219

Coetzee T, Li X, Fujita N, Marcus J, Suzuki K, Francke U, Popko B (1996b) Molecular cloning, chromosomal mapping, and characterization of the mouse UDP-galactose:ceramide galactosyltransferase gene. Genomics 35:215–222

Gossen M, Bujard H (1992) Tight control of gene expression in mammalian cells by tetracyclin-responsive promoters. Proc Natl Acad Sci USA 89:5547–5551

Kopaczyk KC, Radin NS (1965) In vivo conversion of cerebrosides and ceramide in rat brain. J Lipid Res 6:140–148

Morell P, Radin NS (1969) Synthesis of cerebroside by brain from uridine diphosphate galactose- and ceramide-containing hydroxy fatty acid. Biochemistry 8:506–513

Moser HM, Karnovsky ML (1959) Studies on the biosynthesis of glycolipids and other lipids of the brain. J Biol Chem 234:1990–1997

Nescovic NM, Sarlieve LL, Mandel P (1974) Purification and properties of UDP-galactose:ceramide galactosyltransferase from rat brain microsomes. Biochem Biophys Acta 334:309–315

Nescovic NM, Mandel P, Gatt S (1981) UDP-Galactose:(2-hydroxyacyl) sphingosine galactosyltransferase from rat brain. Methods Enzymol 71:321–327

Neskovic N, Roussel G, Nussbaum J (1986) UDP-galactose:ceramide galactosyltransferase of rat brain: a new method of purification and production of specific antibodies. J Neurochem 47:1412–1418

Ong D, Brady R (1972) Synthesis of ceramides using N-hydroxysuccinimide esters. J Lipid Res 13:819–822

Sauer B, Henderson N (1988) Site-specific DANN recombination in mammalian cells by the Cre recombinase of bacteriophage P1. Proc Natl Acad Sci USA 85:166–170

Schulte S, Stoffel W (1993) Ceramide UDP-galactosyltransferase from myelinating rat brain: purification, cloning, and expression. Proc Natl Acad Sci USA 90:10265–10269

Schulte S, Stoffel W (1995) UDP-galactose:ceramide galactosyltransferase and glutamate/aspartate transporter. Copurification, separation and characterization of the two glycoproteins. Eur J Biochem 233:947–953

N-Acetylglucosaminyltransferases

N-Acetylglucosaminyltransferase-I

Introduction

Structural analyses of sugars on secreted glycoproteins performed about 30 years ago revealed bi-, tri-, and tetraantennary N-glycans in which GlcNAc residues linked to a conserved trimannosyl core initiated each antenna. These same structures were lectin binding sites on red cell glycoproteins (Kornfeld and Kornfeld 1970), prompting the search for the GlcNAc-transferases that catalyzed the addition of each GlcNAc residue. N-Acetylglucosaminyltransferase-I (GnT-I) was the first N-glycan branching GlcNAc-transferase for which an assay was developed (Gottlieb et al. 1975; Stanley et al. 1975). It is a Type II transmembrane protein of ~447 amino acids (Kumar et al. 1990; Sarkar et al. 1991) that resides in the medial/*trans* Golgi. GnT-I catalyzes the transfer of GlcNAc from UDP-GlcNAc to the terminal α-1,3-linked Man in $Man_5GlcNAc_2Asn$ to initiate the synthesis of hybrid and complex N-linked glycans in multicellular organisms (reviewed in Kornfeld and Kornfeld 1985). It is not found in yeast or bacteria. The human gene encoding GnT-I is termed MGAT1 and resides on chromosome 5q35 (Kumar et al. 1992; Tan et al. 1995), and the mouse gene, Mgat1, is on chromosome 11 (Pownall et al. 1992). Two transcripts of ~2.9 kb and ~3.3 kb are observed in most mammalian tissues, with the shorter transcript predominating in liver, and the longer transcript in brain (Yang et al. 1994; Yip et al. 1997; Fukada et al. 1998). In mammals, the coding region is in a single exon and the Mgat1 gene is ubiquitously expressed. Mutant mice with a targeted Mgat1 gene mutation that inactivates GnT-I die at mid-gestation (Ioffe and Stanley 1994; Metzler et al. 1994). However, cultured cells (Gottlieb et al. 1975; Meager et al. 1975; Stanley et al. 1975) and plants (von Schaewen et al. 1993) lacking GnT-I are viable and healthy.

PAMELA STANLEY

Department of Cell Biology, Albert Einstein College of Medicine, 1300 Morris Park Ave., Bronx, NY 10461, USA
Tel. +1-718-430-3346; Fax +1-718-430-8574
e-mail: stanley@aecom.yu.edu

Databanks

N-Acetylglucosaminyltransferase-I

NC-IUBMB enzyme classification: E.C.2.4.1.101

Species	Gene	Protein	mRNA	Genomic
Homo sapiens	MGAT1	AAA75523.1	M55621	M61829
		AAA52563.1		L77081(5′ exon)
Oryctolagus cuniculus	GNTI	AAA31493	M57301	–
Rattus norvegicus	NAGT	BAA03807	D16302	–
		JC2076	AB012874–AB012878	
Mus musculus	Mgat1	AAA40478	L07037	M73491
		AAA37698	X77487(5′utr)	NM 010794
Cricetulus griseus	Mgat1	AAC52872	U65791	–
Mesocricetus auratus	Mgat1	AAD04130	AF087456	–
Drosophila melanogaster	CG13431	AAF70177	–	AF251495
Caenorhabditis elegans	gly-12	AAD03023	AF082011	–
	gly-13	AAD03022	AF082010	–
	gly-14	AAD03024	AF082012	–
Arabidopsis thaliana	cgl	JC7084	AJ243198	–
Nicotiana tabacum	cgl	CAB70464	–	–
Solanum tuberosum	cgl	CAB70462	–	–

Name and History

N-Acetylglucosaminyltransferase-I (EC 2.4.1.101), also known as UDP-GlcNAc:α-D-mannoside β2-*N*-acetylglucosaminyltransferase-I and abbreviated as GnT-I (or GnT-I or NAGT 1), was so named because it was the first GlcNAc-transferase activity identified in a crude cell extract to transfer GlcNAc from UDP-GlcNAc to glycoproteins containing the trimannoside *N*-linked core (Manα1-3(Manα1-6)-Manβ1-4GlcNAcβ1-4GlcNAcβ1-Asn. An assay for GnT-I was developed in cell mutants that were resistant to the cytotoxicity of several plant lectins (reviewed in Briles 1982; Stanley 1984). *N*-glycan containing acceptors derived by sequential glycosidase digestion of glycoproteins, including α1-acid glycoprotein, fetuin, and IgG, were tested in cell extracts with a variety of nucleotide–sugar donors. Several independent CHO mutants, including PhaR1-1 (Stanley et al. 1975), Clone 15B (Gottlieb et al. 1975), and a BHK mutant Ric[R] 14 (Meager et al. 1975) had a specific reduction in the ability to transfer GlcNAc from UDP-GlcNAc to mannose-terminating acceptors. When a more defined set of biantennary complex *N*-glycan glycopeptide acceptors was used, CHO cells were found to possess at least two GlcNAc-transferase activities (Narasimhan et al. 1977). GnT-I was the activity missing from PhaR1-1 CHO cells assayed with UDP-GlcNAc and the trimannosyl glycopeptide (Manα1-3(Manα1-6)Manβ1-4GlcNAcβ1-4GlcNAcβ1-Asn as acceptor (Narasimhan et al. 1977). CHO cells lacking GnT-I activity were subsequently termed Lec1 (Stanley 1983). Point mutations that inactivate GnT-I are a conversion of the conserved Cys at position 123 to Arg (Puthalakath et al. 1996), and conversion of the conserved Gly at position 320 to Asp (Opat et al. 1998). GnT-I lacking the first 106 amino acids, that include the cytoplasmic, transmembrane, and stem domains, is active but removal of the C-terminal 7 amino acids results in a 40%

reduction in activity (Sarkar et al. 1998). Missense mutations that dramatically increase the apparent Km of GnT-I for both substrates include conversion of Asp 212 to Asn or Arg 303 to Tryp (Chen et al. 2001).

Enzyme Activity Assay and Substrate Specificity

When GnT-I was partially purified from rat liver, it was found to prefer the Man$_5$ acceptor Manα1-3(Manα1-6)Manα1-6(Manα1-3)Manβ1-4GlcNAcβ1-4GlcNAcβ1-Asn over the (Manα1-3(Manα1-6)Manβ1-4GlcNAcβ1-4GlcNAcβ1-Asn acceptor originally used (Oppenheimer and Hill 1981). Since glycoproteins from Lec1 CHO cells have Man$_5$GlcNAc$_2$Asn in place of complex N-glycans (Robertson et al. 1978; Tabas et al. 1978), the preference of GnT-I for Man$_5$GlcNAc$_2$Asn was consistent with it being an in vivo substrate of GnT-I according to the following scheme:

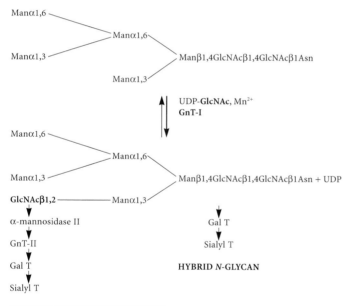

The addition of the β1,2-linked GlcNAc to Man$_5$GlcNAc$_2$Asn generates a substrate for α-mannosidase II and initiates the synthesis of complex N-glycans through the subsequent action of GnT-II (reviewed in Kornfeld and Kornfeld 1985). Alternatively, if α-mannosidase II does not act, a hybrid structure may be formed by the addition of Gal and potentially sialic acid to the β1,2-linked GlcNAc of GlcNAcβ1,2Man$_5$GlcNAc$_2$Asn. In mice lacking α-mannosidase II, complex *N*-glycans are formed in many tissues, although not in red blood cells, revealing an alternative pathway that operates in the absence of α-mannosidase II (Chui et al. 1997). A mannosidase activity termed α-mannosidase III, that is enriched in Golgi fractions and removes mannose residues from Man$_5$GlcNAc$_2$Asn, was identified in mouse tissues and could generate Man$_3$GlcNAc$_2$Asn, which is known from the initial in vitro studies to be a substrate for GnT-I (Gottlieb et al. 1975; Stanley et al. 1975; Narasimhan et al. 1977), thereby allowing α-mannosidase II-deficient mice to synthesize complex *N*-glycans. Acceptor specificity studies show that GnT-I will not transfer GlcNAc to terminal α-mannose residues in a biantennary *N*-glycan in which the α1,6-mannose is substituted at the O-2 position or the β1,4-mannose is substituted at the O-4 position (Nishikawa et al. 1988).

Optimal assay conditions were determined for purified rabbit liver GnT-I (Nishikawa et al. 1988): 0.25 mM Man$_5$GlcNAc$_2$Asn in 100 mM 2-(*N*-morpholino)-ethanesulfonate acid (MES) pH 6.1, 0.5 mM UDP-[^{14}C]-GlcNAc, 20 mM MnCl$_2$, 5 mM AMP (as pyrophosphorylase inhibitor), 1.0% Triton X-100, 100 mM GlcNAc (as hexosaminidase inhibitor), bovine serum albumin at 5 mg/ml (enzyme stabilizer), and enzyme (usually ~0.1 milliunits) in a volume of 50 μl. After 30 min at 37°C, the reaction is terminated by the addition of ice-cold water containing 20 mM EDTA, and the products fractionated on a 1-ml ion exchange column (AG-1 × 8, chloride form) eluted with water. The flow through contains [^{14}C]-GlcNAcMan$_5$GlcNAc$_2$Asn product, unmodified Man$_5$GlcNAc$_2$Asn, and free [^{14}C]-GlcNAc generated by hydrolysis of UDP-[^{14}C]-GlcNAc. Alternatively, products can be fractionated on a 1.5-ml column of concanavalin A (Con A)-Sepharose to which the [^{14}C]-GlcNAcMan$_5$GlcNAc$_2$Asn product binds and may be eluted with 10 mM α-methylmannoside. Neither UDP-GlcNAc nor free GlcNAc bind to Con A, and thus background due to hydrolysis of UDP-GlcNAc is eliminated. The background of the GnT-I assay is determined from labeled products generated by boiled enzyme and from enzyme incubated with UDP-[^{14}C]-GlcNAc in the absence of Man$_5$GlcNAc$_2$Asn acceptor.

To assay GnT-I in cell or tissue extracts, the following conditions, optimized for nonionic detergent extracts of CHO cells, may be used. Cells are washed three times in saline and extracted in 1.5% NP-40 in cold distilled water containing protease inhibitors (75 μl detergent solution per 10^7 packed cells). After 10 min on ice, the extract is centrifuged at low speed to remove nuclei and 5–20 μl extract containing 50–100 μg protein is added to an assay tube on ice containing, in a final volume of 40 μl, 62.5 mM MES buffer pH 6.25, 25 mM MnCl$_2$, 1 mM Man$_5$GlcNAc$_2$Asn, 1 mM UDP-[^{3}H]-GlcNAc (specific activity ~10 000 cpm per nmole), and 50–100 μg protein. After 30–90 min at 37°C, the reaction is terminated by the addition of 1 ml cold Con A buffer (1 M Na acetate, 1 mM MnCl$_2$, 1 mM MgCl$_2$ and 1 mM CaCl$_2$, pH 7.0). The mixture is subsequently passed over a 1.5-ml column of Con A-Sepharose, washed with 10 column volumes of Con A buffer to remove UDP-[^{3}H]-GlcNAc and [^{3}H]-GlcNAc, and the radiolabeled product [^{3}H]-GlcNAcMan$_5$GlcNAc$_2$Asn is eluted with a 6-ml aliquot of 200 mM α-methylmannoside in Con A buffer. The specific activity of GnT-I in CHO cell extracts is ~5–10 nmole/h/mg protein. CHO mutants in the Lec1A group have a point mutation that increases the Km for both substrates, and have barely detectable GnT-I activity under these assay conditions. Their activity becomes normal, however, if the pH of the assay is increased to 7.5, the UDP-GlcNAc concentration to 15 mM, and the $Man_5GlcNAc_2Asn$ concentrattion to 5 mM (Chaney and Stanley 1986; Chen et al. 2001).

Preparation

GnT-I was purified ~64 000-fold to apparent homogeneity from a Triton X-100 extract of rabbit liver using three affinity chromaography steps on UDP-hexanolamine, followed by two affinity chromatography steps on 5-Hg-UDP-GlcNAc (Nishikawa et al. 1988). Km values of purified rabbit liver GnT-I for UDP-GlcNAc and $Man_5GlcNAc_2Asn$ substrates were ~0.04 mM and ~2 mM, respectively. The Vmax was ~16 μmol/min/mg protein and the specific activity was ~20 μmol/min/mg protein. SDS–PAGE analysis revealed a major species of 45 kDa and minor species of 50 kDa and 54 kDa. Since cleavage of glycosyltransferases often occurs in the Golgi at a position just beyond the stem region, and since mammalian GnT-I is known to be O-glycosylated but not N-glycosylated (Hoe et al. 1995), the three forms probably represented O-glycosylated and/or proteolyzed forms of GnT-I. GnT-I may also be produced in active form in bacteria, yeast, insect, plant, and mammalian cells. Plant, *Drosophila* and *C. elegans* GnT-I sequences include N-glycan Asn-X-Ser/Thr consensus sites, although their positions in the protein are not conserved and it is not known whether they are utilized.

Biological Aspects

Cultured cells that lack GnT-I grow well in monolayer and suspension culture, and produce secreted and membrane glycoproteins in apparently normal amounts (Stanley 1989). Plants that lack GnT-I are also unaffected by its absence (von Schaewen et al. 1993; Wenderoth and von Schaewen 2000). Since a lack of GnT-I does not alter the synthesis or processing of N-glycans in the endoplasmic reticulum (ER), nor in the *cis* Golgi, glycoproteins are not compromised in their interactions with ER chaperones such as calnexin and calreticulin, and lysosomal enzymes acquire their usual complement of Man-6-phosphate residues for targeting to lysosomes. Thus, a lack of GnT-I affects N-glycan structures later in the secretory pathway, leading to a dramatic alteration in the array of N-glycans expressed at the cell surface because all complex and hybrid N-glycans are replaced by $Man_5GlcNAc_2Asn$. While this change may not significantly affect the biological or structural properties of a glycoprotein, it has a major effect on the tissue targeting of recombinant glycoproteins. Glycoproteins with oligomannosyl N-glycans are targeted to the reticuloendothelial system.

   In contrast to cells and plants, mammals have an absolute requirement for GnT-I during early embryogenesis. Mice with a null mutation in the *Mgat1* gene die at E9.5 (Ioffe and Stanley 1994; Metzler et al. 1994). They are underdeveloped, with fewer somites, a tube-like heart, an open neural tube, and some are altered in left–right symmetry. However, the cause of death of $Mgat1^{-/-}$ embryos is not known. Maternal *Mgat1* gene transcripts rescue the earliest embryos, and thus it is still not known whether hybrid or complex *N*-glycans are required for blastocyst formation or for implantation (Campbell et al. 1995; Ioffe et al. 1997).

In order to identify a cell type that requires GnT-I to develop or differentiate, $Mgat1^{-/-}$ embryonic stem (ES) cells with an inert transgene were developed and tracked in E10 to E16.5 chimeric embryos by DNA:DNA in situ hybridization (Ioffe et al. 1996). These experiments showed that complex and/or hybrid *N*-glycans are essential for the formation of the organized layer of bronchial epithelium. Since heterozygote $Mgat1^{+/-}$ WW6 cells also contributed very poorly to organized bronchial epithelium (Ioffe et al. 1996), it is possible that some form of lung disorder could arise in humans with only one active *Mgat1* allele.

Future Perspectives

Crystal structures of the catalytic domain of rabbit GnT-I in the presence and absence of UDP-GlcNAc were recently reported (Unligil et al. 2000). Key residues that coordinate manganese and bind UDP were identified. The missense mutations identified in GnT-I from Lec1A CHO mutants alter residues conserved in GnT-I from plants through lower organisms and mammals (Chen et al. 2001) that are important in metal binding and catalysis (Asp212), or stabilization of a structural element involved in UDP-GlcNAc binding and catalysis (R303). It is now important to obtain a crystal structure with both UDP-GlcNAc and $Man_5GlcNAc_2Asn$ bound to GnT-I, and to model point mutations that weaken or inactivate GnT-I onto each crystal structure. Crystal structures of GnT-Is from lower organisms will provide insight into the enzyme mechanism as they are only about 30%–40% identical to mammalian GnT-I in amino acid sequence.

Another key question for the future is whether mammals have additional GnT-I transferases. Schachter and colleagues have shown that there are three genes with homology to *Mgat1* in *C. elegans*, termed *gly-12*, *gly-13*, and *gly-14* (Chen et al. 1999). All have cytoplasmic, transmembrane, stem, and catalytic domains typical of Golgi glycosyltransferases. However, only *gly-12* and *gly-14* gave GnT-I activity when expressed in insect cells. Whereas each gene is expressed in many cell types during development, and *gly-12* and *gly-13* are expressed ubiquitously in the adult, *gly-14* is expressed only in gut cells of the adult. While it is clear that no other gene product rescues $Mgat1^{-/-}$ mouse embryos from death at E9.5 during embryogenesis, it is possible that other GnT-I genes are expressed in adult mammals. Tissue-specific knockout of a floxed *Mgat1* gene may reveal such an activity.

Conditional knockout of *Mgat1* in specific tissues will identify cell types that require complex or hybrid *N*-glycans for development or differentiation. Chimera experiments with $Mgat1^{-/-}$ ES cells in $Rag2^{-/-}$ blastocysts will determine whether T and/or B cells require complex or hybrid *N*-glycans to be generated or to function in immunity. While humans with a mutation in one *MGAT1* allele would not be expected to have developmental problems typical of congenital disorders of glycosylation (CDG) patients, they may have an altered susceptibility to lung disease (Ioffe et al. 1996) or other subtle problems.

Further Reading

Stanley P, Ioffe E (1995) Glycosyltransferase mutants: key to new insights in glycobiology. FASEB J 9:1436–1444

Schachter H, Chen SH, Zhou S, Tan J, Yip B, Sarkar M, Spence A (1997) Structure and function of the genes encoding N-acetylglucosaminyltransferases which initiate N-glycan antennae. Biochem Soc Trans 25:875–880

References

Briles EB (1982) Lectin-resistant cell surface variants of eukaryotic cells. Int Rev Cytol 75:101–165

Campbell RM, Metzler M, Granovsky M, Dennis JW, Marth JD (1995) Complex asparagine-linked oligosaccharides in Mgat1-null embryos. Glycobiology 5:535–543

Chaney W, Stanley P (1986) Lec1A Chinese hamster ovary cell mutants appear to arise from a structural alteration in N-acetylglucosaminyltransferase I. J Biol Chem 261: 10551–10557

Chen W, Unligil UM, Rini JM, Stanley P (2001) Independent Lec1A CHO glycosylation mutants arise from point mutations in N-acetylglucosaminyltransferase I that reduce affinity for both substrates. Molecular consequences based on the crystal structure of GlcNAc-TI. Biochem 40:8765–8772

Chen S, Zhou S, Sarkar M, Spence AM, Schachter H (1999) Expression of three Caenorhabditis elegans N-acetylglucosaminyltransferase I genes during development. J Biol Chem 274:288–297

Chui D, Oh-Eda M, Liao YF, Panneerselvam K, Lal A, Marek KW, Freeze HH, Moremen KW, Fukuda MN, Marth JD (1997) α-mannosidase-II deficiency results in dyserythropoiesis and unveils an alternate pathway in oligosaccharide biosynthesis. Cell 90:157–167

Fukada T, Kioka N, Nishiu J, Sakata S, Sakai H, Yamada M, Komano T (1998) Different response to inflammation of the multiple mRNAs of rat N-acetylglucosaminyltransferase I with variable 5'-untranslated sequences. FEBS Lett 436:228–232

Gottlieb C, Baenziger J, Kornfeld S (1975) Deficient uridine diphosphate-N-acetylglucosamine:glycoprotein N-acetylglucosaminyltransferase activity in a clone of Chinese hamster ovary cells with altered surface glycoproteins. J Biol Chem 250: 3303–3309

Hoe MH, Slusarewicz P, Misteli T, Watson R, Warren G (1995) Evidence for recycling of the resident medial/trans Golgi enzyme, N-acetylglucosaminyltransferase I, in ldlD cells. J Biol Chem 270:25057–25063

Ioffe E, Stanley P (1994) Mice lacking N-acetylglucosaminyltransferase I activity die at mid-gestation, revealing an essential role for complex or hybrid N-linked carbohydrates. Proc Natl Acad Sci USA 91:728–732

Ioffe E, Liu Y, Stanley P (1996) Essential role for complex N-glycans in forming an organized layer of bronchial epithelium. Proc Natl Acad Sci USA 93:11041–11046

Ioffe E, Liu Y, Stanley P (1997) Complex N-glycans in Mgat1 null preimplantation embryos arise from maternal Mgat1 RNA. Glycobiology 7:913–919

Kornfeld R, Kornfeld S (1970) The structure of a phytohemagglutinin receptor site from human erythrocytes. J Biol Chem 245:2536–2545

Kornfeld R, Kornfeld S (1985) Assembly of asparagine-linked oligosaccharides. Annu Rev Biochem 54:631–664

Kumar R, Yang J, Larsen RD, Stanley P (1990) Cloning and expression of N-acetylglucosaminyltransferase I, the medial Golgi transferase that initiates complex N-linked carbohydrate formation. Proc Natl Acad Sci USA 87:9948–9952

Kumar R, Yang J, Eddy RL, Byers MG, Shows TB, Stanley P (1992) Cloning and expression of the murine gene and chromosomal location of the human gene encoding *N*-acetyl-glucosaminyltransferase I. Glycobiology 2:383–393. (erratum Glycobiology (1999) 9(8):ix)

Meager A, Ungkitchanukit A, Nairn R, Hughes RC (1975) Ricin resistance in baby hamster kidney cells. Nature 257:137–139

Metzler M, Gertz A, Sarkar M, Schachter H, Schrader JW, Marth JD (1994) Complex asparagine-linked oligosaccharides are required for morphogenic events during post-implantation development. EMBO J 13:2056–2065

Narasimhan S, Stanley P, Schachter H (1977) Control of glycoprotein synthesis. Lectin-resistant mutant containing only one of two distinct *N*-acetylglucosaminyltransferase activities present in wild-type Chinese hamster ovary cells. J Biol Chem 252: 3926–3933

Nishikawa Y, Pegg W, Paulsen H, Schachter H (1988) Control of glycoprotein synthesis. Purification and characterization of rabbit liver UDP-*N*-acetylglucosamine:α-3-D-mannoside β-1,2-*N*-acetylglucosaminyltransferase I. J Biol Chem 263:8270–8281

Opat AS, Puthalakath H, Burke J, Gleeson PA (1998) Genetic defect in *N*-acetylgluco-saminyltransferase I gene of a ricin-resistant baby hamster kidney mutant. Biochem J 336:593–598

Oppenheimer CL, Hill RL (1981) Purification and characterization of a rabbit liver α,1,3-mannoside β1,2 *N*-acetylglucosaminyltransferase. J Biol Chem 256:799–804

Pownall S, Kozak CA, Schappert K, Sarkar M, Hull E, Schachter H, Marth JD (1992) Molecular cloning and characterization of the mouse UDP-*N*-acetylglucosamine:α-3-D-mannoside β-1,2-*N*-acetylglucosaminyltransferase I gene. Genomics 12:699–704

Puthalakath H, Burke J, Gleeson PA (1996) Glycosylation defect in *Lec1* Chinese hamster ovary mutant is due to a point mutation in *N*-acetylglucosaminyltransferase I gene. J Biol Chem 271:27818–27822

Robertson MA, Etchison JR, Robertson JS, Summers DF, Stanley P (1978) Specific changes in the oligosaccharide moieties of VSV grown in different lectin-resistant CHO cells. Cell 13:515–526

Sarkar M, Hull E, Nishikawa Y, Simpson RJ, Moritz RL, Dunn R, Schachter H (1991) Molecular cloning and expression of cDNA encoding the enzyme that controls conversion of high-mannose to hybrid and complex *N*-glycans: UDP-*N*-acetylglucosamine: α-3-D-mannoside β-1,2-*N*-acetylglucosaminyltransferase I. Proc Natl Acad Sci USA 88:234–238

Sarkar M, Pagny S, Unligil U, Joziasse D, Mucha J, Glossl J, Schachter H (1998) Removal of 106 amino acids from the *N*-terminus of UDP-GlcNAc: α-3-D-mannoside β-1,2-*N*-acetylglucosaminyltransferase I does not inactivate the enzyme. Glycoconj J 15: 193–197

Stanley P (1983) Selection of lectin-resistant mutants of animal cells. Methods Enzymol 96:157–184

Stanley P (1984) Glycosylation mutants of animal cells. Annu Rev Genet 18:525–552

Stanley P (1989) Chinese hamster ovary cell mutants with multiple glycosylation defects for production of glycoproteins with minimal carbohydrate heterogeneity. Mol Cell Biol 9:377–383

Stanley P, Narasimhan S, Siminovitch L, Schachter H (1975) Chinese hamster ovary cells selected for resistance to the cytotoxicity of phytohemagglutinin are deficient in a UDP-*N*-acetylglucosamine–glycoprotein *N*-acetylglucosaminyltransferase activity. Proc Natl Acad Sci USA 72:3323–3327

Tabas I, Schlesinger S, Kornfeld S (1978) Processing of high mannose oligosaccharides to form complex-type oligosaccharides on the newly synthesized polypeptides of the vesicular stomatitis virus G protein and the IgG heavy chain. J Biol Chem 253:716–722

Tan J, D'Agostaro AF, Bendiak B, Reck F, Sarkar M, Squire JA, Leong P, Schachter H (1995) The human UDP-*N*-acetylglucosamine: α-6-D-mannoside-β-1,2-*N*-acetylglucosaminyltransferase II gene (*MGAT2*). Cloning of genomic DNA, localization to chromosome 14q21, expression in insect cells and purification of the recombinant protein. Eur J Biochem 231:317–328

Unligil UM, Zhou S, Yuwaraj S, Sarkar M, Schachter H, Rini JM (2000) X-ray crystal structure of rabbit *N*-acetylglucosaminyltransferase I, a key enzyme in the biosynthesis of *N*-linked glycans. EMBO J 19:5269–5280

von Schaewen A, Sturm A, O'Neill J, Chrispeels MJ (1993) Isolation of a mutant *Arabidopsis* plant that lacks *N*-acetyl glucosaminyl transferase I and is unable to synthesize Golgi-modified complex *N*-linked glycans. Plant Physiol 102:1109–1118

Wenderoth I, von Schaewen A (2000) Isolation and characterization of plant *N*-acetyl glucosaminyltransferase I (GntI) cDNA sequences. Functional analyses in the *Arabidopsis cgl* mutant in antisense plants. Plant Physiol 123:1097–1108

Yang J, Bhaumik M, Liu Y, Stanley P (1994) Regulation of *N*-linked glycosylation. Neuronal cell-specific expression of a 5′-extended transcript from the gene encoding *N*-acetylglucosaminyltransferase I. Glycobiology 4:703–712

Yip B, Chen SH, Mulder H, Hoppener JW, Schachter H (1997) Organization of the human β-1,2-*N*-acetylglucosaminyltransferase I gene (*MGAT1*), which controls complex and hybrid *N*-glycan synthesis. Biochem J 321:465–474

10

N-Acetylglucosaminyltransferase-II

Introduction

The synthesis of complex *N*-glycans can be divided into three distinct stages. The first stage occurs primarily in the cytoplasm and rough endoplasmic reticulum, and involves the synthesis of $Glc_3Man_9GlcNAc_2$-pyrophosphate-dolichol. The second stage begins with the transfer of $Glc_3Man_9GlcNAc_2$ from $Glc_3Man_9GlcNAc_2$-pyrophosphate-dolichol to an Asn residue of the nascent glycoprotein followed by processing to $Man_5GlcNAc_2$-Asn-X. The third stage occurs primarily in the Golgi apparatus and starts with the action of GnT-I on $Man_5GlcNAc_2$-Asn-X followed by the removal of two mannose residues by mannosidase II to form the substrate for GnT-II (Fig. 1). GnT-II transfers GlcNAc from UDP-α-GlcNAc in the β1,2-linkage to the Manα1-6 arm of the *N*-glycan core, and is essential for normal complex *N*-glycan formation (Fig. 1).

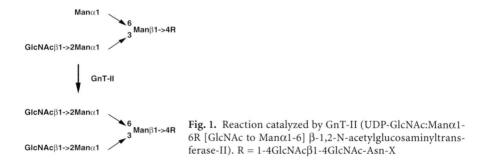

Fig. 1. Reaction catalyzed by GnT-II (UDP-GlcNAc:Manα1-6R [GlcNAc to Manα1-6] β-1,2-N-acetylglucosaminyltransferase-II). R = 1-4GlcNAcβ1-4GlcNAc-Asn-X

HARRY SCHACHTER

Program in Structural Biology and Biochemistry, Hospital for Sick Children, 555 University Avenue, Toronto, Ontario M5G 1X8, Canada
Tel. +1-416-813-5915; Fax +1-416-813-5022
e-mail: harry@sickkids.on.ca

Databanks

N-Acetylglucosaminyltransferase-II

NC-IUBMB enzyme classification: E.C.2.4.1.143

Species	Gene	Protein	mRNA	Genomic
Homo sapiens	*MGAT2*	NP_002399	NM_002408.2	U15128.1
		AAA86956	Hs.172195 (Unigene)	L36537
		Q10469		AL139099.2
		S66256		
Mus musculus	*Mgat2*	–	Mm.24293 (Unigene)	–
Rattus norvegicus	–	AAA86721	U21662.1	–
		Q09326		
Sus scrofa	*SSNAGATII*	CAA70732		Y09537.1
Drosophila melanogaster	–	AAF56991	AY055120	AE003772.1
				(CG7921)
Caenorhabditis elegans	–	CAB03823	AF251126.1	Z81458.1
		AAF71273		(C03E10.4)
Arabidopsis thaliana	–	AAD29068	–	AC007018.6

Name and History

There are six β-N-acetylglucosaminyltransferases (GnT-I–VI) known to incorporate GlcNAc into one of the three mannose residues of the N-glycan core, Manα1-6(Manα1-3)Manβ1-4GlcNAcβ1-4GlcNAc-Asn-X (Schachter 1986). GnT-I and -II were numbered according to their sequential order of action in the biosynthetic pathway leading to complex N-glycans (Schachter 1991). The full name of GnT-II is UDP-GlcNAc:α-6-D-mannoside β2-N-acetylglucosaminyltransferase II. Other names in the literature are mannosyl (α1,6-)-glycoprotein β2-N-acetylglucosaminyltransferase and α1,6-mannosyl-glycoprotein β2-N-acetylglucosaminyltransferase. Abbreviated names are β2-N-acetylglucosaminyltransferase II, GnT-II and GnT-II. The human gene is named *MGAT2*.

The enzyme has been described in humans (Tan et al. 1995), Chinese hamster ovary (CHO) cells (Narasimhan et al. 1977; Stanley et al. 1975b), rat (D'Agostaro et al. 1995), pig (Leeb et al. 1997), mouse (Y. Wang et al. Glycobiology, 2001, in press; Campbell et al. 1997), frog (Mucha et al. 1995), insects (*Spodoptera frugiperda*: Altmann et al. 1993; März et al. 1995; *Drosophila melanogaster*: J. Tan and H. Schachter unpublished results, 2001), nematodes (*Caenorhabditis elegans*: J. Tan et al. unpublished results, 2001); *Lymnaea stagnalis*: (Mulder et al. 1995), and plants (*Arabidopsis thaliana*: R. Strasser et al. unpublished results, 2001). The GnT-II gene has been cloned from many of these species (see Databanks).

Johnston et al. (1966) first reported the presence in goat colostrum of an enzyme activity catalyzing the incorporation of GlcNAc from UDP-GlcNAc into α₁-acid-glycoprotein pretreated with sialidase, β-galactosidase, and β-N-acetylglucosaminidase to uncover the mannose residues of the N-glycan core. It was subsequently shown (Gottlieb et al. 1975; Narasimhan et al. 1977; Stanley et al. 1975a, 1975b) that extracts from a CHO cell line (Lec1) resistant to the toxic action of

Phaseolus vulgaris phytohemagglutinin (PHA) can transfer GlcNAc to Manα1-6(GlcNAcβ1-2Manα1-3)Manβ1-R but not to Manα1-6(Manα1-3)Manβ1-R (R = 4GlcNAcβ1-4GlcNAc-Asn-X), whereas wild-type CHO cell extracts can transfer GlcNAc to both glycopeptide acceptors. It was concluded that the previously described GlcNAc-transferase activity was due to at least two separate enzymes. The enzyme missing from Lec1 cells was called GnT-I because it clearly had to act before the other enzyme (GnT-II). It was later shown (Schachter 1986; Schachter et al. 1983) that GnT-I action is a prerequisite to several enzymes in the complex *N*-glycan synthesis pathway (GnT-II–IV; mannosidase-II; core α6-fucosyltransferase). The resolution of GnT-I and -II depended on two novel (at that time) tools, lectin-resistant mutant somatic cell lines (Stanley et al. 1975a) and well-defined (by nuclear magnetic resonance spectroscopy) low molecular weight glycopeptide acceptors prepared from human multiple myeloma IgG (Narasimhan et al. 1977, 1980).

Enzyme Activity Assay and Substrate Specificity

The acceptor substrate now used for routine GnT-II assays is Manα1-6(GlcNAcβ1-2Manα1-3)Manβ-octyl (GnM$_3$-octyl) synthesized from Manα1-6(Manα1-3)Manβ-octyl (Toronto Research Chemicals, Toronto, Ontario, Canada) using recombinant GnT-I (Reck et al. 1994a). The enzyme assay reaction mixture (Bendiak and Schachter 1987a,b) contains, in a total volume of 0.025 ml, 0.5 mM GnM$_3$-octyl, 0.1 M MES, pH 6.1, 0.1 M NaCl, 20 mM MnCl$_2$, 0.5 mM UDP-[^{14}C]GlcNAc (2000–5000 dpm/nmole) or UDP-[^{3}H]GlcNAc (5000–50 000 dpm/nmole), 0.1% Triton X-100, 10 mM AMP, 0.2 M GlcNAc, and enzyme extract. If the enzyme has been purified and is soluble and free of hydrolases, bovine serum albumin (1 mg/ml) is added and Triton X-100, AMP, and GlcNAc are omitted. After incubation at 37°C for 30–120 min, the reaction is stopped either by adding 0.5 ml water and freezing, or by adding 0.4 ml 0.02 M sodium tetraborate/2 mM disodium EDTA. Purification of the product can be done in two ways. (i) The Dowex method, in which the reaction mixture is passed through a 1 ml column of AG 1-X8 (chloride form) equilibrated with water. The column is washed with 2 ml water, and the radioactivity in the eluate is counted after addition of suitable amounts of scintillation fluid. (ii) The SepPak method (Palcic et al. 1988), which is used if the substrate contains a hydrophobic group like octyl. Then product formation can be assayed, with little background, by adsorption to and elution from Sep-Pak C$_{18}$ reverse-phase cartridges. The Sep-Pak cartridge is activated with 20 ml methanol followed by 30 ml water, the reaction mixture is added to the cartridge, the cartridge is washed with 30 ml water, and the product is eluted with 3.0 ml methanol and counted. The values obtained are corrected against radioactivity obtained in control incubations lacking acceptor substrate.

GnT-II is an inverting glycosyltransferase. Detailed kinetic analysis of the purified rat liver enzyme (Bendiak and Schachter 1987a,b) has shown that catalysis is by an ordered sequential Bi–Bi mechanism in which UDP-GlcNAc binds first ($K_{ia} = k_{-1}/k_1 = 0.80$ mM) and UDP leaves last ($K_{iq} = k_4/k_{-4} = 0.85$ mM) (see Segel 1975 for definitions of the kinetic constants). Mn^{2+} is essential for activity, with an optimum at 10–20 mM. The minimal substrate is Manα1-6(GlcNAcβ1-2Manα1-3)Manβ1-R where R can be 4GlcNAcβ1-4GlcNAc-Asn-X, *N*-acetylglucosaminitol or a hydrophobic aglycone such as n-octyl. Galβ1-4 on the GlcNAcβ1-2Manα1-3 arm or a bisecting GlcNAcβ1-4 on the β-Man residue abolish catalytic activity.

UDP and analogues of UDP and UTP in which the hydrogen at the 5-position of the uracil is substituted with methyl, bromine, or mercury are good reversible inhibitors of the enzyme at 1.4 mM concentrations, but substitution at other sites lowers the inhibitory potency (Bendiak and Schachter 1987b). UDP-hexanolamine is a poor inhibitor even at 14 mM. These findings were used to design a suitable affinity matrix for purification of the enzyme (see below). UDP-Glc, UDP-Gal, and UDP-GalNAc inhibit the enzyme at 14 mM concentrations, but the enzyme cannot use UDP-Gal or UDP-GalNAc as donor substrates (UDP-Glc was not tested). The 2-deoxyManα1-6(GlcNAcβ1-2Manα1-3)Manβ1-R analogue is a competitive inhibitor ($K_i = 0.13$ mM), but the other hydroxyl groups of the Manα1-6 residue are not essential for activity (Reck et al. 1994b). Substitution of the C-4 hydroxyl of the Manβ residue by a GlcNAc residue or by a methyl group, but not its replacement by H, leads to an inactive substrate. The 3-deoxyManα1-3 and 3-deoxyGlcNAcβ1-2 derivatives are not active as substrates, but the 4-deoxy and 6-deoxy derivatives of both these residues are active. GlcNAcβ1-2Manα1-3Manβ1-octyl is a good inhibitor of the enzyme ($K_i = 0.9$ mM), indicating that this trisaccharide moiety is required for substrate binding to the enzyme.

Preparation

GnT-II has been partially purified from bovine colostrum (Harpaz and Schachter 1980), pig liver (Oppenheimer et al. 1981), and pig trachea (Mendicino et al. 1981; Oppenheimer et al. 1981). The rat liver enzyme was purified 60 000-fold to near homogeneity using affinity chromatography on a column containing UDP-GlcNAc linked at the 5-position of the uracil to thiopropyl-Sepharose via a 5-mercuri mercaptide bond (Bendiak and Schachter 1987a). The recombinant human enzyme was purified on a similar column containing UDP-GlcNAc linked at the 5-position of the uracil to cyanogen bromide-activated Sepharose via a 5-propylamine linker (Reck 1995).

Highly purified recombinant human GnT-II produced in the baculovirus/Sf9 insect cell system (Tan et al. 1995) had a specific activity of 20 International Units (IU)/mg (1 IU = 1 µmole/min). The recombinant enzyme was used to convert various derivatives of Manα1-6(GlcNAcβ1-2Manα1-3)Manβ1-octyl to derivatives of GlcNAcβ1-2Manα1-6(GlcNAcβ1-2Manα1-3)Manβ1-octyl (Reck et al. 1995); these compounds are substrates and potential inhibitors of GnT-III–V. Active recombinant GnT-II has been obtained by expression of cDNA from all the species shown in the Databanks table, and also from frog (Mucha et al. 1995).

Biological Aspects

Tissue Distribution

The gene is widely expressed in many mammalian tissues, i.e., blood, brain, and other parts of the central nervous system, breast, colon, ear, germ cells, heart, kidney, liver, lung, lymph, mammary gland, muscle, ovary, pancreas, parathyroid, placenta, prostate, spleen, stomach, testis, thymus, tonsil, uterus, and cervix, and whole embryo (Unigenes Hs.172195 and Mm.24293; D'Agostaro et al. 1995; Tan et al. 1995).

The GnT-II Gene

The human (Tan et al. 1995), mouse (Campbell et al. 1997), pig (Leeb et al. 1997), rat (D'Agostaro et al. 1995), frog (Mucha et al. 1995), *C. elegans* (J. Tan et al. unpublished results, 2001), *D. melanogaster* (J. Tan and H. Schachter, unpublished results, 2001), and *A. thaliana* (R. Strasser et al. unpublished results, 2001) GnT-II genes have been cloned and expressed (Databanks table). The protein sequence shows the type-2 domain structure typical of all previously cloned Golgi-bound glycosyltransferases, e.g., human GnT-II has cytoplasmic *N*-terminal, and hydrophobic and lumenal C-terminal domains of 9, 20, and 418 residues, respectively. There is a very limited sequence homology to GnT-I but not to any other previously cloned glycosyltransferase. The human amino acid sequence shows identities of 89%/446 residues for rat and 40%/310 residues for *C. elegans*. The human, pig, *C. elegans*, *D. melanogaster*, and *A. thaliana* GnT-II genes are on chromosomes *14q21*, *1q23-q27*, *V*, *3R*, and *II*, respectively. The open reading frame and 3′-untranslated region of the human, rat, and pig genes are on a single exon. The *C. elegans* and *D. melanogaster* GnT-II genes, however, contain at least six and five exons, respectively.

5′-RACE (rapid amplification of cDNA ends) and RNase protection analyses of the human GnT-II gene showed multiple transcription initiation sites at −440 to −489 bp relative to the ATG translation start codon (+1), proving that the entire gene is on a single exon (Chen et al. 1998). The gene has three AATAAA polyadenylation sites downstream of the translation stop codon, all of which are used for transcription termination. The gene has a CCAAT but lacks a TATA box, and the 5′-untranslated region is GC-rich and contains consensus sequences suggestive of multiple binding sites for Sp1; these properties are typical of a housekeeping gene. The pig GnT-II gene also has a GC-rich TATA-less promoter with multiple transcription start sites (Leeb et al. 1997). A series of chimeric constructs containing different lengths of the 5′-untranslated region of the human gene fused to the chloramphenicol acetyltransferase (CAT) reporter gene were tested in transient transfection experiments using HeLa cells (Chen et al. 1998). The CAT activity of the construct containing the longest insert (−1076 bp relative to the ATG start codon) showed a ~38-fold increase as compared to that of the control. Removal of the region between −636 and −553 bp caused a dramatic decrease in CAT activity, indicating that this is the main promoter region of the gene.

Oncogenic transformation of fibroblasts by the *src* oncogene has long been known to cause an increase in the size of cell-surface protein-bound oligosaccharides, owing primarily to increased *N*-glycan branching mediated by GnT-V. The *src*-responsive element of the GnT-V promoter was localized to Ets-binding sites (Buckhaults et al. 1997; Kang et al. 1996). GnT-V requires the prior action of GnT-II, and the human GnT-II promoter contains four putative Ets-binding sites (Chen et al. 1998), suggesting that GnT-II, although a housekeeping enzyme, may in some circumstances be under the control of Ets transcription factor. Co-transfection into HepG2, HeLa or Cos-1 cells of either *ets*-1 or *ets*-2 expression plasmids, together with chimeric GnT-II promoter-CAT plasmids (Zhang et al. 2000), results in a 2–4-fold stimulation of promoter activity. Gel mobility shift assays and South-Western blots localized the functional Ets-binding site to one of the four sites in the GnT-II promoter. Unlike the GnT-V promoter, which is activated by both Ets and Src, the GnT-II promoter is not activated by Src, indicating that the functional role of Ets is different for the two genes.

The Role of Complex N-Glycans in Mammalian Development—Men and Mice with Null Mutations in the GnT-II Gene

Complex *N*-glycans are absent from bacteria and yeast, are present in small amounts in some protozoa, but are major components of all multi-cellular invertebrate and vertebrate animals which have been analyzed, i.e., nematodes, schistosomes, molluscs, insects, fish, birds, and mammals (see Chen et al. 1999 for references). This indicates that complex *N*-glycan synthesis appeared in evolution just prior to the development of multicellular organisms, probably because complex *N*-glycans are required for the cell–cell interaction process and normal development of multicellular animals. The study of mutations in man and mouse provided support for this concept.

Congenital disorders of glycosylation (CDG, formerly called carbohydrate-deficient glycoprotein syndromes) are a group of autosomal recessive diseases with multisystemic abnormalities, including a severe disturbance of nervous system development (Jaeken et al. 1993). Fibroblasts and mononuclear cells from two children with CDG Type II (CDG-II) have been shown to lack GnT-II activity due to homozygous point mutations in the GnT-II gene, and are therefore unable to synthesize normal complex *N*-glycans (Charuk et al. 1995; Jaeken et al. 1994, 1996; Tan et al. 1996). CDG-II presents with failure to thrive, psychomotor retardation, facial dysmorphism, hypotonia, kyphoscoliosis, cardiac murmur associated with a ventricular septal defect, osteopenia, frequent infections, gastrointestinal problems, coagulopathy, and abnormal stereotypic behavior. Laboratory tests showed low serum values for many glycoproteins (including several coagulation factors) and other abnormalities. The dominant serum transferrin glycoform from CDG-II patients has only two sialic acids per mole rather than the normal value of four due to truncation of the Manα1-6Manβ1-4 *N*-glycan antennae.

Mouse embryos lacking a functional GnT-I gene die prenatally at about 10 days of gestation (Ioffe and Stanley 1994; Metzler et al. 1994) with multisystemic abnormalities and no obvious cause of death. Over 60% of mouse embryos with null mutations in the GnT-II gene (Y. Wang et al. Glycobiology, 2001, in press; Campbell et al. 1997) survive to term, but 99% of newborns die during the first week of postnatal development. Survivors lacking *Mgat2* function are runts and exhibit facial dysmorphism, kyphoscoliosis, muscular atrophy, tremors, and osteopenia. Pathohistologic findings indicate reduced ossification at the growth plates, increased prevalence of skeletal cartilage, and a calcified bone density that is reduced by over 30%. Although female mice are fertile, males are infertile, with a block in spermatogenesis. Various serum coagulation factors show reduced levels. The majority of *Mgat2*-deficient mice die with gastrointestinal blockage. These and other findings indicate that the surviving *Mgat2*-deficient mice and human CDG-II patients are strikingly similar in pathological and biochemical features.

Future Perspectives

Although the above findings clearly establish that complex *N*-glycans are essential for normal postimplantation embryogenesis and development in both man and mouse, the detailed mechanisms involved are not understood. There is strong evidence to suggest that an underlying factor is the interaction between cell surface *N*-glycans and receptors on nearby cells and in the noncellular environment. Classic examples are (a) the interactions between leukocytes and the endothelial lining of small blood vessels mediated by the binding of selectins to their carbohydrate ligands, and (b) the role of highly branched complex *N*-glycans in metastasis (Dennis et al. 1999; McEver 1997). Since embryogenesis and development involve a large number of *N*-glycan-carrying glycoproteins, it is a daunting task to unravel the specific receptor-ligand systems that are essential to these processes in organisms such as man or mouse. Several groups have therefore initiated glycobiology studies on other models of development, such as *C. elegans* (Arata et al. 1997; Chen et al. 1999; DeBose-Boyd et al. 1998; Drickamer and Dodd 1999; Hagen and Nehrke 1998; Herman and Horvitz 1999; Hirabayashi et al. 1996) and *D. melanogaster* (Foster et al. 1995; Jurado et al. 1999; Kerscher et al. 1995; Rabouille et al. 1999; Roberts et al. 1998; Selleck 2000). These organisms are hardly "simple," and undergo rather complex developmental processes, but the availability of the complete genomic sequences and sophisticated genetic and developmental biological tools may provide answers that are not readily obtained in mammalian systems.

References

Altmann F, Kornfeld G, Dalik T, Staudacher E, Glossl J (1993) Processing of asparagine-linked oligosaccharides in insect cells: N-acetylglucosaminyltransferase I and II activities in cultured lepidopteran cells. Glycobiology 3:619–625

Arata Y, Hirabayashi J, Kasai K (1997) Structure of the *32-kDa* galectin gene of the nematode *Caenorhabditis elegans*. J Biol Chem 272:26669–26677

Bendiak B, Schachter H (1987a) Control of glycoprotein synthesis. XII. Purification of UDP-GlcNAc:α-D-mannoside β1-2-N-acetylglucosaminyltransferase II from rat liver. J Biol Chem 262:5775–5783

Bendiak B, Schachter H (1987b) Control of glycoprotein synthesis. XIII. Kinetic mechanism, substrate specificity, and inhibition characteristics of UDP-GlcNAc:α-D-mannoside β1-2-N-acetylglucosaminyltransferase II from rat liver. J Biol Chem 262:5784–5790

Buckhaults P, Chen L, Fregien N, Pierce M (1997) Transcriptional regulation of N-acetylglucosaminyltransferase V by the *src* oncogene. J Biol Chem 272:19575–19581

Campbell R, Tan J, Schachter H, Bendiak B, Marth J (1997) Targeted inactivation of the murine UDP-GlcNAc:alpha-6-D-mannoside beta-1,2-N-acetylglucosaminyltransferase II gene. Glycobiology 7:1050

Charuk JHM, Tan J, Bernardini M, Haddad S, Reithmeier RAF, Jaeken J, Schachter H (1995) Carbohydrate-deficient glycoprotein syndrome type II. An autosomal recessive

N-acetylglucosaminyltransferase II deficiency different from typical hereditary erythroblastic multinuclearity, with a positive acidified-serum lysis test (HEMPAS). Eur J Biochem 230:797–805

Chen SH, Zhou SH, Tan J, Schachter H (1998) Transcriptional regulation of the human UDP-GlcNAc: alpha-6-D-mannoside beta-1-2-N-acetylglucosaminyltransferase II gene (MGAT2) which controls complex N-glycan synthesis. Glycoconjugate J 15:301–308

Chen SH, Zhou SH, Sarkar M, Spence AM, Schachter H (1999) Expression of three Caenorhabditis elegans N-acetylglucosaminyltransferase I genes during development. J Biol Chem 274:288–297

D'Agostaro GAF, Zingoni A, Moritz RL, Simpson RJ, Schachter H, Bendiak B (1995) Molecular cloning and expression of cDNA encoding the rat UDP-N-acetylglucosamine:alpha-6-D-mannoside beta-1,2-N-acetylglucosaminyltransferase II. J Biol Chem 270:15211–15221

DeBose-Boyd RA, Nyame AK, Cummings RD (1998) Molecular cloning and characterization of an alpha-1,3-fucosyltransferase, CEFT-1, from Caenorhabditis elegans. Glycobiology 8:905–917

Dennis JW, Granovsky M, Warren CE (1999) Protein glycosylation in development and disease. Bioessays 21:412–421

Drickamer K, Dodd RB (1999) C-type lectin-like domains in Caenorhabditis elegans: predictions from the complete genome sequence. Glycobiology 9:1357–1369

Foster JM, Yudkin B, Lockyer AE, Roberts DB (1995) Cloning and sequence analysis of GmII, a Drosophila melanogaster homologue of the cDNA encoding murine Golgi alpha-mannosidase II. Gene 154:183–186

Gottlieb C, Baenziger J, Kornfeld S (1975) Deficient uridine diphosphate-N-acetylglucosamine:glycoprotein N-acetylglucosaminyltransferase activity in a clone of Chinese hamster ovary cells with altered surface glycoproteins. J Biol Chem 250:3303–3309

Hagen FK, Nehrke K (1998) cDNA cloning and expression of a family of UDP-N-acetyl-D-galactosamine:polypeptide N-acetylgalactosaminyltransferase sequence homologs from Caenorhabditis elegans. J Biol Chem 273:8268–8277

Harpaz N, Schachter H (1980) Control of glycoprotein synthesis. IV. Bovine colostrum UDP-N-acetylglucosamine: α-D-mannoside β-2-N-acetylglucosaminyltransferase I. Separation from UDP-N-acetylglucosamine: α-D-mannoside β-2-N-acetylglucosaminyltransferase II, partial purification and substrate specificity. J Biol Chem 255:4885–4893

Herman T, Horvitz HR (1999) Three proteins involved in Caenorhabditis elegans vulval invagination are similar to components of a glycosylation pathway. Proc Natl Acad Sci USA 96:974–979

Hirabayashi J, Ubukata T, Kasai K (1996) Purification and molecular characterization of a novel 16-kDa galectin from the nematode Caenorhabditis elegans. J Biol Chem 271:2497–2505

Ioffe E, Stanley P (1994) Mice lacking N-acetylglucosaminyltransferase I activity die at mid-gestation, revealing an essential role for complex or hybrid N-linked carbohydrates. Proc Natl Acad Sci USA 91:728–732

Jaeken J, Carchon H, Stibler H (1993) The carbohydrate-deficient glycoprotein syndromes—pre-Golgi and Golgi disorders? Glycobiology 3:423–428

Jaeken J, Schachter H, Carchon H, Decock P, Coddeville B, Spik G (1994) Carbohydrate deficient glycoprotein syndrome type II: a deficiency in Golgi localised N-acetylglucosaminyltransferase II. Arch Dis Child 71:123–127

Jaeken J, Spik G, Schachter H (1996) Carbohydrate-deficient glycoprotein syndrome Type II: an autosomal recessive disease due to mutations in the N-acetylglucosaminyl transferase II gene. In: Montreuil J, Vliegenthart JFG, Schachter H (eds) Glycoproteins and disease. Elsevier, Amsterdam, pp 457–467

Johnston IR, McGuire EJ, Jourdian GW, Roseman S (1966) Incorporation of *N*-acetyl-D-glucosamine into glycoproteins. J Biol Chem 241:5735–5737

Jurado LA, Coloma A, Cruces J (1999) Identification of a human homolog of the *Drosophila* rotated abdomen gene (*POMT1*) encoding a putative protein *O*-mannosyl-transferase, and assignment to human chromosome *9q34.1*. Genomics 58:171–180

Kang R, Saito H, Ihara Y, Miyoshi E, Koyama N, Sheng Y, Taniguchi N (1996) Transcriptional regulation of the *N*-acetylglucosaminyltransferase V gene in human bile duct carcinoma cells (*HuCC-T1*) is mediated by Ets-1. J Biol Chem 271:26706–26712

Kerscher S, Albert S, Wucherpfennig D, Heisenberg M, Schneuwly S (1995) Molecular and genetic analysis of the *Drosophila mas-1* (mannosidase-1) gene which encodes a glycoprotein processing alpha-1,2-mannosidase. Dev Biol 168:613–626

Leeb T, Kriegesmann B, Baumgartner BG, Klett C, Yerle M, Hameister H, Brenig B (1997) Molecular cloning of the porcine beta-1,2-*N*-acetylglucosaminyltransferase II gene and assignment to chromosome *1q23-q27*. BBA Gen Subj 1336:361–366

März L, Altmann F, Staudacher E, Kubelka V (1995) Protein glycosylation in insects. In: Montreuil J, Vliegenthart JFG, Schachter H (eds) Glycoproteins. Elsevier, Amsterdam, pp 543–563

McEver RP (1997) Selectin–carbohydrate interactions during inflammation and metastasis. Glycoconj J 14:585–591

Mendicino J, Chandrasekaran EV, Anumula KR, Davila M (1981) Isolation and properties of α-D-mannose:β-1,2-*N*-acetylglucosaminyltransferase from trachea mucosa. Biochemistry 20:967–976

Metzler M, Gertz A, Sarkar M, Schachter H, Schrader JW, Marth JD (1994) Complex asparagine-linked oligosaccharides are required for morphogenic events during post-implantation development. EMBO J 13:2056–2065

Mucha J, Kappel S, Schachter H, Hane W, Glössl J (1995) Molecular cloning and characterization of cDNAs coding for *N*-acetylglucosaminyltransferases I and II from *Xenopus laevis* ovary. Glycoconj J 12:473

Mulder H, Dideberg F, Schachter H, Spronk BA, De Jong-Brink M, Kamerling JP, Vliegenthart JFG (1995) In the biosynthesis of *N*-glycans in connective tissue of the snail *Lymnaea stagnalis*, incorporation of GlcNAc by beta-2-GlcNAc-transferase I is an essential prerequisite for the action of beta-2-GlcNAc-transferase II and beta-2-Xyl-transferase. Eur J Biochem 232:272–283

Narasimhan S, Stanley P, Schachter H (1977) Control of glycoprotein synthesis. II. Lectin-resistant mutant containing only one of two distinct *N*-acetylglucosaminyltransferase activities present in wild type Chinese hamster ovary cells. J Biol Chem 252:3926–3933

Narasimhan S, Harpaz N, Longmore G, Carver JP, Grey AA, Schachter H (1980) Control of glycoprotein synthesis. The purification by preparative high voltage electrophoresis in borate of glycopeptides containing high mannose and complex oligosaccharide chains linked to asparagine. J Biol Chem 255:4876–4884

Oppenheimer CL, Eckhardt AE, Hill RL (1981) The non-identity of porcine *N*-acetylglucosaminyltransferases I and II. J Biol Chem 256:11477–11482

Palcic MM, Heerze LD, Pierce M, Hindsgaul O (1988) The use of hydrophobic synthetic glycosides as acceptors in glycosyltransferase assays. Glycoconj J 5:49–63

Rabouille C, Kuntz DA, Lockyer A, Watson R, Signorelli T, Rose DR, van den Heuvel M, Roberts DB (1999) The *Drosophila GMII* gene encodes a Golgi α-mannosidase II. J Cell Sci 112:3319–3330

Reck F (1995) Synthesis of uridine-5-propylamine derivatives and their use in affinity chromatography of *N*-acetylglucosaminyltransferases I and II. Carbohydr Res 276:321–335

Reck F, Springer M, Paulsen H, Brockhausen I, Sarkar M, Schachter H (1994a) Synthesis of tetrasaccharide analogues of the *N*-glycan substrate of beta-(1→2)-*N*-acetylglucosaminyltransferase II using trisaccharide precursors and recombinant beta-(1→2)-*N*-acetylglucosaminyltransferase I. Carbohydr Res 259:93–101

Reck F, Meinjohanns E, Springer M, Wilkens R, Vandorst JALM, Paulsen H, Moller G, Brockhausen I, Schachter H (1994b) Synthetic substrate analogues for UDP-GlcNAc:Man alpha-1-6-R beta(1-2)-n-acetylglucosaminyltransferase II. Substrate specificity and inhibitors for the enzyme. Glycoconj J 11:210–216

Reck F, Meinjohanns E, Tan J, Grey AA, Paulsen H, Schachter H (1995) Synthesis of pentasaccharide analogues of the N-glycan substrates of N-acetylglucosaminyltransferases III, IV and V using tetrasaccharide precursors and recombinant beta-(1→2)-N-acetylglucosaminyltransferase II. Carbohydr Res 275:221–229

Roberts DB, Mulvany WJ, Dwek RA, Rudd PM (1998) Mutant analysis reveals an alternative pathway for N-linked glycosylation in Drosophila melanogaster. Eur J Biochem 253:494–498

Schachter H (1986) Biosynthetic controls that determine the branching and microheterogeneity of protein-bound oligosaccharides. Biochem Cell Biol 64:163–181

Schachter H (1991) The "yellow brick road" to branched complex N-glycans. Glycobiology 1:453–461

Schachter H, Narasimhan S, Gleeson P, Vella G (1983) Control of branching during the biosynthesis of asparagine-linked oligosaccharides. Can J Biochem Cell Biol 61:1049–1066

Segel IH (1975) Enzyme kinetics. Wiley, New York

Selleck SB (2000) Proteoglycans and pattern formation. Sugar chemistry meets developmental genetics. Trends in Genetics 16:206–212

Stanley P, Caillibot V, Siminovitch L (1975a) Selection and characterization of eight phenotypically distinct lines of lectin-resistant Chinese hamster ovary cells. Cell 6:121–128

Stanley P, Narasimhan S, Siminovitch L, Schachter H (1975b) Chinese hamster ovary cells selected for resistance to the cytotoxicity of phytohemagglutinin are deficient in a UDP-N-acetylglucosamine: glycoprotein N-acetylglucosaminyltransferase activity. Proc Natl Acad Sci USA 72:3323–3327

Tan J, D'Agostaro GAF, Bendiak B, Reck F, Sarkar M, Squire JA, Leong P, Schachter H (1995) The human UDP-N-acetylglucosamine: alpha-6-D-mannoside-beta-1,2-N-acetylglucosaminyltransferase II gene (MGAT2)—cloning of genomic DNA, localization to chromosome 14q21, expression in insect cells and purification of the recombinant protein. Eur J Biochem 231:317–328

Tan J, Dunn J, Jaeken J, Schachter H (1996) Mutations in the MGAT2 gene controlling complex N-glycan synthesis cause carbohydrate-deficient glycoprotein syndrome type II, an autosomal recessive disease with defective brain development. Am J Hum Genet 59:810–817

Zhang W, Revers L, Pierce M, Schachter H (2000) Regulation of expression of the human beta-1,2-N-acetylglucosaminyltransferase II gene (MGAT2) by Ets transcription factors. Biochem J 347:511–518

11

N-Acetylglucosaminyltransferase-III

Introduction

N-Acetylglucosaminyltransferase-III (β-1,4-mannosyl-glycoprotein β1,4-N-acetylglucosaminyltransferase: EC 2.4.1.144) catalyzes the formation of a unique structure, bisecting GlcNAc, and is involved in the biosynthesis of complex and hybrid types of N-glycans. The addition of the bisecting GlcNAc residue to a core β-mannose by the enzyme prevents the actions of other GlcNAc transferases involved in the biosynthesis of multiantennary sugar chains, leading to a decrease in the branch formation of N-glycans. Because of this regulatory role, the enzyme has been considered to be a key glycosyltransferase in N-glycan biosynthesis. Relatively high levels of the activity were found in kidney and brain of mammals (Nishikawa et al. 1988a). Expression of GnT-III is enhanced during hepatocarcinogenesis, while the activity is nearly undetectable in normal liver (Narasimhan et al. 1988; Nishikawa et al. 1988b). Since expression of the enzyme appears to lead to a remarkable structural alteration of the sugar chains on the cell surface, it seems that the enzyme is associated with various biological events such as differentiation and carcinogenesis.

YOSHITAKA IKEDA and NAOYUKI TANIGUCHI

Department of Biochemistry, Osaka University Medical School, 2-2 Yamadaoka, Suita, Osaka 565-0871, Japan
Tel. +81-6-6879-3420; Fax +81-6-6879-3429
e-mail: proftani@biochem.med.osaka-u.ac.jp

Databanks

N-Acetylglucosaminyltransferase-III

NC-IUBMB enzyme classificantion: E.C.2.4.1.144

Species	Gene	Protein	mRNA	Genomic
Homo sapiens	–	–	D13789	D85377–D85379
			L48489	AL022312
	MGAT3	–	NM_002409	–
			Hs.112 (Unigene)	–
Rattus norvegicus			D10852	–
			E13193	–
Mus musculus	*Mgat3*	–	NM_010795	L39373
			Mm.57077 (Unigene)	U66844

Name and History

β4-*N*-Acetylglucosaminyltransferase-III is frequently abbreviated GnT-III or GlcNAcT-III. The activity of this enzyme was first observed in hen oviduct by Narasimhan (1982). GnT-III was purified from rat kidney, and the cDNAs for the rat, human, and mouse enzymes have been cloned (Nishikawa et al. 1992; Ihara et al. 1993). A primary structure of mouse enzyme was determined via isolation of a genomic clone (Bhaumik et al. 1995). Genomic analysis revealed that the coding region is encoded by a single exon, and the human gene was localized to chromosome 22q.13.1 using fluorescence in situ hybridization (Ihara et al. 1993).

Enzyme Activity Assay and Substrate Specificity

GnT-III catalyzes the transfer of GlcNAc from UDP-GlcNAc to a core β-Man of an *N*-glycan via β1,4-linkage (Fig. 1). This transferred β-GlcNAc residue is referred to as a bisecting GlcNAc. The assay for GnT-III activity was carried out using radiolabeled UDP-GlcNAc as a donor substrate (Narasimhan 1982), as has frequently been performed for many other glycosyltransferases. Another sensitive and convenient

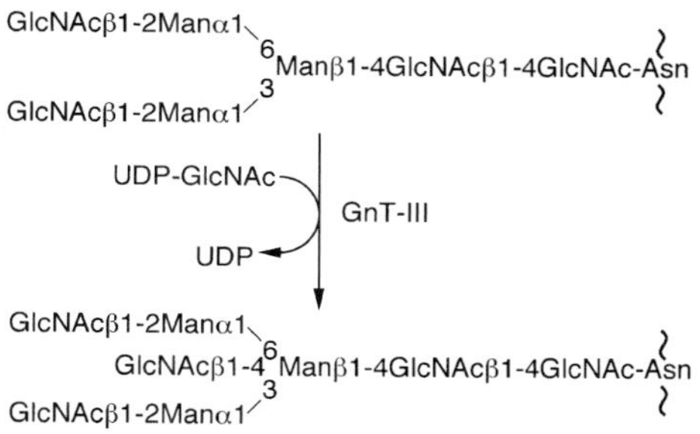

Fig. 1. A reaction catalyzed by GnT-III. A GlcNAc residue linked to an α1,6Man may be substituted by a Man, and the presence of α1,6fucosyl residue at the innermost GlcNAc does not affect the action of GnT-III. An acceptor oligosaccharide is active even when its reducing end is modified by reductive amination, e.g., labeling with 2-aminopyridine

method involving a fluorescence-labeled oligosaccharide acceptor, pyridylaminated agalacto-biantennary sugar chain, is also available (Nishikawa et al. 1988a; Taniguchi et al. 1989). In this assay system, a product, pyridylaminated bisected agalacto-biantennary, can be separated from the unreacted acceptor and quantitated by reversed phase HPLC equipped with a fluorescence detector. This enzyme requires a divalent cation, typically a manganese ion, for its reaction, and therefore approximately 10 mM Mn should be included in the reaction.

The substrate specificity of GnT-III toward the acceptor has been investigated (Schachter et al. 1983; Gleeson et al. 1983; Allen et al. 1984; Bendiak and Schachter 1987; Brockhausen et al. 1988). The specificity studies showed that β1,2GlcNAc linked to the α1,3Man residue, whose reaction is catalyzed by GnT-I, is absolutely required, and that β-galactosylation of this GlcNAc inhibits the action of GnT-III. GnT-III is capable of transferring GlcNAc to agalacto forms of tri- and tetra-antennae as well as bi-antennary. α1,6Fucosylated oligosaccharides are also active as the acceptor. The specificity with respect to the donor was also investigated using a purified recombinant enzyme (Ikeda et al. 2000). A kinetic study with various UDP-sugars revealed that the enzyme can transfer GalNAc and Glc with 0.1%–0.2% catalytic efficiency as compared with GlcNAc, although the transfer of Gal was not observed. In addition, it appears that GnT-III utilizes UDP-sugars but not ADP-, GDP-, CDP-, and TDP-sugars.

Structural Chemistry

GnT-III is a typical type-II membrane glycoprotein which consists of an *N*-terminal short cytoplasmic tail, a transmembrane domain, a stem region and a large catalytic domain, and the molecular mass of the protein portion is approximately 53 kDa, as calculated on the basis of the nucleotide sequence of cDNA. In the case of rat GnT-III, the enzyme has three potential sites for *N*-glycosylation, all of which are actually glycosylated, and furthermore it was found that these *N*-linked sugar chains are essential for the fully active enzyme and its Golgi localization (Nagai et al. 1997). The amino acid sequence of GnT-III contains a D–X–D motif, which is conserved in certain divalent cation-requiring glycosyltransferases and is thought to be involved in the coordination of the metal.

Preparation

GnT-III has been purified from rat kidney microsomal fraction by the use of several column chromatographic techniques, including ligand-coupled matrices based on the affinities to the donor and acceptor. Bacterial expression of the active recombinant enzyme was unsuccessful, probably because of the requirement of *N*-glycosylation for the activity, and thus use of the eukaryotic expression system is essential for the preparation of the active recombinant enzyme. An SV40-based vectors/COS cell system or β-actin promoter/CMV-enhancer allows overexpression of active GnT-III in mammalian cells (Nishikawa et al. 1992; Tanemura et al. 1997). A secretable form of rat GnT-III was expressed by a baculovirus/insect cell system, and the recombinant enzyme was purified from the culture medium (Ikeda et al. 2000).

Biological Aspects

The addition of the bisecting GlcNAc to acceptor oligosaccharides by GnT-III prevents the subsequent actions of other GlcNAc-transferases, GnTs-IV and -V, both of which are involved in the biosynthesis of tri- and tetra-antennary sugar chains, and thereby the action of GnT-III leads to the inhibition of further branch formation (Schachter 1986). It has therefore been suggested that GnT-III plays a regulatory role in the biosynthesis of N-linked oligosaccharides. As has been shown by DNA transfection experiments with GnT-III cDNA, the resulting increase in the level of sugar chains bearing bisecting GlcNAc residues (bisected sugar chains) gives rise to some significant alterations in cells. Overexpression of GnT-III in highly metastatic melanoma cells reduced the $\beta 1,6$-branch in cell-surface N-glycans, in conjunction with the increase in the bisecting GlcNAc, and this structure has been considered to be associated with metastatic potential. This structural alteration was found to suppress the lung metastasis of the melanoma cells (Yoshimura et al. 1995). It was also shown that GnT-III transfected K562 cells are resistant to the cytotoxicity of the natural killer cell and develop spleen colonization in athymic mice (Yoshimura et al. 1996). In addition, it has been reported that increased levels of GnT-III result in the reduction of gene expression of hepatitis B virus (Miyoshi et al. 1995) and the altered sorting of glycoproteins in cells (Sultan et al. 1997). It was also suggested that overexpression of GnT-III impairs the functions of the receptors for epidermal growth factor and nerve growth factor (Ihara et al. 1997; Rebbaa et al. 1997). Thus, GnT-III could potentially be responsible for a variety of biological events, although the detailed molecular bases are not known.

Transgenic mice in which GnT-III was expressed specifically in the liver by use of a serum amyloid P component gene promoter were established to investigate the biological significance of the enzyme in hepatocytes (Ihara et al. 1998). Histological examination revealed that the hepatocytes in the transgenic mice had a swollen oval-like morphology, and an abnormal lipid accumulation was also observed within the cells. This lipid storage appeared to be associated with impairment of apolipoprotein B secretion, as revealed by biochemical analyses, and thus it was suggested that aberrant glycosylation caused by the ectopic expression of GnT-III disrupts the function of apolipoprotein B. Furthermore, a study using another transgenic mouse which systemically expresses GnT-III suggested that expression of the bisecting GlcNAc suppresses stroma-dependent hemopoiesis (Yoshimura et al. 1998).

In order to explore the physiological roles of GnT-III, on the other hand, GnT-III-deficient mice have also been established by gene targeting (Priatel et al. 1997; Bhaumik et al. 1998). No bisected sugar chains were found in the deficient mice, as shown by lectin blot analysis, and this indicated that mice have no other isoenzymes which are capable of forming the bisecting GlcNAc. Although the studies based on overexpression of GnT-III by the transgene, i.e., GnT-III transfected cells and transgenic mice, have shown the involvement of GnT-III and the bisected sugar chains in a variety of biological events, the growth and development of the GnT-III-deficient mice were apparently normal (Priatel et al. 1997). Because of the absence of apparent physical abnormalities, the physiological roles of GnT-III remain unclear. In the treatment with diethylnitrosamine, however, the suppression of hepatocarcinogenesis was observed in the deficient mice (Bhaumik et al. 1998). A further study suggested that this suppression of carcinogenesis involves a serum glycoprotein(s) which is biosynthesized in other tissues, and is possibly due to a structural alteration of the oligosaccharide moiety of the glycoprotein responsible (Yang et al. 2000).

Future Perspectives

It has been believed that GnT-III plays a regulatory role in the biosynthesis of *N*-linked oligosaccharides via the inhibitory effects of the bisecting GlcNAc on the formation of tri- and tetra-antennary sugar chains. Although the biosyntheses of biologically functional sugar chains may be modified by the enzyme, the loss of GnT-III activity and the bisecting GlcNAc led to no apparent abnormalities in mice, and thus their biological significance is still unknown. However, it seems likely that GnT-III plays an important role in diseases such as carcinogenesis and cancer metastasis. In order to understand the role of GnT-III in more detail, it is necessary to identify the phenotypes of the GnT-III-deficient mice under pathological conditions, such as infection and carcinogenesis, rather than in a healthy state.

References

Allen SD, Tsai D, Schachter H (1984) Control of glycoprotein synthesis. The in vitro synthesis by hen oviduct membrane preparations of hybrid asparagine-linked oligosaccharides containing five mannose residues. J Biol Chem 259:6984–6990

Bendiak B, Schachter H (1987) Control of glycoprotein synthesis. Kinetic mechanism, substrate specificity, and inhibition characteristics of UDP-*N*-acetylglucosamine:α-D-mannoside β1–2 *N*-acetylglucosaminyltransferase II from rat liver. J Biol Chem 262:5784–5790

Bhaumik M, Seldin MF, Stanley P (1995) Cloning and chromosomal mapping of the mouse *Mgat3* gene encoding *N*-acetylglucosaminyltransferase III. Gene 164:295–300

Bhaumik M, Harris T, Sundaram S, Johnson L, Guttenplan J, Rogler C, Stanley P (1998) Progression of hepatic neoplasms is severely retarded in mice lacking the bisecting *N*-acetylglucosamine on *N*-glycans: evidence for a glycoprotein factor that facilitates hepatic tumor progression. Cancer Res 58:2881–2887

Brockhausen I, Carver JP, Schachter H (1988) Control of glycoprotein synthesis. The use of oligosaccharide substrates and HPLC to study the sequential pathway for *N*-acetylglucosaminyltransferases I, II, III, IV, V, and VI in the biosynthesis of highly branched *N*-glycans by hen oviduct membranes. Biochem Cell Biol 66:1134–1151

Gleeson PA, Schachter H (1983) Control of glycoprotein synthesis. J Biol Chem 258: 6162–6173

Ihara Y, Nishikawa A, Tohma T, Soejima H, Niikawa N, Taniguchi N (1993) cDNA cloning, expression, and chromosomal localization of human *N*-acetylglucosaminyltransferase III (GnT-III). J Biochem 113:692–698

Ihara Y, Sakamoto Y, Mihara M, Shimizu K, Taniguchi N (1997) Overexpression of *N*-acetylglucosaminyltransferase III disrupts the tyrosine phosphorylation of Trk with resultant signaling dysfunction in PC12 cells treated with nerve growth factor. J Biol Chem 272:9629–9634

Ihara Y, Yoshimura M, Miyoshi E, Nishikawa A, Sultan AS, Toyosawa S, Ohnishi A, Suzuki M, Yamamura K, Ijuhin N, Taniguchi N (1998) Ectopic expression of *N*-acetylglucosaminyltransferase III in transgenic hepatocytes disrupts apolipoprotein

B secretion and induces aberrant cellular morphology with lipid storage. Proc Natl Acad Sci USA 95:2526–2530

Ikeda Y, Koyota S, Ihara H, Yamaguchi Y, Korekane H, Tsuda T, Sasai K, Taniguchi N (2000) Kinetic basis for the donor nucleotide-sugar specificity of β-1,4-N-acetylglucosaminyltransferase III. J Biochem 128:609–619.

Miyoshi E, Ihara Y, Hayashi N, Fusamoto H, Kamada T, Taniguchi N (1995) Transfection of N-acetylglucosaminyltransferase III gene suppresses expression of hepatitis B virus in a human hepatoma cell line, HB611. J Biol Chem 270:28311–28315

Nagai K, Ihara Y, Wada Y, Taniguchi N (1997) N-glycosylation is a requisite for the enzyme activity and Golgi retention of N-acetylglucosaminyltransferase III. Glycobiology 7:769–776

Narasimhan S (1982) Control of glycoprotein synthesis. UDP-GlcNAc:glycopeptide β4-N-acetylglucosaminyltransferase III, an enzyme in hen oviduct which adds GlcNAc in β1-4 linkage to the β-linked mannose of the trimannosylcore of N-glycosyl oligosaccharides. J Biol Chem 257:10235–10242

Narasimhan S, Schachter H, Rajalakshmi S (1988) Expression of N-acetylglucosaminyltransferase III in hepatic nodules during rat liver carcinogenesis promoted by orotic acid. J Biol Chem 263:1273–1281

Nishikawa A, Fujii S, Sugiyama T, Taniguchi N (1988a) A method for the determination of N-acetylglucosaminyltransferase III activity in rat tissues involving HPLC. Anal Biochem 170:349–354

Nishikawa A, Fujii S, Sugiyama T, Hayashi N, Taniguchi N (1988b) High expression of an N-acetylglucosaminyltransferase III in 3′-methyl DAB-induced hepatoma and ascites hepatoma. Biochem Biophys Res Commun 152:107–112

Nishikawa A, Ihara Y, Hatakeyama M, Kangawa K, Taniguchi N (1992) Purification, cDNA cloning, and expression of UDP-N-acetylglucosamine: β-D-mannoside β-1,4N-acetylglucosaminyltransferase III from rat kidney. J Biol Chem 267:18199–181204

Priatel JJ, Sarkar M, Schachter H, Marth JD (1997) Isolation, characterization and inactivation of the mouse *Mgat3* gene: the bisecting N-acetylglucosamine in asparagine-linked oligosaccharides appears dispensable for viability and reproduction. Glycobiology 7:45–56

Rebbaa A, Yamamoto H, Saito T, Meuillet E, Kim P, Kersey DS, Bremer EG, Taniguchi N, Moskal JR (1997) Gene transfection-mediated overexpression of β-1,4-N-acetylglucosamine bisecting oligosaccharides in glioma cell line U373 MG inhibits epidermal growth factor receptor function. J Biol Chem 272:9275–9279

Schachter H (1986) Biosynthetic controls that determine the branching and microheterogeneity of protein-bound oligosaccharides. Biochem Cell Biol 64:163–181

Schachter H, Narasimhan S, Gleeson P, Vella G (1983) Control of branching during the biosynthesis of asparagine-linked oligosaccharides. Can J Biochem Cell Biol 61:1049–1066

Sultan AS, Miyoshi E, Ihara Y, Nishikawa A, Tsukada Y, Taniguchi N (1997) Bisecting GlcNAc structures act as negative sorting signals for cell surface glycoproteins in forskolin-treated rat hepatoma cells. J Biol Chem 272:2866–2872

Tanemura M, Miyagawa S, Ihara Y, Matsuda H, Shirakura R, Taniguchi N (1997) Significant downregulation of the major swine xenoantigen by N-acetylglucosaminyltransferase III gene transfection. Biochem Biophys Res Commun 235:359–364.

Taniguchi N, Nishikawa A, Fujii S, Gu JG (1989) Glycosyltransferase assays using pyridylaminated acceptors: N-acetylglucosaminyltransferase III, IV, and V. Methods Enzymol 179:397–408

Yang X, Bhaumik M, Bhattacharyya R, Gong S, Rogler CE, Stanley P (2000) New evidence for an extra-hepatic role of N-acetylglucosaminyltransferase III in the progression of diethylnitrosamine-induced liver tumors in mice. Cancer Res 60:3313–3319

Yoshimura M, Nishikawa A, Ihara Y, Taniguchi S, Taniguchi N (1995) Suppression of lung metastasis of B16 mouse melanoma by *N*-acetylglucosaminyltransferase III gene transfection. Proc Natl Acad Sci USA 92:8754–8758

Yoshimura M, Ihara Y, Ohnishi A, Ijuhin N, Nishiura T, Kanakura Y, Matsuzawa Y, Taniguchi N (1996) Bisecting *N*-acetylglucosamine on K562 cells suppresses natural killer cytotoxicity and promotes spleen colonization. Cancer Res 56:412–418

Yoshimura M, Ihara Y, Nishiura T, Okajima Y, Ogawa M, Yoshida H, Suzuki M, Yamamura K, Kanakura Y, Matsuzawa Y, Taniguchi N (1998) Bisecting GlcNAc structure is implicated in suppression of stroma-dependent haemopoiesis in transgenic mice expressing *N*-acetylglucosaminyltransferase III. Biochem J 331:733–742

N-Acetylglucosaminyltransferase-IV

Introduction

Complex-type Asn-linked oligosaccharides do not always have the same number of branches, although there are normally from two to five. Each branch is formed by a specific *N*-acetylglucosaminyltransferase (GnT). *N*-Acetylglucosaminyltransferase-IV (GnT-IV) is essential in producing multiantennary sugar chains cooperatively with GnT-V. Much evidence has been accumulated which indicates the profound effect of multiantennary sugar chains on the physiological functions of bioactive glycoproteins. While there are many reports on the expression of GnT-V products related to tumorigenesis and organ differentiation, little information is available about when and where GnT-IV products are expressed. This is because of a lack of probes specific for GnT-IV products. However, the GnT-IV gene family is unique among GnTs for *N*-glycan biosynthesis. Although the biological meaning of each member is not well understood, GnT-IV may have functional roles which are no less important than those of GnT-V.

Mari T. Minowa[1]*, Suguru Oguri[2], Aruto Yoshida[1], and
Makoto Takeuchi[1]

[1] Glycotechnology Group, Central Laboratories for Key Technology, Kirin Brewery Co., Ltd.,
1-13-5 Fuku-ura, Kanazawa-ku, Yokohama 236-0004, Japan
Tel. +81-45-788-7228; Fax +81-45-788-4047
e-mail: mminowa@kirin.co.jp, makotot@kirin.co.jp
[2] Department of Bioproduction, Faculty of Bioindustry, Tokyo University of Agriculture, 196
Yasaka, Abashiri, Hokkaido 099-2493, Japan
* Present address: Annotation Center, Life Science Group, Hitachi, Ltd., 1-3-1 Minamidai,
Kawagoe, Saitama 350-1165, Japan

Databanks

N-Acetylglucosaminyltransferase-IV

NC-IUBMB enzyme classification: E.C.2.4.1.145

Species	Gene	Protein	mRNA	Genomic
GnT-IVa				
Homo sapiens	*Mgat4a*	BAA86388	AB000616	–
Bos taurus	–	BAA31162	AB000628	–
GnT-IVb				
Homo sapiens	*Mgat4b*	BAA83464	AB000624	–

Name and History

UDP-N-acetylglucosamine:α1,3-D-mannoside β1,4-N-acetylglucosaminyltransferase (N-acetylglucosaminyltransferase-IV) is frequently abbreviated to GnT-IV or MGAT4. The activity of this enzyme was first identified in the hen oviduct by Gleeson and Schachter (1983). The enzyme was first purified from bovine small intestine (Oguri et al. 1997a). After the cloning of bovine cDNA (Minowa et al. 1998), two homologous cDNAs were cloned from human, and named GnT-IVa (which is highly homologous to bovine enzyme, Yoshida et al. 1999) and GnT-IVb (which is less homologous to bovine enzyme, Yoshida et al. 1998). Recently, a possible third gene in human was reported and named hGnT-IV-H by Furukawa et al. (1999), although its enzymatic character had not yet been examined.

Enzyme Activity Assay and Substrate Specificity

GnT-IV catalyzes the transfer of GlcNAc from UDP-GlcNAc to the GlcNAcβ1-2Manα1-3 arm of the core structure of Asn-linked oligosaccharides (Man$_3$GlcNAc$_2$-Asn) through β1-4 linkage. The activity of GnT-IV was assayed using a pyridylaminated asialoagalacto–biantennary sugar chain as the acceptor substrate according to the method of Nishikawa et al. (1990), with some modifications (Tokugawa et al. 1996).

The kinetic properties of the purified bovine GnT-IVa enzyme were examined (Oguri et al. 1997a), and transiently expressed human GnT-IVa examined later showed almost the same properties (Oguri et al. 1999). To purify the enzyme, 10% glycerol and 1% BSA are added for stability. The addition of 10 mM ME does not affect the activity. The enzyme is most active between pH 7.0 and pH 8.0, with an optimum at pH 7.3. The activity of the bovine enzyme depends on divalent cations, especially Mn^{2+}, and could be inactivated by ethylenediaminetetraacetate (EDTA) treatment. The optimum concentration of Mn^{2+} is 10 mM. Other divalent cations, such as Co^{2+} and Mg^{2+}, are also able to activate the enzyme to a lesser extent than Mn^{2+}. Ca^{2+} and Fe^{2+} had no effect, while Cu^{2+} abolished the activity completely. Since this enzyme requires UDP-GlcNAc as a donor substrate, UDP and some UDP-sugars, such as UDP-glucose and UDP-N-acetylgalactosamine, are potent inhibitors.

The acceptor specificity of bovine GnT-IVa was examined with various PA-labeled oligosaccharides (Fig. 1). GlcNAc branches transferred by the other GnTs affect the GnT-IVa activity. GnT-IVa activity needs the GlcNAcβ1-2 branch to be transferred by GnT-I onto α1-3 Man in the core Asn-linked oligosaccharide (compare acceptors 1 and 2). Enzyme reactivity increases in accordance with the number of terminal GlcNAc (acceptors 2–4), and a triantennary sugar chain having GlcNAcs transferred by GnT-I, II, and V (acceptor 4) is the best acceptor examined. However, bisecting the GlcNAc structure prevents enzyme activity completely (acceptor 5). Galactosylation of GlcNAc and α1-6 Fuc on GlcNAc next to Asn inhibits the activity (acceptor 6–8). The purified enzyme is confirmed by its ability to transfer GlcNAc to the asialoa-galacto-oligosaccharides bound to the protein (Oguri et al. 1997b). The enzymatic properties of GnT-IV isozymes are compared using transiently expressed hGnT-IVa and hGnT-IVb (Oguri et al. 1999). Both enzymes are similar in optimum pH and optimum Mn^{2+} concentration, but there is a difference in cation dependency. Co^{2+} could activate hGnT-IVb to the same extent as Mn^{2+}. Kinetic analysis showed that both enzymes have the higher affinity to the oligosaccharide with more GlcNAcs (Table 1).

Acceptor	Relative activity (%)
1. $\begin{array}{l}M\alpha1\searrow_6 \\ \quad\quad M\beta1\text{-}4GN\beta1\text{-}4GN\text{-}PA \\ M\alpha1\nearrow^3\end{array}$	0
2. $\begin{array}{l}M\alpha1\searrow_6 \\ \quad\quad M\beta1\text{-}4GN\beta1\text{-}4GN\text{-}PA \\ GN\beta1\text{-}2M\alpha1\nearrow^3\end{array}$	54
3. $\begin{array}{l}GN\beta1\text{-}2M\alpha1\searrow_6 \\ \quad\quad\quad M\beta1\text{-}4GN\beta1\text{-}4GN\text{-}PA \\ GN\beta1\text{-}2M\alpha1\nearrow^3\end{array}$	100
4. $\begin{array}{l}GN\beta1\searrow_6 \\ GN\beta1\text{-}2M\alpha1\searrow_6 \\ \quad\quad\quad M\beta1\text{-}4GN\beta1\text{-}4GN\text{-}PA \\ GN\beta1\text{-}2M\alpha1\nearrow^3\end{array}$	164
5. $\begin{array}{l}GN\beta1\text{-}2M\alpha1\searrow \\ \quad GN\beta1\text{-}4\text{-}_3^6M\beta1\text{-}4GN\beta1\text{-}4GN\text{-}PA \\ GN\beta1\text{-}2M\alpha1\nearrow\end{array}$	0
6. $\begin{array}{l}G\beta1\text{-}4GN\beta1\text{-}2M\alpha1\searrow_6 \\ \quad\quad\quad\quad M\beta1\text{-}4GN\beta1\text{-}4GN\text{-}PA \\ GN\beta1\text{-}2M\alpha1\nearrow^3\end{array}$	16
7. $\begin{array}{l}GN\beta1\text{-}2M\alpha1\searrow_6 \\ \quad\quad\quad\quad M\beta1\text{-}4GN\beta1\text{-}4GN\text{-}PA \\ G\beta1\text{-}4GN\beta1\text{-}2M\alpha1\nearrow^3\end{array}$	0
8. $\begin{array}{l}\quad\quad\quad\quad\quad F\alpha1 \\ \quad\quad\quad\quad\quad\mid^6 \\ GN\beta1\text{-}2M\alpha1\searrow_6 \\ \quad\quad\quad M\beta1\text{-}4GN\beta1\text{-}4GN\text{-}PA \\ GN\beta1\text{-}2M\alpha1\nearrow^3\end{array}$	46

Fig. 1. Acceptor substrate specificity of bovine N-acetylglucosaminyltransferase IVa. Relative activity is expressed as the percentage of β1-4 GlcNAc transfer observed in the presence of acceptor 3. Abbreviations: GN, GlcNAc; M, Man; G, Gal; F, Fuc. (Revised from Oguri et al. 1997a, with permission)

Table 1. Km of hGnT-IVa and hGnT-IVb toward donor and major acceptor substrates

Substrate		hGnT-IVa (mM)	hGnT-IVb (mM)
Donor UDP-GlcNAc		0.12	0.24
Acceptor	2	3.19	10.5
	3	0.97	5.72
	4	0.53	3.35

The structure of each acceptor is shown in Fig. 1
hGnT-IVa, human N-acetylglucosaminyltransferase-IVa; hGnT-IVb, human N-acetylglucosaminyltransferase-IVb
all values are for recombinant enzymes

hGnT-IVb has higher Km values for a donor or acceptor compared with hGnT-IVa, which may suggest that the hGnT-IVa is more active under physiological conditions and primarily contributes to the biosynthesis of oligosaccharides.

Preparation

In the rat, GnT-IV acitivity is found in most tissues, and is relatively high in spleen and small intestine (Nishikawa et al. 1990). Enzyme purification was performed using bovine small intestine as the source. The procedure includes solubilization of the enzyme from a microsome fraction of tissue by Triton X-100, Q-Sepharose, Cu-chelate Sepharose, two affinity (UDP-hexanolamine agarose) chromatographies under different elution conditions, and Superdex 200 gel filtration. Each of full-length enzyme has been successfully expressed in COS7 cells. Enzymes were also expressed as secreted forms by replacing transmembrane and stem regions with an appropriate signal sequence in both animal cells and insect cells. In the *E. coli* system, the active enzyme could be expressed, but the proteins often formed inclusion bodies and were not fully active.

Biological Aspects

The structure of the GnT-IV enzyme is a typical type-II transmembrane protein. The bovine and human GnT-IVa enzymes have two potential *N*-glycosylation sites, and the human GnT-IVb enzyme has three in the peptide portion located in Golgi lumen. One *N*-glycosylation site close to the C-terminus is confirmed to be glycosylated in bovine purified enzyme, although this glycosylation is not essential for its activity.

Because of the homology to the bovine cDNA, a human counterpart was subcloned and, with help of an EST database, another human cDNA was isolated. The former, which has >90% identity to the bovine enzyme, is GnT-IVa, and the latter, which is approximately 60% identical, is GnT-IVb. Genomic Southern analysis proved that most of the animals had both genes. Northern analysis of normal tissues and cell lines of human using specific probes for each gene revealed that GnT-IVb is widely expressed, while the expression of GnT-IVa is restricted to a few tissues and has multiple transcripts (Fig. 2). The latter observation may suggest the presence of complex regulation of GnT-IVa expression. The third GnT-IV gene, hGnT-IV-H, was identified, and the amino acid sequence has 27% identity to bovine protein. It is expressed in multiple sizes and in a regulated manner, but in different tissues from those for GnT-IVa. The chromosome localizations of the three genes are 2q12 for GnT-IVa, 5q35 for GnT-IVb, and 12q21 for hGnT-IV-H.

The products of the GnT-IV enzyme are normally found in many glycoproteins. In erythropoietin, the ratio of tetra/bi-antennary sugar chains is positively correlated with its in vivo activity (Takeuchi et al. 1989) because the bulky structure of the sugar chains prevents rapid clearance of protein from the blood stream. On the other hand, asialoglycoprotein receptor prefers three Gal residues on triantennary sugar chains, with rigid recognition of the Gal on the GlcNAc transferred by GnT-IV (Rice et al. 1990). These observations together suggested that branching of the oligosaccharides is important in regulating the lifetime of glycoprotein hormones in the blood stream.

Fig. 2. Northern analysis of GnT-IVa and Gnt-IVb mRNA expression in human tissues and cancer cells. 32P-labeled cDNA fragments of GnT-IVa (*top four panels*) and GnT-IVb (*middle four panels*) were hybridized to mRNA from various human tissues and cell lines. Glyceraldehyde 3-phosphate dehydrogenase (G3PDH)-specific probe was used as a control (*bottom*). The sizes of the RNA marker bands are indicated on the *left*. (From Yoshida et al. 1998, with permission)

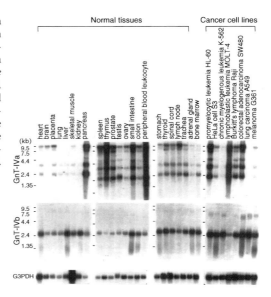

GnT-IV activity is also elevated in cells treated by 1α,25-dihydroxyvitamine D3 and interleukin-6, which suggests that the expression of GnT-IV is regulated during differentiation (Koenderman et al. 1987; Nakao et al. 1990). Upregulation of GnT-IV activity was also observed during oncogenesis. Several hepatoma cell lines have higher GnT-IV activity than normal tissue (Nishikawa et al. 1990), and γ-glutaminyl-transferase from human hepatoma tissue has more GlcNAc branches by GnT-IV than from normal tissue (Yamashita et al. 1989). Mizuochi et al. (1983) showed that human chorionic gonadotropin from choriocarcinoma patients has aberrant biantennary sugar chains which have two GlcNAc branches on α1-3Man. The molecular basis of the formation of this abnormal structure was revealed as an elevated GnT-IVa, but not GnT-IVb, expression in choriocarcinoma cell lines (Takamatsu et al. 1999).

Future Perspectives

GnT-IV is the only GnT for *N*-glycosylation that forms a gene family in higher vertebrates. It will be very interesting to clarify the role of each gene. A comparison of the characteristics of these enzymes is important to elucidate the molecular mechanism of enzymatic function. Since the expression profile of each gene is quite different from the others, each gene has some particular function. In addition, there should be a complex regulation in transcription/translation of GnT-IVa and hGnT-IV-H, which leads to the change in oligosaccharide structures under some physiological conditions.

To elucidate the biological functions of GnT-IVa and GnT-IVb, knockout mouse experiments are now in progress.

Further Reading

For the enzyme preparation and characterization, see Oguri et al. 1997a. For GnT-IV gene family members, see Yoshida et al. 1998 and 1999.

References

Furukawa T, Youssef EM, Yatsuoka T, Yokoyama T, Makino N, Inoue H, Fukushige S, Hoshi M, Hayashi Y, Sunamura M, Horii A (1999) Cloning and characterization of the human UDP-*N*-acetylglucosamine: alpha-1,3-D-mannoside beta-1,4-*N*-acetylglucosaminyl-transferase IV-homologue (hGnT-IV-H) gene. J Hum Genet 44:397–401

Gleeson PA, Schachter H (1983) Control of glycoprotein synthesis. J Biol Chem 258: 6162–6173

Koenderman AH, Wijermans PW, van den Eijnden DH (1987) Changes in the expression of *N*-acetylglucosaminyltransferase III, IV, V associated with the differentiation of HL-60 cells. FEBS Lett 222:42–46

Minowa MT, Oguri S, Yoshida A, Hara T, Iwamatsu A, Ikenaga H, Takeuchi M (1998) cDNA cloning and expression of bovine UDP-*N*-acetylglucosamine: α1,3-D-mannoside-β1,4-*N*-acetylglucosaminyltransferase-IV. J Biol Chem 273:11556–11562

Misaizu T, Matsuki S, Strickland TW, Takeuchi M, Kobata A, Takasaki S (1995) Role of antennary structure of *N*-linked sugar chains in renal handling of recombinant human erythropoietin. Blood 86:4097–4104

Mizuochi T, Nishimura R, Derappe C, Taniguchi T, Hamamoto T, Mochizuki M, Kobata A (1983) Structures of the asparagine-linked oligosaccharides of human chorionic gonadotropin produced in choriocarcinoma. Appearance of triantennary sugar chains and unique biantennary sugar chains. J Biol Chem 258:14126–14129

Nakao H, Nishikawa A, Karasuno T, Nishiura T, Iida M, Kanayama Y, Yonezawa T, Tarui S, Taniguchi N (1990) Modulation of *N*-acetylglucosaminyltransferase III, IV and V activities and alteration of the surface oligosaccharide structure of a myeloma cell line by interleukin 6. Biochem Biophys Res Commun 172:1260–1266

Nishikawa A, Gu J, Fujii S, Taniguchi N (1990) Determination of *N*-acetylglucosaminyl-transferase III, IV and V in normal and hepatoma tissues of rats. Biochim Biophys Acta 1035:313–318

Oguri S, Minowa MT, Ihara Y, Taniguchi N, Ikenaga H, Takeuchi M (1997a) Purification and characterization of UDP-*N*-acetylglucosamine:α1,3-D-mannoside β1,4-*N*-acetylglucosaminyltransferase. J Biol Chem 272:22721–22727

Oguri S, Minowa MT, Ihara Y, Taniguchi N, Ikenaga H, Takeuchi M (1997b) Characteriza-tion of *N*-acetylglucosaminyltransferase IV (GnT-IV) purified from bovine small intestine. Glycoconj J 14:S45

Oguri S, Minowa MT, Yoshida A, Takamatsu S, Hara T, Takeuchi M (1999) Two isozymes of human *N*-acetylglucosaminyltransferase IV (GnT-IV)—what are the differences between them?. Glycoconj J 16:S74

Rice KG, Weisz OA, Barthel T, Lee RT, Lee YC (1990) Defined geometry of binding between triantennary glycopeptide and the asialoglycoprotein receptor of rat hepatocytes. J Biol Chem 265:18429–18434

Takamatsu S, Oguri S, Minowa MT, Yoshida A, Nakamura K, Takeuchi M, Kobata A (1999) Unusually high expression of *N*-acetylglucosaminyltransferase-IVa in human chorio-carcinoma cell lines: a possible enzymatic basis of the formation of abnormal bianten-nary sugar chain. Cancer Res 59:3949–3953

Takeuchi M, Inoue N, Strickland TW, Kubota M, Wada M, Shimizu R, Hoshi S, Kozutsumi H, Takasaki S, Kobata A (1989) Relationship between sugar chain structure and biological activity of recombinant human erythropoietin produced in Chinese hamster ovary cells. Proc Natl Acad Sci USA 86:7819–7822

Tokugawa K, Oguri S, Takeuchi M (1996) Large-scale preparation of PA-oligosaccharides from glycoproteins using an improved extraction method. Glycoconj J 13:53–56

Yamashita K, Totani K, Iwaki Y, Takamisawa I, Tateishi N, Higashi T, Sakamoto Y, Kobata A (1989) Comparative study of the sugar chains of gamma-glutamyltranspeptidases purified from human hepatocellular carcinoma and from human liver. J Biochem (Tokyo) 105:728–735

Yoshida A, Minowa MT, Hara T, Takamatsu S, Ikenaga H, Takeuchi M (1998) A novel second isoenzyme of the human UDP-N-acetylglucosamine:α1,3-D-mannoside β1,4-N-acetylglucosaminyltransferase family. Glycoconj J 15:1115–1123

Yoshida A, Minowa MT, Takamatsu S, Hara T, Oguri S, Ikenaga H, Takeuchi M (1999) Tissue-specific expression and chromosomal mapping of a human UDP-N-acetylglucosamine:α1,3-D-mannnoside β1,4-N-acetylglucosaminyltransferase. Glycobiology 9:303–310

13

N-Acetylglucosaminyltransferase-V

Introduction

N-Acetylglucosaminyltransferase-V (GnT-V, GnT-V or Mgat5) catalyzes the transfer of GlcNAc from UDP-GlcNAc to the OH-6 position of the α-linked Man residue in GlcNAcβ1-2Manα1-6Manβ1-4GlcNAc. This acceptor sequence is found in *N*-glycan intermediates at the medial Golgi stage of glycoprotein production (Fig. 1). Maturation of the *N*-glycans with passage of glycoproteins through the *trans*-Golgi produces the tri (2,2,6)- and tetra (2,4,2,6)-antennary complex-type *N*-glycans. A variety of oligosaccharide sequences are added to complete these glycans, comprising various combinations of *N*-acetyllactosamine and poly-*N*-acetyllactosamine capped with sialic acid and fucose. GnT-V is a rate-limiting enzyme for the addition of poly-*N*-acetyllactosamine to *N*-glycans. In vitro assays indicate that tri (2,2,6)- and tetra (2,4,2,6)-antennary glycans are preferentially elongated by β3GnT(i) and β4GalT to produce poly-*N*-acetyllactosamine (van den Eijnden et al. 1988). Furthermore, the mutant lymphoma cell lines BW5147-PHA^{R}2.1 (Cummings and Kornfeld 1984), and KBL-1 (Yousefi et al. 1991) are GnT-V-deficient and severely depleted of poly-*N*-acetyllactosamine on *N*-glycans, but not the *O*-glycans. These somatic cell mutants were selected for resistance to the toxic effects of leukoagglutinin (L-PHA) in culture. L-PHA binds preferentially to mature GnT-V products, notably the Galβ1-4GlcNAcβ1-6 (Galβ1-4GlcNAcβ1-2)Manα1-6 portion of tri- and tetraantennary *N*-glycans (Cummings and Kornfeld 1982) (Fig. 1). GnT-V-modified *N*-glycans detected by L-PHA lectin histochemistry are often increased in human breast and colorectal carcinomas, and correlate with poor prognosis and reduced patient survival time (Fernandes et al. 1991; Seelentag et al. 1998).

JAMES W. DENNIS

Samuel Lunenfeld Research Institute, Mount Sinai Hospital, 600 University Avenue, Room 988, Toronto, Ontario M5G 1X5, and Department of Molecular and Medical Genetics, University of Toronto, Canada
Tel. +1-416-586-8233, Fax +1-416-586-8857
e-mail: Dennis@MSHRI.ON.CA

Fig. 1. Schematic diagram of the Golgi N-glycan biosynthesis pathway showing GnT-V (*TV*) in the production of tri (2,2,6)- and tetra (2,4,2,6)-antennary (the numbers in parentheses refer to the linkages of the antennae from left to right). *Open diamonds*, sialic acid; *solid circles*, galactose; *solid squares*, GlcNAc; *open circles*, mannose. *TI, TII, TIV, TV*, and *T(i)* refer to the GlcNAc-transferases; *GalT* is β1,4-galactosyltransferase; *ST* is α-sialyltransferase. The *shaded ovals* mark the minimal acceptor for GnT-V, and the *shaded boxes* mark the L-PHA binding site in the mature N-glycans

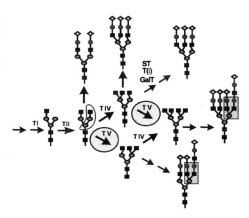

GnT-V has a type-II, single-pass transmembrane organization common to Golgi luminal enzymes. The human and rat GnT-V sequences are 740 amino acids and share 97% identity (Shoreibah et al. 1993; Saito et al. 1994). The peptide sequence $S_{213-740}$ of GnT-V is essential for the catalytic activity, leaving a 183 amino acids stem region, 17 amino acids transmembrane domain, and 12 amino acids in the cytosol. Deletion of as few as five amino acids from the C-terminal end destroys catalytic activity. Similar K_m and V_{max} values for donor and acceptor were observed for $S_{213-740}$, the minimal catalytic domain, and Q_{39-740}, which included the stem region. Secondary structure predictions suggest a high frequency of turns in the stem region, and more contiguous stretches of alpha helix in the catalytic domain $S_{213-740}$ (Korczak et al. 2000)

The GnT-V cDNA from the Lec4 CHO cell mutant cell line possesses two insertions that result in a truncated protein missing 585 amino acids from the catalytic domain. The Lec4A CHO mutant cell line retains full GnT-V enzyme activity, but the cells are deficient in cell surface L-PHA reactive glycans. GnT-V cDNA from Lec4A cells has an L188R point mutation which may either prevent association with exit complexes from the ER, or cause retrograde transport from a Golgi compartment (Weinstein et al. 1996). Leu188 is in the neck region joining the transmembrane domain and the catalytic domain of GnT-V. The 8–10 amino acid sequence flanking Leu188 is notably more conserved than the rest of the neck region when comparing the mammalian and *Caenorhabditis elegans* sequences (Warren and Dennis 2001). The *C. elegans* gene, designated *gly-2*, is a functional homologue of the mammalian gene *Mgat5*, as it complements the Lec4 mutation restoring wild-type levels of L-PHA-reactive glycans at the cell surface. The corresponding L188R mutation in the *gly-2* sequence does not rescue the Lec4 defect and is misslocalized, confirming that this region of the stem is required for Golgi localization. Therefore, mammalian and *C. elegans* GnT-V share significant sequence homology, catalytic specificity, and the Golgi localization signal defined by the region containing Leu188.

Databanks

N-Acetylglucosaminyltransferase-V

NC-IUBMB enzyme classification: E.C.2.4.1.155

Species	Gene	Protein	mRNA	Genomic
Homo sapiens	*MGAT5*	Q09328	D17716	X91652–X91653
			Hs.121502 (Unigene)	
Cricetulus griseus	Mutation Lec4A	–	U62587	–
	Mutation Lec4	–	U62588	–
Rattus norvegicus	*Mgat5*	Q08834	L14284	–
Caenorhabditis elegans	*gly-2*	–	AF154122	–

Name and History

N-Acetylglucosaminyltransferase-V is also referred to as UDP-*N*-acetylglucosamine: α-6-D-mannosidase β1,6-*N*-acetylglucosaminyltransferase V, abbreviated to GnT -V, or GnT-V, or the protein product of the *MGAT5* gene. The activity of the enzyme was first described in lysates of lymphoma cells (Cummings et al. 1982) and inferred by earlier structural analysis of *N*-glycans.

Enzyme Activity Assay and Substrate Specificity

Radiolabeled UDP-GlcNAc and the synthetic acceptor GlcNAcβ1-2Manα1-6Man-βoctylCH₃ or GlcNAcβ1-2Manα1-6GlcβoctylCH₃ are used to measure enzyme activity in biological samples. Inversion of the GlcNAc configuration at its anomeric centers occurs with transfer from UDP-GlcNAc to the acceptor. Removal of the 6-OH on the α-linked mannose converts the synthetic trisaccharide acceptor into a competitive inhibitor (Khan et al. 1993). The 3,4,6-OH positions in the terminal GlcNAc residue of the acceptor are key polar groups recognized by the GnT-V enzyme, and must be unsubstituted for efficient acceptor activity (Kanie et al. 1994). In this regard, when glycopeptides from natural glycoproteins are employed as acceptors for GnT-V, sialic acid and β1,4-galactose must first be removed (Schachter 1986). This may be accomplished by mild acid treatment, 0.1 M HCl 80°C for 1 h, dialysis, and β-galactosidase digestion. The action of GnT-III also inhibits the subsequent use of glycans as GnT-V acceptors. GnT-V activity does not require divalent cations, the pH optimum of the enzyme reaction is ~6.5, and it can be done in the presence of 0.5% to 1.0% Triton-X100 or NP-40 detergents. The kinetics of the GnT-V reaction have been examined using partially purified enzyme from rat kidney and from tumor cell lysates, and the Kms for UDP-GlcNAc and synthetic acceptor are ~2 mM and ~50 μM, respectively (Khan et al. 1993; Yousefi et al. 1991).

GnT-V reactions typically contain 30 mM MES buffer, pH 6.5, 1 mM UDP-GlcNAc, 0.5 μCi UDP-[³H]-GlcNAc, 0.5 mM GlcNAcβ1-2Manα1-6Manβ1octylCH₃ acceptor and an enzyme source. If a soluble recombinant enzyme is used, detergent is unnecessary, but the addition of 1 mg/ml of BSA prevents loss of enzyme. Reactions are incubated at 37°C for 1–4 h in a total volume of 30 μl, and stopped by adding 0.5 ml cold water.

Tubes are processed immediately or stored at −20°C. Mixtures are diluted to 5 ml in water applied to a Sep-Pak C-18 column and washed with 20 ml H_2O. The product is eluted with 5 ml 100% methanol and counted in a β-scintillation counter. The reaction volumes and radioactivity can be reduced using recombinant enzyme, and processing of the products has been adapted to a robotic platform using Sep Pak C-18-packed 200-μl pipette tips (Donovan et al. 2000). A solid-phase radioactive assay using 96-well plastic plates coated with acceptor linked to a polymer has also been developed to screen for inhibitors of GnT-V and other glycosyltransferases (Donovan et al. 2000).

Compounds that can enter the cell and block the biosynthesis of β1,6GlcNAc-branched glycan, either as a competitive inhibitor of GnT-V or upstream in the pathway, may be useful anticancer agents. A simple homogeneous cell-based assay to screen for compounds with these properties has been developed (Datti et al. 2000). The assay detects compounds that, when preincubated for 24 h with MDAY-D2 tumor cells, block β1,6GlcNAc-branched glycan expression, and thereby protect the cells from the subsequent addition of L-PHA. The viable MDAY-D2 cell number is directly proportional to the level of endogenous alkaline phosphatase activity measured by A_{405} in the cultures following the addition of substrate, and the assay is readily adapted for high-throughput screening. Compounds with overt toxicity will not confound the analysis, as they will register as negative on the screen. Finally, reagents to detect regulators of *Mgat5* transcription in cells and mice are also available. We designed a gene replacement vector to produce GnT-V-deficient mice, and the vector contained the reporter bacterial β-galactosidase gene (*LacZ*), which replaced the first exon of *Mgat5*. *LacZ* is expressed with the same tissue-specificity as *Mgat5* transcripts in *Mgat5*$^{+/-}$ and *Mgat5*$^{-/-}$ mice (Granovsky et al. 2000). Therefore, *LacZ* expression can be monitored in cells and tissues to determine physiological conditions and drugs that regulate *Mgat5* gene expression.

Preparation

GnT-V has been measured in human serum and freshly prepared detergent lysates from cells and tissues. Extracts made from cell lines and tissue homogenates in saline, 1% Triton-X100 plus protease inhibitor, and diluted to protein concentrations of 10–20 mg/ml can be assayed directly for GnT-V activity. The enzyme has been partially purified from acetone extracts of kidney, and up to 450000-fold by sequential affinity chromatography using UDP–hexanolamine–agarose, and a synthetic oligosaccharide-inhibitor agarose column (Shoreibah et al. 1992). The pPROTA vector has been used to express recombinant GnT-V by fusing the IgG-binding domains of *staphylococcal* Protein A to the N-terminal catalytic domain of GnT-V (Korczak et al. 2000). ProtA-GnT-V is produce as a secreted protein from CHO cells transfected with the expression vector. Stable CHO transfected lines can be generated that express up to 1 mg/l of enzyme in the culture medium, and 30–60 nmol/mg/min after affinity purification. The recombinant enzyme is isolated from the cell culture medium using IgG–Sepharose affinity chromatography, as described (Korczak et al. 2000). The enzyme is stable in 0.1 MES pH 6.5, 1 mg/ml BSA stored at either 4°C or −20°C.

Biological Aspects

In 1970, it was reported that the *N*-glycans pool increases in size following malignant transformation (Buck et al. 1970), which was subsequently found to be due to increased GlcNAc-branching and poly-*N*-acetyllactosamine extension. GnT-V activity increases in cells transformed by polyomavirus (Yamashita et al. 1985) and Rous sarcoma virus (Pierce and Arango 1986), and by transfection with activated Ras (Dennis et al. 1987). Further analysis has revealed that *Mgat5* gene transcription is positively regulation by signaling downstream of these oncogenes, notably the Ras–Raf–Ets pathway (Kang et al. 1996; Chen et al. 1998). Other interesting changes in *Mgat5* gene expression in human cancers have been observed. An intron of the *Mgat5* gene encodes a transcript detected in tumor cells but not in normal tissues, producing a widely occurring tumor-associated antigen (Guilloux et al. 1996).

The *Mgat5* message is low in the mouse embryo at day 7 of gestation (E7), but widely expressed by E9.5. In late-stage embryos, *Mgat5* expression becomes restricted to regions of the developing central nervous system and to specialized epithelia of skin, intestine, kidney, endocrine tissues, and respiratory tract (Granovsky et al. 1995). The *Mgat5* gene knockout mice lack both detectable enzyme activity and L-PHA reactive glycoproteins, suggesting a single gene encoding GnT-V activity. Although *Mgat5* is expressed early in the embryo, *Mgat5*$^{-/-}$ mice are normal at birth. However, adult mice differed in their responses to certain extrinsic conditions (Granovsky et al. 2000). The polyomavirus middle T oncogene (*PyMT*) induces multifocal breast tumors in mice when expressed from a transgene in mammary epithelium. We observed that *PyMT*-induced tumor growth and metastasis are markedly suppressed in GnT-V-deficient mice compared with their *PyMT* transgenic littermates expressing GnT-V. *Mgat5* gene expression is induced by the *PyMT* oncogene, possibly through both Ras–Raf–Ets and also PI3K–PKB pathways. GnT-V-modified glycans promote focal adhesion turnover, which amplifies PI3 kinase/PKB signaling to promote tumor growth and metastasis. GnT-V selectively substitutes only a subset of *N*-glycan intermediates, presumably specified by the structural features of the glycoprotein substrates (Do et al. 1994). However, there is limited information on the location of GnT-V-modified glycans to specific glycoproteins and glycosylation sites. These glycans have been found on the lysosomal-associated membrane glycoproteins LAMP-1 and LAMP-2 (Heffernan and Dennis 1989; Carlsson and Fukuda 1990) and integrins LFA-1 and $\alpha_1\beta_5$ (Nakayama et al. 1999; Asada et al. 1991), and L-PHA binding suggests their presence on T cell receptor (TCR) (Hubbard et al. 1986). GnT-V-modified glycans increase on integrin subunits α_5, α_v, and β_1 in cells transfected with *Mgat5*, and this was associated with loss of contact inhibition, morphological transformation, and tumor formation in mice (Demetriou et al. 1995).

GnT-V-deficient mice displayed increased incidence of kidney autoimmune disease in mice over 1 year of age. The mutant mice display enhanced delayed-type hypersensitivity and increased susceptibility to experimental autoimmune encephalomyelitis, a model of multiple sclerosis (Demetriou et al. 2001). T cells from *Mgat5*$^{-/-}$ mice are hypersensitive to TCR agonists. Recruitment of TCR to sites of contact with agonist-coated beads and downstream signaling is increased in *Mgat5*$^{-/-}$ compared with *Mgat5*$^{+/+}$ T cells, suggesting that *Mgat5*-modified glycans regulate TCR clustering, and thereby the T cell activation threshold.

Future Perspectives

The α5β1 integrin adhesion signaling pathway and T cell receptor signaling are both regulated by GnT-V glycosylation, with profound effects on cancer growth and T cell function. The GnT-V-modified glycans appear to destabilize focal adhesions, which accelerates membrane ruffling and the associated downstream intracellular signaling. A similar destabilizing effect on T cell receptor clustering by GnT-V-modified glycans slows clustering, increases the threshold for agonist, and acts as a negative regulator of T cell activation. The large size of *Mgat5* glycans may limit the geometry and spacing of receptor clusters in the plane of the membrane. Alternatively, GnT-V-modified glycans can bind to galectins or other lectins, and this has recently been shown to slow antigen-induced TCR clustering (Demetriou et al. 2001). The galectins are defined operationally as *N*-acetyllactosamine-binding proteins, sequences that are enriched in GnT-V-modified glycans. It is possible that the avidity of a multivalent galectin—glycoprotein lattice at the cell surface may negatively regulate receptor clustering. Although the mechanism is unclear, the posttranslational modification of Notch receptor by Fringe, a fucose-specific β3GnT, provides another example of differential receptor glycosylation as a regulator of receptor signaling (Moloney et al. 2000). Structural information on glycosyltransferases and their target glycoproteins, combined with genetic analyses in mice and other model organisms, will soon provide exciting new insights into the function of glycans in development and diseases processes.

Acknowledgments

This research is supported by the National Science and Engineering Research Council of Canada, GlycoDesign Inc., and NCI of Canada.

Further Reading

Reviews by J.W. Dennis, M. Granovsky, and C.E. Warren: Biochim Biophys Acta (1999) 1473:21–34, and Bioessays (1999) 5:412–421.

References

Asada M, Furukawa K, Kantor C, Gahmberg CG, Kobata A (1991) Structural study of the sugar chains of human leukocyte cell adhesion molecules CD11/CD18. Biochemistry 30:1561–1571

Buck CA, Glick MC, Warren L (1970) A comparative study of glycoproteins from the surface of control and Rous sarcoma virus transformed hamster cells. Biochemistry 9: 4567–4576

Carlsson SR, Fukuda M (1990) The polylactosaminoglycans of human lysosomal membrane glycoproteins lamp-1 and lamp-2. J Biol Chem 265:20488–20495

Chen L, Zhang W, Fregien N, Pierce M (1998) The *her-2/neu* oncogene stimulates the transcription of *N*-acetylglucosaminyltransferase V and expression of its cell surface oligosaccharide products. Oncogene 17:2087–2093

Cummings RD, Kornfeld S (1982) Characterization of the structural determinants required for the high-affinity interaction of asparagine-linked oligosaccharides with immobilized *Phaseolus vulgaris* leukoagglutinating and erythroagglutinating lectin. J Biol Chem 257:11230–11234

Cummings RD, Kornfeld S (1984) The distribution of repeating Gal β1-4GlcNAc β1-3 sequences in asparagine-linked oligosaccharides of the mouse lymphoma cell line BW5147 and PHAR 2.1. J Biol Chem 259:6253–6260

Cummings RD, Trowbridge IS, Kornfeld S (1982) A mouse lymphoma cell line resistant to the leukoagglutinating lectin from *Phaseolus vulgaris* is deficient in UDP-GlcNAc:α-D-mannoside β1,6-*N*-acetylglucosaminyltransferase. J Biol Chem 257:13421–13427

Datti A, Donovan RS, Korczak B, Dennis JW (2000) A homogeneous cell-based assay to identify *N*-linked carbohydrate processing inhibitors. Anal Biochem 280:137–142

Demetriou M, Nabi IR, Coppolino M, Dedhar S, Dennis JW (1995) Reduced contact-inhibition and substratum adhesion in epithelial cells expressing GlcNAc-transferase V. J Cell Biol 130:383–392

Demetriou M, Granovsky M, Quaggin S, Dennis JW (2001) Negative regulation of T cell receptor and autoimmunity by *Mgat5* *N*-glycosylation. Nature 409:733–739

Dennis JW, Laferte S, Waghorne C, Breitman ML, Kerbel RS (1987) β 1-6 branching of Asn-linked oligosaccharides is directly associated with metastasis. Science 236:582–585

Do K-Y, Fregien N, Pierce M, Cummings RD (1994) Modification of glycoproteins by *N*-acetylglucosaminyltransferase V is greatly influenced by accessibility of the enzyme to oligosacharide acceptors. J Biol Chem 269:23456–23464

Donovan RS, Datti A, Baek M, Wu Q, Sas IJ, Korczak B, Berger EG, Roy R, Dennis JW (2000) A solid-phase glycosyltransferase assay for high-throughput screening in drug discovery research. Glycoconj J 16:607–615

Fernandes B, Sagman U, Auger M, Demetriou M, Dennis JW (1991) β-1-6 branched oligosaccharides as a marker of tumor progression in human breast and colon neoplasia. Cancer Res 51:718–723

Granovsky M, Fode C, Warren CE, Campbell RM, Marth JD, Pierce M, Fregien N, Dennis JW (1995) GlcNAc-transferase V and core 2 GlcNAc-transferase expression in the developing mouse embryo. Glycobiology 5:797–806

Granovsky M, Fata J, Pawling J, Muller WJ, Khokha R, Dennis JW (2000) Suppression of tumor growth and metastasis in *Mgat5*-deficient mice. Nat Med 6:306–312

Guilloux Y, Lucas S, Brichard VG, Van Pel A, Viret C, De Plaen E, Brasseur F, Lethe B, Jotereau F, Boon T (1996) A peptide recognized by human cytolytic T lymphocytes on HLA-A2 melanomas is incoded by an intron sequence of the *N*-acetylglucosaminyltransferase V gene. J Exp Med 183:1173–1183

Heffernan M, Dennis J (1989) Molecular characterization of P2B/LAMP-1: a major protein target of a metastasis-associated oligossacharide structure. Cancer Res 49:6077–6084

Hubbard SC, Kranz DM, Longmore GD, Sitkovsky MV, Eisen HN (1986) Glycosylation of the T cell antigen-specific receptor and its potential role in lectin-mediated cytotoxicity. Proc Natl Acad Sci USA 83:1852–1856

Kang R, Saito H, Ihara Y, Miyoshi E, Koyama N, Sheng Y, Taniguchi N (1996) Transcriptional regulation of the *N*-acetylglucosaminyltranserase V gene in human bile duct carcinoma cells (HuCC-T1) is mediated by Ets-1. J Biol Chem 271:26706–26712

Kanie O, Crawley SC, Palcic MM, Hindsgaul O (1994) Key involvement of all three GlcNAc hydroxyl groups in the recognition of β-D-GlcpNAc-(1→2)-α-D-Manp-(1→6)-β-D-Glcp-OR by *N*-acetylglucosaminyltransferase-V. Bioorg Med Chem 2:1234–1241

Khan SH, Crawley SC, Kanie O, Hindsgaul O (1993) A trisaccharide acceptor analog for *N*-acetylglucosaminyltransferase V which binds to the enzyme but sterically precludes the transfer reaction. J Biol Chem 268:2468–2473

Korczak B, Le T, Elowe S, Datti A, Dennis JW (2000) Minimal catalytic domain of *N*-acetylglucosaminyltransferase V. Glycobiology 10:595–599

Moloney DJ, Panin VM, Johnston SH, Chen J, Shao L, Wilson R, Wang Y, Stanley P, Irvine KD, Haltiwanger RS, Vogt TF (2000) Fringe is a glycosyltransferase that modifies Notch. Nature 406:369–375

Nakayama J, Yeh JC, Misra AK, Ito S, Katsuyama T, Fukuda M (1999) Expression cloning of a human alpha1, 4-*N*-acetylglucosaminyltransferase that forms GlcNAcα1→ 4Galβ→R, a glycan specifically expressed in the gastric gland mucous cell-type mucin. Proc Natl Acad Sci USA 96:8991–8996

Pierce M, Arango J (1986) Rous sarcoma virus-transformed baby hamster kidney cells express higher levels of asparagine-linked tri- and tetraantennary glycopeptides containing [GlcNAc-β(1,6)Man-α(1,6)Man] and poly-*N*-acetyllactosamine sequences than baby hamster kidney cells. J Biol Chem 261:10772–10777

Saito H, Nishikawa A, Gu J, Ihara Y, Soejima H, Wada Y, Sekiya C, Niikawa N, Taniguchi N (1994) cDNA cloning and chromosomal mapping of human *N*-acetylglucosaminyltransferase V. Biochem Biophys Res Commun 198:318–327

Schachter H (1986) Biosynthetic controls that determine the branching and microheterogeneity of protein-bound oligosaccharides. Biochem Cell Biol 64:163–181

Seelentag WK, Li WP, Schmitz SF, Metzger U, Aeberhard P, Heitz PU, Roth J (1998) Prognostic value of β1,6-branched oligosaccharides in human colorectal carcinoma. Cancer Res 58:5559–5564

Shoreibah MG, Hindsgaul O, Pierce M (1992) Purification and characterization of rat kidney UDP-*N*-acetylglucosamine: α-6-D-mannosidase β-1,6-*N*-acetylglucosaminyltransferase. J Biol Chem 267:2920–2927

Shoreibah M, Perng G-S, Adler B, Weinstein J, Basu R, Cupples R, Wen D, Browne JK, Buckhaults P, Fregien N, Pierce M (1993) Isolation, characterization, and expression of cDNA encoding *N*-acetylglucosaminyltransferase V. J Biol Chem 268:15381–15385

van den Eijnden DH, Koenderman AHL, Schiphorst WECM (1988) Biosynthesis of blood group i-active polylactosaminoglycans. J Biol Chem 263:12461–12465

Warren CE, Krizus A, Roy PJ, Culotti JG, Dennis JW (2001) The non-essential *C. elegans* gene, *gly-2*, can rescue the *N*-acetylglucosaminyltransferase V mutation of Lec4 cells. J Biol Chem (in press)

Weinstein J, Sundaram S, Wang X, Delgado D, Basu R, Stanley P (1996) A point mutation causes mistargeting of golgi GnT V in the Lec4A Chinese hamster ovary glycosylation mutant. J Biol Chem 271:27462–27469

Yamashita K, Tachibana Y, Ohkura T, Kobata A (1985) Enzymatic basis for the structural changes of asparagine-linked sugar chains of membrane glycoproteins of baby hamster kidney cells induced by polyoma transformation. J Biol Chem 260:3963–3969

Yousefi S, Higgins E, Doaling Z, Hindsgaul O, Pollex-Kruger A, Dennis JW (1991) Increased UDP-GlcNAc:Gal β1-3GalNAc-R (GlcNAc to GalNAc) β1-6 *N*-acetylglucosaminyltransferase activity in transformed and metastatic murine tumor cell lines: control of polylactosamine synthesis. J Biol Chem 266:1772–1783

14

N-Acetylglucosaminyltransferase-VI

Introduction

The biological roles of asparagine-linked oligosaccharides (N-glycans) on glycoproteins are thought to take place through the interaction of terminal glycan structures and their receptors. The diversity and avidity of the terminal structures, however, are regulated by the core structure of N-glycans (Schachter 1991). In vertebrates, six different N-acetylglucosaminyltransferases (GnT-I–VI) are involved in initiating the synthesis of highly branched N-glycan core structures (Fig. 1). GnT-VI catalyzes the transfer of GlcNAc to position four of the Manα1,6 arm of the core structure of N-glycan, forming the most highly branched pentaantennary glycans with a bisecting GlcNAc. These glycans have been found in hen ovomucoid (Yamashita et al. 1982; Paz Parente et al. 1982) and fish egg glycoprotein (Taguchi et al. 1995).

Very recently, this enzyme was purified (Taguchi et al. 2000) and its cDNA was cloned (Sakamoto et al. 2000). The primary structure was found to be significantly similar to human GnT-IV-homologue (hGnT-IVh) (Furukawa et al. 1999), which was cloned from the deleted region in pancreatic cancer, and to human and bovine GnT-IVs (Minowa et al. 1998; Yoshida et al. 1998; Yoshida et al. 1999). The sequence of hGnT-IVh is more similar to that of GnT-VI than that of GnT-IV. However, a recombinant hGnT-IVh has neither GnT-VI nor GnT-IV activity (Sakamoto et al. 2000).

Databanks

N-Acetylglucosaminyltransferase-VI

NC-IUBMB enzyme classification: E.C.2.4.1.201

Species	Gene	Protein	mRNA	Genomic
Gallus gallus	–	–	AB040608	–

KOICHI HONKE and NAOYUKI TANIGUCHI

Department of Biochemistry, Osaka University Medical School, 2-2 Yamadaoka, Suita, Osaka 565-0871, Japan
Tel. +81-6-6879-3421; Fax +81-6-6879-3429
e-mail: khonke@biochem.med.osaka-u.ac.jp

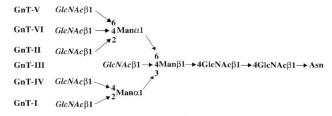

Fig. 1. Structure of pentaantennary *N*-linked oligosaccharide with the bisecting GlcNAc

Name and History

In 1974, Montreuil named the characteristic trisaccharides, sialic acid-Gal-GlcNAc-, at the nonreducing terminal of complex *N*-glycans "antennae" to emphasize a possible functional role of these terminal structures in the transmission of biological recognition signals (Montreuil 1974). Five different types of antennae have been described based on the site of initiation (Kobata 1984). The antennae are initiated by the *N*-acetylglucosaminyltransferases (GnTs) which transfer GlcNAc from UDP-GlcNAc to Man residues in the *N*-glycan core. The six GnTs involved in the synthesis of the five antennae and of the bisecting GlcNAc residue are numbered as Fig. 1 (Schachter et al. 1985).

The pentaantennary complex-type *N*-glycan structures were originally reported in hen ovomucoid (Yamashita et al. 1982; Paz Parente et al. 1982). The GnT-VI activity in hen oviduct was first demonstrated using synthetic substrates (Brockhausen et al. 1989), and the enzyme was shown to transfer GlcNAc in β1,4-linkage to the mannose residue of GlcNAcβ1-6(GlcNAcβ1-2)Manα1-6Manβ-(CH$_2$)$_8$COOCH$_3$. Very recently, this enzyme was purified to apparent homogeneity from hen oviduct (Taguchi et al. 2000), and its cDNA was cloned from a hen oviduct cDNA library based on the partial amino acid sequences of the purified enzyme (Sakamoto et al. 2000).

Enzyme Activity Assay and Substrate Specificity

This enzyme catalyzes the transfer of GlcNAc from UDP-GlcNAc to tetraantennary oligosaccharides and produces pentaantennary oligosaccharide with the β1,4-linked GlcNAc residue on the Manα1-6 arm. A simple and sensitive assay method for GnT-VI activity was developed using a pyridylaminated tetraantennary oligosaccharide as the acceptor substrate (Taguchi et al. 1998). The agalacto-tetraantennary oligosaccharide is prepared by hydrazinolysis of human α$_1$-acid glycoprotein and labeled with 2-aminopyridine. The standard reaction mixture contains 130 mM HEPES (pH 8.0), 25 mM MnCl$_2$, 75 mM GlcNAc, 0.5% Triton X-100, 25 mM UDP-GlcNAc, 0.2 nmol of the pyridylaminated tetraantennary oligosaccharide, and enzyme source in a total volume of 12 μl. After incubation at 37°C for 2 h, the reaction is stopped by adding 40 μl water, followed by heating at 100°C for 2 min. The sample is centrifuged at 13 000 rpm for 5 min, and then the supernatant is injected onto a TSK-gel ODS-80TM column (4.6 × 75 mm, Tosoh, Tokyo). Elution is effected at 55°C with 20 mM ammonium acetate/0.02% *n*-butanol, pH 4.0, at a flow rate of 1.0 ml/min, and monitored with a fluorescence spectrophotometer.

Hen oviduct GnT-VI requires Mn^{2+} and shows a broad pH optimum around pH 8.0 (Taguchi et al. 2000; Brockhausen et al. 1989).

GnT-VI does not act on biantennary oligosaccharide (GnT-I and GnT-II product), and β1,6-*N*-acetylglucosaminylation of the Manα1,6 arm (GnT-V product) is a prerequisite for its activity (Taguchi et al. 2000; Brockhausen et al. 1989). It has no GnT-VI' activity (Brockhausen et al. 1992), which is defined as making the GlcNAcβ1-2(GlcNAcβ1-4)Manα1-6-linkage irrespective of the β1,6-*N*-acetylglucosaminylation of the Manα1,6 arm. A bisecting GlcNAc residue does not inhibit GnT-VI activity (Brockhausen et al. 1989). In this respect, GnT-VI is different from GnT-IV and V, which cannot act on bisected substrates.

Preparation

GnT-VI activity has been demonstrated in hen oviduct (Brockhausen et al. 1989; Taguchi et al. 1998) and fish ovary (Taguchi et al. 1997). Like other glycosyltransferases, GnT-VI activity is found in the microsome fraction and solubilized by extraction with neutral detergent such as Triton X-100. The specific activity of hen oviduct homogenates was approximately 30 pmol/h/mg protein (Taguchi et al. 2000).

GnT-VI was purified 64 000-fold in a 16% yield from a homogenate of hen oviduct by column chromatographic procedures using Q-Sepharose FF, Ni^{2+}-chelating Sepharose FF, and UDP-hexanolamine-agarose (Taguchi et al. 2000). The purified enzyme shows a single band with an apparent molecular weight of 72 000 on nonreducing SDS-PAGE, and 60 000 on reducing SDS-PAGE.

Recombinant chicken GnT-VI is obtained from COS-1 cells transfected with the pSV-GnT-VI plasmid (Sakamoto et al. 2000). The cDNA-introduced COS-1 cells showed an enzyme activity of 26.8 pmol/h/mg protein, which is almost the same level as that of hen oviduct homogenates.

Biological Aspects

The GnT-VI gene is relatively highly expressed in hen oviduct, spleen, lung, and colon (Sakamoto et al. 2000). However, GnT-VI activity in chicken tissues was reported to be high in hen oviduct, and low but significant in liver and colon, but with none in heart and spleen (Brockhausen et al. 1989). Although the reason for the discrepancy between the amount of mRNA and enzymatic activity is unknown, translational regulation and posttranslational modification may be responsible.

Pentaantennary *N*-glycans have not as yet been found in mammalian tissues. This is consistent with the lack of GnT-VI activity in several mammal species (Brockhausen et al. 1989). To determine whether such pentaantennary structures are really absent in mammals, more sensitive and specific assay methods, e.g., immunochemical ones, should be developed. Studies on the presence of a GnT-VI gene and its expression in mammals could prove useful. Multiantennary structures generated by GnT-IV and GnT-V have been associated with phenotypic changes in malignant transformations (Dennis et al. 1999; Kobata 1998). Bulky multiantennary *N*-linked glycan chains are also thought to be involved in communication between cells in conditions relating to fertility, receptor function, and immune responses (Rademacher et al. 1988).

Future Perspectives

The biological function of the pentaantennary *N*-glycans is unknown. The isolation of the GnT-VI gene has enabled us to investigate the expression of this gene in various tissues of diverse species. Such studies and experimental remodeling of GnT-VI expression may clarify the physiological role of highly branched *N*-glycans in various biological processes.

References

Brockhausen I, Hull E, Hindsgaul O, Schachter H, Shah RN, Michnick SW, Carver JP (1989) Control of glycoprotein synthesis. Detection and characterization of a novel branching enzyme from hen oviduct, UDP-*N*-acetylglucosamine:GlcNAcβ1-6(GlcNAcβ1-2)Manα-R(GlcNAc to Man)β-4-*N*-acetylglucosaminyltransferase VI. J Biol Chem 264:11211–11221
Brockhausen I, Möller G, Yang J-M, Khan SH, Matta KL, Paulsen H, Grey AA, Shah RN, Schachter H (1992) Control of glycoprotein synthesis. Characterization of (1m4)-*N*-acetyl-β-D-glucosaminyltransferases acting on the α-D-(1m3)- and α-D-(1m6)-linked arms of *N*-linked oligosaccharides. Carbohydrate Res 236:281–299
Dennis JW, Granovsky M, Warren CE (1999) Glycoprotein glycosylation and cancer progression. Biochim Biophys Acta 1473:21–34
Furukawa T, Youssef EM, Yatsuoka T, Yokoyama T, Makino N, Inoue H, Fukushige S, Hoshi M, Hayashi Y, Sunamura M, Horii A (1999) Cloning and characterization of the human UDP-*N*-actylglucosamine:α-1,3-D-mammoside   β-1,4-*N*-acetylglucosaminyltransferase IV-homologue (*hGnT-IV-H*) gene. J Hum Genet 44:397–401
Kobata A (1984) The carbohydrates of glycoproteins. In: Ginsburg V, Robbins PW (eds) Biology of carbohydrates. Wiley, New York, pp 87–161
Kobata A (1998) A retrospective and prospective view of glycopathology. Glycoconj J 15:323–331
Minowa MT, Oguri S, Yoshida A, Hara T, Iwamatsu A, Ikenaga H, Takeuchi M (1998) cDNA cloning and expression of bovine UDP-*N*-acetylglucosamine:α1,3-D-mannnoside β1,4-*N*-acetylglucoaminyltransferase IV. J Biol Chem 273:11556–11562
Montreuil J (1974) Recent data on the structure of the carbohydrate moiety of glycoproteins. Metabolic and biological implications. Pure Appl Chem 42:431–477
Paz Parente J, Wieruszeski J-M, Strecker G, Montreuil J, Fournet B, van Halbeek H, Dorland L, Vliegenthart JFG (1982) A novel type of carbohydrate structure present in hen ovomucoid. J Biol Chem 257:13173–13176
Rademacher TW, Parekh RB, Dwek RA (1988) Glycobiology. Annu Rev Biochem 57:785–838
Sakamoto Y, Taguchi T, Honke K, Korekane H, Watanabe H, Tano Y, Dohmae N, Takio K, Horii A, Taniguchi N (2000) Molecular cloning and expression of cDNA encoding chicken UDP-GlcNAc: GlcNAcβ1-6(GlcNAcβ1-2)Manα1-R[GlcNAc to Man]β1,4*N*-acetylglucosaminyltransferase VI. J Biol Chem 275:36029–36034
Schachter H (1991) The "yellow brick road" to branched complex *N*-glycans. Glycobiology 1:453–461
Schachter H, Narasimhan S, Gleeson P, Vella G, Brockhausen I (1985) Glycosyltransferases involved in the biosynthesis of protein-bound oligosaccharides of the aspargine-*N*-acetyl-D-glucosamine and serine (threonine)-*N*-acetyl-D-galactosamine types. In:

Martonosi AN (ed) The enzymes of biological membranes. Plenum Press, New York, pp 227–277

Taguchi T, Kitajima K, Muto Y, Inoue S, Khoo K-H, Morris HR, Dell A, Wallace RA, Selman K, Inoue Y (1995) A precise structural analysis of a fertilization-associated carbohydrate-rich glycopeptide isolated from the fertilized eggs of euryhaline killi fish (*Fundulus heteroclitus*). Novel pentaantennary *N*-glycan chains with a bisecting *N*-acetylglucosaminyl residue. Glycobiology 5:611–624

Taguchi T, Kitajima K, Inoue S, Inoue Y, Yang JM, Schachter H, Brockhausen I (1997) Activity of UDP-GlcNAc:GlcNAcβ1-6(GlcNAcβ1-2)Manα1-R[GlcNAc to Man]β1-4*N*-acetylglucosaminyltransferase VI (GnT-VI) from the ovaries of *Oryzias latipes* (Medaka fish). Biochem Biophys Res Commun 230:533–536

Taguchi T, Ogawa T, Kitajima K, Inoue S, Inoue Y, Ihara Y, Sakamoto Y, Nagai K, Taniguchi N (1998) A method for determination of UDP-GlcNAc:GlcNAcβ1-6(GlcNAcβ1-2)Manα1-R[GlcNAc to Man]β1-4*N*-acetylglucosaminyltransferase VI activity using a pyridylaminated tetraantennary oligosaccharide as an acceptor substrate. Anal Biochem 255:155–157

Taguchi T, Ogawa T, Inoue S, Inoue Y, Sakamoto Y, Korekane H, Taniguchi N (2000) Purification and characterization of UDP-GlcNAc:GlcNAcβ1-6(GlcNAcβ1-2)Manα1-R [GlcNAc to Man]β1,4*N*-acetylglucosaminyltransferase VI from hen oviduct. J Biol Chem 275:32598–32602

Yamashita K, Kamerling JP, Kobata A (1982) Structural study of the carbohydrate moiety of hen ovomucoid. Occurrence of a series of pentaantennary complex-type asparagine-linked sugar chains. J Biol Chem 257:12809–12814

Yoshida A, Minowa, MT, Hara T, Takamatsu S, Ikenaga H, Takeuchi M (1998) A novel second isoenzyme of the human UDP-*N*-acetylglucosamine:α1,3-D-mannnoside β1,4-*N*-acetylglucoaminyltransferase family: cDNA cloning, expression, and chromosomal assignment. Glycoconj J 15:1115–1123

Yoshida A, Minowa MT, Takamatsu S, Hara T, Oguri S, Ikenaga H, Takeuchi M (1999) Tissue-specific expression and chromosomal mapping of a human UDP-*N*-acetylglucosamine:α1,3-D-mannnoside β1,4-*N*-acetylglucoaminyltransferase. Glycobiology 9:303–310

β3-*N*-Acetylglucosaminyltransferase (Fringe)

Introduction

Fringe provides a clear example of the role that carbohydrate modifications can play in regulating signal transduction events. Fringe was originally identified for its role in dorsal/ventral boundary formation during *Drosophila* wing development (Irvine and Wieschaus 1994). It functions by altering the response of the Notch receptor to its ligands, potentiating signaling from Delta and inhibiting that from Serrate (Fleming et al. 1997; Panin et al. 1997). Recent results have shown that fringe modulates Notch activity by altering the structure of the *O*-fucose glycans on the EGF repeats in the extracellular domain of Notch (Bruckner et al. 2000; Moloney et al. 2000a). *O*-fucose modifications occur between the second and third conserved cysteines of an EGF repeat at the consensus site C_2-X-X-G-G-$\underline{S/T}$-C_3, where X can be any amino acid and $\underline{S/T}$ is the modification site (Harris and Spellman 1993). Numerous cell surface and secreted proteins have EGF repeats containing these sites. Fringe catalyzes the addition of a β-linked GlcNAc to the 3′-hydroxyl of *O*-fucose, which can be further elongated to a tetrasaccharide with the structure NeuAcα2-3/6Galβ1-4GlcNAcβ1-3Fuc (Harris and Spellman 1993; Moloney et al. 2000a; Moloney et al. 2000b). The glycosyltransferase activity of fringe is essential for its ability to modulate Notch signaling (Bruckner et al. 2000; Moloney et al. 2000a; Munro and Freeman 2000), demonstrating that signal transduction events can be regulated by alterations in the glycosylation state of receptors.

ROBERT S. HALTIWANGER

Department of Biochemistry and Cell Biology, State University of New York at Stony Brook, Stony Brook, NY 11794-5215, USA
Tel. +1-631-632-7336; Fax +1-631-632-8575
e-mail: Robert.Haltiwanger@SUNYSB.EDU

Databanks

β3-*N*-Acetylglucosaminyltransferase (Fringe)

NO EC number has been allocated

Species	Gene	Protein	mRNA	Genomic
Homo sapiens	*MANIC FRINGE*	O00587	U94352	–
	RADICAL FRINGE	O00588	U94353	–
	LUNATIC FRINGE	O00589	U94354	–
Mus musculus	*Manic fringe*	O09008	U94349	–
	Radical fringe	O09009	U94350	–
	Lunatic fringe	Q9DC10	U94351	–
Rattus norvegicus	*Radical fringe*	Q9R1U9	AB016486	–
Gallus gallus	*Lunatic fringe*	O12971	U91849	–
	Radical fringe	O12970	U82088	–
Xenopus laevis	*Lunatic fringe*	P79948	U77640	–
	Radical fringe	P79949	U77641	–
Brachydanio rerio	*Lunatic fringe*	Q9DEV1	AY007434	–
Drosophila melanogaster	*Fringe*	Q24342	L35770	–

Name and History

Fringe derives its name from the fact that it plays an essential role in boundary formation between dorsal and ventral compartments during *Drosophila* development (Irvine and Wieschaus 1994). Three vertebrate homologues have been identified: Lunatic fringe (Lfng); Manic fringe (Mfng); and Radical fringe (Rfng) (Johnston et al. 1997). The vertebrate fringes act similarly to *Drosophila* fringe (D-fng) when expressed in *Drosophila*, suggesting that they work through a common mechanism (Johnston et al. 1997). D-fng, Lfng, and Mfng all have fucose-specific β3-*N*-acetylglucosminyltransferase activity (Moloney et al. 2000a). Although no in vitro assays have been performed with Rfng to date, preliminary data from our laboratory show that transfection of Rfng into CHO cells results in the addition of a β1,3-linked GlcNAc to *O*-fucose on EGF repeats (unpublished observations, L. Shao and R.S. Haltiwanger), suggesting that it shares the same enzymatic activity.

Enzyme Activity and Substrate Specificity

Fringe catalyzes the transfer of GlcNAc from UDP to the 3′-hydroxyl of *L*-fucose in the presence of manganese:

$$\text{UDP-GlcNAc} + \text{Fuc} \xrightarrow{\text{Mn}^{++}} \text{UDP} + \text{GlcNAc}\beta1\text{-}3\text{Fuc}$$

Fringe enzymatic activity has been characterized in vitro using high concentrations of low molecular weight acceptor substrates (Bruckner et al. 2000; Moloney et al. 2000a). D-fng, Lfng, and Mfng transfer GlcNAc to *p*-nitrophenyl-α-*L*-fucose (4 mM), but not to *p*-nitrophenyl-galactose or glucose. Product characterization demonstrated that GlcNAc is β-linked to the 3′-hydroxyl of fucose (Moloney et al. 2000a). D-fng also transfers GlcNAc to very high concentrations of free L-fucose (500 mM), but not to free glucose, galactose, GlcNAc, or GalNAc (Bruckner et al. 2000). Fringe is specific for UDP-GlcNAc as a nucleotide donor, as no transfer from UDP-galactose, UDP-glucose, and UDP-GalNAc to any of these acceptors was detected (Bruckner et al. 2000; Moloney et al. 2000a).

Although fucose is most commonly found as a terminal modification of N-glycans in mammalian systems, no significant transfer of GlcNAc to fucose on N-glycans by fringe has been observed. In contrast, O-fucose in the context of an EGF repeat serves as a highly effective substrate (Moloney et al. 2000a). To generate such a substrate, a recombinant EGF repeat from factor VII (containing an O-fucose consensus site) was expressed in bacteria and modified in vitro with O-fucose. As an acceptor substrate for fringe, factor VII EGF-O-fucose is nearly 1000-fold better than p-nitrophenyl-fucose (Moloney et al. 2000a). Interestingly, reduction and alkylation of the disulfide bonds in the EGF repeat abolish its ability to serve as a substrate (L. Shao and R.S. Haltiwanger, unpublished observation). These results strongly suggest that the in vivo substrates for the fringe proteins are O-fucose residues in the context of EGF repeats, and that the fringe proteins recognize features of folded EGF repeats in substrate recognition.

Fringe also modifies O-fucose on EGF repeats of proteins other than factor VII (Bruckner et al. 2000; Moloney et al. 2000a). The Notch receptor contains 36 tandem EGF repeats in its extracellular domain, many of which contain consensus sequences for O-fucose modification (11 for *Drosophila* Notch, nine for mouse Notch1 [Moloney et al. 2000b]). Five of the consensus sites are conserved across species (EGFs 3, 20, 24, 26, and 31). Transfection of Mfng into CHO cells results in modification of O-fucose on endogenous Notch1 in these cells (Moloney et al. 2000a). Similarly, transfection of Mfng causes modification of O-fucose on a fragment of the extracellular domain from mouse Notch1 (EGFs 19–23, single conserved O-fucose site at EGF 20) expressed in the same cells. In vitro assays have demonstrated that D-fng can transfer GlcNAc to O-fucose on either the entire extracellular domain of *Drosophila* Notch (Moloney et al. 2000a) or a fragment containing several O-fucose consensus sites (EGFs 1–3) (Bruckner et al. 2000). Interestingly, preliminary work suggests that fringe can also modify O-fucose on other proteins. Both Delta1 and Jagged1, ligands of Notch that contain EGF repeats with O-fucose consensus sites, are modified by Mfng in CHO cells (L. Shao and R.S. Haltiwanger, unpublished observation). Since numerous other proteins contain O-fucose consensus sites in EGF repeats (Harris and Spellman 1993), these results suggest that fringe may modify O-fucose on a wide variety of proteins.

Preparation

The preparations of fringe that have been assayed in vitro to date have been overexpressed, tagged versions of the proteins. Hexahistidine-tagged D-fng has been expressed in both Schneider cells (Moloney et al. 2000a) and high five cells (Bruckner et al. 2000) and purified for use in assays. Mouse Lfng and Mfng have been expressed in HEK 293T cells as fusion proteins with the Fc portion of human immunoglobulin (Moloney et al. 2000a) and purified for use in assays. Rfng has not yet been successfully overexpressed and assayed in vitro. Interestingly, D-fng-His$_6$,

Lfng-Fc, and Mfng-Fc could all be purified from the medium, suggesting they were secreted proteins (Johnston et al. 1997; Moloney et al. 2000a). Subsequent studies in *Drosophila* have shown that fringe functions in the Golgi apparatus and appears to be a type II transmembrane protein, consistent with what is known about most other Golgi glycosyltransferases (Bruckner et al. 2000; Munro and Freeman 2000). The fact that large amounts of the proteins are secreted suggests that the stem region of the fringe proteins is highly sensitive to proteases, releasing the catalytic domain from the membrane. Rfng does not appear to be secreted to any significant extent (Johnston et al. 1997), suggesting that its stem region may be less susceptible to proteolysis than those of the other fringes.

Biological Aspects

The genetic locus for Notch was identified in one of the early genetic screens in *Drosophila* as an X-linked lethal mutation where the female flies have a notch in their wings (Artavanis-Tsakonas et al. 1999). The gene was shown to code for a large cell surface receptor, and homologues have been identified in all metazoans, with four in mammals. Notch becomes activated upon binding to its ligands, Delta and Serrate/Jagged, which are also cell surface, transmembrane proteins (Artavanis-Tsakonas et al. 1999). Thus, activation of Notch requires expression of ligand on an adjacent cell. Ligand binding activates Notch through a unique signaling pathway termed regulated intramembrane proteolysis, the details of which have only recently been determined (for a recent review see Mumm and Kopan 2000). Ligand binding activates an extracellular metalloprotease (e.g., TACE/ADAM17), cleaving Notch approximately 12 residues outside the membrane and releasing the extracellular domain of the protein. This is followed by a second proteolytic cleavage just inside the membrane catalyzed by a γ-secretase-like protease. The second cleavage releases the cytoplasmic domain of Notch which translocates to the nucleus and associates with a member of the CSL family of transcriptional regulators, converting it from a transcriptional repressor to a transcriptional activator, inducing the expression of downstream gene products. Defects in Notch signaling result in a number of developmental abnormalities in both invertebrates and vertebrates and are associated with several human diseases (Joutel and Tournier-Lasserve 1998).

Fringe functions by regulating the Notch signaling pathway in a wide variety of tissues in both invertebrates and vertebrates (Irvine 1999). Detailed genetic studies in *Drosophila* have shown that fringe acts cell autonomously, potentiating signaling from Delta while inhibiting signaling from Serrate/Jagged (Fleming et al. 1997; Panin et al. 1997). The result of this regulation is the positioning of Notch activation to a particular set of cells which forms a boundary between dorsal and ventral compartments of developing organs. Fringe performs this function in numerous contexts, including wing, eye, and leg development in *Drosophila*, and appears to function similarly in vertebrates (Irvine 1999). For instance, expression of Rfng in the apical ectodermal ridge of the developing chick wing sets up a dorsal–ventral boundary in a manner analogous to the *Drosophila* wing, and genetic ablation of Lfng in mice results in defects in boundary formation between somites. Nonetheless, the fact that three vertebrate fringe molecules exist, coupled with four Notches and a growing number of ligands, suggests that the situation in vertebrates is much more complicated.

Modification of *O*-fucose glycans on Notch by fringe has been linked to its biological function in two ways. First, cell-based signaling assays were performed using glycosylation mutants of CHO cells (see Moloney et al. 2000a for details). These assays demonstrated that complex-type *N*-glycans are not required for fringe to inhibit ligand-dependent Notch activation. Interestingly, these same studies showed that fucosylation is required for fringe to exert its effects. Since *O*-fucose is the major form of protein fucosylation in CHO cells other than fucose found on complex-type *N*-glycans (Lin et al. 1994), these results suggested that fringe alters *O*-fucose modifications.

To demonstrate further that the glycosyltransferase activity of fringe is essential for its biological function, the putative DxD motif in D-fng was mutated (Bruckner et al. 2000; Moloney et al. 2000a; Munro and Freeman 2000). Mutational analysis of several glycosyltransferases has demonstrated that DxD motifs play a key role in catalysis (Wiggins and Munro 1998). All of the fringe molecules have a conserved DDD patch believed to be a DxD motif. Even minor alterations in this sequence in D-fng (e.g., DDD to DEE) abolished in vitro enzymatic activity (Moloney et al. 2000a), confirming that these residues are essential for catalysis. The biological activity of the mutant D-fng was analyzed in an ectopic expression assay in *Drosophila* wing imaginal discs and shown to be inactive. Thus, the glycosyltransferase activity of fringe is essential for it to modulate Notch activity. Taken together with the cell-based signaling assays described above and the known specificity of fringe, these results suggest that fringe alters Notch function by altering the *O*-fucose glycans on the Notch extracellular domain.

Future Perspectives

The discovery that fringe acts by altering the sugar structures on Notch has raised a number of important questions. For instance, the mechanism by which the change in *O*-fucose structure modulates Notch activation is not yet clear. The simplest model is that the change in sugar structure alters the affinity of Notch for its ligands. Several lines of data support this model. Bruckner and coworkers (Bruckner et al. 2000) showed that D-fng increased the binding of a soluble form of Delta to *Drosophila* SL2 cells expressing Notch. Using mammalian cells and proteins, Shimizu and coworkers (Shimizu et al. 2001) showed that both Mfng and Lfng inhibited binding of a soluble form of Jagged1 (Serrate homologue) to CHO cells expressing Notch2. These results are consistent with the in vivo data (Fleming et al. 1997; Panin et al. 1997) and suggest that the alterations in carbohydrate structure somehow affect the binding between Notch and its ligands. Nonetheless, other data obtained by these groups and others are hard to reconcile with the in vivo data. For instance, Bruckner and coworkers (Bruckner et al. 2000) observed no binding of soluble Serrate to SL2 cells expressing Notch, even in the absence of D-fng. Shimizu and coworkers (Shimizu et al. 2001) did not observe significantly enhanced Delta binding in the presence of fringe. Finally, Hicks and coworkers (Hicks et al. 2000), who performed ligand-binding studies using

HEK 293T cells, saw no decrease in binding of Jagged1 to Notch1 caused by Lfng. The differences in these studies may be due to the differences in cells used or to differences between binding of ligands to Notch1 versus Notch2. Nonetheless, the conflicts in the data can only be resolved by further study. Since all of these studies were performed with whole cells, it is impossible to determine whether the change in sugar structure directly or indirectly affects binding of ligand. The question of whether the change in O-fucose structure alters the affinity between Notch and its ligands will only be ultimately resolved using purified components in an in vitro binding assay.

Since fringe alters signaling from Delta and Serrate/Jagged in opposite directions, it is possible that more than one mechanism for regulation of signaling is at work. Two other possibilities that deserve attention are inhibition of cell autonomous inhibition of Notch signaling by ligands, and regulation of TACE cleavage (Moloney et al. 2000a). Cell autonomous inhibition of Notch by ligand refers to the fact that ligand expressed in the same cell as Notch can have an inhibitory effect on Notch activation (De Celis and Bray 2000), and fringe could exert its effects through regulation of this inhibition. Alternatively, the alteration in O-fucose structure could affect the ability of TACE to recognize and cleave the extracellular domain. These models all deserve consideration as further experiments are performed.

In addition to questions about how the change in sugar structure mediates alterations in Notch signaling, the specificity of the three vertebrate fringe proteins must be determined. Preliminary data from our laboratory suggest that the fringe proteins show specificity for individual EGF repeats, and it is likely that each fringe will show a different specificity (L. Shao and R.S. Haltiwanger, unpublished observation). These experiments will also help to clarify the diversity of Notch receptors and ligands as functional partners are determined.

Further Reading

Several reviews on fringe as a glycosyltransferase have been published, including those by Blair (2000) and Haltiwanger (2001), and a recent review on fringe biology by Irvine (1999).

The two papers demonstrating that the fringe proteins are glycosyltransferases are by Moloney et al. (2000a) and Bruckner et al. (2000). Recent reviews on Notch include those by Artavanis-Tsakonas et al. (1999) and Mumm and Kopan (2000).

References

Artavanis-Tsakonas S, Rand MD, Lake RJ (1999) Notch signaling: Cell fate control and signal integration in development. Science 284:770–776

Blair SS (2000) Notch signaling: Fringe really is a glycosyltransferase. Curr Biol 10: R608–R612

Bruckner K, Perez L, Clausen H, Cohen S (2000) Glycosyltransferase activity of Fringe modulates Notch-Delta interactions. Nature 406:411–415

De Celis JF, Bray SJ (2000) The Abruptex domain of Notch regulates negative interactions between Notch, its ligands and Fringe. Development 127:1291–1302

Fleming RJ, Gu SJ, Hukriede NA (1997) *Serrate*-mediated ctivation of *Notch* is specifically blocked by the product of the gene *fringe* in the dorsal compartment of the *Drosophila* wing imaginal disc. Development 124:2973–2981

Haltiwanger RS (2001) Fringe: A glycosyltransferase that modulates Notch signaling. TIGG. 13:157–165

Harris RJ, Spellman MW (1993) O-linked fucose and other post-translational modifications unique to EGF modules. Glycobiology 3:219–224

Hicks C, Johnston SH, DiSibio G, Collazo A, Vogt TF, Weinmaster G (2000) Jagged1 and Delta1 mediated Notch signaling is differentially regulated by Lunatic Fringe. Nat Cell Biol 2:515–520

Irvine KD (1999) Fringe, Notch, and making developmental boundaries. Curr Opin Genet Dev 9:434–441

Irvine KD, Wieschaus E (1994) fringe, a Boundary-specific signaling molecule, mediates interactions between dorsal and ventral cells during Drosophila wing development. Cell 79:595–606

Johnston SH, Rauskolb C, Wilson R, Prabhakaran B, Irvine KD, Vogt TF (1997) A family of mammalian Fringe genes implicated in boundary determination and the Notch pathway. Development 124:2245–2254

Joutel A, Tournier-Lasserve E (1998) Notch signaling pathway and human diseases. Semin Cell Dev Biol 9:619–625

Lin AI, Philipsberg GA, Haltiwanger RS (1994) Core fucosylation of high-mannose-type oligosaccharides in GlcNAc transferase I-deficient (Lec1) CHO cells. Glycobiology 4:895–901

Moloney DJ, Panin VM, Johnston SH, Chen J, Shao L, Wilson R, Wang Y, Stanley P, Irvine KD, Haltiwanger RS, Vogt TF (2000a) Fringe is a glycosyltransferase that modifies Notch. Nature 406:369–375

Moloney DJ, Shair L, Lu FM, Xia J, Locke R, Matta KL, Haltiwanger RS (2000b) Mammalian Notch1 is modified with two unusual forms of O-linked glycosylation found on epidermal growth factor-like modules. J Biol Chem 275:9604–9611

Mumm JS, Kopan R (2000) Notch signaling: from the outside in. Dev Biol 228:151–165

Munro S, Freeman M (2000) The notch signaling regulator fringe acts in the Golgi apparatus and requires the glycosyltransferase signature motif DXD. Curr Biol 10:813–820

Panin VM, Papayannopoulos V, Wilson R, Irvine KD (1997) Fringe modulates notch ligand interactions. Nature 387:908–912

Shimizu K, Chiba S, Saito T, Kumano K, Takahashi T, Hirai H (2001) Manic fringe and lunatic fringe modify different sites of the notch2 extracellular region, resulting in different signaling modulation. J Biol Chem 276:25753–25758

Wiggins CAR, Munro S (1998) Activity of the yeast MNN1 α-1,3-mannosyltransferase requires a motif conserved in many other families of glycosyltransferases. Proc Natl Acad Sci USA 95:7945–7950

16

β3-*N*-Acetylglucosaminyltransferase (iGnT)

Introduction

β3-*N*-Acetylglucosaminyltransferase, i-extension enzyme (iGnT), is a glycosyltransferase that catalyzes the transfer of GlcNAc from UDP-GlcNAc to Gal in the Galβ1-4Glc(NAc) structure with β1,3-linkage. In *N*-glycans, the addition of β1,3-linked GlcNAc is usually followed by galactosylation by β1,4-galactosyltransferase I, the predominant member of the β1,4-galactosyltransferase gene family (Ujita et al. 1999, 2000). It has been shown that poly-*N*-acetyllactosamines can be efficiently modified to form functional oligosaccharides such as sialyl Le[x].

In granulocytes, sialyl Le[x] is attached to both *O*-glycans (Fukuda et al. 1986) and *N*-glycans (Fukuda et al. 1984; Mizoguchi et al. 1984). In *O*-glycans of blood cells, sialyl Le[x] can be attached only to core2 branched oligosaccharides, since core2 branched *O*-glycans can have *N*-acetyllactosamine. Some of those core2 branches can be further extended to have poly-*N*-acetyllactosamines, which then can be further modified to express sialyl Le[x] (Fig. 1). In core2 *O*-glycans, galactosylation is mainly carried out by β4-galactosyltransferase-IV (Ujita et al. 1998). In *N*-glycans, poly-*N*-acetyllactosamines can be preferentially added to side chains (or antennas) derived from C-6 and C-2 of α1,6-linked mannose (van den Eijnden et al. 1988). In general, the amount and length of poly-*N*-acetyllactosamines in *O*-glycans are less than those attached to *N*-glycans. This is mostly due to the fact that β1,4-galactosyltransferase-IV loses its efficiency as the acceptor contains more *N*-acetyllactosamine repeats (Ujita et al. 1998).

It appears that poly-*N*-acetyllactosaminyl backbones are better substrates for other glycosyltransferases such as α3-fucosyltransferases (Niemelä et al. 1998). This is probably the reason why the sialyl Le[x] terminal is more frequently present in side

Minoru Fukuda

Glycobiology Program, The Burnham Institute, 10901 North Torrey Pines Road, La Jolla, CA 92037, USA
Tel. +1-858-646-3144; Fax +1-858-646-3193
e-mail: minoru@burnham-inst.org

Fig. 1. Synthesis of *N*-acetyllactosamine repeats and sialyl Le^x terminal. *N*-acetyllactosamine (Galβ1→4GlcNAcβ1→R) is extended by β1,3-*N*-acetyllactosaminyltransferase (iGnT) and β1,4-galactosyltransferase I (β4GalT I) (*A*). This can be sialylated by α2,3-sialyltransferase (ST3Gal), forming sialyl di-*N*-acetyllactosamine (*B*). The resultant oligosaccharide can be fucosylated with α1,3-fucosyltransferase VII (FucT-VII) forming sialyl Le^x, and then further fucosylated with FucT-IV forming sialyl diLe^x

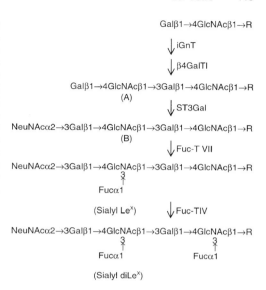

chains derived from the C-6 and C-2 positions of α-mannose, where poly-*N*-acetyllactosamines are frequently present (Fukuda et al. 1984). Recent studies have also demonstrated that a sialyl diLe^x structure such as NeuNAα2-3Galβ1-4(Fucα1-3)GlcNAcβ1-3Galβ1-4(Fucα1-3)GlcNAcβ1-R is a better ligand for C-type lectin than a sialyl Le^x such as NeuNAα2-3Galβ1-4(Fucα1-3)GlcNAcβ1-3Galβ1-4GlcNAcβ1-R (Handa et al. 1997). These two fucose residues are synthesized by FucT-VII (adding to GlcNAc next to a nonreducing terminal) and FucT-IV (adding to GlcNAc far from a nonreducing terminal) (Fig. 1). These results combined indicate that the concerted actions of iGnT, FucT-IV and FucT-VII lead to the synthesis of sialyl diLe^x, expressed on human granulocytes.

Recently, two different i-extension enzymes have been cloned from mammalian cells. One enzyme was cloned by an expression cloning strategy using human anti-i serum to enrich those cells expressing increased amounts of i-antigen (Sasaki et al. 1997). In this cloning, cDNA libraries from human melanoma and colon carcinoma cells were stably transfected into Namalwa KJM-I cells using a pAMo vector containing *ori*P of the Epstein–Barr virus. After enriching the i-expressing Namalwa cells three times, plasmids were recovered by the Hirt procedure (Hirt 1967) from those cells that were highly positive for the i-antigen. After narrowing the plasmids into smaller pools, one of the plasmids directed the expression of i-antigen upon transfection of the pAMo clone, and its cDNA insert was then cloned into pcDNA 3.1, resulting in pcDNA 3.1-iGnT. This enzyme is designated iGnT1, and recently renamed β3GlcNAcT1.

The second enzyme was cloned by identifying an enzyme, which is homologous to the β3-galactosyltransferase gene family. A β3-galactosyltransferase was originally cloned by an expression cloning strategy (Sasaki et al. 1994; Kyowa Hakko Kogyo KK

1994). By using this cloned protein as a probe, three additional β3-galactosyltransferases were cloned by searching the EST database for sequences homologous to this previously cloned β3-galactosyltransferase (Hennet et al. 1998). Further search of the EST database for a sequence homologous to the β3-galactosyltransferase gene family resulted in identification of a cDNA which was found to code for i-GnT (Zho et al. 1999). The enzyme directs the expression of i-antigen when transfected into HeLa cells. This enzyme is designated iGnT2, or β3GlcNAcT2.

In the report by Zhou et al., however, the cDNA sequence encoding β3GlcNAcT2 was mistaken as the cDNA sequence based on EST #AA150140 (Zhou et al. 2000). This β3GlcNAcT2 was independently cloned by Shiraishi et al. (2001) and found to have β1,3-*N*-acetylglucosaminyltransferase activity. Thus, the enzymatic property described by Zhou et al. (1999) and that of β3GlcNAcT2 should be on the same enzyme. The cloned enzyme based on EST #AA150140 is now designated β3GalT6 (see below). Shiraishi et al. (2001) also reported two additional β3GlcNAcTs, β3GlcNAcT3 and β3GlcNAcT4, based on their homology to β3GalT and by screening a phage library harboring human gastric mucosa- and placenta-derived cDNAs or neuroblastoma cell line SK-N-MC-derived single-stranded cDNA. In parallel to these studies, Yeh et al. cloned β3GlcNAcT2, β3GlcNAcT3, and β3GlcNAcT4, based on EST sequences and RT-PCR using human HT29 colonic cell-derived RNAs. In this study, it was discovered that β3GlcNAcT3 can add *N*-acetylglucosamine to core1 *O*-glycans, Galβ1-3GalNAcα1-R, forming GlcNAcβ1-3Galβ1-3GalNAcα1-R. This unique property of β3GlcNAcT3 allows the synthesis of extended core1 containing 6-sulfo GlcNAc, Galβ1-4(sulfo-6)GlcNAcβ1-3Galβ1-3GalNAcα1-R. This structure was found to be the minimum epitope for MECA-79 antibody (Yeh et al. 2001). In addition, β3GlcNAcT5 has recently been cloned (Togayachi et al. 2001). This cDNA was initially identified by random sequencing of rat cDNA and then isolation of full-length cDNA from the human 205 colonic tumor cell line. β3GlcNAcT5 was found to encode for glycolipid extension enzyme, adding *N*-acetylglucosamine to lactosylceramide. Enzymes β3GlcNAcT2 to β3GlcNAcT5 share highly homologous sequences, forming the β3GlcNAcT gene family (Fig. 2).

The major problem in having these two different sets of enzymes (β3GlcNAcT1 versus β3GlcNAcT2-5) that they are dissimilar. Although β3GlcNAcT1-5 are a type-II membrane protein, these two groups of the enzymes are entirely unrelated to each other. At this point, there are thus two entirely different sets of enzymes, which probably evolved from independent evolutionary origins. Recently β3GalT6 that forms Galβ1-3Gal was reported (Bai et al. 2001). The sequence of β3GalT6 is identical to that reported in Zhou et al. (1999), thus the original report described a similarity of two β3GalTs. It was rather fortunate that Zhou et al. (1999) obtained a correct conclusion based on a wrong comparison.

β3-*N*-Acetylglucosaminyltransferase was also cloned from *Neisseria meningitidis* (Blixt et al. 1999). The *lgtA* gene of *N. meningitidis* was found to transfer β1,3-linked *N*-acetylglucosamine to Galβ1-4Glc(NAc), Galβ1-3GalNAc, Galβ1-4Xyl, and Galα1-3Galβ1-4GlcNAc. This acceptor specificity is distinct from mammalian β1,3GlcNAcTs described above, since the mammalian enzymes require Galβ1-4GlcNAc or Galβ1-4Glc as an acceptor. No homology was found between this enzyme and mammalian iGnT. Since this bacterial enzyme has not been shown to form i-antigen, the enzyme is designated β3GnT6.

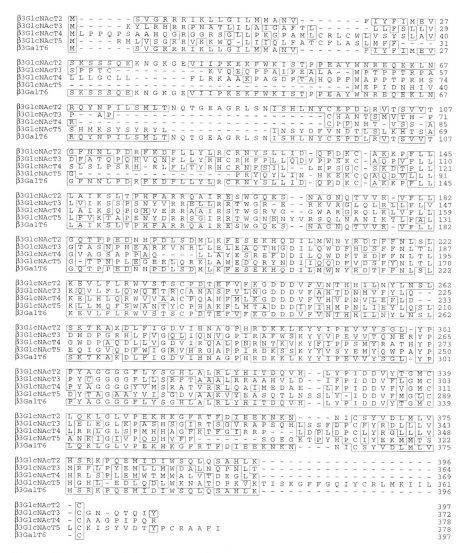

Fig. 2. Comparison of the amino acid sequences of human β3GlcNAcT2, β3GlcNAcT3, β3GlcNAcT4, β3GlcNAcT5, and β3GalT6. Introduced gaps are shown by *hyphens*, and aligned indentical residues are *boxed*

Databanks

β3-*N*-Acetylglucosaminyltransferase (iGnT)

NC-IUBMB enzyme classification: E.C.2.4.149

Species	Gene	Protein	mRNA	Genomic
Homo sapiens	*β3GlcNAcT1 (iGnT1)*	–	AF029893	–
	β3GlcNAcT2		AB049584	
	β3GlcNAcT3		AB049588	
			AF293973	
	β3GlcNAcT4		AB049586	
	β3GlcNAcT5		AB045278	
Neisseria meningitidis	*β3GlcNAcT6*	–	U25389	–

Name and History

β3-*N*-Acetylglucosaminyltransferase was designated as an i-extension enzyme, reflecting that it adds β1,3-GlcNAc to *N*-acetyllactosamine, which is converted to Galβ1-4GlcNAcβ1-3Galβ1-4GlcNAc, i-antigen. β3GlcNAcT1 and β3GlcNAcT2, β3GlcNAcT4, and β3GlcNAcT5 belong to this family, since all of these add β1,3-linked GlcNAc to lactose or *N*-acetyllactosamine. There are additional β3-*N*-acetylglucosaminyltransferases, for example, β3GlcNAcT that adds β1,3-linked GlcNAc to Galβ1-3GalNAc (EC 2.4.1.146), β3GlcNAcT3, or Core1-β3GlcNAcT. β1,3GlcNAcT6 belongs to an entirely different gene that forms lacto-*N*-tetraose in *Neisseria* cell walls. At this point, it was assumed that this enzyme is an entirely different gene product from the mammalian enzymes.

Enzyme Activity Assay and Substrate Specificity

The following reaction can be catalyzed by β3GlcNAcT1–β3GlcNAcT5:

$$\text{UDP-GlcNAc} + \text{Gal}\beta1\text{-}4\text{Glc(NAc)}\beta\text{-R}$$
$$\rightarrow \text{GlcNAc}\beta1\text{-}3\text{Gal}\beta1\text{-}4\text{Glc(NAc)}\beta\text{-R}$$

When both β3GlcNAcT and β4-galactosyltransferase are present, *N*-acetyllactosamine repeats are synthesized as

$$\text{UDP-GlcNAc} + \text{Gal}\beta1\text{-}4\text{Glc(NAc)}\beta\text{-R}$$
$$\rightarrow \text{GlcNAc}\beta1\text{-}3\text{Gal}\beta1\text{-}4\text{Glc(NAc)}\beta\text{-R} + \text{UDP-Gal}$$
$$\rightarrow \text{Gal}\beta1\text{-}4\text{GlcNAc}\beta1\text{-}3\text{Gal}\beta1\text{-}4\text{Glc(NAc)}\beta1\text{-R}$$

The repetition of these reactions leads to $(\text{Gal}\beta1\text{-}4\text{GlcNAc}\beta1\text{-}3)_n\text{Gal}\beta1\text{-}4\text{Glc(NAc)}\beta1\text{-}$ R, poly-*N*-acetyllactosamine. The enzyme assay and acceptor specificity were carried out using a partially purified iGnT from human and bovine sera (Piller and Cartron 1983; van den Eijnden et al. 1988; Kawashima et al. 1993). iGnT is more efficient when it acts on tri- and tetraantennary glycans (van den Eijnden et al. 1988). This unique property is related to the fact that the amount of poly-*N*-acetyllactosamine is increased when *N*-glycans with a GlcNAcβ1-6Man branch is present (see below). In addition, iGnT was found not to act on the Galβ1-4(Fucα1-3)GlcNAc-R structure.

β3GlcNAcT3 (or Core1-β3GlcNAcT) can act on core1 *O*-glycans as shown below (Yeh et al. 2001).

$$\text{UDP-GlcNAc} + \text{Gal}\beta1\text{-}3\text{GalNAc}\alpha1\text{-R}$$

$$\rightarrow \text{GlcNAc}\beta1\text{-}3\text{Gal}\beta1\text{-}3\text{GalNAc}\alpha1\text{-R}$$

The product will be converted to *N*-acetyllactosaminyl core1 (or extended core1), Galβ1-4GlcNAcβ1-3Galβ1-3GalNAcα1-R. Moreover, β3GlcNAcT3 was found to act on core2 branched *O*-glycans as shown below.

$$\pm\text{Gal}\beta1\text{-}4\text{GlcNAcb}1\text{-}6(\text{Gal}\beta1\text{-}3)\text{GalNAc}\alpha1\text{-R}$$

$$\rightarrow \pm\text{Gal}\beta1\text{-}4\text{GlcNAc}\beta1\text{-}6(\text{GlcNAc}\beta1\text{-}3\text{Gal}\beta1\text{-}3)\text{GalNAc}\alpha1\text{-R}$$

This reaction thus allows the formation of biantennary *O*-glycans, Galβ1-4GlcNAcβ1-6(Galβ1-4GlcNAcβ1-3Galβ1-3)GalNAcα1-R. Two side chains of the biantennary *O*-glycans can be modified to express sialyl Le^x or 6-sulfo sialyl Le^x, resulting in the formation of oligosaccharide ligands with high affinity.

Preparation

Native iGnT was partially purified from bovine serum using many steps of column chromatography (Kawashima et al. 1993). However, iGnT is more readily available in recombinant form. iGnT1 (or β3GlcNAcT1) was prepared as a soluble chimeric protein fused with protein A (Sasaki et al. 1997; Ujita et al. 1998). A mammalian expression vector, pcDNAI encoding iGnT1 fused with protein A, was transfected into COS-1 cells. Spent medium was incubated with IgG–Sepharose, and the chimeric enzyme adsorbed to IgG–Sepharose was incubated with 5 mM Galβ1-4Glcβ1-*p*-nitrophenol in 100 μM cacodylate buffer, pH 7.0, containing 0.1 mM UDP-[^{3}H]GlcNAc (1×10^6 cpm/nmol), 20 mM MnCl$_2$, and 5 mM ATP. 10 mM *N*-acetylglucosaminolactone was added to inhibit breakdown of products by β-hexosaminidase when poly-*N*-acetyllactosamine formation is assayed. The reaction product was adsorbed to a Sep-Pak(C18) column and eluted with methanol. In this assay, the cultured medium from COS-1 cells was extensively concentrated and thus some activity may be derived from β3GlcNAcT endogenously expressed in COS-1 cells.

β3GlcNAcT2–β3GlcNAcT4 were expressed in SF9 insect cells using a baculovirus system, Bac-to-Bac system (a transposition-mediated recombination system) purchased from Invitrogen (Carlsbad, CA; Luckow et al. 1993). The lysates from transfected insect cells were used as an enzyme source (Zhou et al. 1999).

Alternatively, β3GlcNAcT3 and other enzymes were cloned in expression vector pcDNA3.1/HSH. This pcDNA3.1/HSH vector harbors cDNAs encoding a signal peptide for the pcDNA3.1·A vector (Ujita et al. 1998), and six histidine residues, which was constructed as described previously (Angata et al. 2001). CHO cells were transfected with pcDNA3.1/HSH harboring cDNA for the soluble form of β3GlcNAcT.

Twenty-four hours after transfection, the medium was changed to serum-free OptiMEM medium (Invitrogen, Carlsbad, CA). The culture medium obtained after an additional 48 h of culture was filtered and concentrated with Microcon 30 (Millipore Bedford, MA) and used as the enzyme source.

Biological Aspects

The results so far obtained indicate that poly-*N*-acetyllactosamines synthesized by iGnT provide critical backbone structures for the addition of functional oligosaccharides.

In cancer cells, the amounts of poly-*N*-acetyllactosamines and *N*-acetylglucosaminyltransferase V (GnT-V) increase (Yamashita et al. 1984; Pierce and Arango 1986; Dennis et al. 1987; Saitoh et al. 1992). It has been assumed that a side chain derived from C-6 of an outer α-mannose core is elongated with poly-*N*-acetyl-lactosamines, since GnT-V forms the side chain. More recent studies, however, revealed that the addition of a 6-linked side chain to the outer mannose makes the dibranched side chains much better acceptors for iGnT and β4GalT-I (Ujita et al. 1999). This change is probably brought about by conformational changes of the dibranched acceptor (Bock et al. 1992). Because of this change, approximately eight times more poly-*N*-acetyllactosamine is added to the branched acceptor than to an unbranched acceptor. In addition, the complementary branch specificity of iGnT and β4GalT-I results in an elongation of poly-*N*-acetyllactosamines which is equal to both side chains derived from C-6 and C-2 (Fig. 3). Such an equal extension is consistent with poly-*N*-acetyllactosaminyl structures found in nature (Sasaki et al. 1987; Watson et al. 1994; Hokke et al. 1995). The increase in poly-*N*-acetyllactosamines, due to the increase in GnT-V, then results in an increase in functional oligosaccharides such as sialyl Le^x and sialyl diLe^x. Those cancer cells expressing poly-*N*-acetyllactosamines and disialyl Le^x may adhere efficiently to C-type lectins, including selectins present on microvascular endothelial cells (Trunen et al. 1995), thereby making cancer cells adhere efficiently to metastatic sites (Ohyama et al. 1999).

It has been demonstrated that functional oligosaccharides such as sialyl Le^x can be synthesized in core2 branched *O*-glycans (see Fig. 1, Chapter 18). The presence of Core2GlcNAcT is thus essential for synthesizing sialyl Le^x and other functional oligosaccharides. A more recent discovery, however, demonstrated that those functional oligosaccharides can be synthesized in extended core1 oligosaccharides, which are formed by β3GlcNAcT3 (Yeh et al. 2001). Moreover, the presence of both β3GlcNAcT3 and Core2GlcNAcT results in the formation of biantennary *O*-glycans, which can be sulfated, sialylated, and fucosylated as follows.

$$\text{NeuNAc}\alpha 2\text{-3Gal}\beta 1\text{-4}[\text{Fuc}\alpha 1\text{-3(sulfo-6)}]\text{GlcNAc}\beta 1 \diagdown \atop {}_{3}^{6}\text{GalNAc}$$
$$\text{NeuNAc}\alpha 2\text{-3Gal}\beta 1\text{-4}[\text{Fuc}\alpha 1\text{-3(sulfo-6)}]\text{Gal}\beta 1 \diagup$$

Such twin 6-sulfo sialyl Le^x in biantennary *O*-glycans serve as very efficient L-selectin ligands (Yeh et al. 2001).

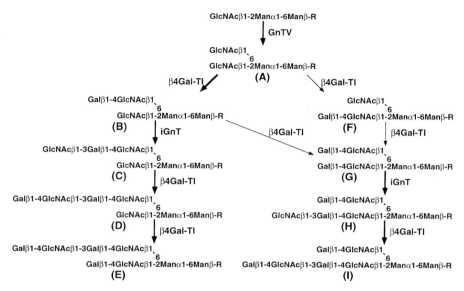

Fig. 3. Proposed biosynthetic steps of poly-*N*-acetyllactosamines in dibranched GlcNAcβ1-6(GlcNAcβ1-2)Manα1-6Manβ-R. β1,6-linked *N*-acetylglucosamine is first added by GnT-V, forming GlcNAcβ1-6(GlcNAcβ1-2)Manα1-6Manβ-R (*A*). This is followed mainly by galactosylation of β1,6-linked GlcNAc (*B*), addition of β1,3-linked GlcNAc and β1,4-linked galactose (*C* and *D*), and galactosylation of the GlcNAcβ1-2Manα-R side chain (*E*), forming poly-*N*-acetyllactosamine in the GlcNAcβ1-6 branch. As a minor biosynthetic pathway, *B* is converted to *G* by galactosylation of the GlcNAcβ1-2Manα-R side chain in *B*. This is followed by addition of *N*-acetylglucosamine (*H*) and galactose (*I*), forming poly-*N*-acetyllactosamine extension on the Galβ1-4GlcNAcβ1-2Manα-R side chain (*I*). As a minor biosynthetic pathway, *G* is also formed from *A* through *F*, leading to the formation of *I*. Although it is not shown in the study, *E* and *I* are most likely converted to those containing *N*-acetyllactosamine repeats in both Galβ1-4GlcNAcβ1-6 and Galβ1-4GlcNAcβ1-2 branches. (From Ujita et al. 1999)

Future Perspectives

In the near future, it is important to identify which enzyme is actually responsible for poly-*N*-acetyllactosamine synthesis in each cell. Among β3GlcNAcT and related enzymes, it is possible that one enzyme is predominant in poly-*N*-acetyllactosamine synthesis in a given cell. Is there an iGnT that adds *N*-acetyllactosamine repeats preferentially to *N*-glycans, or an iGnT that adds *N*-acetyllactosamine repeats preferentially to *O*-glycans?

There are already five cDNA sequences that encode β3GlcNAcT (iGnT). Further characterization of the proteins encoded by these cDNAs is essential to understand those points.

Further Reading

References recommended for further reading are marked by asterisks.

References

Angata K, Yen T-Y, El-Battari A, Macher BA, Fukuda M (2001) Unique disulfide bond structures found in ST8Sia IV polysialyltransferase are required for its activity. J Biol Chem 276:15369–15377

Bai X, Dapeng Z, Brown J, Hennet T, Esko J (2001) Biosynthesis of the linkage region of glycosaminoglycans: Cloning and activity of Glactosyltransferase II, the sixth member of the β1,3galactosyltransferase family (β3GalT6). J Biol Chem in press

Blixt O, van Die I, Norberg T, van den Eijnden DH (1999) High-level expression of the *Neisseria meningitidis lgtA* gene in *Escherichia coli* and characterization of the encoded *N*-acetylglucosaminyltransferase as a useful catalyst in the synthesis of GlcNAcβ-1→3Gal and GalNAcβ-1→3Gal linkages. Glycobiology 9:1061–1071

*Bock K, Duus JO, Hindsgaul O, Lindh I (1992) Analysis of conformationally restricted models for the (1-6)-branch of asparagine-linked oligosaccharides by NMRT spectroscopy and HSEA calculation. Carbohydr Res 228:1–20

Dennis JW, Laferte S, Waghorne C, Breitman ML, Kerbel RS (1987) β1-6 branching of As n-linked oligosaccharides is directly associated with metastasis. Science 236:582–585

*Fukuda M, Spooncer E, Oates JE, Dell A, Klock JC (1984) Structure of sialylated fucosyl lactosaminoglycan isolated from human granulocytes. J Biol Chem 259:10925–10935

Fukuda M, Carlsson SR, Klock JC, Dell A (1986) Structures of O-linked oligosaccharides isolated from normal granulocytes, chronic myelogenous leukemia cells, and acute myelogenous leukemia cells. J Biol Chem 261:12796–12806

Handa K, Stroud MR, Hakomori S (1997) Sialosyl–fucosyl Poly-LacNAc without the sialosyl-Lex epitope as the physiological myeloid cell ligand in E-selectin-dependent adhesion: studies under static and dynamic flow conditions. Biochemistry 36:12412–12420

Hennet T, Dinter A, Kuhnert P, Mattu TS, Rudd PM, Berger EG (1998) Genomic cloning and expression of three murine UDP-galactose: β-*N*-acetylglucosamine β1,3-galactosyltransferase genes. J Biol Chem 273:58–65

Hirt B (1967) Selective extraction of polyoma DNA from infected mouse cell cultures. J Mol Biol 26:365–369

Hokke CH, Bergwerff AA, Van Dedem GW, Kamerling JP, Vliegenthart JF (1995) Structural analysis of the sialylated *N*- and *O*-linked carbohydrate chains of recombinant human erythropoietin expressed in Chinese hamster ovary cells. Sialylation patterns and branch location of dimeric *N*-acetyllactosamine units. Eur J Biochem 228:981–1008

Kawashima H, Yamamoto K, Osawa T, Irimura T (1993) Purification and characterization of UDP-GlcNAc:Gal-β-1-4Glc(NAc) β-1,3-*N*-acetylglucosaminyltransferase (poly-*N*-acetyllactosamine extension enzyme) from calf serum. J Biol Chem 268:27118–27126

Kyowa Hakko Kogyo KK (July 5, 1994) Japanese Patent JP6181759

Luckow VA, Lee SC, Barry GF, Olins PO (1993) Efficient generation of infectious recombinant baculoviruses by site-specific transposon-mediated insertion of foreign genes into a baculovirus genome propagated in Escherichia coli. J Virol 67:4566–4579

Mizoguchi A, Takasaki S, Maeda S, Kobata A (1984) Changes in asparagine-linked sugar chains of human promyelocytic leukemic cells (HL-60) during monocytoid differentiation and myeloid differentiation. Appearance of high mannose-type oligosaccharides in neutral fraction. J Biol Chem 259:11943–11948

*Niemelä R, Natunen J, Majuri ML, Maaheimo H, Helin J, Lowe JB, Renkonen O, Renkonen R (1998) Complementary acceptor and site specificities of FucT IV and FucT VII allow effective biosynthesis of sialyl-TriLex and related polylactosamines present on glycoprotein counterreceptors of selectins. J Biol Chem 273:4021–4026

Ohyama C, Tsuboi S, Fukuda M (1999) Dual roles of sialyl Lewis X oligosaccharides in tumor metastasis and rejection by natural killer cells. EMBO J 18:1516–1525

Pierce M, Arango J (1986) Rous sarcoma virus-transformed baby hamster kidney cells express higher levels of asparagine-linked tri- and tetraantennary glycopeptides containing [GlcNAc-β (1,6)MaN-α (1,6)Man] and poly-N-acetyllactosamine sequences than baby hamster kidney cells. J Biol Chem 261:10772–10777

Piller F, Cartron JP (1983) UDP-GlcNAc:Gal β 1-4Glc(NAc) β 1-3N-acetylglucosaminyltransferase. Identification and characterization in human serum. J Biol Chem 258: 12293–12299

Saitoh O, Wang WC, Lotan R, Fukuda M (1992) Differential glycosylation and cell surface expression of lysosomal membrane glycoproteins in sublines of a human colon cancer exhibiting distinct metastatic potentials. J Biol Chem 267:5700–5711

Sasaki H, Bothner B, Dell A, Fukuda M (1987) Carbohydrate structure of erythropoietin expressed in Chinese hamster ovary cells by a human erythropoietin cDNA. J Biol Chem 262:12059–12076

*Sasaki K, Kurata-Miura K, Ujita M, Angata K, Nakagawa S, Sekine S, Nishi T, Fukuda M (1997) Expression cloning of cDNA encoding a human β-1,3-N-acetylglucosaminyltransferase that is essential for poly-N-acetyllactosamine synthesis. Proc Natl Acad Sci USA 94:14294–14299

*Shiraishi N, Natsume A, Togayachi A, Endo T, Akashima T, Yamada Y, Imai N, Nakagawa S, Koizumi S, Sekine S, Narimatsu H, Sasaki K (2001) Identification and characterization of three novel β1,3-N-acetylglucosaminyltransferases structurally related to the β1,3-galactosyltransferase family. J Biol Chem 276:3498–3507

*Togayachi A, Akashima T, Ookubo R, Kudo T, Nishihara S, Iwasaki H, Natsume A, Mio H, Inokuchi J-I, Iramura T, Sasaki K, Narimatsu H (2001) Molecular cloning and characterization of UDP-GlcNAc:lactosylceramide β1,3-N-acetylglucosaminyltransferase (β3Gn-T5), an essential enzyme for the expression of HNK-1 and Lewis x epitopes on glycolipids. J Biol Chem 276:22032–22040

*Toppila S, Renkonen R, Penttila L, Natunen J, Salminen H, Helin J, Maaheimo H, Renkonen O (1999) Enzymatic synthesis of α3′sialylated and multiply α3fucosylated biantennary polylactosamines. A bivalent [sialyl diLex]-saccharide inhibited lymphocyte–endothelium adhesion orgaN-selectively. Eur J Biochem 261:208–215

Turunen JP, Majuri M-L, Seppo A, Tiisala S, Paavonen T, Miyasaka M, Lemström K, Penttilä L, Renkonen O, Renokonen R (1995) de Novo expression of sialyl Lewisᵃ and sialyl Lewisˣ during cardiac transplant rejection: superior capacity of a tetravalent sialyl Lewisˣ oligosaccharide in inhibiting L-selectin-dependent lymphocyte adhesion. J Exp Med 182:1133–1142

*Ujita M, McAuliffe J, Schwientek T, Almeida R, Hindsgaul O, Clausen H, Fukuda M (1998) Synthesis of poly-N-acetyllactosamine in core2-branched O-glycans. The requirement of novel β-1,4-galactosyltransferase IV and β-1,3-N-acetylglucosaminyltransferase. J Biol Chem 273:34843–34849

Ujita M, McAuliffe J, Hindsgaul O, Sasaki K, Fukuda MN, Fukuda M (1999) Poly-N-acetyllactosamine synthesis in branched N-glycans is controlled by complemental branch specificity of I-extension enzyme and β1,4-galactosyltransferase I. J Biol Chem 274:16717–16726

Ujita M, Misra AK, McAuliffe J, Hindsgaul O, Fukuda M (2000) Poly-N-acetyllactosamine extension in N-glycans and core2- and core4-branched O-glycans is differentially controlled by i-extension enzyme and different members of the β1,4-galactosyltransferase gene family. J Biol Chem 275:15868–15875 van den Eijnden DH, Koenderman AH, Schiphorst WE (1988) Biosynthesis of blood group i-active polylactosaminoglycans. Partial purification and properties of an UDP-GlcNAc:*N*-acetyllactosaminide β1,3-*N*-acetylglucosaminyltransferase from Novikoff tumor cell ascites fluid. J Biol Chem 263:12461–12471

Watson E, Bhide A, van Halbeek H (1994) Structure determination of the intact major sialylated oligosaccharide chains of recombinant human erythropoietin expressed in Chinese hamster ovary cells. Glycobiology 4:227–237

*Yamashita K, Ohkura T, Tachibana Y, Takasaki S, Kobata A (1984) Comparative study of the oligosaccharides released from baby hamster kidney cells and their polyoma transformant by hydrazinolysis. J Biol Chem 259:10834–10840

*Yeh J-C, Hiraoka N, Petryniak B, Nakayama J, Ellies LG, Rabuka D, Hindsgaul O, Marth JD, Lowe JB, Fukuda M (2001) Novel sulfated lymphocyte homing receptors and their control by a core1 extension β1,3-*N*-acetylglucosaminyltransferase Cell 105:957–969

*Zhou D, Dinter A, Gutierrez Gallego R, Kamerling JP, Vliegenthart JF, Berger EG, Hennet T (1999) A β-1,3-*N*-acetylglucosaminyltransferase with poly-*N*-acetyllactosamine synthase activity is structurally related to β1,3-galactosyltransferases. Proc Natl Acad Sci U.S.A 96:406–411

Zhou D, Dinter A, Guiterrez Gallego R, Kamerling JP, Vliegenthart JF, Berger EG, Hennet T (2000) Correction for vol. 96, p. 406. Proc Natl Acad Sci USA 97:11673–11675

β6-*N*-Acetylglucosaminyltransferase (IGnT)

Introduction

I-branching β6-*N*-acetylglucosaminyltransferase (IGnT) is a glycosyltransferase that catalyzes the transfer of GlcNAc from UDP-GlcNAc to β1,4-linked Gal residue in a linear poly-*N*-acetyllactosamine, ±Galβ1-4GlcNAcβ1-3Galβ1-4Glc(NAc)-R, forming ±Galβ1-4GlcNAcβ1-3(GlcNAcβ1-6)Galβ1-4Glc(NAc)-R. The formation of the I branch is usually followed by galactosylation with β4-galactosyltransferase I, resulting in the I antigen. The I-antigen can be further modified to express functional oligosaccharides, such as sialyl Lewis X (Fig. 1).

During human development, fetal and neonatal erythrocytes express i-antigen, while adult erythrocytes express I-antigen. The epitope of these i- and I-antigens was found to be linear poly-*N*-acetyllactosamines, Galβ1→4GlcNAcβ1→3Galβ1→4GlcNAcβ1→R and branched poly-*N*-acetyllactosamines, Galβ1→4GlcNAcβ1→3(Galβ1→4GlcNAcβ1→6)Galβ1→4GlcNAcβ1→R, respectively (Feizi et al. 1979; Watanabe et al. 1979). The expression of I-antigen is entirely dependent on IGnT, and it is assumed that the expression of IGnT is developmentally regulated. In addition, I-antigen is apparently the major carbohydrate antigen present on both human and mouse embryonic cells (Muramatsu et al. 1979; Kapadia et al. 1981; Fukuda et al. 1985).

There are at least two, and possibly three, different I β6-*N*-acetylglucosaminyltransferases. The first IGnT acts on central galactose residues in Galβ1-4GlcNAcβ1-3<u>Gal</u>β1-4Glc(NAc)-R (the acceptor galactose is underlined), forming Galβ1-4GlcNAcβ1-3(GlcNAcβ1-6)<u>Gal</u>β1-4Glc(NAc)-R (Fig. 1). This enzyme, called centrally acting IGnT or cIGnT, is the dominant form of IGnT present in PA-1 human teratocarcinoma cells, hog small intestine, and most likely human erythroid precursors (Fukuda et al. 1985; Mattila et al. 1998; Sakamoto et al. 1998).

The second IGnT, called dIGnT, acts on a pre<u>d</u>istal galactose residue in GlcNAcβ1-3<u>Gal</u>β1-4Glc(NAc)-R (the acceptor galactose is underlined), forming GlcNAcβ1-3(GlcNAcβ1-6)<u>Gal</u>β1-4Glc(NAc)-R (Ropp et al. 1991; Gu et al. 1992).

Minoru Fukuda

Glycobiology Program, The Burnham Institute, 10901 North Torrey Pines Road, La Jolla, CA 92037, USA
Tel. +1-858-646-3144; Fax +1-858-646-3193
e-mail: minoru@burnham-inst.org

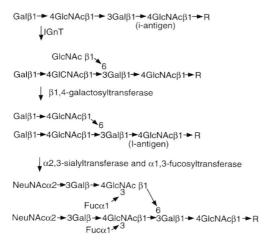

Fig. 1. Biosynthetic steps of I-branches containing sialyl Lewis X termini. A linear poly-*N*-acetyllactosamine (i-antigen) is converted to a branched poly-*N*-acetyllactosamine (I-antigen) by the actions of cIGnT and β4-galactosyltransferase. By dIGnT, a branch is added to GlcNAcβ1-3Galβ1-4GlcN(Ac)β1-R acceptor. I-branched poly-*N*-acetyllactosamines can be further modified to express sialyl Lewis X by the actions of α3-sialyltransferase and α3-fucosyltransferase (Fuc-TVII)

The third enzyme apparently adds *N*-acetylglucosamine to Galβ1-4Glc(NAc)-R, forming GlcNAcβ1-6Galβ1-4Glc(NAc)-R (van der Eijnden et al. 1983). Although the presence of this enzyme has not been confirmed by a purified IGnT or cloned cDNA, it may be responsible for forming [Galβ1-4GlcNAcβ1-6]$_n$(Galβ1-3)GalNAc in human milk mucins (Hanisch et al. 1989).

The cDNA encoding IGnT has been cloned from human PA-1 cells (Bierhuizen et al. 1993). The recombinant IGnT derived from PA-1 cells mainly acts as cIGnT (Mattila et al. 1998). However, it was also noted that this IGnT can act as dIGnT, corresponding to 6%–30% of cIGnT activity, depending on the assays employed (Ujita et al. 1999; Yeh et al. 1999). Most recent studies also demonstrated that the IGnT cloned from PA-1 cells can act on galactose residues distant from the nonreducing terminal, resulting in multiple short I branches along a poly-*N*-acetyllactosaminyl side chain (Ujita et al. 1999) (Fig. 2).

dIGnT was originally discovered in hog gastric mucosa, adding *N*-acetylglucosamine to GlcNAcβ1-3Galβ1-4Glc to form GlcNAcβ1-3(GlcNAcβ1-6)Galβ1-4Glc (Piller et al. 1984). To date, there is no report on the presence of an enzyme that contains only dIGnT activity. As described above, dIGnT activity is preset as a secondary activity in the IGnT cloned from PA-1 cells. Similarly, core 2 β1,6-*N*-acetylglucosaminyltransferase 2 (or core2GlcNAcT-mucin type, see Chap. 19) contains approximately 10% of dIGnT activity compared with core2GlcNAcT activity. It is possible that an enzyme that exclusively contains dIGnT is only present in very few tissues or cells.

As described above, an enzyme containing mainly cIGnT activity was cloned by expression cloning from PA-1 human tetracarcinoma cells (Bierhuizen et al. 1993). The enzyme is a typical type II membrane protein composed of 400 amino acids (Mr 45 860). This enzyme is highly homologous to core2GlcNAcT-I, having 41.6% identity at the amino acid level (see next chapter). The transcript of the IGnT is strongly expressed in adult prostate and cerebellum, and the frontal lobe of adult brain, moderately in adult heart, small intestine, and colon, and fetal brain, kidney, and lung also express the transcript (Sasaki et al. 1997).

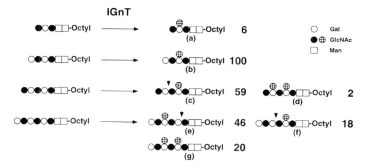

Fig. 2. Schematic representation of oligosaccharide products obtained after incubation with recombinant IGnT derived from PA-1 cells. *Open circles*, Gal; *solid circles*, GlcNAc; *open squares*, Man; *hatched circles*, I-branched β1,6-linked GlcNAc. The *numbers* indicate the relative molar ratio of the products. *Arrowheads* indicate the positions where end-β-galactosidase cleaves. (From Ujita et al. 1999)

Databanks

β6-*N*-Acetylglucosaminyltransferase (IGnT)

NC-IUBMB enzyme classification: E.C.2.4.1.150 (tentative)

Species	Gene	Protein	mRNA	Genomic
Homo sapiens	*IGnT*	–	Z19550	–
Mus musculus	*IGnT*	–	U68182	–

Name and History

I-branching β6-*N*-acetylglucosaminyltransferase (IGnT) was originally described in hog gastric mucosa (Piller et al. 1984). This enzyme acted as dIGnT, forming GlcNAcβ1-3(GlcNAcβ1-6)Galβ1-3GalNAcα1-R from GlcNAcβ1-3Galβ1-3GalNAcα1-R. On the other hand, core2GlcNAcT purified from bovine trachea also had core4GlcNAcT and dIGnT activities (Ropp et al. 1991). The presence of such C2/C4/dIGnT was proven when core2GlcNAcT-II was cloned (Yeh et al. 1999; Schwientek et al. 1999). To date, there has been no report that any enzyme contains only dIGnT activity. All of the enzymes which contain dIGnT activity, either a purified or a recombinant enzyme derived from a cloned cDNA, contain C2/C4/IGnT activities. The presence of cIGnT was first demonstrated in rat liver (Gu et al. 1992). IGnT cloned from PA-1 human teratocarcinoma cells (Bierhuizen et al. 1993) contains mostly cIGnT activity (Mattila et al. 1998; Yeh et al. 1999). The mouse orthologue of this IGnT has been cloned (Magnet and Fukuda 1997). However, the dIGnT activity of the same human IGnT cannot be ignored (Yeh et al. 1999). Similarly, IGnT purified from hog small intestine contains mainly cIGnT activity (Sakamoto et al. 1998).

It is possible that there is yet another IGnT that contains almost exclusively dIGnT activity.

Enzyme Activity Assay and Substrate Specificity

The following reaction can be catalyzed by dIGnT:

$$\text{UDP-GlcNAc} + \text{GlcNAc}\beta1\text{-}3\text{Gal}\beta1\text{-}4\text{Glc(NAc)}\beta1\text{-R}$$
$$\longrightarrow \text{GlcNAc}\beta1\text{-}3(\text{GlcNAc}\beta1\text{-}6)\text{Gal}\beta1\text{-}4\text{Glc(NAc)}\beta1\text{-R}$$

The following reaction was also reported for the action of dIGnT (Piller et al. 1984):

$$\text{UDP-GlcNAc} + \text{GlcNAc}\beta1\text{-}3\text{Gal}\beta1\text{-}3\text{GalNAc}\alpha1\text{-R}$$
$$\longrightarrow \text{GlcNAc}\beta1\text{-}3(\text{GlcNAc}\beta1\text{-}6)\text{Gal}\beta1\text{-}3\text{GalNAc}\alpha1\text{-R}$$

The following reaction can be catalyzed by cIGnT:

$$\text{UDP-GlcNAc} + \text{Gal}\beta1\text{-}4\text{GlcNAc}\beta1\text{-}3\text{Gal}\beta1\text{-}4\text{GlcNAc}\beta1\text{-}3\text{Gal}\beta1\text{-}$$
$$4\text{Glc(NAc)}\beta1\text{-R} \longrightarrow \text{Gal}\beta1\text{-}4\text{GlcNAc}\beta1\text{-}3(\text{GlcNAc}\beta1\text{-}6)\underline{\text{Gal}}\beta1\text{-}$$
$$4\text{GlcNAc}\beta1\text{-}3(\text{GlcNAc}\beta1\text{-}6)\underline{\underline{\text{Gal}}}\beta1\text{-}4\text{Glc(NAc)}\beta1\text{-R}$$

Galactose residue close to a nonreducing terminal (single underlining above) serves more efficiently as an acceptor than that close to a reducing terminal (double underlining, Ujita et al. 1999, see Fig. 2). The action of cIGnT results in poly-*N*-acetyllactosamines with multiple short I branches, each of which consists of only one *N*-acetyllactosamine. In contrast, the action of dIGnT results in a complex highly branched poly-*N*-acetyllactosamine.

Preparation

To date, cIGnT has only been purified from PA-1 human embryocarcinoma cells (Leppanen et al. 1998) and hog small intestine (Sakamoto et al. 1998). A recombinant form of cIGnT, its catalytic domain fused with protein A, is available and has been used as an enzyme source (Mattila et al. 1998; Ujita et al. 1999; Yeh et al. 1999). This recombinant cIGnT was found to have the same substrate specificity as the full-length cIGnT, which contains the transmembrane domain and also the enzyme purified from PA-1 cells. Using this recombinant cIGnT sialyl. Lewis X in I-branch was synthesized (Turunen et al. 1995).

The cDNA encoding a human cIGnT has been cloned by an expression cloning strategy (Bierhuizen et al. 1993). In this cloning, CHO-Py·leu cells were transiently transfected with a cDNA library constructed from poly (A)$^+$ RNA of PA-1 human teratocarcinoma cells. It was shown previously that PA-1 cells contain I-branched poly-*N*-acetyllactosamines (Fukuda et al. 1985). Because CHO-Py·leu cells stably express polyoma large T antigens, pcDNA I plasmids containing the polyoma replication origin can be amplified in the CHO-Py·leu cell line. The transfected CHO-Py·leu cells expressing I antigen were enriched by incubation with human anti-I serum (Ma), followed by panning to goat antihuman IgM antibody coated on bacterial dishes. Plasmids were recovered by a Hirt procedure, and the plasmids were sequentially divided into smaller active pools, resulting in a single plasmid that directed the I antigen expression at the cell surface (for details of the cloning strategy, see Chap. 18).

Biological Aspects

The action mode of cIGnT was examined using various synthetic oligosaccharides and a recombinant IGnT cloned from PA-1 cells. As expected, almost all contained I-branches at centrally located galactose residues (see Fig. 2). By incubating the I branch formed with β4GalT I, it became clear that galactosylation of GlcNAcβ1-3(GlcNAcβ1-6)Galβ1-4Glc(NAc) (where the acceptor GlcNAc is underlined) is much less efficient than the addition of N-acetyllactosamine to form Galβ1-4GlcNAcβ1-3(GlcNAcβ1-6)Galβ1-4Glc(NAc) (Ujita et al. 1999).

This is probably due to competition between UDP-Gal and the terminal Gal residue in the acceptor. Because of this inefficient galactosylation, an I branch is short and contains only one N-acetyllactosamine (Fig. 3). This result is entirely consistent with the structural analysis of I-branched N-glycans present in human Band 3 (Fukuda

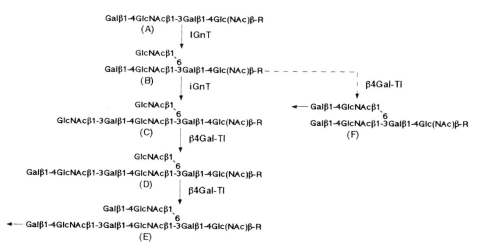

Fig. 3. Proposed biosynthetic steps of I-branched poly-N-acetyllactosamine by cIGnT. β1,6-Linked N-acetylglucosamine is first added to a central galactose by cIGnT (B). This is followed by the addition of of β1,3-linked N-acetylglucosamine by iGnT (C) and β-1,4-linked galactose by β4GalT I (D), adding N-acetyllactosamine to the linear poly-N-acetyllactosamine side chain. This is followed by galactosylation of β1,6-linked N-acetylglucosamine, forming an I-branch (E). As a minor biosynthetic pathway (*dashed arrow*), galactosylation of the I-branch may take place as soon as β1,6-linked N-acetylglucosamine is preferentially added to the I-branch by IGnT, potentially leading to more complex poly-N-acetyllactosamines. (From Ujita et al. 1999)

et al. 1984) and human PA-1 teratocarcinoma cells (Fukuda et al. 1985), and rabbit glycolipids (Dabrowski et al. 1984).

In contrast, a branched structure formed by dIGnT, such as GlcNAcβ1-3(GlcNAcβ1-6)Galβ1-4Glc(NAc)-R, can become extremely complex since both β1,6-linked GlcNAc and β1,3-linked GlcNAc have an equal chance to elongate (Blanken et al. 1982). It is thus expected that multiple branched poly-*N*-acetyllactosamines such as [Galβ1-4GlcNAcβ1-3(Galβ1-4GlcNAcβ1-6)Galβ1-4GlcNAcβ1-6](Galβ1-4GlcNAcβ1-3)Gal can be formed in the presence of dIGnT (Zdebska et al. 1983).

I-branches at the terminal of a carbohydrate chain can provide two functional groups in a close neighborhood (see Fig. 1). This allows, for example, binding to two binding sites of IgG or a dimer of carbohydrate binding proteins such as selectins. Such a binding (termed "monogamous bivalency") significantly increases the affinity between I-branched poly-*N*-acetyllactosamine and carbohydrate-binding proteins or antibodies (Romans et al. 1980). In fact, it has been demonstrated that sialyl Lewis X termini in two I-branched oligosaccharides are much more efficient at inhibiting the binding of L-selectin to high endothelial venules (Turunen et al. 1995). On the other hand, a linear poly-*N*-acetyllactosamine has much less affinity toward antibodies, and thus allows the suppression of agglutination of fetal erythrocytes by antibodies when the mother and fetus have incompatible blood group antigens (Romans et al. 1980).

Multiple presentation of functional carbohydrates or carbohydrate ligands would be extremely powerful if we could use dIGnT to form backbone structures. Such studies are worth pursuing in the future.

Future Perspectives

cIGnT cloned from PA-1 cells is expressed only in very few tissues. It is thus necessary to determine if the cloned IGnT is responsible for I branch formation in various cells, including embryonic stem cells. It is still possible that additional dIGnT exists which plays a critical role in I-branch synthesis in mucin-producing cells. Such a dIGnT should produce highly complex, branched structures (Schachter and Brockhausen 1992).

Future studies should also include the inactivation of the IGnT gene by homologous recombination. Such a study should provide a clear understanding of how I branches play roles in various ligand presentations.

Further Reading

References recommended for further reading are marked by asterisks.

References

*Bierhuizen MF, Mattei MG, Fukuda M (1993) Expression of the developmental I antigen by a cloned human cDNA encoding a member of a β1,6-*N*-acetylglucosaminyltransferase gene family. Gene Dev 7:468–478

Blanken WM, Hooghwinkel GJ, Van Den Eijnden DH (1982) Biosynthesis of blood-group I and i substances. Specificity of bovine colostrum β-*N*-acetyl-D-glucosaminide β 1 leads to 4-galactosyltransferase. Eur J Biochem 127:547–552

Dabrowski U, Hanfland P, Egge H, Kuhn S, Dabrowski J (1984) Immunochemistry of I/i-active oligo- and polyglycosylceramides from rabbit erythrocyte membranes. Determination of branching patterns of a ceramide pentadecasaccharide by 1H nuclear magnetic resonance. J Biol Chem 259:7648–7651

Feizi T, Childs RA, Watanabe K, Hakomori SI (1979) Three types of blood group I specificity among monoclonal anti-I autoantibodies revealed by analogues of a branched erythrocyte glycolipid. J Exp Med 149:975–980

*Fukuda M, Dell A, Oates JE, Fukuda MN (1984) Structure of branched lactosaminoglycan, the carbohydrate moiety of band 3 isolated from adult human erythrocytes. J Biol Chem 259:8260–8273

*Fukuda MN, Dell A, Oates JE, Fukuda M (1985) Embryonal lactosaminoglycan. The structure of branched lactosaminoglycans with novel disialosyl (sialyl α 2,9 sialyl) terminals isolated from PA1 human embryonal carcinoma cells. J Biol Chem 260:6623–6631

*Gu J, Nishikawa A, Fujii S, Gasa S, Taniguchi N (1992) Biosynthesis of blood group I and i antigens in rat tissues. Identification of a novel β1-6-N-acetylglucosaminyltransferase. J Biol Chem 267:2994–2999

Hanisch FG, Uhlenbruck G, Peter-Katalinic J, Egge H, Dabrowski J, Dabrowski U (1989) Structures of neutral O-linked polylactosaminoglycans on human skim milk mucins. A novel type of linearly extended poly-N-acetyllactosamine backbones with Gal β(1-4)GlcNAc β(1-6) repeating units. J Biol Chem 264:872–883

Kapadia A, Feizi T, Evans MJ (1981) Changes in the expression and polarization of blood group I and i antigens in post-implantation embryos and teratocarcinomas of mouse associated with cell differentiation. Exp Cell Res 131:185–195

Leppanen A, Zhu Y, Maaheimo H, Helin J, Lehtonen E, Renkonen O (1998) Biosynthesis of branched polylactosaminoglycans. Embryonal carcinoma cells express midchain β1,6-N-acetylglucosaminyltransferase activity that generates branches to preformed linear backbones. J Biol Chem 273:17399–17405

Magnet AD, Fukuda M (1997) Expression of the large I antigen-forming β-1,6-N-acetylglucosaminyltransferase in various tissues of adult mice. Glycobiology 7:285–295

*Mattila P, Salminen H, Hirvas L, Niittymaki J, Salo H, Niemela R, Fukuda M, Renkonen O, Renkonen R (1998) The centrally acting β1,6N-acetylglucosaminyltransferase (GlcNAc to gal). Functional expression, purification, and acceptor specificity of a human enzyme involved in midchain branching of linear poly-N-acetyllactosamines. J Biol Chem 273:27633–27639

Muramatsu T, Gachelin G, Damonneville M, Delarbre C, Jacob F (1979) Cell surface carbohydrates of embryonal carcinoma cells: polysaccharidic side chains of F9 antigens and of receptors to two lectins, FBP and PNA. Cell 18:183–191

*Piller F, Cartron JP, Maranduba A, Veyrieres A, Leroy Y, Fournet B (1984) Biosynthesis of blood group I antigens. Identification of a UDP-GlcNAc:GlcNAc β1-3Gal(-R) β1-6(GlcNAc to Gal) N-acetylglucosaminyltransferase in hog gastric mucosa. J Biol Chem 259:13385–13390

Romans DG, Tilley CA, Dorrington KJ (1980) Monogamous bivalency of IgG antibodies. I. Deficiency of branched ABHI-active oligosaccharide chains on red cells of infants causes the weak antiglobulin reactions in hemolytic disease of the newborn due to ABO incompatibility. J Immunol 124:2807–2811

Ropp PA, Little MR, Cheng PW (1991) Mucin biosynthesis: purification and characterization of a mucin β6 N-acetylglucosaminyltransferase. J Biol Chem 266:23863–23871

Sakamoto Y, Taguchi T, Tano Y, Ogawa T, Leppanen A, Kinnunen M, Aitio O, Parmanne P, Renkonen O, Taniguchi N (1998) Purification and characterization of UDP-GlcNAc:Galβ1-4GlcNAcβ1-3*Galβ1-4Glc(NAc)-R(GlcNAc to *Gal) β1,6N-acetylglucosaminyltransferase from hog small intestine. J Biol Chem 273:27625–27632

Sasaki K, Kurata-Miura K, Ujita M, Angata K, Nakagawa S, Sekine S, Nishi T, Fukuda M (1997) Expression cloning of cDNA encoding a human β-1,3-N-acetylglucosaminyltransferase that is essential for poly-*N*-acetyllactosamine synthesis. Proc Natl Acad Sci USA 94:14294–14299

Schachter H, Brockhausen I (1992) In: Allen HJ, Kisailus EC (eds) Glycoconjugates: composition, structure, and function, 1st edn. Marcel Dekker, New York, pp 263–332

Schwientek T, Nomoto M, Levery SB, Merkx G, van Kessel AG, Bennett EP, Hollingsworth MA, Clausen H (1999) Control of *O*-glycan branch formation. Molecular cloning of human cDNA encoding a novel β1,6-*N*-acetylglucosaminyltransferase forming core 2 and core 4. J Biol Chem 274:4504–4512

*Turunen JP, Majuri ML, Seppo A, Tiisala S, Paavonen T, Miyasaka M, Lemstrom K, Penttila L, Renkonen O, Renkonen R (1995) De novo expression of endothelial sialyl Lewis(a) and sialyl Lewis(x) during cardiac transplant rejection: superior capacity of a tetravalent sialyl Lewis(x) oligosaccharide in inhibiting L-selectin-dependent lymphocyte adhesion. J Exp Med 182:1133–1141

*Ujita M, McAuliffe J, Suzuki M, Hindsgaul O, Clausen H, Fukuda MN, Fukuda M (1999) Regulation of I-branched poly-*N*-acetyllactosamine synthesis. Concerted actions by I-extension enzyme, I-branching enzyme, and β1,4-galactosyltransferase I. J Biol Chem 274:9296–9304

van den Eijnden DH, Winterwerp H, Smeeman P, Schiphorst WE (1983) Novikoff ascites tumor cells contain *N*-acetyllactosaminide β 1 → to 3 and β 1 → to 6 *N*-acetylglucosaminyltransferase activity. J Biol Chem 258:3435–3437

Watanabe K, Hakomori SI, Childs RA, Feizi T (1979) Characterization of a blood group I-active ganglioside. Structural requirements for I and i specificities. J Biol Chem 254:3221–3228

*Yeh JC, Ong E, Fukuda M (1999) Molecular cloning and expression of a novel β-1,6-*N*-acetylglucosaminyltransferase that forms core 2, core 4, and I branches. J Biol Chem 274:3215–3221

Zdebska E, Krauze R, Koscielak J (1983) Structure and blood-group I activity of poly(glycosyl)-ceramides. Carbohydr Res 120:113–130

18

Core 2 β6-*N*-Acetylglucosaminyltransferase-I and -III

Introduction

Core 2 β6-*N*-acetylglucosaminyltransferase (Core2GlcNAcT) is a glycosyltransferase that transfers GlcNAc from UDP-GlcNAc to αGalNAc residues in core 1, Galβ1-3GalNAcα1-Ser/Thr with β1,6-linkage, forming Galβ1-3(GlcNAcβ1-6)GalNAcα1-Ser/Thr. The formation of the core 2 branch is usually followed by galactosylation by β4-galactosyltransferase-IV, a member of the β4-galactosyltransferase gene family (Ujita et al. 1998), resulting in the formation of *N*-acetyllactosamines in *O*-glycans. Such *N*-acetyllactosamines can be modified to form functional oligosaccharides such as sialyl Lewis X (Fig. 1).

Sialyl Lewis X in core 2-branched *O*-glycans have been discovered in human granulocytes (Fukuda et al. 1986), and shown to be critical structures for selectin recognition (Wilkins et al. 1996). Moreover, sulfated forms of sialyl Lewis X in core 2-branched *O*-glycans were found in L-selectin ligand present in high endothelial venules (HEV) of lymph nodes (Hemmerich et al. 1995) and a sulfotransferase that forms 6-sulfo sialyl Lewis X, NeuNAcα2-3Galβ1-4[Fucα1-3(sulfo-6)]GlcNAcβ1-R in HEV, has a strong preference for core 2-branched oligosaccharides (Hiraoka et al. 1999) (see also Chap. 63 for LSST).

During the early development of T lymphocytes, Core2GlcNAcT is significantly expressed in immature T lymphocytes present in cortical thymus. After maturation and entering into medullary thymus, T lymphocytes express only low levels of Core2GlcNAcT activity (Baum et al. 1995). However, Core2GlcNAcT activity is increased once T lymphocytes are activated during immune responses (Piller et al. 1988; Priatel et al. 2000). These results indicate that Core2GlcNAcT plays an important role in a broad range of biological systems.

Minoru Fukuda[1], Tilo Schwientek[2], and Henrik Clausen[2]

[1] Glycobiology Program, The Burnham Institute, 10901 North Torrey Pines Road, La Jolla, CA 92037, USA
Tel. +1-858-646-3144; Fax +1-858-646-3193
e-mail: minoru@burnham-inst.org
[2] School of Dentistry, University of Copenhagen, 2200 Copenhagen N, Denmark

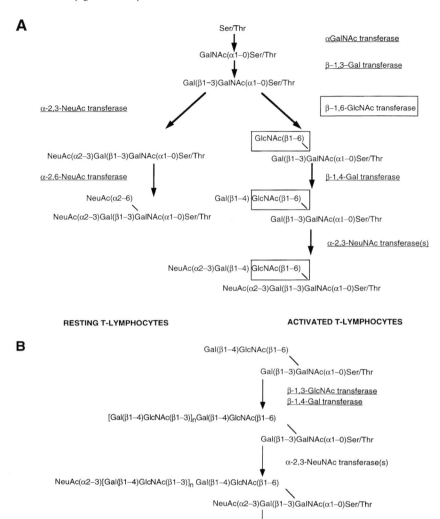

Fig. 1. The proposed biosynthetic pathways of *O*-glycans (**A**) and poly-*N*-acetyllactosaminyl *O*-glycans (**B**). **A** It has been shown that the tetrasaccharide (*bottom left*) is formed by the sequential action of α2-3-sialyltransferase followed by α2-6-sialyltransferase. When β6 *N*-acetylglucosaminyltransferase, Core2GlcNAcT, is present, the branched hexasaccharide (*bottom right*) is formed. **B** Poly-*N*-acetyllactosaminyl chain can be extended from the GlcNAcβ1,6-linkage synthesized in core 2. Poly-*N*-acetyllactosaminyl extension can be further modified by α3 fucosyltransferase, forming sialyl Le^x termini. (From Maemura and Fukuda 1992)

Since Core2GlcNAcT was initially recognized as a critical enzyme in blood cell differentiation, the cDNA encoding Core2GlcNAcT was first cloned from a cDNA library constructed from human promyelocytic leukemia HL-60 cells (Bierhuizen and Fukuda 1992). This enzyme is a typical type II membrane protein composed of 428 amino acids (Mr 49790). Immediately after cloning of Core2GlcNAcT, another β1,6-N-acetylglucosaminyltransferase, I-branching enzyme (IGnT), was also cloned (Bierhuizen et al. 1993, see also Chap. 17). A comparison of these two enzymes demonstrated that the two β6-N-acetylglucosaminyltransferases are highly homologous to each other (41.6% identity at the amino acid level).

Using such a homologous sequence as a probe, two additional cDNAs encoding Core2GlcNAcTs have recently been cloned. One of them almost exclusively contains Core2GlcNAcT activity and is termed Core2GlcNAcT-III (Schwientek et al. 2000). The enzyme cloned previously was termed Core2GlcNAcT-leukocyte type, but is now renamed Core2GlcNAcT-I. Another recently cloned Core2GlcNAcT (Core2GlcNAcT-II) contains additional β6-N-acetylglucosaminyltransferase activities such as core 4-branching activity (Schwientek et al. 1999; Yeh et al. 1999) and will be described separately (see next chapter, Chap. 19). Core2GlcNAcT-II has 48.2% and 33.8% identity with Core2GlcNAcT-I and IGnT, respectively, while Core2GlcNAcT-III has 42% identity with Core2GlcNAcT-I and Core2GlcNAcT-II, and 39% identity with IGnT (Fig. 2).

Core2GlcNAcT-III is also a typical type-II membrane protein containing 453 amino acids (MR 53028). This enzyme is highly expressed in the thymus, but barely present in other tissues (Fig. 3). In contrast, Core2GlcNAcT-I is widely expressed, including in leukocytes (see next chapter). These different expression patterns in the two enzymes is likely to provide a clue about how the two enzymes function in different tissues.

Databanks

Core 2 β6-N-Acetylglucosaminyltransferase-I and -III

NC-IUBMB enzyme classification: E.C.2.4.1.102

Species	Gene	Protein	mRNA	Genomic
C2GlcNAcT-I				
Homo sapiens	*Core2GlcNAcT-I*	–	M97347	–
Mus musculus	*Core2GlcNAcT-I*	–	U19265	–
Bos taurus	*Core2GlcNAcT-I*	–	U41320	–
Rattus sp.	*Core2GlcNAcT-I*	–	S79797	–
C2GlcNAcT-III				
Homo sapiens	*Core2GlcNAcT-III*	–	AF132035	–

Name and History

Core 2 β1,6-N-acetylglucosaminyltransferase, Core2GlcNAcT, was originally detected in canine submaxillary gland microsomes (Williams et al. 1980). Advances in biosynthesis and the structures of O-glycans made it necessary to understand O-glycans by classifying core structures (Schachter and Williams 1982). This important concept is now widely used.

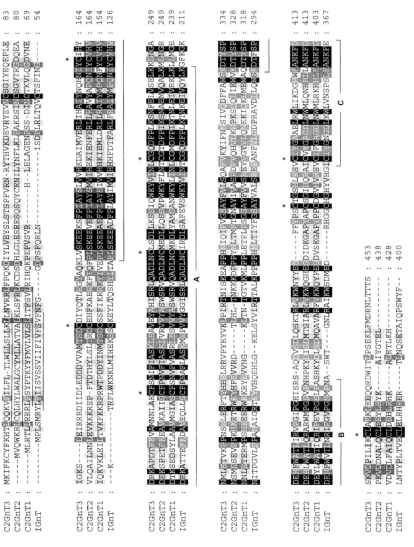

Fig. 2. Multiple amino acid sequence analysis (ClustalW) of human Core2GlcNAcTIII-TI (C2GnT3, C2GnT2, C2GnT1), and IGnT. Introduced gaps are shown as *hyphens*, and aligned identical residues are *boxed* (*black* for all sequences, *dark gray* for three sequences, and *light gray* for two sequences). The putative transmembrane domains are *underlined* with a *single line*. Highly homologous *A*, *B*, and *C* regions are shown (Bierhuizen et al. 1993). The positions of conserved cysteines are indicated by *asterisks*. One conserved *N*-glycosylation site is indicated by an *open*

Fig. 3. Northern analysis of Core2GlcNAcT-III in human tissues. A multiple human tissue Northern blot (MTN II from CLONETECH) was hybridized with a ^{32}P-labeled probe corresponding to the soluble form of Core2GlcNAcT-III. (From Shwientek et al. 2000)

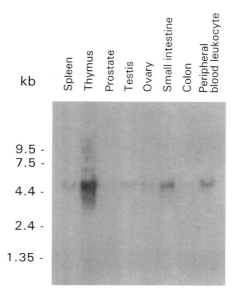

Core 2-branched O-glycans have been found in various mucins. The major breakthrough, however, took place when the expression of core 2 oligosaccharides and an increase of Core2GlcNAcT activity were found to be associated with T cell activation (Piller et al. 1988). That finding clearly placed O-glycans as cell-type specific oligosaccharides, displaying structural change in various biological processes. Although Core2GlcNAcT was partially purified from various sources, the isolation of Core2GlcNAcT was achieved by the cloning of cDNA encoding Core2GlcNAcT-I (Bierhuizen and Fukuda 1992). Subsequently, the cDNAs encoding Core2GlcNAcT-I were cloned from mouse (Granovsky et al. 1995; Sekine et al. 1997), rat (Koya et al. 1999), and cattle (Li et al. 1998) cells.

Enzyme Activity Assay and Substrate Specificity

The following reaction can be catalyzed by Core2GlcNAcT-I or -III:

$$UDP\text{-}GlcNAc + Gal\beta1\text{-}3GalNAc\alpha1\text{-}Ser/Thr$$

$$\rightarrow Gal\beta1\text{-}3(GlcNAc\beta1\text{-}6)GalNAc\alpha1\text{-}Ser/Thr$$

The enzyme assay and acceptor specificity of Core2GlcNAcT were performed using various partially purified enzyme preparations, and more recently using a soluble chimeric Core2GlcNAcT fused with a protein A. The results show the absolute requirement of the Galβ1-3GalNAcα1-R structure, and that α1,2-fucose substitution or α2,3-sialylation at the terminal galactose inhibits the activity (Williams et al. 1980).

This specificity dictates that Core2GlcNAcT acts before the galactose is modified by other sugars or sulfates. Despite the presence of α6-sialyltransferase that adds α2,6-linked sialic acid to GalNAc, the expression of Core2GlcNAcT usually leads into core 2-branched oligosaccharides. This is possible if Core2GlcNAcT is a more efficient enzyme than α6-sialyltransferase. Alternatively, the same result is obtained if Core2GlcNAcT is present in earlier Golgi compartments than α3-sialyltransferase (see Skrincosky et al. 1997; Whitehouse et al. 1997). Thus, an understanding of how core 2 oligosaccharide synthesis is regulated will provide an insight about the critical regulation taking place in relation to localization and enzyme reactions on actual polypeptide acceptors.

Preparation

Core2GlcNAcT-I and -III proteins are available only in recombinant forms. The cDNA encoding a human Core2GlcNAcT-I has been cloned by an expression cloning strategy (Bierhuizen and Fukuda 1992). Briefly, CHO cells were first stably transfected with vectors encoding leukosialin (CD 43) and polyoma large T antigen. Expression of leukosialin allows the enrichment of CHO cells expressing core 2 oligosaccharides by T305, since this antibody preferentially reacts with core 2 oligosaccharides attached to leukosialin (Piller et al. 1991). The expression of polyoma large T antigen in CHO cells allows the amplification of plasmids in the transfected CHO cells, when plasmids have a replication origin of polyoma virus. It was judged that polyoma large T is better than SV 40 large T since CHO cells, which are rodent cells, should have a better replication capacity for the polyoma virus that infects rodent cells than the SV 40 virus that infects monkey cells.

This CHO cell line, CHO-Py·leu, was transiently transfected with an HL-60 cDNA library constructed in pcDNA I. The transfected CHO cells expressing core 2 *O*-glycans were enriched by panning through binding to T305 antibody followed by anti-mouse IgG coated on plates. Plasmids were recovered by the Hirt procedure (Hirt 1967) and then transiently transfected into the same CHO-Py·leu cells. The plasmids recovered were then divided into eight pools of ~500 plasmids each and tested for their ability to direct the expression of T305 reactivity on CHO-Py·leu cells. One plasmid pool, which was strongly positive for T305 expression, was sequentially divided into smaller active pools, resulting in a single plasmid that directed the expression of the T305 antigen at the cell surface (Bierhuizen et al. 1992).

The isolated cDNA in pcDNA I has been used worldwide to express Core2GlcNAcT in transfected cells. A soluble chimeric form of the enzyme fused with protein A has been expressed in CHO cells or COS-1 cells and used for in vitro Core2GlcNAcT assay (Bierhuizen and Fukuda 1992).

Core2GlcNAcT-III was cloned using tBLASTn analysis of the human genome survey sequence for a sequence homologous to human Core2GlcNAcT-II (see Schwientek et al. 2000) (see Chap. 19 for Core2GlcNAcT-II). One clone, 1H-HSP-2288BIY (GenBank accession number A2005888), was found to contain a novel open reading frame sequence in one exon. The coding regions of Core2GlcNAcT-I and Core2GlcNAcT-II were previously shown to be in one exon (Bierhuizen et al. 1995; Schwientek et al. 1999), thus suggesting that a new gene, Core2GlcNAcT-III, may also be coded in one exon. To confirm this and obtain the 3′-sequence of the open reading frame, a genomic P1 clone containing the determined sequence was isolated, and then the sequences were extended to complete the coding sequence. The obtained nucleotide sequence encoding an open reading frame of Core2GlcNAcT-III was confirmed by PCR using reverse transcribed poly (A)+ RNA from human thymus as a template.

Using P1 DNA as a template, PCR was carried out to amplify the whole coding region, and the cDNA obtained was cloned into BamH I and EcoR I sites of pcDNA 3. Similarly, cDNA encoding amino acid residues 39–453 of Core2GlcNAcT-III was prepared by PCR and cloned into pACGp67A (Pharmingen). This vector was used for expression in Sf9 cells using the baculovirus system (Schwientek et al. 2000).

Biological Aspects

Since Core2GlcNAcT-I was cloned in 1992 (Bierhuizen and Fukuda 1992), a large body of results has accumulated on the expression of the enzyme and its effect on biological functions.

Among these, it has been demonstrated that cortical thymocytes contain a significant amount of the Core2GlcNAcT-I transcripts, consistent with expression of T305 antigen, while medulla thymocytes barely expressed the transcripts and were negative for T305 antigen (Baum et al. 1995).

Core2GlcNAcT-I is minimally present in resting T cells under normal conditions, while its activity is substantially increased in leukemia, colon and lung carcinomas, and immunodeficient syndromes such as AIDS and Wiskott–Aldrich (Brockhausen et al. 1991; Higgins et al. 1991; Piller et al. 1991; Saitoh et al. 1991; Yousefi et al. 1991; Shimodaira et al. 1997; Machida et al. 2001). Transgenic mice were made to express Core2GlcNAcT-I under the influence of lck promoter, resulting in core 2 oligosaccharide expression in resting T cells. T lymphocytes from the transgenic mice adhered less to fibronectin or ICAM-1 than T lymphocytes from wild-type mice. As a result, the immune response through the interaction between T lymphocytes and antigen-presenting cells was reduced, and this resulted in weaker delayed-type hypersensitivity (Tsuboi and Fukuda 1997). Similarly, the immune response due to T–B lymphocyte interaction was also impaired. This was observed in a diminished expression of the germinal center in spleen and impaired immunoglobulin isotype switching (Tsuboi and Fukuda 1998). These results indicate that cell–cell interaction in immune cells is highly attenuated by the expression of core 2 oligosaccharides, suggesting that overexpression of core 2 oligosaccharides may be a cause for immunodeficient diseases. In a separate study, core 2 oligosaccharides were overexpressed in cardiac muscle of transgenic mice (Koya et al. 1999). This study was initiated because Core2GlcNAcT is elevated in the heart of diabetic animals and is regulated by hyperglycemia and insulin (Nishio et al. 1995). The hearts of the transgenic mice showed an increase of c-fos gene expression and AP-1 activity indicative of cardiac stress. It is thus possible that hyperglycemia induces Core2GlcNAcT gene expression, which in turn, induced c-fos expression, presumably due to a change in the cell surface glycoproteins (Koya et al. 1999).

Core2GlcNAcT-I was inactivated by homologous recombination, and mice defective for Core2GlcNAcT-I were established. Interestingly, no impairment in the development of lymphocytes or intestinal mucosa was detected, suggesting that additional Core2GlcNAcT is present in these tissues (Ellies et al. 1998). In fact, Core2GlcNAcT-II (Yeh et al. 1999) and Core2GlcNAcT-III (Schwientek et al. 2000) are present in mucin-producing tissues and thymus, respectively, compensating for the loss of Core2GlcNAcT-I in these tissues.

A phenotype of Core2GlcNAcT-I-deficient mice was analyzed extensively for selectin ligand expression. Leukocytes from Core2GlcNAcT-I-deficient mice exhibited significantly reduced binding to L-, P-, and E-selectins, indicating that the majority of selectin ligands in leukocytes are present in core 2-branched *O*-glycans. In contrast, lymphocyte homing was barely impaired in Core2GlcNAcT-I-deficient mice (Ellies et al. 1998). This unexpected finding indicates that L-selectin ligands in high endo-thelial venules (HEV) are synthesized by another Core2GlcNAcT or synthesized in oligosaccharides which are different from core 2-branched *O*-glycans. Alternatively, *O*-glycans other than core 2-branched oligosaccharide may be present in HEV of Core2GlcNAcT-I-deficient mice fact, recent studies revealed that core 1 is extended by β3GlcNAcT-III, forming 6-sulfo sialyl Lewis X in extended core 1 structure, NeuNAcα2-3Galβ1-4[Fucα1-3(sulfo-6)]GlcNAcβ1-3Galβ1-3GalNAcα1-R. This struc-ture is responsible for L-selectin ligand in HEV of Core2GlcNAcT-I deficient mice, and extended core 1 oligosaccharides containing 6-sulfo GlcNAc serve as the minimum epitope for MECA-79 antibody (Yeh et al. 2001).

Future Perspectives

Since three different Core2GlcNAcTs have been identified, the roles of core 2-branched oligosaccharides can be addressed in various tissues. This can be carried out by indi-vidually inactivating each enzyme, as was done for Core2GlcNAcT-I. In addition, mating of mice with one deficient gene with mice with another deficient gene will result in double knockout mice. These studies will reveal the roles of core 2-branched oligosaccharides, in particular those tissues which did not display overt phenotype in mice deficient in only the Core2GlcNAcT-I gene.

Further Reading

References recommended for further reading are marked by asterisks.

References

Baum LG, Pang M, Perillo NL, Wu T, Delegeane A, Uittenbogaart CH, Fukuda M, Seilhamer JJ (1995) Human thymic epithelial cells express an endogenous lectin, galectin-1, which binds to core 2 *O*-glycans on thymocytes and T lymphoblastoid cells. J Exp Med 181:877–887

*Bierhuizen MF, Fukuda M (1992) Expression cloning of a cDNA encoding UDP-GlcNAc:Gal β1-3-GalNAc-R (GlcNAc to GalNAc) β1-6GlcNAc transferase by gene transfer into CHO cells expressing polyoma large tumor antigen. Proc Natl Acad Sci USA 89:9326–9330

Bierhuizen MF, Mattei MG, Fukuda M (1993) Expression of the developmental I antigen by a cloned human cDNA encoding a member of a β1,6-*N*-acetylglucosaminyltransferase gene family. Gene Dev 7:468–478

Bierhuizen MFA, Maemura K, Kudo S, Fukuda M (1995) Genomic organization of core 2 and I-branching β-1,6-*N*-acetylglucosaminyltransferases. Implication for evolution of the β-1,6-*N*-acetylglucosaminyltransferase gene family. Glycobiology 5:417–425

Brockhausen I, Kuhns W, Schachter H, Matta KL, Sutherland DR, Baker MA (1991) Biosynthesis of O-glycans in leukocytes from normal donors and from patients with leukemia: increase in O-glycan core 2 UDP-GlcNAc:Gal β3 GalNAc α-R (GlcNAc to GalNAc) β1,6-N-acetylglucosaminyltransferase in leukemic cells. Cancer Res 51: 1257–1263

*Ellies LG, Tsuboi S, Petryniak B, Lowe JB, Fukuda M, Marth JD (1998) Core 2 oligosaccharide biosynthesis distinguishes between selectin ligands essential for leukocyte homing and inflammation. Immunity 9:881–890

*Fukuda M, Carlsson SR, Klock JC, Dell A (1986) Structures of O-linked oligosaccharides isolated from normal granulocytes, chronic myelogenous leukemia cells, and acute myelogenous leukemia cells. J Biol Chem 261:12796–12806

Granovsky M, Fode C, Warren CE, Campbell RM, Marth JD, Pierce M, Fregien N, Dennis JW (1995) GlcNAc-transferase V and core 2 GlcNAc-transferase expression in the developing mouse embryo. Glycobiology 5:797–806

Hemmerich S, Leffler H, Rosen SD (1995) Structure of the O-glycans in GlyCAM-1, an endothelial-derived ligand for L-selectin. J Biol Chem 270:12035–12047

Higgins EA, Siminovitch KA, Zhuang DL, Brockhausen I, Dennis JW (1991) Aberrant O-linked oligosaccharide biosynthesis in lymphocytes and platelets from patients with the Wiskott–Aldrich syndrome. J Biol Chem 266:6280–6290

*Hiraoka N, Petryniak B, Nakayama J, Tsuboi S, Suzuki M, Yeh JC, Izawa D, Tanaka T, Miyasaka M, Lowe JB, Fukuda M (1999) A novel, high endothelial venule-specific sulfotransferase expresses 6-sulfo sialyl Lewis(x), an L-selectin ligand displayed by CD34. Immunity 11:79–89

Hirt B (1967) Selective extraction of polyoma DNA from infected mouse cell cultures. J Mol Biol 26:365–369

Koya D, Dennis JW, Warren CE, Takahara N, Schoen FJ, Nishio Y, Nakajima T, Lipes MA, King GL (1999) Overexpression of core 2 N-acetylglycosaminyltransferase enhances cytokine actions and induces hypertrophic myocardium in transgenic mice. FASEB J 13:2329–2337

Li CM, Adler KB, Cheng PW (1998) Mucin biosynthesis: molecular cloning and expression of bovine lung mucin core 2 N-acetylglucosaminyltransferase cDNA. Am J Respir Cell Mol Biol 18:343–352

Machida E, Nakayama J, Amano J, Fukada M (2001) Clinicopathological Significance of Core2 β1,6-N-Acetylglucosaminyltransferase Messenger RNA Expressed in the Pulmonary Adenocarcinoma Determined by in Situ Hybridization. Cancer Res 61:2226–2231

Maemura K, Fukuda M (1992) Poly-N-acetyllactosaminyl O-glycans attached to leukosialin. The presence of sialyl Le^x structures in O-glycans. J Biol Chem 267:24379–24386

*Nishio Y, Warren CE, Buczek-Thomas JA, Rulfs J, Koya D, Aiello LP, Feener EP, Miller TB Jr, Dennis JW, King GL (1995) Identification and characterization of a gene regulating enzymatic glycosylation which is induced by diabetes and hyperglycemia specifically in rat cardiac tissue. J Clin Invest 96:1759–1767

*Piller F, Piller V, Fox RI, Fukuda M (1988) Human T-lymphocyte activation is associated with changes in O-glycan biosynthesis. J Biol Chem 263:15146–15150

Piller F, Le Deist F, Weinberg KI, Parkman R, Fukuda M (1991) Altered O-glycan synthesis in lymphocytes from patients with Wiskott–Aldrich syndrome. J Exp Med 173: 1501–1510

*Priatel JJ, Chui D, Hiraoka N, Simmons CJ, Richardson KB, Page DM, Fukuda M, Varki NM, Marth JD (2000) The ST3Gal-I sialyltransferase controls CD8$^+$ T lymphocyte homeostasis by modulating O-glycan biosynthesis [In Process Citation]. Immunity 12:273–283

Saitoh O, Piller F, Fox RI, Fukuda M (1991) T-lymphocytic leukemia expresses complex, branched O-linked oligosaccharides on a major sialoglycoprotein, leukosialin. Blood 77:1491–1499

Schachter H, Williams D (1982) Biosyntheses of mucus glycoproteins. In: Chantler EN, Elder JB, Elstein M (eds) Mucus in health and disease, vol. II. Plenum Press, New York, pp 3–28

Schwientek T, Nomoto M, Levery SB, Merkx G, van Kessel AG, Bennett EP, Hollingsworth MA, Clausen H (1999) Control of *O*-glycan branch formation. Molecular cloning of human cDNA encoding a novel β1,6-*N*-acetylglucosaminyltransferase forming core 2 and core 4. J Biol Chem 274:4504–4512

*Schwientek T, Yeh JC, Levery SB, Keck B, Nerkx G, van Kessel AD, Fukuda M, Clausen H (2000) Control of *O*-glycan branch formation: molecular cloning and characterizaiton of a novel thymus-associated core 2 β1,6-*N*-acetylglucosaminyltransferase. J Biol Chem 275:11106–11113

Sekine M, Nara K, Suzuki A (1997) Tissue-specific regulation of mouse core 2 β-1,6-*N*-acetylglucosaminyltransferase. J Biol Chem 272:27246–27252

Shimodaira K, Nakayama J, Nakamura N, Hasebe O, Katsuyama T, Fukuda M (1997) Carcinoma-associated expression of core 2 β1,6-*N*-acetylglucosaminyltransferase gene in human colorectal cancer: role of *O*-glycans in tumor progression. Cancer Res 57:5201–5206

Skrincosky D, Kain R, El-Battari A, Exner M, Kerjaschki D, Fukuda M (1997) Altered Golgi localization of core 2 β1,6-*N*-acetylglucosaminyltransferase leads to decreased synthesis of branched *O*-glycans. J Biol Chem 272:22695–22702

*Tsuboi S, Fukuda M (1997) Branched *O*-linked oligosaccharides ectopically expressed in transgenic mice reduce primary T-cell immune responses. EMBO J 16:6364–6373

Tsuboi S, Fukuda M (1998) Overexpression of branched *O*-linked oligosaccharides on T cell surface glycoproteins impairs humoral immune responses in transgenic mice. J Biol Chem 273:30680–30687

*Ujita M, McAuliffe J, Schwientek T, Almeida R, Hindsgaul O, Clausen H, Fukuda M (1998) Synthesis of poly-*N*-acetyllactosamine in core 2 branched *O*-glycans. The requirement of novel β-1,4-galactosyltransferase IV and β1,3-*N*-acetylglucosaminyltransferase. J Biol Chem 273:34843–34849

Whitehouse C, Burchell J, Gschmeissner S, Brockhausen I, Lloyd KO, Taylor-Papadimitriou J (1997) A transfected sialyltransferase that is elevated in breast cancer and localizes to the medial/trans-Golgi apparatus inhibits the development of core-2-based *O*-glycans. J Cell Biol 137:1229–1241

Wilkins PP, McEver RP, Cummings RD (1996) Structures of the *O*-glycans on *P*-selectin glycoprotein ligand-1 from HL-60 cells. J Biol Chem 271:18732–18742

Williams D, Longmore G, Matta KL, Schachter H (1980) Mucin synthesis. II. Substrate specificity and product identification studies on canine submaxillary gland UDP-GlcNAc:Gal β1-3GalNAc(GlcNAc leads to GalNAc) β6-*N*-acetylglucosaminyltransferase. J Biol Chem 255:11253–11261

*Yeh JC, Ong E, Fukuda M (1999) Molecular cloning and expression of a novel β1,6-*N*-acetylglucosaminyltransferase that forms core 2, core 4, and I branches. J Biol Chem 274:3215–3221

Yeh JC, Hiraoka N, Petryniak B, Nakayama J, Ellies L, Rabuka D, Hindsgaul O, Marth J, Lowe J, Fukuda M (2001) Novel Sulfated Lymphocyte Homing Receptors and Their Control by a Core1 Extension β1,3-*N*-Acetylglucosaminyltransferase. Cell 105:957–969

Yousefi S, Higgins E, Daoling Z, Pollex-Kruger A, Hindsgaul O, Dennis JW (1991) Increased UDP-GlcNAc:Gal β1-3GaLNAc-R (GlcNAc to GaLNAc) β1,6-*N*-acetylglucosaminyltransferase activity in metastatic murine tumor cell lines. Control of polylactosamine synthesis. J Biol Chem 266:1772–1782

Core 2 β6-*N*-Acetylglucosaminyltransferase-II

Introduction

Core 2 β6-*N*-acetylglucosaminyltransferase-II (Core2GlcNAcT-II) is unique in having all of the core 2, core 4, and I branching activities (Yeh et al. 1999). Core2, core4, and I branching activities are related to each other, but distinguished by substrate acceptors (Fig. 1). Core2 and core4 branching enzymes add β1,6-*N*-acetylglucosamine to the GalNAc residue of either Galβ1-3GalNAcα1-Ser/Thr (for Core2GlcNAcT) or GlcNAcβ1-3GalNAcα1-Ser/Thr (for Core4GlcNAcT). I-branching enzyme adds β1,6-linked *N*-acetylglucosamine to Galβ1-4GlcNAcβ1-3<u>Gal</u>β1-4Glc(NAc)β1-R (for centrally acting cIGnT) or GlcNAcβ1-3<u>Gal</u>β1-4Glc(NAc)β1-R (for distally acting dIGnT), where the acceptor galactose is underlined.

Core2GlcNAcT has been described in many tissues and is often expressed in a cell-type specific manner. Details of these enzymes (Core2GlcNAcT-I and Core2GlcNAcT-III) can be found in Chap. 18. Core4GlcNAcT activity has mainly been found in mucin-producing tissues, and was originally discovered in sheep gastric mucosa (Hounsell et al. 1980). The formation of core4-branched oligosaccharides is entirely dependent on the for-mation of core3 oligosaccharide, which is synthesized by core3 forming β3-*N*-acetylglucosaminyltransferase, core3 β3GlcNAcT (Fig. 2). The determination of core1, core2, core3, and core4-forming activities to date has demonstrated that the presence of Core4GlcNAcT activity is always associated with the presence of Core2GlcNAcT (Schachter and Brockhausen 1992). On the other hand, core4-branched oligosaccharides are often absent in human gastrointestinal tumor cell lines, mainly because core3 β3GlcNAcT (see Fig. 2) is diminished in these tissues (Yang et al. 1994; Vavasseur et al. 1995). In these tissues, Core4GlcNAcT remains the same or even increases compared with normal tissues. Core2GlcNAcT enzyme preparation purified from bovine trachea had as much as 40% dIGnT activity and ~100% Core4GlcNAcT activity compared with Core2GlcNAcT activity (Ropp et al. 1991).

Minoru Fukuda and Jiunn-Chern Yeh

Glycobiology Program, The Burnham Institute, 10901 North Torrey Pines Road, La Jolla, CA 92037, USA
Tel. +1-858-646-3144; Fax +1-858-646-3193
e-mail: minoru@burnham-inst.org

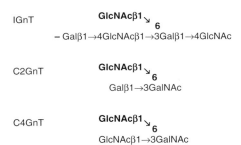

Fig. 1. Acceptors and resultant products by three different β6-*N*-acetyllactosaminyltransferases. I β6-*N*-acetylglucosaminyltransferase (cIGnT), core2 β6-*N*-acetylglucosaminyltransferase (Core2GlcNAcT), and core4 β6-*N*-acetylglucosaminyltransferase (Core4GlcNAcT) are shown. For distally acting IGnT (dIGnT), GlcNAcβ1-3Galβ1-4 Glc(NAC)β-1-R is utilized as an acceptor

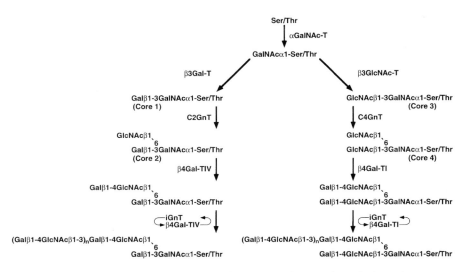

Fig. 2. Proposed biosynthetic pathways of core2- and core4-branched *O*-glycans. *N*-Acetylgalactosamines are transferred to serine or threonin residues in a polypeptide by α-*N*-acetylgalactosaminyltransferase (GalNAcT). This is followed by the action of core1 β1,3-galactosyltransferase (β3GalT), forming Galβ1-3GalNAcα1-R (core1). Core1 is then converted to GlcNAcβ1-6(Galβ1-3)GalNAcα1-R (core2) by core2 β6-*N*-actylglucosaminyltransferase (C2GnT), and then Galβ1-4GlcNAcβ1-6(Galβ1-3)GalNAcα1-R by β4GalT-IV (Ujita et al. 1998). Poly-*N*-acetyllactosamines will be added on galactosylated core 2 by alternate actions of iGnT and β4GalT-IV. Alternatively, GalNAcα1-R can be extended by core3 β3-*N*-acetylglucosaminyltransferase (β3GlcNAcT), forming GlcNAcβ1-3GalNAα1-R (core3). This will be followed by the action of core4 β6-*N*-acetylglucosaminyltransferase (Core4GlcNAcT) to form GlcNAcβ1-6(GlcNAcβ1-3)GalNAcα1-R (core4). Core4 is galactosylated by β4GalT-I, resulting in Galβ1-4GlcNAcβ1-6(Galβ1-4GlcNAcβ1-3)GalNAcα1-R. Poly-*N*-acetyllactosamines will be added by alternate actions of iGnT and β4GalT-I. (Adapted from Ujita et al. 2000)

As shown in Chap. 18, Core2GlcNAcT-I and IGnT were found to be highly homologous to each other. Recent progress in the EST (expressed sequence tag) database allowed the screening of genes related to Core2GlcNAcT-I or IGnT using a probe which corresponds to nucleotide sequences where Core2GlcNAcT and IGnT are strongly homologous to each other. Such a screening of the cDNA database resulted in the isolation of a cDNA, which is more homologous to Core2GlcNAcT-I (48.2% identity) than to IGnT (33.8% identity) at amino acid levels.

This enzyme, Core2GlcNAcT-II, is a typical type II membrane protein composed of 438 amino acids (50 863 Da). Core2GlcNAcT-II is highly homologous in three regions of the catalytic domain, sharing 72.1%, 57.7%, and 75.0% identity in the A, B, and C regions, respectively (see Fig. 2 in Chap. 18). When full-length Core2GlcNAcT-II was expressed in CHO cells, core 2-branched oligosaccharides on leukosialin and I antigen were detected by T305 antibody and human anti-I serum (Ma), respectively. A soluble chimeric Core2GlcNAcT-II fused with protein A exhibits 34.1% of Core4GlcNAcT activity, 7% of dIGnT activity, and 3% of cIGnT activity compared with Core2GlcNAcT activity. Core4GlcNAcT activity is usually lower than that of Core2GlcNAcT in those tissues expressing Core4GlcNAcT activity. These results strongly suggest that the Core4GlcNAcT activity in these tissues is due to Core2GlcNAcT-II.

Databanks

Core 2 β6-*N*-Acetylglucosaminyltransferase-II

NC-IUBMB enzyme classification: E.C.2.4.1.102 (Core2GlcNAcT) and 2.4.1.248 (Core4GlcNAcT)

Species	Gene	Protein	mRNA	Genomic
Homo sapiens	*Core2GlcNAcT-II*	–	AF03865	–
		–	AF102542	–
Bovine herpes virus type 4 (Gammaherpesvirinae)	*Core2GlcNAcT-II*	–	AF231105	–

Name and History

In this chapter, we mainly discuss Core4GlcNAcT activity, since Core2GlcNAcT and IGnT activities are described separately in other chapters (Chaps. 18 and 17) and respectively). Core 4-branched *O*-glycans were originally discovered in sheep gastric mucosa (Hounsell et al. 1980). They were then found in many other tissues, but the majority of tissues expressing core 4-branched *O*-glycans are those producing mucins.

The enzyme expressing core4 activity was partially purified from bovine trachea (Ropp et al. 1991), and the results on that enzyme provided the first clue that Core4GlcNAcT activity may be a secondary activity of Core2GlcNAcT activity. Moreover, the same study suggested that dIGnT activity may be associated with C2/C4/IGnT enzyme preparation. dIGnT and cIGnT were first distinguished in an enzyme preparation of rat liver (Gu et al. 1992). However, no attention was paid to the association between dIGnT and Core4GlcNAcT in that study. The two studies combined suggested the presence of Core2GlcNAcT that contains Core4GlcNAcT as well as dIGnT activities. This was proven once cDNA encoding Core2GlcNAcT-II

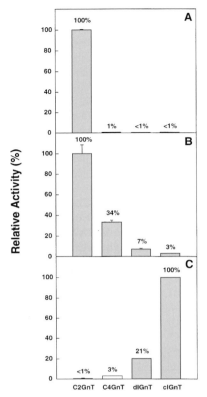

Fig. 3. Incorporation of ^{3}H-GlcNAc from UDP-[^{3}H]GlcNAc to acceptors by **A** Core2GlcNAcT-I, **B** Core2GlcNAcT-II, and **C** IGnT. The soluble chimeric enzyme in the spent medium (**A** and **C**) or that bound to IgG-Sepharose (**B**) were assayed for Core2GlcNAcT, Core4GlcNAcT, dIGnT, and cIGnT activities. The highest activity of Core2GlcNAcT-I, Core2GlcNAcT-II, and IGnT were 30 nmoles/h/ml (for Core2GlcNAcT activity), 130 pmoles/h/ml (for Core2GlcNAcT activity), and 894 pmoles/h/ml (for cIGnT activity), respectively, and they were taken as 100%. The standard errors are shown by bars. (From Yeh et al. 1999)

was cloned (Yeh et al. 1999). As described above, Core2GlcNAcT-II contains Core4GlcNAcT, dIGnT activities, and minor cIGnT activity (Fig. 3), in addition to Core2GlcNAcT activity.

Enzyme Activity Assay and Substrate Specificity

The reactions given below can be catalyzed by Core2GlcNAcT-II. Acceptor Gal or GalNAc residues are underlined.

UDP-GlcNAc + Galβ1-3GalNAcα1-Ser/Thr

    → Galβ1-3(GlcNAcβ1-6)GalNAcα1-Ser/Thr   (Core2GlcNAcT activity)   (1)

UDP-GlcNAc + GlcNAcβ1-3GalNAcα1-Ser/Thr

    → GlcNAcβ1-3(GlcNAcβ1-6)GalNAcα1-Ser/Thr   (Core4GlcNAcT activity)   (2)

UDP-GlcNAc + GlcNAcβ1-3Galβ1-4Glc(NAc)β1-R

    → GlcNAcβ1-3(GlcNAcβ1-6)Galβ1-4Glc(NAc)β1-R   (dIGnT activity)   (3)

$$UDP\text{-}GlcNAc + Gal\beta1\text{-}4GlcNac\beta1\text{-}3\underline{Gal}\beta1\text{-}4Glc(NAc)\beta1\text{-}R$$

$$\rightarrow Gal\beta1\text{-}4GlcNAc\beta1\text{-}3(GlcNAc\beta1\text{-}6)\underline{Gal}\beta1$$

$$\text{-}4Glc(NAc)\beta1\text{-}R \quad \text{(cIGnT activity)} \tag{4}$$

Although the ratio of these activities in the enzyme might differ from species to species, the ratio was found to be 100 (Core2GlcNAcT), 34 (Core4GlcNAcT), 7 (dIGnT), and 3 (cIGnT) for the enzyme cloned from human fetal brain (Fig. 3). On the other hand, an enzyme sample purified from bovine trachea was found to contain these activities in a ratio of 100 (Core2GlcNAcT), ~100 (Core4GlcNAcT), and ~35 (dIGnT). It is not clear if Core4GlcNAcT activity in this trachea enzyme preparation contains additional Core4GlcNAcT which is yet to be identified. Core2GlcNAcT-II is not activated by Mn^{++} or inactivated in the presence of 10 mM EDTA.

Preparation

Although this enzyme was partially purified from bovine trachea (Ropp et al. 1991), it is not certain if the enzyme was homogenous. Core2GlcNAcT-II protein is thus available only in a recombinant form. The cDNA encoding Core2GlcNAcT-II was cloned by identifying cDNA that is highly homologous to Core2GlcNAcT-I and IGnT. This attempt was possible because the homology between Core2GlcNAcT-I and IGnT had already been revealed (Bierhuizen and Fukuda 1992; Bierhuizen et al. 1993). Two laboratories thus cloned the same cDNA encoding Core2GlcNAcT-II. The full-length cDNA was cloned into *Hind* III and *Xba* I sites of pcDNA 3.1/Zeo.

The cDNA encoding a soluble chimeric Core2GlcNAcT-II harboring amino acid residues 34–834 was cloned into *Bam*H I and *Xba* I sites of pcDNA3-A harboring a DNA fragment encoding a signal peptide and IgG binding domain of staphylococcal protein A (Angata et al. 1998). This plasmid can be transfected into COS-1 or CHO cells, and the cultured medium can be used as an enzyme source after absorbing to IgG-Sepharose (Yeh et al. 1999). Alternatively, the cultured medium can be used directly after concentration by Centriprep 30 (Millipore, Bedford, MA). Similarly, the cDNA encoding amino acid residues 31–438 was cloned into the *Eco*R I site of pAcGpb7A (Pharmingen) and expressed in SF9 insect cells (Schwientek et al. 1999).

Biological Aspects

Core4GlcNAcT activity can be detected in a wide variety of tissues, including human colon, ovary, rat colon, stomach, small intestine, and submaxillary (Schachter and Brockhausen 1992). Core4GlcNAcT activity, however, is not detected in human granulocytes, and acute and chronic myelogenous leukemia cells. This distribution is consistent with the distribution of Core2GlcNAcT-II transcripts shown in Fig. 4. It is also important to point out that the product, core4-branched oligosaccharides, is not necessarily present in those tissues containing Core4GlcNAcT. This is because the synthesis of core4 oligosaccharides is entirely dependent on the presence of core3 oligosaccharide (see Fig. 2), which is formed by core3 β1,3-*N*-acetylglucosaminyl-transferase. Apparently, this core3 β1,3-*N*-acetylglucosaminyltransferase is present in

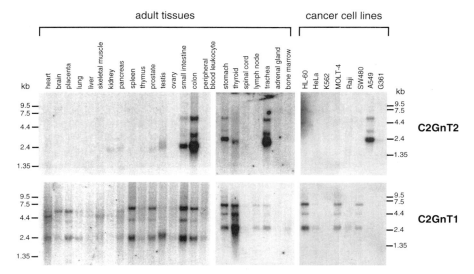

Fig. 4. Northern analysis of Core2GlcNAcT-II and -I (C2GnT2 and C2GnT1). Each lane contained 2 μg poly(A)⁺RNA. The blots were hybridized with ³²P-labeled Core2GlcNAcT-II cDNA followed by ³²P-labeled Core2GlcNAcT-I cDNA. Each blot contained eight lanes and was run separately. The migration positions of the molecular markers are the same in the two blots (shown on the *right*). (From Yeh et al. 1999)

a restricted manner, and is also present in human bronchial and bovine gastric mucins and human meconium.

One striking result is the differential expression of Core2GlcNAcT-II and Core2GlcNAcT-I in various cancer cell lines. For example, the human A549 lung carcinoma cell line expresses Core2GlcNAcT-II but not Core2GlcNAcT-I, while HL-60 promyelocytic, MOLT-4, T cell leukemia, and Raji B cell leukemia cells express Core2GlcNAcT-I but not Core2GlcNAcT-II (Fig. 4). It appears that this differential expression reflects differential cell-type expression. The coding regions of Core2GlcNAcT-I, II, and III are encoded in one exon (Bierhuizen et al. 1995; Schwientek et al. 2000).

Using radiation hybrids of the Stanford Human Genome Center G3 RH panel, the genes of Core2GlcNAcT-I, Core2GlcNAcT-II, Core2GlcNAcT-III, and IGnT were found to be at q13 of chromosome 9, q22.1 of chromosome 15, q12 of chromosome 5, and p24 of chromosome 6, respectively (Yeh et al. 1999; Schwientek et al. 2000).

Future Perspectives

Since the cDNA encoding Core2GlcNAcT-II was cloned only recently, further information is expected to be obtained using this gene. Knockouts of this gene are in progress at Jamey Marth's laboratory in collaborative efforts with the laboratory of Minoru Fukuda. It is also possible that additional Core4GlcNAcT will be identified, which contains more Core4GlcNAcT activity than Core2GlcNAcT activity, as compared with Core2GlcNAcT-II.

Very recently, it was discovered that bovine herpes virus 4 contains a gene encoding Core2GlcNAcT-II (Vanderplasschen et al. 2000). Its predicted amino acid sequence is highly homologous (81.1% identity at amino acid levels) to that of human Core2GlcNAcT-II. This gene is transcribed during virus infection. Future studies are important to determine the roles of this enzyme during viral infection and proliferation.

References

Angata K, Suzuki M, Fukuda M (1998) Differential and cooperative polysialylation of the neural cell adhesion molecule by two polysialyltransferases, PST and STX. J Biol Chem 273:28524–28532

Bierhuizen MF, Fukuda M (1992) Expression cloning of a cDNA encoding UDP-GlcNAc:Gal β1-3-GalNAc-R (GlcNAc to GalNAc) β1-6GlcNAc transferase by gene transfer into CHO cells expressing polyoma large tumor antigen. Proc Natl Acad Sci USA 89: 9326–9330

Bierhuizen MF, Mattei MG, Fukuda M (1993) Expression of the developmental I antigen by a cloned human cDNA encoding a member of a β1,6-N-acetylglucosaminyltransferase gene family. Gene Dev 7:468–478

Bierhuizen MFA, Maemura K, Kudo S, Fukuda M (1995) Genomic organization of core 2 and I-branching β-1,6-N-acetylglucosaminyltransferases. Implication for evolution of the β-1,6-N-acetylglucosaminyltransferase gene family. Glycobiology 5:417–425

Gu J, Nishikawa A, Fujii S, Gasa S, Taniguchi N (1992) Biosynthesis of blood group I and i antigens in rat tissues. Identification of a novel β1-6-N-acetylglucosaminyltransferase. J Biol Chem 267:2994–2999

Hounsell EF, Fukuda M, Powell ME, Feizi T, Hakomori S (1980) A new O-glycosidically linked tri-hexosamine core structure in sheep gastric mucin: a preliminary note. Biochem Biophys Res Commun 92:1143–1150

Ropp PA, Little MR, Cheng PW (1991) Mucin biosynthesis: purification and characterization of a mucin β6 N-acetylglucosaminyltransferase. J Biol Chem 266:23863–23871

Schachter H, Brockhausen I (1992) Glycoconjugates: composition, structure, and function, 1st edn. In: Allen HJ, Kisailus EC (eds) Marcel Dekker, New York, pp 263–332

Schwientek T, Nomoto M, Levery SB, Merkx G, van Kessel AG, Bennett EP, Hollingsworth MA, Clausen H (1999) Control of O-glycan branch formation. Molecular cloning of human cDNA encoding a novel β,6-N-acetylglucosaminyltransferase forming core 2 and core 4. J Biol Chem 274:4504–4512

Schwientek T, Yeh JC, Levery SB, Keck B, Nerkx G, van Kessel AD, Fukuda M, Clausen H (2000) Control of O-glycan branch formation: molecular cloning and characterization of a novel thymus associated core 2 β1,6-N-acetylglucosaminyltransferase. J Biol Chem 275:11106–11113

Ujita M, McAuliffe J, Schwientek T, Almeida R, Hindsgaul O, Clausen H, Fukuda M (1998) Synthesis of poly-N-acetyllactosamine in core 2-branched O-glycans. The requirement of novel β1,4-galactosyltransferase IV and β1,3-N-acetylglucosaminyltransferase. J Biol Chem 273:34843–34849

Ujita M, Anup K, Misra AK, McAuliffe J, Hindsgaul O, Fukuda M (2000) Poly-N-acetyllactosamine extension in N-glycans and core 2- and core 4-branched O-glycans is differentially controlled by i-extension enzyme and different members of the β1,4-galactosyltransferase gene family. J Biol Chem 275:15868–15875

Vanderplasschen A, Markine-Goriaynoff N, Lomonte P, Suzuki M, Hiraoka N, Yeh JC, Bureau F, Willems L, Etienne T, Fukuda M, Pastoret PP (2000) A multipotential β1,6-N-acetylglucosaminyltransferase is encoded by bovine herpes virus type 4. Proc Natl Acad Sci USA 97:5756–5761

Vavasseur F, Yang JM, Dole K, Paulsen H, Brockhausen I (1995) Synthesis of O-glycan core 3: characterization of UDP-GlcNAc: GalNAc-R β3-*N*-acetylglucosaminyltransferase activity from colonic mucosal tissues and lack of the activity in human cancer cell lines. Glycobiology 5:351–357

Yang JM, Byrd JC, Siddiki BB, Chung YS, Okuno M, Sowa M, Kim YS, Matta KL, Brockhausen I (1994) Alterations of O-glycan biosynthesis in human colon cancer tissues. Glycobiology 4:873–884

Yeh JC, Ong E, Fukuda M (1999) Molecular cloning and expression of a novel β1,6-*N*-acetylglucosaminyltransferase that forms core 2, core 4, and I branches. J Biol Chem 274:3215–3221

20

α4-N-Acetylglucosaminyltransferase

Introduction

α4-N-Acetylglucosaminyltransferase (α4GnT) is a glycosyltransferase that mediates transfer of GlcNAc with α1,4-linkage from UDP-GlcNAc to βGal residues preferentially present in O-glycans, forming GlcNAcα1-4Galβ-R (Nakayama et al. 1999). In the human, glycoproteins having GlcNAcα1-4Galβ-R at nonreducing terminals are exclusively limited to the mucins secreted from gland mucous cells (cardiac gland cell, mucous neck cell, pyloric gland cell) of the stomach, Brunner's glands of the duodenum, accessory glands of the pancreaticobiliary tract, and pancreatic ducts showing gastric metaplasia (Nakamura et al. 1998) (Fig. 1). Thus, these mucous glycoproteins as a whole have been regarded as gastric gland mucous cell-type mucin, and α4GnT plays a key role in synthesizing this particular mucin which has GlcNAcα1-4Galβ-R structures. In the gastric mucin, the oligosaccharides having GlcNacα1-4Galβ-R structures were found to be attached to not only MUC6 but also MUC5AC (Zhang et al. 2001).

Recently, cDNA encoding a human α4GnT was cloned by expression cloning (Nakayama et al. 1999). The deduced amino acid sequence of the human α4GnT reveals that this enzyme is a typical type II membrane protein composed of 340 amino acids with a molecular weight of 39 497 Da. This enzyme consists of a very short cytoplasmic NH_2-terminal segment of three amino acid residues, a transmembrane/signal anchoring domain of 22 amino acid residues, and a large COOH-terminal segment of stem and catalytic domains. Four potential N-glycosylation sites are found in the catalytic domain. Among the glycosyltransferases cloned to date, significant homology at the amino acid level (35% similarity) is found with human α4-galactosyltransferase responsible for Gb_3 (Galα1-4Galβ1-4Glcβ-Cer), which is also known as blood group P^k (Furukawa et al 2000; Keusch et al. 2000; Steffensen et al. 2000).

JUN NAKAYAMA

Institute of Organ Transplants, Reconstructive Medicine and Tissue Engineering, Shinshu University Graduate School of Medicine and Central Clinical Laboratories, Shinshu University Hospital, Asahi 3-1-1, Matsumoto 390-8621, Japan
Tel. +81-263-37-2802; Fax +81-263-34-5316
e-mail: jun@hsp.md.shinshu-u.ac.jp

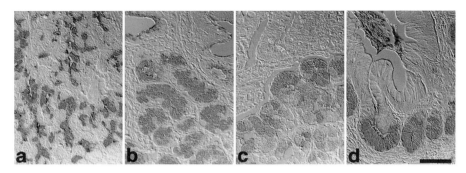

Fig. 1. Expression of the gastric gland mucous cell-type mucin carrying GlcNAcα1-4Galβ-R in the human gastric mucosa, duodenal mucosa, and pancreatic duct. Mucous neck cells (**a**) and pyloric gland cells (**b**) of the gastric mucosa, Brunner's gland cells of the duodenum (**c**), and pancreatic duct showing gastric metaplasia (**d**) produce gastric gland mucous cell-type mucin carrying GlcNAcα1-4Galβ-R. Immunostaining with HIK1083 antibody specific for GlcNAcα1-4Galβ-R; *bar* = 50 μm

Databanks

α4-*N*-Acetylglucosaminyltransferase

No EC number has been allocated

Species	Gene	Protein	mRNA	Genomic
Homo sapiens	*α4GnT*	–	AF141315	–

Name and History

α4-*N*-Acetylglucosaminyltransferase is abbreviated to α4GnT. No other synonyms for α4GnT have been used. cDNA encoding a human α4GnT was isolated from stomach (Nakayama et al. 1999). Glycoprotein having αGlcNAc-substituents was first isolated from the hog gastric mucin (Lloyd et al. 1969). Later, it was demonstrated that the GlcNAcα1-4Galβ-R structure was a characteristic constituent of the pig gastric mucin (Kochetkov et al. 1976), rat gastric mucin (Ishihara et al. 1996), and duodenal glands of rat and pig (Van Halbeek et al. 1983). However, the glycosyltransferase, α4GnT, responsible for the biosynthesis of GlcNAcα1-4Galβ-R was not isolated before the cloning of this gene.

Enzyme Activity Assay and Substrate Specificity

The following reaction can be catalyzed by α4GnT:

$$\text{UDP-GlcNAc} + \text{Gal}\beta\text{1-R} \rightarrow \text{GlcNAc}\alpha\text{1-4Gal}\beta\text{-R} + \text{UDP}$$

Fig. 2. Substrate specificity of α4GnT. Radioactivity of incorporated [³H]GlcNAc to various acceptors by a soluble α4GnT is compared with that obtained when Galβ1-4GlcNAcβ1-6(Galβ1-3)GalNAcα-pNP was used as an acceptor. The concentration of acceptors was 1.0 mM, except for 0.7 mM for Galβ1-4GlcNAcβ1-6(Galβ1-3)GalNAcα-pNP. pNP; p-nitrophenol (From Nakayama et al. 1999, with permission)

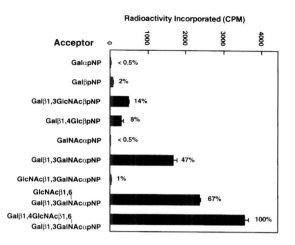

The enzyme assay and acceptor specificity of α4GnT were performed using a soluble chimeric α4GnT fused with protein A as an enzyme source (Nakayama et al. 1999). A mammalian expression vector, pcDNAI, encoding the catalytic domain of α4GnT fused with protein A was transfected into CHO cells. Then the chimeric protein was purified by using IgG-Sepharose and incubated with 1.0 mM or 0.7 mM of various synthetic acceptors, together with 1.0 mM UDP-GlcNAc containing 0.5 μCi of UDP-[³H]GlcNAc in the presence of 5 mM MnCl₂, pH 7.0. After incubation at 37°C for 1 h, the reaction products, purified by a C18 reversed-phase column, were subjected to high-performance liquid chromatography.

As shown in Fig. 2, α4GnT incorporates GlcNAc most efficiently to core 2 branched oligosaccharides, Galβ1-4GlcNAcβ1-6(Galβ1-3)GalNAcα-pNP. In addition, analysis with NMR demonstrated that the incorporated GlcNAc was actually attached with α1,4-linkage to Galβ at a nonreducing terminal of the acceptor. Interestingly, α4GnT acts on Galβ1-3(GlcNAcβ1-6)GalNAcα-pNP better than a core 1 acceptor, Galβ1-3GalNAcα-pNP, and only GlcNAcα1-4Galβ1-3(GlcNAcβ1-6)GalNAcα-pNP was formed from Galβ1-3(GlcNAcβ1-6)GalNAcα-pNP. These results suggest that the addition of β1,6-linked GlcNAc may change the conformation of the acceptor, resulting in a better acceptor for α4GnT. The α4GnT acted on Galβ1-3GlcNAcβ-pNP slightly better than Galβ1-4Glcβ1-pNP, but acted much better on Galβ1-4GlcNAcβ1-6(Galβ1-3)GalNAcα-pNP or Galβ1-3GalNAcα-pNP. These results indicate that this enzyme prefers β1,4-linked or β1,3-linked Gal residues on O-glycans. α4GnT did not utilize UDP-GalNAc when these acceptors were tested.

Preparation

α4GnT protein is available only in a recombinant form. The cDNA encoding a human α4GnT has been cloned by an expression cloning strategy (Nakayama et al. 1999) (Fig. 3). Briefly, COS-1 cells were co-transfected with human stomach cDNA library

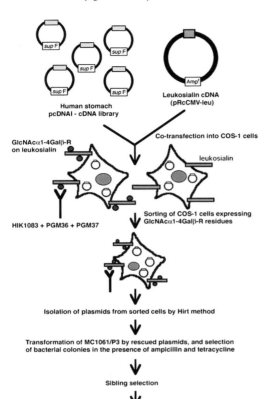

Fig. 3. Strategy of expression cloning of a human α4GnT. Since COS-1 cells do not express GlcNAcα1-4Galβ-R, they were co-transfected with human stomach cDNA library constructed in pcDNAI and a leukosialin vector, pRcCMV-leu. After 60 h, COS-1 cells expressing GlcNAcα1-4Galβ-R residues were isolated by cell sorting using an antibody mixture of HIK1083, PGM36, and PGM37. Plasmid DNA isolated from the sorted cells by the Hirt procedure was amplified in *E. coli* MC1061/P3 cells in the presence of ampicillin and tetracycline. Sibling selection was performed by dividing the plasmids into small pools, and they were separately co-transfected into COS-1 cells together with the leukosialin vector. Then the transfectants stained with the same antibody mixture were screened under immunofluorescence microscopy to isolate a single plasmid encoding a human α4GnT

constructed by a mammalian expression vector, pcDNAI, together with a leukosialin cDNA. Leukosialin is a major membrane-bound sialoglycoprotein of leukocytes and contains 80 *O*-glycans in its extracellular domain (Fukuda and Tsuboi 1999). Although polypeptides carrying GlcNAcα1-4Galβ-R was not identified, we envisaged that leukosialin could be a candidate for the preferred substrate of α4GnT, since leukosialin was reported to be present in COLO 205 cells (Baeckström et al. 1995). In fact, it was proven later that co-transfection with leukosialin cDNA was critical for this cloning, because the expression level of GlcNAcα1-4Galβ-R is much stronger when COS-1 cells co-transfected with α4GnT and leukosialin cDNAs than when COS-1 cells are transfected with α4GnT cDNA alone (Nakayama et al. 1999).

The transfected COS-1 cells were then screened by using monoclonal antibodies specific for GlcNAcα1-4Galβ-R, consisting of HIK1083 (Ishihara et al. 1996), PGM36, and PGM37 (Kurihara et al. 1998), and COS-1 cells positive for these antibodies were enriched by fluorescence-activated cell sorting. The plasmid cDNA was rescued from the sorted cells and amplified in bacteria MC1061/P3 cells. As pcDNAI vector encodes a *sup* F gene that corrects the defect of both ampicillin (Amp)- and tetracycline (Tet)-resistant genes in the P3 episome, MC1061/P3 cells transformed by pcDNAI become resistant to both antibiotics. In contrast, MC1061/P3 cells transformed by the leukosialin cDNA alone are resistant to Amp but not to Tet. Because of this difference, only plasmids derived from the library were selectively amplified. This procedure, followed by several rounds of sibling selection, leads to the isolation of a cDNA encoding human α4GnT. The recombinant enzyme has been expressed in COS-1 cells as well as in human gastric adenocarcinoma AGS cells. A soluble form of this enzyme fused with protein A has been expressed in CHO cells and used for in vitro GlcNAc transferase assay, as described above.

Biological Aspects

α4GnT plays a key role to form GlcNAcα1-4Galβ-R structure in the gastric gland mucous cell-type mucin. This mucin, also termed class III mucin, was originally identified by paradoxical ConA staining (PCS), a sequential histochemical method composed of periodate oxidation, sodium borohydride reduction, ConA binding, and horseradish perioxidase reaction (Katsuyama and Spicer 1978), and molecular cloning of α4GnT allows us to establish that the carbohydrate moiety recognized by PCS is GlcNAcα1-4Galβ-R structure (Nakayama et al. 1999).

In human tissues, transcripts of α4GnT are distributed only in stomach and pancreas (Nakayama et al. 1999). Immunohistochemistry using anti-α4GnT antibody also demonstrates that α4GnT can be detectable in the Golgi region of the mucous cells secreting gastric gland mucous cell-type mucin (Fig. 4; Zhang et al. 2001). Thus, the expression of α4GnT is strictly regulated in a cell-specific manner. The gene of this enzyme is mapped to chromosome 3p14.3. The genomic organization and promoter analysis of this gene have not been reported.

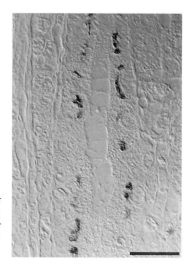

Fig. 4. Expression of α4GnT in the mucous neck cells of the gastric mucosa. α4GnT is expressed in the Golgi region of the mucous neck cells of the fundic mucosa of the stomach. Immunostaining with anti-human α4GnT antibody; bar = 20 μm

Future Perspectives

Recently it was demonstrated that gastric gland mucous cell-type mucin having GlcNAcα1-4Galβ-R is found in the mucous neck cells and pyloric gland cells of a variety of vertebrates, including amphibians, reptiles, and mammals, but not fish and birds (Ota et al. 1998). This result strongly suggests that α4GnT plays an important role in the development of the gastric mucosa of vertebrates. Extensive analysis of transgenic or knockout mice for the α4GnT gene will address this problem.

It is also noteworthy that the gastric gland mucous cell-type mucin is frequently expressed in gastric and pancreatic cancer cells (Nakamura et al. 1998). In addition, this mucin is also detectable in the mucinous bronchioloalveolar cell carcinoma of the lung (Honda et al. 1998) as well as the so-called adenoma malignum of the uterine cervix (Ishii et al. 1998). It will be of significance to determine the role of this particular mucin carrying GlcNAcα1-4Galβ-R in the pathogenesis of these cancers by using the α4GnT cDNA as a tool.

Acknowledgments

I would like to thank Drs. Minoru Fukuda, Tsutomu Katsuyama, Jiunn-Chern Yeh, Anup K. Misra, Susumu Ito, and Mu Xia Zhang for their productive collaboration, suggestions, and encouragement.

References

Baeckström D, Zhang K, Asker N, Rüetschi U, Ek M, Hansson GG (1995) Expression of the leukocyte-associated sialoglycoprotein CD43 by a colon carcinoma cell line. J Biol Chem 270:13688–13692

Fukuda M, Tsuboi S (1999) Mucin-type *O*-glycans and leukosialin. Biochim Biophys Acta 1455:205–217

Furukawa K, Iwamura K, Uchikawa M, Sojka BN, Wiels J, Okajima T, Urano T, Furukawa K (2000) Molecular basis for the phenotype: Identification of distinct and multiple mutations in the α1,4-galactosyltransferase gene in Swedish and Japanese individuals. J Biol Chem 275:37752–37756

Honda T, Ota H, Ishii K, Nakamura N, Kubo K, Katsuyama T (1998) Mucous bronchioloalveolar carcinoma with organized differentiation simulating the pyloric mucosa of the stomach. Clinicopathologic, histochemical, and immunohistochemical analysis. Am J Clin Pathol 109:423–430

Ishihara K, Kurihara M, Goso Y, Urata T, Ota H, Katsuyama T, Hotta K (1996) Peripheral α-linked *N*-acetylglucosamine on the carbohydrate moiety of mucin derived from mammalian gastric mucous cells: epitope recognized by a newly characterized monoclonal antibody. Biochem J 318:409–416

Ishii K, Hosaka N, Toki T, Momose M, Hidaka E, Tsuchiya S, Katsuyama T (1998) A new view of the so-called adenoma malignum of the uterine cervix. Virchows Arch 432:315–322

Katsuyama T, Spicer SS (1978) Histochemical differentiation of complex carbohydrates with variants of the concanavalin A—horseradish peroxidase method. J Histochem Cytochem 26:233–250

Keusch JJ, Manzella SM, Nyame KA, Cummings RD, Baenziger JU (2000) Cloning of Gb₃ synthase, the key enzyme in globo-series glycosphingolipid synthesis, predicts a family of α1,4-glycosyltransferases conserved in plants, insects, and mammals. J Biol Chem 275:25315–25321

Kochetkov NK, Derevitskaya VA, Arbatsky NP (1976) The structure of pentasaccharides and hexasaccharides from blood group substance H. Eur J Biochem 67:129–136

Kurihara M, Ishihara K, Ota H, Katsuyama T, Nakano T, Naito M, Hotta K (1998) Comparison of four monoclonal antibodies reacting with gastric gland mucous cell-derived mucins of rat and frog. Comp Biochem Physiol 121B:315–321

Lloyd KO, Kabat EA, Beychok S (1969) Immunochemical studies on blood groups. XLIII. The interaction of blood group substances from various sources with a plant lectin, Concanavalin A. J Immunol 102:1354–1362

Nakamura N, Ota H, Katsuyama T, Akamatsu T, Ishihara K, Kurihara K, Hotta K (1998) Histochemical reactivity of normal, metaplastic, and neoplastic tissues to α-linked N-acetylglucosamine residue-specific monoclonal antibody HIK1083. J Histochem Cytochem 46:793–801

Nakayama J, Yeh J-C, Misra AK, Ito S, Katsuyama T, Fukuda M (1999) Expression cloning of a human α1,4-N-acetylglucosaminyltransferase that forms GlcNAcα1-4Galβ-R, a glycan specifically expressed in the gastric gland mucous cell-type mucin. Proc Natl Acad Sci USA 96:8991–8996

Ota H, Nakayama J, Momose M, Kurihara M, Ishihara K, Hotta K, Katsuyama T (1998) New monoclonal antibodies against gastric gland mucous cell-type mucins: a comparative immunohistochemical study. Histochem Cell Biol 110:113–119

Steffensen R, Carlier K, Wiels J, Levery SB, Stroud M, Cedergren B, Nilsson Sojka B, Bennett EP, Jersild C, Clausen H (2000) Cloning and expression of the histo-blood group Pᵏ UDP-galactose: Galβ1-4Glcβ1-Cer α1,4-galactosyltransferase. Molecular genetic basis of the p phenotype. J Biol Chem 275:16723–16729

Van Halbeek H, Gerwig GJ, Vliegenthart JG, Smits HL, Van Kerkhof PJ, Kramer MF (1983) Terminal α(1→4)-linked N-acetylglucosamine: a characteristic constituent of duodenal-gland mucous glycoproteins in rat and pig. A high-resolution ¹H-NMR study. Biochim Biophys Acta 747:107–116

Zhang MX, Nakayama J, Hidaka E, Kubota S, Yan J, Ota H, Fukuda M (2001) Immunohistochemical demonstration of α1,4-N-acetylglucosaminyltransferase that forms GlcNAcα1,4Galβ residues in human gastrointestinal mucosa. J Histochem Cytochem 49:587–596

21

O-GlcNAc Transferase

Introduction

Glycosylation of nuclear and cytoplasmic proteins by the addition of a single *N*-acetylglucosamine monosaccharide (*O*-GlcNAc) is a major posttranslational modification in higher eukaryotes (Hart 1997). *O*-GlcNAcylation is characterized by the addition of single GlcNAc residues from a UDP-GlcNAc sugar donor to the hydroxyl groups of serine/threonine residues via an *O*-glycosidic bond in a β-linkage. Unlike some other forms of *O*-glycosylation, this sugar is generally unmodified and is not usually further elongated. *O*-GlcNAcylation of proteins has been shown to be both highly abundant and dynamic, often in response to cellular stimuli (Hart 1997), and it has been suggested that it is a form of regulation alternative to, but working in concert with, serine/threonine phosphorylation. Proteins modified by *O*-GlcNAc are myriad in form and function, with classes which include nuclear pore proteins, transcription factors, RNA polymerase II, oncoproteins, protein kinases and phosphatases, steroid receptors, and cytoskeletal proteins such as MAPs, neurofilaments, and α-crystallins (Hart 1997). Recently, an enzyme that catalyzes this modification was purified to homogeneity from rat liver (Haltiwanger et al. 1992), and the rat and human cDNAs were cloned (Kreppel et al. 1997; Lubas et al. 1997). This chapter examines the characteristics of this enzyme in detail.

SAI PRASAD N. IYER[1,2] and GERALD W. HART[2]

[1] Graduate Program, Department of Biochemistry and Molecular Genetics, University of Alabama at Birmingham, Birmingham, AL 35205, USA
[2] Department of Biological Chemistry, Johns Hopkins University School of Medicine, 725 North Wolfe Street, Baltimore, MD 21205-2185, USA
Tel. +1-410-614-5993; Fax +1-410-614-8804
e-mail: gwhart@bs.jhmi.edu

Databanks

O-GlcNAc transferase

NC-IUBMB enzyme classification: E.C. 2.4.1

Species	Gene	Protein	mRNA	Genomic
Homo sapiens	OGT	AAB63466.1	U77413	–
Rattus norvegicus	OGT	AAC53121.1	U76557	–
Drosophila melanogaster (hypothetical)	–	–	AAF57338	–
Caenorhabditis elegans	–	AAB63465.1	U77412	–
Arabidopsis thaliana (putative homolog)	SPY	g1589778	U62135.1	–

Name and History

In 1984, Torres and Hart made the serendipitous observation that numerous lymphocyte proteins appeared to contain *N*-glycanase-resistant, *O*-glycosidically linked GlcNAc monosaccharides, and that the bulk of these protein-bound GlcNAc residues were found on the inside of the cell (Torres and Hart 1984). Over the next 5 years, this led to a rapid identification of intracellular proteins that were modified by *O*-GlcNAc by several investigators. It now appears that thousands of nuclear and cytosolic proteins are dynamically modified by *O*-GlcNAc. The enzymatic activity that attaches *O*-GlcNAc to proteins is defined as uridine diphospho-*N*-acetylglucosamine:peptide β-*N*-acetylglucosaminyltransferase or *O*-GlcNAc transferase (OGT), and an activity assay was developed by Haltiwanger et al. (Haltiwanger et al. 1990). This assay facilitated purification of the OGT enzyme to homogeneity from rat liver (Haltiwanger et al. 1992). Sequences derived from tryptic OGT peptides subsequently led to cloning of the novel rat and human cDNA through conventional cloning techniques in 1996 (Kreppel et al. 1997; Lubas et al. 1997). Expression of recombinant OGT enzyme and activity has recently been established in heterologous systems such as baculovirus and *E. coli* (Kreppel and Hart 1999; Lubas and Hanover 2000). Current research is focused on regulation of the OGT and elucidation of the function of *O*-GlcNAc.

Enzyme Activity Assay and Substrate Specificity

O-GlcNAc transferase activity is defined as the transfer of GlcNAc from a UDP-GlcNAc sugar donor on to an acceptor peptide (protein):

$$UDP\text{-}GlcNAc + protein \xrightarrow{\text{OGT}} UDP + protein\text{-}O\text{-}GlcNAC$$

The standard assay, as originally defined by Haltiwanger et al. (1990), utilized synthetic peptides as acceptor substrates for the enzyme, whose sequences were based on sequences of *O*-GlcNAcylation sites mapped from three proteins: a nuclear pore protein (D'Onofrio et al. 1988), human erythrocyte band 4.1, and a 65-kDa erythrocyte cytosolic protein (Haltiwanger et al. 1990). A peptide incorporating the common features of these sites of glycosylation, namely an acidic amino acid followed by a serine, a proline, and then a run of serines and threonines gave rise to the YSDSPSTST acceptor substrate.

Specificity for the YSDSPSTST peptide of the enzyme was demonstrated by assaying either YSDSP or YSDSGSTST peptides as substrates (Haltiwanger et al. 1990). Neither of these peptides served as substrates, indicating that the OGT enzyme was highly selective for its substrates. These studies demonstrated that a stretch of serines and threonines alone was not sufficient for activity (Haltiwanger et al. 1990). It also indicated that a proline residue needed to be present for the peptide to be active. This was an important observation, since many of the O-GlcNAc sites on proteins that have been mapped seem to contain a proline at either the −1 or −2 positions relative to the modified serine or threonine (Hart 1997).

The standard assay used in the original purification of the OGT from rat liver (Haltiwanger et al. 1992) was based on the original assay (Haltiwanger et al. 1990), with a few minor modifications. Assays[1] are performed in a final volume of 50 µl, and the reaction mixture contains 50 mM sodium cacodylate pH 6.0, 3.2 mM YSDSPSTST peptide (150 µg), 2.5 mM 5′-adenosine monophosphate, and 0.5–1 µCi of UDP-[6-^{3}H]GlcNAc (26.8 Ci/mM). When assaying crude tissue fractions (such as rat liver or rat brain fractions), sodium fluoride and 1-amino-GlcNAc is added to the reaction mixture to a final concentration of 25 mM and 1 mM, respectively, to inhibit endogenous phosphatases and O-GlcNAcases. In addition, bovine serum albumin (BSA) is added to a final concentration of 1 mg/ml for enzyme stability. It should be noted that a 10× assay buffer cocktail solution can be made, composed of 0.5 M cacodylate acid pH 6.0, 10 mg/ml BSA, 0.25 M NaF, and 0.01 M 1-amino-GlcNAc, which should be diluted to 1× prior to use in an assay reaction. This solution can be stored at 4°C and is stable for months. Standard assay reactions are then started by the addition of enzyme (0–25 µl) and performed at 20°C for 30 min. It should be noted that the assay can also be performed at ambient temperature (22–24°C) for 60 min with good reproducibility.

Reactions are stopped by the addition of 450 µl 50 mM formic acid, and labeled peptides can be separated from unincorporated label by either one of two means. The first method involves loading the reaction mixtures directly onto a 0.5-ml SP-Sephadex (SP-C25-120, Sigma) column equilibrated in 50 mM formic acid. The column is then washed with 10 ml 50 mM formic acid, and the labeled peptides are eluted with 1 ml 0.5 M NaCl. Incorporation of the [^{3}H]GlcNAc label on the peptides is then measured directly by liquid scintillation counting. Alternatively, with the second method, reactions can be stopped by the addition of 450 µl 50 mM formic acid and loaded onto a C_{18} cartridge (Sep-Pak, Waters) equilibrated in 50 mM formic acid. The cartridge is then washed sequentially with 10 ml 50 mM formic acid, 10 ml 50 mM formic acid containing 1 M NaCl, and 10 ml MilliQ water. Peptides are then eluted with 5 ml 50% methanol directly, and the radioactivity incorporated is measured as described above. Units of activity are defined as nanomoles of GlcNAc transferred per minute.

The quality of the enzymatic source is critical for the assay. OGT activity is very sensitive to salt and the presence of UDP, as expected for a glycosyltransferase using

---

[1] The assay conditions described are performed under nonsaturating conditions. When performed under saturating conditions (5 µM UDP-GlcNAc, 15 mM YSDSPSTST), activity is significantly higher (up to 8–10-fold) than under the standard assay conditions. The cost of the reagents, especially the donor and acceptor substrates, may make it unfeasible to perform the assay under saturating conditions

UDP-GlcNAc as the donor sugar. Free UDP is a very potent inhibitor of activity, with 50% loss of activity in the presence of less than 0.5 µM. Therefore, to obviate such potential endogenous inhibitors,[2] enzyme samples are desalted over a 1-ml G-50-80 (Sigma) column, into desalt buffer (20 mM tris pH 7.8, 20% glycerol, 0.02% NaN$_3$) at 4°C. V$_o$ fractions containing enzyme are collected and used for assays.

Variations of the standard assay have been reported in a number of cases (Haltiwanger and Philipsberg 1997; Lubas et al. 1997; Yki-Jarvinen et al. 1997). The most notable variation is the substituted use of the major O-GlcNAcylation[3] site of casein kinase II as the acceptor peptide substrate (PGGSTVPV*S*SANMM), as it has been shown that the OGT has a much higher affinity for this site than the standard peptide (Kreppel and Hart 1999), which increases the sensitivity of the assay by about two orders of magnitude. Other variations include the use of recombinant Nup62 produced in *E. coli* as a protein substrate for recombinant *E. coli*-expressed OGT (Lubas and Hanover 2000).

Preparation

OGT activity and expression has been shown to be ubiquitously distributed in all mammalian tissues examined, including brain, pancreas, thymus, and muscle (Kreppel et al. 1997; Lubas et al. 1997). Interestingly, the activity in brain and thymus is extremely high, suggesting the possibility of tissue-specific isoforms (Kreppel et al. 1997). Purification from rat liver involved standard and affinity chromatography techniques (Haltiwanger et al. 1992). Since then, cloning of the OGT cDNA has facilitated the expression of the recombinant enzyme in baculovirus and *E. coli* (Kreppel and Hart 1999; Lubas and Hanover 2000) as his-tagged and S-tagged N-term fusion proteins, respectively. Purification of recombinant protein is then carried out through IMAC and S-protein affinity chromatography for the his-tagged and S-tagged fusion proteins, respectively. The purified enzyme is then stored indefinitely at −20°C in 20%–40% glycerol and 1 mM DTT.

Biological Aspects

Native O-GlcNAc transferase, purified from rat liver, is a large holoenzyme, consisting of two 110 kDa (α subunit) and one 78 kDa (β subunit) in a α2β configuration, with an apparent M$_r$ of 340 kDa (Haltiwanger et al. 1992). The α or p110 subunit was shown to contain the active site of the enzyme. The p110 cDNA was cloned from rat, human, and *C. elegans,* and its deduced amino acid sequence revealed no homology to any known glycosyltransferase (Field and Wainwright 1995; Joziasse 1992; Paulson and Colley 1989). The cloned enzymes are highly conserved throughout evolution from human to worm. Southern blot analysis revealed that the enzyme is not a member of a multigene family, and northern blots show the presence of at least four

---

[2] Columns are equilibrated in desalt buffer prior to loading the sample

[3] The site of O-GlcNAcylation in the CKII acceptor peptide is in bold and italicized

transcripts ranging in size from 1.7 to 8.0 kb. The 78-kDa subunit was subsequently shown to be immunologically related to the p110 subunit, suggesting its origin either as a proteolytic fragment or an alternatively spliced variant. Interestingly, both subunits are posttranslationally modified by tyrosine phosphorylation and by *O*-GlcNAcylation. Subcellular localization of the OGT showed that it is nuclear and cytoplasmic. Active recombinant OGT from rat, expressed in baculovirus, is a homotrimer of three p110 subunits (M_r ~360 kDa) and exhibits properties similar those of to native OGT (Kreppel and Hart 1999).

While the carboxyl terminus of the OGT shows no homology to any protein in the database, analysis of the amino acid sequence reveals the presence of a tetratricopeptide repeat (TPR) protein–protein interaction domain in the amino terminal, comprising half of the protein. Comprised of individual tandem 34 amino acid repeats, TPRs have been shown to many mediate intra- and intermolecular protein–protein interactions (Goebl and Yanagida 1991; Lamb et al. 1995). Systematic deletions of the 11 TPRs of the recombinant rat OGT revealed that TPRs 4–6 were responsible for p110 intrasubunit trimerization (Kreppel and Hart 1999). Yeast two hybrid screens for interacting proteins have led to the cloning of two novel proteins, OIP98 and OIP102, that interact with the OGT via its TPR domain (Iyer and Hart, manuscripts in preparation). OIP98 and OIP102 may be involved in targeting OGT to transcriptional complexes and to active sites of nascent protein synthesis in neuronal growth cones, respectively (Iyer and Hart, manuscripts in preparation).

The carboxy terminal half of OGT contains the catalytic portion of the enzyme. Photoprobe labelling data indicate that this region contains the UDP-GlcNAc binding site (G.W. Hart, personal communication, 2000). This is of particular importance since it has been shown that levels of donor substrate can influence acceptor substrate specificity (Kreppel and Hart 1999). This can be of particular relevance in studying disease conditions such as diabetes, where glucose metabolic flux is often directly coupled to intracellular pools of UDP-GlcNAc, thereby directly influencing the *O*-GlcNAcylation of key nuclear and cytosolic proteins. Very recently, there has been evidence of protein phosphatases complexed with subpopulations of OGT, and this interaction is thought to occur via the carboxy terminus (Wells and Hart, manuscript in preparation).

Recently, the gene for OGT has been targeted in the mouse. Mutagenesis of the gene leads to loss of cell viability at the single cell level (Shafi et al. 2000). OGT gene deletions in mice resulted in loss of embryonic stem cell viability, which is not surprising since the OGT is a single-copy gene, localized to the X chromosome (Xq13 in humans). These studies support the notion of a key role for *O*-GlcNAcylation in the regulation of cellular metabolism.

Future Perspectives

Since the recent cloning of the OGT cDNA, the focus has now shifted to understanding the regulation of this unique enzyme. One can expect to gain valuable insight into its function in the next few years by studying proteins that interact with the OGT, and also through the posttranslational modifications of the enzyme. Combined with the ongoing effort to expand the list of *O*-GlcNAcylated proteins, the functions of *O*-GlcNAcylation can be expected to be unraveled in the very near future.

Further Reading

For a recent review, see Hart 1997.

References

D'Onofrio M, Starr CM, Park MK, Holt GD, Haltiwanger RS, Hart GW, Hanover JA (1988) Partial cDNA sequence encoding a nuclear pore protein modified by O-linked N-acetylglucosamine. PNAS 85:9595–9599

Field MC, Wainwright LJ (1995) Molecular cloning of eukaryotic glycoprotein and glycolipid glycosyltransferases: a survey [published erratum appears in Glycobiology (1996) 6(1):5]. Glycobiology 5:463–472

Goebl M, Yanagida M (1991) The TPR snap helix: a novel protein repeat motif from mitosis to transcription. Trends Biochem Sci 16:173–177

Haltiwanger RS, Philipsberg GA (1997) Mitotic arrest with nocodazole induces seletive changes in the level of O-linked N-acetylglucosamine and accumulation of incompletely processed N-glycans on proteins from HT29 cells. J Biol Chem 272:8752–8758

Haltiwanger RS, Holt GD, Hart GW (1990) Enzymatic addition of O-GlcNAc to nuclear and cytoplasmic proteins. Identification of a uridine diphospho-N-acetylglucosamine:peptide β-N-acetylglucosaminyltransferase. J Biol Chem 265:2563–2568

Haltiwanger RS, Blomberg MA, Hart GW (1992) Glycosylation of nuclear and cytoplasmic proteins. Purification and characterization of a uridine diphospho-N-acetylglucosamine:peptide β-N-acetylglucosaminyltransferase. J Biol Chem 267:9005–9013

Hart GW (1997) Dynamic O-linked glycosylation of nuclear and cytoskeletal Proteins. Annu Rev Biochem 66:315–335

Joziasse DH (1992) Mammalian glycosyltransferases: genomic organization and protein structure. Glycobiology 2:271–277

Kreppel LK, Hart GW (1999) Regulation of a cytosolic and nuclear O-GlcNAc transferase. Role of the tetratricopeptide repeats. J Biol Chem 274:32015–32022

Kreppel LK, Blomberg BA, Hart GW (1997) Dynamic glycosylation of nuclear and cytosolic proteins. Cloning and characterization of a unique O-GlcNAc transferase with multiple tetratricopeptide repeats. J Biol Chem 272:9308–9315

Lamb JR, Tugendreich S, Heiter P (1995) Tetratricopeptide repeat interactions: to TPR or not to TPR? Trends Biochem Sci 20:257–259

Lubas WA, Hanover JA (2000) Functional expression of O-linked GlcNAc transferase. Domain structure and substrate specificity. J Biol Chem 275:10983–10988

Lubas WA, Frank DW, Krause M, Hanover JA (1997) O-linked GlcNAc transferase is a conserved nucleocytoplasmic protein containing tetratricopeptide repeats. J Biol Chem 272:9316–9324

Paulson JC, Colley KJ (1989) Glycosyltransferases. Structure, localization, and control of cell type-specific glycosylation. J Biol Chem 264:17615–17618

Shafi R, Iyer SN, Ellies LG, O'Donnell N, Marek KW, Chui D, Hart GW, Marth JD (2000) The O-GlcNAc transferase gene resides on the X chromosome and is essential for embryonic stem cell viability and mouse ontogeny. PNAS 97:5735–5739

Torres C-R, Hart GW (1984) Topography and polypeptide distribution of terminal N-acetylglucosamine residues on the surfaces of intact lymphocytes. Evidence for O-linked GlcNAc. J Biol Chem 259:3308–3317

Yki-Järvinen H, Vogt C, Iozzo P, Pipek R, Daniels MC, Virkamäki A, Mäkimattila S, Mandarino L, DeFronzo RA, McClain D, Gottschalk WK (1997) UDP-N-acetylglucosaminetransferase and glutamine:fructose 6-phosphate amidotransferase activities in insulin-sensitive tissues. Diabetologia 40:76–81

N-Acetylgalactosaminyltransferases

Polypeptide
N-Acetylgalactosaminyltransferases

Introduction

The initiation of (mucin-type) *O*-glycosylation is catalyzed by a family of UDP-GalNAc:polypeptide *N*-acetylgalactosaminyltransferases (ppGaNTases). These enzymes catalyze the transfer of GalNAc from the nucleotide sugar UDP-GalNAc to the hydroxyl group of either serine or threonine.

O-Glycans impart unique structural features to mucin-glycoproteins and numerous membrane receptors (Jentoft 1990). In addition, *O*-linked carbohydrates function as ligands for receptors (e.g., leukocyte trafficking, Leppänen et al. 1999) and as intracellular sorting signals (e.g., sucrase–isomaltase, Alfalah et al. 1999).

Databanks

Polypeptide *N*-Acetylgalactosaminyltransferases

NC-IUBMB enzyme classification: E.C.2.4.1.41
Glycosyltransferase classification: family 27 of retaining nucleotide-disphospho-sugar glycosyltransferases (Campbell et al. 1998)

Species	Gene	Protein	mRNA	Genomic
ppGaNTase-T1				
Bos taurus	–	–	L07780	–
			L17437	
Rattus norvegicus	Galnt1	–	U35890	–
Homo sapiens	GALNT1	–	X85018	–
Mus musculus	Galnt1	–	U73820	–
Sus scrofa		–	D85389	–

Fred K. Hagen*, Kelly G. Ten Hagen*, and Lawrence A. Tabak[†]

Center for Oral Biology, Aab Institute of Biomedical Sciences, University of Rochester, 601 Elmwood Avenue, Medical Center Box 611, Rochester, NY 14642, USA
Tel. +1-716 275-0770; Fax +1-716 473-2679
e-mail: Lawrence.Tabak@nih.gov
* The first two authors contributed equally to this work
[†] Present address: National Institute of Dental and Craniofacial Research. NIH, 31 Center Drive MSC 2290, Building 31, Room 2C39, Bethesda, MD 20892-2290, USA

(Continued)

Species	Gene	Protein	mRNA	Genomic
ppGaNTase-T2				
Homo sapiens	*GALNT2*	–	X85019	–
ppGaNTase-T3				
Homo sapiens	*GALNT3*	–	X92689	–
Mus musculus	*Galnt3*	–	U70538	–
ppGaNTase-T4				
Mus musculus	*Galnt4*	–	U73819	–
Homo sapiens	*GALNT4*	–	Y08564	–
ppGaNTase-T5				
Rattus norvegicus	*Galnt5*	–	AF049344	–
ppGaNTase-T6				
Homo sapiens	*GALNT6*	–	Y08565	–
ppGaNTase-T7				
Rattus norvegicus	*Galnt7*	–	AF076167	–
Homo sapiens	*GALNT7*	–	AJ002744	–
ppGaNTase-T9				
Rattus norvegicus	*Galnt9*	–	AF241241	–
Caenorhabditis elegans	*gly-3*	–	AF031833	–
	gly-4	–	AF031834	–
	gly-5a	–	AF031835	–
	gly-5b	–	AF031836	–
	gly-5c	–	AF031837	–

Name and History

ppGaNTase is also referred to as polypeptide GalNAc transferase, or as UDP-protein N-acetylgalactosaminyltransferase. Work over the past decade has demonstrated that there are a family of ppGaNTases which display subtle differences in both their substrate specificity and their patterns of expression. The first detailed analysis of this enzymatic activity was described using a microsomal fraction derived from ovine submandibular glands and enzymatically deglycosylated ovine submandibular gland mucin as an acceptor (McGuire and Roseman 1967). A ppGaNTase was purified from an ascites hepatoma (Sugiura et al. 1982); it had an apparent molecular mass of 55 000 daltons. Subsequently, Elhammer and Kornfeld (1986) reported the purification of a ppGaNTase from bovine colostrum and a murine lymphoma which had an apparent molecular mass of 70 000. Although it was not fully appreciated at the time, this represented the first evidence for a family of closely related ppGaNTase isoforms.

The great similarity among ppGaNTase isoforms has made it necessary to clone and functionally express each one for characterization. Further studies, principally from the Clausen laboratory in Denmark and the Tabak laboratory in the USA, have led to the identification of nine distinct mammalian isoforms to date (see Ten Hagen et al. 2001 and references therein). Each isoform has been numbered in order of its cloning and expression; i.e., ppGaNTase-T1 was the first isoform to be cloned and functionally expressed. Independently, Hagen has characterized the ppGaNTase gene family in *C. elegans*, thus demonstrating that this group of enzymes has been conserved for much of evolution (Hagen and Nehrke 1998). Analyses of EST and genome data bases make it apparent that additional forms of ppGaNTases remain to be characterized.

Enzyme Activity Assay and Substrate Specificity

Only a subset of hydroxyamino acids are substituted with GalNAc in the mammalian proteome. Thus, rules must exist which specify which serines and threonines will become decorated with *O*-glycans. Despite intense investigation, no consensus sequon has emerged which is both necessary and sufficient for *O*-glyosylation to proceed. Evidence has been presented that implicates the charge distribution of amino acids flanking a potential site (Nehrke et al. 1997) or the tertiary structure of the acceptor site (see Kirnarsky et al. 1998 and references therein for a review). These are not mutually exclusive, and thus what emerges is that ppGaNTase isoforms have overlapping specificities and each enzyme may accommodate a rather broad range of substrates (for a review see Elhammer et al. 1999). Prior glycosylation of specific serines and threonines in peptide substrates can influence the acquisition of sugar to vicinal positions (Hanisch et al. 1999). Most recently, Ten Hagen et al. (1999; 2001) have demonstrated that ppGaNTase-T7 and -T9 act as glycopeptide transferases, strongly suggesting that *O*-glycosylation of multisite substrates proceeds in a specific hierarchical manner.

A series of isoform-selective peptides have emerged, but none of these are absolutely specific for any one isoform (Table 1). Enzyme activity requires the presence of divalent cations with Mn^{2+} the preferred cofactor (Sugiura et al. 1982; Elhammer and Kornfeld 1986). ppGaNTase-T1 functions over the pH range 6.5–8.6 (Elhammer and Kornfeld 1986). The only inhibitor that has been reported is a dead-end peptide analogue, PPDAAGAAPLR, which acts as a noncompetitive inhibitor of ppGaNTase-T1 (Wragg et al. 1995).

Preparation

ppGaNTases have been purified to homogeneity from several sources, including bovine colostrum (Elhammer and Kornfeld 1996; Hagen et al. 1993). Recombinant ppGaNTases have been successfully expressed using COS-7 cells; the expression construct was engineered to substitute an insulin secretion signal for the transmembrane

Table 1. Preferred peptide substrates for ppGaNTases

Peptide sequence	Isoform preference	Reference
LSESTTQLPGGGPGCA	ppGaNTase-T1	Wandall et al. 1997
PRFQDSSSSKAPPPLPSPSRLPG	ppGaNTase-T2	Wandall et al. 1997
CIRIQRGPGRAFVTIGKIGNMR	ppGaNTase-T3, -T6	Bennett et al. 1996
		Nehrke et al. 1998
		Bennett et al. 1999
Ac-QATEYEYLDYDFLPETEPPEM	ppGaNTase-T4	Bennett et al. 1998
GTT·PSPVPTTSTT·SAP	PPGaNTase-T7, T-9	Ten Hagen et al. 1999;
(where T· = GalNAc α 1-0-T)		Ten Hagen et al. 2001

domain (e.g., see Hagen et al. 1993), allowing secreted transferase to be separated from endogenous enzyme. Sf-9 insect cells, transfected with the baculovirus system, have also proven to be a good source of recombinant ppGaNTases (e.g., see White et al. 1995). We have recently expressed mg quantities of ppGaNTase-T1, -T2, and -T3 using *Pichia pastoris* (Mao et al. in preparation).

Structure

ppGaNTases are type-II membrane proteins, characterized by a short (4–24 aa) *N*-terminal cytoplasmic tail, followed by a small (15–25 aa) transmembrane anchor, which is tethered to a large (>450 aa) segment in the lumen of the Golgi by a stem region of variable length. Glycosyltransferases which retain the anomeric configuration of the sugar nucleotide bond are thought to work via a double displacement mechanism which would require at least two dicarboxylic acid-containing aa. Either histidine or dicarboxylate acid-containing amino acids would be expected to coordinate the preferred Mn^{2+} co-factor which is suggested to assist in the binding of UDP-GalNAc.

Using a combined bioinformatics/site-directed mutagenesis approach, Hagen et al. (1999) have proposed that ppGaNTases share a common structural fold that consists of two domains, each of which contains a parallel β-sheet, flanked by α-helices. Site-directed mutagenesis of conserved and invariant aspartate, glutamate, and histidine residues supports the view that these residues line the proposed enzyme-active site. The C-terminus of all known ppGaNTases end with a ricin-like motif of unknown function; the mutation of key residues in this region does not compromise the enzymatic activity of ppGaNTase-T1 (Hagen et al. 1999). However, Hassan et al. (2000) have reported that point mutations in this region alter the glycopeptide activity of ppGaNTase-T4.

Biological Aspects

Each ppGaNTase displays a unique pattern of transcript expression across tissues derived from adult rodents or humans as ascertained by Northern blot analysis; transcripts encoding ppGaNTase-T1 and -T2 are expressed in most rodent and human tissues (Hagen et al. 1997; White et al. 1995). ppGaNTase-T3 is found predominantly in the human and murine testis, kidney, and digestive and reproductive tracts (Bennett et al. 1996; Zara et al. 1996). ppGaNTase-T4 transcript is highly expressed in rodent sublingual gland, stomach, and colon, with lower levels in the small intestine, urogenital track, and lung (Hagen et al. 1997). ppGaNTase-T7 expression is similar to that of ppGaNTase-T4; transcript levels are highest in rodent sublingual gland and colon, with lower levels in the small intestine and urogenital tract (Ten Hagen et al. 1999). ppGaNTase-T6 is highly expressed in human placenta and trachea, and weakly expressed in brain and pancreas (Bennett et al. 1999). ppGaNTase-T5 transcript is most abundant in murine sublingual gland and colon, with lower levels of expression in stomach and small intestine (Ten Hagen et al. 1998). ppGaNTase T-9 transcript is found mainly in the murine sublingual gland, testes, and digestive and urogenital tracts (Ten Hagen et al. 2001). The family of ppGaNTases is expressed in a unique spatial and temporal manner during murine development (Kingsley et al. 2000).

The intracellular localizations of ppGaNTases-T1, -T2, and -T3 have been compared (Röttger et al. 1998); they are present throughout the Golgi stack, suggesting that initiation may not be confined to the *cis*-Golgi compartment. This fits well with biochemical data which demonstrates that *O*-glycosylation of peptide substrates proceeds in a specific hierarchical manner (Ten Hagen et al. 1999; 2001).

Cre-lox recombination has been used to generate mice lacking exon 3 of ppGaNTase-T1 (Westerman et al. 1999). Homozygous nulls are fertile and develop normally. However, they lack significant ppGaNTase-T1 activity in the spleen, kidney, thymus, liver, submandibular gland, and lung. Variation in the *O*-glycan lectin binding profile of lymphocyte subsets was observed, and the consequence of this is currently under study. Hennet et al. (1995) have ablated expression of a highly homologous form of ppGaNTase-T1, which has been termed ppGaNTase-T8. Homozygous nulls are fertile and develop normally. Preliminary evidence suggests that the transcript for this uncharacterized isoform is most highly expressed within the brain.

Future Perspectives

As the human genome project reaches its conclusion, we will have a clearer understanding of the full complement of mammalian ppGaNTases. The next challenge will be to define the hierarchical network of glycosyltransferases interaction in vivo, and to identify the repertoire of substrates which are glycosylated by each isoform. Only then will a complete understanding of the biological function played by this glycosyltransferase family be achieved.

Further Reading

Hagen FK, Hazes B, Raffo R, deSa D, Tabak LA (1999) Structure–function analysis of the UDP-*N*-acetyl-D-galactosamine:polypeptide *N*-acetylgalactosaminyltransferase. J Biol Chem 274:6797–6803

Hanisch F-G, Müller S, Hassan H, Clausen H, Zachara N, Gooley AA, Paulsen H, Alving K, Peter-Katalinic J (1999) Dynamic epigenetic regulation of initial *O*-glycosylation by UDP-*N*-acetylgalactosamine:peptide *N*-acetylgalactosaminyltransferases. J Biol Chem 274:9946–9954

Ten Hagen KG, Tetaert D, Hagen FK, Richet C, Beres TM, Gagnon J, Balys MM, VanWuyckhuyse B, Bedi GS, Degand P, Tabak LA (1999) Characterization of a UDP-GalNAc:polypeptide *N*-acetylgalactosaminyltransferase which displays glycopeptide *N*-acetylgalactosaminyltransferase activity. J Biol Chem 274:27867–27874

References

Alfalah M, Jacob R, Preuss U, Zimmer K-P, Naim H, Naim HY (1999) *O*-linked glycans mediate apical sorting of human intestinal sucrase–isomaltase through association with lipid rafts. Curr Biol 9:593–596

Bennett EP, Hassan H, Clausen H (1996) cDNA cloning and expression of a novel human UDP-*N*-acetyl-α-D-galactosamine. J Biol Chem 271:17006–17012

Bennett EP, Hassan H, Mandel U, Hollingsworth MA, Akisawa N, Ikematsu Y, Merkx G, van Kessel AG, Olofsson S, Clausen H (1999) Cloning and characterization of a close homologue of human UDP-*N*-acetyl-D-galactosamine:polypeptide *N*-acetylgalactosaminyltransferase-T3, designated GalNAc-T6. J Biol Chem 274:25362–25370

Campbell JA, Davies GJ, Bulone V, Henrissat B (1998) A classification of nucleotide-diphospho-sugar glycosyltransferases based on amino acid sequence similarities. Biochem J 329:719

Elhammer A, Kornfeld S (1996) Purification and characterization of UDP-*N*-acetylgalactosamine:polypeptide *N*-acetylgalactosaminyltransferase from bovine colostrum and murine lymphoma BW5147 cells. J Biol Chem 261:5249–5255

Elhammer AP, Kézdy FJ, Kurosaka A (1999) The acceptor specificity of UDP-GalNAc:polypeptide *N*-acetylgalactosaminyltransferases. Glyco J 16:171–180

Hagen FK, Nehrke K (1998) cDNA cloning and expression of a family of UDP-*N*-acetyl-D-galactosamine:polypeptide *N*-acetylgalactosaminyltransferase sequence homologs from *Caenorhabditis elegans*. J Biol Chem 272:8268–8277

Hagen FK, VanWuyckhuyse B, Tabak LA (1993) Purification, cloning, and expression of a bovine UDP-GalNAc:polypeptide *N*-acetylgalactosaminyltransferase. J Biol Chem 268:18960–18965

Hagen FK, Ten Hagen KG, Beres T, Balys MM, VanWuyckhuyse BC, Tabak LA (1997) cDNA cloning and expression of a novel UDP-*N*-acetyl-D-galactosamine:polypeptide *N*-acetylgalactosaminyltransferase. J Biol Chem 272:13843–13848

Hagen FK, Hazes B, Raffo R, deSa D, Tabak LA (1999) Structure–function analysis of the UDP-*N*-acetyl-D-galactosamine:polypeptide *N*-acetylgalactosaminyltransferase. J Biol Chem 274:6797–6803

Hanisch F-G, Müller S, Hassan H, Clausen H, Zachara N, Gooley AA, Paulsen H, Alving K, Peter-Katalinic J (1999) Dynamic epigenetic regulation of initial *O*-glycosylation by UDP-*N*-acetylgalactosamine: peptide *N*-acetylgalactosaminyltransferases. J Biol Chem 274:9946–9954

Hassan H, Reis CA, Bennett EP, Mirgorodskaya E, Roespstorff P, Hollingsworth MA, Burchell J, Taylor-Papadimitriou J, Clausen H (2000) The lectin domain of UDP-GalNAc:polypeptide N-acetylgalactosaminyltransferase-T4 directs its glycopeptide specificities. J Biol Chem 275:38197–38205

Hennet T, Hagen FK, Tabak LA, Marth JD (1995) T-cell-specific deletion of a polypeptide *N*-acetylgalactosaminyltransferase gene by site-directed recombination. Proc Natl Acad Sci USA 92:12070–12074

Jentoft N (1990) Why are proteins *O*-glycosylated? Trends Biochem Sci 15:291–294

Kinarsky L, Nomoto M, Ikematsu Y, Hassan H, Bennett EP, Cerny RL, Clausen H, Hollingsworth MA, Sherman S (1998) Structural analysis of peptide substrates for mucin-type *O*-glycosylation. Biochemistry 37:12811–12817

Kingsley PD, Ten Hagen KG, Maltby KM, Zara J, Tabak LA (2000) Diverse spatial expretession patterns of UDP-GalNAc:polypeptide N-acetylgalactosaminyltransferase family member RNAs during mouse development. Glycobiology 10:1317–1323

Leppänen A, Mehta P, Ouyang Y-B, Ju T, Helin J, Moore KL, van Die I, Canfield WM, McEver RP, Cummings RD (1999) A novel glycosulfopeptide binds to P-selectin and inhibits leukocyte adhesion to P-selectin. J Biol Chem 274:24838–24848

McGuire EJ, Roseman S (1967) Enzymatic synthesis of the protein–hexosamine linkage in sheep submaxillary mucin. J Biol Chem 242:3745–3755

Nehrke K, Ten Hagen KG, Hagen FK, Tabak LA (1997) Charge distribution of flanking amino acids inhibits *O*-glycosylation of several single-site acceptors in vivo. Glycobiology 7:1053–1060

Nehrke K, Hagen FK, Tabak LA (1998) Isoform-specific *O*-glycosylation by murine UDP-GalNAc:polypeptide *N*-acetylgalactosaminyltransferase-T3, in vivo. Glycobiology 8:367–371

Röttger S, White J, Wandall HH, Olivo J-C, Stark A, Bennett EP, Whitehouse C, Berger EG, Clausen H, Nilsson T (1998) Localization of three human polypeptide GalNAc-transferases in HeLa cells suggests initiation of O-linked glycosylation throughout the Golgi apparatus. J Cell Sci 111:45–50

Sugiura M, Kawasaki T, Yamashina I (1982) Purification and characterization of UDP-GalNAc:polypeptide N-acetylgalactosamine transferase from an ascites hepatoma, AH 66. J Biol Chem 257:9501–9507

Ten Hagen KG, Hagen FK, Balys MM, Beres TM, VanWuyckhuyse BC, Tabak LA (1998) Cloning and expression of a novel, tissue specifically expressed member of the UDP-GalNAc:polypeptide N-acetylgalactosaminyltransferase family. J Biol Chem 273: 27749–27754

Ten Hagen KG, Tetaert D, Hagen FK, Richet C, Beres TM, Gagnon J, Balys MM, VanWuyckhuyse B, Bedi GS, Degand P, Tabak LA (1999) Characterization of a UDP-GalNAc:polypeptide N-acetylgalactosaminyltransferase which displays glycopeptide N-acetylgalactosaminyltransferase activity. J Biol Chem 274:27867–27874

Ten Hagen KG, Bedi GS, Tetaert D, Kingsley PD, Hagen FK, Balys MM, Beres TM, Degand P, Tabak LA (2001) Cloning and characterization of a ninth member of the UDP-GalNAc:polypeptide N-acetylgalactosaminyltransferase family, ppGaNTase-T9. J Biol Chem 276:17395–17404

Wandall HH, Hassan H, Mirgorodskaya E, Kristensen AK, Roepstorff P, Bennett EP, Nielsen PA, Hollingsworth MA, Burchell J, Taylor-Papadimitriou J, Clausen H (1997) Substrate specificities of three members of the human UDP-N-acetyl-α-D-galactosamine: polypeptide N-acetylgalactosaminyltransferase family, GalNAc-T1, -T2, and -T3. J Biol Chem 272:23503–23514

Westerman EL, Ellies LG, Hagen FK, Marek KW, Sutton-Smith M, Dell A, Tabak LA, Marth JD (1999) Selective loss of O-glycans in mice lacking polypeptide GalNAc-T1. Glyco-biology 9:1121 (abstract No. 80)

White T, Bennett EP, Takio K, Sørensen T, Bonding N, Clausen H (1995) Purification and cDNA cloning of a human UDP-N-acetyl-α-D-galactosamine:polypeptide N-acetylgalactosaminyltransferase. J Biol Chem 270:24156–24165

Wragg S, Hagen FK, Tabak LA (1995) Kinetic analysis of a recombinant UDP-N-acetyl-D-galactosamine:polypeptide N-acetylgalactosaminyltransferase. J Biol Chem 270: 16947–16954

Zara J, Hagen FK, Ten Hagen KG, VanWuyckhuyse BC, Tabak LA (1996) Cloning and expression of mouse UDP-GalNAc: polypeptide N-acetylgalactosaminyltransferase. Biochem Biophys Res Comm 228:38–44

23

β4-N-Acetylgalactosaminyltransferase

Introduction

β4-N-Acetylgalactosaminyltransferase (β4GalNAcT) is a key enzyme to catalyze the conversion of GM3, GD3, and lactosylceramide (LacCer) to GM2, GD2, and asialo-GM2 (GA2), respectively (Yamashiro et al. 1995). This step is critical for the synthesis of all complex gangliosides such as GM1, GD1a, GD1b, GT1b, and GQ1b which are enriched in the nervous system of vertebrates. Therefore, all complex gangliosides are synthesized via the direct products of this enzyme. The cDNAs of β4GalNAcT were isolated by a eularyocyte expression cloning system for the first time as a glycosyl-transferase gene responsible for the ganglioside synthesis in 1992 (Nagata et al. 1992). This enzyme utilizes only glycolipid acceptors, not glycoproteins, and no other enzymes (genes) catalyzing similar functions have been detected to date.

Databanks

β4-N-Acetylgalactosaminyltransferase

NC-IUBMB enzyme classification: E.C.2.4.1.92

Species	Gene	Protein	mRNA	Genomic
Homo sapiens	–	–	M83651 NM_001478	L76079
Mus musculus	–	–	L25885	–
			NM_008080	–
Rattus norvegicus	–	–	D17809	–

KOICHI FURUKAWA

Department of Biochemistry II, Nagoya University School of Medicine, 65 Tsurumai, Showa-ku, Nagoya 466-0065, Japan
Tel. +81-52-744-2070; Fax +81-52-744-2069
e-mail: koichi@med.nagoya-u.ac.jp

Name and History

GM2 synthase.
β1,4-*N*-Acetylgalactosaminyltransferase.
β4GalNAcT.
GM2/GD2 synthase.

Since we succeeded in the cDNA cloning of the gene, the identities of GM2 synthase and GD2 synthase (and GA2 synthase) have been confirmed by experimental results (Yamashiro et al. 1993) (Fig. 1). Consequently, this enzyme is now called GM2/GD2 synthase.

Enzyme Activity Assay and Substrate Specificity

Enzyme activity is measured by the incorporation of UDP-[^{14}C] GalNAc onto acceptors (Yamashiro et al. 1993). The reaction (50 μl) mixture contains 100 mM sodium cacodylate-HCl (pH 7.2), 10 mM MnCl$_2$, 0.3% Triton CF-54, 325 μM GM3 (for GM2 synthesis), 400 μM UDP-GalNAc, UDP-[^{14}C] GalNAc (3.5×10^5 dpm), 10 mM CDP-choline, and membranes containing 200 (100) μg protein. The mixture was incubated at 37°C for 2 h. The products were isolated by C18 Sep-Pak cartridge, and analyzed by thin-layer chromatography and fluorography.

The substrate structures utilized are listed below (measured with extracts from a transfectant line) (Yamashiro et al. 1995).

(NeuAc)GM3, 100.0%; (NeuGc)GM3, 87.5%; GD3, 40.5%; GD1a, 1.0%; LacCer, 2.0%; GalCer, 0.0%; GlcCer, 0.0%; GT1b, 0.0%; (NeuAc)SPG, 2.2%; (NeuGc)SPG, 4.2%.

Enzyme kinetic factors

Acceptor	K_m	V_{max}	V_{max}/K_m
GM3	500 μM	79124 units	158.2
LacCer	40 μM	692 units	17.6

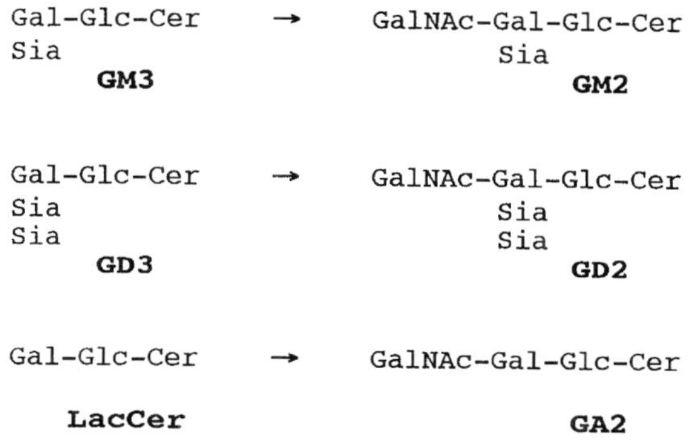

```
Gal-Glc-Cer      →    GalNAc-Gal-Glc-Cer
Sia                        Sia
       GM3                        GM2


Gal-Glc-Cer      →    GalNAc-Gal-Glc-Cer
Sia                        Sia
Sia                        Sia
       GD3                        GD2


Gal-Glc-Cer      →    GalNAc-Gal-Glc-Cer
     LacCer                       GA2
```

Fig. 1. GM2/GD2 synthase responsible for the synthesis of GM2, GD2, and GA2 from GM3, GD3, and LacCer, respectively

Preparation

Source (Biological/Commercial)

Human, melanoma cell line SK-MEL-31 (Yamashiro et al. 1993); rat, cell line AH7974F (ascites hepatoma) (Hidari et al. 1994); proteinA fusion (soluble) protein, (Yamashiro et al. 1995).

Expression Systems

pMIKneo/M2T1-1 expression vector, DEAE-dextran or lipofectin (Nagata et al. 1992).

Isolation/Purification

β4GalNAcT cDNA was isolated by a eukaryotic cell expression cloning system using KF3027 (B16 melanoma expressing polyoma T antigen) and a cDNA library from a human NK-like cell line YT17 (Nagata et al. 1992). For the purification, see Hashimoto et al. 1993.

Biological Aspects

Gene Promoters

Three transcription units (promoters and transcription initiating sites) were defined (Furukawa et al. 1996). Three alternative exon usages at 5′up-stream of the uncoding region were demonstrated. The third promoter was found in the first intron.

Trafficking

β4GalNAcT protein is present as a homodimer in later Golgi apparatus (Jaskiewicz et al. 1996).

Activation

The β4GalNAcT gene is *trans*-activated by the p40[tax] protein encoded by human T lymphotropic virus type I (Furukawa et al. 1993). This gene is also activated with the stage advancement of malignant melanomas. In mice, the β4GalNAcT gene is up-regulated in functionally differentiated T cell subsets, e.g. in cytotoxic T cells with CD8 or CD4 markers (Yoshimura et al. 1994). The β4GalNAcT gene is also activated in mouse thymocytes stimulated by antigen mimicking agents (Takamiya et al. 1995).

Distribution in Tissues

The β4GalNAcT gene is expressed in the late stage of neurodevelopment in mouse brain (Yamamoto et al. 1995, 1996). The expression level gradually reduces after birth. The most strongly expressing sites in the mouse brain are the hippocampus, dentate gyrus, cortex, Purkinje cells in the cerebellum, and in mitral cells of the olfactory nerve. This gene is also expressed in the retina (Daniotti et al. 1997).

Normal Function/Substrates

Significance of N-Glycosylation

Human β4GalNAcT has three N-glycosylation sites at 79, 179, and 274. Replacement of D with Q at some or all of these N-glycosylation sites revealed that all three N-glycosylations occurred, and that this was needed for the full activity of the enzyme (Haraguchi et al. 1995).

Disease Involvement

Among human leukemia cells, only adult T cell leukemia (ATL) cells express GD2/GM2 (Furukawa et al. 1993). Activated T cells or acute T lymphocytic leukemia cells do not express GM2/GD2. These expression patterns corresponded well with the gene expression of β4GalNAcT. Among solid tumors, neuroblastoma cell lines, glioma cell lines, and some malignant melanoma cell lines express very high levels of β4GalNAcT gene (Yamashiro et al. 1993). Moreover, gastric cancer tissues highly express β4GalNAcT gene, while normal stomach does not (Yuyama et al. 1995).

Using transfectant cells of β4GalNAcT cDNA, alterations of tumor phenotypes such as growth rate or adhesion to extracellular matrix were elucidated (Tsurifune et al. 2000; Hyuga et al. 1999). Newly expressed gangliosides, GM2 or GD1a, may modulate the structure/function of integrins. β4GalNAcT gene expression certainly determines the profile of ganglioside expression in tumor cells (Yamashiro et al. 1993; Ruan et al. 1995).

KO Mice

Knock-out mice with the β4GalNAcT gene showed no apparent abnormal morphology of the brain (Takamiya et al. 1996) or of other tissues except for the testis (Takamiya et al. 1998). However, homozygotes showed reduced nerve conductivity. After long-term observation, marked nerve degeneration was found in the sciatic nerve, dorsal root ganglion, and spinal cord (unpublished data, 1999). In the testis, spermatocytes did not fully differentiate, resulting in a lack of sperm and the formation of giant multinuclear cells in the seminiferous tubules (Takamiya et al. 1998). This seems to be due to the disrupted transport of testosterone from Leydig cells to the seminiferous tubules or the vascular system. In homozygotes, the spleen and thymus are slightly smaller than those in the wild type. The T cell response to IL-2 was reduced when spleen T cells were analyzed based on the dysfunction of the IL-2 receptor complex in the mutant mice (Zhao et al. 1999).

Tg Mice

β4GalNAcT Tg mice showed a shift of ganglioside components from b series to a series. In skin-graft experiments, Tg mice showed an enhanced response to foreign substances, mainly by the proliferation of multinuclear polymorphonuclear cells (Fukumoto et al. 1997).

Future Perspectives

Complex gangliosides synthesized via the action of β4GalNAcT are important in the functions and maintenance of nervous systems. However, the molecular mechanisms of the functions of complex gangliosides are still unknown. Molecules interacting with gangliosides remain to be identified.

References

Daniotti JL, Rosales Fritz VM, Martina JA, Furukawa K, Maccioni HJ (1997) Expression of beta 1-4 *N*-acetylgalactosaminyltransferase gene in the developing rat brain and retina: mRNA, protein immunoreactivity and enzyme activity. Neurochem Int 31: 11–19

Fukumoto S, Yamamoto A, Hasegawa T, Abe K, Takamiya K, Okada M, Zhao J, Furukawa K, Miyazaki H, Tsuji Y, Goto G, Suzuki M, Shiku H, Furukawa K (1997) Genetic remodeling of gangliosides resulted in enhanced reactions to the foreign substances in skin. Glycobiology 7:1111–1120

Furukawa K, Akagi T, Nagata Y, Yamada Y, Shimotohno K, Cheung NK, Shiku H, Furukawa K (1993) GD2 ganglioside on human T-lymphotropic virus type I-infected T cells: possible activation of beta-1,4-*N*-acetylgalactosaminyltransferase gene by p40^{tax}. Proc Natl Acad Sci USA 90:1972–1976

Furukawa K, Soejima H, Niikawa N, Shiku H (1996) Genomic organization and chromosomal assignment of the human beta1,4-*N*-acetylgalactosaminyltransferase gene. Identification of multiple transcription units. J Biol Chem 271:20836–20844

Haraguchi M, Yamashiro S, Furukawa K, Takamiya K, Shiku H, Furukawa K (1995) The effects of the site-directed removal of *N*-glycosylation sites from beta-1,4-*N*-acetylgalactosaminyltransferase on its function. Biochem J 312:273–280

Hashimoto Y, Sekine M, Iwasaki K, Suzuki A (1993) Purification and characterization of UDP-*N*-acetylgalactosamine GM3/GD3 *N*-acetylgalactosaminyltransferase from mouse liver. J Biol Chem 268:25857–25864

Hidari JK, Ichikawa S, Furukawa K, Yamasaki M, Hirabayashi Y (1994) Beta 1-4*N*-acetylgalactosaminyltransferase can synthesize both asialoglycosphingolipid GM2 and glycosphingolipid GM2 in vitro and in vivo: isolation and characterization of a beta-1-4*N*-acetylgalactosaminyltransferase cDNA clone from rat ascites hepatoma cell line AH7974F. Biochem J 303:957–965

Hyuga S, Yamagata S, Takatsu Y, Hyuga M, Nakanishi H, Furukawa K, Yamagata T (1999) Suppression by ganglioside GD1a of migration capability, adhesion to vitronectin and metastatic potential of highly metastatic FBJ-LL cells. Int J Cancer 83:685–691

Jaskiewicz E, Zhu G, Bassi R, Darling DS, Young WW Jr (1996) Beta1,4-*N*-acetylgalactosaminyltransferase (GM2 synthase) is released from Golgi membranes as a neuraminidase-sensitive, disulfide-bonded dimer by a cathepsin D-like protease. J Biol Chem 271:26395–26403

Nagata Y, Yamashiro S, Yodoi J, Lloyd KO, Shiku H, Furukawa K (1992) Expression cloning of beta 1,4 *N*-acetylgalactosaminyltransferase cDNAs that determine the expression of GM2 and GD2 gangliosides. J Biol Chem 267:12082–12089

Ruan S, Raj BK, Furukawa K, Lloyd KO (1995) Analysis of melanoma cells stably transfected with beta 1,4GalNAc transferase (GM2/GD2 synthase) cDNA: relative glycosyltransferase levels play a dominant role in determining ganglioside expression. Arch Biochem Biophys 323:11–18

Takamiya K, Yamamoto A, Yamashiro S, Furukawa K, Haraguchi M, Okada M, Ikeda T, Shiku H, Furukawa K (1995) T cell receptor-mediated stimulation of mouse thymocytes induces up-regulation of the GM2/GD2 synthase gene. FEBS Lett 358:79–83

Takamiya K, Yamamoto A, Furukawa K, Yamashiro S, Shin M, Okada M, Fukumoto S, Haraguchi M, Takeda N, Fujimura K, Sakae M, Kishikawa M, Shiku H, Furukawa K, Aizawa S (1996) Mice with disrupted GM2/GD2 synthase gene lack complex gangliosides but exhibit only subtle defects in their nervous system. Proc Natl Acad Sci USA 93:10662–10667

Takamiya K, Yamamoto A, Furukawa K, Zhao J, Fukumoto S, Yamashiro S, Okada M, Haraguchi M, Shin M, Kishikawa M, Shiku H, Aizawa S, Furukawa K (1998) Complex gangliosides are essential in spermatogenesis of mice: possible roles in the transport of testosterone. Proc Natl Acad Sci USA 95:12147–12152

Tsurifune T, Ito T, Li X-J, Yamashiro S, Okada M, Kanematsu T, Shiku H, Furukawa K (2000) Alteration of tumor phenotypes of B16 melanoma after genetic remodeling of the ganglioside profile. Int J Oncol 17:159–165

Yamamoto A, Yamashiro S, Takamiya K, Atsuta M, Shiku H, Furukawa K (1995) Diverse expression of beta 1,4-N-acetylgalactosaminyltransferase gene in the adult mouse brain. J Neurochem 65:2417–2424

Yamamoto A, Haraguchi M, Yamashiro S, Fukumoto S, Furukawa K, Takamiya K, Atsuta M, Shiku H, Furukawa K (1996) Heterogeneity in the expression pattern of two ganglioside synthase genes during mouse brain development. J Neurochem 66:26–34

Yamashiro S, Ruan S, Furukawa K, Tai T, Lloyd KO, Shiku H, Furukawa K (1993) Genetic and enzymatic basis for the differential expression of GM2 and GD2 gangliosides in human cancer cell lines. Cancer Res 53:5395–5400

Yamashiro S, Haraguchi M, Furukawa K, Takamiya K, Yamamoto A, Nagata Y, Lloyd KO, Shiku H, Furukawa K (1995) Substrate specificity of beta 1,4-N-acetylgalactosaminyl-transferase in vitro and in cDNA-transfected cells. GM2/GD2 synthase efficiently generates asialo-GM2 in certain cells. J Biol Chem 270:6149–6155

Yoshimura A, Takamiya K, Kato I, Nakayama E, Shiku H, Furukawa K (1994) GD2 ganglioside-specific monoclonal antibody reacts with murine cytotoxic T lymphocytes reactive with FBL-3N erythroleukaemia. Scand J Immunol 40:557–563

Yuyama Y, Dohi T, Morita H, Furukawa K, Oshima M (1995) Enhanced expression of GM2/GD2 synthase mRNA in human gastrointestinal cancer. Cancer 75:1273–1280

Zhao J, Furukawa K, Fukumoto S, Okada M, Furugen R, Miyazaki H, Takamiya K, Aizawa S, Shiku H, Matsuyama T, Furukawa K (1999) Attenuation of interleukin 2 signal in the spleen cells of complex ganglioside-lacking mice. J Biol Chem 274:13744–13747

24

Histoblood Group A and B Transferases, Their Gene Structures, and Common O Group Gene Structures

Introduction

Histoblood group A and transferases are UDP-GalNAc: H-α3GalNAc transferase and UDP-Gal: H-α3Gal transferase, respectively (Fig. 1). Depending on the structure of H, four types of A antigen and three types of B antigen have been distinguished (Table 1). Type-3 chain A, i.e., repetitive A, is present only as a glycosphingolipid in A_1- but not in A_2-erythrocytes (see Chapter 25). The corresponding "repetitive B" is absent or unknown. Type-4 chain A (globo-A) is also expressed in A_1- but not in A_2-erythrocytes. Type-4 chain B (globo-B) is not known in erythrocytes; however, both globo-A and globo-B are highly expressed in other organs, particularly kidney and urogenital epithelia. The distribution patterns of these isotypes of A and B antigens, and their patterns during development, have been reviewed (Hakomori 1981; Oriol et al. 1986; Clausen and Hakomori 1989; Oriol 1995). In this chapter, only allelic structures of A and B transferases, and the allele for the common O blood group gene, are described. Alleles for variants are described in Chapter 25.

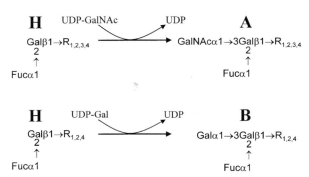

Fig. 1. Enzymatic reactions for the synthesis of A and B antigens from various types of H. For the synthesis of A, four types of H structure with different R residues ($R_{1,2,3,4}$) are involved. For the synthesis of B, only three types of H structure with different R residues ($R_{1,2,4}$) are involved. The structures of R are shown in Table 1

SEN-ITIROH HAKOMORI

Pacific Northwest Research Institute, 720 Broadway, Seattle, WA 98122-4327, USA, and Departments of Pathobiology and Microbiology, University of Washington, Seattle, WA, USA
Tel. +1-206-726-1222; Fax +1-206-726-1212
e-mail: hakomori@u.washington.edu

Table 1. The four types of R structure shown in Fig. 1

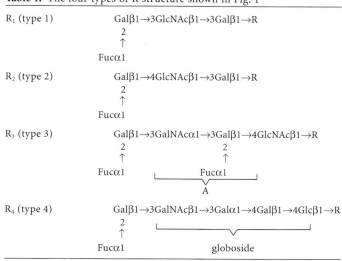

R₁ (type 1)	Galβ1→3GlcNAcβ1→3Galβ1→R

Databanks

Databanks

Histoblood groups A and B transferases, their gene structure, and common O group gene structure

NC-IUBMB enzyme classification: E.C.2.4.1.40 and 2.4.1.37

Species	Gene	Protein	mRNA	Genomic
Histoblood group A transferase				
Homo sapiens	–	–	J05175	–
Histoblood group B transferase				
Homo sapiens	–	AAD26581	AF134429	–
O group gene product				
Homo sapiens	–	AAD26584	AF134439	–

Name and History

Following the discovery of the ABO status of erythrocytes by Landsteiner in 1900, a great effort was made to elucidate the chemical properties of antigens ("blood group substances") on erythrocytes and in mucins. Because of the abundance, solubility, and ease of preparation of antigens in mucin, the essential structures of A and B determinants (see Fig. 1) were elucidated in mucin by Morgan and Watkins (e.g., Morgan and Watkins 1969), and by Kabat and associates (Kabat 1973), during the 1950s and early 1960s. In contrast, antigens on erythrocytes are insoluble in water, and they can only be obtained in very small quantities. They were eventually identified as being carried by glycosphingolipids (Yamakawa and Iida 1953; Koscielak 1963; Hakomori and Strycharz 1968) and transmembrane proteins (Fukuda et al. 1979). However, their characterization was only possible after methodologies for the separation of glycosphingolipids and the solubilization of transmembrane proteins became established in the 1970s (for a review, see Hakomori 1981). The characterization of four isotypes of A and B antigens became possible when techniques for monoclonal antibodies, nuclear magnetic resonance spectroscopy, and mass spectrometry were fully developed in the 1980s (for a review, see Clausen and Hakomori 1989).

Soon after the essential structures of A and B determinants were established in the 1960s, the enzymatic basis for genetically defined synthesis of the determinants was studied, mainly by Watkins and Ginsburg, and their associates. A concept for the genetic basis of the expression of ABO, H, and Lewis antigens, and their secretor status, was proposed in 1966 (Watkins 1966), and fully developed during the 1970s (for a review, see Watkins 1980). The concept was not substantiated until the genes for each glycosyltransferase responsible for the synthesis of ABH and Lewis antigens were cloned during the early 1990s, when the recombinant DNA technique became fully developed.

Enzyme Activity Assay and Substrate Specificity

The enzymatic activity of A and B transferase was determined using radiolabeled UDP-GalNAc or UDP-Gal as the sugar donor and H-glycolipid or 2-fucosyllactose as the acceptor. The activity was expressed in "units" defined as 1 μmole of product formed per minute from a defined substrate, as above. Examples of the procedure, using glycolipid or 2-fucosyllactose as substrate (Clausen et al. 1990), are described below.

Glycolipids

The α-GalNAc transferase activity was determined in reaction mixtures containing 10 mM Tris buffer (pH 7.4), 25 μg H_1 or H_2 type-2 chain substrate glycolipid, 2 μmol $MnCl_2$, 0.5 μmol CDP-choline, 40 μg Cutscum, 11 nmol UDP[^{14}C]GalNAc (22 816 cpm/nmole), and enzyme preparations as described below, in a total volume of 100 μl. Radioactive glycolipid products were located by autoradiography, scraped from the plate, and counted using a liquid scintillation counter. Identification of the reaction product was assessed by high-pressure thin-layer chromatography (HPTLC) immunostaining using anti-A mAbs with well-characterized specificity.

2-Fucosyllactose

Transferase activity was determined in the same reaction mixture as for the glycolipid assay, but with the omission of Cutscum and a lower specific activity of the sugar nucleotide (4000 cpm/nmole). The acceptor substrate 2′FL was used in concentrations of 5–10 nM, and product determination was by scintillation counting after Dowex-1 formic acid cycle chromatography.

Preparation

The cloning of the group A gene was based on the sequence of six peptides released by endopeptidase digestion of isolated A enzyme extracted from human lung, which was a glycoprotein with Mr 40 kDa (34 kDa for the nonglycosylated form). The glycoprotein was isolated by affinity binding of the enzyme to Sepharose 4B and elution with UDP, followed by repeated cation exchange chromatography on a mono-S column, and finally by reversed-phase chromatography on a C-8 column. After mono-S column chromatography, the purified A enzyme had 5.7 units/mg (~630 000× higher than crude extract) (Clausen et al. 1990). Based on the sequence information, degenerate oligodeoxynucleotides were used for a polymerase chain reaction (PCR) to detect, identify, and clone cDNA from A$^+$ gastric cancer cell line MKN45 initially, and subsequently from other A$^+$ cell lines. The cDNA clones encoding A enzyme, having a coding region of 1065 bp, had a common nucleotide sequence. Based on the nucleotide sequence, the A enzyme is a typical type-2 glycoprotein having a transmembrane domain close to the N-terminal region, a proline-rich stalk region, and a long C-terminal catalytic domain (Yamamoto et al. 1990b).

Several cDNAs encoding B transferase from various B cell lines were cloned and sequenced, based on their high homology with the A transferase gene (Yamamoto et al. 1990a). Four consistent differences in nucleotide substitution were found between cDNAs for A vs. B transferase (Fig. 2). cDNA almost identical to those encoding A or B transferase, except for the deletion of nucleotide 261G, was cloned from histoblood group O cells. The deleted nucleotide was originally reported as 258G for the reason described in the caption to Fig. 2. This deletion causes a shift of the reading frame which creates a new termination codon (TAA) at nucleotide number 353–355, and another termination codon TGA at nucleotide number 386–388, resulting in a polypeptide which is inactive and much shorter than the normal A or B enzyme (Yamamoto et al. 1990a) (Fig. 3).

Among the four nucleotide differences which create differences in four amino acid substitutions between A and B transferases, we need to know which differences are predominant in defining A and B activity. To answer this question, we constructed various cDNA chimeras with different combinations of substitutions. These constructs, in plasmids, were transfected to HeLa cells (genotype OO), and the expression of blood group A and B antigens was compared. The results indicate that the third and fourth nucleotide differences (i.e., nucleotides 796 and 803) and the amino acids encoded thereby (i.e., 266 and 268) are predominant in determining A or B transferase activity (Yamamoto and Hakomori 1990). This is presumably because amino acids 266 and 268 provide a binding site for UDP-GalNAc or UDP-Gal.

Biological Aspects

Expression of histoblood group A/B antigens is high in epithelial cells of gastrointestinal, esophageal, bronchopulmonary, oral, and urogenital tissues, absent in brain and muscle, and minimal in parenchymatous cells of liver, spleen, and kidney. Expression also changes in cancer tissue (Hakomori 1984) and during human ontogenesis

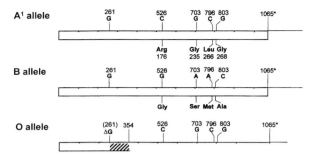

Fig. 2. Basic structural differences between A^1, B, and O alleles. A^1 and B alleles have a termination codon at nucleotide numbers 1063–65 counting from the initiation codon. The differences in nucleotide substitution and amino acid substitution between A^1 and B alleles are shown in the *top* and *bottom lines*, respectively. The differences in amino acid substitution in A and B enzymes are 176 Arg for A vs. 176 Gly for B, 235 Gly (A) vs. Ser (B), 266 Leu (A) vs. Met (B), and 268 Gly (A) vs. Ala (B). The common O allele, O^1, is characterized by the deletion of nucleotide 261 G, causing a frame shift resulting in a termination codon at 352–4, thus transcribing a much shorter and unrelated protein. For a further explanation, see Fig. 3. The *boxed areas* denote the translated A^1 consensus, Δ indicates the deletion of a nucleotide, and a *straight line* indicates nontranslated DNA, with its nucleotide number counting from the initiation codon. The nucleotide number reported originally, based on the sequence in clone FY-59-5 (isolated from human gastric cancer cell line MKN45) (Yamamoto et al. 1990b), was three less than in common A or B genes, since FY-59-5 is uniquely devoid of three nucleotides at the splicing acceptor site of exon 6. This does not modify the frame of the codons, but replaces two amino acids (Cys and Arg) with a single amino acid (Trp). Thus, all amino acid numbers of typical A and B enzymes are one higher than originally reported for FY-59-5 (Yamamoto et al. 1990b), as explained in our more recent paper (Yamamoto et al. 1995)

```
          234                    252           261           270
A/B    ACA CCG TGT AGG AAG GAT GTC CTC GTG GTG ACC CCT TGG CTG GCT
        T   P   C   R   K   D   V   L   V   V   T   P   W   L   A

O       *   *   *   *   *   *   *   *   *  GTA CCC CTT GGC TGG CTC
        *   *   *   *   *   *   *   *   *   V   P   L   G   W   L


          279         288         297         306         315
A/B    CCC ATT GTC TGG GAG GGC ACA TTC AAC ATC GAC ATC CTC AAC GAG
        P   I   V   W   E   G   T   F   N   I   D   I   L   N   E

O       CCA TTG TCT GGG AGG GCA CAT TCA ACA TCG ACA TCC TCA ACG AGC
        P   L   S   G   R   A   H   S   T   S   T   S   S   T   S


          324         333         342         351         359
A/B    CAG TTC AGG CTC CAG AAC ACC ACC ATT GGG TTA ACT GTG TTT GCC ---
        Q   F   R   L   Q   N   T   T   I   G   L   T   V   F   A
                                              354
O       AGT TCA GGC TCC AGA ACA CCA CCA TTG GGT TAA CTG TGT TTG CCA ---
        S   S   G   S   R   T   P   P   L   G   =  ──no translation──
```

Fig. 3. Deletion of 261 G in the most common O allele, causing a shift of the reading frame and transcription of an entirely different amino acid sequence (indicated by bold letters), with the termination codon at 352–4. An *asterisk* indicates the same nucleotide and amino acid sequence as for A and B alleles

(Szulman 1971). These findings indicate that gene expression is regulated differently in different types of cells, and that this regulation becomes aberrant under pathological conditions. To understand what regulates gene expression, under both normal and pathological conditions, it is essential to identify the regulatory mechanism. Often, expression is regulated at the transcription initiation step and the responsible *cis*-regulatory DNA elements as well as *trans*-acting factors can be identified.

Genomic Organization of ABO Genes

We studied the organization of the ABO gene, which spans ~19 kb of genomic DNA on chromosome 9, band q34 (Yamamoto et al. 1995). The locations of exons and introns were mapped, and the nucleotide sequences of the exon–intron boundaries were determined. The sequences of the boundaries are in accordance with the GT–AG rule (Reed and Maniatis 1986), and this rule seems to apply to all the splicing junctions of human ABO genes. Human ABO genes consist of at least 7 exons, and the coding sequences of these exons span over 18 kbp of the genomic DNA. Exons 1–5 range in size from 35 to 69 bp. Most of the coding sequence is present in exon 7, which has 688 bp. The hydrophobic region representing the transmembrane domain is present mostly in exon 2, the nucleotide G 261 whose deletion represents the O allele is located in exon 6, and catalytic domains are located in exon 7 (Yamamoto et al. 1995). Essentially the same genomic organization of the ABO locus, consisting of 7 exons, was identified by characterization of a single clone obtained by screening the P1 phage library using PCR primers FY57 and FY46 (Bennett et al. 1995). A typical genomic structure and an expression pattern of the transcript in various human organs are shown in Fig. 4. Interestingly, the transcripts of the ABO gene detected by reverse transcriptase–polymerase chain reaction (RT–PCR) of various human organs show significant variation. Highly heterogeneous transcripts were found in testis, homogeneous transcripts were found in pancreas and salivary glands, and homogeneous but minimal level of transcripts were found in liver. No transcripts were detectable in muscle, thymus, heart, brain, or placenta. The heterogeneity of transcripts found in testis was due to alternative exon usages. One transcript was missing only exon 6; others were missing exons 2, 3, 6, or exons 2, 3, 4, 5, 6 (Bennett et al. 1995).

Expression of Gene Regulation Through the Promoter Region

The 5' upstream sequence of the ABO gene was analyzed extensively in the gastric carcinoma cell line KATO III by Kominato et al. (1997). The sequence just upstream of the transcription initiation site (cap site), and the enhancer element, which is located further upstream between −3899 and −3618 bp from the transcription initiation site and contains four tandem copies of the 43-bp repeat unit, were responsible for the transcriptional activity of the ABO gene. DNA binding studies demonstrated that transcription factor CBF/NF-Y binds to the 43-bp repeat unit in the minisatellite. The functional importance of the CBF/NF-Y binding site in enhancer activity was confirmed by transfection of reporter plasmid with mutated binding sites. Thus, expression of the ABO gene is regulated by the binding of this transcription factor to the minisatellite (Kominato et al. 1997). Further studies indicated that the ABO gene promoter contains a CpG island whose methylation status is correlated with gene

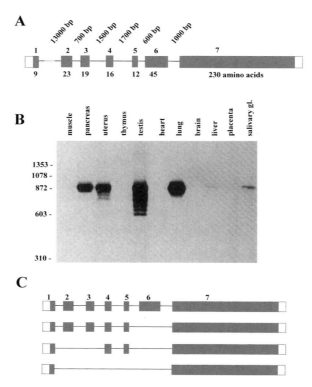

Fig. 4A–C. Genomic organization of the human *ABO* locus, and reverse transcriptase–polymerase chain reaction (RT–PCR) analysis of the *ABO* gene transcript in human organs. **A**, The exon and intron organization of a mature, common *ABO* locus. The size of the intron (bp) is shown between each exon, except between exons 1 and 2. The *number* beneath each exon indicates the number of amino acids encoded. The *ABO* gene has a single base excess in the first two small exons (1 and 2) and in the last coding exon (7). The size of an intron is based on the genomic clone (AC000397, deposited in GenBank). **B**, RT–PCR products probed with an exon 7-specific probe. Note the complete absence of transcript in muscle, thymus, heart, brain, and placenta, and the degree of heterogeneity expressed in each organ. **C**, Heterogeneity of testis transcripts, showing the exclusion or inclusion of some exons. Based on RT–PCR of testis transcripts

expression in various cell lines. CpG islands were hypomethylated in some cell lines showing high expression of A and B transcript, but were hypermethylated in others showing suppression of A and B transcript expression (Kominato et al. 1999), although the cell lines used in this study were genetically unrelated and their comparison is not justified. In a study using genetically comparable A$^+$ vs. A$^-$ variants from SW480 and HT29 colonic carcinoma cell lines, the deletion or reduction of A transcript in A$^-$ variants was correlated with reduced promoter activity of the CBF/NF-Y binding region and with enhanced DNA methylation of CpG islands at certain (but not all) A transferase promoter regions (Iwamoto et al. 2000).

References

Bennett EP, Steffensen R, Clausen H, Weghuis DO, van Kessel AG (1995) Genomic cloning of the human histo-blood group ABO locus. Biochem Biophys Res Commun 206:318–325

Clausen H, Hakomori S (1989) ABH and related histo-blood group antigens: immunochemical differences in carrier isotypes and their distribution. Vox Sang 56:1–20

Clausen H, White T, Takio K, Titani K, Stroud MR, Holmes EH, Karkov J, Thim L, Hakomori S (1990) Isolation to homogeneity and partial characterization of a histo-blood group A defined Fucα1→2Galα1→3-N-acetylgalactosaminyltransferase from human lung tissue. J Biol Chem 265:1139–1145

Fukuda MN, Fukuda M, Hakomori S (1979) Cell surface modification by endo-β-galactosidase: change of blood group activities and release of oligosaccharides from glycoproteins and glycosphingolipids of human erythrocytes. J Biol Chem 254:5458–5465

Hakomori S (1981) Blood group ABH and Ii antigens of human erythrocytes: chemistry, polymorphism, and their developmental change. Semin Hematol 18:39–62

Hakomori S (1984) Blood group glycolipid antigens and their modifications as human cancer antigens. Am J Clin Pathol 82:635–648

Hakomori S, Strycharz GB (1968) Investigations on cellular blood group substance. I. Isolation and chemical composition of blood group ABH and Le^b isoantigens of sphingoglycolipid nature. Biochemistry 7:1279–1286

Iwamoto S, Withers DA, Handa K, Hakomori S (1999) Deletion of A-antigen in a human cancer cell line is associated with reduced promoter activity of CBF/NF-Y binding region, and possibly with enhanced DNA methylation of A transferase promoter. Glycoconj J 16:659–666

Kabat EA (1973) Immunochemical studies on the carbohydrate moiety of water-soluble blood group A, B, H, Le^a, and Le^b substances and their precursor I antigens. In: H. Isbell (ed) Carbohydrates in solution. American Chemical Society, Washington, pp 334–361 (Adv Chemistry Series 117)

Kominato Y, Tsuchiya T, Hata N, Takizawa H, Yamamoto F (1997) Transcription of human ABO histo-blood group genes is dependent upon binding of transcription factor CBF/NF-Y to minisatellite sequence. J Biol Chem 272:25890–25898

Kominato Y, Hata Y, Takizawa H, Tsuchiya T, Tsukada J, Yamamoto F (1999) Expression of human histo-blood group ABO genes is dependent upon DNA methylation of the promoter region. J Biol Chem 274:37240–37250

Koscielak J (1963) Blood-group A-specific glycolipids from human erythrocytes. Biochim Biophys Acta 78:313–328

Morgan WTJ, Watkins WM (1969) Genetic and biochemical aspects of human blood group A-, B-, H-, Le^a- and Le^b-specificity. Br Med Bull 25:30–34

Oriol R (1995) ABO, Hh, Lewis, and secretion: serology, genetics, and tissue distribution. In: Cartron JP, Rouger P (eds) Molecular basis of human blood group antigens. Plenum Press, New York, pp 37–73 (Blood cell biochemistry, vol 6)

Oriol R, Le Pendu J, Mollicone R (1986) Genetics of ABO, H, Lewis, X and related antigens. Vox Sang 51:161–171

Reed R, Maniatis T (1986) A role for exon sequences and splice-site proximity in splice-site selection. Cell 46:681–690

Szulman AE (1971) The histological distribution of the blood group substances in man as disclosed by immunofluorescence. IV. The ABH antigens in embryos at the fifth week post-fertilization. Hum Pathol 2:575–585

Watkins WM (1966) Blood-group substances. Science 152:172–181

Watkins WM (1980) Biochemistry and genetics of the ABO, Lewis, and P blood group systems. In: Harris H, Hirschhorn K (eds) Advances in human genetics, vol 10. Plenum Press, New York, pp 1–136

Yamakawa T, Iida T (1953) Immunochemical study on the red blood cells. I. Globoside, as the agglutinogen of the ABO system on erythrocytes. Jpn J Exp Med 23:327–331

Yamamoto F, Hakomori S (1990) Sugar–nucleotide donor specificity of histo-blood group A and B transferases is based on amino acid substitutions. J Biol Chem 265:19257–19262

Yamamoto F, Clausen H, White T, Marken J, Hakomori S (1990a) Molecular genetic basis of the histo-blood group ABO system. Nature 345:229–233

Yamamoto F, Marken J, Tsuji T, White T, Clausen H, Hakomori S (1990b) Cloning and characterization of DNA complementary to human UDP-GalNAc:Fucα1→2Galα1→3GalNAc transferase (histo-blood group A transferase) mRNA. J Biol Chem 265:1146–1151

Yamamoto F, McNeill PD, Hakomori S (1995) Genomic organization of human histo-blood group ABO genes. Glycobiology 5:51–58

Histoblood Group A Variants, O Variants, and Their Alleles

Introduction

Histoblood group A phenotypes showing weaker agglutination, by anti-A antibodies or lectin (e.g., *Dolichos biflorus*), than regular A phenotypes (A_1) are called collectively "weak A," and include A_2, A_3, A_x, A_m, A_{el}, etc. (Race and Sanger 1975). The A_2 phenotype is common in Caucasians (average incidence 10%–15%), and particularly in Scandinavians (up to 25%), and very rare in Asian populations (to date, only studied in Japanese). It is sometimes difficult to distinguish the A_2 from the O phenotype when A_2 agglutination by anti-A antibodies is very weak. A_3, A_x, A_m, and A_{el} are very rare, even in Caucasians. "Weak B" phenotypes are also very rare. B_x, which is comparable to A_x, is extremely rare in Caucasians, and slightly more common (but still rare) in Japanese (Yamaguchi et al. 1970). The presence of O variants has become clear since the O gene was sequenced, and the genetic structures of the variants have been elucidated (Olsson et al. 1998).

Databanks

Histoblood group A variants, O variants and their alleles

NC-IUBMB enzyme classification: E.C.2.4.1.40 and 2.4.1.37

Species	Gene	Protein	mRNA	Genomic
Histoblood group A2 transferase				
Homo sapiens	–	–	S44054	AH007588
ABO A201 allele				
Homo sapiens	–	–	–	AF134421 (exon 6)
				AF134422 (exon 7)
				(AF170893)

Sen-itiroh Hakomori

Pacific Northwest Research Institute, 720 Broadway, Seattle, WA 98122-4327, USA, and Departments of Pathobiology and Microbiology, University of Washington, Seattle, WA, USA
Tel. +1-206-726-1222; Fax +1-206-726-1212
e-mail: hakomori@u.washington.edu

(Continued)

Species	Gene	Protein	mRNA	Genomic
O^{lv} allele				
Homo sapiens	–	–	–	AF016623
Ax ($B-O^{lv}$ allele)				
Homo sapiens	–	–	–	AF016624
				AF016625
A^{el} allele				
Homo sapiens	–	–	–	AF170889 (exon 7)
O^{l} allele				
Homo sapiens	–	–	–	AF170892 (exons 6 and 7)

Name and History

The A_1/A_2 distinction was discovered in 1911 (von Dungern and Hirszfeld 1911) and the A^2 gene was assumed to be distinguishable from the A^1 gene based on family studies (Race and Sanger 1975). A^1 and A^2 genes are considered to define the expression rate of A antigen and A transferase, since A_1 individuals express a larger quantity of A antigen in erythrocytes and secretions than A_2 individuals. There was a longstanding debate about whether the distinction is qualitative or simply quantitative, i.e., whether A_1- or A_2-specific antigens do or do not exist (Moreno et al. 1971; Kisailus and Kabat 1978). However, A_1 transferase is quantitatively and qualitatively different from A_2 transferase in terms of optimal pH, metal requirement, and pI value, and the two enzymes can be separated by isoelectric focusing (Schachter et al. 1973). A clear structural distinction between A_1 and A_2 antigens was possible only after the introduction of the monoclonal antibody approach, and the full development of carbohydrate analysis by ^{1}H-NMR and mass spectrometry/methylation analysis. A_1-specific antigen was discovered as "repetitive A" (type 3 chain A) defined by mAb TH1 (Clausen et al. 1985), and/or globo A (type 4 chain A; A^x) defined by mAb HH5 (Clausen et al. 1984), although the latter is an extremely minor component in erythrocytes. A-associated H (type 3 chain H) (Clausen et al. 1986b) and/or globo H (type 4 chain H) (Bremer et al. 1984) are considered to be A_2-specific. Type 3 chain A and H, and type 4 chain A and H are associated only with glycosphingolipids, not glycoprotein.

Enzyme Activity Assay and Substrate Specificity

The A_1 enzyme is capable of catalyzing the conversion of A-associated H (type 3 chain H) to "repetitive A" (type 3 chain A) (Fig. 1, left), or the conversion of globo H (type 4 chain H) to globo A (type 4 chain A; A^x) (Fig. 1, right), under the conditions described below, whereas the A_2 enzyme is incapable of catalyzing these conversions (Clausen et al. 1986a).

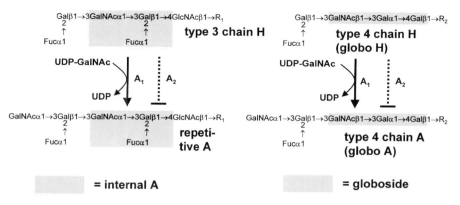

Fig. 1. Distinction between A_1 and A_2 transferase activity. A_1 transferase converts type 3 chain A to repetitive A, and type 4 chain H (globo H) to type 4 chain A (globo A). In contrast, A_2 transferase is incapable of catalyzing these conversions (*dotted line*). Type 3 chain H and repetitive A are characterized by the presence of an internal A epitope (indicated by the *shaded box*)

The reaction mixture contained 10 mM Tris buffer (pH 7.4), 25 or 12 μg glycolipid as above, 2 μmol $MnCl_2$, 0.5 μmol CDP-choline, 40 μg Cutscum, 11 nmol UDP-[^{14}C]GalNAc (22 816 cpm/nmol), and ~5 mg protein (with enzyme activity) in a total volume of 100 μl. In order to obtain a glycolipid substrate, a chloroform-methanol solution of glycolipid and Cutscum was mixed and evaporated to dryness under a nitrogen stream in a conical tube. A Tris buffer solution containing CDP-choline and $MnCl_2$ as above was added and sonicated extensively, followed by the addition of UDP-[^{14}C]GalNAc and enzyme protein. The reaction mixture was incubated for 3 h at 37°C, the reaction was terminated by the addition of 100 μl chloroform-methanol 2:1, and the mixture was quantitatively transferred to Whatman 3MM paper as a streak, dried, and irrigated with water to eliminate water-soluble components. Glycolipids remained at the original streak and were extracted from paper by chloroform –methanol 2:1 and separated by high-performance thin-layer chromatography. Radioactivity corresponding to repetitive A or globo A was counted.

Preparation

Weak A Variants

A^2 Variant

A comparison of nucleotide sequences and deduced amino acid sequences between A_1 and A_2 transferases, illustrating the essential differences between them, are shown in Fig. 2. The A^1 allele has a termination codon, TGA, at bases 1063–65; the C-terminal amino acid is P-encoded by the last codon CCG. In the A^2 allele, any C at 1059–1061

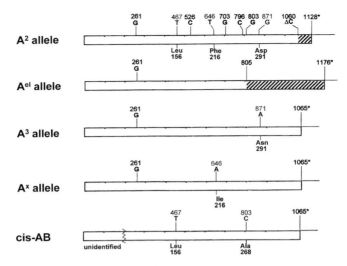

Fig. 2. Allelic structures of weak A (A^2, A^{el}, A^3, A^x) and *cis-AB*. Weak A alleles, A^2 and A^{el}, both show extension of the C-terminal region. The A^2 allele has the same nucleotide and amino acid sequence as A^1, but shows a single nucleotide deletion at 1060, which extends the termination codon to 1128. In contrast, the A^{el} allele has a nucleotide insertion at 805, which creates a different codon frame and extends the termination codon to nucleotide 1176. The extended C-terminal peptide may inhibit the function of two crucial amino acids, 266Leu and 268Gly, located close to the C-terminus of A_1 transferase. The A^3, A^x, and *cis-AB* alleles are character-ized by mutations at specific nucleotides, as shown. A^3 shows 871A as compared with 871G in the A^1 or A^2 allele, resulting in amino acid substitution Asn 291 instead of Asp 291. A^x has nucleotide 646A instead of 646T, and amino acid Ile 216 instead of Phe 216. *cis-AB* has nucleotides 467T (as in A^2) and 803C (as in *B*), resulting in coding of amino acids Leu 156 and Ala 268, respectively. Nucleotide 261G is not deleted in any of these alleles. An *asterisk* indicates translated A^1 consensus. Nucleotide and amino acid substitutions are given in the *top* and *bottom lines*, respectively. *Hatching* indicates an extended C-terminal peptide

(C–C–C; e.g., C1060) is deleted, thereby shifting the reading frame, deleting the ter-minal codon, and extending the C-terminus for 22 amino acids. Thus, the A_2 enzyme may be extended up to residue 375, with C-terminal F (Fig. 3) (Yamamoto et al. 1992). This suggested difference between A_1 and A_2 enzymes needs to be verified by direct comparison of the activities of the enzymes with and without the 22 amino acid exten-sion. If the hypothesis is correct, how does extension of the C-terminal region affect the substrate specificity of the enzyme? It is suggested that the extension may affect the catalytic function of three amino acids (235, 266, and 268). This idea is compati-ble with the structure for the A^{el} allele (see below).

Other Weak A Variants

The weak A variant phenotype A_x showed negative or weak agglutination with anti-A, but good agglutination with a mixture of anti-A and anti-B antibodies (for a review, see Race and Sanger 1975). The A^x allele has a single nucleotide substitution (T→A

Fig. 3. A possible mechanism for the creation of the A^2 allele. The A^1 allele has a termination codon (*underlined*) at 1063–1065. In contrast, the A^2 allele shows the deletion of C1060 or 1061, causing a shift of the reading frame, abolishing the termination codon at 1063–65, and introducing a new termination codon at 1126–28. This causes a 21-amino-acid extension at the C-terminal end, which may inhibit the function (binding to UDP-GalNAc) of amino acids at 266 and 268

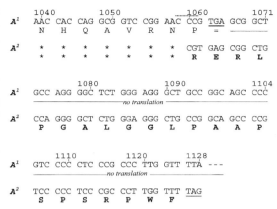

at nucleotide 646), resulting in an amino acid substitution (Phe→Ile) at position 216 (Yamamoto et al. 1993c). Another weak A subgroup, the A^3 allele, was identified as having a single base substitution (G 871 A) resulting in Asp 291→Asn (Yamamoto et al. 1993d). In these weak A and B phenotypes, it is not clear how a single substitution of the allele causes decreased A or B transferase activity and A or B antigen expression.

A very weak A (A_{el}) does not cause agglutination by anti-A antibody, but can be demonstrated only by elution of anti-A antibody bound to cells; the saliva of A_{el} secretors contains H but not A (for a review, see Race and Sanger 1975). Recently, the A^{el} allele was sequenced and found to contain a single nucleotide insertion at nucleotide 805, which alters the amino acid sequence of the A enzyme immediately after its essential catalytic site (a.a. 235–268, encoded by nucleotides 703–803), and extends the translated protein by 37 amino acids, i.e., 16 more amino acids than the A_2 enzyme at the C-terminal end (see Fig. 2) (Olsson et al. 1995).

Weak B Variants

Weak B variants B_3, B_x, B_m, B_{el} are known, and are comparable to A_3, A_x, A_{end}, and A_{el}; however, these B variants are all extremely rare (Issit 1985), and their genes are not elucidated. As an exception, weak B subgroup B_3 was studied and characterized by a single nucleotide substitution at 930A (Yamamoto et al. 1993d). How such a substitution causes weak B expression is unknown.

O Allele Variants

O variants were found only recently, in terms of allelic difference, by sequencing the O gene. Two types of O allele variants are known: $O^{1\ variant}$ and O^2 (Fig. 4).

$O^{1\ variant}$ Allele

A frequently occurring variant of the O^1 allele was recently described by Olsson and Chester (1996b) and termed $O^{1\ v}$ or $O^{1\ variant}$. This allele shows 9 nucleotide substitu-

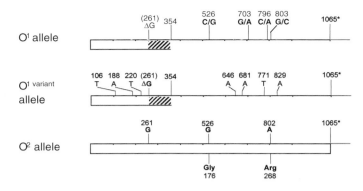

Fig. 4. Allelic structures of *O* variants (*O*[1 variant] allele and *O*[2] allele). The *O*[1 variant] allele showed 9 nucleotide substitutions in addition to the deletion of 261 G: these are 646 A, 681 A, 771 T, and 829 A in exon 7; 297 G in exon 6; 220 T in exon 5; 188 A and 189 T in exon 4; 106 T in exon 3. The substitutions at 189 and 297 are omitted in the drawing of the *O*[1 variant] allele. The mutation at 297 G is also found in the *B* and *O*[2] alleles. The *O*[2] allele shows the characteristic substitution of nucleotide 802 encoding Arg 268

tions in addition to consistent 261 G deletion (i.e., four in exon 7, one in exon 6, one in exon 5, two in exon 4, and one in exon 3). The positions of these substitutions, except for 189 T in exon 4 and 297 G in exon 6, are shown and explained in Fig. 2 and its legend. In addition to *O*[1], *O*[2], and *O*[1 variant] alleles, a new type of *O* allele was found resulting from a combination of *A*[2] nucleotide deletion and *A*[el] nucleotide insertion (see the following section regarding weak *A* alleles) (Olsson and Chester 1996a). In the indigenous populations of Mexico, the West Indies, and South America, >95% are O (e.g., Oriol 1995). These populations were recently claimed to have *O*[1 variant] rather than *O*[1], although the number of specimens examined was still limited (Olsson et al. 1998).

O[2] Allele

Nucleotide G 802 A mutation causes the replacement of Gly by Arg, resulting in a complete blocking of A and B enzymatic activity, since these amino acids may provide the site for the binding of the sugar nucleotide. There is no deletion of 261 G in the *O*[2] allele (see Fig. 2). This allele comprises about 5% of all *O* alleles (Yamamoto et al. 1993b; Grunnet et al. 1994).

cis-AB

Erythrocyte phenotypes showing weak A and weak B and their genes (e.g., *A*[2]*B*[3]) are inherited simultaneously on one chromosome, and called "*cis*-AB" in order to distinguish them from the ordinary AB phenotype and genotype (i.e., *A* and *B* genes are independent; "*trans*-AB") (Yamaguchi et al. 1965). Recently, two *cis*-AB alleles from two *cis*-AB individuals were analyzed, and characterized by the substitution of C567 (*A*[1]) by T, similar to the *A*[2] allele, and the presence of C503, similar to the *B* allele (Yamamoto et al. 1993a) (see Fig. 2).

References

Bremer EG, Levery SB, Sonnino S, Ghidoni R, Canevari S, Kannagi R, Hakomori S (1984) Characterization of a glycosphingolipid antigen defined by the monoclonal antibody MBr1 expressed in normal and neoplastic epithelial cells of human mammary gland. J Biol Chem 259:14773–14777

Clausen H, Watanabe K, Kannagi R, Levery SB, Nudelman ED, Arao-Tomono Y, Hakomori S (1984) Blood group A glycolipid (A^x) with globo-series structure which is specific for blood group A^1 erythrocytes: one of the chemical bases for A^1 and A^2 distinction. Biochem Biophys Res Commun 124:523–529

Clausen H, Levery SB, Nudelman ED, Tsuchiya S, Hakomori S (1985) Repetitive A epitope (type 3 chain A) defined by blood group A_1-specific monoclonal antibody TH-1: chemical basis of qualitative A_1 and A_2 distinction. Proc Natl Acad Sci USA 82: 1199–1203

Clausen H, Holmes E, Hakomori S (1986a) Novel blood group H glycolipid antigens exclusively expressed in blood group A and AB erythrocytes (type 3 chain H). II. Differential conversion of different H substrates by A_1 and A_2 enzymes, and type 3 chain H expression in relation to secretor status. J Biol Chem 261:1388–1392

Clausen H, Levery SB, Kannagi R, Hakomori S (1986b) Novel blood group H glycolipid antigens exclusively expressed in blood group A and AB erythrocytes (type 3 chain H). I. Isolation and chemical characterization. J Biol Chem 261:1380–1387

Grunnet N, Steffenson R, Bennett EP, Clausen H (1994) Evaluation of histo-blood group ABO genotyping in a Danish population: frequency of a novel O allele defined as O^2. Vox Sang 67:210–215

Issit PD (1985) Applied blood group serology. Montgomery Scientific, Miami, FL

Kisailus EC, Kabat EA (1978) Immunochemical studies on blood groups. LXVI. Competitive binding assays of A_1 and A_2 blood group substances with insolubilized anti-A serum and insolubilized A agglutinin from Dolichos biflorus. J Exp Med 147: 830–843

Moreno C, Lundblad A, Kabat EA (1971) Immunochemical studies on blood groups. LI. A comparative study of the reaction of A_1 and A_2 blood group glycoproteins with human anti-A. J Exp Med 134:439–457

Olsson ML, Chester MA (1996a) Evidence for a new type of O allele at the ABO locus, due to a combination of the A^2 nucleotide deletion and the A^{el} nucleotide insertion. Vox Sang 71:113–117

Olsson ML, Chester MA (1996b) Frequent occurrence of a variant O^1 gene at the blood group ABO locus. Vox Sang 70:26–30

Olsson ML, Thuresson B, Chester MA (1995) An A^{EL} allele-specific nucleotide insertion at the blood group ABO locus and its detection using a sequence-specific polymerase chain reaction. Biochem Biophys Res Commun 216:642–647

Olsson ML, Santos SE, Guerreiro JF, Zago MA, Chester MA (1998) Heterogeneity of the O alleles at the blood group ABO locus in Amerindians. Vox Sang 74:46–50

Oriol R (1995) ABO, Hh, Lewis, and secretion: serology, genetics, and tissue distribution. In: Cartron JP, Rouger P (eds) Blood cell biochemistry, vol 6. Molecular basis of human blood group antigens. Plenum Pres, New York, pp 37–73

Race RR, Sanger R (1975) Blood groups in man. Blackwell Scientific, Oxford

Schachter H, Michaels MA, Tilley CA, Crookston MC, Crookston JH (1973) Qualitative differences in the N-acetyl-D-galactosaminyltransferases produced by human A1 and A2 genes. Proc Natl Acad Sci USA 70:220–224

von Dungern E, Hirszfeld L (1911) Über gruppenspezifische Strukturen des Blutes: III Mitteilung. Z Immunitatsforsch Exp Ther 8:526–538

Yamaguchi H, Okubo Y, Hazama F (1965) An A_2B_3 phenotype blood showing atypical mode of inheritance. Proc Jpn Acad Sci 41:316–320

Yamaguchi H, Okubo Y, Tanaka M (1970) A rare blood B_x analogous to A_x in a Japanese family. Proc Jpn Acad Sci 46:446–449

Yamamoto F, McNeill PD, Hakomori S (1992) Human histo-blood group A^2 transferase coded by A^2 allele, one of the A subtypes, is characterized by a single base deletion in the coding sequence, which results in an additional domain at the carboxyl terminal. Biochem Biophys Res Commun 187:366–374

Yamamoto F, McNeill PD, Kominato Y, Yamamoto M, Hakomori S, Ishimoto S, Nishida S, Shima M, Fujimura Y (1993a) Molecular genetic analysis of the ABO blood group system. 2. *cis-AB* alleles. Vox Sang 64:120–123

Yamamoto F, McNeill PD, Yamamoto M, Hakomori S, Bromilow IM, Duguid JKM (1993b) Molecular genetic analysis of the ABO blood group system. 4. Another type of *O* allele. Vox Sang 64:175–178

Yamamoto F, McNeill PD, Yamamoto M, Hakomori S, Harris T (1993c) Molecular genetic analysis of the ABO blood group system. 3. A^x and $B^{(A)}$ alleles. Vox Sang 64:171–174

Yamamoto F, McNeill PD, Yamamoto M, Hakomori S, Harris T, Judd WJ, Davenport RD (1993d) Molecular genetic analysis of the ABO blood group system. 1. Weak subgroups: A^3 and B^3 alleles. Vox Sang 64:116–119

Forssman Glycolipid Synthase

Introduction

More than 200 distinct glycosphingolipids have been reported from a wide variety of eukaryotic sources. Variability in glycosphingolipid structure occurs primarily in the carbohydrate moiety owing to differences in the number, sequence, or way of linkage between monosaccharide residues. Forssman glycolipid (GalNAcα1-3GalNAcβ1-3Galα1-4Galβ1-4Glc-Cer, globopentaosylceramide) is a member of the globo series glycosphingolipid family, and is formed by the addition of GalNAc in α1,3-linkage to the terminal GalNAc residue of globoside (globotetraosylceramide). This reaction is catalyzed by Forssman glycolipid synthase (UDP-N-acetylgalactosamine:globoside α3-N-acetylgalactosaminyltransferase).

The cDNA of Forssman glycolipid synthase was cloned from an MDCK-cell cDNA library using an expresson cloning method (Haslam and Baenziger 1996). The isolated Forssman glycolipid synthase shows 42% identity in the amino acid sequence to the histoblood group ABO transferase (Yamamoto et al. 1990), and 35% identity to α3 galactosyltransferase (Joziasse et al. 1989). The A- or B-transferases transfer GalNAc or Gal in α1-3-linkage to the histo-H acceptor, respectively, and the α3 galactosyltransferase transfers Gal to the terminal Galβ1-4GlcNAc structure on glycoproteins as well as glycolipids. The close sequence identity and the similar enzyme reaction suggested that these glycosyltransferase genes have the same evolutionary origin (Haslam and Baenziger 1996). Furthermore, in humans, all these related glycosyltransferase genes are located on chromosome 9q34, supporting the hypothesis that they arose by gene duplication and subsequent divergence (Joziasse et al. 1992; Yamamoto et al. 1995; Xu et al. 1999).

KOICHI HONKE

Department of Biochemistry, Osaka University Medical School, 2-2 Yamadaoka, Suita, Osaka 565-0871, Japan
Tel. +81-6-6879-3421; Fax +81-6-6879-3429
e-mail: khonke@biochem.med.osaka-u.ac.jp

Databanks

Forssman glycolipid synthase

NC-IUBMB enzyme classification: E.C.2.4.1.88

Species	Gene	Protein	mRNA	Genomic
Homo spaiens	*FS*	–	AF163572	AC002319
			NM_021996	AC00164
			Hs.130783 (Unigene)	–
Canis familiaris	*FS*	–	U66140	–

Name and History

Forssman antigen (Forssman 1911), which is known as a heterophile antigen (Buchbinder 1935), is one of the most potent haptenic glycosphingolipids (Papirmeister and Mallette 1955). Its strucure was proven to be GalNAcα1-3GalNAcβ1-3Galα1-4Galβ1-4Glc-Cer (Siddiqui and Hakomori 1971). Forssman glycolipid synthase (EC 2.4.1.88) catalyzes the transfer of *N*-acetylgalactosamine in α1,3-linkage to the terminal GalNAc residue of globoside. Forssman glycolipid synthase activity has been demonstrated in various mammalian tissues (Kijimoto et al. 1974; Taniguchi et al. 1981).

A cDNA clone of canine Forssman glycolipid synthase was isolated through an expression cloning method using a monoclonal antibody against Forrsman antigen (Haslam and Baenziger 1996).

Enzyme Activity Assay and Substrate Specificity

A typical enzyme reaction is performed in a 100-μl reaction mixture containing 100 mM MES (pH 6.7), 10 mM $MnCl_2$, 5 μM UDP-[^{3}H]GalNAc, 20 μM globoside, and membrane extract as an enzyme source (Haslam and Baenziger 1996). After incubation at 37°C for 2 h, the reaction is terminated by the addition of 1 ml ice-cold water containing EDTA. Glycolipid products are separated from unincorporated sugar nucleotide using a Sep-Pak C18 catridge. The glycolipid fraction is then applied onto a TLC plate and developed with authentic Forssman glycolipid. The radioactivity corresponding to the standard glycolipid is determined by fluorography or a liquid scintillation technique. A unique assay method using anti-Forssman antibody is also employed (Taniguchi et al. 1982).

Canine spleen Forssman glycolipid synthase has a pH optimum at 6.7–6.9 and requires Mn^{2+} (Taniguchi et al. 1982).

Studies on substrate specificity for canine spleen Forssman glycolipid synthase indicate that the enzyme recognizes GalNAcβ1-3Gal-R structure (Taniguchi et al. 1982). The recombinant canine Forssman glycolipid synthase did not act on the histo-H acceptor, *N*-acetyllactosamine, LacCer, globotraiaosylceramide, or GM3 (Haslam and Baenziger 1996). In addition, it had no galactosyltransferase activity toward globoside, indicating that Forssman glycolipid synthase and SSEA-3 (Gal-Gb4Cer) synthase are distinct enzymes (Haslam and Baenziger 1996).

Preparation

Forssman glycolipid synthase was purified over 3500-fold in a 4% yield from a Triton X-100 extract of canine spleen microsomes by affinity chromatography on globoside acid–agarose (Taniguchi et al. 1982). The purified enzyme preparation showed two major bands with apparent molecular weights of 56000 and 66000 on SDS-PAGE under reduced conditions. Since these proteins are too large for the 347 amino acid protein predicted from the cloned cDNA (Haslam and Baenziger 1996), the genuine Forssman glycolipid synthase might have been hidden behind the co-purified proteins.

Recombinant canine Forssman glycolipid synthase was obtained from COS-1 cells transfected with the pFS-7 plasmid (Haslam and Baenziger 1996). The cDNA-introduced COS cells are homogenized by sonication. After the removal of nuclear fractions, membrane fractions are solubilized with 1% Triton X-100 and used as an enzyme source.

Biological Aspects

Forssman glycolipid is expressed in a tissue-specific and developmentally regulated fashion in many mammals (Willison and Stern 1978). However, a biological function for Forssman glycolipid has not been identified. Unlike many other mammalian species, humans do not normally produce Forssman glycolipid, but produce the precursor globoside, suggesting that human tissues lacks Forssman synthase activity. Although the human Forssman glycolipid synthase gene is indeed expressed ubiquitously and generates a highly homologous protein with the canine enzyme, the gene product shows no α1-3 N-acetylgalactosaminyltransferase activity (Xu et al. 1999). This finding confirmed the belief that human cells do not synthesize Forssman antigen, and indicates that Forssman glycolipid is dispensable in terms of physiological functions. Since human Forssman glycolipid synthase retains 83% amino acid sequence identity with the canine orthologue, and is expressed widely in human tissues, the human gene product may have another as yet unknown biochemical function which is different from glycosyltransferase activity. Glycosphingolipids serve as receptors for several pathogenic organisms (Karlsson 1989). Therefore, differences in Forssman glycolipid expression between humans and other species may contribute to variable host susceptibility to microbial pathogens (Xu et al. 1999). In fact, the globo series glycosphingolipids have been reported to be an attachment site for bacteria, viruses, and bacterial toxins (Strömberg et al. 1990; Brown et al. 1993; Lingwood 1993; Jacewicz et al. 1994).

Despite the absence of Forssman glycolipid synthase activity in normal human tissues, Forssman antigen is detected in human tumors (Kawanami 1972; Yoda et al. 1980; Mori et al. 1982; Fredman 1993). The molecular mechanism of how Forssman glycolipid emerges in human tumors is unknown. Some factors may activate the dormant Forssman glycolipid synthase in cancer tissues. Alternatively, other N-acetylgalactoaminyltransferases that are activated in tumors may synthesize Forssman glycolipid as a result of the looseness of substrate specificity. Future studies will be directed to this issue.

References

Brown KE, Anderson SM, Young NS (1993) Erythrocyte P antigen: cellular receptor for B19 parvovirus. Science 262:114–117

Buchbinder L (1935) Heterophile phenomena in immunology. Arch Pathol 19:841–880

Forssman J (1911) Die Herstellung hochwertiger spezifischer Schafhämolysine ohne Verwendung von Schafblut. Biochem Z 37:78–115

Fredman P (1993) Glycosphingolipid tumor antigens. Adv Lipid Res 25:213–234

Haslam DB, Baenziger JU (1996) Expression cloning of Forssman glycolipid synthase: a novel member of the histo-blood group ABO gene family. Proc Natl Acad Sci USA 93:10697–10702

Jacewicz MS, Mobassaleh M, Gross SK, Balasubramanian KA, Daniel PF, Raghavan S, McCluer RH, Keusch GT (1994) Pathogenesis of Shigella diarrhea. XVII. A mammalian cell membrane glycolipid, Gb3, is required but not sufficient to confer sensitivity to Shiga toxin. J Infect Dis 169:538–546

Joziasse DH, Shaper JH, Van den Eijnden DH, Van Tunen AJ, Shaper NL (1989) Bovine alpha-1-3-galactosyltransferase: isolation and characterization of a cDNA clone. Identification of homologous sequences in human genomic DNA. J Biol Chem 264:14290–14297

Joziasse DH, Shaper NL, Kim D, Van den Eijnden DH, Shaper JH (1992) Murine alpha 1,3-galactosyltransferase. A single gene locus specifies four isoforms of the enzyme by alternative splicing. J Biol Chem 267:5534–5541

Karlsson KA (1989) Animal glycosphingolipids as membrane attachment sites for bacteria. Annu Rev Biochem 58:309–350

Kawanami J (1972) The appearence of Forssman hapten in human tumor. J Biochem 72:783–785

Kijimoto S, Ishibashi T, Makita A (1974) Biosynthesis of Forssman hapten from globoside by α-*N*-acetylgalactosaminyltransferase of guinea pig tissues. Biochem Biophys Res Commun 56:177–184

Lingwood CA (1993) Verotoxins and their glycolipid receptors. Adv Lipid Res 25:189–211

Mori E, Mori T, Sanai Y, Nagai Y (1982) Radioimmuno-thin-layer chromatographic detection of Forssman antigen in human carcinoma cell lines. Biochem Biophys Res Commun 108:926–932

Papirmeister B, Mallette MF (1955) The isolation and some properties of the Forssman hapten from sheep erythrocytes. Arch Biochem Biophys 57:94–105

Siddiqui B, Hakomori S (1971) A revised structure for the Forssman glycolipid hapten. J Biol Chem 246:5766–5769

Strömberg N, Marklund BI, Lund B, Ilver D, Hamers A, Gaastra W, Karlsson KA, Normark S (1990) Host-specificity of uropathogenic *Escherichia coli* depends on differences in binding specificity to Gal alpha-1-4-Gal-containing isoreceptors. EMBO J 9:2001–2010

Taniguchi N, Yokosawa N, Narita M, Mitsuyama T, Makita A (1981) Expression of Forssman antigen synthesis and degradation in human lung cancer. J Natl Cancer Inst 67:577–583

Taniguchi N, Yokosawa N, Gasa S, Makita A (1982) UDP-*N*-acetylgalactosamine:globoside α-3-*N*-acetylgalactosaminyltransferase. Purification, characterization, and some properties. J Biol Chem 257:10631–10637

Willison KR, Stern PL (1978) Expression of a Forssman antigenic specificity in the preimplantation mouse embryo. Cell 14:785–793

Xu H, Storch T, Yu M, Elliott SP, Haslam DB (1999) Charcterization of the human Forssman synthase gene. An evolving association between glycolipid synthesis and host–microbial interactions. J Biol Chem 274:29390–29298

Yamamoto F, Marken J, Tsuji T, White T, Clausen H, Hakomori S (1990) Cloning and characterization of DNA complementary to human UDP-GalNAc:Fuc-α-1-2-Gal-α-1-3-GalNAc transferase (histoblood group A transferase) mRNA. J Biol Chem 265:1146–1151

Yamamoto F, McNeill PD, Hakomori S (1995) Genomic organization of human histoblood group ABO genes. Glycobiology 5:51–58

Yoda Y, Ishibashi T, Makita A (1980) Isolation, characterization, and biosynthesis of Forssman antigen in human lung and lung carcinoma. J Biochem 88:1887–1890

Fucosyltransferases

α2-Fucosyltransferases (FUT1, FUT2, and Sec1)

Introduction

The human FUT1 and FUT2 α2-fucosyltransferases corresponding to the H and Se enzymes transfer Fuc in α1,2-linkages onto the terminal galactose of lactosamine. Individuals of histoblood-group "O" express the H antigen under control of the *FUT1* gene on red cells and vascular endothelium, and the H antigen under control of the *FUT2* gene in exocrine secretions.

Databanks

α2-Fucosyltransferases (FUT1, FUT2 and Sec1)

NC-IUBMB enzyme classification: E.C.2.4.1.69

Species	Gene	Protein	mRNA	Genomic	Clone/strain
FUT1					
Mus musculus	–	O09160	U90553	–	–
Rattus norvegicus	–	BAA31130	AB015637	–	–
Oryctolagus cuniculus	–	Q10979	X80226	–	–
Sus scrofa	–	O19101	U70883	–	–
Bos taurus	–	AAF07933	AF186465	–	–
Eulemur fulvus	–	AAF14063	AF045546	–	–
Callithrix jacchus	–	AAF42965	AF111936	–	–
Saimiri sciureus	–	AAF25584	AF136647	–	–
Macaca fascicularis	–	AAF42967	AF112474	–	–
Macaca mulatta	–	AAF14069	AF080607	–	–
Chlorocebus aethiops	–	O77711	D87932	–	–
Hylobates lar/agilis	–	AAF14062	AF045545	–	–
Pongo pygmeus	–	AAF42964	AF111935	–	–
Gorilla gorilla	–	AAF14067	AF080605	–	–

RAFAEL ORIOL and ROSELLA MOLLICONE

INSERM U504, 16 Av. Paul Vaillant-Couturier, 94807 Villejuif, France
Tel. +33145595041; Fax +33146770233
e-mail: oriol@infobiogen.fr

(Continued)

Species	Gene	Protein	mRNA	Genomic	Clone/strain
Pan troglodytes	–	AAF14065	AF080603	–	–
Homo sapiens	–	P19526	M35531	–	–
FUT2					
Mus musculus	–	AAF45146	AF214656	–	–
Rattus norvegicus	–	O35087	AB006138	–	–
Oryctolagus cuniculus	–	Q29505	X91269	–	–
Sus scrofa	–	O19100	U70881	–	–
Bos taurus	–	Q28113	X99620	–	–
Eulemur fulvus	–	AAF25583	AF136646	–	–
Callithrix jacchus	–	AAF25582	AF136645	–	–
Macaca mulatta	–	AAF25581	AF136644	–	–
Chlorocebus aethiops	–	O77712	D87934	–	–
Hylobates lar/agilis	–	AAF25585	AF136648	–	–
Pongo pygmeus	–	O77487	AB015636	–	–
Gorilla gorilla	–	AAF14068	AF080606	–	–
Pan troglodytes	–	O77485	AB015634	–	–
Homo sapiens	–	Q10981	U17894	–	–
Sec1					
Mus musculus	–	AAF45147	AF214657	–	–
Rattus norvegicus	–	AAD24470	AF131239	–	–
Oryctolagus cuniculus	–	Q10983	X80225	–	–
Sus scrofa	–	CAB36074	U70882	–	–
Bos taurus	–	AAF03411	AF187851	–	–
Callithrix jacchus	–	AAF42966	AF111938	–	–
Saimiri sciureus	–	AAF42966	AF111937	–	–
Macaca fascicularis	–	AAF42968	AF112475	–	–
Macaca mulatta	–	AAF14070	AF080608	–	–
Chlorocebus aethiops	–	CAB20642	D87933	–	–
Hylobates lar/agilis	–	BAA21879	AB006609	–	–
Pongo pygmeus	–	BAA21880	AB006610	–	–
Gorilla gorilla	–	(*pseudogene*)	AB006611	–	–
Pan troglodytes	–	(*pseudogene*)	AB006612	–	–
Homo sapiens	–	(*pseudogene*)	U17895	–	–
(invertebrates and bacteria)					
Caenorhabditis elegans	–	O61739	AF067211_1	–	B0205.4
C. elegans	–	P34302	L6559_6	–	C06E1.7
C. elegans	–	O44669	AF039051_3	–	C14C6.3
C. elegans	–	O16533	AF016654_3	–	C17A2.4
C. elegans	–	O61922	AF068708_6	–	C18G1.8
C. elegans	–	P91200	U80026_1	–	EGAP9.2
C. elegans	–	P91201	U80026_2	–	EGAP9.3
C. elegans	–	O62141	Z99710_6	–	F08A8.5
C. elegans	–	O17784	Z92830_5	–	F11A5.5
C. elegans	–	O45376	Z81066_4	–	F17B5.4
C. elegans	–	O17129	AF024503_7	–	F31F4.11
C. elegans	–	O17127	AF024503_5	–	F31F4.17
C. elegans	–	O45510	Z81537_6	–	F41D3.6
C. elegans	–	O17107	AF024500_4	–	K06H6.6
C. elegans	–	Q10017	U28409_2	–	T25D10.1

(Continued)

Species	Gene	Protein	mRNA	Genomic	Clone/strain
C. elegans	–	O45837	Z81132_3	–	T26E4.3
C. elegans	–	O45839	Z81132_6	–	T26E4.4
C. elegans	–	CAB3440	Z81132_5	–	T26E4.5
C. elegans	–	CAB04857	Z82056_6	–	T26H5.8
C. elegans	–	CAB07284	Z92813_2	–	T28A8.2
C. elegans	–	O01660	AF000198_4	–	T28F2.1
C. elegans	–	Q23217	Z78018_2	–	W07G4.2
C. elegans	–	AAF59635	AC006810	–	Y5H2B.i
Leishmania major	–	O60971	AC003011_11	–	L549.10
Yersinia enterocolitica	–	Q56870	U46859_13	–	ORF11.8
Lactococcus lactis cremoris	–	O06036	U93364_11	–	EPSH
Helicobacter pylori	–	O24919	AE000531_3	–	0093/0094
H. pylori	–	AAC99764	AF07679	–	UA802
H. pylori	–	AAD29865	AF093829_1	–	UA1182
H. pylori	–	AAD29868	AF093832_1	–	UA1207
H. pylori	–	AAD29869	AF093833_1	–	UA1210
H. pylori	–	AAD29867	AF093831_1	–	UA1218
H. pylori	–	AAD29863	AF093828	–	UA1234
H. pylori	–	AAD05663	AE001447	–	0086/0094
Vibrio cholerae	–	O87157	AB012957_18	–	ORF22-18
Bacterioides fragilis	–	AAD40713	AF048749	–	wcfW
(plants, expected to use xyloglucans as acceptors)					
Arabidopsis Thaliana	–	AAF15914.1	AC011765_10	–	F1M20.10
A. Thaliana	–	AAD22285.1	AC006920_9	–	F26H6.9
A. Thaliana	–	AAD22287.1	AC006920_11	–	F26H6.11
A. Thaliana	–	AAD22289.1	AC006920_13	–	F26H6.13
A. Thaliana	–	AAD39292.1	AC007576_15	–	F7A19.15
A. Thaliana	–	AAD39293.1	AC007576_16	–	F7A19.16
A. Thaliana	–	AAD39294.1	AC007576_18	–	F7A19.18
A. Thaliana	–	Q9XI77	AC007576_19	–	F7A19.19
A. Thaliana	–	AAD41092.1	AF154111_1	–	T18E12
A. Thaliana	–	O81053	AC005313.2	–	T18E12.12
Pisum sativum	–	AAF62896.1	AF223643	–	taxon 3888

Name and History

In the original definition of the ABO major blood groups, Karl Landsteiner called the blood groups of the individuals with one of these antigens on red cells "A" and "B," and the blood groups of the individuals devoid of A and B antigens "O" or zero (Landsteiner 1901). Later, AB individuals with both A and B antigens, inherited in a co-dominant way, were identified (Decastello and Sturli 1902), but no specific antigen structure was ascribed to the O individuals for some time. Half a century later, the antigen expressed by the red cells of the individuals with this null ABO phenotype was called the H antigen when it was established that it is the fucosylated precursor of the A and B oligosaccharide structures (Morgan and Watkins 1948; Watkins and Morgan 1955). The existence of a human *H-h* genetic polymorphism was established by the discovery, in India (Bombay), of an individual devoid of the H antigen on red cells who had antibodies in plasma reacting with all the normal red cell ABO pheno-types (Bhende et al. 1952).

In the early 1920s, the ABO red cell antigens were also identified in exocrine secre-tions, such as saliva, of 80% of the Caucasian population, and this saliva secretory character of ABH was found to be inherited as a dominant Mendelian trait (Schiff and Sasaki 1932). The remaining 20% of nonsecretor individuals have very low levels of ABH antigens in saliva (less than 5% of the normal amounts found in saliva of secre-tors), but they express normal ABH antigens on red cells. In order to explain the fact that ABH antigens in saliva are independent of the expression of ABH antigens on red cells, a genetic model was proposed by Morgan and Watkins, who postulated the exis-tence of three genes: a single *H* structural gene controlled by two regulatory genes, *Se* in saliva and *Z* in red cells (reviewed in Watkins 1980). This regulatory model was confirmed by the pedigree of an *H*-deficient child of a red-cell-*H*-deficient, salivary ABH nonsecretor mother and a father with normal *H* on red cells, but who was also a nonsecretor of ABH in saliva (Levine et al. 1955).

In 1981, we suggested that the *H* and *Se* genes are both structural genes each encod-ing for a different α2-fucosyltransferase (Oriol et al. 1981a). This two-gene model was incompatible with the *H*-deficient family mentioned above, and we suspected that this family was a nonpaternity case. In our structural genes model, we postulated that the *H* gene encodes for an α2-fucosyltransferase expressed in red cells, and the *Se* gene encodes for another α2-fucosyltransferase expressed in exocrine secretions (Oriol et al. 1981b), and we showed that both enzymes have different acceptor specificity patterns and different kinetics (Le Pendu et al. 1985).

Ten and 5 years ago, the elegant work of the team of J.B. Lowe led to the cloning and expression of the two predicted α2-fucosyltransferase genes: *H* or *FUT1* (Rajan et al. 1989; Larsen et al. 1990) encoding for the α2-fucosyltransferase expressed on red cells, and *Se* or *FUT2* (Kelly et al. 1995; Rouquier et al. 1995) encoding for the α2-fucosyltransferase expressed in exocrine secretions. These two structural genes have been localized in the long arm of chromosome 19 (19q13.3) (Reguigne-Arnould et al. 1995) closely linked to an inactive pseudogene that has been called *Sec1* (Rouquier et al. 1995). Anthropoid apes have the same *FUT1* and *FUT2* genes plus the *Sec1* pseudo-gene, which has probably been inactivated by a nonsense mutation in the codon cor-responding to the amino acid 325 of the ancestor gene of man, gorilla, and chimpanzee *Sec1*, inducing a premature stop codon and the loss of the 23 C-terminal amino acids of the enzyme. The human *Sec1* pseudogene has in addition a deletion of two bases (GG between positions 668 and 671), and the gorilla *Sec1* pseudogene has in addition an insertion of one base (C in position 612 or 613). Both of these two species-specific, and therefore more recent, mutations induce premature frameshifts that mask the common original nonsense inactivating mutation which occurred in the ancestor of chimpanzee, gorilla, and man *Sec1* (Apoil et al. 2000).

The other primates and lower mammals have an active *Sec1* gene. Therefore, three structural genes encoding for three different α2-fucosyltransferases have been found in the animal kingdom. Genes encoding for putative enzymes with the three con-served peptide motifs specific to α2-fucosyltransferases (Breton et al. 1998) have also been found in bacteria, invertebrates (Oriol et al. 1999), and plants (Perrin et al. 1999; Faik et al. 2000), but the plant enzymes use a xyloglucan as the acceptor substrate (Galβ1,2Xylα1,6Glcβ-R).

Enzyme Activity Assay and Substrate Specificity

The α2-fucosyltransferases are inverting enzymes using the same donor substrate GDP-β-L-Fuc to add an α1-2 fucose onto the terminal galactose of the disaccharide acceptor substrates, and they use as cofactor divalent cations such as Mn^{2+} or Mg^{2+}. There are two main natural lactosamine acceptor substrates: type 1 (Galβ1-3GlcNAcβ-R) and type 2 (Galβ-4GlcNAcβ-R). However, these enzymes can also work on type 3 (Galβ1-3GalNAcα-R) or type 4 (Galβ1-3GalNAcβ-R), although less efficiently.

The H type 1 (Fucα1-2Galβ1-3GlcNAcβ-R) and the H type 2 (Fucα1-2Galβ1-4GlcNAcβ-R) glycotopes are the main products of FUT1 and FUT2 enzymes, and they constitute the natural acceptor substrates for the A (αGalNAc transferase) and the B (αGal transferase) making the major A and B histoblood group oligosaccharide antigens.

The α2-fucosyltransferases can be assayed in biological fluids such as plasma or milk, in tissue extracts, or in COS cells transfected with the corresponding *FUT1* and *FUT2* cDNA clones. Different acceptors can be used, but the more common one is phenyl-β-D-galactoside (Sigma–Aldrich USA), which is a good acceptor for most α2-fucosyltransferases and gives different K_m values for FUT1 and FUT2. The type 1 or type 2 disaccharide acceptor substrates (Galβ1,3/4GlcNAc) can also be used, but these acceptors, in contrast to phenyl-β-D-galactoside, give similar K_m values for FUT1 and FUT2, either partially purified or in crude extracts. Some α2-fucosyltransferases which do not use the above-mentioned acceptors can be assayed with the asialo-GM1 glycolipid as acceptor (Barreaud et al. 2000).

The α2-fucosyltransferase assays are performed on ice in a final volume of 50 μl and contain 25 μl plasma or 5–50 μg protein cell extracts in 1% Triton X-100, 5 mM of TRIS/HCl 50 mM pH 7.2 (plasma) or 25 mM cacodylate buffer pH 6 (cell extracts), 5 mM ATP, 10 mM α-L-fucose, 3.5 μM of GDP-[C^{14}]-fucose (300 mCi/mmol, 80 000 cpm, Amersham-Pharmacia-Biotech UK), and the oligosaccharide acceptors. With phenyl-β-D-galactoside or the above-mentioned disaccharides synthesized with a hydrophobic aglycone such as biotin (Syntesome, Munich, Germany) or the methylcarbonyloctyl aglycone-$(CH_2)_8$-COOCH$_3$ (linking arm of Prof. R.U. Lemieux), hydrophobic interaction chromatography can be used to separate the reaction products. In this method, the product of the reaction is retained by, and then eluted from, a Sep-Pack C_{18} reverse chromatography cartridge (Waters, Milford, USA), due to the hydrophobicity of the aglycone (Palcic et al. 1988). The mixtures are incubated for 1 or 2 h at 37°C, and the reactions are stopped by the addition of 3 ml cold water, centrifuged, and the supernatant is applied to the conditioned Sep-Pak C_{18} cartridge. The elution of the radiolabeled reaction product is performed with 2 × 5 ml methanol, collected directly into scintillation vials, and counted in a liquid scintillation counter with one volume of Instagel-Plus (Packard, Illinois, USA).

When nonhydrophobic free oligosaccharide acceptors are used (lacto-N-bioseI or N-acetyllactosamine), the reaction mixture is applied to a small column containing the anion-exchange resin Dowex 1 × 2 (400 mesh, formate form) (Weston et al. 1992). The final product is recovered in the column flow-through fractions and quantified by liquid scintillation counting.

Table 1. Comparative apparent K_m of human α2-fucosyltransferase activities

Substrate	K_m (mM)		References
	H or FUT1	Se or FUT2	
Phenyl-β-galactoside	2.7; 3.0	8.4; 11.5	Sarnesto et al. 1990, 1992
	3.1	15.1	Rajan et al. 1989
	1.4	10	Le Pendu et al. 1985
	4.6; 6.4	46	Kumazaki and Yoshida 1984
	2.4[a]	ND	Larsen et al. 1990
	3[a]	ND	Fernandez-Mateos et al. 1998
	ND	10.5[a]	Henry et al. 1996b
	ND	11.5[a]	Kelly et al. 1995
Lacto-N-bioseI	2	1	Le Pendu et al. 1985
(Galβ1,3GlcNAc)	3.5-O-Me	1.4	Sarnesto et al. 1992
	ND	3.6[a]	Kelly et al. 1995
Lacto-N-tetraose	2.2	1.4	Sarnesto et al. 1992
(Galβ1,3GlcNAc)	ND	1.6	Kumazaki and Yoshida 1984
N-acetyllactosamine	1.9	5.7	Sarnesto et al. 1992
(Galβ1,4GlcNAc)	2.5	36	Kumazaki and Yoshida 1984
	ND	3.8[a]	Kelly et al. 1995
GDP-fucose	0.016; 0.018	0.123	Rajan et al. 1989
	0.027	0.108	Sarnesto et al. 1992
	ND	0.197[a]	Kelly et al. 1995
	ND	0.125[a]	Henry et al. 1996b
	0.029[a]	ND	Fernandez-Mateos et al. 1998

[a] Measured in extracts of COS cells transfected with *FUT1* or *FUT2* cDNA constructs

For kinetic assays, the concentration ranges of the acceptors are: phenyl-β-D-galactoside, 0–170 mM; lacto-N-bioseI, 0–10 mM; N-acetyllactosamine, 0–12.5 mM. These assays are performed at pH 6 using 3.5 µM of GDP-[C^{14}]-fucose and a saturation concentration of total GDP-fucose. To determine the apparent K_m for GDP-fucose, the 3.5 µM of GDP-[C^{14}]-fucose is supplemented with different amounts of unlabeled GDP-fucose to achieve final concentrations from 3.5 to 400 µM. The GDP-fucose K_m determination is evaluated in the presence of 25 mM phenyl-β-D-galactoside (Table 1). Apparent Michaelis constants are derived from Lineweaver–Burk plots of substrate concentration-rate determinations.

Preparation

Individuals with the *H/–, se/se* genotype express only the H or FUT1 enzyme in plasma, whereas individuals with the *h/h, Se/–* genotype express only the Se or FUT2 enzyme in plasma and saliva. These differential expression patterns have been used to test for FUT1 and FUT2 activities in plasma and saliva (Le Pendu et al. 1985).

Both FUT1 (Sarnesto et al. 1992) and FUT2 (Sarnesto et al. 1990) have been partially purified from human serum (5–10l) by hydrophobic chromatography on phenyl-Sepharose, ion exchange chromatography on sulfopropyl-Sepharose, and affinity chromatography on GDP-hexanolamine-Sepharose. The final purification step was achieved by high-pressure liquid chromatography gel-filtration and sodium-dodecyl-sulfate polyacrylamide gel electrophoresis of the radiolabeled proteins. The purified native FUT1 enzyme had an apparent molecular weight of about 200 kDa, and the native FUT2 150 kDa, while under reducing conditions both had an apparent molecular weight of about 50 kDa. However, the overall purification procedure is rather complex, the yields are low, and the purified enzymes have poor stability. Now that the corresponding genes have been cloned, the most promising methodology for the preparation of FUT1, FUT2, or animal Sec1 α2-fucosyltransferases is the expression of the recombinant enzymes in transfected cells.

Biological Aspects

Type 2 ABH glycoproteins made by FUT1 are intrinsic antigens of the human red cell membrane, while type 1 ABH glycolipids made by FUT2 are adsorbed at the surface of erythrocytes from plasma. In epithelial cells making the exocrine secretions both type 1 and type 2 H glycotopes are present, but the H structures of type 1 are predominant. Lower monkeys and mammals are devoid of H antigens on red cells, but express H antigens in exocrine secretions (Oriol et al. 1992).

Comparison of the protein sequences of FUT1, FUT2, and Sec1 enzymes of different animal species allowed us to identify three conserved peptide motifs which are present in all the α2-fucosyltransferases and constitute a signature of these types of enzyme (Breton et al. 1998; Oriol et al. 1999). The presence of these highly conserved sequence motifs and their highly conserved relative positions within the protein sequences of α2-fucosyltransferase enzymes suggested that these enzymes have appeared by sequential gene duplications of a common ancestor gene, followed by divergent evolution (Fig. 1). This phylogenetic tree also illustrates that the FUT2 and Sec1 families of enzymes belong to the same branch, and are more closely related to each other than to FUT1. The distance analysis of the same enzymes from different animal species within each of the three subfamilies of FUT1, FUT2, and Sec1 shows that each of the three families has its own evolutionary rate, i.e., Sec1 has a higher evolutionary rate than FUT1 and FUT2, which is in good accordance with the fact that the Sec1 of anthropoid apes have already incorporated several inactivating mutations transforming them into pseudogenes (Apoil et al. 2000; Barreaud et al. 2000).

Twenty-three inactivating mutations have been described in the FUT1 gene with a rather low incidence; most of them are sporadic nonprevalent mutations (Wagner and Flegel 1997) implicating nonsense, missense, and frame shift mutations (Table 2). Since *FUT1* and *FUT2* are closely linked in a small area of less than 100 kb in 19q13.3, alleles at these two loci tend to co-segregate. Two different H-deficient phenotypes were found on Reunion Island: the Indian red cell H-null *Bombay* phenotype, and the red cell H-weak *Reunion* phenotype, which depended on two distinct point mutations of the *FUT1* gene, the T725→G and the C349→T, respectively (Fernandez-Mateos et al. 1998). Both phenotypes were also nonsecretors of ABH in saliva, but the molecular bases of their *FUT2* gene inactivation are different. The Indian *Bombay* H-deficient phenotype had a total deletion of the *FUT2* gene (Koda et al. 1997), while

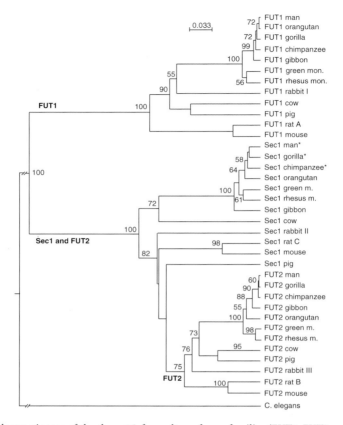

Fig. 1. Phylogenetic tree of the three α2-fucosyltransferase families (FUT1, FUT2, and Sec1) of twelve mammalian species. An α2-fucosyltransferase from *Caenorhabditis elegans* (O17784) has been added as an outgroup. Each family has a similar overall branching pattern, but the Sec1 enzymes have, on average, longer branches than FUT1 and FUT2 in each particular animal species, showing that Sec1 has a higher evolutionary rate (Apoil et al. 2000; Barreaud et al. 2000). The apparent branching out of all the FUT2 family of enzymes from the pig Sec1 is an artifact due to the higher evolutionary rate of Sec1, since the Fitch–Margoliash least-squares method with an evolutionary clock does not correct branch lengths for differences due to evolutionary rates. The bootstrap values (>50%) of 100 sets of data are represented at the divergence point of each branch. *The mutations on the sequences of the corresponding pseudogenes of these three hypothetical enzymes were corrected in order to get comparative complete proteins for the construction of the phylogenetic tree (Apoil et al. 2000)

the *Reunion* H-deficient phenotype had the punctual nonsense mutation G428→A which is also found, in a double dose, in the 20% of Caucasian salivary ABH nonsecretors (Kelly et al. 1994). Another *FUT2*-inactivating mutation, A385→T, is present in about 20% of Orientals. It only partially inactivates the enzyme, and results in the Le(a+b+) red cell phenotype owing to the incomplete formation of Le[b] antigen by the partial lack of α-2-fucosyltransferase activity (Kudo et al. 1996; Henry et al. 1996b;

Table 2. Summary of the inactivating mutations of the human *FUT1* gene, impairing the expression of the red cell H antigen

Nucleotide changes	Amino acid changes	References
349 C→T	117 His→Tyr	Fernandez-Mateos et al. 1998
442 G→T	148 Asp→Tyr	Kaneko et al. 1997
460 T→C	154 Tyr→His	Kaneko et al. 1997; Wang et al. 1997; Yu et al. 1997
461 A→G	154 Tyr→Cys	Wagner and Flegel 1997
491 T→A	164 Leu→His	Kelly et al. 1994
513 G→C	171 Trp→Cys	Wagner and Flegel 1997
547 deletion AG	182 frame shift	Yu et al. 1997
658 C→T	220 Arg→Cys	Yu et al. 1997
695 G→A	232 Trp→stop	Kaneko et al. 1997
721 T→C	241 Tyr→His	Kaneko et al. 1997
725 T→G	242 Leu→Arg	Fernandez-Mateos et al. 1998; Koda et al. 1997
776 T→A	259 Val→Glu	Wagner and Flegel 1997
801 G→C/T	267 Trp→Cys	Johnson et al. 1994
826 C→T	276 Ser→stop	Kelly et al. 1994
832 G→A	278 Asp→Asn	Johnson et al. 1994
880 deletion TT	294 frame shift	Yu et al. 1997
944 C→T	315 Ala→Val	Wagner and Flegel 1997
948 C→G	316 Tyr→stop	Kelly et al. 1994
969 deletion C	323 frame shift	Wagner and Flegel 1997
980 A→C	327 Asn→Thr	Yu et al. 1997
990 deletion G	330 frame shift	Johnson et al. 1994; Kaneko et al. 1997
1042 G→A	348 Glu→Lys	Kaneko et al. 1997; Wang et al. 1997
1047 G→C	349 Trp→Cys	Wagner and Flegel 1997

Koda et al. 1997). The gene mutations resulting in a partial inactivation of the FUT1 and FUT2 enzymes gave similar K_m values to those obtained for their wild-type counterparts. In addition, reduced V_{max} are obtained, suggesting a diminution of the active enzyme due to a decrease of the mutated enzyme stability. Other inactivating point mutations and a fusion gene *Sec1/FUT2* in the Japanese population have also been described (Koda et al. 1996) (Tables 2 and 3).

Future Perspectives

Most of the inactivating mutations found for *FUT1* and *FUT2* genes are located within the coding region of the genes. However, modifications of the upstream regulatory region of these genes may also play a role in their expression. It has recently been postulated that the large deletion of the *FUT2* gene in the Indian *Bombay* phenotype is probably mediated by the insertion of an *Alu* repetitive sequence (Koda et al. 2000a). Another *Alu*-mediated modification might be responsible for the appearance of the red cell expression of *FUT1* in higher apes (Apoil et al. 1999).

The nonprevalent sporadic character of some *FUT1* and *FUT2* mutations might be related to their time of appearance. Mutations found in all ethnic groups are probably more ancient than mutations prevalent only in certain ethnic groups (Koda et al. 2000b), and the nonprevalent sporadic mutations might be the more recent ones. In

Table 3. Summary of the inactivating mutations of the human *FUT2* gene, impairing the expression of the salivary H antigen

Nucleotide changes	Amino acid changes	References
Gene deletion	No enzyme	Fernandez-Mateos et al. 1998; Koda et al. 1997, 2000a
Fusion gene	No enzyme	Koda et al. 1996; Liu et al. 1999
385 A→T	129 Ile→Phe	Henry et al. 1996a,b; Koda et al. 1996; Kudo et al. 1996; Peng et al. 1999; Yu et al. 1995
428 G→A	143 Trp→stop	Fernandez-Mateos et al. 1998; Kelly et al. 1994; Liu et al. 1998; Peng et al. 1999
571 C→T	191 Arg→stop	Henry et al. 1996c; Koda et al. 1996; Peng et al. 1999; Yu et al. 1996
628 C→T	210 Arg→stop	Koda et al. 1996
658 C→T	220 Arg→stop	Liu et al. 1999
685 GTGGT→GT	230 Val deletion	Yu et al. 1999
778 deletion C	259 frame shift	Liu et al. 1998
849 G→A	283 Trp→stop	Peng et al. 1999; Yu et al. 1996

fact, all the inactivating mutations described are relatively recent events compared with the gene duplications at the origin of the present paralogous α-fucosyltransferase genes, which occurred either before the great mammalian radiation (*FUT1*, *FUT2*, *Sec1*, *FUT4*, *FUT7*, *FUT8*, and *FUT9*) (Oulmouden et al. 1997) or after it (*FUT3*, *FUT5*, and *FUT6*), but before the separation of Old World monkeys from the common evolutionary pathway (Costache et al. 1997).

None of the known fucosyltransferases has yet been crystallized, but an α/β barrel structure, with a Rossman fold, has been suggested (Breton et al 1996), and we can reasonably expect that their crystal three-dimensional structures will be soon elucidated, since the structure of a similar type II transmembrane glycosyltransferase, the bovine β4-galactosyltransferase, has recently been solved with a resolution of 2.4 Å (Gastinel et al. 1999).

Further Reading

The first cloning of a *FUT* gene (Larsen et al. 1990) opened a new era for the study of fucosyltransferases. The overall inactivating mutations of α2-fucosyltransferases are reviewed by Fernandez-Mateos et al. (1998), the Japanese alleles by Kaneko et al. (1997), the European alleles by Wagner and Flegel (1997), and the American alleles by Kelly et al. (1994). The conserved peptide motifs and the phylogeny of α2-fucosyltransferases are reviewed by Oriol et al. (1999). The different evolutionary rates of *FUT1*, *FUT2*, and *Sec1* are analyzed by Barreaud et al. (2000).

References

Apoil PA, Roubinet F, Despiau S, Mollicone R, Oriol R, Blancher A (2000) Evolution of α2-fucosyltransferase genes in primates: relation between intronic *Alu*-Y element and red cell expression of ABH antigens. Mol Biol Evol 17:337–351

Barreaud JP, Saunier K, Souchaire J, Delourme D, Oulmouden A, Oriol R, Leveziel H, Julien R, Petit FM (2000) Three bovine α2-fucosyltransferase genes encode enzymes that preferentially transfer fucose on Galβ1,3GalNAc acceptor substrates. Glycobiology 10:611–621

Bhende YM, Deshpande CK, Bhatia HM, Sanger R, Race RR, Morgan WTJ, Watkins WM (1952) A new blood group character related to the ABO system. Lancet i:903–904

Breton C, Oriol R, Imberty A (1996) Sequence alignment and fold recognition of fucosyltransferases. Glycobiology 6:vii–xii

Breton C, Oriol R, Imberty A (1998) Conserved structural features in eukaryotic and prokaryotic fucosyltransferases. Glycobiology 8:87–94

Costache M, Apoil P, Cailleau A, Elmgren A, Larson G, Henry S, Blancher A, Iordachescu D, Oriol R, Mollicone R (1997) Evolution of fucosyltransferase genes in vertebrates. J Biol Chem 272:29721–29728

Decastello AV, Sturli A (1902) Uber die isoagglutinine im serum gesunder und kranker menschen. Munchen Med Wschr II:1090–1095

Faik A, Bar-Peled M, DeRocher AE, Zeng W, Perrin RM, Wilkersson C, Raikhel NV, Keegstra K (2000) Biochemical characterization and molecular cloning of an α-1,2-fucosyltransferase that catalyzes the last step of cell wall xyloglucan biosynthesis in pea. J Biol Chem 275:15082–15089

Fernandez-Mateos P, Cailleau A, Henry S, Costache M, Elmgren A, Svenson L, Larson G, Samuelsson BE, Oriol R, Mollicone R (1998) Point mutations and deletion responsible for the *Bombay* H null and the *Reunion* H weak blood groups. Vox Sang 75:37–46

Gastinel LN, Cambillau C, Bourne Y (1999) Crystal structures of the bovine β4-galactosyltransferase catalytic domain and its complex with uridine diphosphogalactose. EMBO J 18:3546–3557

Henry S, Mollicone R, Fernandez P, Samuelsson B, Oriol R, Larson G (1996a) Homozygous expression of a missense mutation at nucleotide 385 in the *FUT2* gene associates with the Le(a+b+) partial-secretor phenotype in an Indonesian family. Biochem Biophys Res Com 219:675–678

Henry S, Mollicone R, Fernandez P, Samuelsson B, Oriol R, Larson G (1996b) Molecular basis for erythrocyte Le(a+b+) and salivary ABH partial-secretor phenotypes. Expression of a FUT2 secretor allele with an A-T point mutation at nucleotide 385 correlates with reduced α(1,2)fucosyltransferase activity. Glycoconj J 13:985–993

Henry S, Mollicone R, Lowe J, Samuelsson BE, Larson GA (1996c) A second nonsecretor allele of the blood group α(1,2)fucosyltransferase gene (*FUT2*). Vox Sang 70:21–25

Johnson PH, Mak MK, Leong S, Broadberry R, Duraisamy G, Gooch A, Lin CM, Makar I, Okubo Y, Swart E, Kolpsall E, Ewess M (1994) Analysis of mutations in the blood group H gene in donors with H-deficient phenotypes. Vox Sang 67:Suppl 2, 25 (abstract)

Kaneko M, Nishihara S, Shinya N, Kudo T, Iwasaki H, Seno T, Okubo Y, Narimatsu H (1997) Wide variety of point mutations in the *H* gene of *Bombay* and *para-Bombay* individuals that inactivate H enzyme. Blood 90:839–849

Kelly RJ, Ernst LK, Larsen RD, Bryant JG, Robinson JS, Lowe JB (1994) Molecular basis for H blood group deficiency in *Bombay* (Oh) and *para-Bombay* individuals. Proc Natl Acad Sci USA 91:5843–5847

Kelly RJ, Rouquier S, Giorgi D, Lennon GG, Lowe JB (1995) Sequence and expression of a candidate for the human secretor blood group α(1,2)fucosyltransferase gene (*FUT2*). J Biol Chem 270:4640–4649

Koda Y, Soejima M, Liu Y, Kimura H (1996) Molecular basis for secretor type α(1,2)fucosyltransferase gene deficiency in the Japanese population: a fusion gene generated by unequal crossover responsible for a nonsecretor phenotype. Am J Hum Genet 59:343–350

Koda Y, Soejima M, Johnson PH, Smart E, Kimura H (1997) Missense mutation of *FUT1* and deletion of *FUT2* are responsible for the Indian *Bombay* phenotype of the ABO blood group system. Biochem Biophys Res Commun 238:21–25

Koda Y, Soejima M, Johnson P, Smart E, Kimura H (2000a) An *Alu*-mediated large deletion of the *FUT2* gene in individuals with the ABO-*Bombay* phenotype. Hum Genet 106:80–85

Koda Y, Tachida H, Soejima M, Takenaka O, Kimura H (2000b) Ancient origin of the null allele *se(428)* of the human ABO-secretor locus (*FUT2*). J Mol Evol 50:243–248

Kudo T, Iwasaki H, Nishihara S, Shinya N, Ando T, Narimatsu I, Narimatsu H (1996) Molecular genetic analysis of the human Lewis histoblood group system. II. Secretor gene inactivation by a novel single missense mutation A385-T in Japanese nonsecretor individuals. J Biol Chem 271:9830–9837

Kumazaki T, Yoshida A (1984) Biochemical evidence that the secretor gene, *Se*, is a structural gene encoding a specific fucosyltransferase. Proc Natl Acad Sci USA 81: 4193–4197

Landsteiner K (1901) Uber agglutinationserscheinungen normalen menschlichen blutes. Klin Wochenschr 14:1132–1134

Larsen RD, Ernst LK, Nair RP, Lowe JB (1990) Molecular cloning, sequence and expression of a human GDP-L-fucose: β-D-galactoside α2-L-fucosyltransferase cDNA that can form the H blood group antigen. Proc Natl Acad Sci USA 87:6674–6678

Le Pendu J, Cartron JP, Lemieux RU, Oriol R (1985) The presence of at least two different H-blood group related β-D-Gal α-2-L-fucosyltransferases in human serum and the genetics of blood group H substances. Am J Hum Genet 37:749–760

Levine P, Robinson E, Celano M, Briggs O, Falkinburg L (1955) Gene interactions resulting in suppression of blood group substance B. Blood 10:1100–1108

Liu YH, Koda Y, Soejima M, Pang H, Schlaphoff T, duToit ED, Kimura H (1998) Extensive polymorphism of the *FUT2* gene in an African (Xhosa) population of South Africa. Hum Genet 103:204–210

Liu YH, Koda Y, Soejima M, Pang H, Wang BJ, Kim DS, Oh HB, Kimura H (1999) The fusion gene at the ABO-secretor locus (*FUT2*): absence in Chinese populations. J Hum Genet 44:181–184

Morgan WTJ, Watkins WM (1948) The detection of a product of the blood group O gene and the relationship of the so-called O substance to the agglutinogens A and B. Br J Exp Pathol 29:159–173

Oriol R, Danilovs J, Hawkins BR (1981a) A new genetic model proposing that the *Se* gene is a structural gene closely linked to the *H* gene. Am J Hum Genet 33:421–431

Oriol R, Le Pendu J, Sparkes RS, Sparkes MC, Crist M, Gale RP (1981b) Insights into the expression of ABH and Lewis antigens through human bone marrow transplantation. Am J Hum Genet 33:551–560

Oriol R, Mollicone R, Couillin P, Dalix AM, Candelier JJ (1992) Genetic regulation of the expression of ABH and Lewis antigens in tissues. APMIS Suppl 27, 100:28–38

Oriol R, Mollicone R, Cailleau A, Balanzino L, Breton C (1999) Divergent evolution of fucosyltransferase genes from vertebrates, invertebrates, and bacteria. Glycobiology 9:323–334

Oulmouden A, Wierinckx A, Petit J, Costache M, Palcic MM, Mollicone R, Oriol R, Julien R (1997) Molecular cloning and expression of a bovine α(1,3)fucosyltransferase gene homologous to the ancestor of the human *FUT3–FUT5–FUT6* cluster. J Biol Chem 272:8764–8773

Palcic MM, Heerze LD, Pierce M, Hindsgaul O (1988) The use of hydrophobic synthetic glycosides as acceptors in glycosyltransferase assays. Glycoconj J 5:49–63

Peng CT, Tsai CH, Lin TP, Peng LI, Kao MC, Yang TY, Wang NM, Liu TC, Lin SF, Chang FG (1999) Molecular characterization of secretor type α(1,2)-fucosyltransferase gene deficiency in the Philippine population. Ann Hematol 78:463–467

Perrin RM, DeRocher AE, BarPeled M, Zeng WQ, Norambuena L, Orellana A, Raikhel NV, Keegstra K (1999) Xyloglucan fucosyltransferase, an enzyme involved in plant cell wall biosynthesis. Science 284:1976–1979

Rajan VP, Larsen RD, Ajmera S, Ernst LK, Lowe JB (1989) A cloned human DNA restriction fragment determines expression of a GDP-L-fucose: β-D-galactoside 2-α-L-fucosyltransferase in transfected cells. Evidence for isolation and transfer of the human H blood group locus. J Biol Chem 264:11158–11167

Reguigne-Arnould I, Couillin P, Mollicone R, Faure S, Fletcher A, Kelly RJ, Lowe JB, Oriol R (1995) Relative positions of two clusters of human α-L-fucosyltransferases in 19q (FUT1–FUT2) and 19p (FUT6–FUT3–FUT5) within the microsatellite genetic map of chromosome 19. Cytogenet Cell Gen 71:158–162

Rouquier S, Lowe JB, Kelly RJ, Fertitta AL, Lennon GG, Giorgi D (1995) Molecular cloning of a human genomic region containing the H blood group α(1,2)fucosyltransferase gene and two H locus-related DNA restriction fragments. J Biol Chem 270:4632–4639

Sarnesto A, Köhlin T, Thurin J, Blaszczyk-Thurin M (1990) Purification of gene-encoded β-galactoside α1-2-fucosyltransferase from human serum. J Biol Chem 267:15067–15075

Sarnesto A, Kölin T, Hindsgaul O, Thurin J, Blaszczyk-Thurin M (1992) Purification of the secretor-type β-galactoside α1-2 fucosyltransferase from human serum. J Biol Chem 267:2734–2744

Schiff F, Sasaki H (1932) Der ausscheidungstypus, ein auf serologischem wege nachweisbares mendelndes merkmal. Klin Wochenschr II:1426–1429

Wagner FF, Flegel WA (1997) Polymorphism of the h allele and the population frequency of sporadic nonfunctional alleles. Transfusion 37:284–290

Wang BJ, Koda Y, Soejima M, Kimura H (1997) Two missense mutations of H type α(1,2)fucosyltransferase gene (FUT1) responsible for para-Bombay phenotype. Vox Sang 72:31–35

Watkins WM (1980) Biochemistry and genetics of the ABO, Lewis and P blood group systems. Adv Hum Genet 10:1–136

Watkins WM, Morgan WTJ (1955) Some observations on the O and H characters of human blood and secretions. Vox Sang 5:1–14

Weston BW, Nair RP, Larsen RD, Lowe JB (1992) Isolation of a novel human α(1,3)fucosyltransferase gene and molecular comparison to the human Lewis blood group α(1,3/1,4)fucosyltransferase gene. J Biol Chem 267:4152–4160

Yu LC, Yang YH, Broadberry RE, Chen YH, Chan YS, Lin M (1995) Correlation of a missense mutation in the human Secretor α1,2-fucosyltransferase gene with the Lewis(a+b+) phenotype: a potential molecular basis for the weak secretor allele (Se-W). Biochem J 312:329–332

Yu LC, Broadberry RE, Yang YH, Chen YH, Lin M (1996) Heterogeneity of the human secretor α(1,2)fucosyltransferase gene among Lewis(a+b–) nonsecretors. Biochem Biophys Res Com 222:390–394

Yu LC, Yang YH, Broadberry RE, Chen YH, Lin M (1997) Heterogeneity of the human H blood group α(1,2)fucosyltransferase gene among para-Bombay individuals. Vox Sang 72:36–40

Yu LC, Lee HL, Chu CC, Broadberry RE, Lin M (1999) A newly identified nonsecretor allele of the human histoblood group α(1,2)fucosyltransferase gene (FUT2). Vox Sang 76:115–119

28

α3/4-Fucosyltransferase (FUT3, Lewis enzyme)

Introduction

α3/4-Fucosyltransferase was cloned for the first time among members of the α3-fucosyltransferases (Kukowska Latallo et al. 1990), and named Fuc-TIII by J. Lowe et al. or FUT3 by Oriol et al. Later, FUT3 (Fuc-TIII) was demonstrated to be the Lewis enzyme that determines the expression of Lewis histoblood group antigen, i.e., Lewis a (Le^a) and Lewis b (Le^b) (Mollicone et al. 1994; Nishihara et al. 1994). FUT3 catalyzes a transfer of Fuc from GDP-Fuc to the GlcNAc residue of type-2 chain, Galβ1-4GlcNAc, with an α1,3-linkage, or to the GlcNAc residue of type-1 chain, Galβ1-3GlcNAc, with an α1,4-linkage. The biosynthetic pathways of Lewis antigen epitopes in relation to ABH antigens are summarized in Fig. 1.

FUT3 exhibits the broadest specificity synthesizing all Lewis antigens, Le^a, Le^b, sLe^a, Le^x, Le^y, and sLe^x. Two antigens, Le^a and Le^b, among the Lewis antigens in Fig. 1 are histoblood group antigens which are present not only on erythrocytes, but also in gastrointestinal, pancreatic, and other tissues. FUT3 is the only enzyme responsible for the synthesis of Le^a, Le^b, and sLe^a in vivo. FUT5, as described in an other chapter, can synthesize the type-1 Lewis antigens, Le^a and Le^b, but it seems to be a silent enzyme which is not expressed in any tissues in the body.

The sequence of FUT3 is highly homologous to those of FUT5 and FUT6. Three α1,3FUT genes, FUT3, FUT5, and FUT6, form a gene cluster in close proximity to 19p13.3, which probably diverged after the primates by gene-duplication (Nishihara et al. 1993; Cameron et al. 1995). Chimpanzee and gorilla possess three distinct orthologous homologues, each one corresponding to the human FUT3, FUT5, and FUT6 genes (Costache et al. 1997). Bovines possess a single-copy gene, named the futb gene, which is an orthologous homologue of an ancestral gene from which the present

Hisashi Narimatsu

Laboratory of Gene Function Analysis, Institute of Molecular and Cell Biology (IMCB), National Institute of Advanced Industrial Science and Technology (AIST), Central-2, 1-1-1 Umezono, Tsukuba, Ibaraki 305-8568, Japan
Tel. +81-298-61-3200; Fax +81-298-61-3201
e-mail: h.narimatsu@aist.go.jp

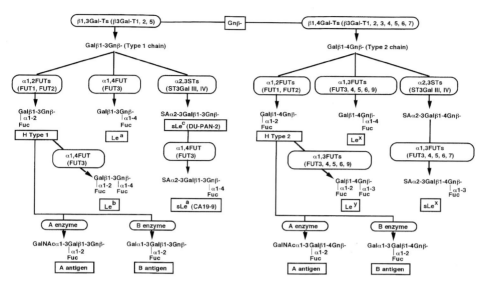

Fig. 1. Biosynthetic pathways of type-1 to and type-2 Lewis antigens in correlation with the synthesis of ABO antigens. Lewis antigens with a type-1 chain are on the *left*, and those with a type-2 chain are on the *right*

human *FUT3–FUT5–FUT6* gene cluster evolved (Oulmouden et al. 1997; Costache et al. 1997). The mouse orthologous homologue corresponding to the *futb* gene, the ancestral gene for the human *FUT3–FUT5–FUT6* gene cluster, seems to be a pseudogene (accession number, U33458). Thus, mouse cells and tissues do not exhibit FUT3 activity, and are considered not to express type-1 Lewis antigens, such as Le^a, Le^b, and sLe^a.

Databanks

α3/4-Fucosyltransferase (FUT3, Lewis enzyme)

NC-IUBMB enzyme classification: E.C.2.4.1.65

Species	Gene	Protein	mRNA	Genomic
Homo sapiens	FUT3, Fuc-TIII	–	X53578	–
		–	U27326	–
		–	U27327	–
		–	U27328	–
		–	M81485	–
Pan troglodytes	FUT3	–	Y14033	–
Mus musculus	Fut3, mFuc-TIII	–	–	U33458
Bos taurus	futb	–	X87810	–
Sus scrofa	FUT3	–	AF130972	–

Name and History

Fuc-TIII is also called FUT3 or a Lewis enzyme. Fuc-TIII (FUT3) was first cloned as a candidate enzyme synthesizing a stage-specific embryonal antigen-1 (SSEA-1) (Kukowska Latallo et al. 1990). SSEA-1, which has a carbohydrate epitope which is defined as Le^x, Galβ1-4(Fucα1-3)GlcNAc, appears to be involved in mouse embryo compaction (Solter and Knowls 1978). However, FUT3 was not expressed in the early embryo, and mice do not possess an active *FUT3* gene. Thus, the real SSEA-1 synthesizing enzyme is not FUT3. FUT9, which is described in another chapter, may be a real SSEA-1 synthesizing enzyme (Narimatsu et al. manuscript in preparation).

   The Lewis histoblood group system has been thoroughly investigated by hematologists in the field of blood transfusion. Individuals are divided into four groups based on their Lewis blood group phenotype, i.e., three Lewis-positive groups, Le(a+,b−), Le(a−,b+), and Le(a+,b+), and a Lewis-negative group, Le(a−,b−). The activity of α1,4-fucosyltransferase (Lewis enzyme) has been found in saliva and some intestinal tissues of Lewis-positive individuals, whereas it was not detected in any tissues of Lewis-negative individuals. The alleles encoding FUT3 were analyzed in relation to Lewis blood group phenotype in Caucasian ethnic groups (Mollicone et al. 1994; Elmgren et al. 1997) and in the Japanese population (Nishihara et al. 1994; Narimatsu et al. 1998). Lewis-negative individuals were found to be homozygotes for the null *FUT3* alleles. Thus, FUT3 (Fuc-TIII) was identified as the Lewis enzyme. Point mutations in the open reading frame of the *FUT3* gene inactivate the Lewis enzyme, and these mutations are ethnic-group-specific. The alleles of the *FUT3* gene found in Japanese and Caucasians are summarized in Fig. 2.

Enzyme Activity Assay and Substrate Specificity

The fine specificity of FUT3 has been examined towards various oligosaccharides (de Vries et al. 1995, 1997) and *N*-glycans on glycoproteins (Grabenhorst et al. 1998). As shown in Fig. 1, FUT3 can synthesize all Lewis antigens, Le^a, Le^b, sLe^a, Le^x, Le^y, and sLe^x. Recently, we found that FUT3 preferentially fucosylates the inner GlcNAc residue of neutral polylactosamine (Nishihara et al. 1999b). As can be seen in a phylogenetic tree of the α1,3-FUT family (Fig. 3), there are four clusters of subfamilies, FUT3–FUT5–FUT6, FUT4, FUT7, and FUT9.

   FUT4, FUT3, FUT5, and FUT6 exhibit a similar specificity for preferential fucose transfer to the innner GlcNAc residue towards neutral polylactosamine, while only FUT9 preferentially transfers a fucose to the distal GlcNAc residue. This is rational given that the FUT9 amino acid sequence is quite different from those of the other five α1,3-FUTs, which share highly homologous sequences.

Preparation

The open reading frame (ORF) of the *FUT3* gene is encoded in a single exon. Therefore, it is easy to clone the ORF of the *FUT3* gene by polymerase chain reaction (PCR) in order to express it as a recombinant enzyme. The primers for obtaining the ORFs of six human α1,3-FUTs by PCR amplification are listed in Table 1.

	Species of Allele	Name of allele	Enzyme activity when transfected into COS cells	Frequency of each allele in the Japanese population
FUT3 alleles found in Japanese	──────────	Le	100%	551/800 (68.9%)
	T59G G508A ●──●──────	le1	0%	198/800(24.8%)
	T59G T1067A ●────●───	le2	0%	47/800(5.8%)
	T59G ●────────	le3	80%	4/800(0.2%)
FUT3 alleles found in Swedes	T202C ──●─────── C314T ──●───────			

Fig. 2. Schematic diagram of the mutant *FUT3* alleles found in the Japanese and Swedish populations. The Japanese *FUT3* alleles are divided into *Le*, *le1*, *le2*, and *le3* (Narimatsu et al. 1998; Nishihara et al. 1999a). The point mutations in each allele are presented schematically. The enzyme activity directed by each allele is also presented. The frequencies with which each allele occurrs in the Japanese population are on the *right*

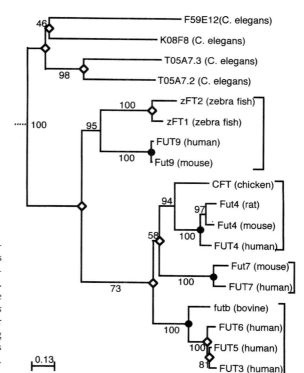

Fig. 3. A phylogenetic tree of α-1,3-FUTs. The branch lengths indicate the evolutionary distances between different genes. *Open diamonds* indicate gene duplication, and *solid circles* indicate speciation. Four cosmid clone names containing *Caenorhabditis elegans* genes are presented (Kaneko et al. 1999)

Table 1. Oligonucleotide primers used for cloning of six human α1,3-fucosyltransferase genes

Gene	Forward primer	Reverse primer
FUT3,5,6	5′-ACCCATGGATCCCCTGGGTGCAGC-3′	5′-CTCTCAGGTGAACCAAGCCGCTATG-3′
FUT4	5′-CAGCGCTGCCTGTTCGCGCCATGG-3′	5′-TCACCGCTCGAACCAGCTGGCCAA-3′
FUT7	5′-AATCTCGGGTCTCTTGGCTG-3′	5′-GGTGGTTTGATTTCGACACC-3′
FUT9	5′-ATGACATCAACATCCAAAGGAATTC-3′	5′-ATCCTCAATACTTGGATGATATCTC-3′

A genomic DNA or cDNAs from the cell lines or tissues derived from human gastrointestinal organs can be used for the templates for PCR to obtain the ORF of the *FUT3* gene. However, care should be taken not to clone the null *FUT3* alleles which are widely distributed in the population (see Fig. 2). The frequency of each null *FUT3* allele distributed in the Japanese population has been reported (Narimatsu et al. 1998). Three inactive *FUT3* alleles, named *le1*, *le2*, and *le3*, are found at a frequency of 27%, 7%, and 0.5%, respectively, in the Japanese population.

The monoclonal antibody which specifically recognizes the human FUT3 is available (Kimura et al. 1995).

Biological Aspects

The activity of the Lewis enzyme (FUT3) has been detected and analyzed in saliva, milk, and other secretions (de Vries et al. 1993; Nishihara et al. 1999a). Lewis-negative individuals, who are approximately 10% of the population, never show Lewis enzyme activity in their secretions. The expression of type-1 Lewis antigens, Le^a, Le^b, and sLe^a, in a body is solely determined by FUT3. Lewis-negative individuals, who genetically lack the Lewis enzyme (FUT3) because they are homozygotes of the null *FUT3* alleles, never express type-1 Lewis antigens in any body tissues.

Lewis blood types are determined by two fucosyltransferases, FUT2 (Se enzyme) and FUT3 (see Fig. 1). Le(a−,b+) individuals, who have the Le^b antigen on erythrocytes, possess at least one active *FUT2* and *FUT3* allele. Le(a+,b−) individuals, who express the Le^a antigen but not the Le^b antigen on erythrocytes, possess at least one active *FUT3* allele, but they are homozygotes of the null *FUT2* alleles. Le(a+,b+) individuals possess at least one active *FUT3* allele, but they are homozygotes of the weakened *FUT2* alleles (Henry et al. 1996). Le(a−,b−) Lewis-negative individuals are homozygotes of null *FUT3* alleles regardless of their *FUT2* alleles. In many Asian countries, including Japan, the weakened *FUT2* allele, named *sej*, is widely distributed in the population, so that about 18% of the Japanese people are Le(a+,b+). However, Le(a+,b−) individuals are very rare in Asian countries, including Japan (Kudo et al. 1996; Henry et al. 1996; Narimatsu et al. 1998). In contrast, the null *FUT2* allele, which is completely inactivated by a nonsense mutation, is widely distributed in Caucasians. Le(a+,b−) individuals are very common, but Le(a+,b+) Caucasians are very rare.

Although FUT3 is essential for the expression of Le^a and Le^b antigens on erythrocytes, FUT3 is not expressed in hematopoietic cells. FUT3 is widely expressed in epithelial cells of the digestive tract, including salivary gland, mammary gland, esophagus, stomach, pancreas, small intestine, colon, and rectum, etc. (Kaneko et al. 1999;

Nishihara et al. 1999a). Le^a- and Le^b-active glycolipid antigens on erythrocytes are not synthesized by erythrocytes themselves, and are probably synthesized elsewhere in the digestive organs by FUT3. They are secreted into serum and adsorbed to erythrocytes as blood-group antigens (Yazawa et al. 1995).

Regardless of *FUT2* alleles, the homozygotes of null *FUT3* alleles (Le(a−,b−) individuals) cannot express type-1 Lewis antigens either on erythrocytes or in other tissues (Nishihara et al. 1999a; Narimatsu et al. 1996; Ikehara et al. 1998). Le^a and Le^b antigens are abundantly expressed in colorectal tissues, and intermediately expressed in stomach and pancreas. However, Lewis-negative individuals never express Le^a and Le^b antigens in any tissues. CA19-9 is a well-known tumor marker which is frequently used for the clinical diagnosis of colorectal, stomach, and pancreatic cancers. CA19-9 recognizes the sLe^a epitope on mucins (see Fig. 1). Lewis-negative individuals cannnot produce the sLe^a epitope on mucins recgnized by CA19-9. DU-PAN-2 is also a well-known tumor marker recognizing the sLe^c epitope, a precursor structure of sLe^a. We suggested that CA19-9 measurements would be useful for cancer diagnosis in Lewis-positive individuals, but not in Lewis-negative ones. DU-PAN-2 measurements should be performed for cancer diagnosis in Lewis-negative patients (Narimatsu et al. 1998).

sLe^a and sLe^x are well-known cancer-associated antigens, and are known to be ligands for selectins (Takada et al. 1993). sLe^a expression definitely requires FUT3, although the amount of sLe^a antigen is not determined by the level of FUT3 activity. The expression of sLe^x antigen in cancer cells is directed by two enzymes, FUT3 and FUT6. The amount of sLe^x antigen in lung cancer cells is determined by the cooperative activity of FUT3 and FUT6 (Togayachi et al. 1999). The cancer cells of Lewis-negative individuals do not express the sLe^a antigen at all, but can express the sLe^x antigen, which is synthesized by FUT6.

Future Perspectives

The point mutations which inactivate the *FUT3* gene differ between Caucasians and Japanese. This demonstrates that inactivation of FUT3 and the Lewis blood group system occurred after the divergence of two ethnic groups, Caucasians and Japanese. It is an interesting question as to why only two or three inactivated *FUT3* alleles are common in each ethnic group. This finding indicates that the number of individuals lacking FUT3 (Lewis-negative individuals) has been increasing in the population, but for unknown reasons. There are very few studies on the biological advantages or disadvantages for Lewis-negative individuals. One study found a biological disadvantage in Lewis-negative individuals, i.e., a significantly restricted lung function as well as a higher prevalence of wheezing and asthma compared with Lewis-positive individuals (Kauffmann et al. 1996). Clinical statistical analysis demonstrated that cancer patients who express abundant sLe^a antigens have a worse prognosis for liver metastasis than patients who do not express sLe^a antigens (Nakayama et al. 1995). Lewis-negative individuals, who cannot produce sLe^a antigens, may have an advantage regarding the prognosis for colorectal cancer.

The tissue distribution patterns of two α1,3-FUTs, FUT3 and FUT6, are very similar. Both enzymes are mainly expressed in the epithelial cells of gastrointestinal and pancreatic tissues, and appear to complement each other in producing type-2 Lewis antigens, Le^x, Le^y, and sLe^x, while type-1 Lewis antigens, Le^a, Le^b, and sLe^a, are solely synthesized by FUT3. The essential roles of such Lewis antigens in those tissues are not yet clear and need to be elucidated.

References

Cameron HS, Szczepaniak D, Weston BW (1995) Physical maps of human alpha(1,3)fucosyltransferase genes FUT3–FUT6 on chromosomes 19p13.3 and 11q21. J Biol Chem 270:20112–20221

Costache M, Apoil PA, Cailleau A, Elmgren A, Larson G, Henry S, Blancher A, Iordachescu D, Oriol R, Mollicone R (1997) Evolution of fucosyltransferase genes in vertebrates. J Biol Chem 272:29721–29728

de Vries T, Norberg T, Lonn H, van den Eijnden DH (1993) The use of human milk fucosyltransferase in the synthesis of tumor-associated trimeric determinants. Eur J Biochem 216:769–777

de Vries T, Srnka CA, Palcic MM, Swiedler SJ, van den Eijnden DH, Macher BA (1995) Acceptor specificity of different length constructs of human recombinant α1,3/4-fucosyltransferases. Replacement of the stem region and the transmembrane domain of fucosyltransferase V by protein A results in an enzyme with GDP-fucose hydrolysing activity. J Biol Chem 270:8712–8722

de Vries T, Palcic MM, Schoenmakers PS, van den Eijnden DH, Joziasse DH (1997) Acceptor specificity of GDP-Fuc: Galβ1-4GlcNAc-R α3-fucosyltransferase VI (Fuc-TVI) expressed in insect cells as soluble, secreted enzyme. Glycobiology 7:921–927

Elmgren A, Mollicone R, Costache M, Borjeson C, Oriol R, Harrington J, Larson G (1997) Significance of individual point mutations, T202C and C314T, in the human Lewis (FUT3) gene for expression of Lewis antigens by the human α(1,3/1,4)-fucosyltransferase, Fuc-TIII. J Biol Chem 272:21994–21998

Grabenhorst E, Nimtz M, Costa J, Conradt HS (1998) In vivo specificity of human α1,3/4-fucosyltransferases III–VII in the biosynthesis of Lewis^x and sialyl Lewis^x motifs on complex-type N-glycans. J Biol Chem 273:30985–30994

Henry S, Mollicone R, Fernandez P, Samuelsson B, Oriol R, Larson G (1996) Molecular basis for erythrocyte Le(a+,b+) and salivary ABH partial-secretor phenotypes: expression of an FUT2 secretor allele with an A→T mutation at nucleotide 385 correlates with reduced alpha(1,2) fucosyltransferase activity. Glycoconj J 13:985–993

Ikehara Y, Nishihara S, Kudo T, Hiraga T, Morozumi K, Hattori T, Narimatsu H (1998) The aberrant expression of Lewis a antigen in intestinal metaplastic cells of gastric mucosa is caused by augmentation of Lewis enzyme expression. Glycoconj J 15:799–807

Kaneko M, Kudo T, Iwasaki H, Ikehara Y, Nishihara S, Nakagawa S, Sasaki K, Shiina T, Inoko H, Saitou N, Narimatsu H (1999) α1,3-fucosyltransferase IX (Fuc-TIX) is very highly conserved between human and mouse: molecular cloning, characterization and tissue distribution of human Fuc-TIX. FEBS Lett 452:237–242

Kauffmann F, Frette C, Pham QT, Nafissi S, Bertrand JP, Oriol R (1996) Associations of blood group-related antigens to FEV1, wheezing, and asthma. Am J Respir Crit Care Med 153:76–82

Kimura H, Kudo T, Nishihara S, Iwasaki H, Shinya N, Watanabe R, Honda H, Takemura F, Narimatsu H (1995) Murine monoclonal antibody recognizing human α(1,3/1,4)fucosyltransferase. Glycoconj J 12:802–812

Kudo T, Iwasaki H, Nishihara S, Shinya N, Ando T, Narimatsu I, Narimatsu H (1996) Molecular genetic analysis of the human Lewis histo-blood group system. II. Secretor gene inactivation by a novel single missense mutation A385T in Japanese nonsecretor individuals. J Biol Chem 271:9830–9837

Kukowska Latallo JF, Larsen RD, Nair RP, Lowe JB (1990) A cloned human cDNA determines expression of a mouse stage-specific embryonic antigen and the Lewis blood group α(1,3/1,4)fucosyltransferase. Genes Dev 4:1288–1303

Mollicone R, Reguigne I, Kelly RJ, Fletcher A, Watt J, Chatfield S, Aziz A, Cameron HS, Weston BW, Lowe JB, Oriol R (1994) Molecular basis for Lewis α(1,3/1,4)fucosyltransferase gene deficiency (FUT3) found in Lewis-negative Indonesian pedigrees. J Biol Chem 269:20987–20994

Nakayama T, Watanabe M, Katsumata T, Teramoto T, Kitajima M (1995) Expression of sialyl Lewis(a) as a new prognostic factor for patients with advanced colorectal carcinoma. Cancer 75:2051–2056

Narimatsu H, Iwasaki H, Nishihara S, Kimura H, Kudo T, Yamauchi Y, Hirohashi S. (1996) Genetic evidence for the Lewis enzyme, which synthesizes type-1 Lewis antigens in colon tissue, and intracellular localization of the enzyme. Cancer Res 56:330–338

Narimatsu H, Iwasaki H, Nakayama F, Ikehara Y, Kudo T, Nishihara S, Sugano K, Okura H, Fujita S, Hirohashi S (1998) Lewis and secretor gene dosages affect CA19-9 and DU-PAN-2 serum levels in normal individuals and colorectal cancer patients. Cancer Res 58:512–518

Nishihara S, Narimatsu H, Iwasaki H, Yazawa S, Akamatsu S, Ando T, Seno T, Narimatsu I (1994) Molecular genetic analysis of the human Lewis histo-blood group system. J Biol Chem 269:29271–29278

Nishihara S, Nakazato M, Kudo T, Kimura H, Ando T, Narimatsu H (1993) Human alpha-1,3 fucosyltransferase (Fuc-TVI) gene is located at only 13 kb 3' to the Lewis type fucosyltransferase (Fuc-TIII) gene on chromosome 19. Biochem Biophys Res Commun 190:42–46

Nishihara S, Hiraga T, Ikehara Y, Iwasaki H, Kudo T, Yazawa S, Morozumi K, Suda Y, Narimatsu H (1999a) Molecular behavior of mutant Lewis enzyme. Glycobiology 9:373–382

Nishihara S, Iwasaki H, Kaneko M, Tawada A, Ito M, Narimatsu H (1999b) α1,3-fucosyltransferase 9 (FUT9; Fuc-TIX) preferentially fucosylates the distal GlcNAc residue of polylactosamine chain while the other four α1,3FUT members preferentially fucosylate the inner GlcNAc residue. FEBS Lett 462:289–294

Oulmouden A, Wierinckx A, Petit JM, Costache M, Palcic MM, Mollicone R, Oriol R, Julien R (1997) Molecular cloning and expression of a bovine α(1,3)-fucosyltransferase gene homologous to a putative ancestor of the human FUT3–FUT5–FUT6 cluster. J Biol Chem 272:8764–8773

Solter D, Knowls BB (1978) Monoclonal antibody defining a stage-specific mouse embryonic antigen (SSEA-1). Proc Natl Acad Sci USA 75:5565–5569

Takada A, Ohmori K, Yoneda T, Tsuyuoka K, Hasegawa A, Kiso M, Kannagi R (1993) Contribution of carbohydrate antigens sialyl Lewis A and sialyl Lewis X to adhesion of human cancer cells to vascular endothelium. Cancer Res 53:354–361

Togayachi A, Kudo T, Ikehara Y, Iwasaki H, Nishihara S, Andoh T, Higashiyama M, Kodama K, Nakamori S, Narimatsu H (1999) Up-regulation of Lewis enzyme (Fuc-TIII) and plasma-type α1,3fucosyltransferase (Fuc-TVI) expression determines the augmented expression of sialyl Lewis X antigen in non-small cell lung cancer. Int J Cancer 83:70–79

Yazawa S, Nishihara S, Iwasaki H, Asao T, Nagamachi Y, Matta KL, Narimatsu H (1995) Genetic and enzymatic evidence for Lewis enzyme expression in Lewis-negative cancer patients. Cancer Res 55:1473–1478

29

α3-Fucosyltransferase-IV (FUT4)

Introduction

α3-Fucosyltransferase-IV (Fuc-TIV, FUT4) was cloned independently by two groups at almost the same time (Goelz et al. 1990; Lowe et al. 1991). It was first cloned as an α1,3-fucosyltransferase determining the expression of the ligand for the endothelial-leukocyte adhesion molecule-1 (ELAM-1) by Goelz et al., and named ELAM-1 ligand fucosyltransferase (ELFT). Lowe et al. cloned it by cross-hybridization with a probe of FUT3 cDNA. However, the terms ELAM-1 and ELFT are not used at the present time. They have been renamed E-selectin and α3-fucosyltransferase-IV(Fuc-TIV, FUT4), respectively. The *Fut4* gene is one of three active *Fut* genes, *Fut4*, *7*, and *9*, in rodents, and have been cloned from several species, i.e., mouse (Gersten et al. 1995, D63380), rat (U58860), and chicken (U73678). FUT4 is ubiquitously expressed in many tissues (Gersten et al. 1995; Kaneko et al. 1999). The chromosomal localization of human *FUT4* and mouse *Fut4* genes have been mapped on human chromosome 11q22–q23 and mouse choromosome 9, respectively (Gersten et al. 1995).

The substrate specificity of FUT4 in comparison to that of the other FUTs has been studied by several investigators (de Vries et al. 1995; Gersten et al. 1995; Niemela et al. 1998; Grabenhorst et al. 1998; Nishihara et al. 1999).

HISASHI NARIMATSU

Laboratory of Gene Function Analysis, Institute of Molecular and Cell Biology (IMCB), National Institute of Advanced Industrial Science and Technology (AIST), Central-2, 1-1-1 Umezono, Tsukuba, Ibaraki 305-8568, Japan
Tel. +81-298-61-3200; Fax +81-298-61-3201
e-mail: h.narimatsu@aist.go.jp

Databanks

α3-Fucosyltransferase-IV (FUT4)

NC-IUBMB enzyme classification: E.C.2.4.1.152

Species	Gene	Protein	mRNA	Genomic
Homo sapiens	*FUT4, Fuc-TIV*	NP_002024	M65030	NT_008984
Mus musculus	*Fut4, mFuc-TIV*	NP_034372	D63380	–
			U33457	–
Rattus norvegicus	*Fut4, rFuc-TIV*	NP_071555	U58860	–
Gallus gallus	*Fut4, CFT1*	AAC60060	U73678	–

Name and History

Fuc-TIV is also called FUT4 or a myeloid-type α3-fucosyltransferase. Before the molecular cloning of α3-fucosyltransferase genes, Mollicone et al. (1990) had predicted that there are at least three α3-fucosyltransferases based on their distinct substrate-specificity patterns. Myeloid cell homogenates showed the myeloid α3-fucosyltransferase pattern, in which leukocyte enzyme(s) transfer fucose onto H type-2 acceptor, and poorly onto sialylated *N*-acetyllactosamine. Later, FUT4 was found to be abundantly expressed in myeloid cells, and its substrate specificity was consistent with that of myeloid cell homogenates. Thus, the myeloid-type enzyme was identified as FUT4. Very recently, a new α3-fucosyltransferase, FUT9, was cloned and its tissue distribution was examined together with FUT4 and FUT7 (Kaneko et al. 1999; Nakayama et al. 2001). Peripheral blood leukocytes (PBL) expressed three α3-fucosyltransferases, FUT4, FUT7, and FUT9. FUT4 and FUT7 were expressed ubiquitously in all subpopulations of PBL, including granulocytes, monocytes, and lymphocytes. In contrast, FUT9 expression was restricted to some subpopulations such as granulocytes, natural killer (NK) cells, and B cells.

Enzyme Activity Assay and Substrate Specificity

Niemela et al. (1998) analyzed the detailed substrate specificity of FUT4 using various oligosaccharide acceptors of sialylated and neutral polylactosamines in a comparison with the FUT7 specificity. FUT4 can transfer fucose effectively to all *N*-acetyllactosamine (LN) units in neutral polylactosamines. Our recent data demonstrated that FUT4 transfers fucose more effectively to the inner LN units than to the distal LN units of neutral polylactosamine (Nishihara et al. 1999). This is consistent with Niemela's data. In contrast, FUT9 preferentially transfers fucose to the distal LN units rather than the inner LN units. FUT4 can also transfer fucose to the inner LN units of sialylated polylactosamine acceptor, resulting in the synthesis of the VIM-2 (CDw65) epitope, but it is ineffective in transferring it to the distal α2,3-sialylated LN units in α 2,3-sialylated acceptors. FUT7, by contrast, effectively fucosylates only the distal LN units in α2,3-sialylated acceptors. The differential site specificities of the three enzymes FUT4, FUT7, and FUT9 are summarized in Fig. 1.

Fig. 1. Preferential fucose transfer to the differential sites of neutral and acidic polylactosamine chains by three α1,3FUTs. **A** FUT4 preferentially transfers Fuc to the inner GlcNAc residue of the neutral polylactosamine chain, while FUT9 preferentially transfers Fuc to the distal GlcNAc residue of the chain. FUT7 cannot transfer Fuc to the neutral chain. **B** FUT7 preferentially transfers Fuc to the distal GlcNAc residue of the sialylated polylactosamine chain, while FUT4 preferentially transfers Fuc to the inner GlcNAc residue. FUT9 exhibits weak activity for Fuc transfer to the inner GlcNAc residue of the sialylated polylactosamine chain

Preparation

Although FUT4 is ubiquitously expressed in almost all tissues, it is most abundantly expressed in myeloid cells. In particular, HL-60 cells, human promyelocytic leukemia cells, express a large amount of FUT4 (Marer et al. 1997; Nakayama et al. 2001). FUT4 protein is available in a recombinant form. The *FUT4* gene was stably expressed in Namalwa cells (Nishihara et al. 1999), Chinese hamster ovary (CHO) cells (Niemela et al. 1998), COS-1 cells (Lowe et al. 1991), and BHK-21 cells (Grabenhorst et al. 1998). Homogenates of these cells were used for enzyme assays. The recombinant forms, i.e., a soluble form expressed as a fusion protein in a mammalian expression vector, and a soluble form expressed in a baculoviral system, are also available.

Biological Aspects

The biological function of FUT4 is unclear. FUT4 is ubiquitously expressed in all tissues, although the expression level is different dependent on the tissues (Gersten et al. 1995; Kaneko et al. 1999). FUT4 transcript is most abundant in myeloid cells, stomach, colon, jejunum, uterus, and kidney; substantial amounts are also detected in the lung, testes, spleen, and liver. FUT4 was believed to be the only enzyme responsible for Le^x (CD15) expression in a body before the discovery of FUT9. It was based on experiments showing that overexpression of FUT4 by transfection experiments resulted in Le^x expression on the cell surface (Goelz et al. 1990; Lowe et al. 1991), whereas FUT7 transfection resulted in sialyl Le^x (sLe^x) expression, but not Le^x expression. The tissue distribution of FUT4 does not necessarily correlate with Le^x expression. It is known that FUT4 and FUT7 are expressed in hematopoietic cells, but FUT3, 5, and 6 are not expressed (Clarke and Watkins 1996; Marer et al. 1997; Kaneko 1999). The CD15 epitope of hematopoietic cells has been defined as the Le^x carbohydrate structure. The expression of FUT4 and FUT7 during hematopoiesis has been investigated in order to correlate the two enzymes with the expression of CD15 and CD15s

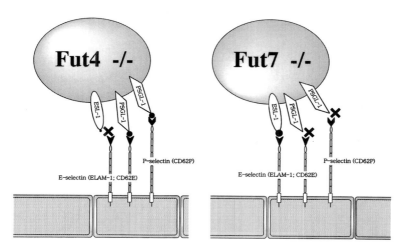

Fig. 2. Glycoforms on PSGL-1 and ESL-1 are differentially fucosylated by Fut7 and Fut4, respectively, determining the binding activity of neutrophils to endothelial cells. Fut4-deficient neutrophils can bind to E-selectin and P-selectin through glycoforms on PSGL-1, which is synthesized by Fut7, but cannot bind to E-selectin through glycoforms on ESL-1, which is synthesized by Fut4. Fut7-deficient neutrophils cannot bind to E- and P-selectins through glycoforms on PSGL-1, but can bind to E-selectin through glycoforms on ESL-1, which is synthesized by Fut4

(sialylated CD15; sLe^x) epitopes during hematopoiesis. Clarke and Watkins (1996) reported that FUT7 definitely determines the expression of CD15s epitope, which is consistent with the other results, but the CD15 expression does not correlate well with the FUT4 expression, which predicted that an unknown α1,3-FUT is involved in CD15 expression in hematopoietic cells.

It was demonstrated that the Le^x expression is determined by FUT9, not by FUT4, except for some tissues (Nakayama et al. 2001). FUT4 mainly participates in VIM-2 (CDw65) expression, not in Le^x (CD15) expression.

During the preparation of this manuscript, some very interesting papers were published regarding the biological function of FUT4 (Weninger et al. 2000; Huang et al. 2000). By analysis of neutrophils of Fut4-deficient mice and Fut7-deficient mice, these authors observed normal binding of E- or P-selectin to P-selectin glycoprotein ligand-1 (PSGL-1) expressed by Fut4-deficient neutrophils, but PSGL-1 expressed by Fut7-deficient neutrophils is not bound by E- or P-selectin. By contrast, E-selectin binds with normal efficiency to E-selectin ligand-1 (ESL-1) on Fut7-deficient neutrophils, but E-selectin exhibits dramatically reduced binding to ESL-1 on Fut4-deficient neutrophils. Thus, glycoforms on PSGL-1 and ESL-1 are differentially fucosylated by Fut7 and Fut4, respectively, which determines the binding activity of neutrophils. Figure 2 shows schematically the recent findings that Fut4 is responsible for glycoforms on ESL-1, which is bound by E-selectin, while Fut7 forms glycoforms on PSGL-1, which is bound by both P-selectin and E-selectin.

Future Perspectives

Fut4-knockout mice have been established. Recent studies demonstrated the essential role of glycoforms on neutrophils, which is directed by Fut4 (Weninger et al. 2000; Huang et al. 2000). However, the biological functions of Fut4 in other tissues are still unknown. Three α3-fucosyltransferases, FUT4, FUT7, and FUT9, are expressed in leukocytes. The complementary in vitro substrate specificities of the three enzymes imply that they cooperate in vivo in the biosynthesis of neutral and sialylated polylactosamine chains which are monofucosylated or multifucosylated. The knockout mice lacking multiple α3-fucosyltransferase genes, i.e., Fut4- and Fut7-deficient mice, Fut4- and Fut9-deficient mice, or Fut7- and Fut9-deficient mice, will be very interesting tools for an analysis of the biological functions of α1,3-fucosylation.

References

Clarke JL, Watkins WM (1996) α1,3-L-fucosyltransferase expression in developing human myeloid cells. J Biol Chem 271:10317–10328

de Vries T, Srnka CA, Palcic MM, Swiedler SJ, van den Eijnden DH, Macher BA (1995) Acceptor specificity of different length constructs of human recombinant α1,3/1, 4-fucosyltransferases. J Biol Chem 270:8712–8722

Gersten KM, Natsuka S, Trinchera M, Petryniak B, Kelly RJ, Hiraiwa N, Jenkins NA, Gilbert DJ, Copeland NG, Lowe JB (1995) Molecular cloning, expression, chromosomal assignment, and tissue-specific expression of a murine α-(1,3)-fucosyltransferase locus corresponding to the human ELAM-1 ligand fucosyltransferase. J Biol Chem 270:25047–25056

Goelz SE, Hession C, Goff D, Griffiths B, Tizard R, Newman B, Chi-Rosso G, Lobb R (1990) ELFT: a gene that directs the expression of an ELAM-1 ligand. Cell 63: 1349–1356

Grabenhorst E, Nimtz M, Costa J, Conradt HS (1998) In vivo specificity of human α1,3/1, 4-fucosyltransferases III–VII in the biosynthesis of Lewis^x and sialyl Lewis^x motifs on complex-type N-glycans. J Biol Chem 273:30985–30994

Huang M-C, Zollner O, Moll T, Maly P, Thall AD, Lowe JB, Vestweber D (2000) P-selectin glycoprotein ligand-1 and E-selectin ligand-1 are differentially modified by fucosyltransferases Fuc-TIV and Fuc-TVII in mouse neutrophils. J Biol Chem 275:31353–31360

Kaneko M, Kudo T, Iwasaki H, Ikehara Y, Nishihara S, Nakagawa S, Sasaki K, Shiina T, Inoko H, Saitou N, Narimatsu H (1999) α1,3-fucosyltransferase IX (Fuc-TIX) is very highly conserved between human and mouse: molecular cloning, characterization, and tissue distribution of human Fuc-TIX. FEBS Lett 452:237–242

Lowe JB, Kukowska-Latallo JF, Nair RP, Larsen RD, Marks RM, Macher BA, Kelly RJ, Ernst LK (1991) Molecular cloning of a human fucosyltransferase gene that determines expression of the Lewis x and VIM-2 epitopes but not ELAM-1-dependent adhesion. J Biol Chem 266:17467–17477

Marer NL, Palcic MM, Clarke JL, Davies D, Skacel PO (1997) Developmental regulation of α1,3-fucosyltransferase expression in CD34-positive progenitors and maturing myeloid cells isolated from normal bone marrow. Glycobiology 7:357–365

Mollicone R, Gibaud A, Francois A, Ratcliffe M, Oriol R (1990) Acceptor specificity and tissue distribution of three human α-3-fucosyltransferases. Eur J Biochem 191: 169–176

Nakayama F, Nishihara S, Iwasaki H, Okubo R, Kaneko M, Kudo T, Nakamura M, Karube M, Narimatsu H (2001) CD15 expression in mature granulocytes is determined by α1,3-fucosyltransferase IX (FUT9), but in promyelocytes and monocytes by α1, 3-fucosyltransferase IV (FUT4) J Biol Chem 276:16100–16106

Niemela R, Natunen J, Majuri M-L, Maaheimo H, Helin J, Lowe JB, Renkonen O, Renkonen R (1998) Complementary acceptor and site specificity of Fuc-TIV and Fuc-TVII allow effective biosynthesis of sialyl-triLe^x and related polylactosamines present on glyco-protein counterreceptors of selectins. J Biol Chem 273:4021–4026

Nishihara S, Iwasaki H, Kaneko M, Tawada A, Ito M, Narimatsu H (1999) α1,3-fucosyltransferase 9 (FUT9; Fuc-TIX) preferentially fucosylates the distal GlcNAc residue of polylactosamine chain while the other four α1,3FUT members preferentially fucosylate the inner GlcNAc residue. FEBS Lett 462:289–294

Weninger W, Ulfman LH, Cheng G, Souchkova N, Quackenbush EJ, Lowe JB, von Andrian UH (2000) Specialized contributions by α(1,3)-fucosyltransferase-IV and Fuc-TVII during leukocyte rolling in dermal microvessels. Immunity 12:665–676

30

α3-Fucosyltransferase-V (FUT5)

Introduction

In 1992, Weston et al. isolated a human α3 fucosyltransferase gene homologous to, but distinct from, previously reported genes for two fucosyltransferases, Fuc-TIII (FUT3) and Fuc-TIV (FUT4), and named it Fuc-TV (FUT5). FUT5 shared 91% amino acid sequence identity with FUT3, and the *FUT5* gene encoding the enzyme was located on chromosome 19, which is similar to that for FUT3. The gene is only sparsely expressed in cells and tissues, and its physiological function remains largely unknown, while the enzymatic characteristics of the recombinant enzyme have been studied extensively.

Databanks

α3-Fucosyltransferase-V (FUT5)

NC-IUBMB enzyme classification: E.C.2.4.1.152

Species	Gene	Protein	mRNA	Genomic
Homo sapiens	*FUT5*	Q11128	NM_002034	M81485
		A42270	U27329	NC_001099
		I39046	U27330	–
		I39047	–	–
Pan troglodytes	*FUT5*	–	Y14034	–

Reiji Kannagi

Department of Molecular Pathology, Aichi Cancer Center Research Institute, 1-1 Kanokoden, Chikusaku, Nagoya 464-8681, Japan
Tel. +81-52-762-6111-int. 7050; Fax +81-52-723-5347
e-mail: rkannagi@aichi-cc.jp

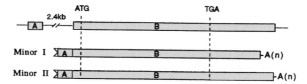

Fig. 1. Genomic structure of *FUT5* and its transcripts. *FUT5* transcripts were isolated by cross-hybridization screenings. Minor transcript I was present at higher levels (>10-fold) than Minor transcript II in all polymerase chain reaction assays (From Cameron et al. 1995)

Name and History

Fucosyltransferase-V is also called FUT5 or Fuc-TV. The amino acid sequence of FUT5 is 91% homologous to that of FUT3. The *FUT5* gene is composed of two exons, and exon B contains an uninterrupted coding sequence for FUT5 protein (Fig. 1), which encodes an enzyme protein of 374 amino acids with four potential *N*-glycosylation sites. The transcripts have no alternative splicing, but use two polyA signal sequences in liver and colon (Cameron et al. 1995). The *FUT5* gene is clustered with two other Lewis-type fucosyltransferases in 19p13.3 in the gene order cen–*FUT5*–*FUT3*–*FUT6*–tel, where *FUT5* and *FUT3* are separated by 23 kb (McCurley et al. 1995).

This enzyme used to be referred to as a "plasma-type" enzyme in earlier literature (Weston et al. 1992; Chandrasekaran et al. 1996). Then, the plasma-type fucosyltransferase had been assumed to be either FUT5 or FUT6, but it has since been clarified, in a study on plasma-type fucosyltransferase deficiency, that FUT6, but not FUT5, is the real plasma-type enzyme (Brinkman-Van der Linden et al. 1996).

The gene encoding this enzyme in rodents has not yet been described. A homologue of *FUT5* is known to be present in chimpanzee (Costache et al. 1997), and is found in COS cells (Ishikawa-Mochizuki et al. 1999).

Enzyme Preparation, Activity Assay, and Substrate Specificity

Recombinant FUT5 enzyme is optimum at pH 5.0–7.0, is sensitive to *N*-ethylmaleimide as well as FUT3, and requires Mn^{2+} for full activity (Holmes et al. 1995; Murray et al. 1996, 1997). A K_m of $9 \pm 2\,\mu M$ for GDP-fucose was reported for FUT5, which is almost the same as that of FUT3 (de Vries et al. 1995). Human recombinant FUT5 enzyme prepared from *Spodoptera frugiperda* is available commercially (Calbiochem Co. Ltd. San Diego, CA) (Murray et al. 1996, 1997). The successful in vitro synthesis of sialyl Lewis x tetrasaccharide has been reported using FUT5 (de Vries et al. 1993).

FUT5 has predominantly α1,3-FUT activity with sialylated or nonsialylated type-2 chain acceptors, and only a low α1,4-FUT activity with type-1 chain acceptors. This is in strong contrast to FUT3, which has predominantly α1,4-FUT activity, although it has some activity with selected type-2 chain acceptor substrates, such as H type-2. Because it shares a high level of amino acid sequence similarity with FUT3, and yet has acceptor substrate specificity distinct from that of FUT3, FUT5 has been a target of extensive study into the relationship between protein structure and the enzymatic activity, catalytic activity, and substrate specificity of fucosyltransferases through domain-swapping and site-directed mutagenesis experiments (Xu et al. 1996; Holmes et al. 1995; Legault et al. 1995; Nguyen et al. 1998; Vo et al. 1998; Sherwood et al. 1998). Interestingly, replacement of the stem and transmembrane domain of FUT5 by protein A is known to result in an enzyme with GDP-fucose hydrolyzing activity (de Vries et al. 1995). N-glycosylation at the two conserved C-terminal N-glycosylation sites is known to be required for full enzyme activity (Christensen et al. 2000).

Biological Aspects

The physiological functions of the *FUT5*-encoded enzyme are not yet known. Transcripts of *FUT5* have not been detected in human tissues by Northern blotting, but expression in liver, colon, testis, and brain was detected by reverse transcriptase-polymerase chain reaction (RT-PCR) (Cameron et al. 1995). A small amount of transcripts, detectable only by RT-PCR, was also found in a few cultured melanoma cells and epithelial cancer cells (Kunzendorf et al. 1994; Yago et al. 1993; Wittig et al. 1996), cancer tissues (Mas et al. 1998), and even in leukocytes and related cell lines (Yago et al. 1993; Cameron et al. 1995) such as T-lymphocytes and HL-60 cells. However, it is not known whether these transcripts lead to the production of a significant amount of FUT5 enzyme in these cells, or merely represent so-called "ectopic" or "illegitimate" transcripts (Kaplan et al. 1992). The antibody specific to FUT5 failed to detect the enzyme protein except in the cells transfected with FUT5 cDNA, where the enzyme was shown to be concentrated at Golgi apparatus and at cell surfaces (Borsig et al. 1996).

Besides FUT3, FUT5 is the only enzyme which, although much lower than FUT3, can utilize type-1 chain substrate and is capable of synthesizing Lewis a and related substances. If the enzyme happens to be really functional somewhere in the human body, it might explain sporadic reports describing the ectopic appearance of Lewis a and/or Lewis a-related antigen in Lewis[a−/b−] cancer patients and healthy individuals (Mandel et al. 1991; Orntoft et al. 1991; Henry et al. 1994). Otherwise, these reports could be the result of false blood typing of the patients either by epigenetic masking of Lewis-related antigens on erythrocytes due to aberrant plasma lipoprotein metabolism, or by inappropriate reagents for blood group typing.

Future Perspectives

Obviously further investigation is awaited to elucidate the physiological significance of this enzyme. Since the enzymatic characteristics of FUT5 are well elucidated using recombinant enzyme, it can be widely applied for industrial purposes such as large-scale synthesis of selectin ligand carbohydrates, and other areas requiring a large amount of fucosyltransferase preparation.

Further Reading

For reviews, see Lowe (1997), Oriol et al. (1999), and Costache et al. (1997)

I thank Dr. Brent W. Weston in University of North Carolina for pertinent advice on the manuscript and allowing reproduction of figures.

References

Borsig L, Kleene R, Dinter A, Berger EG (1996) Immunodetection of α1-3-fucosyltransferase (Fuc-TV). Eur J Cell Biol 70:42–53

Brinkman-Van der Linden ECM, Mollicone R, Oriol R, Larson G, van den Eijnden DH, Van Dijk W (1996) A missense mutation in the *FUT6* gene results in total absence of α3-fucosylation of human α1-acid glycoprotein. J Biol Chem 271:14492–14495

Cameron HS, Szczepaniak D, Weston BW (1995) Expression of human chromosome 19p α(1,3)-fucosyltransferase genes in normal tissues. Alternative splicing, polyadenylation, and isoforms. J Biol Chem 270:20112–20122

Chandrasekaran EV, Jain RK, Larsen RD, Wlasichuk K, DiCioccio RA, Matta KL (1996) Specificity analysis of three clonal and five nonclonal α1,3-L-fucosyltransferases with sulfated, sialylated, or fucosylated synthetic carbohydrates as acceptors in relation to the assembly of 3′-sialyl-6′-sulfo Lewis^x (the L-selectin ligand) and related complex structures. Biochemistry 35:8925–8933

Christensen LL, Jensen UB, Bross P, Orntoft TF (2000) The C-terminal *N*-glycosylation sites of the human α1,3/4-fucosyltransferase III, -V, and VI (hFucTIII, -V and -VI) are necessary for the expression of full enzyme activity. Glycobiology 10:931–939

Costache M, Cailleau A, Fernandez-Mateos P, Oriol R, Mollicone R (1997) Advances in molecular genetics of α2- and α3/4-fucosyltransferases. Transfus Clin Biol 4:367–382

de Vries T, van den Eijnden DH, Schultz J, O'Neill R (1993) Efficient enzymatic synthesis of the sialyl-Lewis^x tetrasaccharide. A ligand for selectin-type adhesion molecules. FEBS Lett 330:243–248

de Vries T, Srnka CA, Palcic MM, Swiedler SJ, van den Eijnden DH, Macher BA (1995) Acceptor specificity of different length constructs of human recombinant α1,3/4-fucosyltransferases. Replacement of the stem region and the transmembrane domain of fucosyltransferase V by protein A results in an enzyme with GDP-fucose hydrolyzing activity. J Biol Chem 270:8712–8722

Henry SM, Oriol R, Samuelsson BE (1994) Detection and characterization of Lewis antigens in plasma of Lewis-negative individuals. Evidence of chain extension as a result of reduced fucosyltransferase competition. Vox Sang 67:387–396

Holmes EH, Xu Z, Sherwood AL, Macher BA (1995) Structure–function analysis of human α1→3fucosyltransferases. A GDP-fucose-protected, *N*-ethylmaleimide-sensitive site in Fuc-TIII and Fuc-TV corresponds to Ser178 in Fuc-TIV. J Biol Chem 270:8145–8151

Ishikawa-Mochizuki I, Kitaura M, Baba M, Nakayama T, Izawa D, Imai T, Yamada H, Hieshima K, Suzuki R, Nomiyama H, Yoshie O (1999) Molecular cloning of a novel CC chemokine, interleukin-11 receptor alpha-locus chemokine (ILC), which is located on chromosome 9p13 and a potential homologue of a CC chemokine encoded by *Molluscum contagiosum* virus. FEBS Lett 460:544–548

Kaplan JC, Kahn A, Chelly J (1992) Illegitimate transcription: its use in the study of inherited disease. Hum Mutat 1:357–360

Kunzendorf U, Krueger-Krasagakes S, Notter M, Hock H, Walz G, Diamantstein T (1994) A sialyl-Le^x-negative melanoma cell line binds to E-selectin but not to P-selectin. Cancer Res 54:1109–1112

Legault DJ, Kelly RJ, Natsuka Y, Lowe JB (1995) Human alpha(1,3/1,4)-fucosyltransferases discriminate between different oligosaccharide acceptor substrates through a discrete peptide fragment. J Biol Chem 270:20987–20996

Lowe JB (1997) Selectin ligands, leukocyte trafficking, and fucosyltransferase genes. Kidney Int 51:1418–1426

Mandel U, Orntoft TF, Holmes EH, Sorensen H, Clausen H, Hakomori S, Dabelsteen E (1991) Lewis blood group antigens in salivary glands and stratified epithelium: lack of regulation of Lewis antigen expression in ductal and buccal mucosal lining epithelia. Vox Sang 61:205–214

Mas E, Pasqualini E, Caillol N, El Battari A, Crotte C, Lombardo D, Sadoulet MO (1998) Fucosyltransferase activities in human pancreatic tissue: comparative study between cancer tissues and established tumoral cell lines. Glycobiology 8:605–613

McCurley RS, Recinos A, Olsen AS, Gingrich JC, Szczepaniak D, Cameron HS, Krauss R, Weston BW (1995) Physical maps of human α(1,3)fucosyltransferase genes FUT3–FUT6 on chromosomes 19p13.3 and 11q21. Genomics 26:142–146

Murray BW, Takayama S, Schultz J, Wong CH (1996) Mechanism and specificity of human α-1,3-fucosyltransferase V. Biochemistry 35:11183–11195

Murray BW, Wittmann V, Burkart MD, Hung SC, Wong CH (1997) Mechanism of human α-1,3-fucosyltransferase V: glycosidic cleavage occurs prior to nucleophilic attack. Biochemistry 36:823–831

Nguyen AT, Holmes EH, Whitaker JM, Ho S, Shetterly S, Macher BA (1998) Human α1,3/4-fucosyltransferases. I. Identification of amino acids involved in acceptor substrate binding by site-directed mutagenesis. J Biol Chem 273:25244–25249

Oriol R, Mollicone R, Cailleau A, Balanzino L, Breton C (1999) Divergent evolution of fucosyltransferase genes from vertebrates, invertebrates, and bacteria. Glycobiology 9:323–334

Orntoft TF, Holmes EH, Johnson P, Hakomori S, Clausen H (1991) Differential tissue expression of the Lewis blood group antigens: enzymatic, immunohistologic, and immunochemical evidence for Lewis a and b antigen expression in Le[a–/b–] individuals. Blood 77:1389–1396

Sherwood AL, Nguyen AT, Whitaker JM, Macher BA, Stroud MR, Holmes EH (1998) Human α1,3/4-fucosyltransferases. III. A Lys/Arg residue located within the α1,3-FucT motif is required for activity but not substrate binding. J Biol Chem 273:25256–25260

Vo L, Lee S, Marcinko MC, Holmes EH, Macher BA (1998) Human α1,3/4-fucosyltransferases. II. A single amino acid at the COOH terminus of Fuc-TIII and -V alters their kinetic properties. J Biol Chem 273:25250–25255

Weston BW, Nair RP, Larsen RD, Lowe JB (1992) Isolation of a novel human α(1,3)fucosyltransferase gene and molecular comparison to the human Lewis blood group α(1,3/1,4)fucosyltransferase gene. Syntenic, homologous, nonallelic genes encoding enzymes with distinct acceptor substrate specificities. J Biol Chem 267:4152–4160

Wittig BM, Thees R, Kaulen H, Gott K, Bartnik E, Schmitt C, Meyer zum Buschenfelde KH, Dippold W (1996) α(1,3)Fucosyltransferase expression in E-selectin-mediated binding of gastrointestinal tumor cells. Int J Cancer 67:80–85

Xu Z, Vo L, Macher BA (1996) Structure–function analysis of human α-1,3-fucosyltransferase. Amino acids involved in acceptor substrate specificity. J Biol Chem 271:8818–8823

Yago K, Zenita K, Ginya H, Sawada M, Ohmori K, Okuma M, Kannagi R, Lowe JB (1993) Expression of α(1,3)fucosyltransferases which synthesize sialyl Le[x] and sialyl Le[a], the carbohydrate ligands for E- and P-selectins, in human malignant cell lines. Cancer Res 53:5559–5565

α3-Fucosyltransferase-VI (FUT6)

Introduction

Fuc-TVI (FUT6) was cloned as a fourth member of the human $\alpha1\rightarrow3/4$ fucosyltransferase gene family by Weston et al. in 1992. The enzyme shared 85% and 89% amino acid sequence identity with previously cloned Fuc-TIII (FUT3) and Fuc-TV (FUT5), respectively, and the *FUT6* gene encoding the enzyme was shown to be located on chromosome 19, where the genes for FUT3 and FUT5 were known to be located. FUT6 could synthesize Lewis X and sialyl Lewis X, but not Lewis A or sialyl Lewis A. The enzyme turned out to be a "plasma-type" fucosyltransferase, and is known to synthesize the sialyl Lewis X determinant carried by plasma proteins and epithelial cancer cells.

Databanks

α3-Fucosyltransferase-VI (FUT6)

NC-IUBMB enzyme classification: E.C.2.4.1.152

Species	Gene	Protein	mRNA	Genomic
Homo sapiens	*FUT6*	P51993	M98825	L01698
		A45156	NM_000150	NC_001099
		I39048–I39054	U27331–U27337	–
		JC1228	–	–
Pan troglodytes	*FUT6*	–	Y14035	–
Bos taurus	*FUTb*	–	X87810	–

REIJI KANNAGI

Department of Molecular Pathology, Aichi Cancer Center Research Institute, 1-1 Kanokoden, Chikusaku, Nagoya 464-8681, Japan
Tel. +81-52-762-6111-int. 7050; Fax +81-52-723-5347
e-mail: rkannagi@aichi-cc.jp

Name and History

The gene encoding FUT6 was cloned as the fourth member of the human α3-fucosyltransferase family by screening the human genomic DNA phage library at low stringency using FUT3 cDNA (Weston et al. 1992), or by polymerase chain reaction (PCR) using primers derived from the DNA sequence of FUT3 (Koszdin and Bowen 1992). The human *FUT6* gene is located approximately 70 kb telomeric of *FUT5*, and is located 14 kb telomeric of *FUT3* in 19p13.3 (McCurley et al. 1995; Nishihara et al. 1993). The *FUT6* gene is composed of seven exons. The first exon transcribed can be either exon A or exon C, suggesting that at least two transcriptional start sites are used, in addition to alternative splicing in 5'- and 3'-UT regions (Cameron et al. 1995). Exon E contains an uninterrupted coding sequence for FUT6 protein, which encode an enzyme protein of 358 amino acids with four potential *N*-glycosylation sites. The nucleic acid sequence of FUT6 is 91% homologous to that of FUT3, and its amino acid sequence is 85% and 89% identical to that of FUT3 and FUT5, respectively (Weston et al. 1992; Koszdin and Bowen 1992).

FUT6 was found to correspond to a "plasma-type" α3-fucosyltransferase from a study of plasma fucosyltransferase deficiency found in Indonesia (Mollicone et al. 1994; Brinkman-Van der Linden et al. 1996). In these individuals, either a missense mutation (G739A, Glu-247 → Lys) in the catalytic domain, or a nonsense mutation (C945A, Tyr-315 → stop codon) truncating the COOH terminus of the enzyme, leads to the total loss of the enzymatic activity of FUT6 (Mollicone et al. 1994). As many as 9% of individuals on the island of Java, Indonesia, were found to lack the plasma-type enzymic activity. The extensive sequence diversity of the *FUT6* gene was also noted in Caucasian, European–African, and Japanese populations (Larson et al. 1996; Pang et al. 1999).

Enzyme Activity Assay and Substrate Specificity

Plasma-type α3-fucosyltransferase is known to be optimum at pH 7.2–8.0, sensitive to *N*-ethylmaleimide, heat labile, and requires a metal ion such as 20 mM Mn^{2+} for full activity (Mollicone et al. 1990). As well as *N*-ethylmaleimide-sensitive cysteine, some histidine residue(s) is suggested to be essential for the catalytic activity, since the diethylpyrocarbonate (DEPC) treatment abrogates the activity of the recombinant enzyme (Britten and Bird 1997). The K_m for GDP-fucose is reported to be in the range 6–10 μM when determined under saturating conditions (Mollicone et al. 1990; Sherwood and Holmes 1999; Cailleau-Thomas et al. 2000). FUT6 is suggested to be co-localized in *trans*-Golgi networks with β4-GalT-I, and actively released into the cell culture medium by proteolytic cleavage (Borsig et al. 1998, 1999).

Both human and chimpanzee FUT6 utilizes only type-2 chain polylactosamines (and not type-1 chain polylactosamines) as substrates (Weston et al. 1992; Costache et al. 1997a). Sialylated oligosaccharides are good substrates as well as nonsialylated ones, and the reaction catalyzed by the enzyme produces sialyl Lewis X and Lewis X determinants. As to sialylated oligosaccharides, only NeuAcα2-3-substituted lactosamine serves as substrate, whereas NeuAcα2-6-substituted lactosamine does not. FUT6 is also active on α1-2 fucosylated type 2 chain substrates. Its Km for an H-type-2 chain substrate was reported to be 17 μM, while Km of FUT3 for the same substrate was 1330 μM (Cailleau-Thomas et al. 2000). The enzyme also has the capability to transfer fucose residues to internal GlcNAc moiety. Preference for the internal and terminal GlcNAc residues is highly dependent on the assay conditions applied for the enzymatic activity (Sherwood and Holmes 1999).

The enzyme was shown to be active on the polylactosamine substrate having 6-sulfate modification at the GlcNAc moiety, and gives rise to sialyl and nonsialyl 6-sulfo Lewis X (Izawa et al. 2000), a potent ligand for L-selectin (Mitsuoka et al. 1998), while Fuc-T-III is less active on 6-sulfated type-2 chain substrates. Polylactosamine substrates having 3'-sulfate modification at the terminal Gal moiety are good substrate for both FUT6 and FUT3 (Ikeda et al. 2001).

Preparation

Plasma-type α1-3 fucosyltransferase was identified in plasma, milk, saliva, and liver. In accordance with this, transcripts of *FUT6* were detected by Northern blotting in liver, kidney, colon, small intestine, bladder, uterus, and salivary gland (Cameron et al. 1995). A 3.5-kb transcript containing exons A, B, C, D, E, and G (Fig. 1) was found in liver, kidney, and small intestine, and presumably in cultured hepatocellular carcinoma cell lines such as HepG2 (Cameron et al. 1995; Yago et al. 1993). A 2.5-kb transcript composed of exons C, D, E, and F (Fig. 1) was widely distributed in kidney, liver, colon, bladder, uterus, and salivary gland, and probably in most epithelial cancer cell lines (Cameron et al. 1995; Yago et al. 1993; Matsuura et al. 1998).

The *FUT6* coding sequence of chimpanzees is highly homologous to human counterparts (Costache et al. 1997a), and bovine *futb* gene is assumed to be the orthologous homolog of human *FUT6* (Oulmouden et al. 1997). The gene encoding the FUT6 homologue has not yet been described in rodents. No α3-fucosyltransferase is detected in murine liver, and the acute-phase protein in mice is known to lack the sialyl Lewis X epitope (Havenaar et al. 1998), while there are sporadic reports describing the presence of plasma-type-like enzymatic activity in certain organs of rodents, such as rat epididymis (Raychoudhury and Millette 1995) or colon (Karaivanova et al. 1996).

The purification of FUT6 to near homogeneity has been achieved by successive chromatography on CM-Sephadex, Phenyl Sepharose, GDP-hexanolamine-Sepharose,

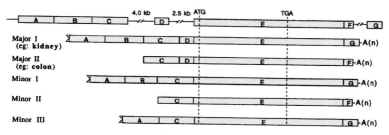

Fig. 1. Genomic and transcript structures of *FUT6* and position of exons A–G. Major transcripts are detectable on Northern analyses and in at least three cDNA clones obtained by hybridization (Major I, six clones; Major II, five clones; Minor I, two clones; Minor II, two clones; Minor III, one clone). The 3.5-kb FUT6 transcript visible on Northern blot analyses (e.g., kidney and liver) is mostly comprised of the Major I species. The 2.5-kb transcript (e.g., colon and liver) corresponds to Major II. (From Cameron et al. 1995)

and HPLC (Johnson et al. 1995). Plasma-type α3-fucosyltransferase was also separated from the Lewis-type enzyme by a bovine IgG glycopep-Sepharose column, and highly purified by Sephacryl S-100 HR column chromatography from human milk (Chandrasekaran et al. 1994). Recombinant enzyme was produced using either insect cells (de Vries et al. 1997) or yeast (Malissard et al. 2000). The recombinant enzyme is commercially available from Calbiochem (San Diego, CA), and was used successfully in the recent enzymological synthesis of selectin ligands, including mucin-type ligand for P-selectin (Leppänen et al. 1999).

Specific antibodies directed to FUT6 were generated, and utilized to study localization and trafficking of the enzyme in transfectant cells (Borsig et al. 1998).

Biological Aspects

It has been shown that the sialyl Lewis X determinant, a carbohydrate ligand for well-known cell adhesion molecules, selectins (Lowe 1997), is expressed on some acute-phase proteins such as α_1-acid glycoprotein, α_1-antichymotripsin, and haptoglobin, and the FUT6 enzyme in the liver is mainly responsible for the determinant (De Graaf et al. 1993; Brinkman-Van der Linden et al. 1998). The increase in soluble plasma proteins carrying sialyl Lewis X in sera of patients with inflammatory diseases is believed to have a regulatory role in leukocyte chemotaxis (Brinkman-Van der Linden et al. 1998). Some inflammatory cytokines, including IL-6, are believed to be responsible for the increased synthesis of the determinant in inflammatory diseases. FUT6 enzyme protein was shown to be released to the cell culture medium by using CHO cells transfected with cDNA for this enzyme (Borsig et al. 1998).

Cancer cells of various origins, such as the colon, are known to express sialyl Lewis X and sialyl Lewis A, which serve as ligands for the cell adhesion molecule, E-selectin, expressed on vascular endothelial cells (Takada et al. 1991, 1993). This cell adhesion system is believed to be involved in the mechanism of hematogenous metastasis of cancer (Kannagi 1997). Since FUT6 is widely expressed in these cancer cells, it is assumed to be at least partly responsible for the synthesis of sialyl Lewis X, but not sialyl Lewis A, in these cancer cells. Another candidate enzyme that has an ability to catalyze sialyl Lewis X synthesis in these cells is FUT3. There is a certain disagreement among researchers as to the relative importance of FUT3 and FUT6 in the synthesis of sialyl Lewis X in these cells, while there is no doubt that FUT3, but not FUT6, synthesizes sialyl Lewis A in these cells. It is not easy to decide which one of the two Fuc-Ts plays a major role in the synthesis of sialyl Lewis X, since most epithelial cancer cells co-express the two enzymes. It is known that a type-2 chain substrate is much less preferred by FUT3 than a type-1 chain substrate when both substrates are available for the enzyme (Nimtz et al. 1998), while FUT6 exclusively utilizes the type-2 chain substrates. Expression of sialyl Lewis X and A on cancer cells was abolished when the antisense cDNA that suppresses transcription of both enzymes was introduced into the cells (Weston et al. 1999; Hiller et al. 2000).

FUT6 is widely assumed to be a major enzyme responsible for the synthesis of sialyl Lewis X in epithelial cells as well as in cancer cells. However, there is no concrete evidence that the α3-fucosyltransferases play a role as a late-limiting enzyme in the synthesis of sialyl Lewis X/A in epithelial cancer cells. The specific activity of α3-fucosyltransferases in most malignant or nonmalignant cells of epithelial origin is usually one or two orders of magnitude higher than that of sialyltransferases, which would provide the appropriate substrates to the enzyme for the synthesis of sialyl Lewis X. It is true that expression of sialyl Lewis X is increased in cancer compared with its normal cell counterpart, but with few exceptions no consistent increase in α3-fucosyltransferase activity or mRNA in cancer cells, nor its significant correlation to the degree of sialyl Lewis X expression, has been observed in several well-controlled studies (Ito et al. 1997; Petretti et al. 1999, 2000; Dohi et al. 1994; Hada et al. 1995; Hutchinson et al. 1991; Mas et al. 1998).

Plasma α3-fucosyltransferase is known to be elevated in patients with a variety of malignant disorders, including liver cancer (Hada et al. 1995; Hutchinson et al. 1991), and the enzyme detected in the plasma must at least partly consist of the plasma-type fucosyltransferase. It is noteworthy, however, that enzyme activity in the homogenates prepared from liver cancer was reported to be significantly lower than that in non-malignant liver (Hutchinson et al. 1991). The plasma enzymatic activity is also reportedly elevated in heavy alcohol drinkers (Thompson et al. 1991). On the other hand, a significant reduction in the plasma α3-fucosyltransferase level was reported in patients with schizophrenia (Yazawa et al. 1999).

A peculiar historical aspect of FUT6 is that a small amount of its transcript was repeatedly detected in human leukocytes (Cameron et al. 1995; Yago et al. 1993). One of the earlier cloning studies even successfully employed a cDNA library prepared from HL60 cells to obtain FUT6 cDNA (Koszdin and Bowen 1992). The transcript was not detectable in leukocytes by Northern blotting, but it is sometimes detected by well-controlled reverse transcriptase-PCR (RT-PCR) studies (Cameron et al. 1995; Yago et al. 1993). This had led to the earlier suggestion that FUT6 might be a candidate for the synthesis of sialyl Lewis X in leukocytes, before the discovery and cloning of FUT7 in 1994 (Sasaki et al. 1994; Natsuka et al. 1994). At present, it is well established that sialyl Lewis X in leukocytes is synthesized through the action of FUT7, but not that of FUT6. The appearance of FUT6 transcript in RT-PCR in leukocytes could be a typical example of so-called "illegitimate transcripts" (Kaplan et al. 1992), and may reflect a certain leakiness in the regulation of the *FUT6* gene transcription.

Stage-specific expressions of α1→3 fucosyltransferases are noted in the developing kidney: myeloid-type (FUT4) appears at weeks 5–8 in the embryo kidney, followed by the plasma-type (FUT6) in the second trimester in proximal convoluted tubules and Henle's loops, and the Lewis-type enzyme (FUT3) in the calyx and collecting tubules (Candelier et al. 1993). Recently FUT6 was found to be located in Weibel-Parade bodies of human endothelial cells (Schnyder-Candrian et al. 2000). This may be a clue to glycosylation-independent functions of FUT6.

Future Perspectives

Since the sialyl Lewis X determinant synthesized by the enzyme plays an important role in cancer metastasis, suppression of its activity would be beneficial for cancer treatment. It is reported that human colon carcinoma cell lines transfected with anti-sense cDNA, which suppresses both FUT3 and FUT6, had much reduced proliferative activity both in vitro and in vivo, suggesting some growth-promoting effects for the products of these enzymes in addition to cell-adhesive activity (Hiller et al. 2000). A lack of description on the gene encoding this enzyme in rodents so far indicates that the knockout-mice approach will not provide any insight into the physiological significance of this enzyme in the very near future. The finding that individuals deficient in FUT6 are apparently healthy would indicate that the enzyme is not indispensable for supporting life.

Acknowledgments

I thank Dr. Brent W. Weston in University of North Carolina for pertinent advice on the manuscript and allowing reproduction of figures.

Further Reading

For reviews, see Lowe (1997), Oriol et al. (1999) and Costache et al. (1997b).

References

Borsig L, Katopodis AG, Bowen BR, Berger EG (1998) Trafficking and localization studies of recombinant α1,3-fucosyltransferase VI stably expressed in CHO cells. Glycobiology 8:259–268

Borsig L, Imbach T, Höchli M, Berger EG (1999) α1,3Fucosyltransferase VI is expressed in HepG2 cells and codistributed with β1,4galactosyltransferase I in the Golgi apparatus and monensin-induced swollen vesicles. Glycobiology 9:1273–1280

Brinkman-Van der Linden ECM, Mollicone R, Oriol R, Larson G, van den Eijnden DH, Van Dijk W (1996) A missense mutation in the FUT6 gene results in total absence of α3-fucosylation of human α1-acid glycoprotein. J Biol Chem 271:14492–14495

Brinkman-Van der Linden EC, de Haan PF, Havenaar EC, Van Dijk W (1998) Inflammation-induced expression of sialyl Lewis X is not restricted to α1-acid glycoprotein, but also occurs to a lesser extent on α1-antichymotrypsin and haptoglobin. Glycoconj J 15:177–182

Britten CJ, Bird MI (1997) Chemical modification of an α3-fucosyltransferase: definition of amino acid residues essential for enzyme activity. Biochim Biophys Acta 1334: 57–64

Cailleau-Thomas A, Coullin P, Candelier JJ, Balanzino L, Mennesson B, Oriol R, Mollicone R (2000) FUT4 and FUT9 genes are expressed early in human embryogenesis. Glycobiology 10:789–802

Cameron HS, Szczepaniak D, Weston BW (1995) Expression of human chromosome 19p α(1,3)-fucosyltransferase genes in normal tissues—alternative splicing, polyadenylation, and isoforms. J Biol Chem 270:20112–20122

Candelier J-J, Mollicone R, Mennesson B, Bergemer A-M, Henry S, Coullin P, Oriol R (1993) α-3-fucosyltransferases and their glycoconjugate antigen products in the developing human kidney. Lab Invest 69:449–459

Chandrasekaran EV, Rhodes JM, Jain RK, Matta KL (1994) Lactose as affinity eluent and a synthetic sulfated copolymer as inhibitor, in conjunction with synthetic and natural acceptors, differentiate human milk Lewis-type and plasma-type α-L-fucosyltransferases. Biochem Biophys Res Commun 198:350–358

Costache M, Apoil PA, Cailleau A, Elmgren A, Larson G, Henry S, Blancher A, Iordachescu D, Oriol R, Mollicone R (1997a) Evolution of fucosyltransferase genes in vertebrates. J Biol Chem 272:29721–29728

Costache M, Cailleau A, Fernandez-Mateos P, Oriol R, Mollicone R (1997b) Advances in molecular genetics of α2- and α3/4-fucosyltransferases. Transfus Clin Biol 4:367–382

De Graaf TW, Van der Stelt ME, Anbergen MG, Van Dijk W (1993) Inflammation-induced expression of sialyl Lewis X-containing glycan structures on α1-acid glycoprotein (orosomucoid) in human sera. J Exp Med 177:657–666

de Vries T, Palcic MP, Schoenmakers PS, van den Eijnden DH, Joziasse DH (1997) Acceptor specificity of GDP-Fuc:Galβ1→4GlcNAc-R α3-fucosyltransferase VI (Fuc-T VI) expressed in insect cells as soluble, secreted enzyme. Glycobiology 7:921–927

Dohi T, Hashiguchi M, Yamamoto S, Morita H, Oshima M (1994) Fucosyltransferase-producing sialyl Le^a and sialyl Le^x carbohydrate antigen in benign and malignant gastrointestinal mucosa. Cancer 73:1552–1561

Ikeda N, Eguchi H, Nishihara S, Narimatsu H, Kannagi R, Irimura T, Ohta M, Matsuda H, Taniguchi N, Honke K (2001) A remodeling system of the 3′-sulfo Lewis a and 3′-sulfo Lewis X epitopes. J Biol Chem, in press ([doi] 10.1074/jbc. M107390200)

Hada T, Fukui K, Ohno M, Akamatsu S, Yazawa S, Enomoto K, Yamaguchi K, Matsuda Y, Amuro Y, Yamanaka N (1995) Increased plasma α(1→3)-L-fucosyltransferase activities in patients with hepatocellular carcinoma. Glycoconj J 12:627–631

Havenaar EC, Hoff RC, van den Eijnden DH, Van Dijk W (1998) Sialyl Lewis^x epitopes do not occur on acute-phase proteins in mice: relationship to the absence of α3-fucosyltransferase in the liver. Glycoconj J 15:389–395

Hiller KM, Mayben JP, Bendt KM, Manousos GA, Senger K, Cameron HS, Weston BW (2000) Transfection of α(1,3)fucosyltransferase antisense sequences impairs the proliferative and tumorigenic ability of human colon carcinoma cells. Mol Carcinog 27:280–288

Hutchinson WL, Du M-Q, Johnson PJ, Williams R (1991) Fucosyltransferases: differential plasma and tissue alterations in hepatocellular carcinoma and cirrhosis. Hepatology 13:683–688

Ito H, Hiraiwa N, Sawada-Kasugai M, Akamatsu S, Tachikawa T, Kasai Y, Akiyama S, Ito K, Takagi H, Kannagi R (1997) Altered mRNA expression of specific molecular species of fucosyl- and sialyltransferases in human colorectal cancer tissues. Int J Cancer 71:556–564

Izawa M, Kumamoto K, Mitsuoka C, Kanamori A, Ohmori K, Ishida H, Nakamura S, Kurata-Miura K, Sasaki K, Nishi T, Kannagi R (2000) Expression of sialyl 6-sulfo Lewis X is inversely correlated with conventional sialyl Lewis^x expression in human colorectal cancer. Cancer Res 60:1410–1416

Johnson PH, Donald AS, Clarke JL, Watkins WM (1995) Purification, properties and possible gene assignment of an α1,3-fucosyltransferase expressed in human liver. Glycoconj J 12:879–893

Kannagi R (1997) Carbohydrate-mediated cell adhesion involved in hematogenous metastasis of cancer. Glycoconj J 14:577–584

Kaplan JC, Kahn A, Chelly J (1992) Illegitimate transcription: its use in the study of inherited disease. Hum Mutat 1:357–360

Karaivanova V, Mookerjea S, Hunt D, Nagpurkar A (1996) Characterization and purification of fucosyltransferases from the cytosol of rat colon. Int J Biochem Cell Biol 28:165–174

Koszdin KL, Bowen BR (1992) The cloning and expression of a human α-1,3-fucosyltransferase capable of forming the E-selectin ligand. Biochem Biophys Res Commun 187:152–157

Larson G, Borjeson C, Elmgren A, Kernholt A, Henry S, Fletcher A, Aziz A, Mollicone R, Oriol R (1996) Identification of a new plasma α(1,3)fucosyltransferase (FUT6) allele requires an extended genotyping strategy. Vox Sang 71:233–241

Leppänen A, Mehta P, Ouyang YB, Ju T, Helin J, Moore KL, van Die I, Canfield WM, McEver RP, Cummings RD (1999) A novel glycosulfopeptide binds to P-selectin and inhibits leukocyte adhesion to P-selectin. J Biol Chem 274:24838–24848

Lowe JB (1997) Selectin ligands, leukocyte trafficking, and fucosyltransferase genes. Kidney Int 51:1418–1426

Malissard M, Zeng S, Berger EG (2000) Expression of functional soluble forms of human β-1,4-galactosyltransferase I, α-2,6-sialyltransferase, and α-1,3-fucosyltransferase VI in the methylotrophic yeast *Pichia pastoris*. Biochem Biophys Res Commun 267: 169–173

Mas E, Pasqualini E, Caillol N, El Battari A, Crotte C, Lombardo D, Sadoulet MO (1998) Fucosyltransferase activities in human pancreatic tissue: comparative study between cancer tissues and established tumoral cell lines. Glycobiology 8:605–613

Matsuura N, Narita T, Hiraiwa N, Hiraiwa M, Murai H, Iwase T, Funahashi H, Imai T, Takagi H, Kannagi R (1998) Gene expression of fucosyl- and sialyltransferases which synthesize sialyl Lewis[x], the carbohydrate ligands for E-selectin, in human breast cancer. Int J Oncol 12:1157–1164

McCurley RS, Recinos A, Olsen AS, Gingrich JC, Szczepaniak D, Cameron HS, Krauss R, Weston BW (1995) Physical maps of human α(1,3)fucosyltransferase genes FUT3–FUT6 on chromosomes 19p13.3 and 11q21. Genomics 26:142–146

Mitsuoka C, Sawada-Kasugai M, Ando-Furui K, Izawa M, Nakanishi H, Nakamura S, Ishida H, Kiso M, Kannagi R (1998) Identification of a major carbohydrate capping group of the L-selectin ligand on high-endothelial venules in human lymph nodes as 6-sulfo sialyl Lewis X. J Biol Chem 273:11225–11233

Mollicone R, Gibaud A, Francois A, Ratcliffe M, Oriol R (1990) Acceptor specificity and tissue distribution of three human α-3-fucosyltransferases. Eur J Biochem 191: 169–176

Mollicone R, Reguigne I, Fletcher A, Aziz A, Rustam M, Weston BW, Kelly RJ, Lowe JB, Oriol R (1994) Molecular basis for plasma α(1,3)-fucosyltransferase gene deficiency (FUT6). J Biol Chem 269:12662–12671

Natsuka S, Gersten KM, Zenita K, Kannagi R, Lowe JB (1994) Molecular cloning of a cDNA encoding a novel human leukocyte α-1,3-fucosyltransferase capable of synthesizing the sialyl Lewis X determinant. J Biol Chem 269:16789–16794

Nimtz M, Grabenhorst E, Gambert U, Costa J, Wray V, Morr M, Thiem J, Conradt HS (1998) In vitro α1-3 or α1-4 fucosylation of type I and II oligosaccharides with secreted forms of recombinant human fucosyltransferases III and VI. Glycoconj J 15:873–883

Nishihara S, Nakazato M, Kudo T, Kimura H, Ando T, Narimatsu H (1993) Human α-1,3-fucosyltransferase (Fuc-T VI) gene is located at only 13 kb 3′ to the Lewis type fucosyltransferase (Fuc-T III) gene on chromosome 19. Biochem Biophys Res Commun 190:42–46

Oriol R, Mollicone R, Cailleau A, Balanzino L, Breton C (1999) Divergent evolution of fucosyltransferase genes from vertebrates, invertebrates, and bacteria. Glycobiology 9:323–334

Oulmouden A, Wierinckx A, Petit JM, Costache M, Palcic MM, Mollicone R, Oriol R, Julien R (1997) Molecular cloning and expression of a bovine α(1,3)-fucosyltransferase gene homologous to a putative ancestor gene of the human FUT3–FUT5–FUT6 cluster. J Biol Chem 272:8764–8773

Pang H, Koda Y, Soejima M, Schlaphoff T, Du Toit ED, Kimura H (1999) Allelic diversity of the human plasma α(1,3)fucosyltransferase gene (FUT6). Ann Hum Genet 63:277–284

Petretti T, Schulze B, Schlag PM, Kemmner W (1999) Altered mRNA expression of glycosyltransferases in human gastric carcinomas. Biochim Biophys Acta 1428:209–218

Petretti T, Kemmner W, Schulze B, Schlag PM (2000) Altered mRNA expression of glycosyltransferases in human colorectal carcinomas and liver metastases. Gut 46:359–366

Raychoudhury SS, Millette CF (1995) Glycosidic specificity of fucosyltransferases present in rat epididymal spermatozoa. J Androl 16:448–456

Sasaki K, Kurata K, Funayama K, Nagata M, Watanabe E, Ohta S, Hanai N, Nishi T (1994) Expression cloning of a novel α1,3-fucosyltransferase that is involved in biosynthesis of the sialyl Lewis X carbohydrate determinants in leukocytes. J Biol Chem 269: 14730–14737

Sherwood AL, Holmes EH (1999) Analysis of the expression and enzymatic properties of $\alpha 1 \rightarrow 3$ fucosyltransferase from human lung carcinoma NCI-H69 and PC9 cells. Glycobiology 9:637–643

Schnyder-Candrian S, Borsig L, Moser R, Berger EG (2000) Localization of $\alpha 1,3$-fucosyltransferase VI in Weibel-Palade bodies of human endothelial cells. Proc Natl Acad Sci USA 97:8369–8374

Takada A, Ohmori K, Takahashi N, Tsuyuoka K, Yago K, Zenita K, Hasegawa A, Kannagi R (1991) Adhesion of human cancer cells to vascular endothelium mediated by a carbohydrate antigen, sialyl Lewis A. Biochem Biophys Res Commun 179:713–719

Takada A, Ohmori K, Yoneda T, Tsuyuoka K, Hasegawa A, Kiso M, Kannagi R (1993) Contribution of carbohydrate antigens sialyl Lewis A and sialyl Lewis X to adhesion of human cancer cells to vascular endothelium. Cancer Res 53:354–361

Thompson S, Matta KL, Turner GA (1991) Changes in fucose metabolism associated with heavy drinking and smoking: a preliminary report. Clin Chim Acta 201:59–64

Weston BW, Smith PL, Kelly RJ, Lowe JB (1992) Molecular cloning of a fourth member of a human $\alpha(1,3)$fucosyltransferase gene family. Multiple homologous sequences that determine expression of the Lewis X, sialyl Lewis X, and difucosyl sialyl Lewis X epitopes. J Biol Chem 267:24575–24584

Weston BW, Hiller KM, Mayben JP, Manousos GA, Bendt KM, Liu R, Cusack JC Jr (1999) Expression of human $\alpha(1,3)$fucosyltransferase antisense sequences inhibits selectin-mediated adhesion and liver metastasis of colon carcinoma cells. Cancer Res 59:2127–2135

Yago K, Zenita K, Ginya H, Sawada M, Ohmori K, Okuma M, Kannagi R, Lowe JB (1993) Expression of $\alpha(1,3)$fucosyltransferases which synthesize sialyl Le^x and sialyl Le^a, the carbohydrate ligands for E- and P-selectins, in human malignant cell lines. Cancer Res 53:5559–5565

Yazawa S, Tanaka S, Nishimura T, Miyanaga K, Kochibe N (1999) Plasma $\alpha 1,3$-fucosyltransferase deficiency in schizophrenia. Exp Clin Immunogenet 16:125–130

32

α3-Fucosyltransferase-VII (FUT7)

Introduction

α1,3-Fucosyltransferase-VII (Fuc-TVII, FUT7) was cloned independently by two groups at almost the same time (Sasaki et al. 1994; Natsuka et al. 1994). Sasaki et al. first cloned the FUT7 cDNA from a THP-1 cell cDNA library by an expression cloning method using an anti-sLe^x mAb KM93. Natsuka et al. employed a cross-hybridization method for cloning the FUT7 cDNA. Flow cytometric analysis on the cells stably expressing FUT7 demonstrated that FUT7 can form sLe^x but not Le^x on the cell surface. FUT7 can transfer fucose only to the GlcNAc residue of the sialylated type-2 chain but not to that of type-1 chain. Thus, FUT7 can synthesize only the sLe^x epitope and not the other Lewis antigens such as Le^x, Le^a, Le^b, and sLe^a. The tissue distribution of FUT7 is limited to leukocytes and high-endothelial venules (HEV) (Maly et al. 1996; Clarke and Watkins 1996; Marer et al. 1997; Kaneko et al. 1999).

Databanks

α3-Fucosyltransferase-VII (FUT7)

NC-IUBMB enzyme classification: E.C.2.4.1.152

Species	Gene	Protein	mRNA	Genomic
Homo sapiens	*FUT7, Fuc-TVII*	NP_004470	X78031	NP_024000
		–	U08112	–
Mus musculus	*Fut7, mFuc-TVII*	NP_038552	NM_013524	U45980

HISASHI NARIMATSU

Laboratory of Gene Function Analysis, Institute of Molecular and Cell Biology (IMCB), National Institute of Advanced Industrial Science and Technology (AIST), Central-2, 1-1-1 Umezono, Tsukuba, Ibaraki 305-8568, Japan
Tel. +81-298-61-3200; Fax +81-298-61-3201
e-mail: h.narimatsu@aist.go.jp

Name and History

Before the FUT7 cloning, four α3-fucosyltransferase genes, i.e., *FUT3*, *FUT4*, *FUT5*, and *FUT6*, had been cloned and characterized. Leukocytes express the sLe^x determinant which is E- and P-selectin ligands. FUT3, FUT5, and FUT6 had been confirmed to direct the appearance of the sLe^x epitope on the transfected cells. However, they were not endogenously expressed in leukocytes. Only FUT4 was found to be expressed in leukocytes, but the capability of FUT4 to generate the sLe^x epitope was very weak. Thus, the existence of unknown α1,3FUT which can form the sLe^x epitope in leukocytes had been predicted. Two groups independently succeeded in the cloning of the *FUT7* gene from cDNA libraries of leukocytes.

Enzyme Activity Assay and Substrate Specificity

The substrate specificity of FUT7 is very strict in comparison with those of other FUTs. FUT7 can form only sLe^x, and cannot synthesize the other Lewis antigens such as Le^x, Le^y, Le^a, Le^b, and sLe^a, as can be seen in Fig. 1 of the chapter on FUT3. Niemela et al. (1998) compared the substrate specificity of two enzymes, FUT4 and FUT7, both of which are expressed in leukocytes. They proposed the complementary FUT4 and FUT7 reactions for the multifucosylation of the sialylated-polylactosamine chain, as shown in Fig. 1 of the chapter on FUT4. FUT7 effectively fucosylates only the distal α2,3-sialylated lactosamine (LN) unit in the α2,3-sialylated polylactosamine chain, resulting in the sLe^x epitope. In contrast, FUT4 preferentially fucosylates the inner LN unit, resulting in the synthesis of the VIM-2 (CDw65) epitope. Later, leukocytes were found to express the third enzyme, FUT9, that is also involved in the fucosylation of neutral and sialylated polylactosamine chains. Three α1,3FUTs are involved in the multifucosylation of polylactosamine chains in leukocytes. Shinoda et al. (1998) analyzed the acceptor specificity of FUT7 toward various analogs of a 2-(trimethylsilyl)ethyl α2,3-sialyllacto-*N*-neotetraose. They demonstrated that FUT7 requires three portions of α2,3-sialylated type-2 oligosaccharide structures, i.e., the hydroxyl group at C-4 of Gal, the hydroxyl group at C-3 of GlcNAc, and the carbonylamino group at C-2 of GlcNAc.

Preparation

FUT7 is expressed in a series of hematopoietic cells and HEV. In many laboratories, the *FUT7* gene was transfected into various cells, i.e., COS-1 cells (Natsuka et al. 1994), Namalwa cells (Kimura et al. 1997), BHK-21 cells (Grabenhorst et al. 1998). FUT7 is also available in a soluble protein A chimeric form (Shinoda et al. 1997).

Biological Aspects

FUT7 has been proven to be a key enzyme in the synthesis of E-, P-, and L-selectin ligands by a series of experiments. The treatment of Jurkat cells with phorbol myristate acetate increased the expression of sLe^x epitopes in correlation with the increase

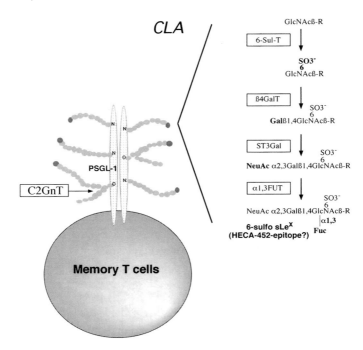

Fig. 1. The biosynthetic pathways of the CLA carbohydrate component, 6-sulfo-sLe^x. Multiple glycosyltransferases are sequentially involved in the CLA synthesis, such as C2GnT, 6-sulfotransferase(s), β4-galactosyltransferase(s), α2,3-sialyltransferase(s), and α1,3FUTs

in FUT7 transcripts, and induced the synthesis of E-selectin ligands (Knibbs et al. 1996). A panel of hematopoietic cell lines were examined for the expression of FUT4 and FUT7. The level of FUT7 expression correlated with E-selectin binding (Wagers et al. 1996). The level of FUT7 expression in the lining of endothelial cells in the HEV of peripheral lymph nodes, mesenteric lymph nodes, and Payer's patches correlated precisely with the L-selectin ligands (Smith et al. 1996). The antisense cDNA of the *FUT7* gene suppressed the expression of the sLe^x epitope, leading to reduced activity of the E-selectin binding (Hiraiwa et al. 1996) Finally, the essential role for FUT7 in E-, P-, and L-selectin ligand biosynthesis has been proven by the knockout mice lacking the *Fut7* gene (Maly et al. 1996). The Fut7-deficient mice exhibited a leukocyte adhesion deficiency characterized by absent leukocyte E- and P-selectin ligand activity and deficient HEV L-selectin ligand activity. This was again clearly demonstrated in recent studies which showed that PSGL-1 expressed by Fut7-deficient neutrophils is not bound by E- or P-selectin (Weninger et al. 2000; Huang et al. 2000) (see Fig. 2 in the chapter on FUT4).

In the human, cutaneous lymphocyte-associated antigen (CLA) plays a key role in skin-homing of CD4$^+$ memory T cells via CLA/E-selectin binding (Fuhlbrigge et al. 1997). The CLA antigenic epitope, which is defined by a monoclonal antibody HECA452, was identified as a 6-sulfo-sLe^x determinant (Mitsuoka et al. 1998) (Fig. 1).

In CD4[+] T cells, interleukin (IL)-12 upregulates FUT7, resulting in the increased activity of E- and P-selectin binding, whereas IL-4 causes their downregulation (Wagers et al. 1998). Figure 1 shows the sequential steps of the biosynthetic pathways of the CLA epitope on O-glycan, in which multiple glycosyltransferases are involved in CLA synthesis. We recently demonstrated that CLA expression in memory T cells is essentially regulated by two glycosyltransferases, FUT7 and β4GalT-I (Nakayama et al. 2000). The results suggested that the *FUT7* gene is transcriptionally regulated directly by IL-12 and IL-4.

However, this is not the case in B cells. Human pre-B cells express a certain amount of sLe[x] antigen and FUT7 transcripts. The amount of sLe[x] antigen in pre-B cells is determined by core 2 β6-*N*-acetylglucosaminyltransferase (C2GnT), not by FUT7 (Nakamura et al. 1998).

Little is known about the transcriptional regulation of the *FUT7* gene. Hiraiwa et al. (1997) demonstrated that Tax protein produced by human T-cell leukemia virus type 1 (HTLV-1) transactivates the *FUT7* gene, leading to the strong expression of sLe[x] antigen in leukemic cells. This indicated that the strong expression of sLe[x] through the upregulation of the *FUT7* gene by Tax protein in adult T-cell leukemic (ATL) cells accelerated the extravascular infiltration of ATL cells.

Future Perspectives

FUT7 expression is restricted to specific cells, i.e., leukocytes and HEV endothelial cells, in which FUT7 is responsible for sLe[x] expression. However, sLe[x] expression in other tissues, such as epithlial cells of the digestive tract, is directed by the enzymes FUT3 and FUT6. The development of chemical compounds which specifically inhibit FUT7 activity will lead to treatments for inflammation and allergy.

Recently, the null alleles of the *FUT7* gene in humans were found in patients with chronic inflammation (Bengtson et al. 2001). The frequency of null alleles of the *FUT7* gene seems to be very low compared with that of null alleles of the *FUT3* and *FUT6* genes, probably because FUT7 is essential to maintain health.

References

Bengtson P, Larson C, Lundblad A, Larson G, Pahlsson P (2001) Identification of a missense mutation (G329A; Arg[110]–Gln) in the human *FUT7* gene. J Biol Chem 276:31575–31582

Clarke JL, Watkins WM (1996) α1,3-L-Fucosyltransferase expression in developing human myeloid cells. J Biol Chem 271:10317–10328

Fuhlbrigge RC, Kieffer JD, Armerding D, Kupper TS (1997) Cutaneous lymphocyte antigen is a specialized form of PSGL-1 expressed on skin-homing T cells. Nature 389:978–981

Grabenhorst E, Nimtz M, Costa J, Conradt HS (1998) In vivo specificity of human α1,3/1,4-fucosyltransferases III–VII in the biosynthesis of Lewis[x] and sialyl Lewis[x] motifs on complex-type *N*-glycans. J Biol Chem 273:30985–30994

Hiraiwa N, Dohi T, Kawakami-Kimura N, Yumen M, Ohmori K, Maeda M, Kannagi R (1996) Suppression of sialyl Lewis X expression and E-selectin-mediated cell adhesion in cultured human lymphoid cells by transfection of antisense cDNA of an α1,3fucosyltransferase (Fuc-TVII). J Biol Chem 271:31556–31561

Hiraiwa N, Hiraiwa M, Kannagi R (1997) Human T-cell leukemia virus-1 encoded Tax protein transactivates α1,3-fucosyltransferase Fuc-TVII, which synthesizes sialyl Lewis X, a selectin ligand expressed on adult T-cell leukemia cells. Biochem Biophys Res Commun 231:183–186

Huang M-C, Zollner O, Moll T, Maly P, Thall AD, Lowe JB, Vestweber D (2000) P-selectin glycoprotein ligand-1 and E-selectin ligand-1 are differentially modified by fucosyltransferases Fuc-TIV and Fuc-TVII in mouse neutrophils. J Biol Chem in press

Kaneko M, Kudo T, Iwasaki H, Ikehara Y, Nishihara S, Nakagawa S, Sasaki K, Shiina T, Inoko H, Saitou N, Narimatsu H (1999) α1,3-fucosyltransferase IX (Fuc-TIX) is very highly conserved between human and mouse; molecular cloning, characterization and tissue distribution of human Fuc-TIX. FEBS Lett 452:237–242

Kimura H, Shinya N, Nishihara S, Kaneko M, Irimura T, Narimatsu H (1997) Distinct substrate specificities of five human α-1,3-fucosyltransferases for in vivo synthesis of the sialyl Lewis X and Lewis X epitopes. Biochem Biophys Res Commun 237:131–137

Knibbs RN, Craig AC, Natsuka S, Chang A, Cameron M, Lowe JB, Stoolman LM (1996) The fucosyltransferase Fuc-TVII regulates E-selectin ligand synthesis in human T cells. J Cell Biol 133:911–920

Maly P, Thall AD, Petryniak B, Rogers CE, Smith PL, Marks RM, Kelly RJ, Gersten KM, Cheng G, Saunders TL, Camper SA, Camphausen RT, Sullivan FX, Isogai Y, Hindsgall O, von Andrian UH, Lowe JB (1996) The α(1,3)fucosyltransferase Fuc-TVII controls leukocyte trafficking through an essential role in L-, E-, and P-selectin ligand biosynthesis. Cell 86:643–653

Marer NL, Palcic MM, Clarke JL, Davies D, Skacel PO (1997) Developmental regulation of α1,3-fucosyltransferase expression in CD34 positive progenitors and maturing myeloid cells isolated from normal bone marrow. Glycobiology 7:357–365

Mitsuoka C, Sawada-Kasugai M, Ando-Fukui K, Nakanishi H, Nakamura S, Ishida H, Kiso M, Kannagi R (1998) Identification of a major carbohydrate capping group of the L-selectin ligand on high-endothelial venules in human lymph nodes as 6-sulfo sialyl Lewis X. J Biol Chem 273:11225–11233

Nakamura M, Kudo T, Narimatsu H, Furukawa Y, Kikuchi J, Asakura S, Yang W, Iwase S, Hatake K, Miura Y (1998) Single glycosyltransferase, Core 2 β1-6-N-acetylglucosaminyltransferase, regulates cell surface sialyl-Lex expression level in human pre-B lymphocytic leukemia cell line KM3 treated with phorbolester. J Biol Chem 273:26779–26789

Nakayama F, Teraki Y, Kudo T, Togayachi A, Iwasaki H, Tamatani T, Nishihara S, Mizukawa Y, Shiohara T, Narimatsu H (2000) Expression of cutaneous lymphocyte-associated antigen regulated by a set of glycosyltransferases in human T cells: involvement of α1,3-fucosyltransferase VII and β1,4-galactosyltransferase I. J Invest Dermatol 115:299–306

Natsuka S, Gersten KM, Zenita K, Kannagi R, Lowe JB (1994) Molecular cloning of a cDNA encoding a novel human leukocyte alpha-1,3-fucosyltransferase capable of synthesizing the sialyl Lewis X determinant. J Biol Chem 269:16789–16794

Niemela R, Natunen J, Majuri M-L, Maaheimo H, Helin J, Lowe JB, Renkonen O, Renkonen R (1998) Complementary acceptor and site specificity of Fuc-TIV and Fuc-TVII allow effective biosynthesis of sialyl-triLe^x and related polylactosamines present on glycoprotein counterreceptors of selectins. J Biol Chem 273:4021–4026

Sasaki K, Kurata K, Funayama K, Nagata M, Watanabe E, Ohta S, Hanai N, Nishi T (1994) Expression cloning of a novel alpha 1,3-fucosyltransferase that is involved in biosynthesis of the sialyl Lewis X carbohydrate determinants in leukocytes. J Biol Chem 269:14730–14737

Shinoda K, Morishita Y, Sasaki K, Matsuda Y, Takahashi I, Nishi T (1997) Enzymatic characterization of human α1,3-fucosyltransferase Fuc-TVII synthesized in a B cell lymphoma cell line. J Biol Chem 272:31992–31997

Shinoda K, Tanahashi E, Fukunaga K, Ishida H, Kiso M (1998) Detailed acceptor specificities of human α-1,3-fucosyltransferases, Fuc-TVII and Fuc-TVI. Glycoconj J 15:
969–974

Smith PL, Gersten KM, Bronislawa P, Kelly RJ, Rogers C, Natsuka Y, Alford III JA,
Scheidegger EP, Natsuka S, Lowe JB (1996) Expression of the α(1,3)fucosyltransferase
Fuc-TVII in lymphoid aggregate high-endothelial venules correlates with expression
of L-selectin ligands. J Biol Chem 271:8250–8259

Wagers A, Lowe JB, Kansas GS (1996) An important role for the α1,3fucosyltransferase,
Fuc-TVII, in leukocyte adhesion to E-selectin. Blood 88:2125–2132

Wagers AJ, Waters CM, Stoolman LM, Kansas GS (1998) Interleukin 12 and interleukin
4 control T cell adhesion to endothelial selectins through opposite effects on α1,3-
fucosyltransferase VII gene expression. J Exp Med 188:2225–2231

Weninger W, Ulfman LH, Cheng G, Souchkova N, Quackenbush EJ, Lowe JB, von Andrian
UH (2000) Specialized contributions by α(1,3)-fucosyltransferase-IV and Fuc-T-VII
during leukocyte rolling in dermal microvessels. Immunity 12:665–676

33

α3-Fucosyltransferase-IX (FUT9)

Introduction

To date, six human α3-fucosyltransferases (α1,3-FUTs) have been cloned. α3-fucosyltransferase-IX (Fuc-TIX, FUT9) is the newest member of the six human α3-FUTs, and has been cloned by an expression cloning method from a mouse brain cDNA library (Kudo et al. 1998). All the α1,3-FUT members, including those of *Helicobacter pylori* (Ge et al. 1997), possess the conserved amino acid stretch FxL/VxFENS/TxxxxYxTEK as a motif of α1,3-FUT. The phylogenetic tree of α1,3-FUTs indicates that the *α1,3-FUT* genes evolved by independent gene duplication between vertebrates and *Caenorhabditis elegans*. There are four clusters in the vertebrate *α1,3-FUT* gene family, corresponding to the *FUT3–FUT5–FUT6* gene cluster, *FUT4*, *FUT7*, and *FUT9* gene subfamilies. The *FUT9* gene family seems to be the first to diverge from the ancestral gene (Kaneko et al. 1999a). The *FUT9* gene does not cross-hybridize with the other five *α1,3-FUT* genes. The FUT9 amino acid sequence is quite different from those of the other five α1,3-FUTs which share highly homologous sequences. More interestingly, the amino acid sequence of FUT9 (Fut9) is very highly conserved between mouse and human. The degree of conservation is almost equivalent to that of the *α-actin* gene. This indicates that FUT9 is under a strong selective pressure for preservation during evolution. Mouse has only three functional *α1,3-Fut* genes, *Fut4*, *Fut7*, and *Fut9*, corresponding to human *FUT4*, *FUT7*, and *FUT9* genes. FUT9 (Fut9) transfers Fuc from GDP-Fuc to the GlcNAc residue of type-2 chain with an α1,3-linkage, resulting in the synthesis of Lewis X (Le^x) epitope, Galβ1-4(Fucα1-3)GlcNAc-R. The *FUT9* gene is localized at 6q16 in human (Kaneko et al. 1999b), and the mouse *Fut9* gene was mapped at chromosome 4 (unpublished data, 1998).

HISASHI NARIMATSU

Laboratory of Gene Function Analysis, Institute of Molecular and Cell Biology (IMCB), National Institute of Advanced Industrial Science and Technology (AIST), Central-2, 1-1-1 Umezono, Tsukuba, Ibaraki 305-8568, Japan
Tel. +81-298-61-3200; Fax +81-298-61-3201
e-mail: h.narimatsu@aist.go.jp

Databanks

α3-Fucosyltransferase-IX (FUT9)

NC-IUBMB enzyme classification: E.C.2.4.1.152

Species	Gene	Protein	mRNA	Genomic
Homo sapiens	FUT9, Fuc-TIX	NP_006572	AB023021	NT_019424
		–	AJ238701	–
Mus musculus	Fut9, mFuc-TIX	NP_034373	AB015426	–
Rattus norvegicus	rFut9	BAB40953	AB049819	–

Name and History

α3-Fucosyltransferase-IX is abbreviated to FucT-IX or FUT9 in human, or Fut9 in species other than human. A cDNA encoding mouse Fut9 was first isolated from brain by an expression cloning method (Kudo et al. 1998). It has two conserved stretches of amino acid motifs among the α1,3-FUT members of vertebrates and *C. elegans*. Mollicone et al. (1990) reported that human brain homogenates exhibited activity for Le^x synthesis but not for sLe^x synthesis. The α1,3-FUTs expressed in brain and leuko-cytes had similar activity features, but they differed in enzyme activity when activated by dications (Mollicone et al. 1990). FUT3, 5, 6, and 7 can synthesize sLe^x, but they are not expressed in brain. FUT4, which can synthesize Le^x and exhibit very weak activity for sLe^x synthesis, is expressed in brain and leukocytes. Therefore, the activity detected in brain and leukocytes had been considered to be directed by FUT4. We recently examined the tissue distribution of six human α1,3-FUTs in a whole body (Kaneko et al. 1999a). FUT9 was found to be the most abundant α1,3-FUT in brain, and was responsible for Le^x expression in brain (manuscript in preparation).

Enzyme Activity Assay and Substrate Specificity

The acceptor specificity of Fut9 (FUT9) was examined using cell homogenates of Namalwa cells stably expressing mouse Fut9 (Kudo et al. 1998) or human FUT9 (Kaneko et al. 1999a; Nishihara et al. 1999). A mammalian expression vector, pAMo, was inserted with full-length cDNAs of FUT9 or Fut9. The vectors containing the full-length cDNAs were stably expressed in Namalwa cells. First, the protein concentration of cell lysates was determined, and the cell lysates were used to measure the α1,3-FUT activity in 50 mM cacodylate buffer (pH 6.8), 5 mM ATP, 75 µM GDP-Fuc, 10 mM L-fucose, and 25 µM of the each acceptor substrate. Oligosaccharides were pyridylaminated (PA) or labeled with 2-aminobenzamide (2AB) and used as acceptor substrates. After incubation at 37°C for 2 h, the enzyme reactions were terminated by boiling followed by the addition of water. After centrifugation of the reaction mixtures, each supernatant was filtrated and subjected to reverse-phase high-performance liquid chromatography (HPLC) analysis.

FUT9 exhibited activity for Le^x synthesis towards LNnT, but not for Le^a, Le^b, or sLe^x synthesis towards LNT or sialyl-LNnT (Kudo et al. 1998). FUT9 activity was not activated in the presence of Mn^{2+} or Co^{2+}, but was markedly suppressed (Kaneko et al.

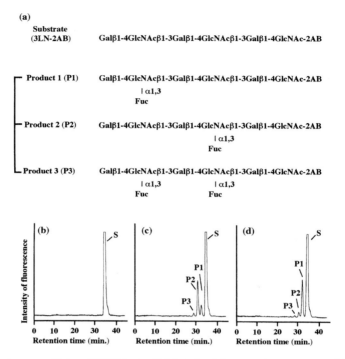

Fig. 1. A unique specificity of FUT9 for 2-AB labeled polylactosamine acceptor. **a** 3LN-2AB, an oligosaccharide structure of an acceptor substrate, and (*below*) three products of the enzyme reaction presented as Products 1, 2, and 3. **b–d** Reverse-phase HPLC analysis of the reaction products after incubation with α1,3FUTs. **b** Mock reaction; **c** FUT4; **d** FUT9. *S*, 3LN-2AB; *P1*, Product 1; *P2*, Product 2; *P3*, Product 3

1999a). In contrast, FUT4 and FUT6 were activated (Kaneko et al. 1999). In the recent study, we found that FUT9 exhibits a unique specificity against polylactosamine, i.e., FUT9 preferentially fucosylates the distal GlcNAc residue, resulting in Le^x synthesis, while FUT3, 4, 5, and 6 preferentially fucosylate the inner GlcNAc residue of polylactosamine, resulting in internal Le^x synthesis (Nishihara et al. 1999). This is rational given that the FUT9 amino acid sequence is quite different from those of the other five FUTs.

In Fig. 1, the P1, P2, and P3 peaks indicate the activities synthesizing Le^x, internal Le^x, and dimeric Le^x, respectively, on polylactosamine. The relative activity of FUT9 for Le^x synthesis is approximately 15 times stronger than that of FUT4, whereas the internal Le^x synthesizing activity of FUT9 is half that of FUT4. This is consistent with the flow cytometric analysis of Namalwa cells transfected stably with each gene (Fig. 2). Namalwa-FUT9 cells, stably expressing FUT9, show strong positive staining with CD15 (Le^x) and very weak positive staining with CDw65 (VIM2), which recognizes the sialylated internal Le^x. In contrast, Namalwa-FUT4 cells, stably expressing FUT4, expressed less CD15 and more CDw65 than Namalwa-FUT9 cells (Fig. 2).

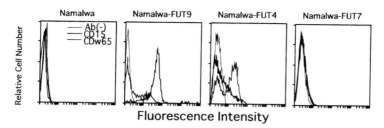

Fig. 2. Flow cytometric analysis of Namalwa cells stably transfected with FUT9 (Namalawa-FUT9), FUT4 (Namalwa-FUT4), and FUT7 (Namalwa-FUT7) with anti-CD15 and anti-CDw65 antibodies

Table 1. Quantitative analysis of transcripts of six α $1,3FUT$ genes in various human tissues by competitive reverse transcriptase-polymerase chain reaction

	Expression levels of α 1,3FUTs relative to the level of β-actin (%)					
	FUT3	FUT4	FUT5	FUT6	FUT7	FUT9
Forebrain	<0.001	0.58	<0.001	0.01	<0.001	0.53
Stomach (antrum)	2.42	1.83	<0.001	0.27	<0.001	0.73
(corpus)	1.87	1.07	<0.001	0.06	<0.001	0.37
Jejunum	13.66	2.33	<0.001	0.86	<0.001	<0.001
Colon	2.90	0.43	<0.001	0.41	<0.001	<0.001
Liver	<0.001	0.39	<0.001	0.14	<0.001	<0.001
Spleen	0.02	0.25	0.02	0.004	0.13	0.004
Lung	0.01	0.01	<0.001	<0.001	<0.001	<0.001
Kidney	0.34	1.13	<0.001	3.18	<0.001	<0.001
Adrenal cortex	<0.001	0.35	<0.001	<0.001	<0.001	<0.001
Uterus, cervix	0.12	5.14	<0.001	0.07	<0.001	<0.001
PBL	<0.001	0.29	<0.001	<0.001	0.15	0.01

Preparation

FUT9 is available only in a recombinant form. FUT9 is expressed in some expression systems, such as Namalwa cells with pAMo vector, many cancer cell lines with pcDNA3 vector, and insect cells with a baculo-expression system. No native FUT9 enzyme is available. However, some human cell lines derived from gastric cancer are known to endogenously express FUT9.

Biological Aspects

In a previous study, we examined the tissue distribution patterns of six human α1,3-FUTs (Kaneko et al. 1999a) (Table 1). FUT9 was found to be expressed in brain, stomach, spleen, and peripheral blood cells (PBL) (Table 1). The Le^x epitope is mainly detected in neuronal and glial cells in brain, proximal tubules in kidney, granulocytes and monocytes in PBL, and stomach subglandular cells in stomach mucosa by immunohistochemical analysis. FUT9 plays a key role to form the Le^x epitope Galβ1-4(Fucα1-3)GlcNAcα1-R in some tissues such as brain, stomach, hematopoietic cells, kidney, and early embryo. FUT9 can synthesize the Le^x epitopes on both glycoproteins and glycolipids (manuscript in preparation). Both neuronal and glial cells in brain expressed FUT9 in a manner of developmental regulation. The Le^x epitope in brain is mainly carried on neutral glycolipids and synthesized by FUT9. The Le^x expression in stomach is restricted to glandular compartments and is well correlated with the FUT9 expression (Kaneko et al. 1999a). FUT9 is expressed in the epithelial cells of proximal tubes in mouse kidney, where the abundant Le^x antigen exists. In hematopoietic cells, FUT9 is mainly expressed in mature granulocytes, resulting in the CD15 (Le^x) expression (Nakayama et al. 2001). In each subpopulation of peripheral blood mononuclear cells, all subpopulations express FUT9 except for monocytes. Monocytes and promyelocytes, which are also CD15-positive cells, do not express FUT9, but they express a large amount of FUT4, which is enough for CD15 expression (Nakayama et al. 2000). In brief, we found that CD15 expression in mature granulocytes is determined by FUT9, but in promyelocytes and monocytes it is determined by FUT4. Figure 3 shows the complementary FUT4, FUT7, and FUT9 reactions for multiple fucosylations on the neutral or sialylated polylactosamine chains traced by previous experiments. FUT9 preferentially fucosylates the distal lactosamine (LN) unit of polylactosamine to form Le^x-LN, whereas FUT4 preferentially fucosylates inner LN units on both the neutral and α2,3-sialylated polylactosamine chains. FUT7 prefers to fucosylate distal LN units on α2,3-sialylated, but not on neutral, polylactosamine chains.

The early embryo of mice, eight cells to morullae, expresses SSEA-1 (stage-specific embryonal antigen-1) antigen (Solter and Knowles 1978), and both Fut9 and Fut4 are expressed during this stage of mouse embryo (unpublished data, 2000). The antigenic epitope of SSEA-1 antigen is also defined as the Le^x structure. In consideration of the stronger activity of Fut9 for Le^x synthesis and the transcript level expressed, Fut9 must be responsible for SSEA-1 expression during the early embryogenesis of mouse.

The biological functions of Le^x (CD15, SSEA-1) antigen are still unclear. SSEA-1 is believed to be essential for mouse early embryogenesis because the blocking experiments of SSEA-1 demonstrated that compaction, which is an essential phenomenon occuring after the eight-cell stage of mouse embryo, is inhibited by the additon

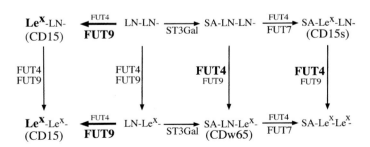

Fig. 3. Complementary pathways of multiple fucosylations by three α1,3FUTs on neutral and sialylated polylactosamine chains. The relative activity of the enzymes is indicated by the *letter size*

of anti-Le^x antibodies. Neurobiologists suggested that the Le^x (CD15) antigen is important for synaptic formation during brain development (Streit et al. 1996; Allendoerfer et al. 1995, 1999). CD15 is considered to be involved in neutrophil functions, i.e., cell–cell interactions, phagocytosis, stimulation of degranulation, and respiratory burst (Melnick et al. 1986; Warren et al. 1996).

Thus, the expression of FUT9 is strictly regulated in a cell-specific manner and developmentally stage-specific manner. The *FUT9* gene is encoded in three exons, and the open reading frame is encoded in a single third exon (unpublished data, 1999). The promoter analysis of this gene has not been reported.

Future Perspectives

Fut9-knockout mice have been established in our laboratory (unpublished data, 1999). The eight-cell and mollulae embryos of Fut9-knockout mice were negatively stained with anti-SSEA-1 antibody, demonstrating that Fut9 is the enzyme responsible for SSEA-1 expression during embryogenesis (manuscript in preparation). However, the Fut9$^{-/-}$ mice were born normally and looked healthy. This suggested that the Le^x epitope on SSEA-1 antigen does not function as an important molecule for cell–cell interaction during embryogenesis. We also observed that the compaction phenomenon occurs in Fut9$^{-/-}$ embryos without the Le^x epitope of SSEA-1 antigen. The Le^x epitope in brain is definitely synthesized by Fut9, because it disappeared completely in the brain of Fut9$^{-/-}$ mice. The Le^x epitopes also disappeared in the proximal tubules of the kidney and the subgrandular compartment of the stomach of Fut9$^{-/-}$ mice.

We are now investigating the functional abnormality of the tissues from which the Le^x reactivity disappeared.

Acknowledgments

I would like to thank Drs. Shoko Nishihara, Takashi Kudo, Hiroko Iwasaki, Mika Kaneko, and Akira Togayachi, from our laboratory, for their productive FUT9 work, and Dr. Katsutoshi Sasaki, of Kyowa Hakko Co., Ltd., for his excellent collaboration.

References

Allendoerfer KL, Magnani JL, Patterson PH (1995) FORSE-1, an antibody that labels regionally restricted subpopulations of progenitor cells in the embryonic central nervous system, recognizes the Lex carbohydrate on a proteoglycan and two glycolipid antigens. Mol Cell Neurosci 6:381–395

Allendoerfer KL, Durairaj A, Matthews GA, Patterson PH (1999) Morphological domains of Lewis-X/FORSE-1 immunolabeling in the embryonic neural tube are due to developmental regulation of cell surface carbohydrate expression. Dev Biol 211:208–219

Ge Z, Chan NW, Palcic MM, Taylor DE (1997) Cloning and heterologous expression of an alpha1,3-fucosyltransferase gene from the gastric pathogen *Helicobacter pylori*. J Biol Chem 272:21357–21363

Kaneko M, Kudo T, Iwasaki H, Ikehara Y, Nishihara S, Nakagawa S, Sasaki K, Shiina T, Inoko H, Saitou N, Narimatsu H (1999a) α1,3-fucosyltransferase IX (Fuc-TIX) is very highly conserved between human and mouse: molecular cloning, characterization and tissue distribution of human Fuc-TIX. FEBS Lett 452:237–242

Kaneko M, Kudo T, Iwasaki H, Shiina T, Inoko H, Kozaki T, Saitou N, Narimatsu H (1999b) Assignment of the human α1,3-fucosyltransferase IX (FUT9) gene to chromosome band 6q16 by in situ hybridization. Cytogenet Cell Genet 86:329–330

Kudo T, Ikehara Y, Togayachi A, Kaneko M, Hiraga T, Sasaki K, Narimatsu H (1998) Expression cloning and characterization of a novel murine α1,3-fucosyltransferase, mFuc-TIX, that synthesizes the Lewis^x (CD15) epitope in brain and kidney. J Biol Chem 273:26729–26738

Melnick DA, Meshulam T, Manto A, Malech HL (1986) Activation of human neutrophils by monoclonal antibody PMN7C3: cell movement and adhesion can be triggered independently from the respiratory burst. Blood 67:1388–1394

Mollicone R, Gibaud A, Francois A, Ratcliffe M, Oriol R (1990) Acceptor specificity and tissue distribution of three human α-3-fucosyltransferases. Eur J Biochem 191: 169–176

Nakayama F, Nishihara S, Iwasaki H, Okubo R, Kaneko M, Kudo T, Nakamura M, Karube M, Narimatsu H (2001) CD15 expression in mature granulocytes is determined by α1,3-fucosyltransferase IX (FUT9), but in promyelocytes and monocytes by α1,3-fucosyltransferase IV (FUT4). J Biol Chem 276:16100–16106

Nishihara S, Iwasaki H, Kaneko M, Tawada A, Ito M, Narimatsu H (1999) α1,3-fucosyltransferase 9 (FUT9; Fuc-TIX) preferentially fucosylates the distal GlcNAc residue of polylactosamine chain while the other four α1,3FUT members preferentially fucosylate the inner GlcNAc residue. FEBS Lett 462:289–294

Solter D, Knowles BB (1978) Monoclonal antibody defining a stage-specific mouse embryonic antigen (SSEA-1) Proc Natl Acad Sci USA 75:5565–5569

Streit A, Yuen CT, Loveless RW, Lawson AM, Finne J, Schmitz B, Feizi T, Stern CD (1996) The Le^x carbohydrate sequence is recognized by antibody to L5, a functional antigen in early neural development. J Neurochem 66:834–844

Warren HS, Altin JG, Waldron JC, Kinnear BF, Parish CR (1996) A carbohydrate structure associated with CD15 (Lewis X) on myeloid cells is a novel ligand for human CD2. J Immunol 156:2899–2873

α6-Fucosyltransferase (FUT8)

Introduction

GDP-L-Fuc:*N*-acetyl-β-D-glucosaminide α1-6fucosyltransferase (FUT8) catalyzes the transfer of fucose from GDP-Fuc to *N*-linked-type complex glycoproteins, as shown in Fig. 1. The enzymatic products, α1,6-fucosylated (core fucosylated) *N*-glycans, are commonly observed in many glycoproteins, and are especially abundant in brain tissue. It is well known that the sugar chains in α-fetoprotein (AFP), a well-known tumor marker of hepatocellular carcinoma, are microheterogenous in nature to sugar chains. The oligosaccharide structures of transferrin as well as AFP, synthesized by hepatocellular carcinoma cells, are highly fucosylated (Champion et al. 1989). In contrast, FUT8 is released from platelets during blood clotting (Koscielak et al. 1987), suggesting that this enzyme might play a role in blood coagulation. An increase in fucosylated carbohydrates in pathological conditions has also been reported in other types of cancer cells (Tatsumura 1977).

FUT8 has been purified from four different sources (Voynow et al. 1991; Uozumi et al. 1996b; Yanagidani et al. 1997; Kaminska et al. 1998). Several studies have been reported on the FUT8 gene. A high expression of FUT8 mRNA was observed in rat hepatoma tissues, but not in the surrounding tissues (Noda et al. 1998). In the case of human liver diseases, expression of FUT8 was observed both in the hepatoma tissues and in the surrounding tissues with liver cirrhosis (Noda et al. 1998). When the FUT8 gene was transfected into hepatoma cells, experimental metastasis was dramatically suppressed (Miyoshi et al. 1999). More than 20 glycoproteins were severely fucosylated. One of those proteins is α5β1-integrin, which functions were downregulated by the fucosylation.

Eiji Miyoshi and Naoyuki Taniguchi

Department of Biochemistry, Osaka University Graduate School of Medicine (Osaka University Medical School), Room B-1, 2-2 Yamadaoka, Suita, Osaka 565-0871, Japan
Tel. +81-6-6879-3421; Fax +81-6-6879-3429
e-mail proftani@biochem.med.osaka-u.ac.jp

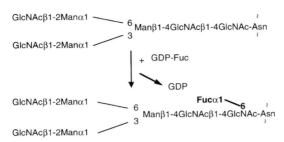

Fig. 1. Reaction pathway of FUT8. *GlcNAc* indicates *N*-acetyl-glucosamine, *Man* indicates mannose, *Fuc* indicates fucose, *GDP-Fuc* indicates guanosinedi-phosphofucopyranoside, and *Asn* indicates asparagine

Databanks

α6-Fucosyltransferase (FUT8)

NC-IUBMB enzyme classification: E.C.2.4.1.68

Species	Gene	Protein	mRNA	Genomic
Homo sapiens	FUT8	–	D89289	SEG_AB032567S
		–	Y17979	–
Sus scrofa		–	D86723	–
Mus musculus	Fut8	–	AB025198	–

Name and History

FUT8 from porcine liver was the first enzyme of this class to be characterized (Wilson et al. 1976). Alterations in fucosylation have been reported in cystic fibrosis glyco-proteins from several sources (Scanlin et al. 1985). Cystic fibrosis is a major heredi-tary disease whose causative gene has recently been identified. It is significant, in this respect, that purified glycoproteins from fibroblasts derived from patients with cystic fibrosis contain much higher levels of α1,6fucose residues compared with normal control fibroblasts (Wang et al. 1990). Based on this background and in relation to α1,6-fucosylated proteins, Voynow et al. (1991) first succeeded in purifying and char-acterizing FUT8 from cultured human skin fibroblasts of cystic fibrosis. While several glycosyltransferases form a gene family, no homology to other fucosyltransferases such as α2FucT, α3FucT, and α4FucT was found except for a region consisting of nine amino acids.

Enzyme Activity Assay and Substrate Specificity

FUT8 assay using radioisotope (RI) methods was reported by Voynow et al. (1991). The reaction is carried out at 37°C for 1 h in the incubation of enzyme sources with 100 μM GnGn-Gp (Gp, glycoprotein), 2.5 mM CDP-choline, 20 μg bovine serum albumin, 75 μl enzyme cacodylate buffer in 20 mM Tris-HCl, pH 7.5, containing 4 mM $MgCl_2$, 15% glycerol, and 0.075% Triton CF54 and 100 mM cacodylate buffer, pH 5.2, and 26.6 μM GDP-[^{14}C] fucose (840 000 cpm). The final volume of the reaction was 100 μl. The ^{14}C-fucosylated product formed is detected by lentil lectin–Sepharose chromatography.

FUT8 assay using non-radioisotope methods was reported by Uozumi et al. (1996a). The reaction is carried out at 37°C for 2h in the incubation of enzyme sources with 200 mM MES–NaOH buffer (pH 5.6 or 7.0; there are two peaks of optimal pH), 1% Triton X-100 + 50 μM GnGn-Asn-PABA [Gn, *N*-acetylglucosamine; Asn, asparagine; PABA, 4-(2-pyridylamino) butylamine], and 500 μM GDP-Fuc (guanosine diphosphofucopyranoside). After stopping the reaction by heating at 100°C for 1 min, the products are analyzed by HPLC.

The substrate specificity of FUT8 was investigated by Voynow et al. (1991). FUT8 is permissible for Gn-M (Gn-M) M-Gn-Gn-Asn and Gn-M (Gn-M) M-Gn-Gn at 66.0 μM and 55.0 μM of K_m value, respectively (Gn indicates *N*-acetylglucosamine and Man indicates mannose) However, α6-FucT is nonpermissible for asialo-transferin-glycopeptide, asialo-agalacto-aglucosaminyltransferrin glycopeptide, $Man_5GlcNAc_2$-O, chitriose (tritely), chitobiose (diacetyl), and GlcNAc-Asn.

pH Optimum

The optimum pH of FUT8 was investigated by Uozumi et al. (1996). The optimum pH of crude enzymes from serum, brain, and spleen is 7.0. In contrast, the optimum pH of crude enzymes from liver is 5.6.

Ionic Strength

The ionic strength of FUT8 was investigated by several groups who purified FUT8 from difference sources (Voynow et al. 1991; Uozumi et al. 1996b; Yanagidani et al. 1997; Kaminska et al. 1998). Both Ni and Cu from all sources downregulate FUT8 activity. Ca, Mg, and Mn have negligible effects on FUT8 activity from porcine brain and a human gastric cancer cell, MKN45. FUT8 purified from fibroblast requires Mg for its enzymatic activity, and the activity of FUT8 purified from platelets is downregulated by Ca and Mg.

Redox Effects

While α1-6 fucosylation of glycoproteins in the serum of patients with hepatocellular carcinoma is increased as compared with that of patients with liver cirrhosis alone, the levels of FUT8 mRNA expression are equal or higher in liver tissues of patients with cirrhosis.

Activators or Inhibitors

Not known.

Preparation

Source

Cultured human skin fibroblasts, porcine brain, and the medium of human gastric cancer cells are biological sources of FUT8. Commercial recombinant FUT8 is available from Toyobo Corp. (Shiga, Japan).

Expression Systems

Human/porcine FUT8 cDNA is available in our laboratory, and the expression system in *E. coli* is available from Toyobo Corp., which has a patent for this system.

Isolation and Purification

Porcine FUT8 was purified in a 12% final yield with a 440000-fold increase in specific activity from brain, and its cDNA was cloned by Uozumi et al. (1996b). The purification and cDNA cloning of human FUT8 were achieved by Yanagidani et al. (1997).

Biological Aspects

The chromosomal mapping of FUT8 was completed by Yamaguchi et al. in 1999. The FUT8 gene (*FUT8*) is located on14q24.3 where no other fucosyltransferases are located. While expression of FUT8 is quite low in normal liver, it is dramatically increased in chronic liver diseases such as hepatitis and liver cirrhosis (Noda et al. 1998). When the FUT8 gene was transfected into a human hepatocellular carcinoma (HCC) cell line, Hep3B, which originally showed a low level of FUT8 expression, the secretion of fucosylated AFP was dramatically increased, suggesting that FUT8 catalyzes the fucosylation of AFP in hepatomas. These results suggest that the increment of $\alpha 6$ fucosylated AFP in the serum of patients with HCC is dependent not solely on the direct upregulation of FUT8, but also on the abnormal secretion system in hepatoma cells. Experimental metastasis of FUT8 transfectants via splenic injection was dramatically suppressed due to the dysfunction of integrin, suggesting that FUT8 might alter the phenotypes of hepatomas which have, biologically speaking, low-malignant characteristics. A project of FUT8 knockout mice is ongoing, and FUT8 transgenic mice showed abnormal lipid accumulation in their liver and kidney. The levels of FUT8 activity in the serum of patients with pancreatic cancer were upregulated (unpublished data, 2000).

Future Perspectives

Since the mRNA expression and enzymatic activity of FUT8 are not correlated in some cases, a family member of FUT8 will be found in the future as well as other fucosyltransferases such as $\alpha 2$FucTs or $\alpha 3$FucTs. The target protein, which is easily fucosylated, will be found because certain glycoproteins are selectively fucosylated in FUT8 transfectants.

Further Reading

Miyoshi E, Noda K, Yamaguchi Y, Inoue S, Ikeda Y, Wang W, Ko JH, Uozumi N, Li W, Taniguchi N (1999) The $\alpha 1$-6 fucosyltransferase gene and its biological significance. Biochim Biophys Acta 1473:9–20.
Many topics relating to FUT8 are reviewed in this paper.

References

Champion B, Leger D, Wieruszeski J-M, Montreuil J, Spik G (1989) Presence of fucosylated triantennary, tetraantennary and pentaantenary glycans in transferrin synthesized by the human hepatocarcinoma cell line HepG2. Eur J Biochem 184:405–413

Kaminska J, Glick MC, Koscielak J (1998) Purification and characterization of GDP-L-Fuc:N-acetyl β-D-glucosaminide α1→6-fucosyltransferase from human blood platelets. Glycoconj J 15:783–788

Koscielak J, Pacuszka T, Kubin J, Zdziechowska H (1987) Serum α-6-L-fucosyltransferase is released from platelets during clotting of blood. Glycoconj J 4:43–49

Miyoshi E, Noda K, Ko JH, Ekuni A, Kitada T, Uozumi N, Ikeda Y, Matsuura N, Sasaki Y, Hayashi N, Hori M, Taniguchi N (1999) Overexpression of α1-6 fucosyltransferase in hepatoma cells suppresses intrahepatic metastasis after splenic injection in athymic mice. Cancer Res 59:2237–2243

Noda K, Miyoshi E, Uozumi N, Ikeda Y, Gao C-X, Suzuki K, Yoshihara H, Yoshikawa K, Kawano K, Hayashi N, Hori M, Taniguchi N (1998) Gene expression of α1-6 fucosyltransferase in human hepatoma tissues: a possible implication for increased fucosylation of α fetoprotein. Hepatology 28:944–952

Scanlin TF, Wang YM, Glick MC (1985) Altered fucosylation of membrane glycoproteins from cystic fibrosis fibroblasts. Pediatr Res 19:368–374

Tatsumura T, Sato H, Mori A, Komori Y, Yamamoto K, Fukutani G, Kuno S (1977) Clinical significance of fucose level in glycoprotein fraction of serum in patients with malignant tumors. Cancer Res 37:4101–4103

Uozumi N, Teshima T, Yamamoto T, Nishikawa A, Gao YE, Miyoshi E, Gao CX, Noda K, Islam KN, Ihara Y, Fujii S, Shiba T, Taniguchi N (1996a) A fluorescent assay method for GDP-L-Fuc:N-acetyl-β-D-glucosaminide α1-6fucosyltransferase activity, involving high-performance liquid chromatography. J Biochem 120:385–392

Uozumi N, Yanagidani S, Miyoshi E, Ihara Y, Sakuma T, Gao CX, Teshima T, Fujii S, Shiba T, Taniguchi N (1996b) Purification and cDNA cloning of porcine brain GDP-L-Fuc:N-acetyl-β-D-glucosaminide:α1-6-fucosyltransferase. J Biol Chem 271:27810–27817

Voynow JA, Kaiser RS, Scanlin TF, Glick MC (1991) Purification and characterization of GDP-L-Fuc: N-acetyl-β-D-glucosaminide α1-6-fucosyltransferase from cultured human skin fibroblasts. J Biol Chem 266:21572–21577

Wang YM, Hare TR, Won B, Stowell CP, Scanlin TF, Glick MC, Hard K, van Kuik JA, Vliegenthart JF (1990) Additional fucosyl residues on membrane glycoproteins but not a secreted glycoprotein from cystic fibrosis fibroblasts. Clin Chim Acta 188:193–210

Wilson JR, Willams D, Schachter H (1976) The control of glycoprotein synthesis: N-acetylglucosamine linkage to α-mannose residue as a signal of L-fucose to the asparagine-linked N-acetylglucosamine residue of glycopeptide from α1-acid glycoprotein. Biochem Biophys Res Commun 72:909–916

Yamaguchi Y, Fujii J, Inoue S, Uozumi N, Yanogidani S, Ikeda Y, Egashira M, Miyoshi O, Niikawa N, Taniguchi N (1999) Mapping of the α1,6-fucosyltransferase gene, FUT8, to human chromosome 14q24.3. Cytogenet. Cell Genet 84, 58–60

Yanagidani S, Uozumi N, Ihara Y, Miyoshi E, Yamaguchi N, Taniguchi N (1997) Purification and cDNA cloning of GDP-L-Fuc:N-acetylglucosaminide:α1-6 fucosyltransferase (FUT8) from human stomach carcinoma MKN45 cells. J Biochem 121:626–632

Sialyltransferases

ST3Gal-I

Introduction

ST3Gal-I was originally purified from porcine liver, and the amino acid sequences derived were used for polymerase chain reaction (PCR)-directed cDNA cloning (Gillespie et al. 1992). ST3Gal-I is at present one of six members of the α2,3-sialyltransferase gene family (Ishii et al. 1993; Tsuji et al. 1996; Okajima et al. 1999), and one of the few sialyltransferases that have relatively strict acceptor specificity. Only the core1 O-glycan Galβ1→3GalNAc→Ser/Thr can be utilized by ST3Gal-I as an acceptor. The resulting oligosaccharide product is a sialylated core1 O-glycan with the structure NeuAcα2→3Galβ1→3GalNAcα1→Ser/Thr, which may also be further sialylated by certain α2,6-sialyltransferases, resulting in a fully sialylated tetrasaccharide: NeuAcα2→3Galβ1→3(NeuAcα2→6)GalNAcα1→Thr/Ser.

On the other hand, the ST3Gal-I product cannot serve as an acceptor for the core 2 β1,6-N-acetylglucosaminyltransferase (core2GlcNAcT) (Bierhuizen and Fukuda 1992). Because of this, and since core2GlcNAcT also uses Galβ1→3GalNAcα1→Ser/Thr as an acceptor, these two glycosyltransferases may compete and affect the structural outcome in O-glycan biosynthesis. Moreover, if core2GlcNAcT wins over ST3Gal-I, the resultant oligosaccharide is a poor acceptor for ST3Gal-I and the resultant O-glycans are primarily NeuAcα2→3Galβ1→4GlcNAcβ1→6(Galβ1→3) GalNAcα1→Ser/Thr.

These properties explain several observations on the expression of O-glycans in vivo. In the thymus, thymocytes in the cortical layer representing immature thymocytes express core2 branched O-glycans (Baum et al. 1995) and at the same time are stained strongly with peanut agglutinin (PNA) (Reisner et al. 1976). This PNA staining is due to the presence of unsialylated core 1 O-glycans, and can include those bearing a core2 O-glycan branch: NeuAcα2→3Galβ1→4GlcNAcβ1→6(Galβ1→3)

Minoru Fukuda[1] and Jamey D. Marth[2]

[1] Glycobiology Program, The Burnham Institute, 10901 North Torrey Pines Road, La Jolla, CA 92037, USA
Tel. +1-858-646-3144; Fax +1-858-646-3193
e-mail: minoru@burnham-inst.org
[2] Howard Hughes Medical Institute, University of California, San Diego, La Jolla, CA 92093, USA

GalNAcα1→Ser/Thr. After the majority of thymocytes die and those positively selected enter the medullary compartment, these medullary thymocytes no longer contain high levels of core2GlcNAcT and have induced ST3Gal-I, such that the effect of ST3Gal-I is pronounced (Gillespie et al. 1993), leading to the core1 O-glycan: NeuAcα2→3Galβ1→3(NeuAcα2→6)GalNAcα1→Ser/Thr. This sialylation of the core1 O-glycan branch results in loss of PNA binding (Reisner et al. 1976).

In peripheral compartments following thymic emigration, resting T lymphocytes contain almost exclusively sialylated core1 O-glycans, while T lymphocytes acquire core2 O-glycans once immune-activated, with an increase in core2GlcNAcT (Piller et al. 1988). The increase in core2 O-glycans may also occur by a significant decrease in the expression of ST3Gal-I RNA (unpublished data, 2000), while α2,6-sialyltransferases such as ST6GalNAc-III appear unchanged (Piller et al. 1988). This increase on human T lymphocytes can be detected by T305 antibody (Fox et al. 1983). To a first approximation, this change in O-glycan biosynthesis can be ascribed to the fact that the core2GlcNAcT enzyme resides in "earlier" compartments of the Golgi apparatus than does ST3Gal-I or other sialyltransferases. In fact, the core2GlcNAcT has been localized to the *medial* Golgi compartment (Skrincosky et al. 1997). Most siayltransferases have been localized to the *trans*-Golgi, although transfection experiments producing high levels of ST3Gal-I have been found to inhibit core2 O-glycan biosynthesis, with evidence that ST3Gal-I molecules in this situation can be found within both the *medial* and *trans*-Golgi (Whitehouse et al. 1997).

Evidence that a competition normally exists in vivo among ST3Gal-I and the core2GlcNAcT has recently been demonstrated in mice rendered genetically deficient in ST3Gal-I (Priatel et al. 2000). T lymphocytes lacking ST3Gal-I were found to have induced core 2 O-glycan biosynthesis but without an increase in core2GlcNAcT activity. Therefore, at least among primary T lymphocytes, ST3Gal-I must co-localize with core2GlcNAcT in order to compete effectively and inhibit the formation of core2 O-glycans until T cell activation occurs.

These findings indicate that a balance exists between core1 sialylation and core2 O-glycan biosynthesis, and that this balance can be controlled by altering the expression levels of core2GlcNAcT and ST3Gal-I (Fig. 1). However, it is also possible that changes in the expression of these glycosyltransferases may affect where these enzymes are localized within the Golgi apparatus. Alterations in the balance of O-glycan branch formation occurs in normal lymphocyte development and activation, and under certain conditions among transformed cells, which may explain some of the changes in O-glycan biosynthesis associated with cancer.

These results are entirely consistent with the cell surface marker of human and mouse CD8[+] memory T cells (Mukasa et al. 1999; Harrington et al. 2000). In both studies, reduced expression of core2 branched O-glycans has been found to distinguish memory CD8[+] T cells from activated/effector CD8[+] T cells expressing an increased amount of core2 branched O-glycans. The above findings are also consistent with the fact that the expression of core2 O-glycans upon T cell activation is a late event, coinciding more with the impending induction of apoptosis or memory cell differentiation than with the primary functions associated with immune activation (Priatel et al. 2000).

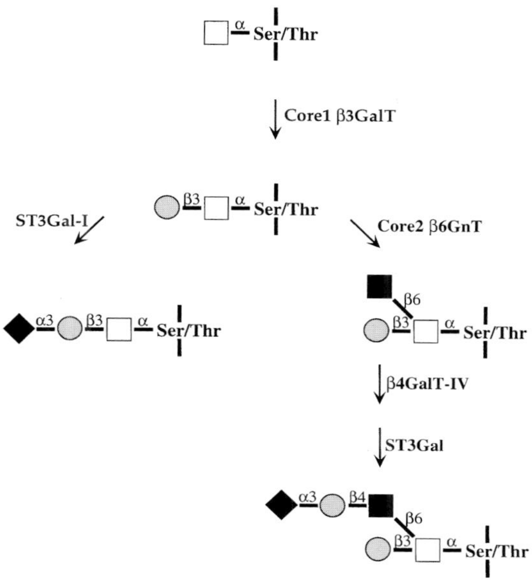

Fig. 1. Synthesis of core1 and core2 O-glycans. Core1 O-glycans are synthesized by the addition of β1,3-linked galactose to N-acetylgalactosamine. The resultant Galβ1→3GalNAcα1→ Ser/Thr can be sialylated by ST3Gal I, forming NeuAcα2→Galβ1→3GalNAcα1→Ser/Thr (*left*). This oligosaccharide is not an acceptor for core2GlcNAcT and is negative for PNA staining. If core2GlcNAcT acts first on Galβ1→3GalNAcα1→Ser/Thr, the resultant oligosaccharide is converted to NeuNAcα2→Galβ1→4GalNAcβ1→6(Galβ1→3)GalNAcα1→Ser/Thr. The oligosaccharide is positive toward PNA staining. □, N-acetylgalactosamine; ■, N-acetylglucosamine; ◯, galactose; ◆, sialic acid

Databanks

ST3Gal-I

NC-IUBMB enzyme classification: E.C.2.4.99.4

Species	Gene	Protein	mRNA	Genomic
Homo sapiens	–	Q11201	L29555	–
Sus scrofa	–	Q02745	M97753	–
Mus musculus	Siat4	g402215	X73523	–
Gallus galuus	–	Q11200	X80503	–

Name and History

ST3Gal-I was originally described as ST3O. This enzyme is apparently very similar, or identical, to ST3GalA.1 and SiaT-4a previously described. Since ST3GalA.1 and SiaT-4a were the names for the partially purified enzymes, they were not necessarily a homogenous protein. Since ST3Gal-I is designated for the enzyme coded by an isolated cDNA, the definition of ST3Gal-I within the context of sequence analyses has been accepted at this time. Moreover, the amino acid sequences of ST3Gal-I among different mammalian species are highly homologous to each other (more than 95% identity). Such a high homology also indicates the designation of ST3Gal-I among different species. To date, cDNAs for ST3Gal-I have been cloned and sequenced from various vertebrates, including porcine (Gillespie et al. 1992), human (Kitagawa and Paulson 1994), mouse (Lee et al. 1993), and chicken (Kurosawa et al. 1995).

Enzyme Activity Assay and Substrate Specificity

The enzyme assay and acceptor specificity of ST3Gal-I was examined using a soluble ST3Gal-I. The insulin signal sequence was fused with the catalytic domain of ST3Gal I and transfected into COS-1 cells. The soluble ST3Gal-I secreted into the medium was used as an enzyme source. ST3Gal-I catalyzes the following reaction (Gillespie et al. 1992):

$$CMP\text{-}NeuAc + Gal\beta1 \rightarrow 3GalNAc\alpha1 \rightarrow Ser/Thr \rightarrow$$
$$NeuAc\alpha2 \rightarrow 3Gal\beta1 \rightarrow 3GalNAc\alpha1 \rightarrow Ser/Thr$$

The enzyme incubation mixture (50 μl) contains 100 mM sodium cacodylate buffer, pH 6.5 containing 1% Triton X-100, 1 mg/ml bovine serum albumin, and 0.5 nmol CMP-[^{14}C]NeuAc. The acceptor used is Galβ1→3GalNAcα→p-nitrophenol (Sigma) at 1.6 mM concentration.

Preparation

ST3Gal-I was purified from porcine liver (Sadler et al. 1979; Gillespie et al. 1992). However, a secreted ST3Gal-I is more readily available as a recombinant protein. After transfection of soluble ST3Gal-I (prepared as above) into COS-1 cells, the cell medium was concentrated 15-fold in a Centricon 30 filter (Millipore, Bedford, MA) and used as an enzyme source (Gillespie et al. 1992).

Biological Aspects

The mouse ST3Gal-I gene is composed of at least three exons (Priatel et al. 2000). In humans, the enzyme is highly expressed in heart, placenta, skeletal muscle, kidney, pancreas, liver, spleen, and peripheral blood leukocytes (Kitagawa and Paulson 1994). Interestingly, mRNA expression in the mouse may be more restricted, with high levels apparent only among lung, oviduct, and hematopoietic tissues of over 24 tissue types surveyed (unpublished data). Mice lacking ST3Gal I were developed by gene-targeting approaches and using Cre-loxP recombination (Priatel et al. 2000). The lymphocytes from ST3Gal-I-deficient mice exhibited a significant increase in unsialylated core1 O-glycans, noted by the induction of PNA binding, and only 15% of normal ST3Gal activity compared with that from wild-type mice when assayed using Galβ1→3GalNAcα1→p-nitrophenol as an acceptor. The enzyme assay indicates that most of the α2,3-sialylation on core1 O-glycans is carried out by ST3Gal-I in lymphocytes, and perhaps even more completely among CD8$^+$ T cells because MAL-II lectin reactivity was abrogated specifically in this lymphoid subpopulation. ST3Gal-I-deficient mice had lost the majority of their peripheral CD8$^+$ T cells, while no substantial effect was observed among thymocytes. The decrease in CD8$^+$ cells was most prominent among the naïve T cells with a 90% decrease, and a 50% reduction was noted among the memory CD8$^+$ T-cell population. These findings were repeated in mutant mice, where ST3Gal-I was inactivated only in T cells by Cre-loxP recombination, indicating that the apoptotic loss of CD8$^+$ T cells in vivo was due to ST3Gal-I deficiency specifically in T cells. These results can also be considered in relation to the binding of PNA. While T cells from wild-type mice express a minimum amount of PNA-positive CD45 or CD43, T cells from knockout mice display CD43 or CD45 with a strong reactivity toward PNA. This indicates that the majority of α2,3-linked sialic acid to Galβ1→3GalNAcα1→Ser/Thr is added by ST3Gal-I in wild-type mice.

CD43 in ST3Gal-I deficient mice contained a significant amount of core2 branched O-glycans despite the fact that core2GlcNAcT activity remained unaffected. These results, combined with the induction of apoptotic cells, suggest that overexpression of core2 branched O-glycans in CD8$^+$ T cells, or perhaps the loss of sialic acid from core 1 O-glycans, render them sensitive to cell death in the absence of immune stimulation. Furthermore, CD8$^+$ memory T cells normally downregulate core2 O-glycans, apparently enabling them to survive for a long time, as ST3Gal-I-deficient memory CD8$^+$ T cells were mostly apoptotic compared with wild-type cells. The most intriguing aspect of this relationship was demonstrated by the significantly increased apoptosis of CD8$^+$ T cells that resulted when an antibody (1B11) specific to core2 O-glycans attached to CD43 (Jones et al. 1994) was applied (Priatel et al. 2000). This suggests the possibility that a lectin specific for either desialylated core1 O-glycans, core2 O-glycans, or both, may drive the apoptotic program in postactivated CD8$^+$ T cells as necessary to reduce cell numbers in the periphery subsequent to an effective immune response.

Future Perspectives

It is now evident that oligosaccharide formation is a dynamic process as directed by ST3Gal-I, and encompasses a competitive balance with core2GlcNAcT function. Since the roles of core2 O-glycans have been implicated in many biological systems, it is critical to understand how ST3Gal-I may regulate O-glycan biosynthesis in various contexts. Further studies should also be aimed at adding to our understanding of the roles of the various core1 O-glycan branch structures themselves.

Further Reading

Useful texts are marked with asterisks in the reference list.

References

*Baum LG, Pang M, Perillo NL, Wu T, Delegeane A, Uittenbogaart CH, Fukuda M, Seilhamer JJ (1995) Human thymic epithelial cells express an endogenous lectin, galectin-1, which binds to core 2 O-glycans on thymocytes and T lymphoblastoid cells. J Exp Med 181:877–887

Bierhuizen MFA, Fukuda M (1992) Expression cloning of cDNA encoding UDP-GlcNAc:Galβ1-3GalNAc-R(GlcNAc to GalNAc) β1,6-N-acetylglucosaminyltransferase by gene transfer into CHO cells expressing polyoma large T antigen. Proc Natl Acad Sci USA 89:9326–9330

Fox RI, Hueniken M, Fong S, Behar S, Royston I, Singhal SK, Thompson L (1983) A novel cell surface antigen (T305) found in increased frequency on acute leukemia cells and in autoimmune disease states. J Immunol 131:762–767

Gillespie W, Kelm S, Paulson JC (1992) Cloning and expression of Gal β1,3GalNAc α 2,3-sialyltransferase. J Biol Chem 267:21004–21010

Gillespie W, Paulson JC, Kelm S, Pang M, Baum LG (1993) Regulation of α2,3-sialyltransferase expression correlates with conversion of peanut agglutinin (PNA)+ to PNA– phenotype in developing thymocytes. J Biol Chem 268:3801–3804

Harrington LE, Galvan M, Baum LG, Altman JD, Ahmed R (2000) Differentiating between memory and effector CD8 T cells by altered expression of cell surface O-glycans. J Exp Med 191:1241–1246

Ishii A, Ohta M, Watanabe Y, Matsuda K, Ishiyama K, Sakoe K, Nakamura M, Inokuchi J, Sanai Y, Saito M (1998) Expression cloning and functional characterization of human cDNA for ganglioside GM3 synthase. J Biol Chem 273:31652–31655

Jones AT, Federsppiel B, Ellies LG, Williams MJ, Burgener R, Duronio V, Smith CA, Takei F, Ziltener HJ (1994) Characterization of the activation-associated isoform of CD43 on murine T lymphocytes. J Immunol 153:3426–3439

Kitagawa H, Paulson JC (1994) Differential expression of five sialyltransferase genes in human tissues. J Biol Chem 269:17872–17878

Kurosawa N, Hamamoto T, Inoue M, Tsuji S (1995) Molecular cloning and expression of chick Gal β1,3GalNAc α2,3-sialyltransferase. Biochim Biophys Acta 1244:216–222

Lee YC, Kurosawa N, Hamamoto T, Nakaoka T, Tsuji S (1993) Molecular cloning and expression of Gal β1,3GalNAc α2,3-sialyltransferase from mouse brain. Eur J Biochem 216:377–385

*Mukasa R, Homma T, Ohtsuki T, Hosono O, Souta A, Kitamura T, Fukuda M, Watanabe S, Morimoto C (1999) Core 2-containing O-glycans on CD43 are preferentially expressed in the memory subset of human CD4 T cells. Int Immunol 11:259–268

Okajima T, Fukumoto S, Miyazaki H, Ishida H, Kiso M, Furukawa K, Urano T (1999) Molecular cloning of a novel α2,3-sialyltransferase (ST3Gal VI) that sialylates type II lactosamine structures on glycoproteins and glycolipids. J Biol Chem 274:11479–11486

*Piller F, Piller V, Fox RI, Fukuda M (1988) Human T-lymphocyte activation is associated with changes in O-glycan biosynthesis. J Biol Chem 263:15146–15150

*Priatel JJ, Chui D, Hiraoka N, Simmons CJ, Richardson KB, Page DM, Fukuda M, Varki NM, Marth JD (2000) The ST3Gal-I sialyltransferase controls CD8+ T lymphocyte homeostasis by modulating O-glycan biosynthesis [in process citation]. Immunity 12:273–283

Reisner Y, Linker-Israeli M, Sharon N (1976) Separation of mouse thymocytes into two subpopulations by the use of peanut agglutinin. Cell Immunol 25:129–134

Sadler JE, Rearick JI, Paulson JC, Hill RL (1979) Purification to homogeneity of a β-galactoside α2→3-sialyltransferase and partial purification of an α-N-acetylgalactosaminide α2→6-sialyltransferase from porcine submaxillary glands. J Biol Chem 254:4434–4442

*Skrincosky D, Kain R, El-Battari A, Exner M, Kerjaschki D, Fukuda M (1997) Altered Golgi localization of core 2 β1,6-N-acetylglucosaminyltransferase leads to decreased synthesis of branched O-glycans. J Biol Chem 272:22695–22702

Tsuji S, Datta AK, Paulson JC (1996) Systematic nomenclature for sialyltransferases [letter]. Glycobiology 6:v–vii

*Whitehouse C, Burchell J, Gschmeissner S, Brockhausen I, Lloyd KO, Taylor-Papadimitriou J (1997) A transfected sialyltransferase that is elevated in breast cancer and localizes to the medial/trans-Golgi apparatus inhibits the development of core-2-based O-glycans. J Cell Biol 137:1229–1241

36

ST3Gal-II (SAT-IV)

Introduction

ST3Gal-II is a β-galactoside α2-3-sialyltransferase which is expressed in brain, liver, and striated muscle, and is presumed to be mainly involved in the synthesis of GD1a and GT1b. The cDNA sequences included an open reading frame coding for 350 amino acids, and the primary structure of this enzyme suggested a type II membrane protein topology with a putative domain structure consisting of four regions, as in other glycosyltransferases. ST3Gal-II exhibits characteristic motifs for the sialyltransferases called sialylmotifs L, S, and VS. ST3Gal-II also has a Kurosawa motif, as seen in the ST3Gal family and two members of the ST6GalNAc family (Tsuji 1999). The deduced amino acid sequence of ST3Gal-II (mouse) showed 76% identity in the active domain with that of ST3Gal-I (Lee et al. 1993, 1994). Northern blotting indicated that it is prominent in brain, liver, skeletal muscle, and heart. This enzyme expressed in COS-7 cells exhibited transferase activity only toward the disaccharide moiety of Gal-β-1,3-GalNAc of glycolipids, as well as glycoproteins and oligosaccharides such as ST3Gal-I, but asialo-GM1 and GM1 were much more suitable substrates for ST3Gal-II than for ST3Gal-I.

TOSHIRO HAMAMOTO[1] and SHUICHI TSUJI[2]

[1] Department of Biochemistry, Jichi Medical School, Minamikawachi, Tochigi 329-0498, Japan
Tel. +81-285-58-7322; Fax +81-285-44-1827
e-mail: thamamot@jichi.ac.jp
[2] Department of Chemistry, Faculty of Science, Ochanomizu University, Otsuka, Bunkyo-ku, Tokyo 112-8610, Japan
Tel. +81-3-5978-5345; Fax +81-3-5978-5344
e-mail: stsuji@cc.ocha.ac.jp

Databanks

ST3Gal-II (SAT-IV)

NC-IUBMB enzyme classification: E.C.2.4.99.4

Species	Gene	Protein	mRNA	Genomic
Homo sapiens	*SIAT4B*	Q16842	U63090	–
		JC5251	–	–
Mus musculus	*Siat5*	Q11204	X76989	–
		A54420	–	–
Rattus norvegicus	*Siat5*	Q11205	X76988	–
		B54420	–	–

Name and History

A sialyltransferase which synthesizes GD1a form GM1 was named monosialoganglio-side sialyltransferase (EC 2.4.99.2, systematic name: CMP-*N*-acetylneuraminate:D-galactosyl-*N*-acetyl-D-galactosaminyl-(*N*-acetylneuraminyl)-D-galactosyl-D-glucosylc eramide). Later, it was shown that GM1b, GD1a, and GT1b synthases are identical, and the enzyme was called SAT-IV (Pohlentz et al. 1988) or SAT-4 (Basu et al. 1987). However, without purification of the authentic enzyme, ST3GalA.2 was cloned as homologous cDNA to ST3Gal-I. This enzyme showed a strict preference for the disaccharide moiety of Galβ1-3GalNAc of glycolipids as well as glycoproteins and oligosaccharides such as ST3Gal-I, but asialo-GM1 and GM1 were much more suit-able substrates than ST3Gal-I. In 1996, it was designated ST3Gal-II according to the abbreviated nomenclature system for cloned sialyltransferases (Tsuji et al. 1996). Comparative analysis of human ST3Gal-II cDNA with mouse ST3Gal-II indicates 89% and 94% homologies in the nucleotide and amino acid levels, respectively, between the two sequences in the predicted coding region (Kim et al. 1996). This enzyme expressed in COS cells showed a similar activity to that of mouse ST3Gal-II. Currently, ST3Gal-II is regarded as a β-galactoside α2-3-sialyltransferase (EC 2.4.99.4, systematic name: CMP-*N*-acetylneuraminate:β-D-galactoside α2-3-*N*-acetylneuraminyl-transferase) along with ST3Gal-I, but the tissue distribution and substrate preference make it a primary candidate for a synthase of GM1b, GD1a, and GT1b. Monosialo-ganglioside sialyltransferase (EC 2.4.99.2) may be an obsolete entry. The International System for Gene Nomenclature named this gene *SIAT4B* (human) and *Siat5* (mouse).

Enzyme Activity Assay and Substrate Specificity

Enzyme Activity Assay

CMP-[^{14}C]]NeuAc(11 GBq/nmole) was from Amersham Pharmacia Biotech (Plscat-away, NJ, USA), asialo-GM1 and Triton CF-54 were from Sigma (St. Louis, MO, USA). The reaction mixture, comprising 0.05 M sodium cacodylate buffer (pH 6.0), 0.05 mM CMP-[^{14}C]NeuAc (0.9 Bq/pmol), 0.5% Triton CF-54, 1 µl enzyme solution, and 1 mM acceptor substrates (oligosaccharides and glycolipids), in a total volume of 10 µl, was incubated at 37°C for 1 h (Lee et al. 1994).

Oligosaccharide and glycoprotein products were separated from CMP-[^{14}C]NeuAc by high-pressure thin-layer chromatography (HPTLC) (Silica Gel 60, E. Merck, Darmstadt, Germany) with a solvent system of ethanol/pyridine/n-butanol/acetate/water (100:10:10:3:30). For glycolipid products, the reaction mixture was subjected to HPTLC with a solvent system of chloroform/methanol/0.02% CaCl$_2$ (55:45:10). When asialoglycoproteins were used as acceptors, the substrate concentration was adjusted to 2 mg/ml (asialofetuin, 140 µM as Galβ1,3GalNAc residue). The radioactivity of corresponding products were visualized with a BAS2000 radio image-analyzer (Fuji Film, Japan).

Substrate Specificity

ST3Gal-II utilizes acceptor substrates which contain the Galβ1-3GalNAc sequence. For oligosaccharide substrates, ST3Gal-II showed the highest activity toward Galβ1-3GalNAc (type III), very low activity toward Galβ1-3GlcNAc (type I), and none toward Galβ1-4GlcNAc (type II). Gangliosides (sialo-GM1, GM1, GD1b) as well as glycoproteins (Asialofetuin, asialo-bovine submaxillary mucin) having terminal Galβ1-3GalNAc served as good acceptors. Thus, the substrate specificity of ST3Gal-II is similar to that of ST3Gal-I and different from those of ST3Gal-III and IV, which exhibit high activity toward the type I and II disaccharides, but very low activity toward the type III. Also, ST3Gal-III and IV rarely utilized glycolipids as substrates.

Between ST3Gal-I and II, ST3Gal-I exhibits the highest Km value for asialo-GM1 (Km = 1.25 mM) and the lowest for asialofetuin (Km = 0.10 mM), whereas the Km values of ST3Gal-II for the substrates are very similar (Km approximately 0.5 mM). Furthermore, the synthesis of GM1b from asialo-GM1 by ST3Gal-I was clearly inhibited in the presence of disaccharide Galβ1,3GalNAc or asialofetuin, but that by ST3Gal-II was not inhibited at all. On the other hand, the activity of ST3Gal-II toward disaccharide Galβ1-3GalNAc or asialofetuin was inhibited by asialo-GM1 or GM1. Further study indicated ST3Gal-II exhibits noncompetitive inhibition between asialo-GM1 and Galβ1-3GalNAc or between asialo-GM1 and asialofetuin, whereas ST3Gal-I exhibits competitive inhibition between all kinds of acceptors. In conclusion, gangliosides serve as the predominant acceptors for ST3Gal-II over O-glycosidically linked oligosaccharides of glycoproteins, which are much better acceptors for ST3Gal-I.

Preparation

Purification of authentic ST3Gal-II has not been reported. CMP-N-acetylneuraminate: β-D-galactoside α2-3-N-acetylneuraminyltransferase has been purified 20000-fold from a Triton X-100 extract of human placenta by affinity chromatography on concanavalin-A-sepharose and CDP-hexanolamine-sepharose (Joziasse et al. 1985). Although, Northern blot analysis indicates medium-level expression of ST3Gal-II in human placenta (Kim et al. 1996), whether the purified enzyme was ST3Gal-I, ST3Gal-II, or a mixture of the two was not clear.

ST3Gal-II, free from ST3Gal-I, was obtained as a recombinant soluble enzyme. Recombinant protein in which the truncated form (lacking 55 N-terminal amino acids) of rat or mouse ST3Gal-II fused with IgM signal peptide was constructed with pcDSR-α expression vector and transiently expressed in COS-7 cells. The cell medium was harvested at 48 h posttransfection, concentrated 10-fold on Centricon 30 filters (Amicon Beverly, MA, USA), and used as an enzyme source (Lee et al. 1994).

Biological Aspects

Even though ST3Gal-II prefers a type III sequence in glycolipid and ST3Gal-I prefers that in glycoprotein, they have overlapping acceptor specificities. Northern blot analysis indicated that the expression of human ST3Gal-II mRNA (4.4 kb) is tissue-specific, being prominent in skeletal muscle and heart, and also expressed in liver, placenta, brain, and pancreas, while that in lung and kidney is very low. The mouse ST3Gal-II gene was strongly expressed in brain and liver, while that in spleen and salivary gland was very low. Quantitative analysis of the expression of mouse sialyltransferase genes by competitive polymerase chain reaction (PCR) indicated that the ST3Gal-II gene was strongly expressed in brain, and weakly in colon, thymus, salivary gland, and testis, and developmentally expressed in liver, heart, kidney, and spleen, while the ST3Gal-I gene was strongly expressed in spleen and salivary gland, and weakly in brain, liver, heart, kidney, and thymus (Takashima et al. 1999).

This tissue-specific expression and substrate specificity strongly indicated that ST3Gal-II participates in the synthesis of glycolipids such as GM1b, GD1a, and GT1b. The brain is known to have a particularly high content of gangliosides, as Pohlentz et al. (1988) reported in a competition experiment GM1b, GD1a, and GT1b were synthesized by a single enzyme (sialyltransferase IV) in Golgi vesicles from rat liver. The K_m of sialyltransferase IV for asialo-GM1 and GM1 was much smaller than that of ST3Gal-II (and, of course, ST3Gal-I). The association of the sialyltransferases with Golgi membranes may be important for this activity, particularly for ganglioside synthesis. It is also possible that some activating factors existing in the Golgi apparatus may enhance the activity of sialyltransferase IV.

The mouse genes encoding β-galactoside $\alpha2,3$-sialyltransferases (Siat4, ST3Gal-I; Siat5, ST3Gal-II) were isolated and characterized (Takashima and Tsuji 2000). The Siat4 and Siat5 genes were found to comprise 8.4 and 14 kilobases, respectively, and to be composed of six exons. The genomic structures of the two genes were quite similar. These results suggested that the Siat4 and Siat5 genes arose from common ancestral genes.

Future Perspectives

ST3Gal-I and ST3Gal-II have different expression patterns in various tissues (Kono et al. 1997; Takashima et al. 1999). Identification of the transcription factors that are involved in stage-specific expression of these genes may facilitate our understanding of their different expression patterns and the mechanisms for stage- and tissue-specific expression. Unlike ST3Gal-I, which is dynamically regulated in carcinogensis or in T lymphocyte activation (Priatel et al. 2000), expression controls, except for those which are tissue-specific or developmental, are not reported for ST3Gal-II.

It should be noted that even though ST3Gal-I and ST3Gal-II showed overlapping substrate specificity against O-linked glycan chains in vitro, ST3Gal-I-deficient mice caused an elimination of sialic acid on core 1 O-glycans in T lymphocytes (see the chapter on ST3Gal-I), suggesting that ST3Gal-II does not work on O-glycans of glycoproteins in vivo.

In addition, it has yet to be determined whether ST3Gal-II is the only enzyme that participates in GM1b, GD1a, and GT1b synthesis.

Further Reading

Kojima et al. (1994) reported enzymatic characterization using purified ST3Gal-II protein and compared it in detail with ST3Gal-I.
Kono et al. (1997) compared the enzymatic characterization of four members of the ST3Gal family.
Lee et al. (1994) successfully cloned ST3Gal-II cDNA from mouse brain.

References

Basu M, De T, Das KK, Kyle JW, Chon H-C, Schaeper RJ, Basu S (1987) Glycolipids. Method Enzymol 138:575–607
Joziasse DH, Bergh ML, ter Hart HG, Koppen PL, Hooghwinkel GJ, Van den Eijnden DH (1985) Purification and enzymatic characterization of CMP-sialic acid: galactosylβ1,3-N-acetylgalactosaminide α-2,3-sialyltransferase from human placenta. J Biol Chem 260:941–951
Kim YJ, Kim KS, Kim SH, Kim CH, Ko JH, Choe IS, Tsuji S, Lee YC (1996) Molecular cloning and expression of human Galβ1,3GalNAcα2,3-sialyltransferase (hST3Gal-II). Biochem Biophys Res Commun 228:24–27
Kojima N, Lee YC, Hamamoto T, Kurosawa N, Tsuji S (1994) Kinetic properties and acceptor substrate preferences of two kinds of Galβ1,3GalNAc α2,3-sialyltransferase from mouse brain. Biochemistry 33:5772–5776
Kono M, Ohyama Y, Lee YC, Hamamoto T, Kojima N, Tsuji S (1997) Mouse β-galactoside α2,3-sialyltransferases: comparison of in vitro substrate specificities and tissue-specific expression. Glycobiology 7:469–479
Lee YC, Kurosawa N, Hamamoto T, Nakaoka T, Tsuji S (1993) Molecular cloning and expression of Galβ1,3GalNAc α2,3-Sialyl-transferase from mouse brain. Eur J Biochem 216:377–385
Lee YC, Kojima N, Wada E, Kurosawa N, Nakaoka T, Hamamoto T, Tsuji S (1994) Cloning and expression of cDNA for a new type of Gal β1,3GalNAc α2,3-sialyltransferase. J Biol Chem 269:10028–10033
Pohlentz G, Klein D, Schwarzmann G, Schmitz D, Sandhoff K (1988) Both GA2, GM2, and GD2 synthases and GM1b, GD1a, and GT1b synthases are single enzymes in Golgi vesicles from rat liver. Proc Natl Acad Sci USA 85:7044–7048
Priatel JJ, Chui D, Hiraoka N, Simmons CJ, Richardson KB, Page DM, Fukuda M, Varki NM, Marth JD (2000) The ST3Gal-I sialyltransferase controls CD[8+] T lymphocyte homeostasis by modulating O-glycan biosynthesis. Immunity 12:273–283
Takashima S, Tsuji S (2000) Comparison of genomic structures of four members of mouse β-galactoside α2,3-sialyltransferase genes. Cytogenet Cell Genet in press
Takashima S, Tachida Y, Nakagawa T, Hamamoto T, Tsuji S (1999) Quantitative analysis of expression of mouse sialyltransferase genes by competitive PCR. Biochem Biophys Res Commun 260:23–27
Tsuji S (1999) Molecular cloning and characterization of sialyltransferases. In Inoue Y, Lee YC, Troy FA (eds) Sialobiology and other novel forms of glycosylation. Gakushin Osaka pp 145–154
Tsuji S, Datta AK, Paulson JC (1996) Systematic nomenclature for sialyltransferases. Glycobiology 6(7):v–vii
Yip MCM (1973) The enzymic synthesis of disialoganglioside: rat brain cytidine-5′-monophospho-N-acetylneuraminic acid:monosialoganglioside (GM1) sialyltransferase. Biochim Biophys Acta 306:298–306

ST3Gal-III

Introduction

ST3Gal-III is a CMP-N-acetylneuraminate: β-galactoside α2-3-sialyltransferase (Galβ1-3(4)GlcNAc α2-3-sialyltransferase). Like other sialyltransferases, ST3Gal-III exhibits type II membrane protein topology and has characteristic motifs for sialyltransferases which are called sialylmotifs L, S, and VS. ST3Gal-III also has a Kurosawa motif as seen in the ST3Gal family and two members of the ST6GalNAc family (Tsuji 1999).

The ST3Gal-III enzyme is one of a few sialyltransferases that have been successfully purified to homogeneity (Weinstein et al. 1982a). Even though it is believed that ST3Gal-III is mainly involved in the α2-3-sialylation of the less common type I (Galβ1-3GlcNAc) sequence in N-linked glycan chains, ST3Gal-III can also catalyze the α2-3-sialylation of type II (Galβ1-4GlcNAc) sequence in N-linked glycan chains (Weinstein et al. 1982b). In contrast, ST3Gal-IV (Galβ1-4GlcNAc-α2-3-sialyltransferase), the gene of which was cloned later, prefers the type II sequence to the type I sequence (Sasaki et al. 1993). Both enzymes rarely utilize glycolipids such as paraglanoside (Galβ1-4GlcNAcβ1-3Galβ1-4Glc-Cer).

SHINOBU KITAZUME-KAWAGUCHI[1] and SHUICHI TSUJI[2]

[1] Frontier Research Program, RIKEN Institute, 2-1 Hirosawa, Wako-shi, Saitama 351-0198, Japan
Tel: +81-48-467-9616; Fax +81-48-467-9617
e-mail: shinobuk@postman.riken.go.jp
[2] Department of Chemistry, Faculty of Science, Ochanomizu University, Otsuka, Bunkyo-ku, Tokyo 112-8610, Japan
Tel. +81-3-5978-5345; Fax +81-3-5978-5344
e-mail: stsuji@cc.ocha.ac.jp

Databanks

ST3Gal-III

NC-IUBMB enzyme classification: E.C.2.4.99.6

Species	Gene	Protein	mRNA	Genomic
Homo sapiens	SIAT3	–	L23768	–
Mus musculus	Siat3	P97325	X84234	–
Rattus norvegicus	–	Q02734	M97754	–

Name and History

ST3Gal-III, Galβ1-3(4)GlcNAc α2-3-sialyltransferase, used to be called ST3N because it utilizes *N*-linked glycan chains as acceptor substrates and because it was biochemically identified earlier than ST3Gal-IV, which also utilizes *N*-linked glycan chains as acceptor substrates (Kitagawa and Paulson 1994). In 1996, it was designated ST3Gal-III according to the abbreviated nomenclature system for cloned sialyltransferases (Tsuji et al. 1996). In 1982, ST6Gal-I and ST3Gal-III were purified from rat liver using CDP-hexanolamine-Sepharose, and were extensively characterized (Weinstein et al. 1982a, b). Later, Wen et al. (1992) successfully determined the partial amino acid sequence of ST3Gal-III by means of mass spectrometry using a tiny amount of purified protein. Using the peptide sequence information obtained, short degenerate primers were constructed for polymerase chain reaction (PCR)-based cDNA cloning. An analysis of the genomic structure of the mouse ST3Gal-III gene has been reported (Takashima and Tsuji 2000). The International System for Gene Nomenclature named this gene *Siat3* (sialyltransferase 3).

Enzyme Activity Assay and Substrate Specificity

Enzyme Activity Assay

Initially, enzymatic characterization of ST3Gal-III was extensively studied using purified protein from rat liver (Weinstein et al. 1982b). Briefly, enzyme reaction mixtures (60 µl) containing ST3Gal-III (0–0.2 milliunits), an acceptor substrate, 9 nmol CMP-[^{14}C]NeuAc (6600 cpm/nmol), and 50 µg bovine serum albumin in 50 mM sodium cacodylate, pH 6.0, with 0.5% Triton CF-54, were incubated at 37°C for various lengths of time (5–30 min). One unit here refers to 1 µmole of product formed per minute of incubation at 37°C at saturating substrate concentrations. The reaction products of glycoproteins and glycolipids were isolated by chromatography on a column of Sephadex G-50. Products of neutral oligosaccharide substrates were isolated by passage of the reaction mixture through a Pasteur pipette column of Dowex 1-x8 (PO_4^{2-}, 100–200 mesh), and products of sialyloligosaccharide substrates were isolated by chromatography on a column of Sephadex G-25. In each case, the elution pattern was monitored by scintillation spectrophotomety.

Subsequently, as has been done for most of the cloned sialyltransferases, recombinant protein in which a soluble form of the enzyme was fused with a protein A IgG binding domain was constructed and expressed in COS cells. The enzymes secreted to the media were condensed with IgG-Sepharose and used for sialyltransferase assay (Kitagawa and Paulson 1993; Kono et al. 1997). The reaction mixture, comprising 0.1 M sodium cacodylate buffer (pH 6.4), 10 mM $MgCl_2$, 2 mM $CaCl_2$, 0.1 mM CMP-[^{14}C]NeuAc (8.5 nCi), the enzyme preparation, and different amounts of acceptor substrates, in a total volume of 20 μl, were incubated at 37°C with gentle agitation using a microtube mixer. When glycolipids were used as acceptor substrates, 0.3% Triton CF-54 was added to the reaction mixtures. Oligosaccharide products were separated from CMP-[^{14}C]NeuAc by high-pressure thin-layer chromatography (HPTLC) with a solvent system of ethanol/pyridine/n-butanol/acetate/water (100:10:10:3:30). For glycolipid products, the reaction mixtures were applied on a C-18 column. After the column was washed with water, the glycolipids were eluted with methanol, dried, and then subjected to HPTLC with a solvent system of chloroform/methanol/0.02% $CaCl_2$ (55:45:10). The radioactivity of the oligosaccharides and glycolipids was visualized with a radio image-analyzer.

Substrate Specificity

There was no significant difference in the substrate specificities of purified protein and recombinant protein. ST3Gal-III utilizes acceptor substrates which contain either the Galβ1-3GlcNAc or Galβ1-4GlcNAc sequence. Because a fixed concentration of lacto-N-tetraose (Galβ1-3GlcNAcβ1-3Galβ1-4Glc) added to the reaction mixtures was shown to be a competitive inhibitor (alternate substrate) of the enzyme using asialo-α_1-acid glycoprotein (Galβ1-4GlcNAc) as the acceptor substrate, both substrates are utilized by a single ST3Gal-III enzyme. It was shown that the purified ST3Gal-III prefers lacto-N-tetraose (apparent Km is 0.09 mM) to lacto-N-neotetraose (Galβ1-4GlcNAcβ1-3Galβ1-4Glc) (apparent Km is 4.22 mM). ST3Gal-IV, whose cDNA was cloned later, shows the opposite substrate specificity and prefers a type II sequence to a type I sequence (Sasaki et al. 1993). ST3Gal-III rarely utilized glycolipids such as paragloboside as acceptor substrates (Kono et al. 1997). The purified ST3Gal-III does not require metal ions for enzyme activity, because the addition of 1 mM EDTA to the assay medium inhibited the enzyme activity by less than 10%.

Preparation

ST3Gal-III protein was purified 860 000-fold to homogeneity from Triton CF-54 extracts of rat liver membranes (Weinstein et al. 1982b). ST6Gal I and ST3Gal-III enzymes were concentrated by affinity chromatography on CDP-hexanolamine-agarose and resolved by NaCl gradient solution from the same absorbent. They were successfully separated by specific elution from CDP-agarose with CDP. The final purification of the ST3Gal-III protein required affinity chromatography on an absorbent prepared by coupling asialoprothrombin to cyanogen bromide-activated agarose. Asialoprothrombin contains the terminal sequence Galβ1-3GlcNAc on N-linked glycan chains, and is the best acceptor substrate of the enzyme. ST3Gal-III protein was found to bind to asialoprothrombin-agarose in the presence of CDP, and could be eluted with a solution containing 0.2 M lactose and no CDP. SDS-polyacrylamide gel analysis revealed a single enzyme with an apparent molecular weight of 44 000. The purified enzyme was completely devoid of detectable (<0.1%) ST6Gal I as judged by the production of only the 3′-isomer of sialyllactose with lactose as substrate.

Alternatively, as has been done for most sialyltransferases, recombinant protein in which a truncated form of rat or human ST3Gal-III was fused with an insulin signal peptide plus a protein A IgG binding domain was also constructed and expressed in COS cells (Kitagawa and Paulson 1993).

Biological Aspects

Even though rat ST3Gal-III was successfully purified from livers, Northern blot analysis showed that human and rat ST3Gal-III genes were abundantly expressed in skeletal muscle and fetal tissues, whereas low levels of mRNA were found in placenta, lung, and liver (Kitagawa and Paulson 1994). Two transcripts (2.3 kb and 2.0 kb) of the mouse ST3Gal-III gene were detected in all tissues tested (Kono et al. 1997). It was shown that the level of the mouse ST3Gal-III gene was developmentally regulated in liver, heart, kidney, and spleen. Liver-produced serum glycoproteins with biantennary oligosaccharides, such as transferrin and α_2-macroglobulin, the terminal sialic acids are predominately (>90%) NeuAcα2-6-Gal-linked, while for other types of serum glycoproteins, such as α_1-acid glycoprotein with tri- and tetraantennary chains, the sialic acids are found in both NeuAcα2-6-Gal (60%) and NeuAcα2-3-Gal (40%) linkages. Thus, despite the minor expression in liver, it seems that ST3Gal-III plays a significant role in the sialylation of the tri- and tetraantennary branched structures of liver glycoproteins. It is not yet known why the ST3Gal-III gene is very highly expressed in skeletal muscle.

The genomic structures of mouse ST3Gal-III and -IV genes (*Siat3* and *4c*) were similar to each other. *Siat3* (ST3Gal-III) was found to comprise over 100 kilobases, and to be composed of 12 exons. Although the genome sizes of *Siat3* (ST3Gal-III) and *Siat4c* (ST3Gal-IV) genes were quite different, some of their exon structures were significantly similar. These results suggested that *Siat3* and *Siat4c* arose from common ancestral genes (Takashima and Tsuji 2000).

Future Perspectives

Even though ST3Gal-III prefers a type I sequence in *N*-linked glycan chains, and ST3Gal-IV prefers a type II sequence, they have overlapping acceptor specificity. However, they have different expression patterns in various tissues (Kono et al. 1997; Takashima et al. 1999). The identification of transcription factors that are involved in stage-specific expression of these genes may facilitate our understanding of their different expression patterns and the mechanisms for stage- and tissue-specific expression. The construction and characterization of ST3Gal-III-deficient mice might tell us whether ST3Gal-III and -IV show overlapping enzyme activity in vivo. From this standpoint, it should be noted that even though ST3Gal-I and ST3Gal-II showed overlapping substrate specificity against *O*-linked glycan chains in vitro, ST3Gal-I knockout mice showed an elimination of sialic acid on core 1 *O*-glycans in T lymphocytes

(Priatel et al. 2000) (see the chapter on ST3Gal-I). With such research, we mighty also be able to understand the biological significance of the Siaα2-3Galβ1-3GlcNAc structure.

Further Reading

Weinstein et al. (1982a) reported in detail the enzymatic characterization using purified ST3Gal-III protein.
Weinstein et al. (1982b) successfully purified the ST3Gal-III protein from rat liver by a classical method.
Kono et al. (1997) compared the enzymatic characterization of four members of the ST3Gal family.

References

Kitagawa H, Paulson JC (1993) Cloning and expression of human Galβ1,3(4)GlcNAc α2,3-sialyltransferase. Biochem Biophys Res Commun 194:375–382
Kitagawa H, Paulson JC (1994) Differential expression of five sialyltransferase genes in human tissues. J Biol Chem 269:17872–17878
Kono M, Ohyama Y, Lee YC, Hamamoto T, Kojima N, Tsuji S (1997) Mouse β-galactoside α2,3-sialyltransferases: comparison of in vitro substrate specificities and tissue-specific expression. Glycobiology 7:469–479
Priatel JJ, Chui D, Hiraoka N, Simmons CJ, Richardson KB, Page DM, Fukuda M, Varki NM, Marth JD (2000) The ST3Gal-I sialyltransferase controls CD[8+] T lymphocyte home-ostasis by modulating O-glycan biosynthesis. Immunity 12:273–283
Sasaki K, Watanabe E, Kawashima K, Sekine S, Dohi T, Oshima M, Hanai N, Nishi T, Hasegawa M (1993) Expression cloning of a novel Galβ(1-3/1-4)GlcNAc α2,3-sialyl-transferase using lectin resistance selection. J Biol Chem 268:22782–22787
Takashima S, Tsuji S (2000) Comparison of genomic structures of four members of mouse β-galactoside α2,3-sialyltransferase genes. Cytogenet Cell Genet 89:101–106
Takashima S, Tachida Y, Nakagawa T, Hamamoto T, Tsuji S (1999) Quantitative analysis of expression of mouse sialyltransferase genes by competitive PCR. Biochem Biophys Res Commun 260:23–27
Tsuji S (1999) Molecular cloning and characterization of sialyltransferases. In: Inoue Y, Lee YC, Troy FA (eds) Sialobiology and other novel forms of glycosylation. Gakushin, Osaka, pp 145–154
Tsuji S, Datta AK, Paulson JC (1996) Systematic nomenclature for sialyltransferases. Glycobiology 6(7):v–vii.
Weinstein J, de Souza-s-Silva U, Paulson JC (1982a) Purification of a Galβ1→4GlcNAcα2→6sialyltransferae and a Galβ1→3(4)GlcNAc α2→3sialyltransferase to homogeneity from rat liver. J Biol Chem 257:13835–13844
Weinstein J, de Souza-s-Silva U, Paulson JC (1982b) Sialylation of glycoprotein oligosac-charides N-linked to asparagine. J Biol Chem 257:13845–13853
Wen DX, Livingston BD, Medzihradszky KF, Kelm S, Burlingame AL, Paulson JC (1992) Primary structure of Galβ1,3(4)GlcNAc α2,3-sialyltransferase determined by mass spectrometry sequence analysis and molecular cloning. J Biol Chem 267:21011–21019

38

ST3Gal-IV

Introduction

ST3Gal-IV is a CMP-*N*-acetylneuraminate: β-galacotside α2-3-sialyltransferase (Galβ1-4(3)GlcNAc α2-3-sialyltransferase) (Tsuji et al. 1996). Expression cloning with cytotoxic lectin resistance selection has successfully isolated a cDNA encoding human ST3Gal-IV. ST3Gal-IV is a type II membrane protein, as are most other glycosyltransferases. It has three conserved motifs, sialylmotif L, S, and VS, within the active domain. ST3Gal-IV also has a Kurosawa motif, as seen in the ST3Gal family and two members of the ST6GalNAc family (Tsuji 1999). ST3Gal-IV mainly catalyzes the α2-3-sialylation of type II (Galβ1-4GlcNAc) sequence in *N*-linked glycan chains. Subsequent α1-3-fucosylation to the GlcNAc residue results in the formation of sialyl Lewis X, which is recognized by the selectin family, cell adhesion molecules. ST3Gal-IV also catalyzes the α2-3-sialylation of a type I (Galβ1-3GlcNAc) sequence in *N*-linked glycan chains. In contrast, ST3Gal-III prefers a type I sequence to a type II sequence. Both enzymes rarely utilize a glycolipid such as paragloboside (Galβ1-4GlcNAcβ1-3Galβ1-4Glc-Cer).

SHINOBU KITAZUME-KAWAGUCHI[1] and SHUICHI TSUJI[2]

[1] Frontier Research Program, RIKEN Institute, 2-1 Hirosawa, Wako-shi, Saitama 351-0198, Japan
Tel. +81-48-467-9616; Fax +81-48-467-9617
e-mail: shinobuk@postman.riken.go.jp
[2] Department of Chemistry, Faculty of Science, Ochanomizu University, Otsuka, Bunkyo-ku, Tokyo 112-8610, Japan
Tel. +81-3-5978-5345; Fax +81-3-5978-5344
e-mail: stsuji@cc.ocha.ac.jp

Databanks

ST3Gal-IV

NC-IUBMB enzyme classification: E.C.2.4.99.6

Species	Gene	Protein	mRNA	Genomic
Homo sapiens	*SIAT4C*	Q11206	L23767	–
		–	L29553	–
		–	X74570	–
Mus musculus	*Siat4c*	P97354	X95809	–
	–	Q61325	D28941	–
Mesocricetus auratus	–	–	AJ245700	–
Gallus gallus	–	–	AF035250	–

Name and History

ST3Gal-IV has been called STZ, SAT-3, or SiaT-4c (Basu et al. 1987; Basu 1991; Kitagawa and Paulson 1994a). STZ is the name given to one of the sialyltransferases, the cDNA of which Paulson's group cloned using a polymerase chain reaction (PCR) homology approach employing degenerate synthetic primers to sialylmotif L. SAT-3 is the name of the particular sialyltransferase activities that have been characterized using disaccharide as an acceptor substrate from the mixtures of sialyltransferases in the solubilized tissue homogenates SiaT-4c (Basu et al. 1987; Basu 1991). SiaT-4c is a symbol, and is still used. Because the ST3Gal-III that was purified from rat liver and characterized earlier utilizes *N*-linked glycan chains as an acceptor substrate (Kitagawa and Paulson 1994b), it has been called ST3N. Subsequent studies showed that ST3Gal-III prefers the less common type I sequence to a type II sequence as an acceptor substrate (Weinstein et al. 1982). Later, Sasaki et al. (1993) successfully isolated the ST3Gal-IV cDNA by expression cloning, and showed that the enzyme prefers a type II sequence to a type I sequence.

Enzyme Activity Assay and Substrate Specificity

Sasaki et al. (1993) used a cell line expressing ST3Gal-IV for the enzyme assay. In brief, cells (1×10^7 cells) were harvested, washed with PBS, resuspended in 100 µl 1% Triton X-100, and homogenized. After centrifugation at $900 \times g$ for 10 min, the supernatants were used as enzyme sources. The supernatant (10 µl of each) or the protein A-ST3Gal-IV fusion protein bound to IgG-Sepharose (5 µl of each) were used for sialyltransferase assay with 0.1 M cacodylate buffer (pH 6.8), 0.01 M $MnCl_2$, 0.45% Triton X-100, 5 mM CMP-sialic acid, and 0.1 mM pyridylaminnated lacto-*N*-tetraose of lacto-*N*-tetraose as acceptor substrates (total volume 30 µl). After incubation at 37°C for 2 h, the reaction mixtures were terminated by boiling for 5 min, and centrifuged at 12 000 $\times g$ for 10 min. The supernatant was subjected to HPLC analysis using a TSK-gel ODS-80T$_M$ column monitored with a fluorescence spectrophotometer to identify the reaction products. Table 1 shows the acceptor specificities of intact ST3Gal-IV and protein A-ST3Gal-IV fusion protein derived from expressing Namalwa cells.

Table 1. ST3Gal-IV activities toward oligosaccharides, showing the relative activities of the incorporation of sialic acid into oligosaccharides

	Acceptor substrates	
	Lacto-N-neotetraose	Lacto-N-tetraose
Cells expressing ST3Gal-IV	618 pmol/mg protein/h	317 pmol/mg protein/h
Control cells	Not detectable	114 pmol/mg protein/h
ST3Gal-IV bound to IgG-Sepharose	73.2 pmol/ml of medium/h	23.0 pmol/ml of medium/h

Kono et al. (1997) extensively studied the enzymatic characterization of ST3Gal-IV in vitro. As has been done for most of the cloned sialyltransferases, recombinant protein, in which the enzyme lacking the N-terminal cytoplasmic tail, transmembrane domain, and stem region was fused with a signal peptide and a protein A IgG binding domain, was constructed and expressed in COS cells. The secreted protein was used for the sialyltransferase assay. In brief, 20 µl of the reaction mixture containing 4 µl of the enzyme fraction absorbed to IgG-Sepharose (20 µl beads per 100 ml culture medium), 0.1 M sodium cacodylate buffer (pH 6.4), 10 mM $MgCl_2$, 2 mM $CaCl_2$, 0.1 mM CMP-[^{14}C]NeuAc (8.5 nCi), and different amount of acceptor substrate was incubated at 37°C for various lengths of time. When glycolipids were used as acceptor substrates, 0.3 % Triton CF-54 was added to the reaction mixture. The reaction products were separated from CMP-[^{14}C]NeuAc, by high-pressure thin-layer chromatography (HPTLC) for oligosaccharides, or by C-18 column for glycolipids, and then quantitated by a radio image-analyzer system. ST3Gal-IV bound to IgG-Sepharose exhibited high activity toward type II disaccharide (Galβ1-4GlcNAc) and lacto-N-neotetraose (Galβ1-4GlcNAcβ1-3Galβ1-4Glc) just like wild-type ST3Gal-IV, while paragloboside (Galβ1-4GlcNAcβ1-3Galβ1-4Glc-Cer) was a poor substrate, suggesting that ST3Gal-IV rarely utilizes glycolipids as acceptor substrates.

Preparation

ST3Gal-IV cDNA was successfully isolated by expression cloning (Sasaki et al. 1993). In brief, a cDNA library of the human melanoma cell line WM266-4 was constructed in an Epstein–Barr virus-based cloning vector. The authors used Namalwa KJM-1, a subline of the human Burkitt lympoma cell line Namalwa, for the following reasons: (i) it has been shown that vectors having their replication origin in the Epstein–Barr virus were stably replicated in an episomal state, (ii) this cell line expresses sialyl Lewis X antigen poorly, and (iii) α2,3-sialyltransferase activity for lacto-N-neotetraose is not detectable in this cell line. In contrast, it has been shown that cell homogenates of the human melanoma cell WM266-4 contain more α2,3-sialyltransferase activity for lacto-N-neotetraose than for lacto-N-tetraose. Namalwa cells expressing cDNAs derived from WM266-4 were selected in the presence of the cytotoxic lectin *Ricinus communis* agglutinin (RCA_{120}) that recognizes the Galβ1-4GlcNAc structure (Sasaki et al. 1993). It is known that sialylation at these galactose residues reduces the association constants of RCA_{120} for complex glycoconjugates. From the RCA_{120}-resistant clones, transfected cDNAs were rescued, and a single plasmid was identified that conferred the RCA_{120}-resistance phenotype.

Alternatively, a recombinant protein in which a truncated form of human or mouse ST3Gal-IV was fused with an insulin signal peptide plus a protein A IgG binding domain was constructed and expressed in B-cell line Namalwa (Sasaki et al. 1993) or COS cells (Kono et al. 1997), respectively. Secreted enzyme protein that was purified on IgG-Sepharose was used for sialyltransferase assay.

Biological Aspects

The ST3Gal-IV gene was expressed in various tissues without any remarkable developmental regulation (Kono et al. 1997; Takashima et al. 1999). Northern blot analysis showed that 2.0 kb of mouse ST3Gal-IV transcript was detected in almost all tissues. In salivary gland tissue, a 3.0-kb transcript was also found. It has been reported that several cell lines expressing considerable amounts of sialyl Lewis X antigens also expressed ST3Gal-IV mRNA (Sasaki et al. 1993). ST3Gal-IV I is involved in the formation of sialyl Lewis X antigen that is a ligand for a selectin family protein on endothelial cells, lymphocytes, and platelets. However, α1,3-fucosyltransferase (FucT VII), which acts after α2,3-sialylation of the Galβ1-4GlcNAc structure and is a key enzyme for the biosynthesis of sialyl Lewis X, has received more attention and has been studied extensively.

The ST3Gal-IV gene (*Siat4c*) was found to comprise over 9.7 kb, and to be composed of ten exons. Although the genome sizes of ST3Gal-III and IV genes (*Siat3* and *Siat4c*) were quite different, some of their exon structures were significantly similar. These results suggest that the ST3Gal-III and IV genes (*Siat3* and *Siat4c*) arose from common ancestral genes (Takashima and Tsuji 2000).

Future Perspectives

Even though ubiquitous ST3Gal-IV expression has been reported, a specific probe may facilitate cell-type-specific transcription, because five different mRNAs were produced in human placenta by a combination of tissue-specific alternative splicing and alternative promoter utilization (Kitagawa et al. 1996). The identification of transcription factors that are involved in stage-specific expression of these genes may facilitate our understanding of their different expression patterns, and the mechanisms for stage- and tissue-specific expression. The development and characterization of ST3Gal-IV knockout mice might tell us whether ST3Gal-III and IV show overlapping enzyme activity in vivo. A recent interesting finding is that even though ST3Gal-I and ST3Gal-II showed overlapping substrate specificity against O-linked glycan chains in vitro, ST3Gal-I knockout mice showed an elimination of sialic acid on core 1 O-glycans in T lymphocytes (Priatel et al. 2000).

Further Reading

Sasaki et al. (1993) report an elegant expression cloning of ST3Gal-IV.
Kono et al. (1997) extensively studied the enzymatic characterization of ST3Gal-I, II, III, and IV.

References

Basu M, De T, Das KK, Kyte JW, Chon H, Schaeper RJ, Basu S (1987) Glycolipids. Methods Enzymol 138:575–607

Basu S (1991) The serendipity of ganglioside biosynthesis: pathway to CARS and HY-CARS glycosyltransferases. Glycobiology 1:469–475

Kitagawa H, Paulson JC (1994a) Cloning of a novel α2,3-sialyltransferase that sialylates glycoprotein and glycolipid carbohydrate groups. J Biol Chem 269:1394–1401

Kitagawa H, Paulson JC (1994b) Differential expression of five sialyltransferase genes in human tissues. J Biol Chem 269:17872–17878

Kitagawa H, Mattei M-G, Paulson JC (1996) Genomic organization and chromosomal mapping of the Galβ1,3GlcNAc/Galβ1,4GlcNAc α2,3-sialyltransferase. J Biol Chem 271:931–938

Kono M, Ohyama Y, Lee YC, Hamamoto T, Kojima N, Tsuji S (1997) Mouse β-galactoside α2,3-sialyltransferases: comparison of in vitro substrate specificities and tissue-specific expression. Glycobiology 7:469–479

Priatel JJ, Chui D, Hiraoka N, Simmons CJ, Richardson KB, Page DM, Fukuda M, Varki NM, Marth JD (2000) The ST3Gal-I sialyltransferase controls CD[8+] T lymphocyte homeostasis by modulating O-glycan biosynthesis. Immunity 12:273–283

Sasaki K, Watanabe E, Kawashima K, Sekine S, Dohi T, Oshima M, Hanai N, Nishi T, Hasegawa M (1993) Expression cloning of a novel Galβ(1-3/1-4)GlcNAc-α2,3-sialyltransferase using lectin resistance selection. J Biol Chem 268:22782–22787

Takashima S, Tsuji S (2000) Comparison of genomic structures of four members of mouse β-galactoside α2,3-sialyltransferase genes. Cytogenet Cell Genet 89:101–106

Takashima S, Tachida Y, Nakagawa T, Hamamoto T, Tsuji S (1999) Quantitative analysis of expression of mouse sialyltransferase genes by competitive PCR. Biochem Biophys Res Commun 260:23–27

Tsuji S (1999) Molecular cloning and characterization of sialyltransferases. In: Inoue Y, Lee YC, Troy FA (eds) Sialobiology and other novel forms of glycosylation. Gakushin, Osaka, pp 145–154

Tsuji S, Datta AK, Paulson JC (1996) Systematic nomenclature for sialyltransferases. Glycobiology 6(7):v–vii

Weinstein J, de Souza-s-Silva U, Paulson JC (1982) Sialylation of glycoprotein oligosaccharides N-linked to asparagine. J Biol Chem 257:13845–13853

ST3Gal-V (GM3 Synthase, SAT-I)

Introduction

ST3Gal-V (GM3 synthase) belongs to the sialyltransferase family, and is a unique enzyme among all sialyltransferases so far identified since it is specifically involved in ganglioside GM3 formation. Since GM3 is a common precursor of almost all gangliosides (especially ganglio-series gangliosides), GM3 synthase plays an important role in the biosynthesis of more complex gangliosides, and primarily regulates the amount of GM3. As is found in all members of the sialyltransferase family, the cloned ST3Gal-Vs from different animal origins encode type II membrane proteins with two conserved regions, i.e., sialylmotifs L and S, although an invariant aspartic acid residue in sialylmotif Ls of all other sialyltransferases is characteristically replaced by histidine residue (Ishii et al. 1998; Kono et al. 1998; Fukumoto et al. 1999).

Databanks

ST3Gal-V (GM3 Synthase, SAT-I)

NC-IUBMB enzyme classification: E.C.2.4.99.9

Species	Gene	Protein	mRNA	Genomic
Homo sapiens	*SIAT9*	BAA33950	AB018356	–
		NP_003887	NM_003896	–
		ADD14634	AF105026	–
		AAF66146	AF119415	–
		–	AF119417	–
		–	AF119418	–
		–	AK001340	–

MASAKI SAITO[1,2] and ATSUSHI ISHII[2]

[1] Department of Oncology and Pharmacodynamics, Meiji Pharmaceutical University, 2-522-1 Noshio, Kiyose-shi, Tokyo 204-8588, Japan
Tel. & Fax +81-424-95-8428
e-mail: mssaito@my-pharm.ac.jp
[2] Virology and Glycobiology Division, National Cancer Center Research Institute, 5-1-1 Tsukiji, Chuo-ku, Tokyo 104-0045, Japan

(Continued)

Species	Gene	Protein	mRNA	Genomic
		–	BE349429	–
		–	AI760803	–
		–	AI806884	–
Mus musculus		BAA33491	AB018048	–
	Siat9	CAA75235/75236	Y15003	–
		NP_035505	NM_011375	–
		BAA76467	AB013302	–
		AAF66147	AF119416	–
		–	BB135305	–
		–	AW212080	–
		–	AW211520	–
Ratus norvegicus	Siat9	BAA33492	AB018049	–

Name and History

Ganglioside GM3 synthase (EC 2.4.99.9) is also referred to as either CMP-*N*-acetylneuraminatelactosylceramide α-2,3-sialyltransferase (the name recommended by NC-IUBMB), CMP-*N*-acetylneuraminate:lactosylceramide α-2,3-*N*-acetylneuraminyltransferase (the systematic name), sialyltransferase-1, or lactosylceramide sialyltransferase. Ganglioside GM3 synthetic sialyltransferase-1, and SAT-1 and ST3Gal-V are abbreviations that are frequently used. The activity of the enzyme was originally detected in embryonic chicken brain (Kaufman et al. 1968), and has also been found in various tissues and organs such as calf brain (Preti et al. 1980), rat liver (Keenan et al. 1974), rat kidney (Fleischer 1977), rat intestinal mucosa (Glükman and Bouhours 1976), and mouse neuroblastoma cells (Duffard et al. 1977), and is a key stage in ganglioside biosynthesis leading to either ganglioside GM2 or ganglioside GD3 and their subsequent polysialyl ganglioside derivatives. Experiments with rat liver showed a great enrichment of this sialyltransferase in the Golgi membranes, although some activity was also found in the smooth endoplasmic reticulum (Richardson et al. 1977; Pacuszka et al. 1978).

Enzyme Activity Assay and Substrate Specificity

ST3Gal-V (GM3 synthase) catalyzes the reaction

$$\text{Lactosylceramide} + \text{CMP-NEUAC} \rightarrow \text{Ganglioside GM3} + \text{CMP}$$

Rat brain GM3 synthase shows a relatively narrow pH optimum, and the highest activity for the enzyme is achieved at pH 6.5 in a cacodylate buffer supplemented with manganese ion (10 mM) and a detergent (0.3%–0.4%), such as Triton CF-54, Triton X-100, or Myrj 59, using lactosylceramide as the acceptor substrate (Preuss et al. 1993). The apparent Km values for acceptor substrate lactosylceramide and donor substrate CMP-NeuAc of rat brain GM3 synthase were 80 μM and 210 μM, respectively. Very similar results were reported in rat liver GM3 synthase (110 μM for lactosylceramide and 260 μM for CMP-NeuAc) (Melkerson-Watson and Sweeley 1991) or cloned mouse

ST3Gal-V (270μM for lactosylceramide and 313μM for CMP-NeuAc) (Fukumoto et al. 1999). In contrast with the other α-2,3-sialyltransferases, GM3 synthase shows marked preferences toward glycosphingolipid substrates. Galactosylceramide, asialo-GM1, GM1, and paragloboside, which commonly contain nonreducing terminal Galβ1-structure, could serve as acceptors for the sialylation catalyzed by the enzyme, but to a lesser extent, as well as lactosylceramide (Melkerson-Watson and Sweeley 1991; Preuss et al. 1993; C. Nakamura et al. unpublished data, 2000). However, it has been demonstrated that GM1b, GD1a, and GT1b are mainly synthesized by ST3Gal-II/SAT-IV (Lee et al. 1994), and that ST3Gal-VI is responsible for the production of sialylparagloboside (Okajima et al. 1999). Although Yu and Lee (1976) suggested that the activity of GM4 synthase was indistinguishable from that of GM3 synthase, using mouse microsome fractions from brain, liver, spleen, and kidney as the enzyme sources, it is uncertain, at present, whether GM3 and GM4 are generated with the expression product of a single genetic locus or not. In any case, it has not been proven that the above glycolipids, except for lactosylceramide, may act as substrate for GM3 synthase in vivo.

Preparation

The fact that GM3 is ubiquitous in vertebrates, and detectable in most organs, tissues, and various established cell lines, indicates a wide distribution of GM3 synthase in higher animals.

The expression levels of GM3 synthase in humans, mice, and rats showed both tissue- and species-specific patterns. In humans, brain, skeletal muscle, and testis expressed high levels of the transcript (2.4kb) of the GM3 synthase gene, whereas low levels of the mRNA were found in liver and kidney (Ishii et al. 1998). In mice, in contrast, the highest level of mRNA (2.4kb) was detected in liver, while the lowest was in kidney (A. Ishii et al. unpublished data, 2000). On the other hand, kidney, as well as heart, brain, and spleen, expressed high levels of mRNA (2.4kb) in rats (C. Nakamura et al. unpublished data, 2000).

Because of the relatively high activity, GM3 synthase was purified from rat liver (Melkerson-Watson and Sweeley 1991) and rat brain (Preuss et al. 1993) to 43000-fold and 7200-fold, respectively, with acceptor glycolipid–acid column chromatography. The molecular mass of the purified liver enzymes was 60kDa, which was reduced to 43kDa after N-glycanase treatment, whereas that of brain GM3 synthase was 76kDa. The predicted molecular masses of cloned human, mouse, monkey, and rat ST3Gal-V were 41.7kDa (Ishii et al. 1998), 41.2kDa (Fukumoto et al. 1999), 41.8kDa, and 41.3kDa (unpublished results, 2000), respectively. Very recently, mouse brain, liver, and testes GM3 synthase has been found to be 45kDa by Western blot analysis (Stern et al. 2000).

Biological Aspects

GM3 synthase is one of the key enzymes acting at a metabolic branch point in the biosynthesis of glycosphingolipids. Competition for a common substrate, lactosylceramide, between β1,3-N-acetylglucosaminyltransferase, α1,3(4)-galactosyltransferase, β1,4-N-acetylgalactosaminyltransferase, and GM3 synthase results in a spatially and developmentally regulated expression of glycosphingolipids. In some cases, changes in the glycosphingolipid composition during cell differentiation, oncogenesis, and tissue development have been reported as a consequence of the regulation of enzymes active at this branch point (Nakamura et al. 1992; Holmes et al. 1987; Chou and Jungalwala 1996; Tsunoda et al. 1995).

Different results were reported on the sub-Golgi distribution of GM3 synthase. Biochemical fractionation as well as pharmacological analysis with Brefeldin A (van Echten et al. 1990; Young et al. 1990) has indicated the synthesis of GM3 in the *cis*-compartment (Trinchera and Ghidoni 1989; Trinchera et al. 1991, Iber et al. 1992). Other fractionation studies, however, have demonstrated that GM3 synthase was enriched in the *trans*-Golgi and TGN compartments, coupled with other ganglioside synthetic glycosyltransferases (Lannert et al. 1998). A similar observation was provided from confocal immunochemical localization: mouse ST3Gal-V co-localized with a medial/*trans*-Golgi marker but not with a *cis*-Golgi marker (Stern et al. 2000).

The GM3 synthase gene spans 56 kb of human genomic DNA located on chromosome 2p11.2. Coding sequences for the GM3 synthase protein of the genes are divided into six exons, from the second to the seventh exon. Little is known about the transcriptional regulation of human GM3 synthase. A luciferase reporter assay indicated that the 205 bp region around the transcriptional start site of the gene possessed strong promoter activity. This region contained the GC-rich domain (82%) and two Sp1 binding sites, without TATA box. On the other hand, the mouse ST3Gal-V gene spans 58 kb and is distributed among nine exons. Unlike the human gene, three variants of the transcript containing a different 5′-noncoding region were identified in particular tissues, suggesting that alternative promoter utilization and alternative splicing may be involved in the transcriptional control of mouse ST3Gal-V gene expression in a tissue-specific manner (unpublished results, 2000).

Future Perspectives

Gene targetting experiments in vivo with the mouse sialyltransferase gene and some in an in vitro cell culture system with the human enzyme gene are now in progress internationally, so news of great interest is expected soon.

References

Chou DKH, Jungalwala FB (1996) N-Acetylglucosaminyl transferase regulates the expression of the sulfoglucuronyl glycolipids in specific cell types in cerebellum during development. J Biol Chem 271:28868–28874

Duffard RO, Fishman PH, Bradley RM, Lauter CJ, Brady RO, Trams EG (1977) Ganglioside composition and biosynthesis in cultred cells derived from CNS. J Neurochem 28:1161–1166

Fleischer B (1977) Localization of some glycolipid glycosylating enzymes in the Golgi apparatus of rat kidney. J Supramol Struct 7:79–89

Fukumoto S, Miyazaki H, Goto G, Urano T, Furukawa K, Furukawa K (1999) Expression cloning of mouse cDNA of CMP-NeuAc:lactosylceramide α2,3-sialyltransferase, an enzyme that initiates the synthesis of gangliosides. J Biol Chem 274:9271–9276

Glickman RM, Bouhours JF (1976) Characterization, distribution and biosynthesis of the major ganglioside of rat intestinal mucosa. Biochim Biophys Acta 424:17–25

Holmes EH, Hakomori S, Ostrander GK (1987) Synthesis of type 1 and 2 lacto-series glycolipid antigens in human colonic adenocarinoma and derived cell lines is due to activation of a normally unexpressed β1-3 N-acetylglucosaminyltransferase. J Biol Chem 262:15649–15658

Iber H, van Echten G, Sandhoff K (1992) Fractionation of primary cultured cerebellar neurons: distribution of sialyltransferases involved in ganglioside biosynthesis. J Neuro chem 58:1533–1537

Ishii A, Ohta M, Watanabe Y, Matsuda K, Ishiyama K, Sakoe K, Nakamura M, Inokuchi J, Sanai Y, Saito M (1998) Expression cloning and functional characterization of human cDNA for ganglioside GM3 synthase. J Biol Chem 273:31652–31655

Kaufman B, Basu S, Roseman S (1968) Enzymatic synthesis of disialogangliosides from monosialogangliosides by sialyltransferases from embryonic chicken brain. J Biol Chem 243:5804–5807

Keenan TW, Morre DJ, Basu S (1974) Ganglioside biosynthesis. Concentration of glycosphingolipid glycosyltransferases in Golgi apparatus from rat liver. J Biol Chem 249: 310–315

Kono M, Takashima S, Liu H, Inoue M, Kojima N, Lee Y-S, Hamamoto T, Tsuji S (1998) Molecular cloning and functional expression of a fifth-type α2,3-sialyltransferase (mST3Gal-V: GM3 synthase). Biochem Biophys Res Commun 253:170–175

Lannert H, Gorgas K, Meibner I, Wieland FT, Jeckel D (1998) Functional organization of the Golgi apparatus in glycosphingolipid biosynthesis. J Biol Chem 273:2939–2946

Lee Y-C, Kojima N, Wada E, Kurosawa N, Nakaoka T, Hamamoto T, Tsuji S (1994) Cloning and expression of cDNA for a new type of Galβ1,3GalNAcα2,3-sialyltransferase. J Biol Chem 269:10028–10033

Melkerson-Watson LJ, Sweeley CC (1991) Purification to apparent homogeneity by immunoaffinity chromatography and partial characterization of the GM3 ganglioside-forming enzyme, CMP-sialic acid:lactosylceramide α2,3-sialyltransferase (SAT-I), from rat liver Golgi. J Biol Chem 267:23507–23514

Nakamura M, Tsunoda A, Sakoe K, Gu J, Nishikawa A, Taniguchi N, Saito M (1992) Total metabolic flow of glycosphingolipid biosynthesis is regulated by UDP-GlcNAc:lacosylceramide β1-3 N-acetylglucosaminyltransferase and CMP-NeuAc:lactosylceramide α2-3 sialyltransferase in human hematopoietic cell line HL-60 during differentiation. J Biol Chem 267:23507–23514

Pacuszka T, Duffard RO, Nishimura RN, Brady RO, Fishman PH (1978) Biosynthesis of bovine thyroid gangliosides. J Biol Chem 253:5839–5846

Preti A, Fiorilli A, Lombardo A, Caimi L, Tettamanti G (1980) Occurrence of sialyltransferase activity in the synaptosomal membranes prepared from calf brain cortex. J Neurochem 35:281–296

Preuss U, Gu X, Gu T, Yu RK (1993) Purification and characterization of CMP-N-acetylneuraminic acid:lactosylceramide (α2-3) sialyltransferase (GM3-synthase) from rat brain. J Biol Chem 268:26273–26278

Richardson CL, Keenan TW, Morre DJ (1977) Ganglioside biosynthesis—characterization of CMP-N-acetylneuraminic acid: lactosylceramide sialyltransferase in Golgi apparatus from rat liver. Biochim Biophys Acta 488:88–96

Stern AS, Braverman TR, Tiemeyer M (2000) Molecular identification, tissue distribution and subcellular localization of mST3Gal-V/GM3 synthase. Glycobiology 10:365–374

Trams EG (1977) J Neurochem 28:1161–1166

Trinchera M, Ghidoni R (1989) Two glycosphingolipid sialyltransferases are localized in different sub-Golgi compartments in rat liver. J Biol Chem 264:15766–15769

Trinchera M, Fabbri M, Ghidoni R (1991) Topography of glycosyltransferases involved in the initial glycosylations of gangliosides. J Biol Chem 266:20907–20912

Tsunoda A, Nakamura M, Kirito K, Hara K, Saito M (1995) Interleukin-3-associated expression of gangliosides in mouse myelogenous leukemia NFS60 cells introduced with interleukin-3 gene: expression of ganglioside GD1a and key involvement of CMP-NeuAc:lactosylceramide α-2-3-sialyltransferase in GD1a expression. Biochemistry 34:9356–9367

van Echten G, Iber H, Stotz H, Takatsuki A, Sandhoff (1990) Uncoupling of ganglioside biosynthesis by Brefeldin A. Eur J Cell Biol 51:135–139

Young WW Jr, Lutz MS, Mills SE, Lechler-Osborn S (1990) Use of brefeldin A to define sites of glycosphingolipid synthesis: GA2/GM2/GD2 synthase is trans to the brefeldin A block. Proc Natl Acad Sci USA 87:6838–6842

Yu RK, Lee SH (1976) In vitro biosynthesis of sialylgalactosylceramide (G7) by mouse brain microsomes. J Biol Chem 251:198–203

ST6Gal-I

Introduction

Galβ1-4GlcNAc α2-6-sialyltransferase (EC 2.4.99.1) catalyzes the incorporation of sialic acids at the terminal positions of glycoconjugates with NeuAc α2-6-Gal linkage. cDNA sequences from mouse, rat, human, and chicken, along with the human genomic DNA sequence and-tissue specific alternative splicing in rat have been described. Also, the domains responsible for localization to the Golgi apparatus and the acceptor substrate specificity, with synthetic acceptors, have been reported. Cloned sialyltransferases have been expressed in cultured eukaryotic cells and in *E. coli*. Like other sialyltransferases, ST6Gal-I exhibits type II membrane protein topology and has characteristic motifs for sialyltransferases called sialylmotifs L, S, and VS. ST3Gal-II also has a Kurosawa motif (Cys-Xaa$_{75-82}$-Cys-Xaa-Cys-Ala-Xaa-Val-Xaa$_{150-160}$-Cys; Xaa denotes any amino acid residue) as seen in the ST3Gal family and two members of the ST6GalNAc family (Kurosawa et al. 1996; Tsuji 1999). Currently, ST6Gal-I is the sole member of the ST6Gal subfamily that can also synthesize the ganglioside terminal NeuAc α2-6-Gal glycoconjugates.

Toshiro Hamamoto[1] and Shuichi Tsuji[2]

[1] Department of Biochemistry, Jichi Medical School, Minamikawachi, Tochigi 329-0498, Japan
Tel. +81-285-58-7322; Fax +81-285-44-1827
e-mail: thamamot@jichi.ac.jp
[2] Department of Chemistry, Faculty of Science, Ochanomizu University, Otsuka, Bunkyo-ku, Tokyo 112-8610, Japan
Tel. +81-3-5978-5345; Fax +81-3-5978-5344
e-mail: stsuji@cc.ocha.ac.jp

Databanks

ST6Gal-I

NC-IUBMB enzyme classification: E.C.2.4.99.1

Species	Gene	Protein	mRNA	Genomic
Homo sapiens	*SIAT1*	P15907	X62822	AC007488 (exons 1–6)
		A41734	–	–
Mus musculus	*Siat1*	Q64685	D16106	–
Ratus norvegicus	*Siat1*	P13721	M18769	–
		A28451	–	–
		C42327	M83143 (kidney type)	–
Bos taurus	–	O18974	–	Y15111
Gallus gallus	–	Q92182	X75558	–
		S41114	–	–

Name and History

β-Galactoside α-2-6-sialyltransferase (recommended name; systematic name is CMP-*N*-acetylneuraminate:β-D-galactosyl-1–4-*N*-acetyl-β-D-glucosamine-α-2-6-*N*-acetylneuraminyltransferase) was purified from bovine colostrum and rat liver (Weinstein et al. 1982a). It was cloned from rat liver (Weinstein et al. 1987), and then homologous cDNAs were found in human placenta (Grundnann et al. 1990), mouse liver, and chicken embryo (Kurosawa et al. 1994). The enzyme was called Galβ1-4GlcNAc α2-6sialyltransferase, or abbreviated to ST6N, because the substrates are mainly *N*-linked oligosaccharides. The products of these clones was designated ST6Gal-I according to the abbreviated nomenclature system for cloned sialyltransferases (Tsuji et al. 1996). The recombinant ST6Gal-Is, which lack the *N*-terminal cytosolic and transmembrane regions, were expressed in cultured eukaryotic cells as soluble proteins, and showed sialyltransferase activity which is almost identical to that of authentic β-galactoside α2-6sialyltransferases in their kinetic parameters and their specificity to the Galβ1-4GlcNAc sequence. ST6Gal-Is also prefer Galβ1-4GlcNAc residues on Manα1-3Man branches over Manα1-6Man branches on bi- and triantennary *N*-linked glycoconjugates (Joziasse et al. 1985), and utilize CMP-NeuGc much faster than CMP-NeuAc (Hamamoto et al. 1995). The International System for Gene Nomenclature named this gene *Siat1*.

Enzyme Activity Assay and Substrate Specificity

Enzyme Activity Assay

The enzyme activity of rat ST6Gal-I was measured using the full-length form of this enzyme (Weinstein et al. 1982a). The assay mixture consisted of 150 μM CMP-NeuAc with 56 000 cpm CMP-[^{14}C]NeuAc added as tracer, 0.5% Triton CF-54, 50 mM cacodylate, pH 6.0, 50 μg bovine serum albumin, sialyltransferase (0–0.2 miliunits), and 50 μg enzyme substrate (asialo-α1-acid glycoprotein) in 60 μl of reaction mixture. Radioactive products were separated from CMP-NeuAc by gel filtration on Sephadex G50 after incubation at 37°C (5–30 min), and quantitated.

The enzyme activity of the recombinant, soluble form of this enzyme was measured with 50 mM CMP-[^{14}C]NeuAc (0.9 Bq/pmole) as a donor substrate, 5 mM Galβ1-4GlcNAc (*N*-acetyllactosamine) as an acceptor substrate, 1 mg/ml bovine serum albumin, 1 μl of the enzyme solution, and 50 mM sodium cacodylate, pH 6.0, in a total volume of 10 μl, with incubation at 37°C for 1 h. The samples were then subjected to high-performance thin-layer chromatography (HPTLC) (silica gel 60, E. Merck, Darmstadt, Germany), and development with ethanol/pyridine/*n*-butanol/acetic acid/water (100:10:10:3:30). The radioactivity transferred was determined with a radio image analyzer, BAS2000 (Fuji Film Tokyo, Japan) (Kurosawa 1994). One unit of activity was defined as 1 μmol of sialic acid transferred per minute. When glycoproteins were used as substrates, the reaction was terminated by the addition of SDS-polyacrylamide electrophoresis loading buffer, and the radioactivity incorporated into the proteins was determined by radio image analyzer after the appropriate gel electrophoresis.

Substrate Specificity

ST6Gal-I exhibited remarkable specificity to Galβ1-4GlcNAc-R as an acceptor substrate, and the Michaelis–Menten constants of the purified enzymes were generally 30–50 μM for CMP-NeuAc, 0.3–0.5 mM for the Galβ1-4GlcNAc-R sequence on α1-acid glycoprotein, 2–10 mM for free *N*-acetyllactosamine (Galβ1-4GlcNAc), and >0.1 M for lactose and Galβ1,3GlcNAc (Weinstein et al. 1982b). The enzyme did not require divalent cations and was not inhibited by EDTA. When bi- and triantennary *N*-linked glycoconjugates were used as acceptor substrates, the enzyme preferentially sialylated Gal residues on Manα1,3Man branches (Van den Eijnden et al. 1980). It also utilizes CMP-NeuGc three times faster than CMP-NeuAc as a donor substrate. The difference in acceptor substrates has little effect on the preference of the donor substrates (i.e., NeuAc and NeuGc) (Hamamoto et al. 1995).

Preparation

For the purification of rat liver ST6Gal-I, the tissue was homogenized with 25 mM Na cacodylate (pH 6.0) and 20 mM NaCl, and then a 5000-r.p.m. 60-min pellet was extracted with 1.4% Triton CF-54. Mn^{2+} was chelated by the addition of Na EDTA while keeping the pH at 6.0 by the drop-wise addition of NaOH. The extracts were then applied to a CDP-hexanolamine-Sepharose (Sigma) column, three times. The enzyme was eluted from the column with NaCl (at the first and second steps) and with 0–1 mM CDP-gradient (at the final step) (Weinstein et al. 1982a).

For preparation of the soluble form of chick embryo ST6Gal-I, the DNA fragment encoding a truncated form of ST6Gal-I, lacking the first 53 amino acids of the open reading frame, was prepared, and this fragment was inserted into a pcDSRα vector. The resulting plasmid consisted of the IgM signal peptide sequence, the linker peptide, and a truncated form of ST6Gal-I. This expression vector was transiently transfected into COS-7 cells on a 150-mm plate. The catalytic domain of ST6Gal-I secreted in the medium was used as the soluble enzyme source (Kurosawa et al. 1994).

For the expression in *E. coli*, the recombinant enzyme, which is the c-terminal catalytic domains from mouse ST6Gal-I, was accumulated in the form of insoluble inclusion bodies in *E. coli* cells. The insoluble fraction of the cell lysate was washed with

1% Triton X-100 and then extracted by 8 M urea. The extract was diluted with renaturation buffer (0.5 M NaCl, 10 mM lactose, 0.5 mM EDTA, 20 mM MOPS-NaOH, pH 7.0) and incubated at 4°C for 48 h. The substrate specificity and kinetic parameters of the renatured ST6Gal-I were similar to those of the enzyme obtained from rat liver (Hamamoto et al. 1994).

Biological Aspects

ST6Gal-I transfers sialic acid from CMP-Sia (cytidine 5'-monophospho-sialic acid) to galactose residues, and produces the Siaα2-6Gal linkage. This linkage is found in most N-linked oligosaccharides of glycoproteins as the Siaα2-6Galβ1-4GlcNAc-R sequence. O-linked oligosaccharides of glycoproteins and glycosphingolipids contain this structure, but to lower extents.

Northern blot analysis in rat revealed three different sized mRNAs related to ST6Gal-I (4.7 kb, common to all tissues at low level; 4.3 kb, specific to liver and intestine; 3.6 kb, specific to kidney) with the highest in the liver. Analysis of the gene structure in rat revealed that these mRNAs are the products of the same gene derived through alternative promoter utilization and alternative splicing. Both the common transcript and the liver specific transcript code the whole ST6Gal-I enzyme, differing only in the 5' noncoding region. The kidney-specific transcripts code smaller proteins which share the C-terminal 171 a.a. polypeptide with the ST6Gal-I enzyme, but sialyltransferase activity was not detected (Svensson et al. 1990, 1992). Furthermore, B lymphocyte-specific transcripts which code active ST6Gal-I were reported for human and mouse (Wuensch et al. 2000). ST6Gal-I controls production of the Siaα2-6Galβ1-4GlcNAc sequence, which is the ligand for the lectin CD22 on B lymphocyte. ST6Gal-I expression of B cells was elevated upon activation using the CD40 ligand with repression of the one isoform and the appearance of three new isoforms differing only in their 5'-UT regions.

Investigation of ST6Gal-I knockout mice showed that they are viable, but exhibit severe immunosuppression, reduced serum IgM levels, and impaired B cell proliferation in response to activation (Hennet et al. 1998).

It was reported that the attached carbohydrate chain is important in controlling the activity of the sialyltransferase, and that it is part of the mechanism regulating the compartment-specific expression of the activity in the Golgi apparatus (Fast et al. 1993).

Future Perspectives

A deficiency of β-galactoside α2-6-sialyltransferase in human has not been reported. If it exists, because of a lack of sialic acids on N-linked glycans, nondenaturing polyacrylamide gel electrophoresis of the serum protein from such a patient should show an abnormal migration pattern of multiple proteins, just as in the case of N-acetylglucosaminyltransferase deficiency. Although ST6Gal-I knockout mice were viable, a deficiency of β-galactoside α2-6-sialyltransferase in human could be fatal, one reason being that desialylated glycoproteins are rapidly removed from the circulation by the liver.

ST6Gal-I is the only β-galactoside α2-6-sialyltransferase so far cloned. Other β-galactoside α2-6-sialyltransferases with different substrate specificities or preferences, such as an enzyme specific to glycolipids, one which prefers Gal residues on Manα1-6Man branches, or one which is more active to lactose, may be found in the future. ST6Gal-I can transfer sialic acid to these substrates, but the rate is very slow in vitro assay conditions.

Further Reading

Weinstein et al. (1982a,b). These two articles describe the purification of the authentic enzyme from rat liver and its characterization.
Hennet et al. (1998). A description of the ST6Gal-I knockout mouse.
Wuensch et al. (2000). The activation-regulated expression of ST6Gal-I isoforms in B lymphocytes.

References

Fast DG, Jamieson JC, McCaffrey G (1993) The role of the carbohydrate chains of Gal β-1,4-GlcNAc α 2,6-sialyltransferase for enzyme activity. Biochim Biophys Acta 1202:325–330
Grundnann V, Nerlich C, Rein T, Zettlmeissl G (1990) Complete cDNA sequence encoding human β-galactoside α2,6-sialyltransferase Nucleic Acids Res 18:667
Hamamoto T, Lee Y-C, Kurosawa N, Nakaoka T, Kojima N, Tsuji S (1994) Expression of mouse Gal β 1,4GlcNAc α 2,6-sialyltransferase in an insoluble form in *Escherichia coli* and partial renaturation. Bioorg Med Chem 2:79–84
Hamamoto T, Kurosawa N, Lee Y-C, Tsuji S (1995) Donor substrate specificities of Galβ1,4GlcNAc α2,6-sialyltransferase and Galβ1,3GalNAc α2,3-sialyltransferase: comparison of N-acetyl and N-glycolylneuraminic acids. Biochim Biophys Acta 1244:223–228
Hennet T, Chui D, Paulson JC, Marth JD (1998) Immune regulation by the ST6Gal sialyltransferase. Proc Natl Acad Sci USA 95:4504–4509
Joziasse DH, Schiphorst WECM, Van den Eijnden DH, Van Kuik JA, Van Halbeek H, Vliegenthart JFG (1985) Branch specificity of bovine colostrum CMP-sialic acid: N-acetyllactosaminide α 2-6-sialyltransferase. Interaction with biantennary oligosaccharides and glycopeptides of N-glycosylproteins. J Biol Chem 260:714–719
Kurosawa N, Hamamoto T, Lee YC, Nakaoka T, Kojima N, Tsuji S (1994) Molecular cloning and expression of GalNAc α2,6-sialyltransferase. J Biol Chem 269:1402–1409
Kurosawa N, Kojima N, Inoue M, Hamamoto T, Tsuji S (1994) Cloning and expression of Gal β 1,3GalNAc-specific GalNAc α 2,6-sialyltransferase. J Biol Chem 269:19048–19053
Kurosawa N, Inoue M, Yoshida Y, Tsuji S (1996) Molecular cloning and genomic analysis of mouse Galβ1,3GalNAc-specific GalNAc α2,6-sialyltransferase. J Biol Chem 271:15109–15116
Svensson EC, Soreghan B, Paulson JC (1990) Organization of the β-galactoside α-2,6-sialyltransferase gene. Evidence for the transcriptional regulation of terminal glycosylation. J Biol Chem 265:20863–20868
Svensson EC, Conley PB, Paulson JC (1992) Regulated expression of α 2,6-sialyltransferase by the liver-enriched transcription factors HNF-1, DBP, and LAP. J Biol Chem 267:466–472

Tsuji S (1999) Molecular cloning and characterization of sialyltransferases. In: Inoue Y, Lee YC, Troy FA (eds) Sialobiology and other novel forms of glycosylation. Gakushin, Osaka, pp 145–154

Tsuji S, Datta AK, Paulson JC (1996) Systematic nomenclature for sialyltransferases. Glycobiology 6 (7):v–vii

Van den Eijnden DH, Joziasse DH, Dorland L, Van Halbeek H, Vliegenthart JFG, Schmid K (1980) Specificity in the enzymic transfer of sialic acid to the oligosaccharide branches of bi- and triantennary glycopeptides of α1-acid glycoprotein. Biochem Biophys Res Commun 92:839–845

Weinstein J, de Souza-e-Silva U, Paulson JC (1982a) Purification of a Gal β 1 to 4GlcNAc α 2 to 6 sialyltransferase and a Gal β 1 to 3(4) GlcNAc α 2 to 3 sialyltransferase to homogeneity from rat liver. J Biol Chem 257:13835–13844

Weinstein J, de Souza-e-Silva U, Paulson JC (1982b) Sialylation of glycoprotein oligosaccharides N-linked to asparagine. Enzymatic characterization of a Gal β 1 to 3(4)GlcNAc α 2 to 3 sialyltransferase and a Gal β 1 to 4GlcNAc α 2 to 6 sialyltransferase from rat liver. J Biol Chem 257:13845–13853

Weinstein J, Lee EU, McEntee K, Lai PH, Paulson JC (1987) Primary structure of β-galactoside α 2,6-sialyltransferase. Conversion of membrane-bound enzyme to soluble forms by cleavage of the NH$_2$-terminal signal anchor. J Biol Chem 262:17735–17743

Wen DX, Svensson EC, Paulson JC (1992) Tissue-specific alternative splicing of the β-galactoside α 2,6-sialyltransferase gene. J Biol Chem 267:2512–2518

Wuensch SA, Huang RY, Ewing J, Liang X-L, Lau JTY (2000) Murine B cell differentiation is accompanied by programmed expression of multiple novel β-galactoside α2,6-sialyltransferase mRNA forms. Glycobiology 10:67–75

ST6GalNAc-I

Introduction

ST6GalNAc-I is a CMP-sialic acid: N-acetylgalactosaminide α2-6-sialyltransferase (GalNAc α2-6-sialyltransferase), and a member of the ST6GalNAc subfamily that exhibits activity toward GalNAc-O-Ser/Thr, Galβ1-3GalNAc-O-Ser/Thr, and NeuAcα2-3Galβ1-3GalNAc-O-Ser/Thr (Kurosawa et al. 1994, 2000; Ikehara et al. 1999; Kono et al. 2000). Like other sialyltransferases, ST6GalNAc-I exhibits type II membrane protein topology and has characteristic motifs for sialyltransferases called sialylmotifs L, S, and VS. ST6GalNAc-I also has the Kurosawa motif (Cys-Xaa$_{75-82}$-Cys-Xaa-Cys-Ala-Xaa-Val-Xaa$_{150-160}$-Cys; Xaa denotes any amino acid residue) as seen in the ST3Gal family and two members of the ST6GalNAc family (Kurosawa et al. 1996; Tsuji 1999). ST6GalNAc-I is a relatively large sialyltransferase (600 amino acids in length in *Homo sapiens*, 526 in *Mus musculus*, and 566 in *Gallus gallus*) compared with other sialyltransferases characterized to date. This structural character is attributed to its long stem domain. The putative active domain of mouse ST6GalNAc-I (250 amino acid residues from the C-terminal end) showed a high identity to the corresponding region of human ST6GalNAc-I (85%) and to the chick enzyme (67%), but showed low identity to other members of the mouse ST6GalNAc family: ST6GalNAc-II 48%, ST6GalNAc-III 41%, ST6GalNAc-IV 23%, and ST6GalNAc-V 16%. Human ST6GalNAc-I cDNA has two isoforms (2.46 and 2.23 kb). The former encodes an active enzyme with a predicted 600 amino acid sequence. The latter, a splice-variant of the long form, encodes an inactive enzyme (Ikehara et al. 1999).

Nobuyuki Kurosawa[1] and Shuichi Tsuji[2]

[1] Department of Materials and Biosystem Engineering, Faculty of Engineering, Toyama University, Toyama 930-8555, Japan
Tel. +81-76-445-6982; Fax +81-76-445-6874
e-mail: kurosawa@eng.toyama-u.ac.jp
[2] Department of Chemistry, Faculty of Science, Ochanomizu University, Otsuka, Bunkyo-ku, Tokyo 112-8610, Japan
Tel. +81-3-5978-5345; Fax +81-3-5978-5344
e-mail: stsuji@cc.ocha.ac.jp

ST6GalNAc-I gene showed a tissue-specific expression pattern. In mouse, ST6GalNAc-I was expressed in submaxillary glands, spleen, mammary gland, and colon, although the ST6GalNAc-II gene showed a rather ubiquitous expression pattern (Kurosawa et al. 2000).

Databanks

ST6GalNAc-I

NC-IUBMB enzyme classification: E.C.2.4.99.3

Species	Gene	Protein	mRNA	Genomic
Homo sapiens	–	CAA72179.2	Y11339	Y11340–Y11341
Mus musculus	Siat7a	CAA72137.1	Y11274	Y10294–Y10295
Gallus gallus	–	g453197	X74946	

Name and History

This enzyme cDNA was firstly cloned as a member of the GalNAc α2-6 sialyltransferase family, designated ST6GalNAc-I (Tsuji et al. 1996). The enzyme which exhibits activity towards GalNAc-Ser/Thr was purified to homogeneity from porcine submaxillary glands (Sadler et al. 1979), and was designated ST6O-I. ST6GalNAc-I is a candidate for ST6O-I. ST6GalNAc-I cDNA was firstly cloned from the chick embryo cDNA library by a polymerase chain reaction (PCR)-based approach using degenerate oligonucleotide primers corresponding to sialylmotif L. To date, chick, mouse, and human ST6GlaNac-I cDNA have been cloned (Kurosawa et al. 1994, 2000; Ikehara et al. 1999). Analysis of the genomic structure and transcription regulation of the mouse ST6GalNAc-I gene has been carried out (Kurosawa et al. 2000). The International System for Gene Nomenclature named this gene *Siat7a*.

Enzyme Activity Assay and Substrate Specificity

Enzyme Activity Assay

The enzyme activity of ST6GalNAc-I was measured using the soluble form of this enzyme fused with protein A, or the full-length form (Kurosawa et al. 1994, 2000; Kono et al. 2000; Ikehara et al. 1999). Each reaction mixture comprised 50 mM MES buffer (pH 6.0), 10 mM MgCl$_2$, 2 mM CaCl$_2$, 200 μM CMP-[^{14}C]NeuAc (7.1 kBq), acceptor substrate (5 mg/ml glycoprotein, 0.1 mM glycolipid, and 0.1 mM [Ala-Thr(GalNAc)-Ala]$_n$ polymer), and the enzyme preparation in a total volume of 20 μl. [Ala-Thr(GalNAc)-Ala]$_n$ polymer (n = 2–7) was synthesized by Tsuda and Nishimura (1996). After 4 h incubation at 37°C, the reaction was terminated by the addition of SDS-PAGE loading buffer, and then the mixtures were directly subjected to SDS-PAGE (for glycoprotein acceptors). For glycolipid acceptors, the incubation mixtures were applied on a C-18 column (Sep-Pak Vac, 100 mg; Waters, Milford, MA, USA), as described previously (Kono et al. 1996). When the [Ala-Thr(GalNAc)-Ala]$_n$ polymer was used as the acceptor substrate, the reaction mixture was directly subjected to high-performance thin-layer chromatography (HPTLC) with a solvent system of 1-propanol : 1-butanol : water (3:1:2). The radioactive materials were visualized with a BAS2000 radio image analyzer (Fuji Film, Tokyo, Japan).

Substrate Specificity

The soluble form of chick and mouse, and the intact form of human ST6GalNAc-I showed NeuAc-transferred activity toward fetuin, asialofetuin, agalactoasialofetuin, asialo-BSM, and asialo-OSM, and weak activity toward BSM and [Ala-Thr(GalNAc)-Ala]$_n$ polymer with an α2-6-linkage. However, it was not active toward α1-acid glycoprotein, asialo α1-acid glycoprotein, benzyl-GalNAc, GM1, or asialo-GM1. These results clearly show that ST6GalNAc-I exhibits a broad substrate specificity, transferring CMP-NeuAc with an α2-6-linkage to the GalNAc residues of following structures: GalNAc-O-Ser/Thr, Galβ1-3GalNAc-O-Ser/Thr, and NeuAcα2-3Galβ1-3GalNAc-O-Ser/Thr (Kurosawa et al. 1994, 1996; Ikehara et al. 1999; Kono et al. 2000).

Preparation

Truncated Form of Enzyme

The truncated form of ST6GalNAc-I, lacking the N-terminal putative cytosolic and transmembrane region, was prepared by PCR amplification (Kurosawa et al. 1994, 2000). The amplified fragment was inserted into a pcDSA vector (Kono et al. 1996). The resulting plasmid was designated pcDSA-ST6GalNAc-I, which encodes a fusion protein of the IgM signal peptide sequence, a protein A IgG binding domain, and a truncated form of mST6GalNAc-I.

COS-7 cells on a 100-mm plate (5×10^6) were transiently transfected with pcDSA-ST6GalNAc-I DNAs (10 μg) using LipofectAMINE (GIBCO) reagent. After 5 h and 24 h of transfection, the medium was changed to DMEM containing 2% FCS and macrophage SFM (GIBCO), respectively. From 48 h to 72 h after transfection, the medium was collected and centrifuged at $800 \times g$ for 10 min at 4°C to remove cell contamination. The protein A fusion enzyme, which was expressed in the culture medium of the COS-7 cells, was absorbed to IgG-sepharose gel at 4°C for 16 h, and the IgG-sepharose gel-bound protein A fusion enzyme was used as the enzyme source.

Full-Length Form of Enzyme

The full-length ORF of human cDNA was subcloned into an expression vector, pCXN$_2$ (Niwa et al. 1991). HCT15 cells, a human colorectal cancer cell line, were transfected with 10 μg of the expression plasmid DNAs by the electroporation method (Ikehara et al. 1999). The cells were selected in the presence of geneticin (G418; 0.6 mg/ml) (GIBCO-BRL) in RPMI 1640 (GIBCO-BRL) medium supplemented with 10% heat-inactivated fetal bovine serum. After 10 days exposure to geneticin, an aliquot of the cells was examined by flow cytometric analysis to determine the sialyl-Tn antigen expression, and the rest of the cells were subjected to limiting dilution to obtain single transformant clones. The cell lysate of a transformant clone was used as an enzyme source.

Biological Aspects

A Candidate for Sialyl-Tn-Antigen Synthase

The sialyl-Tn antigen is found on ovine submaxillary mucin and in colorectal cancer as a cancer-associated antigen. The structure of the sialyl-Tn epitope was biochemically determined by the transfer of sialic acid in 2-6-linkage to an O-linked N-acetylglucosamine residue on mucin. Among the substrate specificities of six ST6GalNAc transferases characterized to date, ST6GalNAc-I and -II are the candidates for the synthesis of sialyl-Tn antigens (ST6GalNAc-III–V exhibit restricted substrate specificity, only utilizing the Neuα2-3Galβ1,3GalNAc-sequences as an acceptor). When human ST6GalNAc-I cDNA was stably expressed in HCT15 human colorectal cancer cells, the cell expressed the sialyl-Tn epitope (Ikehara et al. 1999). The human ST6GalNAc-I and -II genes are located close to each other on chromosomes 17 (17q25), suggesting that this gene pair is closely related from an evolutionary standpoint.

Tissue-Specific Expression of ST6GalNAc-I Gene

The ST6GalNAc-I gene showed a tissue-specific expression pattern. In mouse, ST6GalNAc-I was expressed in submaxillary glands, spleen, mammary gland, and colon (Kurosawa et al. 1996). The mouse ST6GalNAc-I gene is a complex genomic locus, the open reading frame of the gene consisting of nine exons distributed over 10 kb, with a single copy in the haploid mouse genome. Exon 1 contains the entire 5′-untranslated region, a cytoplasmic domain, a hydrophobic domain, and a stem domain. Exon 2 contains a stem domain, and exons 3–9 contain a putative active domain. The structure of the mouse ST6GalNAc-I gene is similar to that of the mouse ST6GalNAc-II gene, including intron–exon boundaries between codons. These results suggest that the ST6GlaNAc-I and -II genes were derived from a common ancestral gene. Analysis of the sequence immediately upstream of the transcription initiation site reveal that the mouse ST6GalNAc-I gene has two promoters, and neither of them contain canonical TATA, CAAT, and GC-box. However, they have some putative binding sites for tumor-associated transcription factors such as c-Myb, c-Myc, and cEts, suggesting the possibility of tumor-associated expression of sialyl-Tn antigen by the ST6GalNAc-I gene promoter. These features are quite different from those of a house-keeping gene such as the promoter of the mouse ST6GlanAc-II gene (Kurosawa et al. 2000).

Future Perspectives

ST6GalNAc-I and -II have similar substrate specificities and both enzymes have sialyl-Tn antigen synthase activity. The genes have similar structures but different expression patterns. At present, we cannot distinguish the functions of these enzymes in vivo exactly. The existence of these enzymes is probably important for fine control of the expression of O-glycans, resulting in stage- and tissue-specific varieties. Identification of the transcription factors that are involved in stage- and tissue-specific expression of these genes may facilitate an understanding of the different expression patterns of these genes, and also the mechanisms for stage- and tissue-specific expression. Generating knockout mice of these genes will also help to clarify the exact biological functions of these enzymes.

Further Reading

Y. Ikehara et al. 1999: cDNA cloning and characterization of human ST6GalNAc-I.

References

Ikehara Y, Naoya Kojima N, Kurosawa N, Kudo T, Nishihara S, Morozumi K, Kono M, Tsuji S, Narimatsu H (1999) Cloning and expression of a human gene encoding an N-acetylgalactosamine-α2,6-sialyltransferase (ST6GalNAc-I) which is a candidate for synthesis of cancer-associated sialyl-Tn. Glycobiology 9:1213–1224

Kono M, Yoshida Y, Kojima N, Tsuji S (1996) Molecular cloning and expression of a fifth-type of α2,8-sialyltransferase (ST8Sia V): its substrate specificity is similar to that of SAT-V/III, which synthesize GD1c, GT1a, GQ1b and GT3. J Biol Chem 271: 29366–29371

Kono M, Tsuda T, Ogata S, Takashima S, Liu H, Hamamoto T, Itzkowitz SH, Nishimura S, Tsuji S (2000) Re-defined substrate specificity of ST6GalNAc-II: a second candidate sialyl-Tn synthase. Biochem Biophys Res Commun 272:94–97

Kurosawa N, Hamamoto T, Lee Y-C, Nakaoka T, Kojima N, Tsuji S (1994) Molecular cloning and expression of GalNAc α2,6-sialyltransferase. J Biol Chem 269:1402–1409

Kurosawa N, Inoue M, Yoshida Y, Tsuji S (1996) Molecular cloning and genomic analysis of mouse Galβ1,3GalNAc-specific GalNAc α2,6-sialyltransferase. J Biol Chem 271: 15109–15116

Kurosawa N, Takashima S, Kono M, Ikehara Y, Inoue M, Tachida Y, Narimatsu H, Tsuji S (2000) Molecular cloning and genomic analysis of mouse GalNAc α2,6-sialyltransferase (ST6GalNAc-I). J Biochem 127:845–854

Niwa H, Yamamura K, Miyazaki J (1991) Efficient selection for high-expression transfectants with a novel eukaryotic vector. Gene 108:193–199

Sadler JE, Rearick JI, Hill RL (1979) Purification to homogeneity and enzymatic characterization of an α-N-acetylgalactosaminide α2-6 sialyltransferase from porcine submaxillary glands. J Biol Chem 254:5934–5941

Tsuda T, Nishimura S (1996) Synthesis of an antifreeze glycoprotein analogue: efficient preparation of sequential glycopeptide polymers. Chem Commun 2779–2780

Tsuji S (1999) Molecular cloning and characterization of sialyltransferases. In: Inoue Y, Lee YC, Troy FA (eds) Sialobiology and other novel forms of glycosylation. Gakushin, Osaka, pp 145–154

Tsuji S, Datta AK, Paulson JC (1996) Systematic nomenclature for sialyltransferases. Glycobiology 6(7):v–vii

42

ST6GalNAc-II

Introduction

ST6GalNAc-II is a CMP-sialic acid: N-acetylgalactosaminide α2-6-sialyltransferase (GalNAc α2-6-sialyltransferase), and the second member of the ST6GalNAc subfamily that exhibits activity toward GalNAc-O-Ser/Thr, Galβ1,3GalNAc-O-Ser/Thr, and NeuAcα2,3Galβ1,3GalNAc-O-Ser/Thr (Kurosawa et al. 1994, 1996; Kono et al. 2000). Like other sialyltransferases, ST6GalNAc-II exhibits type II membrane protein topology and has characteristic motifs for sialyltransferases called sialylmotifs L, S, and VS. ST6GalNAc-II also has a Kurosawa motif (Cys-Xaa$_{75-82}$-Cys-Xaa-Cys-Ala-Xaa-Val-Xaa$_{150-160}$-Cys; Xaa denotes any amino acid residue) as seen in the ST3Gal family and two members of the ST6GalNAc family (Kurosawa et al. 1996; Tsuji 1999). Mouse ST6GalNAc-II cDNA encodes a protein of 373 amino acids, showing low identity to chick ST6GalNAc-II (57%) and -I (38%), and mouse ST6GalNAc-I (41%), -III (20%), -IV (22%), and -V (20%).

The mouse ST6GalNAc-II gene is constitutively expressed in various tissues, but is highly expressed in lactating mammary gland and adult testis. The mouse gene contains nine exons spanning about 25 kb of genomic DNA, and encodes a messenger RNA of 1995 nucleotides (Kurosawa et al. 1996).

Nobuyuki Kurosawa[1] and Shuichi Tsuji[2]

[1] Department of Materials and Biosystem Engineering, Faculty of Engineering, Toyama University, Toyama 930-8555, Japan
Tel. +81-76-445-6982; Fax +81-76-445-6874
e-mail: kurosawa@eng.toyama-u.ac.jp
[2] Department of Chemistry, Faculty of Science, Ochanomizu University, Otsuka, Bunkyo-ku, Tokyo 112-8610, Japan
Tel. +81-3-5978-5345; Fax +81-3-5978-5344
e-mail: stsuji@cc.ocha.ac.jp

Databanks

ST6GalNAc-II

NC-IUBMB enzyme classification: E.C.2.4.99.3

Species	Gene	Protein	mRNA	Genomic
Homo sapiens	–	AAA5228.1	U14550	–
Mus musculus	*Siat7b*	CAA63821.1	X93999	X94000
		P70277	–	
Gallus gallus	–	–	X77775	–

Name and History

The enzyme that exhibits the activity towards GalNAc-Ser/Thr was purified to homo-geneity from porcine submaxillary glands (Sadler et al. 1979), and was designated ST6O-I. ST6GalNAc-II is the second candidate for ST6O-I. ST6GalNAc-II cDNA was first isolated from chick cDNA libraries by sequence information of chick ST6GalNAc-I (Kurosawa et al. 1994). To date, chick, mouse, and human ST6GlaNAc-II cDNA have been cloned. Analyses of the genomic structure and transcription regulation of the mouse ST6GalNAc-II gene has been performed (Kurosawa et al. 1996). The International System for Gene Nomenclature named this gene *Siat7b*.

Enzyme Activity Assay and Substrate Specificity

Enzyme Activity Assay

The enzyme activity of ST6GalNAc-II was measured using the soluble form of the enzyme fused with protein A (Kurosawa et al. 1994, 1996; Kono et al. 2000). Each re-action mixture comprised 50 mM MES buffer (pH 6.0), 10 mM $MgCl_2$, 2 mM $CaCl_2$, 200 μM CMP-[^{14}C]NeuAc (7.1 kBq), acceptor substrate (5 mg/ml glycoprotein, 0.1 mM glycolipid, 0.1 mM [Ala-Thr(GalNAc)-Ala]$_n$ polymer), and enzyme preparation in a total volume of 20 μl. [Ala-Thr(GalNAc)-Ala]$_n$ polymer ($n = 2$–7) was synthesized by Tsuda and Nishimura (1996). After 4 h incubation at 37°C, the reaction was terminated by the addition of SDS-PAGE loading buffer, and then the mixtures were directly sub-jected to SDS-PAGE (for glycoprotein acceptors). For glycolipid acceptors, the incu-bation mixtures were applied on a C-18 column (Sep-Pak Vac, 100 mg; Waters, Milford, MA, USA), as described previously (Kono et al. 1996). When [Ala-Thr(GalNAc)-Ala]$_n$ polymer was used as the acceptor substrate, the reaction mixture was directly subjected to high-performance thin-layer chromatography (HPTLC) with a solvent system of 1-propanol:1-butanol:water (3:1:2). The radioactive materials were visu-alized with a BAS2000 radio image analyzer (Fuji Film, Tokyo, Japan).

Substrate Specificity

ST6GalNAc-II efficiently transferred sialic acids to asialo fetuin and asialo BSM. No significant activity was observed toward α1-acid glycoprotein, which has *N*-linked oligosaccharides. In addition, oligosaccharides and glycosphingolipids did not serve as acceptors for ST6GalNAc-II. ST6GalNAc-I and -II have almost the same activity toward asialo-OSM and [Ala-Thr(GalNAc)-Ala]$_n$ polymer. The carbohydrate structures of OSM are reported to be (NeuAcα2-6)GalNAc-O-Ser/Thr (Gottschalk and Bhargava 1972), and thus almost all the carbohydrate structures of asialo-OSM are of GalNAc-O-Ser/Thr structure. These results clearly show that both ST6GalNAc-I and -II exhibit activity toward GalNAc-O-Ser/Thr structure, suggesting that ST6GalNAc-II is also a candidate for the sialyl-Tn synthase.

Preparation

Truncated forms of ST6GalNAc-II, lacking the N-terminal putative cytosolic and transmembrane region, were prepared by polymerase chain reaction (PCR) amplification (Kurosawa et al. 1996). The amplified fragments were inserted into a pcDSA vector (Kono et al. 1996). The resulting plasmid encodes a fusion protein of the IgM signal peptide sequence, a protein A IgG binding domain, and a truncated form of ST6GalNAc-II. COS-7 cells on a 100-mm plate (5×10^6) were transiently transfected with pcDSA-ST6GalNAc-II DNAs (10μg) using LipofectAMINE (GIBCO) reagent. After 5h and 24h of transfection, the medium was changed to DMEM containing 2% FCS and macrophage SFM (GIBCO), respectively. From 48h to 72h after transfection, the medium was collected and centrifuged at $800 \times g$ for 10min at 4°C to remove cell contamination. The protein A fusion enzyme, which was expressed in the culture medium of COS-7 cells, was absorbed to IgG-sepharose gel at 4°C for 16h, and the IgG-sepharose gel bound protein A fusion enzyme was used as the enzyme source.

Biological Aspects

A Second Candidate for Sialyl-Tn-Antigen Synthase

Although ST6GalNAc-I and -II exhibit similar substrate specificities, the expression patterns of these genes were quite different. The expression levels were highest in lactating mammary gland and adult testis, medium levels were found in kidney, and only low signals were detected for other tissues. The gene expression in testis increased with testis maturation (Kurosawa et al. 1996). On the other hand, the expression of the mouse ST6GalNAc-I gene was limited in submaxillary gland, mammary gland, colon, and spleen. In addition, the expression level of the ST6GalNAc-II gene was much higher than that of the ST6GalNAc-I gene (Kurosawa et al. 2000). Siaα2-6GalNAc-O-Ser/Thr structures occur as a cancer-associated carbohydrate antigen, sialyl-Tn. It is known that elevation of sialyl-Tn antigen correlates with a poor prognosis for gastric cancer (Kushima et al. 1993) and colon cancer (Itzkowitz et al. 1990) patients.

Genomic and Promoter Analysis of the ST6GalNAc-II Gene

The gene contains nine exons spanning about 25kb of genomic DNA and encodes a messenger RNA of 1995 nucleotides. Exon 1 contains the entire 5-untranslated region, a cytoplasmic domain, a hydrophobic domain, and a stem domain. Exon 2 contains a stem domain, and exons 3–9 contain a putative active domain. The structure of the mouse ST6GalNAc-II gene is similar to that of the mouse ST6GalNAc-I gene. All exon–intron boundaries within codons of the mouse ST6GalNAc-II gene were identical to those of mouse ST6GalNAc-I. The splice junction in exon 3 occurred after the first nucleotide of the amino acid codon, those in exons 1, 4, 6, and 7 occurred after the second nucleotide of the amino acid codon, and those in the rest of the intron–exon boundaries between codons. In addition, the human ST6GalNAc-I and -II genes are located close to each other on chromosomes 17 (17q25). These results suggest that ST6GlaNAc-I and -II genes were derived from a common ancestral gene.

Analysis of the sequence immediately upstream of the transcription initiation site revealed that the mouse ST6GalNAc-II gene promoter consists of a G+C-rich sequence lacking canonical TATA and CAAT boxes (Kurosawa et al. 2000). This promoter is embedded in a G+C-rich domain (1.1 kb) that extends from nucleotide 900 through the first exon. This domain has the characteristics of a CpG island: it is enriched in G+C (61.3% over 1109 nucleotides) and contains a 60 CpG dinucleotides with a CpG/GpC ratio of 0.65. The TATA- and CAAT-less ST6GalNAc-II gene promoter contains two Sp1 binding sequences at positions 41 and +36. An inverted Sp1 binding site is found at position 54. These structural features suggest that the regulatory region upstream of the ST6GalNAc-II gene functions as a housekeeping promoter. Interestingly, the tissue expression pattern of mouse ST6GalNAc-II is similar to that observed for mouse β1,4-galactosyltransferase gene, and low-level transcription driven by a housekeeping promoter and high level expression driven by germ cell- and mammary cell-specific promoters have been detected. The ST6GalNAc-II gene expression in testis or mammary gland may be achieved through differential utilization of a number of physically distinct promoter regions, as found for the β1,4-galactosyltransferase gene. The low but constitutive expression of the ST6GalNAc-II gene may be driven by the housekeeping promoter. High and developmental gene expression in testis may be involved in the biosynthesis of developmentally regulated, testis-specific O-linked sialylated oligosaccharide structures that have not yet been characterized.

Future Perspectives

ST6GalNAc-I and -II have similar substrate specificities and both enzymes have sialyl-Tn-antigen synthase activity. The genes have similar structures but different expression patterns. At present, we cannot distinguish the functions of these enzymes in vivo exactly. In order to clarify the relationship between sialyl-Tn antigen expression and the biology of cancers, it is important to examine the expression of sialyl-Tn synthases in carcinogenesis. Since the mRNA expression of ST6GalNAc-I was not well correlated with the expression of sialyl-Tn antigen during the maturation of goblet cells in intestinal metaplastic glands (Ikehara et al. 1999), a more precise investigation of ST6GalNAc-II, including the expression patterns and biological functions of sialyl-Tn synthases, is necessary. The existence of these enzymes is probably important for fine control of the expression of O-glycans, resulting in stage- and tissue-specific variety. Identification of the transcription factors that are involved in stage- and tissue-specific expression of these genes may facilitate our understanding of the different expression patterns of these genes, and also the mechanisms for stage- and tissue-specific expression. Developing knockout mice of these genes will also help to establish the exact biological functions of the enzymes.

Further Reading

M. Kono et al., 2000, a short report about a re-evaluation of the substrate specificities of mouse ST6GalNAc-I and -II.

References

Gottschalk A, Bhargava AS (1972) In: Gottschalk A (ed) Glycoproteins: their composition, structure and function. Elsevier, New York

Ikehara Y, Naoya Kojima N, Kurosawa N, Kudo T, Nishihara S, Morozumi K, Kono M, Tsuji S, Narimatsu H (1999) Cloning and expression of a human gene encoding an *N*-acetylgalactosamine-α2,6-sialyltransferase (ST6GalNAc-I) which is a candidate for synthesis of cancer-associated sialyl-Tn. Glycobiology 9:1213–1224

Itzkowitz SH, Bloom EJ, Kokal WA, Modin G, Hakomori S, Kim YS (1990) Sialosyl-Tn: a novel mucin antigen associated with prognosis in colorectal cancer patients. Cancer 66:1960–1966

Kono M, Yoshida Y, Kojima N, Tsuji S (1996) Molecular cloning and expression of a fifth type of α2,8-sialyltransferase (ST8Sia V): its substrate specificity is similar to that of SAT-V/III, which synthesize GD1c, GT1a, GQ1b and GT3. J Biol Chem 271: 29366–29371

Kono M, Tsuda T, Ogata S, Takashima S, Liu H, Hamamoto T, Itzkowitz SH, Nishimura S, Tsuji S (2000) Re-defined substrate specificity of ST6GalNAc-II: a second candidate sialyl-Tn synthase. Biochem. Biophys Res Commun 272:94–97

Kurosawa N, Kojima N, Inoue M, Hamamoto T, Tsuji S (1994) Cloning and expression of Galβ1,3GalNAc-specific GalNAc α2,6-sialyltransferase. J Biol Chem 269:19048–19053

Kurosawa N, Inoue M, Yoshida Y, Tsuji S (1996) Molecular cloning and genomic analysis of mouse Galβ1,3GalNAc-specific GalNAc α2,6-sialyltransferase. J Biol Chem 271: 15109–15116

Kurosawa N, Takashima S, Kono M, Ikehara Y, Inoue M, Tachida Y, Narimatsu H, Tsuji S (2000) Molecular cloning and genomic analysis of mouse GalNAc-α2,6-sialyltransferase (ST6GalNAc-I) J Biochem 127:845–854

Kushima R, Jancic S, Hattori T (1993) Association between expression of sialosyl-Tn antigen and intestinalization of gastric carcinomas. Int J Cancer 55:904–908

Sadler JE, Rearick JI, Hill RL (1979) Purification to homogeneity and enzymatic characterization of an α-*N*-acetylgalactosaminide α2-6 sialyltransferase from porcine submaxillary glands. J Biol Chem 254:5934–5941

Tsuda T, Nishimura S (1996) Synthesis of an antifreeze glycoprotein analogue: efficient preparation of sequential glycopeptide polymers. Chem Commun 2779–2780

Tsuji S (1999) Molecular cloning and characterization of sialyltransferases. In Inoue Y, Lee YC, Troy FA (eds) Sialobiology and other novel forms of glycosylation. Gakushin, Osaka, pp 145–154

ST6GalNAc-III (STY)

Introduction

ST6GalNAc-III is a relatively small sialyltransferase (305 amino acids in length) compared with other sialyltransferases characterized to date. Like other sialyltransferases, ST6GalNAc-III exhibits type II membrane protein topology and has characteristic motifs for sialyltransferases called sialylmotifs L, S, and VS. ST6GalNAc-III–VI are the members of one ST6GalNAc subfamily that can synthesize the ganglioside GD1α from GM1b (Sjoberg et al. 1996; Lee et al. 1999; Okajima et al. 1999, 2000; Ikehara et al. 1999). GD1α has been implicated as a molecular component of a variety of important biological processes. The overall amino acid sequence identity of mouse ST6GalNAc-III is 94.4% to rat ST6GalNAc-III, 43.0% to mouse ST6GalNAc-IV, 42.6% to mouse ST6GalNAc-V, and 41.4% to mouse ST6GalNAc-VI, but ST6GalNAc-III shows no significant similarity to other sialyltransferases except in sialylmotifs. This ST6GalNAc subfamily (ST6GalNAc-III–VI) has been suggested to have a different domain structure to other sialyltransferases (Lee et al. 1999; Okajima et al. 1999, 2000; Ikehara et al. 1999). ST6GalNAc-III, -V, and -VI prefer glycolipids to O-glycans as substrates, while ST6GalNAc-IV prefers O-glycans to glycolipids. Each gene has different tissue-specific expression patterns, suggesting that there may be several tissue-specific ST6GalNAc members capable of synthesizing GD1α.

SHOU TAKASHIMA[1] and SHUICHI TSUJI[2]

[1] Cellular Biochemistry Laboratory, Institute of Physical and Chemical Research (RIKEN), Wako, Saitama 351-0198, Japan
Tel. +81-48-462-1111 (ext. 5465); Fax +81-48-462-4670
e-mail: staka@postman.riken.go.jp
[2] Department of Chemistry, Faculty of Science, Ochanomizu University, Otsuka, Bunkyo-ku, Tokyo 112-8610, Japan
Tel. +81-3-5978-5345; Fax +81-3-5978-5344
e-mail: stsuji@cc.ocha.ac.jp

Databanks

ST6GalNAc-III (STY)

NC-IUBMB enzyme classification: E.C.2.4.99.7

Species	Gene	Protein	mRNA	Genomic
Mus musculus	*Siat7c*	CAA72181	Y11342	Y11343–Y11346 (exons 1–5)
Rattus norvegicus	*Siat7c*	–	L29554	–

Name and History

The polymerase chain reaction (PCR)-based cloning approach has been widely applied to obtain novel sialyltransferase cDNA clones. The ST6GalNAc-III cDNA clone is one such clone, and to date has been cloned from rat and mouse. Using degenerate primers deduced on the conserved sequence in sialylmotif L, PCR was performed with rat brain cDNA as a template, which resulted in the isolation of the 150-bp fragment encoding a new sialylmotif, SMY (Sjoberg et al. 1996). Using this fragment as a probe, positive clones named STY were obtained from a rat brain cDNA library. STY was found to encode a third type of α-2-6-GalNAc sialyltransferase, and was thus designated ST6GalNAc-III according to the abbreviated nomenclature system for cloned sialyltransferases (Tsuji et al. 1996). Mouse ST6GalNAc-III was also cloned by a PCR-based approach (Lee et al. 1999). An analysis of the genomic structure and transcription regulation of the mouse ST6GalNAc-III gene has been performed (Takashima et al. 2000). The International System for Gene Nomenclature named this gene *Siat7c*.

Enzyme Activity Assay and Substrate Specificity

Enzyme Activity Assay

The enzyme activity of rat ST6GalNAc-III was measured using the full-length form of this enzyme (Sjoberg et al. 1996). The assay mixture consisted of 50 μM CMP-NeuAc with 250 000 cpm of CMP-[^{14}C]NeuAc added as tracer, 0.1% Triton CF-54, 20 mM cacodylate, pH 6.0, and 10 μl enzyme extract (containing Mn^{2+}) in 30 μl of reaction mixture. Glycoprotein and glycolipid products were separated from CMP-NeuAc by gel filtration, and the sialic acid content of each glycoconjugate was fixed at 0.5 mM. In the case of oligosaccharides, the disialylated product formed by ST6GalNAc-III was separeted from the monosialylated acceptor and CMP-NeuAc by MonoQ HPLC. The oligosaccharide acceptor concentration was fixed at 0.1 mM, and 20 μM CMP-NeuAc plus 200 000 cpm of CMP-[^{14}C]NeuAc was incubated as a tracer in 30 μl of assay mixture. Then the relative activities were determined. It should be noted that rat ST6GalNAc-III requires the presence of either Mg^{2+} or Mn^{2+} for optimal activity.

The enzyme activity of mouse ST6GalNAc-III was measured using the soluble form of this enzyme fused with protein A or the full-length form of the enzyme (Lee et al. 1999). Enzyme activity was measured in 50 mM MES buffer, pH 6.0, 1 mM $MgCl_2$, 1 mM $CaCl_2$, 0.5% Triton CF-54, 100 μM CMP-[^{14}C]NeuAc (10.2 kBq), an acceptor substrate, and an enzyme preparation, in a total volume of 10 µl. As acceptor substrates, 10 µg proteins, 5 µg glycolipids, or 10 µg oligosaccharides were used. The enzyme reaction was performed at 37°C for 2 h. The reaction was terminated by the addition of SDS-polyacrylamide gel electrophoresis loading buffer (10 µl), and then the incubation mixtures were directly subjected to SDS-polyacrylamide gel electrophoresis (for glycoprotein acceptors). For glycolipid acceptors, the incubation mixtures were applied on a C-18 column (Sep-Pak Vac, 100 mg; Waters, Milford, MA, USA), which had been washed with water. The glycolipids were eluted from the column with methanol, dried, and then subjected to chromatography on a high-performance thin-layer chromatography (HPTLC) plate (E. Merck, Darmstadt, Germany) with a solvent system of chloroform, methanol, and 0.02% $CaCl_2$ (55:45:10). Acceptor substrates were visualized by staining with Coomassie brilliant blue (for glycoproteins) or by the orcinol-H_2SO_4 method (for glycolipids). The radioactive materials in the glycoproteins and glycolipids were visualized with a BAS2000 radio image analyzer (Fuji Film, Tokyo, Japan), and the radioactivity incorporated into the acceptors was counted.

Substrate Specificity

Rat ST6GalNAc-III transferred sialic acid only to sialylated glycoconjugates displaying the NeuAcα2,3Galβ1,3GalNAc sequence including α2-3-sialylated antifreeze glycoprotein, the ganglioside GM1b, fetuin, and sialylated oligosaccharide NeuAcα2-3Galβ1-3GalNAc (Sjoberg et al. 1996). The substrate specificity of mouse ST6GalNAc-III was similar to that of rat ST6GalNAc-III, and mouse ST6GalNAc-III exhibited activity toward fetuin, GM1b, and sialylated oligosaccharide NeuAcα2-3Galβ1-3GalNAc (Lee et al. 1999). These results suggest that ST6GalNAc-III cannot discriminate between α- and β-linked GalNAc. NMR spectroscopy experiments showed that rat ST6GalNAc-III forms NeuAcα2,3Galβ1,3(NeuAc2,6)GalNAc structure (Sjoberg et al. 1996). In contrast, asialo derivatives of these substrates were not acceptors for ST6GalNAc-III. While GM1b was a good acceptor for ST6GalNAc-III, other gangliosides containing the NeuAcα2-3Galβ1-3GalNAc sequence, such as GD1a and GT1b, were not acceptors. Since GD1a and GT1b both contain a terminal NeuAcα2-3Galβ1-3GalNAc sequence, it is clear that the α2,3-linked sialic acid attached to the internal galactose of GD1a and GT1b abolishes the ability of ST6GalNAc-III to transfer sialic acid to this sequence. On the other hand, GD1a and GT1b are acceptors for ST6GalNAc-VI (Okajima et al. 2000).

Preparation

The rat ST6GalNAc-III expression vector was constructed by inserting an *Eco*RI fragment encoding the rat ST6GalNAc-III coding region in its entirety into the *Eco*RI site of the pcDNA3 vector (Invitrogen) (Sjoberg et al. 1996). This plasmid was stably transfected into the human kidney carcinoma cell line 293. Cells in which the plasmid was stably integrated were selected and grown to confluence in 225-cm² tissue culture flasks. After being harvested, cells were solubilized in 1 ml 1% triton X-100, 50 mM NaCl, 5 mM $MnCl_2$, and 25 mM MES, pH 6.0. The crude homogenate was centrifuged for 10 min at $1000 \times g$. The resulting supernatant was removed and used directly as the enzyme source.

To prepare the soluble form of mouse ST6GalNAc-III, the DNA fragment encoding a truncated form of mouse ST6GalNAc-III, lacking the first 28 amino acids of the open reading frame, was prepared by PCR amplification, and this fragment was inserted into a pcDSA vector (Lee et al. 1999). The resulting plasmid consisted of the IgM signal peptide sequence, the protein A IgG binding domain, and a truncated form of ST6GalNAc-III. This expression vector was transiently transfected into COS-7 cells on a 150-mm plate. The protein A-fused ST6GalNAc-III secreted in the medium was adsorbed to an IgG-Sepharose gel (Amersham Pharmacia Biotech, Piscataway, NJ, USA) at 4°C for 16h. The resin was collected by centrifugation, washed three times with phosphate-buffered saline, suspended in 50 μl (final volume) Dulbecco's modified Eagle medium without fetal bovine serum, and used as the soluble enzyme.

To prepare the full-length form of mouse ST6GalNAc-III, the expression vector containing the whole coding region of mouse ST6GalNAc-III was inserted into the pcDL-SRα vector. This vector was transiently transfected into COS-7 cells. After 5h transfection, the culture medium was changed to Dulbecco's modified Eagle's medium containing 10% fetal calf serum. After 48h, the COS-7 cells were collected and the membrane-bound proteins were extracted by sonication in 20 mM MES buffer, pH 6.4, containing 0.3% Triton CF-54. After centrifugation of the cell lysate at $10\,000 \times g$ for 15 min, the resultant supernatant was used as the enzyme source.

Biological Aspects

Northern blot analysis of the rat ST6GalNAc-III gene revealed that the expression of this gene is highest in spleen, followed by kidney and lung in adult tissues (Sjoberg et al. 1996). The transcript size is approximately 3.0 kb. Although the rat ST6GalNAc-III gene is expressed below the detectable limits of Northern blot analysis in adult brain, it is abundantly expressed in both newborn brain and kidney, indicating that the expression of the rat ST6GalNAc-III gene is developmentally regulated. On the other hand, the expression level of the mouse ST6GalNAc-III gene was very low, and it could not be detected on Northern blot analysis. Therefore, the expression of the gene was examined by the competitive PCR method (Lee et al. 1999). Mouse ST6GalNAc-III gene expression is detected in adult brain, lung, and heart, followed by lower levels of expression in kidney, mammary gland, spleen, thymus, and testis. Unlike rat ST6GalNAc-III, no dramatic change of the expression of the ST6GalNAc-III gene is observed during mouse brain development. The discrepancies between rat and mouse ST6GalNAc-III expression suggest that there are species-specific regulation mechanisms for gene expression. The exact biological functions of ST6GalNAc-III in vivo have not yet been demonstrated, but one potential function of ST6GalNAc-III is to synthesize GD1α in tissues where the ST6GalNAc-III transcript is observed. There also remains a possibility that ST6GalNAc-III contributes to the synthesis of O-glycans, although ST6GalNAc-III prefers glycolipids to O-glycans as substrates (Lee et al. 1999).

An analysis of the genomic structure and transcription regulation of the mouse ST6GalNAc-III gene has been performed (Takashima et al. 2000). The ST6GalNAc-III gene spans over 120 kb of genomic DNA with five exons. The exon–intron boundaries of the ST6GalNAc-III gene and those of the ST6GalNAc-IV gene are very similar, suggesting that these genes arose from a common ancestral gene through gene duplication. Measurement of the promoter activity demonstrated that the first 189 bp upstream sequence from the translational initiation codon contains the minimum promoter for expression by a mouse embryonal carcinoma cell line, P19. This region does not contain a TATA or CAAT box, but has three putative Sp1 binding sites. Mobility shift assaying and mutational analysis of the promoter region indicated that two of them are synergistically involved in the transcription regulation of the ST6GalNAc-III gene in P19 cells.

Future Perspectives

ST6GalNAc-III and -IV have similar substrate specificities and both enzymes have GD1α synthase activity. In addition, ST6GalNAc-V and -VI also have GD1α synthase activity, but each gene has characteristic and different expression patterns. At present, we cannot distinguish the functions of these enzymes in vivo exactly. The existence of these enzymes is probably important for fine control of the expression of sialylglycoconjugates, resulting in stage- and tissue-specific variety. Identification of transcription factors that are involved in stage- and tissue-specific expression of these genes may facilitate our understanding of the different expression patterns of these genes, and also of the mechanisms for stage- and tissue-specific expression. Generating knockout mice of these genes will also help to clarify the exact biological functions of these enzymes.

Further Reading

Lee et al. (1999) report on the cDNA cloning of mouse ST6GalNAc-III and -IV, the substrate specificities of these enzymes, and the expression patterns of the ST6GalNAc-III and -IV genes.

Sjoberg et al. (1996) reported the cDNA cloning of rat ST6GalNAc-III, the substrate specificity of the enzyme, and NMR analysis of the sialoside formed by ST6GalNAc-III.

Takashima et al. (2000) compare the genomic structures of mouse ST6GalNAc-III and -IV genes, and report an analysis of promoter activities, and the identification of Sp1 binding sites that are involved in the transcription regulation of these genes.

References

Ikehara Y, Shimizu N, Kono M, Nishihara S, Nakanishi H, Kitamura T, Narimatsu H, Tsuji S, Tatematsu M (1999) A novel glycosyltransferase with a polyglutamine repeat: a new candidate for GD1α synthase (ST6GalNAc-V) FEBS Lett 463:92–96

Lee YC, Kaufmann M, Kitazume-Kawaguchi S, Kono M, Takashima S, Kurosawa N, Liu H, Pircher H, Tsuji S (1999) Molecular cloning and functional expression of two members of mouse NeuAcα2,3Galβ1,3GalNAc GalNAcα2,6-sialyltransferase family, ST6GalNAc-III and -IV. J Biol Chem 274:11958–11967

Okajima T, Fukumoto S, Ito H, Kiso M, Hirabayashi Y, Urano T, Furukawa K, Furukawa K (1999) Molecular cloning of brain-specific GD1α synthase (ST6GalNAc-V) containing CAG/glutamine repeats. J Biol Chem 274:30557–30562

Okajima T, Chen HH, Ito H, Kiso M, Tai T, Furukawa K, Urano T, Furukawa K (2000) Molecular cloning and expression of mouse GD1α/GT1aα/GQ1bα synthase (ST6GalNAc-VI) gene. J Biol Chem 275:6717–6723

Sjoberg ER, Kitagawa H, Glushka J, van Halbeek H, Paulson JC (1996) Molecular cloning of a developmentally regulated N-acetylgalactosamine α2,6-sialyltransferase specific for sialylated glycoconjugates. J Biol Chem 271:7450–7459

Takashima S, Kurosawa N, Tachida Y, Inoue M, Tsuji S (2000) Comparative analysis of the genomic structures and promoter activities of mouse Siaα2,3Galβ1,3GalNAc GalNAcα2,6-sialyltransferase genes (ST6GalNAc-III and -IV): characterization of their Sp1 binding sites. J Biochem 127:399–409

Tsuji S, Datta AK, Paulson JC (1996) Systematic nomenclature for sialyltransferases. Glycobiology 6(7):v–vii

ST6GalNAc-IV

Introduction

Like ST6GalNAc-III, ST6GalNAc-IV is a relatively small sialyltransferase (302 amino acids in length) compared with other sialyltransferases characterized to date. ST6GalNAc-IV has a type II membrane protein topology. The transmembrane domain of mouse ST6GalNAc-IV is 23 amino acids long, from positions 14 to 36. The stem region of mouse ST6GalNAc-IV is very short, and there are only 38 amino acid residues between the transmembrane domain and sialylmotif L, while mouse ST6GalNAc-I, -II, and -III have 261, 123, and 53 amino acid residues, respectively. ST6GalNAc-III–VI are members of one ST6GalNAc subfamily that can synthesize the ganglioside GD1α from GM1b (Sjoberg et al. 1996; Lee et al. 1999; Okajima et al. 1999, 2000; Ikehara et al. 1999). ST6GalNAc-IV prefers O-glycans to glycolipids as substrates, while ST6GalNAc-III, -V, and -VI prefer glycolipids to O-glycans. Therefore, ST6GalNAc-IV may be the main candidate for synthesizing the NeuAcα2-3Galβ1-3(NeuAcα2-6)GalNAc residue, which is usually found in the O-linked glycan chains of glycoproteins. ST6GalNAc-IV is also considered to be involved in the early alteration of the sialylation pattern of cell-surface molecules in activated lymphocytes (Kaufmann et al. 1999). The overall amino acid sequence identity of mouse ST6GalNAc-IV is 43.0% to mouse ST6GalNAc-III, 44.8% to mouse ST6GalNAc-V, and 41.2% to mouse ST6GalNAc-VI, but ST6GalNAc-IV shows no significant similarity to other sialyltransferases except in sialylmotifs.

Shou Takashima[1] and Shuichi Tsuji[2]

[1] Cellular Biochemistry Laboratory, Institute of Physical and Chemical Research (RIKEN), Wako, Saitama 351-0198, Japan
Tel. +81-48-462-1111 (ext. 5465); Fax +81-48-462-4670
e-mail: staka@postman.riken.go.jp
[2] Department of Chemistry, Faculty of Science, Ochanomizu University, Otsuka, Bunkyo-ku, Tokyo 112-8610, Japan
Tel. +81-3-5978-5345; Fax +81-3-5978-5344
e-mail: stsuji@cc.ocha.ac.jp

Databanks

ST6GalNAc-IV

NC-IUBMB enzyme classification: E.C.2.4.99.7

Species	Gene	Protein	mRNA	Genomic
Homo sapiens	–	AAF00102	AF127142	AF162789
		AAF32237	–	–
Mus musculus	*Siat7d*	CAB43507–CAB43508	Y15779–Y15780	Y19053–Y19057 (exons 1–6)
		CAB43514–CAB43515	–	–
		CAA07446	AJ007310	–
		NP_035501	NM_011373	–

Name and History

The mouse ST6GalNAc-IV cDNA has been independently cloned by two groups. Lee et al. (1999) cloned a fragment encoding a new sialylmotif by a polymerase chain reaction (PCR)-based approach using two primers deduced from the sialylmotif L in mouse ST3Gal-I and -II with mouse brain cDNA as a template. A mouse brain cDNA library was then screened with this fragment as a probe. Among the positive clones, one clone included an open reading frame of 906 bp, encoding a protein of 302 amino acids in length. Homology searching and the substrate specificities of the protein revealed that this is the fourth type of GalNAc α2-6-sialyltransferase (and the second type of NeuAcα2-3Galβ1-3GalNAc α2-6-sialyltransferase). Therefore, it was designated ST6GalNAc-IV according to the abbreviated nomenclature system for cloned sialyltransferase (Tsuji et al. 1996).

Kaufmann et al. (1999) used an mRNA differential display PCR to search for genes in activated T cells, and they obtained a 3.6 kb cDNA clone named MK45, encoding a protein having sialylmotifs L and S. The deduced amino acid sequence of MK45 was found to be identical to that of ST6GalNAc-IV.

An analysis of the genomic structure and transcription regulation of the mouse ST6GalNAc-IV gene has been performed (Takashima et al. 2000). The International System for Gene Nomenclature named this gene *Siat7d*.

Enzyme Activity Assay and Substrate Specificity

Enzyme Activity Assay

The enzyme activity of mouse ST6GalNAc-IV was measured using the soluble form of this enzyme fused with protein A (Lee et al. 1999). Enzyme activity was measured in 50 mM MES buffer, pH 6.0, 1 mM $MgCl_2$, 1 mM $CaCl_2$, 0.5% Triton CF-54, 100 μM CMP-[^{14}C]NeuAc (10.2 kBq), an acceptor substrate, and an enzyme preparation, in a total volume of 10 μl. As acceptor substrates, 10 μg proteins, 5 μg glycolipids, or 10 μg oligosaccharides were used. The enzyme reaction was performed at 37°C for 2 h. The reaction was terminated by the addition of SDS-polyacrylamide gel electrophoresis loading buffer (10 μl), and then the incubation mixtures were directly subjected to

SDS-polyacrylamide gel electrophoresis (for glycoprotein acceptors). For glycolipid acceptors, the incubation mixtures were applied on a C-18 column (Sep-Pak Vac, 100 mg; Waters, Milford, MA, USA), which had been washed with water. The glycolipids were eluted from the column with methanol, dried, and then subjected to chromatography on a high-performance thin-layer chromatography (HPTLC) plate (Merck, Germany) with a solvent system of chloroform, methanol, and 0.02% CaCl$_2$ (55:45:10). Acceptor substrates were visualized by staining with Coomassie Brilliant Blue (for glycoproteins) or by the orcinol-H$_2$SO$_4$ method (for glycolipids). The radioactive materials in glycoproteins and glycolipids were visualized with a BAS2000 radio image analyzer (Fuji Film Tokyo, Japan), and the radioactivity incorporated into acceptors was counted.

Substrate Specificity

The substrate specificity of mouse ST6GalNAc-IV is similar to that of ST6GalNAc-III, although ST6GalNAc-IV prefers O-glycans to glycolipids as substrates. Among the glycolipids examined, only GM1b served as an acceptor substrate for ST6GalNAc-IV. ST6GalNAc-IV exhibited higher activity toward fetuin than GM1b, but very low activity toward asialofetuin. These results suggest that ST6GalNAc-IV requires the NeuAcα2-3Galβ1-3GalNAc sequence in fetuin and GM1b just like ST6GalNAc-III. On the other hand, GD1a, which has the NeuAcα2-3Galβ1-3GalNAc sequence and an additional NeuAc residue at the internal galactose, did not serve as an acceptor substrate. Just as for ST6GalNAc-III, the α2-3-linked sialic acid attached to the internal galactose of GD1a seems to abolish the ability of ST6GalNAc-IV to transfer sialic acid to the NeuAcα2-3Galβ1-3GalNAc sequence. It should be noted that the oligosaccharide, NeuAcα2-3Galβ1-3GalNAc was a good acceptor substrate for ST6GalNAc-IV, while such an oligosaccharide was a poor substrate for mouse ST6GalNAc-III. The activities of ST6GalNAc-IV toward nonsialylated Galβ1,3GalNAc and disialylated NeuAcα2-3Galβ1-3(NeuAcα2-6)GalNAc were almost negligible. Thin-layer chromatography analysis revealed that the oligosaccharide alditol derived from [^{14}C]NeuAc-incorporated fetuin synthesized with ST6GalNAc-IV migrated to the same position as NeuAcα2-3Galβ1-3(NeuAcα2-6)GalNAc-ol. From these results, ST6GalNAc-IV is considered to be the second type of NeuAcα2-3Galβ1-3GalNAc α2-6-sialyltransferase which can synthesize the NeuAcα2-3Galβ1-3(NeuAcα2-6)GalNAc structure.

Preparation

To prepare the soluble form of mouse ST6GalNAc-IV, the DNA fragment encoding a truncated form of mouse ST6GalNAc-IV, lacking the first 36 amino acids of the open reading frame, was prepared by PCR amplification, and this fragment was inserted into a pcDSA vector (Lee et al. 1999). The resulting plasmid consisted of the IgM signal peptide sequence, the protein A IgG binding domain, and a truncated form of ST6GalNAc-IV. This expression vector was transiently transfected into COS-7 cells on a 150-mm plate using LipofectAMINE reagent (Life Technologies, Inc. Tokyo, Japan). The protein A-fused ST6GalNAc-IV expressed in the medium was adsorbed to an IgG-

Sepharose gel (Amersham Pharmacia Biotech, Piscataway, NJ, USA; 50 µl resin/50 ml culture medium) at 4°C for 16 h. The resin was collected by centrifugation, washed three times with phosphate-buffered saline, suspended in 50 µl (final volume) of Dulbecco's modified Eagle medium without fetal bovine serum, and used as the soluble enzyme.

Biological Aspects

Northern blot analysis of the mouse ST6GalNAc-IV gene in several tissues revealed that this gene is highly expressed in colon and brain, and moderately expressed in lung, heart, thymus, and spleen (Lee et al. 1999). Three major transcripts (1.6–1.9, 2.0–2.2, and 3.6–3.7 kb), which mainly differed in the length of their 3′-untranslated regions, were generated from this gene (Kaufmann et al. 1999; Lee et al. 1999; Takashima et al. 2000). Expression in embryonal stage (E12) and 1-day-old (P1) mouse brain was relatively low, whereas expression in 3-week-old and 8-week-old mouse brain was high, indicating that the expression in brain is developmentally regulated (Lee et al. 1999). The precise biological functions of ST6GalNAc-IV in vivo have not yet been demonstrated, but one potential function of ST6GalNAc-IV seems to be to synthesize NeuAcα2,3Galβ1,3(NeuAcα2,6)GalNAc residue on glycoproteins.

Interestingly, it has been shown that ST6GalNAc-IV expression is rapidly induced in activated CD8 T cells in vivo (Kaufmann et al. 1999). The expression reached the highest level 4 h after antigen triggering, and then declined rapidly to near base levels within 45 h. It should be noted that the induced expression level is much higher than the normal expression levels in mouse tissues. Moreover, ST6GalNAc-IV expression was also induced in lipopolysaccharide-activated B cells and antigen-triggered CD4 T cells in vitro. The rapidly induced expression in activated lymphocytes is specific for ST6GalNAc-IV, since mRNA expression levels of other sialyltransferases remained largely unchanged during the early stage of lymphocyte activation. Thus, ST6GalNAc-IV is considered to be a potent candidate sialyltransferase that is involved in the early alteration of the sialylation pattern of cell-surface molecules in activated lymphocytes.

Analysis of the genomic structure and transcription regulation of the mouse ST6GalNAc-IV gene has also been carried out (Takashima et al. 2000). The ST6GalNAc-IV gene spans over 12 kb of genomic DNA with six exons. The exon–intron boundaries of the ST6GalNAc-IV gene and those of the ST6GalNAc-III gene are very similar, suggesting that these genes arose from a common ancestral gene through gene duplication. Measurement of the promoter activity demonstrated that the first 441 bp upstream sequence from the translational initiation codon contains the minimum promoter for expression by a mouse fibroblast cell line, NIH3T3. Like the ST6GalNAc-III promoter, the minimum promoter region does not contain a TATA or CAAT box but has three putative Sp1 binding sites. Mobility shift assaying and mutational analysis of the promoter region indicated that all three Sp1 binding sites are independently involved in the transcription regulation of the ST6GalNAc-IV gene in NIH3T3 cells. The difference between the basic transcription regulation of the ST6GalNAc-III and -IV genes by Sp1 may be related to the expression levels of these genes.

Future Perspectives

ST6GalNAc-III–VI are the members of one ST6GalNAc subfamily and some of their enzymatic properties overlap. Each gene has characteristic but different expression patterns, suggesting that the expression of these genes is regulated in a different manner. Although the basic transcription mechanisms of the ST6GalNAc-III and -IV genes regulated by Sp1 have been shown, the mechanisms of cell-, tissue-, and stage-specific expression of these genes have not yet been clarified. In addition, the mechanism of the rapid induction of ST6GalNAc-IV in activated lymphocytes and its biological significance in the immune response have not been elucidated. At present, we cannot distinguish the precise biological functions of these enzymes in vivo. To solve this problem, it is necessary to know the mechanisms of cell-, tissue-, and stage-specific expression of these genes. Identification of transcription factors that are involved in these cell-, tissue-, and stage-specific expressions may clarify our understanding of the different expression patterns of these genes. Generating knock out mice of these genes will also help to reveal the precise biological functions of these enzymes. It would also be interesting to know the genomic organization of the ST6GalNAc-V and -VI genes, and evaluate the evolutional relation among all ST6GalNAc genes.

Further Reading

Kaufmann et al. (1999) report the cDNA cloning of mouse ST6GalNAc-IV by the mRNA differential display approach, and the rapid induction of ST6GalNAc-IV in lymphocytes after activation.

Lee et al. (1999) report the cDNA cloning of mouse ST6GalNAc-III and -IV, the substrate specificities of these enzymes, and the expression patterns of the ST6GalNAc-III and -IV genes.

Takashima et al. (2000) compare the genomic structures of mouse ST6GalNAc-III and -IV genes, and report on an analysis of promoter activities and the identification of Sp1 binding sites that are involved in the transcription regulation of these genes.

References

Ikehara Y, Shimizu N, Kono M, Nishihara S, Nakanishi H, Kitamura T, Narimatsu H, Tsuji S, Tatematsu M (1999) A novel glycosyltransferase with a polyglutamine repeat: a new candidate for GD1α synthase (ST6GalNAc-V) FEBS Lett 463:92–96

Kaufmann M, Blaser C, Takashima S, Schwartz-Albiez R, Tsuji S, Pircher H (1999) Identification of an α2,6-sialyltransferase induced early after lymphocyte activation. Int Immunol 11:731–738

Lee YC, Kaufmann M, Kitazume-Kawaguchi S, Kono M, Takashima S, Kurosawa N, Liu H, Pircher H, Tsuji S (1999) Molecular cloning and functional expression of two members of mouse NeuAcα2,3Galβ1,3GalNAc GalNAcα2,6-sialyltransferase family, ST6GalNAc-III and -IV. J Biol Chem 274:11958–11967

Okajima T, Fukumoto S, Ito H, Kiso M, Hirabayashi Y, Urano T, Furukawa K, Furukawa K (1999) Molecular cloning of brain-specific GD1α synthase (ST6GalNAc-V) containing CAG/glutamine repeats. J Biol Chem 274:30557–30562

Okajima T, Chen HH, Ito H, Kiso M, Tai T, Furukawa K, Urano T, Furukawa K (2000) Molecular cloning and expression of mouse GD1α/GT1aα/GQ1bα synthase (ST6GalNAc-VI) gene. J Biol Chem 275:6717–6723

Sjoberg ER, Kitagawa H, Glushka J, van Halbeek H, Paulson JC (1996) Molecular cloning of a developmentally regulated N-acetylgalactosamine α2,6-sialyltransferase specific for sialylated glycoconjugates. J Biol Chem 271:7450–7459

Takashima S, Kurosawa N, Tachida Y, Inoue M, Tsuji S (2000) Comparative analysis of the genomic structures and promoter activities of mouse Siaα2,3Galβ1,3GalNAc GalNAcα2,6-sialyltransferase genes (ST6GalNAc-III and -IV): characterization of their Sp1 binding sites. J Biochem 127:399–409

Tsuji S, Datta AK, Paulson JC (1996) Systematic nomenclature for sialyltransferases. Glycobiology 6(7):v–vii

ST8Sia-I (GD3 Synthase, SAT-II)

Introduction

GD3 synthase and its biosynthetic products, b-series gangliosides, have been suggested to have significant roles in development because of their unique expression patterns during neuronal differentiation. In extraneural tissues, GD3 is implicated in cell attachment, cell-to-cell interactions during embryogenesis, and signal transduction in the glycolipid-enriched microdomain. These observations suggest that GD3 may play an important role not only in brain development, but also in extraneural tissues. GD3 synthase (CMP-N-acetylneuraminate: GM3 α2,8-sialyltransferase) is a key enzyme of ganglioside synthesis that, in concert with GM2 synthase, regulates the ratio of a- and b-pathway gangliosides. The ganglioside composition of cells and tissues thus reflects the relative expression of these two biosynthetic glycosyltransferases. GD3 synthase is characteristically expressed in the early developmental stage of brain tissues.

Databanks

ST8Sia-I (GD3 synthase, SAT-II)

NC-IUBMB enzyme classification: E.C.2.4.99.8

Species	Gene	Protein	mRNA	Genomic
Homo sapiens	*SIAT8A*	Q92185	L32867	–
		–	D26360	–
		–	L43494	–
		–	X77922	–

YUTAKA SANAI

Department of Biochemical Cell Research, Tokyo Metropolitan Institute of Medical Science, 3-18-22 Honkomagome, Bunkyo-ku, Tokyo 113-8613, Japan
Tel. +81-3-3823-2101; Fax +81-3-3828-6663
e-mail: sanai@rinshoken.or.jp

(Continued)

Species	Gene	Protein	mRNA	Genomic
Mus musculus	Siat8a	Q64687	X84235	–
		–	L38677	–
Mesocricetus auratus		–	AF141657	–
Rattus norvegicus	Siat8a	–	P70554	–
		–	U53833	–
		–	AF016405	–
Gallus gallus	–	–	U73176	–

Name and History

GD3 synthase (CMP-*N*-acetylneuraminate: GM3 α2,8-sialyltransferase) is a member of the α2,8 sialyltransferase family and is referred to as ST8Sia-I (Tsuji et al. 1996), SAT-II (Basu et al. 1987; Pohlentz et al. 2000). GD3 synthase activity was found in embryonic chicken brain. This enzyme was well characterized using rat liver Golgi membrane as an enzyme source because the methods of preparation of the membrane were well established. The cDNA of GD3 synthase has been cloned in human, mouse, rat, and chicken. The chromosomal localization of the GD3 synthase (locus symbol: SIAT8A) was determine in human and mouse. The human GD3 synthase gene was mapped to p12.1–p11.2 of chromosome 12. The mouse homologue was mapped 2.8 cM distal to D6Mit52 and 4.3 cM proximal to D6Mit25 of chromosome 6.

Enzyme Activity Assay and Substrate Specificity

GD3 synthase, like other ganglioside glycosyltransferases, is predicted to type II membrane protein with an active site that faces the lumen of the Golgi apparatus and utilizes a membrane-resident lipid acceptor (GM3) and a water-soluble activated sugar donor (CMP-NeuAc).

$$NeuAc\alpha2\text{-}3Gal\beta1\text{-}4Glc\beta1\text{-}1Cer + CMP\text{-}NeuAc \rightarrow$$
$$NeuAc\alpha2\text{-}8NeuAc\alpha2\text{-}3Gal\beta1\text{-}4Glc\beta1\text{-}1Cer + CMP\text{-}NeuAc$$

The activity is greatest in the presence of certain detergents, and has a pH optimum of around 6.5. Apparent Km values for CMP-NeuAc and GM3 were about 0.8 and 0.2 mM, respectively. Several methods for assaying the enzyme activity have been developed for the treatment of the reaction products, such as a thin-layer chromatography (TLC), a gel filtration column method, a paper chromatographic method, and a reverse-phase chromatographic method (Pohlentz et al. 2000; Basu et al. 1987). A typical assay method for the human GD3 synthase expressed in COS-7 cells is described in the following paragraph (Nara et al. 1996).

The reaction mixture contained 100 mM sodium cacodylate, pH 6.5, 10 mM MgCl$_2$, 0.4% Triton CF-54, 0.3 mM acceptor substrate (GM3), 1.5 mM [^{14}C] CMP-NeuAc (50000 cpm in 10 μl), and enzyme protein. The incubation time depends on the specific activity of the enzyme source. After the addition of an equal volume of methanol to stop the reaction, the reaction mixture was subjected to reverse-phase TLC. The ganglioside-bound radioactivity was measured using a Fujix BAS 2000 bio-imaging analyzer (Fuji Photo Film, Tokyo, Japan) after the TLC plate had been developed with water.

The substrate specificity of GD3 synthase of rat liver Golgi was analyzed using GM3 and its synthetic derivatives (Klein et al. 1987). The activity of GD3 synthase is highly dependent on N-acylation of the sphingoid amino group of the GM3 molecule. GD3 synthase activity is also dependent on the size of the acyl group bound to the amino group of the sialic acid residue of GM3. The substrate specificity of the GD3 synthase was examined using the recombinant protein expressed in COS-7 cells. The cloned GD3 synthase catalyzes the production of GD3 from GM3, and to some extent utilizes GM1b/GD1a/GT1b as the substrate. Furthermore, this enzyme produces a significant amount of GT3 and polysialogangliosides (Nakayama et al. 1996; Watanabe et al. 1996).

Structural Chemistry

The deduced GD3 synthase protein is predicted to have a type II membrane topology and three potential N-glycosylation sites (Asn-X-Ser/Thr) in the chicken and rat enzymes, and four in the human and mouse. SDS-PAGE of the purified GD3 synthase from rat brain revealed a single major protein band with an apparent molecular weight of 55000 (Gu et al. 1990).

GD3 synthase is concentrated in Golgi apparatus fractions. Using a sucrose density gradient fractionation of a highly purified Golgi apparatus from rat liver, it was shown that GD3 synthase activity and GQ1b activity are physically separate. It is suggested that glycosyltransferases are ordered in the order in which they act, i.e., that early reactions occur in early stations (endoplasmic reticulum and cis/medial Golgi) of the pathway, and late reactions occur in late stations (trans-Golgi/trans-Golgi network) (Trinchera et al. 1990). However, it has been demonstrated that lactosylceramide and subsequent glycosphingolipids are formed in the lumen of the late Golgi (Lannert et al. 1998).

The sub-Golgi location of an epitope-tagged chicken GD3 synthase expressed in CHO-K1 cells was studied (Daniotti et al. 2000). Confocal immunofluorescence microscopy and biochemical analysis showed most GD3 synthase localized in the proximal Golgi, and that a functional fraction is also present in the trans-Golgi network.

It has been shown that N-glycosylation of GD3 synthase may affect its activity or subcellular localization (Martina et al. 1998). Elimination of N-linked oligosaccharides reduced the activity of the enzyme to <10% of the activity as determined with cells with fully glycosylated enzyme. The endoplasmic reticulum (ER) retention and increased turnover of GD3 synthase were most likely due to its inability to bind to calnexin upon inhibition of early N-glycoprotein processing.

Preparation

The enzyme is widely distributed in various mammalian tissues such as brain, kidney, heart, lung, pancreas, and liver. The highest activity is found in kidney. GD3 synthase activity was well characterized using the rat liver Golgi membrane as an enzyme source. GD3 synthase was purified 10 000-fold from the Triton X-100 extract of rat brain by affinity chromatography (Gu et al. 1990). The purified recombinant enzyme is obtained in mammalian expression systems as the protein A-fusion protein (Watanabe et al. 1996).

Biological Aspects

GD3 has been identified as a melanoma-associated antigen. Melanoma cell lines showed extremely high levels of GD3 synthase gene expression as well as the enzyme activity (Yamashiro et al. 1995). The upregulation of GD3 synthase gene and GD3 expression during the activation of peripheral T lymphocytes has also been demonstrated. GD3 synthase gene is highly expressed in adult T cell leukemia (ATL) cells (Okada et al. 1996).

GD3 was expressed mainly in the retina and fetal brain. GD3 synthase is characteristically expressed in the early developmental stage of brain tissues. Site-restricted expression of GD3 synthase mRNA has been demonstrated in mouse (Yamamoto et al. 1996a,b) and rat (Watanabe et al. 1996), and in P19-derived neural cells (Osanai et al. 1997). The GD3 synthase gene was strongly expressed in the neural tube at embryonic day 9. The expression level of the GD3 synthase gene was gradually reduced along with the development of whole brain tissue. However, some parts of mouse brain showed persistent expression of the GD3 synthase gene even after birth.

The transfection of GD3 synthase cDNA into Neuro2a cells, a neuroblastoma cell line, caused cell differentiation with neurite sprouting (Kojima et al. 1994). A rat pheochromocytoma cell line (PC12) transfected with GD3 synthase cDNA showed a marked increase in cell growth and an inability to respond to NGF stimulation (Fukumoto et al. 2000).

Treatment of the human promyelocytic leukemia cell line HL-60 cells with antisense oligonucleotides to GM2 synthase and GD3 synthase sequences effectively downregulated the synthesis of complex gangliosides, resulting in a remarkable increase in endogenous GM3. The treated cells underwent monocyte differentiation (Zeng et al. 1995). Suppression of GD3 synthase expression of the neuroblastoma cells with antisense vector against GD3 synthase gene resulted in reduced cell migration, tumor growth, and experimental metastasis (Zeng et al. 2000).

Future Perspectives

Strategies to control glycolipid expression are remarkably well developed, i.e., targeted disruption of the GD3 synthase gene and the expression of the antisense GD3 synthase gene. These technologies will be applied to elucidate the functions of GD3 and b-series gangliosides.

Further Reading

For reviews on GD3 synthase, see Tsuji et al. 1996 and Maccioni et al. 1999.

References

Basu M, De T, Das KK, Kyle JW, Chon HC, Schaeper RJ, Basu S (1987) Glycolipids. Methods Enzymol 138:575–607

Daniotti JL, Martina JA, Giraudo CG, Zurita AR, Maccioni, HJ (2000) GM3 α2,8-sialyltransferase (GD3 synthase): protein characterization and sub-Golgi location in CHO-K1 cells. J Neurochem 74:1711–1720

Fukumoto S, Mutoh T, Hasegawa T, Miyazaki H, Okada M, Goto G, Furukawa K, Urano T (2000) GD3 synthase gene expression in PC12 cells results in the continuous activation of TrkA and ERK1/2 and enhanced proliferation. J Biol Chem 275:5832–5838

Gu XB, Gu TJ, Yu RK (1990) Purification to homogeneity of GD3 synthase and partial purification of GM3 synthase from rat brain. Biochem Biophys Res Commun 166:387–393

Klein D, Pohlentz G, Schwarzmann G, Sandhoff K (1987) Substrate specificity of GM2 and GD3 synthase of Golgi vesicles derived from rat liver. Eur J Biochem 167:417–424

Kojima N, Kurosawa N, Nishi T, Hanai N, Tsuji S (1994) Induction of cholinergic differentiation with neurite sprouting by de novo biosynthesis and expression of GD3 and b-series gangliosides in Neuro2a cells. J Biol Chem 269:30451–30456

Lannert H, Gorgas K, Meissner I, Wieland FT, Jeckel D (1998) Functional organization of the Golgi apparatus in glycosphingolipid biosynthesis. Lactosylceramide and subsequent glycosphingolipids are formed in the lumen of the late Golgi. J Biol Chem 273:2939–2946

Maccioni HJ, Daniotti JL, Martina, JA (1999) Organization of ganglioside synthesis in the Golgi apparatus. Biochim Biophys Acta 1437:101–118

Martina JA, Daniotti JL, Maccioni, HJ (1998) Influence of N-glycosylation and N-glycan trimming on the activity and intracellular traffic of GD3 synthase. J Biol Chem 273:3725–3731

Nakayama J, Fukuda MN, Hirabayashi Y, Kanamori A, Sasaki K, Nishi T, Fukuda M (1996) Expression cloning of a human GT3 synthase. GD3 and GT3 are synthesized by a single enzyme. J Biol Chem 271:3684–3691

Nara K, Watanabe Y, Kawashima I, Tai T, Nagai Y, Sanai Y (1996) Acceptor substrate specificity of a cloned GD3 synthase that catalyzes the biosynthesis of both GD3 and GD1c/GT1a/GQ1b. Eur J Biochem 238:647–652

Okada M, Furukawa K, Yamashiro S, Yamada Y, Haraguchi M, Horibe K, Kato K, Tsuji Y (1996) High expression of ganglioside α-2,8-sialyltransferase (GD3 synthase) gene in adult T-cell leukemia cells unrelated to the gene expression of human T-lymphotropic virus type ICancer Res 56:2844–2848

Osanai T, Watanabe Y, Sanai Y (1997) Glycolipid sialyltransferases are enhanced during neural differentiation of mouse embryonic carcinoma cells, P19. Biochem Biophys Res Commun 241:327–333

Pohlentz G, Kaes C, Sandhoff K (2000) In vitro assays for enzymes of ganglioside synthesis. Methods Enzymol 311:82–94

Trinchera M, Pirovano B, Ghidoni R (1990) Sub-Golgi distribution in rat liver of CMP-NeuAc GM3- and CMP-NeuAc:GT1b α-2-8sialyltransferases and comparison with the distribution of the other glycosyltransferase activities involved in ganglioside biosynthesis. J Biol Chem 265:18242–18247

Tsuji S, Datta AK, Paulson, JC (1996) Systematic nomenclature for sialyltransferases [letter]. Glycobiology 6:v–vii

Watanabe Y, Nara K, Takahashi H, Nagai Y, Sanai Y (1996) The molecular cloning and expression of α-2,8-sialyltransferase (GD3 synthase) in a rat brain. J Biochem (Tokyo) 120:1020–1027

Yamamoto A, Haraguchi M, Yamashiro S, Fukumoto S, Furukawa K, Takamiya K, Atsuta M, Shiku H (1996a) Heterogeneity in the expression pattern of two ganglioside synthase genes during mouse brain development. J Neurochem 66:26–34

Yamamoto A, Yamashiro S, Fukumoto S, Haraguchi M, Atsuta M, Shiku H, Furukawa K (1996b). Site-restricted and neuron-dominant expression of α-2,8-sialyltransferase gene in the adult mouse brain and retina. Glycoconj J 13:471–480

Yamashiro S, Okada M, Haraguchi M, Furukawa K, Lloyd KO, Shiku H (1995) Expression of α-2,8-sialyltransferase (GD3 synthase) gene in human cancer cell lines: high-level expression in melanomas and up-regulation in activated T lymphocytes. Glycoconj J 12:894–900

Yu RK, Macala LJ, Taki T, Weinfield HM, Yu FS (1988) Developmental changes in ganglioside composition and synthesis in embryonic rat brain. J Neurochem 50:1825–1829

Zeng G, Ariga T, Gu XB, Yu RK (1995) Regulation of glycolipid synthesis in HL-60 cells by antisense oligodeoxynucleotides to glycosyltransferase sequences: effect on cellular differentiation. Proc Natl Acad Sci USA 92:8670–8674

Zeng G, Gao L, Yu RK (2000) Reduced cell migration, tumor growth and experimental metastasis of rat F-11 cells whose expression of GD3-synthase is suppressed. Int J Cancer 88:53–57

ST8Sia-II (STX)

Introduction

ST8Sia-II (STX) belongs to a family of α2-8-sialyltransferases, and exhibits polysialic acid synthase activity like ST8Sia-IV (see Chap. 48). The cDNAs encoding ST8Sia-II have been cloned from rat (Livingston and Paulson 1993), mouse (Kojima et al. 1995a), human (Scheidegger et al. 1995; Angata et al. 1997), and *Xenopus* (Kudo et al. 1998). The deduced amino acid sequences of the ST8Sia-II cloned from human, mouse, and rat exhibit the same 375 amino acid lengths, and conserve over 99% among them. The amino acid sequence of ST8Sia-II cloned from mouse showed 34.4% and 56.0% identity with mouse ST8Sia-III and mouse ST8Sia-IV, respectively, but no similarity of the amino acid sequence (less than 30%) of mouse ST8Sia-II was observed toward those of other sialyltransferases cloned from mouse. Like other sialyltransferases, ST8Sia-II exhibits type II membrane protein topology and has characteristic motifs for sialyltransferases called sialylmotifs L, S, and VS.

The mRNA expression of mammalian ST8Sia-II is highly restricted in brain and is regulated during brain development, as seen in the case of the expression of polysialic acid on N-CAM (Kojima et al. 1995a; Kurosawa et al. 1997). *N*-glycans on N-CAM with terminal sialic acid moieties in α2-3 linkage serve as the substrate for the attachment of the initial α2-8-linked sialic acid residue. This initial reaction (initiation) is followed by an elongation reaction in which the α2-8 sialic acid moiety added in the preceding step serves as the attachment site for the next α2-8-linked sialic acid. Only ST8Sia-II, and also ST8Sia-IV, exhibit both the initiation and elongation reactions in vitro (Kojima et al. 1995a,b, 1996; Nakayama and Fukuda 1996).

NAOYA KOJIMA[1] and SHUICHI TSUJI[2]

[1] Department of Applied Chemistry, Tokai University, 1117 Kitakaname, Hiratsuka, Kanagawa 259-1292, Japan
Tel. +81-463-58-1211 (ext 4175); Fax +81-463-50-2012
e-mail: naoyaki@keyaki.cc.u-tokai.ac.jp
[2] Department of Chemistry, Faculty of Science, Ochanomizu University, Otsuka, Bunkyo-ku, Tokyo 112-8610, Japan
Tel. +81-3-5978-5345; Fax +81-3-5978-5344
e-mail: stsuji@cc.ocha.ac.jp

Databanks

ST8Sia-II (STX)

No EC number has been allocated.

Speices	Gene	Protein	mRNA	Genomic
Homo sapiens	*SIAT8B* or *STX*	g522199	L29556	–
		g995770	U33551	–
		Q9218689	U82762	–
		g1916841	–	–
Mus musculus	*Siat8b*	e1001357	X83562	X99645–X99651 (exons 1–6)
Rattus norvegicus	–	g310229	L13445	–
Xenopus laevis	–	BAA32617.1	AB007468	–

Name and History

ST8Sia-II was first cloned from rat, in 1993, as a developmentally regulated member of the sialyltransferase family by a polymerase chain reaction (PCR)-based approach using the homology of sialylmotif L (Livingston and Paulson 1993). At that time, the enzymatic activity of rat STX could not be detected in vitro or in vivo, although the deduced amino acid sequence revealed that STX has a type II membrane topology, and has sialylmotifs L, S, and VS. Therefore, it was named sialyltransferase X (STX). In 1995, cDNA encoding STX was cloned from mouse brain cDNA library, and the recombinant mouse STX was confirmed to exhibit α2-8-sialyltransferase activity toward N-linked oligosaccharides of glycoproteins (Kojima et al. 1995a). Since, GD3 synthase (ST8Sia-I) had been cloned as the first member of the α2-8-sialyltransferase family, STX was renamed ST8Sia-II, i.e., the second member of the α2-8-sialyltransferase family (Tsuji et al. 1996). Subsequently, mouse ST8Sia-II was shown to exhibit polysialic acid synthase activity toward N-CAM in vivo and in vitro (Kojima et al. 1995b, 1996, 1997). To date, the cDNAs encoding ST8Sia-II have been cloned from human, mouse, rat, and *Xenopus* (Livingston and Paulson 1993; Kojima et al. 1995a; Scheidegger et al. 1995; Angata et al. 1997; Kudo et al. 1998). An analysis of the genomic structure and transcription regulation of the mouse ST8Sia-II gene has been performed (Yoshida et al. 1996).

Enzyme Activity Assay and Substrate Specificity

Enzyme Activity Assay

The enzyme activity of mouse ST8Sia-II was measured using the soluble form of this enzyme fused with protein A after being adsorbed to an IgG-Sepharose (15 μl resin per 10 ml culture medium) (Kojima et al. 1995a). The assay mixture consisted of 0.1 M sodium cacodylate buffer (pH 6.0), 10 mM $MgCl_2$, 2 mM $CaCl_2$, 0.5% Triton CF-54, 34 mM CMP-[^{14}C]NeuAc (0.4 mCi), 1 mg/ml acceptor substrate, and 2 μl enzyme preparation in a total volume of 10 μl (Kojima et al. 1995a,b). After 6 h incubation at 37°C, the reaction was terminated by the addition of SDS-PAGE loading buffer (10 μl), and the incubation mixtures were directly subjected to SDS-PAGE. Acceptor substrates were visualized by staining with Coomassie brilliant blue. The radioactive materials in glycoproteins were visualized with a BAS2000 radio image analyzer, and the radioactivity incorporated into acceptor glycoproteins was counted. When N-CAM-Fc was used as the substrate glycoprotein, protein G-Sepharose resin was added to the supernatant, and the protein G-Sepharose-bound fraction was used for SDS-PAGE or enzymatic treatment.

Substrate Specificity

Linkage-specific neuraminidase treatment of substrate glycoproteins indicated that the presence of the α2-3-linked sialic acid residues on substrate glycoproteins is essential for enzymatic activity. However, it was reported that α2-6-linked sialic acid also serves as an acceptor for the enzyme (Angata et al. 1998). Although recombinant ST8Sia-II can synthesize polysialic acid on several glycoproteins, N-CAM served as a 1500-fold better acceptor for mouse ST8Sia-II as compared with other glycoproteins in vitro (Kojima et al. 1996). Transfection of mouse ST8Sia-II cDNA into Neuro2a cells, which express the 180-, 140-, and 120-Kda isoforms of N-CAM, leads to the polysialylation of only the 180- and 140-Kda isoforms, and no other polysialylated glycoproteins were observed in the transfected cells (Kojima et al. 1997). Therefore, ST8Sia-II is specific for N-CAM in vivo.

Preparation

A truncated form of mouse ST8Sia-II, lacking the first 33 amino acids of the open reading frame, was prepared by PCR amplification, and this fragment was inserted into a pcDSA vector (Kojima et al. 1995a). The resulting plasmid consisted of the IgM signal peptide sequence, the protein A IgG binding domain, and a truncated form of ST8Sia-II. This expression vector (10 mg) was transiently transfected into COS-7 cells on a 150-mm plate by the DEAE-dextran method and cultured for 16 h in Dulbecco's modified Eagle medium supplemented with 2% fetal bovine serum. This medium was then replaced with a serum-free medium, and the cells were cultured for another 32 h. After 48 h transfection, the culture medium was collected and the protein A-fused ST8Sia-II secreted in the medium was adsorbed to IgG-Sepharose (15 µl resin per 10 ml culture medium) at 4°C for 16 h. The resin was collected by centrifugation, washed three times with phosphate-buffered saline, suspended in 50 µl (final volume) of Dulbecco's modified eagle medium without fetal bovine serum, and used as the soluble enzyme.

For preparation of the full-length form of mouse ST8Sia-II, the expression vector containing the whole coding region of mouse ST6GalNAc III was inserted into the pRc/CMV.

Biological Aspects

The deduced amino acid sequences of the ST8Sia-II cloned from human, mouse, and rat showed over 99% conservation among them, and ST8Sia-II cloned from *Xenopus* had 80% amino acid similarity to the mammlian ST8Sia-II. The high conservation of

ST8Sia-II among the animal species may suggest the biological importance of this enzyme in animals.

ST8Sia-II transfers the first α2-8-sialic acid residues to the α2-3-linked sialic acids of *N*-glycans (initiation), and also transfers multiple α2-8-sialic acid residues to the resulting Siaα2-8Siaα2-3Galβ-R structure (polymerization), yielding polysialic acid on *N*-linked oligosaccharides of several glycoproteins, such as fetuin, α1-acid glycoprotein, and N-CAM, to form polysialic acid in vitro (Angata et al. 1997; Kojima et al. 1996).

Transfection of ST8Sia-II cDNA leads to polysialic acid on the cell surface of various cell lines (Angata et al. 1997; Kojima et al. 1995b, 1997; Scheidegger et al. 1995). It has been shown that HeLa cells doubly transfected with the ST8Sia-II cDNA and N-CAM cDNA supported neurite outgrowth much better than HeLa cells expressing N-CAM alone (Angata et al. 1997). These characteristics of ST8Sia-II are very similar to those of ST8Sia-IV (Eckhardt et al. 1995; Nakayama and Fukuda 1996). However, several differences are observed between ST8Sia-II and -IV. The expression of mRNA of mammalian ST8Sia-II is highly restricted in brain and highly regulated during brain development, whereas ST8Sia-IV mRNA expresses not only in brain but also in other tissues (Kurosawa et al. 1997). The most striking difference between the two polysialic acid synthases (ST8Sia-II and -IV) was the degree of polysialylation of the polysialic acid synthesized on the glycoproteins, i.e., polysialic acid synthesized by ST8Sia-IV is larger than that synthesized by ST8Sia-II (Kojima et al. 1996, 1997). In addition, ST8Sia-II is more specific for N-CAM than ST8Sia-IV in vivo. Recently, it has been shown that ST8Sia-IV strongly prefers the sixth *N*-glycosylation site, which is the closest to the transmembrane domain, over the fifth *N*-glycosylation site of N-CAM, whereas ST8Sia-II also prefers the sixth site, it utilizes the fifth site more than does ST8Sia-IV. In addition, a mixture of ST8Sia-II and -IV synthesized polysialic acid on N-CAM more efficiently than ST8Sia-II or -IV alone, suggesting that ST8Sia-II and -IV form polysialylated N-CAM in a synergistic manner (Angata et al. 1998). A striking abundance of mouse ST8Sia-II mRNA during the late embryonic to early postnatal stages of brain development has been shown (Kurosawa et al. 1997). The major increase in mRNA is observed between E12 and E14, and the highest level is observed between postnatal day 1 (P1) and P3. Then the signal gradually decreases to a hardly detectable level by P14, indicating that expression of this gene is highly regulated. In situ hybridization of postnatal day 3 mouse brain shows high levels of ST8Sia-II mRNA expression in the cerebral neocortex, striatum, hippocampus, subiculum, medial habenular nucleus, thalamus, pontine nuclei, and inferior colliculus, intermediate levels of expression in the olfactory bulb, hypothalamus, superior colliculus, and cerebellum, and low-level expression in other regions. In the hippocampus, ST8Sia-II mRNA is expressed abundantly in the pyramidal cell layer at an early postnatal stage (P1–5). The cerebellum exhibits high levels of ST8Sia-II mRNA expression in the deep cerebellar nuclei and just below the molecular layer (Kurosawa et al. 1997).

Although the overall structures of ST8Sia-II and -IV genes are very similar, there is no extensive sequence homology between the 5'-flanking regions of both genes, suggesting that these two genes are expressed under different regulatory systems (Takashima et al. 1998). The gene encoding mouse ST8Sia-II is found to span about 80 kb and to be composed of six exons, and the transcription started from 167 nt upstream of the translational initiation site (Yoshida et al. 1996). Minimal promoter activity specific for neuronal cells has been detected within the proximal region

325 bp upstream from the translational initiation codon. The minimal promoter is embedded in a GC-rich domain (74%, GC content), in which two Sp1 binding motifs as well as a long purine-rich region are found, but it lacks TATA- and CAAT boxes (Yoshida et al. 1996). In addition, the locus of the ST8Sia-II gene is also different from that of ST8Sia-IV, i.e., human ST8Sia-II and -IV genes reside at chromosome 15, band q26, and chromosome 5, band p21, respectively (Angata et al. 1997).

Future Perspectives

Polysialic acid is mainly attached to N-CAM and its expression is shown to be highly restricted and regulated. The presence of polysialic acid in developmental brain and metastatic tumors has attracted the attention of many researchers. Recent data imply the importance of polysialic acid in the pathfinding and targeting on innervation of axons, the migration of neuronal cells and tumor cells, and spatial learning and memory (Kiss and Rougon 1997; Rutishauser and Landmesser 1996; Rutishauser 1998). The aim of cloning the sialyltransferase(s) responsible for polysialic acid biosynthesis was to elucidate the regulation mechanisms of polysialic acid expression, and the functions of polysialic acid in biological phenomena. Cloning the enzymes responsible for polysialic acid synthesis and finding that a single enzyme (ST8Sia-II or ST8Sia-IV) is able to confer polysialylation to N-CAM contributed in part to our understanding of the mechanisms of regulation of polysialic acid expression. However, it is not still clear how many enzymes participate in polysialic acid synthesis, or how to control polysialic acid expression. There seem to be several small inconsistencies between the results from the characterization of the cloned enzymes and those from biochemical structural analyses of the polysialylated oligosaccharides on N-CAM. Structural analysis of the *N*-linked oligosaccharides of N-CAM is required for an understanding of the regulation of polysialic acid biosynthesis and expression. In addition, the detailed functions of polysialic acid in biological phenomena are also still unclear. Since ST8Sia-II is one of the key enzymes in the biosynthesis of polysilic acid, and manipulation of the gene encoding ST8Sia-II in cells and animals may help to elucidate polysialic acid functions.

Further Reading

Rutishauser and Landmesser 1996 and Rutishauser 1998 are reviews of the functions of polysialic acid in the nervous system.

References

Angata K, Nakayama J, Fredette B, Chong K, Ranscht B, Fukuda M (1997) Human STX polysialyltransferase forms the embryonic form of the neural cell adhesion molecule. Tissue-specific expression, neurite outgrowth, and chromosomal localization in comparison with another polysialyltransferase, PST. J Biol Chem 272:7182–7190

Angata K, Suzuki M, Fukuda M (1998) Differential and cooperative polysialylation of the neural cell adhesion molecule by two polysialyltransferases, PST and STX. J Biol Chem 273:28524–28532

Eckhardt M, Muhlenhoff M, Bethe A, Koopman J, Frosch M, Gerardy-Schahn R (1995) Molecular characterization of eukaryotic polysialyltransferase-1. Nature 373:715–718

Kiss JZ, Rougon G (1997) Cell biology of polysialic acid. Curr Opin Neurobiol 7:640–646

Kojima N, Yoshida Y, Kurosawa N, Lee YC, Tsuji S (1995a) Enzymatic activity of a developmentally regulated member of the sialyltransferase family (STX): evidence for α2,8-sialyltransferase activity toward N-linked oligosaccharides. FEBS Lett 360:1–4

Kojima N, Yoshida Y, Tsuji S (1995b) A developmentally regulated member of the sialyltransferase family (ST8Sia-II, STX) is a polysialic acid synthase. FEBS Lett 373:119–122

Kojima N, Tachida Y, Yoshida Y, Tsuji S (1996) Characterization of mouse ST8Sia-II (STX) as a neural cell adhesion molecule-specific polysialic acid synthase. Requirement of core α1,6-linked fucose and a polypeptide chain for polysialylation. J Biol Chem 271:19457–19463

Kojima N, Tachida Y, Tsuji S (1997) Two polysialic acid synthases, mouse ST8Sia-II and -IV, synthesize different degrees of polysialic acids on different substrate glycoproteins in mouse neuroblastoma Neuro2a cells. J Biochem (Tokyo) 122:1265–1273

Kudo M, Takayama E, Tashiro K, Fukamachi H, Nakata T, Tadakuma T, Kitajima K, Inoue Y, Shiokawa K (1998) Cloning and expression of an α-2,8-polysialyltransferase (STX) from Xenopus laevis. Glycobiology 8:771–777

Kurosawa N, Yoshida Y, Kojima N, Tsuji S (1997) Polysialic acid synthase (ST8Sia-II/STX) mRNA expression in the developing mouse central nervous system. J Neurochem 69:494–503

Livingston BD, Paulson JC (1993) Polymerase chain reaction cloning of a developmentally regulated member of the sialyltransferase gene family. J Biol Chem 268:11504–11507

Nakayama J, Fukuda M (1996) A human polysialyltransferase directs in vitro synthesis of polysialic acid. J Biol Chem 271:1829–1832

Rutishauser U (1998) Polysialic acid at the cell surface: biophysics in service of cell interactions and tissue plasticity. J Cell Biochem 70:304–312

Rutishauser U, Landmesser L (1996) Polysialic acid in the vertebrate nervous system: a promoter of plasticity in cell–cell interactions. Trends Neurosci 19:422–427

Scheidegger EP, Sternberg LR, Roth J, Lowe JB (1995) A human STX cDNA confers polysialic acid expression in mammalian cells. J Biol Chem 270:22685–22688

Takashima S, Yoshida Y, Kanematsu T, Kojima N, Tsuji S (1998) Genomic structure and promoter activity of the mouse polysialic acid synthase (mST8Sia-IV/PST) gene. J Biol Chem 273:7675–7683

Tsuji S, Datta AK, Paulson JC (1996) Systematic nomenclature for sialyltransferases. Glycobiology 6(7):v–vii

Yoshida Y, Kurosawa N, Kanematsu T, Kojima N, Tsuji S (1996) Genomic structure and promoter activity of the mouse polysialic acid synthase gene (mST8Sia-II). Brain-specific expression from a TATA-less GC-rich sequence. J Biol Chem 271:30167–30173

47

ST8Sia-III

Introduction

The predicted amino acid sequence of mouse ST8Sia-III (mST8Sia-III) shows a protein of 380 amino acids with a type II transmembrane topology, as has been found for all sialyltransferases cloned to date. The sequence shows 16.9, 26.1, 26.2, and 17.2% identity with those of mST8Sia-I, -II, -IV, and -V, respectively. The mST8Sia-III transfers sialic acids toward Siaα2-3Galβ1-4GlcNAc sequences on both N-linked oligosaccharides of glycoproteins and α2-3sialylated glycosphingolipids, such as α2,3-sialylparagloboside and GM3. The mST8Sia-III gene was expressed only in brain and testis, and was well regulated during brain development.

Databanks

ST8Sia-III

NC-IUBMB enzyme classification: E.C.2.4.99.8

Species	Gene	Protein	mRNA	Genomic
Homo sapiens	–	JC5600	AF004668	–
Mus musculus	Siat8c	NP_033208	X80502	X93998
Rattus norvegicus	Siat8c	–	U55938	–

Yukiko Yoshida[1] and Shuichi Tsuji[2]

[1] Department of Tumor Immunology, Tokyo Metropolitan Institute of Medical Science, Honkomagome 3-18-22, Bunkyo-ku, Tokyo 113-8613, Japan
Tel. +81-3-3823-2101 (ext. 5249); Fax +81-3-3823-2965
e-mail: yyosida@rinshoken.or.jp
[2] Department of Chemistry, Faculty of Science, Ochanomizu University, Otsuka, Bunkyo-ku, Tokyo 112-8610, Japan
Tel. +81-3-5978-5345; Fax +81-3-5978-5344
e-mail: stsuji@cc.ocha.ac.jp

Name and History

ST8Sia-III is the third molecule isolated as an α2,8-sialyltransferase (Tsuji et al. 1996). Its cDNA was cloned from mouse brain by means of the polymerase chain reaction (PCR)-based approach involving two degenerate oligonucleotide primer bases on two highly conserved regions, sialylmotifs L and S, of human ST8Sia-I and rat ST8Sia-II (Yoshida et al. 1995a). In 1998, Lee et al. isolated ST8Sia-III cDNA from human brain. The genomic gene of mST8Sia-III has been cloned, and its gene organization and 5′-flanking region have been analyzed (Yoshida et al. 1996). The International System for Gene Nomenclature named this gene *Siat8c*.

Enzyme Activity Assay and Substrate Specificity

Sialyltransferase Assay

The enzyme activity was measured in the presence of 0.1 M sodium cacodylate buffer (pH 6.0), 10 mM MgCl$_2$, 2 mM CaCl$_2$, 0.5% Triton CF-54, 100 mM CMP-[^{14}C]NeuAc (0.25 mCi), 10 μg acceptor substrate, and 2 μl enzyme preparation in a total volume of 10 μl (Sasaki et al. 1994). After 4 h incubation at 37°C, the reaction was terminated by the addition of SDS-PAGE loading buffer (10 μl), and the incubation mixtures were directly subjected to SDS-PAGE (for glycoprotein acceptors). For glycolipid acceptors, the incubation mixtures were applied on a C-18 column (Sep-Pak Vac, 100 mg; Waters, Milford, MA, USA), which had been washed with water. The glycolipids were eluted from the column with methanol, dried, and then subjected to chromatography on a high-performance thin-layer chromatography (HPTLC) plate with a solvent system of chloroform, methanol, and 0.02% CaCl$_2$ (55:45:10). Acceptor substrates were visualized by staining with Coomassie brilliant blue (for glycoproteins) or by the orcinol/H$_2$SO$_4$ method (for glycolipids). The radioactive materials in glycoproteins or glycolipids were visualized with a BAS2000 radio image analyzer (Fuji Film, Tokyo, Japan), and the radioactivity incorporated into acceptor glycoproteins was counted.

Acceptor Substrate Specificity

The acceptor substrate specificity of ST8Sia-III was compared with that of ST8Sia-I and ST8Sia-II, as shown in Table 1. ST8Sia-II exhibited sialyltransfer activity only toward sialylated glycoproteins such as α1-acid glycoprotein or fetuin. A comparison of the substrate specificities of these two α2,8-sialyltransferases showed that that of ST8Sia-III was rather broad. Both sialylated glycoproteins and glycolipids served as acceptors for this sialyltransferase. Although the substrate specificity for glycoproteins of ST8Sia-III was similar to that of ST8Sia-II, fetuin acts as a better acceptor (10-fold) than α1-acid glycoprotein in the case of ST8Sia-III, whereas the incorporation of sialic acids into fetuin was almost the same as that into α1-acid glycoprotein in the case of ST8Sia-II. Thus, the structure of oligosaccharides on glycoproteins acting as acceptors for ST8Sia-III is probably different from that in the case of ST8Sia-II. On the other hand, the substrate specificity of ST8Sia-III toward glycolipids was similar to that of ST8Sia-I (GD3 synthase) because both sialyltransferases were able to synthesize GD3

Table 1. Comparison of the acceptor substrate specificities of the three cloned α2-8-sialyltransferases (pmol/ml medium, h)

Acceptor	ST8Sia-III	ST8Sia-II (STX)	ST8Sia-I[a] (GD3 syntase)
Glycoproteins			
α1-Acid glycoprotein	7.8	7.6	0[b]
Asialo-α1-acid glycoprotein	0	0	0
Fetuin	92.1	8.0	0
Asialofetuin	0	0	0
Ovomucoid	1.7	1.3	0
Transferrin (bovine)	1.3	0.38	0
BSM	0	0	0
Glycolipids			
Lactosylceramide	0	0	0
GM3	2.1	0	0.18
GD3	0.86	0	0
GM1	0	0	0
GD1a	0	0	0
GD1b	0	0	0
GT1b	0	0	0
GQ1b	0	0	0
2,3-SPG	7.5	0	NT[c]
2,6-SPG	0	0	NT

[a] Human ST8Sia-I (GD3 synthase) expressed by Namalwa cells was used (Sasaki et al. 1994)
[b] 0 indicates values under 0.1 pmol/ml medium, h for mouse ST8Sia-II and -III, and values under 0.01 pmol/ml medium, h for human ST8Sia-I
[c] NT, not tested

from GM3. ST8Sia-III, but not ST8Sia-I, could synthesize GT3 from GD3. The kinetic properties of ST8Sia-III for 2,3-SPG, GM3, and GD3 are summarized in Table 2. The Vmax/Km values shown strongly indicate that 2,3-SPG is a much more suitable acceptor for ST8Sia-III than GM3 or GD3. In addition, the Vmax/Km values for fetuin indicate that ST8Sia-III is much more specific for complex-type N-linked oligosaccharides, which contain Siaα2,3Galβ1,4GlcNAc sequences (Yoshida et al. 1995a).

Preparation

A cDNA encoding a truncated form of mST8Sia-III, lacking the first 39 amino acids of the open reading frame, was inserted into a pcDSA vector (Kojima et al. 1995). The resulting plasmid was designated pcDSA-ST8Sia-III, and consisted of an IgM signal peptide sequence, a protein A IgG binding domain, and the truncated form of mST8Sia-III. COS-7 cells on a 100-mm φ plate were transiently transfected with 10 μg pcDSA-ST8Sia-III using the DEAE-dextran procedure and cultured. After 48 h of transfection, the culture medium was collected, and the protein A-ST8Sia-III expressed in the medium was adsorbed to IgG-Sepharose (15 μl resin/10 ml culture medium) at 4°C for 16 h. The resin was collected by centrifugation, washed three times with phosphate-buffered saline, suspended in 50 μl (final volume) of Dulbecco's modified eagle medium without fetal bovine serum, and used as the soluble enzyme (Yoshida et al. 1995a).

Table 2. Kinetic properties of ST8Sia-III

Acceptor	Km (mM)	Vmax (pmol/h, ml)	Vmax/Km
2,3-SPG	0.082	9.2	112.1
GM3	0.588	3.7	6.3
GD3	3.30	6.1	1.8
Fetuin[a]	0.020	424	21 200

[a] The numbers of α2-3-linked sialic acids on *N*-linked oligosaccharides (about 30 nmol/mg) were calculated from the difference between sialic acid residues in fetuin and those in α2-3-specific sialidase-treated fetuin, and the number of *O*-linked oligosaccharides (about 70 nmol/mg)

Biological Aspects

Tissue-Specific and Developmentally Regulated Expression of the ST8Sia-III Gene

The expression pattern of the ST8Sia-III gene is tissue- and stage-specific. The mST8Sia-III gene was expressed only in brain (6.7, 2.2, and 1.7 kb) and testis (3.7 kb). Its transcript was not detected until 14 days postcoitum, but it was highly expressed in the 20-day postcoitum fetal brain. The expression level decreased thereafter during brain development (Yoshida et al. 1995a). The hST8Sia-III gene was expressed in both fetal and adult brain (single 11-kb transcript), while the expression of the 5.5-kb transcript was restricted to fetal liver (Lee et al. 1998).

Genomic Organization and Expression of the mST8Sia-III Gene

The genomic mST8Sia-III gene is found to span about 8 kb and to be composed of only four exons. The mST8Sia-III gene is much smaller and its organization much simpler than other sialyltransferase genes so far reported. In particular, the sialyl motif L of mST8Sia-III is in one exon, while this motif is encoded by discrete exons in other sialyltransferases. The mST8Sia-III gene was highly expressed in the mouse brain and gave rise to at least three transcripts, which differed in the length of their 3′-untranslated regions through the alternative use of different polyadenylation sites. Although the promoter region lacked an apparent TATA or CCAAT box and potential regulatory motifs, a transfection experiment employing neuroblastoma cells expressing mST8Sia-III demonstrated that the minimal promoter activity was mapped in the proximal region 418 bp upstream from the ATG codon, which suggests the presence of tissue-specific enhancer elements (Yoshida et al. 1996).

Future Perspectives

To date, five species of α2,8-sialyltransferases have been cloned. ST8Sia-I and -V exhibit activity toward glycolipids (Sasaki et al. 1994; Kono et al. 1996). On the other hands, ST8Sia-II and -IV has been shown to transfer multiple α2,8-sialic acid residues onto α2,3-sialylated *N*-glycolated glycoproteins, particularly on the N-CAM, yielding polysialic acid chains (Yoshida et al. 1995b; Kojima et al. 1996). Only ST8Sia-III exhibits activity toward the Siaα2,3Galβ1,4GlcNAc sequences of N-linked oligos-accahrides and glycosphingolipids in vitro. Identification of the actual substrates for ST8Sia-III is necessary to understand its biological function. Generating knockout mice of these genes will also help to clarify the exact biological functions of these enzymes. Identification of transcription factors that are involved in tissue (brain and testis)-specific expression of this gene may facilitate our understanding of the mechanisms for tissue-specific expression.

Further Reading

Yoshida et al. (1995a) report on the cDNA cloning and substrate specificities of mouse ST8Sia-III, and the expression pattern of the ST6GalNAc III gene.

Yoshida et al. (1996) report the genomic structure of the mouse ST8Sia-III gene and an analysis of promoter activities.

Tsuji (1999) gives a short review of the characterization of sialyltransferases cloned before 1999.

References

Kojima N, Yoshida Y, Kurosawa N, Lee YC, Tsuji S (1995) Enzymatic activity of a develop-mentally regulated member of the sialyltransferase family (STX): evidence for α2,8-sialyltransferase activity toward N-linked oligosaccharides. FEBS Lett 360:1–4

Kojima N, Tachida Y, Yoshida Y, Tsuji S (1996) Characterization of mouse ST8Sia-II (STX) as a neural cell adhesion molecule-specific polysialic acid synthase. J Biol Chem 271:19457–19463

Kono M, Yoshida Y, Kojima N, Tsuji S (1996) Molecular cloning and expression of a fifth type of α2,8-sialyltransferases (ST8Sia-V). J Biol Chem 271:29366–29371

Lee YC, Kim YJ, Lee KY, Kim KS, Kim BU, Kim HN, Kim CH, Do SI (1998) Cloning and expression of cDNA for a human α2,3 Gal β1,4-GlcNAc α-2,8-sialyltransferase (hST8Sia-III). Arch Biochem Biophys 360:41–46

Sasaki K, Kurata K, Kojima N, Kurosawa N, Ohta S, Hanai N, Tsuji S, Nishi T (1994) Expression cloning of a GM3-specific α2,8-sialyltransferase (GD3 synthase). J Biol Chem 269:15950–15956

Tsuji S (1999) Molecular cloning and characterization of sialyltransferases. In: Inoue Y, Lee YC, Troy FA (eds) Sialobiology and other novel forms of glycosylation. Gakushin, Osaka, pp 145–154

Tsuji S, Datta AK, Paulson JC (1996) Systematic nomenclature for sialyltransferases. Glycobiology 6(7):v–vii

Yoshida Y, Kojima N, Kurosawa N, Hamamoto T, Tsuji S (1995a) Molecular cloning of Siaα2,3Galβ1,4GlcNAc α2,8-sialyltransferase from mouse brain. J Biol Chem 270:14628–14633

Yoshida Y, Kojima N, Tsuji S (1995b) Molecular cloning and characterization of a third type of N-glycan α2,8-sialyltransferase from mouse lung. J Biochem 118:658–664

Yoshida Y, Kurosawa N, Kanematsu T, Taguchi A, Arita M, Kojima N, Tsuji S (1996) Unique genomic structure and expression of the mouse α2,8-sialyltransferase (ST8Sia-III) gene. Glycobiology 6:573–580

48

ST8Sia-IV (PST-1)

Introduction

ST8Sia-IV belongs to a member of the sialyltransferase family and forms polysialic acid (PSA) mainly attached to neural cell adhesion molecules (N-CAM). PSA is a glycan composed of a linear homopolymer of $\alpha2,8$-linked sialic acid, which contains as many as 55 sialic acid residues per chain (Livingston et al. 1988). The expression of PSA in the brain dramatically decreases during development (Fig. 1), suggesting that this unique glycan is implicated in various neural events such as cell migration, neurite outgrowth, and neural plasticity by modulating the adhesive property of N-CAM (for reviews, see: Fryer and Hockfield 1996; Rutishauser and Landmesser 1996; Walsh and Doherty 1996; Kiss and Rougon 1997; Nakayama et al. 1998). In addition, PSA is also involved in the pathogenesis or metastatic potential of certain tumors such as neuroblastoma, small and nonsmall cell lung cancer, Wilm's tumor, and pancreatic cancer (Roth et al. 1988; Scheidegger et al. 1994; Hildebrandt et al. 1998; Kameda et al. 1999; Tanaka et al. 2000; Tanaka et al. 2001).

Recently, cDNA encoding ST8Sia-IV was cloned from five animal species: hamster, human, mouse, rat, and chicken (Eckhardt et al. 1995; Nakayama et al. 1995; Yoshida et al. 1995; Phillips et al. 1997; Bruses et al. 1998; Ong et al. 1998). The deduced amino acid sequence of ST8Sia-IV predicts that this enzyme is composed of 359 amino acid residues having a typical type II membrane topology. The distinctive sialyl motifs of S and L are found in the catalytic domain of the enzyme. ST8Sia-IV is highly conserved among these species; i.e., 97.8% identical between hamster and human, and 99.2% identical between hamster and mouse. In vitro polysialyltransferase assays

Jun Nakayama,[1] Kiyohiko Angata,[2] Misa Suzuki,[2] and Minoru Fukuda[2]

[1] Institutes of Organ Transplants, Reconstructive Medicine and Tissue Engineering, Shinshu University Graduate School of Medicine, and Central Clinical Laboratories, Shinshu University Hospital, Asahi 3-1-1, Matsumoto 390-8621, Japan
Tel. +81-263-37-2802; Fax +81-263-34-5316
e-mail: jun@hsp.md.shinshu-u.ac.jp
[2] Glycobiology Program, Burnham Institute, 10901 North Torrey Pines Road, La Jolla, CA 92037, USA

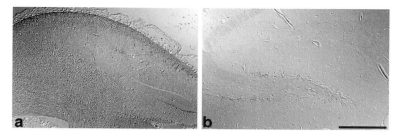

Fig. 1. Developmental expression of PSA in mouse brain. PSA is abundantly expressed in the midbrain at embryonic day 12 (**a**), whereas its expression level is significantly decreased in the adult brain, except for the hippocampus (**b**) and olfactory bulb. Immunostaining with anti-PSA antibody, 12F8; *bar* = 500 μm

using a soluble form of ST8Sia-IV demonstrated that PSA can actually be synthesized by a single enzyme, ST8Sia-IV alone (Nakayama and Fukuda 1996; Mühlenhoff et al. 1996a), or another polysialyltransferase, ST8Sia-II alone (Kojima et al. 1995; Angata et al. 1997). PSA formed by ST8Sia-IV is larger than that formed by ST8Sia-II, and PSA-formed by ST8Sia-IV facilitates neurite outgrowth better than PSA formed by ST8Sia-II (Angata et al. 1997). Moreover, the polysialylation of N-CAM is more efficiently facilitated in the presence of both enzymes, suggesting that ST8Sia-IV and ST8Sia-II synthesize PSA in a cooperative manner (Angata et al. 1998).

Databanks

ST8Sia-IV (PST-1)

No EC number has been allocated.

Species	Gene	Protein	mRNA	Genomic
Homo sapiens	STAT8D or PST	g945221	L41680	–
Mus musculus	Siat8d	e145119	X86000	Y09483-Y09488
		g1223771	AJ223955	–
			AJ223956	–
			NM_009183	–
Rattus norvegicus	–	g1899186	U90215	–
Cricetulus griseus	–	Q64690	Z46801	–
Gallus gallus	–	g2749960	AF008194	–

Name and History

ST8Sia-IV for hamster and human was originally designated PST-1 and PST, respectively. Other synonyms have not been used. The presence of PSA in the vertebrate brain was first described by Finne (1982), and the enzymatic properties of the polysialyltransferase responsible for the polysialylation of N-CAM were characterized by Oka et al. (1995). Regarding the cDNA cloning of ST8Sia-IV, hamster and human

ST8Sia-IV were isolated using the expression cloning approach (Eckhardt et al. 1995; Nakayama et al. 1995), and the mouse and rat homologues were cloned utilizing the PCR-based method (Yoshida et al. 1995; Phillips et al. 1997; Ong et al. 1998).

Enzyme Activity Assay and Substrate Specificity

The following reaction can be catalyzed by ST8Sia-IV:

$$n(CMP\text{-}NeuAc) + NeuAc\alpha2 \rightarrow 3(6)Gal\beta \rightarrow R \rightarrow$$
$$(NeuNAc\alpha2 \rightarrow 8NeuAc\alpha2)_n \rightarrow 3(6)Gal\beta \rightarrow R$$

The enzyme assay and acceptor specificity of human ST8Sia-IV was carried out using a soluble chimeric ST8Sia-IV fused with protein A (Angata et al. 1998). A mammalian expression vector, pcDNAI, encoding ST8Sia-IV fused with protein A was transfected into COS-1 cells. Then the fused enzyme, purified by IgG Sepharose, was incubated with Fc-N-CAM dissolved in 0.1 M cacodylate buffer (pH 6.0) containing 5 mM MnCl$_2$ and 2 mM CaCl$_2$, 1% Triton CF-54, and 2.4 mmol CMP-[^{14}C]NeuAc at 37°C for 18 h. Similarly, N-glycans derived from N-CAM, and synthetic oligosaccharides such as NeuAc$\alpha2 \rightarrow 3$Gal$\beta1 \rightarrow 4$GlcNAc$\beta1 \rightarrow$octyl were incubated with the recombinant enzyme (Angata et al. 2000). The results indicate that N-CAM is at least 50 times more efficient as an acceptor than N-glycans isolated from N-CAM. ST8Sia-IV and ST8Sia-II add PSA almost equally to NeuAc$\alpha2 \rightarrow 3$Gal$\beta1 \rightarrow 4$GlcNAc$\beta1 \rightarrow$octyl and NeuAc$\alpha2 \rightarrow 8$NeuAc$\alpha2 \rightarrow 3$Gal$\beta1 \rightarrow 4$GlcNAc$\beta1 \rightarrow$octyl, suggesting that a disialosyl structure (presumably synthesized by an "initiation enzyme") is not necessary for PSA synthesis (Angata et al. 2000). The unique intramolecular disulfide bonds found in ST8Sia-IV between the COOH terminus and sialylmotif L, and between the sialylmotif S and sialylmotif L are critical for the catalytic activity of this enzyme (Angata et al. 2001).

Preparation

ST8Sia-IV protein is available only in recombinant form. Here, we describe our cloning procedure for human ST8Sia-IV (PST) using an antibody specific for PSA (Fig. 2) (Nakayama et al. 1995). COS-1 cells express neither polysialic acid nor N-CAM. Thus, COS-1 cells were co-transfected by human fetal brain cDNA library constructed by pcDNAI and N-CAM cDNA. The transfected COS-1 cells were then incubated with monoclonal antibody 735 specific for PSA, and COS-1 cells expressing PSA were enriched by fluorescence-activated cell sorting. The plasmid DNA was then rescued from the sorted cells and amplified in bacteria, MC1061/P3 cells in the presence of ampicillin (Amp) and tetracycline (Tet). The pcDNAI encodes a *supF* gene that corrects the defects of Amp- and Tet-resistant genes in the P3 episome, while MC1061/P3 cells transformed by the N-CAM vector alone are resistant to Amp but not to Tet. Because of this difference, only plasmids derived from the library were selectively amplified, and several rounds of sibling selections allowed us to isolate a cDNA encoding ST8Sia-IV. The recombinant ST8Sia-IV has been expressed in COS-1 cells, HeLa cells (Nakayama et al. 1995; Angata et al. 1997), and CHO cells (Angata et al. 1998).

Fig. 2. Strategy of expression cloning of human ST8Sia-IV. COS-1 cells do not express PSA after transfection with N-CAM cDNA, pHβAPr-1-neo-N-CAM. Thus COS-1 cells were co-transfected with pcDNAI-based human fetal brain cDNA library and the N-CAM vector. After 48 h, polysialic acid-positive cells were isolated by cell sorting using PSA-specific antibody, 735. Plasmid DNA isolated from the sorted cells by the Hirt procedure was amplified in *E. coli* MC1061/P3 cells in the presence of ampicillin and tetracycline. Sibling selection was performed by dividing the plasmids into small pools, and they were separately co-transfected into COS-1 cells together with the N-CAM vector. The transfectants were then screened by immunofluorescence with the 735 antibody to isolate a single plasmid harboring a human ST8Sia-IV cDNA (From Nakayama et al. 1998, with permission)

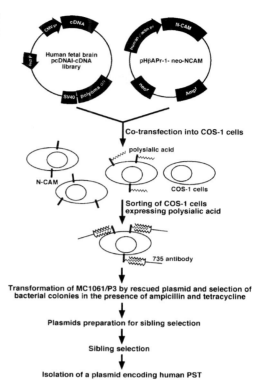

Co-transfection into COS-1 cells

polysialic acid

N-CAM

COS-1 cells

Sorting of COS-1 cells expressing polysialic acid

735 antibody

Transformation of MC1061/P3 by rescued plasmid and selection of bacterial colonies in the presence of ampicillin and tetracycline

↓

Plasmids preparation for sibling selection

↓

Sibling selection

↓

Isolation of a plasmid encoding human PST

Biological Aspects

Mouse ST8Sia-IV genes contain five exons and span over 55 kb (Eckhardt and Gerardy-Schahn 1998; Takashima et al. 1998). The promoter region lacks a TATA box, but Sp1 and NF-Y binding sites are found in this region. ST8Sia-IV itself can be poly-sialylated, and this process is called "autopolysialylation" (Mühlenhoff et al. 1996b). Polysialylation of N-CAM is more significantly facilitated by the polysialylated ST8Sia-IV than the nonpolysialylated ST8Sia-IV, but autopolysialylation is not necessarily required for enzyme activity (Close et al. 2000; Angata et al. 2001).

Regarding the tissue distribution of ST8Sia-IV in human tissues (Nakayama et al. 1995; Angata et al. 1997), the transcripts of this enzyme are strongly expressed in fetal brain, but its expression level is much lower in adult brain. As well as in the nervous system, ST8Sia-IV mRNA is expressed in nonneural tissues such as spleen, thymus, and heart. ST8Sia-IV mRNA is also detectable in human neuroblastoma (Fig. 3). The expression profile of ST8Sia-IV mRNA during mouse development was examined (Ong et al. 1998). The transcripts are strongly expressed at embryonic day 11 and gradually decline during development, but still remain in a moderate amount in the adult brain. The *ST8Sia-IV* gene is mapped to human chromosome 5, band q21 (Angata et al. 1997). For human diseases involving PSA, the expression of PSA in the

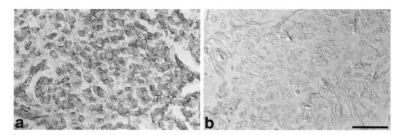

Fig. 3. Expression of ST8Sia-IV mRNA in human neuroblastoma. **a** Transcripts of human ST8Sia-IV are strongly expressed in the neuroblastoma cells by using an antisense probe. **b** No specific signals are detectable using a sense probe. In situ hybridization for human ST8Sia-IV; *bar* = 20 μm

hippocampus of schizophrenic patients decreased significantly compared with that of healthy individuals (Barbeau et al. 1995), but it is not clear whether the ST8Sia-IV gene is involved in the pathogenesis of the disease.

Future Perspectives

N-CAM-deficient mice exhibit a smaller olfactory bulb and hippocampus, and show defects in spatial learning and disruption in circadian rhythmicity (Tomasiewicz et al. 1993; Cremer et al. 1994; Shen et al. 1997), suggesting the involvement of PSA in (1) the migration of olfactory neurons from the anterior subventricular zone to the olfactory bulb, (2) learning and memory through hippocampal formation, and (3) the circadian clock function. Recent results on the mice deficient in ST8Sia-IV gene indicate the age-dependent impairment of long-term potentiation and long-term depression in Schaffer collateral-CA1 synapses of the hippocampus of the adult mutants (Eckhardt et al. 2000). It is important to determine how two polysialyltransferases, ST8Sia-IV and ST8Sia-II, contribute the biosynthesis of PSA in vivo. Transgenic or knockout mice of ST8Sia-IV and/or ST8Sia-II genes will address this problem.

References

Angata K, Nakayama J, Fredette B, Chong K, Ranscht B, Fukuda M (1997) Human STX polysialyltransferase forms the embryonic form of the neural cell adhesion molecule: tissue-specific expression, neurite outgrowth, and chromosomal localization in comparison with another polysialyltransferase, PST. J Biol Chem 272:7182–7190

Angata K, Suzuki M, Fukuda M (1998) Differential and cooperative polysialylation of the neural cell adhesion molecule by two polysialyltransferases, PST and STX. J Biol Chem 273:28524–28532

Angata K, Suzuki M, McAuliffe J, Ding Y, Hindsgaul O, Fukuda M (2000) Differential biosynthesis of polysialic acid on NCAM and oligosaccharide acceptors by three distinct α2,8-sialyltransferases, PST (ST8Sia-IV), STX (ST8Sia-II), and ST8Sia-III. J Biol Chem 275:18594–18601

Angata K, Yen TY, El-Battari A, Macher BA, Fukuda M (2001) Unique disulfide bond structures found in ST8Sia IV polysialyltransferase are required for its activity. J Biol Chem 276:15369–15377

Barbeau D, Liang JJ, Robitalille Y, Quirion R, Srivastava LK (1995) Decreased expression of the embryonic form of the neural cell adhesion molecule in schizophrenic brains. Proc Natl Acad Sci USA 92:2785–2789

Bruses JL, Rollins KG, Rutishauser U (1998) Chicken polysialyltransferase (PST) from embryonic brain. GenBank database; accession No. AF008194

Close BE, Tao K, Colley KJ (2000) Polysialyltransferase-1 autopolysialylation is not requisite for polysialylation of neural cell adhesion molecules. J Biol Chem 275:4484–4491

Cremer H, Lange R, Christoph A, Plomann M, Vopper G, Roes J, Brown R, Baldwin S, Kraemer P, Scheff S, Barthels D, Rajewsky K, Wille W (1994) Inactivation of the N-CAM gene in mice results in size reduction of the olfactory bulb and deficits in spatial learning. Nature 367:455–459

Eckhardt M, Bukalo O, Chazel G, Wang L, Goridis C, Schachner M, Gerardy-Schahn R, Cremer H, Dityatev A (2000) Mice deficient in the polysialyltransferase ST8SiaIV/PST-1 allow discrimination of the roles of neural cell adhesion molecule protein and polysialic acid in neural development and synaptic plasticity. J Neurosci 20:5234–5244

Eckhardt M, Gerardy-Schahn R (1998) Genomic organization of the murine polysialyltransferase gene ST8SiaIV (PST-1). Glycobiology 8:1165–1172

Eckhardt M, Mühlenhoff M, Bethe A, Koopman J, Frosch M, Gerardy-Schahn R (1995) Molecular characterization of eukaryotic polysialyltransferase-1. Nature 373:715–718

Finne J (1982) Occurrence of unique polysialosyl carbohydrate units in glycoproteins of developing brain. J Biol Chem 257:11966–11970

Fryer HJL, Hockfield S (1996) The role of polysialic acid and other carbohydrate polymers in neural structural plasticity. Curr Opin Neurobiol 6:113–118

Hildebrandt H, Becker C, Gluer S, Rosner H, Gerardy-Schahn R, Rahmann H (1988) Polysialic acid on the neural cell adhesion molecule correlates with expression of polysialyltransferases and promotes neuroblastoma cell growth. Cancer Res 58:779–784

Kameda K, Shimada H, Ishikawa T, Takimoto A, Momiyama N, Hasegawa S, Misuta K, Nakano A, Nagashima Y, Ichikawa Y (1999) Expression of highly polysialylated neural cell adhesion molecule in pancreatic cancer neural invasive lesion. Cancer Lett 137:201–207

Kiss JZ, Rougon G (1997) Cell biology of polysialic acid. Curr Opin Neurobiol 7:640–646

Kojima N, Yoshida Y, Tsuji S (1995) A developmentally regulated member of the sialyltransferase family (ST8Sia-II, STX) is a polysialic acid synthase. FEBS Lett 373:119–122

Livingston BD, Jacobs JL, Glick MC, Troy FA (1988) Extended polysialic acid chains (n > 55) in glycoproteins from human neuroblastoma cells. J Biol Chem 263:9443–9448

Mühlenhoff M, Eckhardt M, Bethe A, Frosch M, Gerardy-Schahn R (1996a) Polysialylation of N-CAM by a single enzyme. Curr Biol 6:1188–1191

Mühlenhoff M, Eckhardt M, Bethe A, Frosch M, Gerardy-Schahn R (1996b) Autocatalytic polysialylation of polysialyltransferase-1. EMBO J 15:6943–6950

Nakayama J, Fukuda M (1996) A human polysialyltransferase directs in vitro synthesis of polysialic acid. J Biol Chem 271:1829–1832

Nakayama J, Fukuda MN, Fredette B, Ranscht B, Fukuda M (1995) Expression cloning of a human polysialyltransferase that forms the polysialylated neural cell adhesion molecule present in embryonic brain. Proc Natl Acad Sci USA 92:7031–7035

Nakayama J, Angata K, Ong E, Katsuyama T, Fukuda M (1998) Polysialic acid: a unique glycan that is developmentally regulated by two polysialyltransferases, PST and STX, in the central nervous system. From biosynthesis to function. Pathol Int 48:665–677

Oka S, Bruses JL, Nelson RW, Rutishauser U (1995) Properties and developmental regulation of polysialyltransferase activity in the chicken embryo brain. J Biol Chem 270:19357–19363

Ong E, Nakayama J, Angata K, Reyes L, Katsuyama T, Arai Y, Fukuda M (1998) Developmental regulation of polysialic acid synthesis in mouse directed by two polysialyltransferases, PST and STX. Glycobiology 8:415–424

Phillips GR, Krushel LA, Crossin KL (1997) Developmental expression of two rat sialyl-transferases that modify the neural cell adhesion molecule, N-CAM. Brain Res Dev Brain Res 102:143–155

Roth J, Zuber C, Wagner P, Taatjes DJ, Weisgerber C, Heitz PU, Goridis C, Bitter-Suermann D (1988) Reexpression of poly(sialic acid) units of the neural cell adhesion molecule in Wilms tumor. Proc Natl Acad Sci USA 85:2999–3003

Rutishauser U, Landmesser L (1996) Polysialic acid in the vertebrate nervous system: a promoter of plasticity in cell–cell interactions. Trends Neurosci 19:422–427

Scheidegger EP, Lackie PM, Papay J, Roth J (1994) In vitro and in vivo growth of clonal sublines of human small cell lung carcinoma is modulated by polysialic acid of the neural cell adhesion molecule. Lab Invest 70:95–106

Shen H, Watanabe M, Tomasiewicz H, Rutishauser U, Magnuson T, Glass JD (1997) Role of neural cell adhesion molecule and polysialic acid in mouse circadian clock function. Neuroscience 17:5221–5229

Takashima S, Yoshida Y, Kanematsu T, Kojima N, Tsuji S (1998) Genomic structure and promoter activity of the mouse polysialic acid synthase (mST8Sia-IV/PST) gene. J Biol Chem 273:7675–7683

Tanaka F, Otake Y, Nakagawa T, Kawano Y, Miyahara R, Li M, Yanagihara K, Inui K, Oyanagi H, Yamada T, Nakayama J, Fujimoto I, Ikenaka K, Wada H (2001) Prognostic significance of polysialic acid expression in resected non-small cell lung cancer. Cancer Res 61:1666–1670

Tanaka F, Otake Y, Nakagawa T, Kawano Y, Miyahara R, Li M, Yanagihara K, Nakayama J, Fujimoto I, Ikenaka K, Wada H (2000) Expression of polysialic acid and STX, a human polysialyltransferase, are correlated with tumor progression in non-small cell lung cancer. Cancer Res 60:3072–3080

Tomasiewicz H, Ono K, Yee D, Thompson C, Goridis C, Rutishauser U, Magnuson T (1993) Genetic deletion of a neural cell adhesion molecule variant (N-CAM-180) produces distinct defects in the central nervous system. Neuron 11:1163–1174

Walsh FS, Doherty P (1996) Cell adhesion molecules and neuronal regeneration. Curr Opin Cell Biol 8:707–713

Yoshida Y, Kojima N, Tsuji S (1995) Molecular cloning and characterization of a third type of N-glycan α2,8-sialyltransferase from mouse lung. J Biochem 118:658–664

ST8Sia-V (SAT-V/SAT-III)

Introduction

ST8Sia-V was the fifth molecule cloned as an α2-8-sialyltransferase. This enzyme exists as three isoforms as ST8Sia-V-L, -M, -S in mice. The predicted amino acid sequences of mouse ST8Sia-V-L, -M, -S are 412, 376, and 345, respectively, all of which exhibit a type II transmembrane topology, consisting of an NH_2 terminal cytoplasmic tail, a transmembrane domain, a proline-rich stem region, and a large COOH-terminal active domain. The nucleotide sequences of the three isoforms are identical to each other except in the putative stem region (Kono et al. 1996).

The recombinant protein A-fused ST8Sia-V (-L, -M, -S) expressed in COS-7 cells exhibited an α2-8-sialyltransfarase activity toward GM1b, GD1a, GT1b, and GD3, respectively. The apparent Kms for GM1b, GD1a, GT1b, and GD3 were 1.1, 0.082, 0.070, and 0.28 mM, respectively. The mouse ST8Sia-V gene was mainly expressed in brain and the expression increased with the aging. Gene expression in the brain was first observed in 14 day-embryo, and then increased up to postnatal 49 day (Kono et al. 1996).

Mari Kono[1] and Shuichi Tsuji[2]

[1] Genetics of Development and Disease Branch, National Institutes of Health, Building 10/Room 9D-11, 10 Center DR MSC 1810, Bethesda, MD 20892-1810, USA
Tel. +1-301-496-6774; Fax +1-301-496-9878
e-mail: marikono@helix.nih.gov
[2] Department of Chemistry, Faculty of Science, Ochanomizu University, Otsuka, Bunkyo-ku, Tokyo 112-8610, Japan
Tel. +81-3-5978-5345; Fax +81-3-5978-5344
e-mail: stsuji@cc.ocha.ac.jp

Databanks

ST8Sia-V (SAT-V/SAT-III)

NC-IUBMB enzyme classification: E.C.2.4.99.8

Species	Gene	Protein	mRNA	Genomic
Homo sapiens	–	AAC51727.1	U91641	–
Mus musculus	*Siat8e*	CAA66642.1	X98014	–
		CAA66643.1	–	–
		CAA66644.1	–	–

Name and History

ST8Sia-V was the fifth molecule isolated as an α2-8-sialyltransferase (Tsuji et al. 1996). Its cDNA was cloned from mouse brain cDNA library using a PCR-based approach, which involved two degenerate oligonucleotide primer bases on two highly conserved regions, sialylmotifs L and S, of mouse ST8Sia-I and mouse ST8Sia-III. In 1997, Kim et al. isolated ST8Sia-V cDNA from human brain.

Enzyme Activity Assay and Substrate Specificity

Sialyltransferase Assay

Sialyltransferase activity was measured in the presence of 0.1 M sodium cacodylate buffer (pH 6.0), 10 mM $MgCl_2$, 2 mM $CaCl_2$, 0.3% Triton CF-54, 100 μM CMP-[^{14}C]NeuAc (3.9 Bq), various amounts of glycolipids, and 2 μl enzyme preparation in a total volume of 10 μl (Kono et al. 1996). The enzyme reaction was performed within the time that the reaction proceeded linearly. After 4 h incubation at 37°C, the reaction was terminated by the addition of 200 μl PBS. For the separation of [^{14}C]NeuAc conjugated glycolipids and CMP-[^{14}C]NeuAc, the reaction mixtures were applied to C-18 columns (Sep-Pak Vac, 100 mg; Waters, Milford, MA, USA) which had been washed with water and 0.1 M KCl. The glycolipids were eluted from the column with methanol, dried, and then subjected high-performance thin-layer chromatography (HPTLC) with solvent systems of chloroform/methanol/0.02% $CaCl_2$ (55:45:10) and n-propanol/28% ammonia solution/water (75:5:25). Acceptor substrates were visualized by staining with a resorcinol/HCl reagent. The radioactive materials in glycolipids were visualized with a BAS2000 radio image analyzer (Fuji Film, Tokyo, Japan), and the radioactivity incorporated into the acceptor was counted.

Acceptor Substrate Specificity

The acceptor substrate specificity of ST8Sia-V was compared with those of ST8Sia-I and ST8Sia-III, which exhibit activity toward α2-3-sialylated glycosphingolipids (see Table 1 in the chapter 47 ST8Sia-III). The soluble ST8Sia-I, -III, and -V showed broad activities toward most gangliosides tested. However, the Km value for each substrate was characteristic for each enzyme. The Km values of ST8Sia-V for GD1a and GT1b were very low (70–80 μM), that for GD3 was medium (about 300 μM), and those for

GM1b and 2,3-SPG were relatively high (over 1000 μM), indicating that ST8Sia-V exhibits a substrate preference for GD1a and GT1b rather than GD3, 2,3-SPG, or GM1b. On the other hand, the Km values of ST8Sia-I for GD1a, GT1b, GD3, and 2,3-SPG were very high (2–5 mM) compared with that of GM3 (30 μM). ST8Sia-III exhibited the highest activity toward 2,3-SPG and only very low activity toward GD1a and GT1b. The V_{max}/Km values indicated that GT1b and GD1a served as much better acceptors for ST8Sia-V than ST8Sia-I or III. In addition, the apparent Km values of the three sialyltransferases for GD3 (280 μM for ST8Sia-V, and 3–5 mM for ST8Sia-I and -III) indicated that ST8Sia-V is a candidate for GT3 synthase.

Preparation

Truncated forms of ST8Sia-V-L, -M, and -S lacking the first 35 amino acids of the open reading frame were prepared by PCR amplification using three kinds of ST8Sia-V isoforms as templates with 5′- and 3′-primers containing an *XhoI* site. The resulting amplified and digested 1161-bp, 1053-bp, and 960-bp *XhoI* fragments were inserted into the *XhoI* site of the expression vector pcDSA. The resulting plasmids consisted of the IgM signal peptide sequence, the protein A IgG binding domain, and the truncated forms of ST8Sia-V-L, -M, and -S, respectively. Each expression plasmid (10 μg) was transiently transfected into COS-7 cells on a 100-mm plate by the DEAE-dextran method. After 5 h or 24 h transfection, the culture medium was changed to DMEM containing 2% FCS or macrophage-SFM (GIBCO), respectively. After 48–72 h transfection, the cell culture medium was collected and the protein A-ST8Sia-V expressed in the medium was adsorbed to an IgG-sepharose gel (Pharmacia; 25 μl resin/50 ml culture medium) at 4°C for 16 h. The IgG-Sepharose gel and enzyme fused with protein A complex was used as the enzyme source.

Biological Aspects

Ganglioside Biosynthesis and ST8Sia-V

Ganglioside biosynthesis takes place in the Golgi apparatus, where starting with Glc-Cer, it progresses with the sequential addition of Gal, GalNAc, and Sia to the growing oligosaccharide chain. These reactions are catalyzed by specific glycosyltransferases. Many of these enzyme activities have been studied and partially characterized in rat liver Golgi membrane fractions. These studies suggested a model for ganglioside biosynthesis (Fig. 1, modified from Iber et al. 1991, 1992; Pohlentz et al. 1988). GM3, GD3, and GT3 synthase activities (SAT-I, -II, and -III) can be discriminated on the basis of their enzymatic characteristics. However, a competition-based assay suggested that all galactosyltransferase-II (GalT-II), GalNAc-transferase (GalNAcT), SAT-IV, and -V activities are due to a sole specific enzyme. In addition, SAT-V was reported to synthesize GT3 from GD3. In this model, three α2-8-sialyltransferases are involved in ganglioside biosynthesis. Comparisons of the kinetic parameters of mouse ST8Sia-I, -III, and -V suggested that ST8Sia-V is a most probable candidate for SAT-V and a probable candidate for SAT-III.

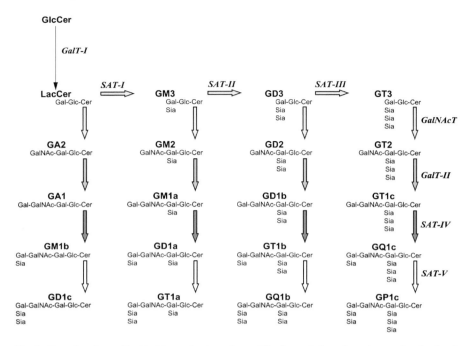

Fig. 1. Postulated ganglioside biosynthesis pathway. The figure is based on the results obtained from rat Golgi membrane fractions. (Modified from Iber et al. 1991, 1992; Pohlentz et al. 1988)

Tissue-Specific Expression and Increase Accompanied by Aging of ST8Sia-V Gene

In adult mouse tissues, a strong signal corresponding to 2.3 kb was observed in brain, and a weak signal of the same size was observed in testis (the probe used in Northern blot analysis can detect three isoforms). The signals for the ST8Sia-V gene were barely detectable in other tissues by Northern blot analysis. However, reocroe transcriptase (RT)-PCR analysis indicated that transcripts were expressed in mouse liver, lung, placenta, and spleen, in addition to brain and testis. Gene expression in the brain was first observed in 14-day embryos, and it then increased up to postnatal day 49.

Population of Three Isoforms

Several overlapping clones were obtained and sequenced. Three isoforms of different lengths were cloned (ST8Sia-V-L, -M, -S). Of nine clones, six were ST8Sia-V-M, two were ST8Sia-V-L, and only one was ST8Sia-V-S.

Future Perspectives

ST8Sia-I, -III, and -V exhibit activity toward glycolipids (Sasaki et al. 1994; Nara et al. 1994; Haraguchi et al. 1994; Yoshida et al. 1995; Kono et al. 1996), and kinetic analyses suggested that they are candidates for GD3 synthase (Fig. 1, SAT-II), Siaα2-3Galβ1-4GlcNAc α2-8-sialyltransferase, and GQ1b/GT1a/GD1c/GT3 synthase (Fig. 1, SAT-V/SAT-III), respectively. However, the substrate specificities of the cloned α2-8-sialyltransferases were broadly overlapping in vitro. To confirm the in vivo substrate specificities of the cloned enzymes, it is important to determine their expressions in vivo and the localization of the cloned enzymes in the Golgi networks. Generating knockout mice of these genes will also help to clarify the exact biological functions of these enzymes.

Further Reading

Tsuji S (1999) Molecular cloning and characterization of sialyltransferases. In: Inoue Y, Lee YC, Troy FA (eds) Sialobiology and other novel forms of glycosylation. Gakushin, Osaka, pp 145–154. This is a short review of the characterization of sialyltransferases cloned before 1999.

References

Haraguchi M, Yamashiro S, Yamamoto A, Furukawa K, Takamiya K, Lloyd KO, Shiku H, Furukawa K (1994), Nucleotide isolation of GD3 synthase gene by expression cloning of GM3 α-2,8-sialyltransferase cDNA using anti-GD2 monoclonal antibody. Proc Natl Acad Sci USA 25:10455–10459

Iber H, van Echten G, Sandhoff K (1991) Substrate specificity of α2-3-sialyltransferases in ganglioside biosynthesis of rat liver Golgi. Eur J Biochem 195:115–120

Iber H, Zacharias C, Sandhoff K (1992) The c-series gangliosides GT3, GT2, and GP1c are formed in rat liver Golgi by the same set of glycosyltransferases that catalyse the biosynthesis of asialo-, a- and b-series gangliosides. Glycobiology 2:137–142

Kim YJ, Kim KS, Do S, Kim CH, Kim SK, Lee YC (1997) Molecular cloning and expression of human α2,8-sialyltransferase (hST8Sia-V). Biochem Biophys Res Commun 18:327–330

Kono M, Yoshida Y, Kojima N, Tsuji S (1996) Molecular cloning and expression of fifth type of α2,8-sialyltransferase (ST8Sia-V). J Biol Chem 271:29366–29371

Nara K, Watanabe Y, Maruyama K, Kasahara K, Nagai Y, Sanai Y (1994) Expression cloning of a CMP-NeuAc:NeuAc α 2-3Gal β 1-4Glc β 1-1′Cer α2,8-sialyltransferase (GD3 synthase) from human melanoma cells. Proc Natl Acad Sci USA 16:7952–7956

Pohlentz G, Klein D, Schwarzmann G, Schmitz D, Sandhoff K (1988) Both GA2, GM2, and GD2 synthases and GM1b, GD1a, and GT1b synthases are single enzymes in Golgi vesicles from rat liver. Proc Natl Acad Sci USA 85:7044–7048

Sasaki K, Kurata K, Kojima N, Kurosawa N, Ohta S, Hanai N, Tsuji S, Nishi T (1994) Expression cloning of a GM3-specific α2,8-sialyltransferase (GD3 synthase). J Biol Chem 269:15950–15956

Tsuji S, Datta AK, Paulson JC (1996) Systematic nomenclature for sialyltransferases. Glycobiology 6(7):v–vii

Yoshida Y, Kojima N, Kurosawa N, Hamamoto T, Tsuji S (1995) Molecular cloning of Siaα2,3Galβ1,4GlcNAc α2,8-sialyltransferase from mouse brain. J Biol Chem 270:14628–14633

50

CMP-NeuAc Hydroxylase

Introduction

Cytidine monophosphate-*N*-acetylneuraminic acid hydroxylase (CMP-NeuAc hydroxylase) participates in the hydroxylation reaction of CMP-NeuAc to form CMP-*N*-glycolylneuraminic acid (CMP-NeuGc). The hydroxylation is carried out by an enzyme complex composed of cytochrome b5, NADH cytochrome b5 reductase, and CMP-NeuAc hydroxylase in the presence of NADH. Other characteristic features of the hydroxylation are that (1) the reaction is a rate-limiting step for the expression of NeuGc; (2) CMP-NeuAc hydroxylase is a soluble and cytosolic protein, but the other enzymes, cytochrome b5 and NADH cytochrome b5 reductase, are endoplasmic reticulum (ER) membrane-bound proteins; and (3) the expression of NeuGc is tissue- and species-specific, and both phenotypes are regulated by CMP-NeuAc hydroxylase.

Databanks

CMP-NeuAc hydroxylase

NC-IUBMB enzyme classification: E.C.1.14.13.45

Species	Gene	Protein	mRNA	Genomic
Homo sapiens	–	–	D86324	AB009668
			AF074480	
Pan troglodytes	–	–	AF074481	–
Mus musculus	–	–	D21826	–

Akemi Suzuki

RIKEN Frontier Research System, 2-1 Hirosawa, Wako-shi, Saitama 351-0198, Japan
Tel. +81-48-467-9615; Fax +81-48-462-4692
e-mail: aksuzuki@postman.riken.go.jp

Name and History

The conversion of NeuAc to NeuGc was suggested by a finding that radiolabeled precursor, ManNAc, was incorporated first into NeuAc and then into NeuGc during an incubation with sliced porcine submandibular gland. Further extensive studies by Schauer's group (Corfield and Schauer 1982) suggested three possible pathways for the biosynthesis of NeuGc: hydroxylation of free NeuAc, CMP-NeuAc, and glycoconjugate-bound NeuAc. The enzyme responsible for the hydxoylation was called monooxygenase because the oxygen of the hydroxyl group comes from O_2 in incubation mixtures. The name, hydroxylase, has been used more frequently. An important finding by Shaw and Schauer (1988) was that the major route is hydroxylation of CMP-NeuAc but not of free NeuAc or NeuAc of glycoconjugates. The second important finding was the involvement of cytochrome b5 in the hydroxylation of CMP-NeuAc (Kozutsimi et al. 1990), leading to successful purification of CMP-NeuAc hydroxylase. Enzyme purification became possible only when we stood on the concept that the hydroxylation is a complex reaction carried out by multiple enzymes including the cytochrome b5 electron transport system (Kawano et al. 1994). It is well known that normal human tissues do not contain NeuGc as a component of glycan chains of glycoconjugates, but the reason for this was not discovered until 1999. cDNA cloning of mouse CMP-NeuAc hydroxylase and then of human homolog revealed that humans have a 92-bp deletion in the hydroxylase gene, which results in production of a protein with no catalytic activities (Kawano et al. 1995; Chou et al. 1998; Irie et al. 1998). The chimpanzee has an intact hydroxylase gene, so the deletion occurred after divergence of the human ancestor from the chimpanzee (Chou et al. 1998).

Enzyme Activity Assay and Substrate Specificity

As mentioned, the hydroxylation requires CMP-NeuAc hydroxylase, cytochrome b5, and NADH cytochrome b5 reductase in the presence of NADH. Therefore, an in vitro assay of the hydroxylation activity requires that all these components be present. The presence of excess amounts of cytochrome b5 and cytochrome b5 reductase in assay mixtures is critical to quantitate CMP-NeuAc hydroxylase activity. CMP-NeuAc hydroxylase is a soluble and cytosolic protein; and the substrate CMP-NeuAc is a soluble substance as well. Thus, soluble forms of cytochrome b5 and cytochrome b5 reductase are ideal for the assay, but they should be purified from erythrocyte lysate or be prepared as recombinant proteins or by tryptic digestion of membrane-bound forms.

A convenient method for preparing these two factors is to obtain a detergent-soluble fraction of microsomes from cells or tissues. Mouse liver contains a reasonably large amount of the hydroxylase in the cytosolic fraction and at the same time is a good source for the preparation of cytochrome b5 and cytochrome b5 reductase from the microsome. A pelleted microsome obtained from 1 g of mouse liver was suspended with 1 ml of 10 mM Tris-HCl buffer pH 7.5 and 1 mM dithiothneitol (DTT). The cytosolic and microsomal fractions were dialyzed against 10 mM Tris-HCl buffer pH 7.5 and 1 mM DTT. The microsomal fraction was adjusted to 1% Triton X-100 solution, and centrifuged at 100000 g for 1 h. Aliquots of supernatants of cytosolic and microsomal fractions were kept at −60°C.

Incubation mixtures contain CMP-NeuAc (2 nmol), NADH (34 nmol), 10 mM Tris-HCl buffer pH 7.5, 1 mM DTT, the dialyzed cytosolic fraction (up to 100 μg protein), the dialyzed and Triton-X 100-solubilized microsomal fraction (<25 μl), and 0.5% Triton X-100 as a final concentration in a final volume of 50 μl (Kozutsumi et al. 1991; Takematsu et al. 1994). Incubation is performed at 37°C for 1 h. The reaction was stopped by mixing with 0.3 ml of ice-cold ethanol. After 15 min on ice, the incubation mixtures were centrifuged at 12 000 g for 5 min at 4°C. An aliquot of supernatant was analyzed by high-performance liquid chromatography (HPLC) on an ODS column (4.6 × 250 mm) with 50 mM ammonium phosphate at a flow rate of 0.5 ml/min. CMP-sialic acids were monitored at 271 nm (Kozutsumi et al. 1990).

The hydroxylation is a reaction sensitive to salt concentration. KCl, at 50 mM, reduces the activity to 20%. The hydroxylase takes CMP-NeuAc as a substrate but not free NeuAc or glycoconjugate containing NeuAc.

Preparation

CMP-NeuAc hydroxylase is a soluble enzyme that does not have a hydrophobic membrane spanning domain. It is considered to be a simple, single polypeptide based on its molecular mass determined by sodium dodecyl sulfate-polyacrylamide gel electrophoresis (SDS-PAGE) (64 kDa), gel filtration (60 kDa), and calculation from an amino acid sequence (66 kDa) deduced from a cDNA sequence (Kawano et al. 1994, 1995). Therefore, routine methods for preparing cytosolic and soluble proteins can be applied to preparation of the hydroxylase. Good enzyme sources are mouse liver, as mentioned above, and pig submandibular glands (Schlenzka et al. 1994). The enzyme has been expressed in COS cells as a native form (Koyama et al. 1996) and a recombinant with FLAG-tag (Irie et al. 1998). The hydroxylation reaction in vitro is rather sensitive to the salt concentration and other conditions we cannot identify. Therefore, salts and other low-molecular-weight materials should be removed from enzyme solutions by dialysis or with a molecular cutoff membrane.

Biological Aspects

CMP-NeuAc hydroxylation is the biosynthetic step that can control the expression of NeuGc in a critical way. Humans lost the expression of NeuGc-containing glyco-chains owing to the loss of CMP-NeuAc hydroxylase activity produced by a 92-bp deletion from the genome, probably during an early stage of establishing the species. We speculate that we changed the susceptibility to several infectious diseases produced by *Escheiechia coli* K99 and viruses including influenza. At the same time we lost a family of carbohydrate chains containing NeuGc, probably about 100 structures. Siglecs recognize structures containing sialic acid at the terminus, and recognition through Siglecs and their lignads must be modified with or without NeuGc. We do not yet understand what roles this deletion plays in the evolution and survival of our species.

CMP-NeuAc hydroxylase gene-targeted mice were established. The mice lost expression of NeuGc in various tissues, but at present abnormal phenotypes are not found.

The highest concentration of sialic acid is found in the brain of most mammals, and sialic acid is composed of NeuAc and a small amount of NeuGc in all mammals except humans. Obiously, suppression of CMP-NeuAc hydroxylase transcription is observed in the brain of mice; but, surprisingly, this suppression is conserved even in humans who have lost the intact hydroxylase gene. Peripheral nerves of mammals (except humans) contain NeuGc, an interesting difference.

Future Perspectives

Humans produce an anti-NeuGc antibody called the H-D antibody when treated with sera prepared from horse, goat, and sheep because these sera contain NeuGc-glycoconjugates and NeuGc is a foreign substance to humans. Antibody production occurs when recombinant proteins produced by cells expressing CMP-NeuAc hydroxylase are used for clinical treatment or if xenotransplantation from pig to human is performed. Several approaches to make cells or pig NeuGc-negative have been tested, and clinical trials will be one of our future projects.

Several human cancers were reported to have detectable amounts of NeuGc. If CMP-NeuAc hydroxylation is the only biosynthetic pathway, how can we explain this result? Production of active CMP-NeuAc hydroxylase in cancer tissues of humans is impossible because of the nature of the defect in the genome. The notion that the major pathway of NeuGc production is CMP-NeuAc is also true because humans do not express NeuGc in normal tissues. One possibility is that there are minor pathways for hydroxylation of free NeuAc or NeuAc-containing glycoconjugates in humans, and one of these two pathways is activated in cancer tissues. One paper reported hydroxylase activity of free NeuAc (Mukuria 1995). This possibility should be subjected to extensive research.

References

Chou H-H, Takematsu H, Diaz S, Iber J, Nickerson E, Wright K, Muchmore EA, Nelson DL, Warren ST, Varki A (1998) A mutation in human CMP-sialic acid hydroxylase occurred after Homo-Pan divergence. Proc Natl Acad Sci USA 95:11751–11756

Corfield AP, Schauer R (1982) Metabolism of sialic acid. In: Schauer R (ed) Sialic acids, chemistry, metabolism and function. Springer Berlin Heidelberg, New York, pp 196–261

Irie A, Koyama S, Kozutsumi Y, Kawasaki T, Suzuki A (1998) The molecular basis for the absence of N-glycolylneuraminic acid in humans. J Biol Chem 273:15866–15871

Kawano T, Koyama S, Takematsu H, Kozutsumi Y, Kawasaki H, Kawashima S, Kawasaki T, Suzuki A (1995) Molecular cloning of cytidine-monophosho-N-acetylneuraminic acid hydroxylase: regulation of species- and tissue-specific expression of N-glycolylneuraminic acid. J Biol Chem 270:16458–16463

Kawano T, Kozutsumi K, Kawasaki T, Suzuki A (1994) Biosynthesis of N-glycolylneuraminic acid-containing glycoconjugates: purification and characterization of the key enzyme of the cytidine monophospho-N-acetylneuraminic acid hydroxylation system. J Biol Chem 269:9024–9029

Koyama S, Yamaji T, Takematsu H, Kawano T, Kozutsumi Y, Suzuki A, Kawasaki T (1996) A naturally occurring 46-amino acid deletion of cytidine monophospho-N-acetylneuraminic acid hydroxylase leads to a change in the intracellular distribution of the protein. Glycoconjugate J 13:353–358

Kozutsumi Y, Kawano T, Kawasaki H, Suzuki K, Yamakawa T, Suzuki A (1991) Reconstitution of CMP-N-acetylneuraminic acid hydroxylation activity using a mouse liver cytosol fraction and soluble cytochrome b5 purified from horse erythrocytes. J Biochem 110:429–435

Kozutsumi Y, Kawano T, Yamakawa T, Suzuki A (1990) Participation of cytochrome b5 in CMP-N-acetylnerminic acid hydroxylation in mouse liver cytosol. J Biochem 108:704–706

Mukuria CJ, Mwangi WD, Noguchi A, Waiyaki GP, Asano T, Naiki M (1995) Evidence for a free N-acetylneuraminic acid-hydroxylating enzyme in pig mandibular gland soluble fraction. Biochem J 305:459–464

Schlenzka W, Shaw L, Schneckenburger P, Schauer R (1994) Purification and characterization of CMP-N-acetylneuraminic acid hydroxylase from pig submandibular glands. Glycobiology 4:675–683

Shaw L, Schauer R (1988) Detection of CMP-N-acetylneuraminic acid hydroxylase activity in fractionated mouse liver. Biochem J 263:355–364

Takematsu H, Kawano T, Koyama S, Kozutsumi Y, Suzuki A, Kawasaki T (1994) Reaction mechanism underlying CMP-N-acetylneuraminic acid hydroxylation in mouse liver: formation of a ternary complex of cytochrome b5, CMP-N-acetylneuraminic acid, and a hydroxylation enzyme. J Biochem 115:381–386

Glucuronyltransferases

HNK-1 Glucuronyltransferase

Introduction

The HNK-1 carbohydrate epitope, recognized by the monoclonal antibody HNK-1 (Abo and Balch 1981), is characteristically expressed on a series of cell adhesion molecules, including neural cell adhesion molecule (NCAM), myelin-associated glycoprotein (MAG), L1, P0, telencephalin, and others (McGarry et al. 1983; Kruse et al. 1984; Bollensen and Schachner 1987; Yoshihara et al. 1994), and on some glycolipids (Ilyas et al. 1984) in the nervous system. The HNK-1 epitope is spatially and temporally regulated during development of the nervous system (Schwarting et al. 1987); and characteristic expression of this epitope is observed in migrating neural crest cells (Bronner-Fraser 1986), rhombomeres (Kuratani 1991), and cerebellum (Eisenman and Hawkes 1993). The structure of the HNK-1 epitope is demonstrated to be the sulfated trisaccharide SO_4-3GlcAβ1-3Galβ1-4GlcNAc, which is shared with glycolipid and glycoprotein epitope (Chou et al. 1986; Ariga et al. 1987; Voshol et al. 1996). The HNK-1 epitope associates with neural crest cell migration (Bronner-Fraser 1987), neuron to glial cell adhesion (Keilhauer et al. 1985), outgrowth of astrocytic processes and migration of the cell body (Künemund et al. 1988), as well as the preferential outgrowth of neurites from motor neurons (Martini et al. 1992). These lines of evidence indicate that the HNK-1 epitope plays important roles in cell–cell and cell–substrate interaction during development of the nervous system.

Shogo Oka and Toshisuke Kawasaki

Department of Biological Chemistry and CREST (Core Research for Educational Science and Technology) Project, Japan Science and Technology Corporation, Graduate School of Pharmaceutical Sciences, Kyoto University, Kyoto 606-8501, Japan
Tel. +81-75-753-4572; Fax +81-75-753-4605
e-mail: kawasaki@pharm.kyoto-u.ac.jp

Databanks

HNK-1 glucuronyltransferase

No EC number has been allocated

Species	Gene	Protein	mRNA	Genomic
Homo sapiens	*GlcAT-P, B3GAT1*	–	AB029396	–
Rattus norvegicus	*GlcAT-P*	–	D88035	–
	GlcAT-S	–	AB010441	–
	GlcAT-D	–	AF106624	–

Name and History

The monoclonal antibody HNK-1 was raised against a membrane fraction of the human HSB-2 T-cell line. The antigen was originally noted as a maker of *h*uman *n*atural *k*iller (HNK) cells (Abo and Balch 1981) and is called CD57 in immunology. The HNK-1 carbohydrate antigen appears to be highly immunogenic because several monoclonal antibodies such as L2 (Kruse et al. 1984), NC-1 (Tucker et al. 1984), 4F4 (Schwarting et al. 1987), VC1.1 (Naegele and Barnstable 1991), M6749 (Obata and Tanaka, 1988), and others with specificities similar to those of the HNK-1 antibody have been obtained. Furthermore, the antigen attracted attention as an autoantigen involved in peripheral demyelinative neuropathy because the HNK-1 carbohydrate also reacts with immunoglobulin M (IgM) M-proteins from patients with plasma cell abnormalities, or paraproteinemia (Ilyas et al. 1984; Shy et al. 1984).

The characteristic structural feature of this epitope is the sulfoglucuronyl residue because the inner structure, Gal β1-4 GlcNAc, is found commonly in various glycoproteins and glycolipids, suggesting that glucuronyltransferase(s) and sulfotransferase(s) are key enzymes in the biosynthesis. Several years ago we and others demonstrated HNK-1-associated glucuronyltransferase activity using neolactotetraosylceramide as a glycolipid acceptor in chicken and rat brain (Chou et al. 1991; Das et al. 1991; Kawashima et al. 1992). A little later we also found glucuronyltransferase activity using asialoorosomucoid (ASOR) as a glycoprotein acceptor in rat brain (Oka et al. 1992). Using these assay systems we demonstrated that respective glucuronyltransferases are involved in biosynthesis of the HNK-1 epitope on glycoproteins (GlcAT-P) and on glycolipids (GlcAT-L) (Kawashima et al. 1992; Oka et al. 1992). We have purified and cloned GlcAT-P from rat brain (Terayama et al. 1997, 1998). Using GlcAT-P cDNA, we and others have cloned a second glucuronyltransferase (GlcAT-S) (Seiki et al. 1999; Shimoda et al. 1999) and sulfotransferase (Sulfo-T) (Bakker et al. 1997; Ong et al. 1998) responsible for the HNK-1 carbohydrate biosynthesis.

Enzyme Activity Assay and Substrate Specificity

UDP-GlcA + Galβ1-4GlcNAc-R → GlcAβ1-3Galβ1-4GlcNAc-R

Glucuronyltransferase Activity Toward Glycoprotein Acceptors

Glucuronyltransferase activity toward glycoprotein acceptors was measured essentially as described previously (Oka et al. 1992; Terayama et al. 1998) with slight modification. Incubation was carried out at 37°C for 3h in an assay mixture comprising 20 µg ASOR, 100 µM UDP-[^{14}C]-GlcA (2×10^5 dmp), 200mM MES buffer pH 6.5, 20 mM MnCl$_2$, 0.5 mM ATP, 0.2% (v/v) Nonidet P-40, and 2 µl of a 2% NP-40 extract of rat forebrain that had been treated at 100°C for 3min, in a final volume of 50 µl. After incubation the assay mixture was spotted onto a 2.5-cm Whatman no. 1 disk. The disks were washed with a 10% (w/v) trichloroacetic acid solution three times, followed by ethanol/ether (2:1, v/v) and then ether. The disks were then air-dried, and the radioactivity of [^{14}C]-GlcA-ASOR on the disks was counted with a liquid scintillation counter.

The enzyme transferred glucuronic acid to various glycoprotein acceptors bearing a terminal N-acetyllactosamine structure such as the asialoorosomucoid, asialofetuin, and asialo-neural cell adhesion molecules (Terayama et al. 1998). This enzyme is highly specific for the terminal N-acetyllactosamine structure of the glycoprotein acceptors because the enzymatic activity is almost completely inhibited by N-acetyllactosamine (Galβ1-4GlcNAc), whereas lacto-N-biose (Galβ1-3GlcNAc) has no effect on the activity (Oka et al. 1992; Terayama et al. 1998). The pH optimum of the enzyme is at 6.5. Divalent cations are essential for the enzymatic activity, and Mn^{2+} at 20 mM gives the highest activity. The activity is completely abolished in the presence of EDTA or 10 mM UDP (Terayama et al. 1998). The primary structure of GlcAT-P revealed that the enzyme has two cysteine residues: one (cys-70) in the stem region and the other (cys-317) in the catalytic region (Terayama et al. 1997). The SH-blocking reagent, N-ethylmaleimide, completely inhibited a soluble form of the GlcAT-P, which is replaced with the first 74 amino acids with protein A, indicating that a free cysteine residue at 317 is associated the catalytic activity of the enzyme (Terayama et al. 1997). During purification of GlcAT-P from rat brain, we noted that the enzyme had lost most of its catalytic activity, although addition of a lipid of a lipid extract of rat brain to the assay mixture restored the activity completely. Further studies indicated that sphingomyelin (SM) is specifically required for expression of the enzyme activity. As for the length of the acyl group, stearoyl-SM (18:0) was most effective, followed by palmitoyl-SM (16:0) and lignoceroyl-SM (24:0). More interestingly, activity was demonstrated only for SM with a saturated fatty acid (i.e., not for that with an unsaturated fatty acid, regardless of the length of the acyl group) (Terayama et al. 1998).

Glucuronyltransferase Activity Toward Glycolipid Acceptors

Glucuronyltransferase activity toward glycolipid acceptors was measured essentially as described previously (Kawashima et al. 1992). The standard assay mixture contained the following components in a total volume of 50 µl: 150 µM nLc-PA14 or 7.5 µg nLc-Cer, 100 µM UDP-[^{14}C] GlcA (2×10^5 cpm), 200mM MES buffer pH 4.5 or 80 mM sodium cacodylate buffer pH 6.0, 10 mM MnCl$_2$, 10 mM ATP, and 0.4% (v/v) NP-40. These reaction mixtures were incubated for 3h at 37°C, and the reactions were terminated by adding 20 volumes of chloroform-methanol (2:1, v/v). The radioactive reaction products were separated from labeled precursors according to the procedure of Fishman et al. (1976). The reaction mixture was applied to a column (0.5 × 5.0 cm) of Sephadex G-25 that had been equilibrated with chloroform-methanol-water (60 : 30 : 4.5). The column was then washed with 5 ml of the same solvent, and the effluent and washings were collected in a scintillation vial and evaporated to dryness by gentle warming. The radioactivity of the sample was counted using a liquid scintillation counter. The glucuronyltransferase activity toward glycolipid acceptors exhibits a dual optimum: at 4.5 (MES or sodium acetate buffer) and at 6.0 (sodium cacodylate buffer). Divalent cations are essential for the enzymatic activity, and Mn^{2+} at 10 mM gives the highest activity (Kawashima et al. 1992). The activity is completely abolished in the presence of EDTA and is stimulated by the addition of phosphatidylinositol (PI) and phosphatidylserine (PS), whereas phosphatidylcholine (PC) affected the reaction negatively.

Preparation

Both of these enzymes are expressed only in the nervous system, with little activity detected in extraneural tissues such as the liver, kidney, thymus, spleen, or lung (Kawashima et al. 1992; Oka et al. 1992). The activity of the cerebral cortex increased rapidly just before birth and reached a maximum 2 weeks after birth; it decreased gradually thereafter. The activity of the cerebellum reached a maximum just after birth and then disappeared rapidly. GlcAT-P was purified to an apparent homogeneity from the NP-40 extract of postnatal 2-week-old rat forebrain by sequential chromatography on CM-Sepharose CL-6B, UDP-GlcA-Sepharose 4B, asialoorosomucoid-Sepharose 4B, Matrex gel Blue A, Mono Q, HiTrap Chelating, and HiTrap Heparin columns (Terayama et al. 1998). The purified enzyme migrated as a 45-kDa protein with sodium dodecyl sulfate polyacrylamide gel electrophoresis (SDS-PAGE) under reducing conditions but eluted as a 90-kDa protein upon Superose gel filtration in the presence of NP-40, suggesting that the enzyme forms homodimers under nondenatured conditions. The primary structure deduced from the cDNA sequence revealed that GlcAT-P is composed of 347 amino acid residues of a predicted molecular weight of 39 706 with three potential N-glycosylation sites (Terayama et al. 1997). This is in agreement with the apparent molecular weight of 45 kDa estimated by SDS-PAGE. Hydropathy analysis indicated that GlcAT-P is a typical type II transmembrane protein and that it has a domain profile similar to that of many other glycosyltransferases. It has a short N-terminal region (19 amino acids), which is a putative cytoplasmic tail, a hydrophobic transmembrane region (17 amino acids), and a large C-terminal catalytic domain. For expression of the soluble form of GlcAT-P, the first 74 amino acids containing a cytoplasmic region and transmembrane region were replaced with protein A and glutathione S-transferase (GST). The expression plasmid (pEF-BOS) containing the GlcAT-P/protein A fusion was expressed in the COS-1 cells (Terayama et al. 1997). The fusion protein expressed in the culture medium was concentrated with rabbit IgG-conjugated Sepharose beads. The expression plasmid (pGEX-4T-1) containing GlcAT-P/GST fusion protein was expressed in *Escherichia coli* BL21. The fusion protein expressed in the bacteria was extracted and purified with glutathione-conjugated Sepharose beads.

Biological Aspects

As described above, there are at least two types of glucuronyltransferase associated with biosynthesis of the HNK-1 carbohydrate epitope. Using the GlcAT-P cDNA as a probe under low stringency conditions, another cDNA encoding a novel glucuronyltransferase was cloned from a rat brain cDNA library (Seiki et al. 1999). The cDNA sequence contained an open reading frame that encoded 324 amino acids with type II transmembrane topology. The amino acid sequence comparison of GlcAT-S and GlcAT-P is shown in Fig. 1. The overall identity was 49%, and the highest identity was found in the COOH-terminal catalytic domain containing the previously identified four highly conserved motifs (I–IV) following the proline-rich region (amino acids from Pro-81 to Val-324 overlap with 58.6% homology). We confirmed that GlcAT-S has glucuronyltransferase activity toward glycoprotein acceptors. Accordingly, we called this newly cloned cDNA GlcAT-S (the second HNK-1-associated GlcAT).

Glucuronyltransferase(s) and sulfotransferase(s) are thought to be key enzymes in biosynthesis of the HNK-1 carbohydrate epitope because of the characteristic structure of the HNK-1 carbohydrate epitope. Northern blot analysis of the GlcAT-P,

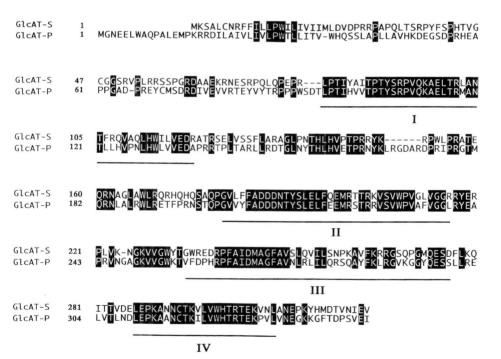

Fig. 1. Comparison of amino acid sequences of two glucuronyltransferases: GlcAT-P and GlcAT-S. *Black background* indicates identical amino acid residues. *Dashes* indicate gaps introduced for maximal alignment. Four highly conserved regions, called modules I–IV (Terayama et al. 1997) are indicated by *bars*

GlcAT-S, and Sulfo-T genes in various adult rat tissues revealed that expression of GlcAT-P and GlcAT-S mRNAs was limited in the nervous system (Terayama et al. 1997; Seiki et al. 1999), and that of sulfotransferase was expressed in various tissues in addition to the nervous system (Ong et al. 1998). These results suggest that glucuronyltransferases are regulatory enzymes in biosynthesis of the HNK-1 carbohydrate. Signals of GlcAT-P mRNA detected in the adult rat brain by Northern blot analysis were stronger than those of GlcAT-S, suggesting that GlcAT-P is the major glucuronyltransferase involved in biosynthesis of the HNK-1 carbohydrate epitope in adult brain or GlcAT-S might be expressed in restricted brain regions (or both).

With regard to the biological function of the HNK-1 epitope, the results of the transfection of full-length GlcAT-P cDNA into COS-1 cells should be noted. Transfection of GlcAT-P or GlcAT-S cDNA in COS-1 cells induced expression of the HNK-1 epitope on their cell surface, suggesting that COS-1 cells contain sulfotransferase transferring sulfate to glycoconjugates bearing GlcA to complete the HNK-1 epitope. More surprisingly, COS-1 cells, which express HNK-1 epitope, exhibited dramatic changes in cell architecture. The HNK-1-expressing cells had long, branched processes with irregular shapes (Fig. 2). A number of microspikes were observed on the soma and processes of cells expressing HNK-1. Such processes and microspikes are rare in cells transfected with vector alone. The HNK-1 epitope itself may have the ability to modulate the cell–substratum interaction and to reorganize cytoskeletal proteins in the cells. This unique function of the HNK-1 epitope may be associated with its presumptive roles during development of the nervous system.

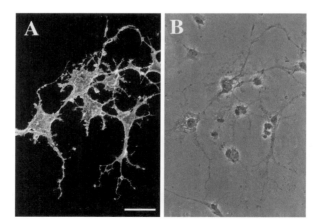

Fig. 2. Transient expression of GlcAT-P cDNA in COS-1 cells. COS-1 cells are transfected with GlcAT-P cDNA in pEF-BOS, a mammalian expression plasmid. Indirect immunofluorescence staining of HNK-1 carbohydrate epitope (**A**) and the corresponding phase-contrast micrograph (**B**) are shown. Transfection of GlcAT-P cDNA into COS-1 cells induces not only expression of the HNK-1 epitope on the cell surface but also marked morphological changes in the cells. Bar 50 μm

Future Perspectives

The most characteristic feature of this epitope is commonly expressed on a series of cell adhesion molecules, including immunoglobulin superfamily molecules. Why is the HNK-1 carbohydrate epitope expressed on some glycoproteins involved in the cell adhesion molecules? This question contains two important issues. One is how GlcAT-P and GlcAT-S recognize these cell adhesion molecules to express the HNK-1 carbohydrate epitope. From this point of view, it is important to elucidate the acceptor specificity of these glucuronyltransferases. The other is how the HNK-1 carbohydrate epitope on cell adhesion molecules expresses its function. It is possible that the HNK-1 carbohydrate epitope plays its role in the following ways. (1) The HNK-1 carbohydrate epitope is directly involved in cell recognition and adhesion mediated via binding proteins (receptors) on the cell surface or in the cell matrix. Several binding proteins have been identified, such as laminin (Mohan et al. 1990), selectins (Needham and Schnaar 1993), and SBP-1 (Nair and Jungalwala 1997). Receptors or binding proteins highly specific for the HNK-1 carbohydrate have not been identified. (2) The HNK-1 carbohydrate epitope on cell adhesion molecules may play a role in modulating the cell adhesion function of its own protein portion, as is the case with polysialic acid on NCAM. In addition, the functional significance of the HNK-1 epitope in the immune systems is still unclear despite its characteristic expression pattern. Further study is required to clarify these issues.

References

Abo T, Balch CM (1981) A differentiation antigen of human NK and K cells identified by a monoclonal antibody (HNK-1). J Immunol 127:1024–1029

Ariga T, Kohriyama T, Freddo L, Latov N, Saito M, Kon K, Ando S, Suzuki M, Hemling ME, Rinehart KL Jr (1987) Characterization of sulfated glucuronic acid-containing glycolipids reacting with IgM M-proteins in patients with neuropathy. J Biol Chem 262:848–853

Bakker H, Friedmann I, Oka S, Kawasaki T, Nifant'ev N, Schachner M, Mantei N (1997) Expression cloning of a cDNA encoding a sulfotransferase involved in the biosynthesis of the HNK-1 carbohydrate epitope. J Biol Chem 272:29942–29946

Bollensen E, Schachner M (1987) The peripheral myelin glycoprotein P0 expresses the L2/HNK-1 and L3 carbohydrate structures shared by neural adhesion molecules. Neurosci Lett 82:77–82

Bronner-Fraser M (1986) Analysis of the early stages of trunk neural crest migration in avian embryos using monoclonal antibody HNK-1. Dev Biol 115:44–55

Bronner-Fraser M (1987) Perturbation of cranial neural crest migration by the HNK-1 antibody. Dev Biol 123:321–331

Chou DK, Flores S, Jungalwala FB (1991) Expression and regulation of UDP-glucuronate: neolactotetraosylceramide glucuronyltransferase in the nervous system. J Biol Chem 266:17941–17947

Chou DKH, Ilyas AA, Evans JE, Costello C, Quarles RH, Jungalwala FB (1986) Structure of sulfated glucuronyl glycolipids in the nervous system reacting with HNK-1 antibody and some IgM paraproteins in neuropathy. J Biol Chem 261:11717–11725

Das KK, Basu M, Basu S, Chou DK, Jungalwala FB (1991) Biosynthesis in vitro of GlcA beta 1-3nLcOse4Cer by a novel glucuronyltransferase (GlcAT-1) from embryonic chicken brain. J Boil Chem 266:5238–5243

Eisenman LM, Hawkes R (1993) Antigenic compartmentation in the mouse cerebellar cortex: zebrin and HNK-1 reveal a complex, overlapping molecular topography. J Comp Neurol 335:586–605

Fishman PH, Bradley RM, Henneberry RC (1976) Butyrate-induced glycolipid biosynthesis in HeLa cells: properties of the induced sialyltransferase. Arch Biochem Biophys 172:618–626

Ilyas AA, Quarles RH, Brady RO (1984) The monoclonal antibody HNK-1 reacts with a human peripheral nerve ganglioside. Biochem Biophys Res Commun 122:1206–1211

Kawashima C, Terayama K, Ii M, Oka S, Kawasaki T (1992) Characterization of a glucuronyltransferase: neolactotetraosylceramide glucuronyltransferase from rat brain. Glycoconj J 9:307–314

Keilhauer G, Faissner A, Schachner M (1985) Differential inhibition of neurone-neurone, neurone-astrocyte and astrocyte-astrocyte adhesion by L1, L2 and N-CAM antibodies. Nature 316:728–730

Kruse J, Mailhammer R, Wernecke H, Faissner A, Sommer I, Goridis C, Schachner M (1984) Neural cell adhesion molecules and myelin-associated glycoprotein share a common carbohydrate moiety recognized by monoclonal antibodies L2 and HNK-1. Nature 311:153–155

Künemund V, Jungalwala FB, Fischer G, Chou DK, Keilhauer G, Schachner M (1988) The L2/HNK-1 carbohydrate of neural cell adhesion molecules is involved in cell interactions. J Cell Biol 106:213–223

Kuratani SC (1991) Alternate expression of the HNK-1 epitope in rhombomeres of the chick embryo. Dev Biol 144:215–219

Martini R, Xin Y, Schmitz B, Schachner M (1992) The L2/HNK-1 carbohydrate epitope is involved in the preferential outgrowth of motor neurons on ventral roots and motor nerves. Eur J Neurosci 4:628–639

McGarry RC, Helfand SL, Quarles RH, Roder JC (1983) Recognition of myelin-associated glycoprotein by the monoclonal antibody HNK-1. Nature 306:376–378

Mohan PS, Chou DK, Jungalwala FB (1990) Sulfoglucuronyl glycolipids bind laminin. J Neurochem 54:2024–2031

Naegele JR, Barnstable CJ (1991) A carbohydrate epitope defined by monoclonal antibody VC1.1 is found on N-CAM and other cell adhesion molecules. Brain Res 559:118–129

Nair SM, Jungalwala FB (1997) Characterization of a sulfoglucuronyl carbohydrate binding protein in the developing nervous system. J Neurochem 68:1286–1297

Needham LK, Schnaar RL (1993) The HNK-1 reactive sulfoglucuronyl glycolipids are ligands for L-selectin and P-selectin but not E-selectin. Proc Natl Acad Sci USA 90:1359–1363

Obata K, Tanaka H (1988) Molecular differentiation of the otic vesicle and neural tube in the chick embryo demonstrated by monoclonal antibodies. Neurosci Res 6:131–142

Oka S, Terayama K, Kawashima C, Kawasaki T (1992) A novel glucuronyltransferase in nervous system presumably associated with the biosynthesis of HNK-1 carbohydrate epitope on glycoproteins. J Biol Chem 267:22711–22714

Ong E, Yeh JC, Ding Y, Hindsgaul O, Fukuda M (1998) Expression cloning of a human sulfotransferase that directs the synthesis of the HNK-1 glycan on the neural cell adhesion molecule and glycolipids. J Biol Chem 273:5190–5195

Schwarting GA, Jungalwala FB, Chou DK, Boyer AM, Yamamoto M (1987) Sulfated glucuronic acid-containing glycoconjugates are temporally and spatially regulated antigens in the developing mammalian nervous system. Dev Biol 120:65–76

Seiki T, Oka S, Terayama K, Imiya K, Kawasaki T (1999) Molecular cloning and expression of a second glucuronyltransferase involved in the biosynthesis of the HNK-1 carbohydrate epitope. Biochem Biophys Res Commun 255:182–187

Shimoda Y, Tajima Y, Nagase T, Harii K, Osumi N, Sanai Y (1999) Cloning and expression of a novel galactoside β1, 3-glucuronyltransferase involved in the biosynthesis of HNK-1 epitope. J Biol Chem 274:17115–17122

Shy ME, Vietorisz T, Nobile-Orazio E, Latov N (1984) Specificity of human IgM M-proteins that bind to myelin-associated glycoprotein: peptide mapping, deglycosylation, and competitive binding studies. J Immunol 133:2509–2512

Terayama K, Oka S, Seiki T, Miki Y, Nakamura A, Kozutsumi Y, Takio K, Kawasaki T (1997) Cloning and functional expression of a novel glucuronyltransferase involved in the biosynthesis of the carbohydrate epitope HNK-1. Proc Natl Acad Sci USA 94:6093–6098

Terayama K, Seiki T, Nakamura A, Matsumori K, Ohta S, Oka S, Sugita M, Kawasaki T (1998) Purification and characterization of a glucuronyltransferase involved in the biosynthesis of the HNK-1 epitope on glycoproteins from rat brain. J Biol Chem 273:30295–30300

Tucker GC, Aoyama H, Lipinski M, Tursz T, Thiery JP (1984) Identical reactivity of monoclonal antibodies HNK-1 and NC-1: conservation in vertebrates on cells derived from the neural primordium and on some leukocytes. Cell Differ 14:223–230

Voshol H, van-Zuylen CW, Orberger G, Vliegenthart JF, Schachner M (1996) Structure of the HNK-1 carbohydrate epitope on bovine peripheral myelin glycoprotein P0. J Biol Chem 271:22957–22960

Yoshihara Y, Oka S, Nemoto Y, Watanabe Y, Nagata S, Kagamiyama H, Mori K (1994) An ICAM-related neuronal glycoprotein, telencephalin, with brain segment-specific expression. Neuron 12:541–553

52

GAG Glucuronyltransferase-I

Introduction

GAG glucuronyltransferase I (GlcAT-I) transfers GlcA from UDP-GlcA to the trisaccharide-serine Galβ1-3Galβ1-4Xylβ1-O-Ser, forming the GAG–protein linkage region (GlcAβ1-3Galβ1-3Galβ1-4Xylβ1-O-Ser) common to various proteoglycans. The enzyme consists of 335 amino acids with one N-glycan and has a type II transmembrane orientation characteristic of many of the other glycosyltransferases (Kitagawa et al. 1998; Wei et al. 1999). The molecular mass of the peptide backbone is 37 kDa, but the mass of the mature protein increases with glycosylation to around 47 kDa. An additional characteristic feature in the amino acid sequence of GlcAT-I is a proline-rich domain (from Pro-30 to Pro-75) next to the transmembrane region, as is seen in several other glycosyltransferases, including glucuronyltransferase-P (GlcAT-P) and glucuronyltransferase-D (GlcAT-D), which synthesize the precursor structure GlcAβ1-3Galβ1-4GlcNAc-R for the HNK-1 carbohydrate epitope GlcA(3-O-sulfate)β1-3Galβ1-4GlcNAc-R (Terayama et al. 1997; Seiki et al. 1999; Shimoda et al. 1999). Database searches indicate that the amino acid sequence of GlcAT-I displays 43% and 46% identity to GlcAT-P and GlcAT-D, respectively. The highest sequence identity is found in the COOH-terminal catalytic domain, which follows the proline-rich region (252 amino acids between Pro-68 and Glu-319 overlap with about 60% identity) (Kitagawa et al. 1998; Wei et al. 1999) and contains the four previously identified highly conserved motifs (I–IV) for putative GlcAT among animal species (Terayama et al. 1997; Shimoda et al. 1999).

Hiroshi Kitagawa and Kazuyuki Sugahara

Department of Biochemistry, Kobe Pharmaceutical University, 4-19-1 Motoyamakita-machi, Higashinada-ku, Kobe 658-8558, Japan
Tel. +81-78-441-7570; Fax +81-78-441-7569
e-mail: k-sugar@kobepharma-u.ac.jp

Databanks

GAG glucuronyltransferase-I

NC-IUBMB enzyme classification: E.C.2.4.1.135

Species	Gene	Protein	mRNA	Genomic
Homo sapiens	GlcAT-1	–	AB009598	AB047194
Mus musculus	GlcAT-1	–	AB019523	–
Cricetulus griseus	GlcAT-1	–	AF113703	–

Name and History

The enzyme transferring GlcA to the GAG linkage trisaccharide has been termed GlcAT-I because the enzyme is thought to be distinct from the glucuronyltransferases involved in formation of the repeating disaccharide units characteristic of various GAG species (Helting and Rodén 1969). GAG GlcAT-I was first detected in an embryonic chick cartilage extract (Helting and Rodén 1969) and was partially purified from embryonic chick brain (Brandt et al. 1969) and mouse mastocytoma cells (Helting 1972). Attempts to purify GlcAT-I to homogeneity have not been successful owing to the low concentrations and the difficulty solubilizing the membrane-bound enzyme. The initial characterization of crude enzyme preparations showed that the enzyme preparations catalyzed the GlcA transfer not only to GAG–protein linkage region fragments with the characteristic structure (e.g., Galβ1-3Gal) but also to disaccharides with analogous structures such as lactose (Galβ1-4Glc) and *N*-acetyllactosamine (Galβ1-4GlcNAc) (Brandt et al. 1969; Helting and Rodén 1969; Helting 1972). Curenton et al. (1991) suggested that these glucuronyl transfer reactions are catalyzed by two distinct β1,3-glucuronyltransferases. This hypothesis has now been confirmed by the cDNA cloning of GlcAT-I and GlcAT-P. The cDNA encoding GlcAT-P was cloned first (Terayama et al. 1997), and the cDNA encoding the GlcAT-I from human placenta was subsequently cloned based on information about the amino acid sequence alignment of rat GlcAT-P with putative proteins in *Caenorhabditis elegans* and *Schistosoma mansoni* (Kitagawa et al. 1998). It is not surprising, however, that a molecular similarity exists between the two enzymes despite the different product glycan types (the GAG linkage region and the HNK-1 epitope on *N*-linked glycoproteins), as these two enzymes exhibit a certain similarity in their substrate recognition (both enzymes recognize the terminal β-linked Gal moiety of their acceptor substrates). In addition, the molecular similarity suggests that the two genes may be evolutionarily related and originate from one primordial gene.

Enzyme Assay and Substrate Specificity

GlcAT-I catalyzes the transfer of a GlcA residue from UDP-GlcA to the GAG–protein linkage region trisaccharide, Galβ1-3Galβ1-4Xyl, through a β1,3-linkage.

$$\text{Gal}\beta1\text{-}3\text{Gal}\beta1\text{-}4\text{Xyl}\beta1\text{-}O\text{-}R \xrightarrow{\text{UDP-GlcA}} \text{GlcA}\beta1\text{-}3\text{Gal}\beta1\text{-}3\text{Gal}\beta1\text{-}4\text{Xyl}\beta1\text{-}O\text{-}R$$

Although GlcAT-I was first identified in 1969, the properties of GlcAT-I had not been described in detail until the cDNA encoding GlcAT-I was cloned (Kitagawa et al. 1998; Wei et al. 1999). The properties and substrate specificities of the recombinant GlcAT-I, which was expressed in COS-1 cells as a soluble protein A chimeric form and was purified using immunoglobulin G (IgG)-Sepharose, have now been determined (Tone et al. 1999; Wei et al. 1999). In addition, the enzyme assay method has been described, although the chemically synthesized acceptor substrates used are not commercially available (Tone et al. 1999; Wei et al. 1999). The standard assay mixture contained 1–10 µl of an enzyme solution, 1 nmol Galβ1-3Galβ1-4Xyl, 20 µM UDP-[^{14}C]GlcA (about 2.0×10^5 dpm), 50 mM MES buffer pH 6.5, and 2 mM MnCl$_2$ in a total volume of 30 µl. Reaction mixtures were incubated at 37°C for 1–4 h, and then ^{14}C-labeled products were separated from UDP-[^{14}C]GlcA by passing the reaction mixture diluted in 1 ml 5 mM sodium phosphate pH 6.8 through Pasteur pipette columns containing Dowex 1-X8 (a PO$_4^{2-}$ form, 100–400 mesh; Bio-Rad) as described previously (Tone et al. 1999). The isolated products were quantified by scintillation spectrophotometry.

Characterization of the purified recombinant GlcAT-I using Galβ1-3Galβ1-4Xyl as an acceptor substrate has revealed an Mn^{2+} requirement for maximal activity (10 mM EDTA completely abolishes the activity) and a pH optimum at pH 6.5 (Tone et al. 1999), basically in accordance with the findings obtained using a particulate enzyme fraction prepared from embryonic chick cartilage (Helting and Rodén 1969). In contrast to the previous findings that the reaction occurred over a wide pH range with a poorly defined maximum at pH 5.4 and reached a saturation level at 15 mM Mn^{2+} (Helting and Rodén 1969), the recombinant GlcAT-I exhibited a relatively sharp maximum at pH 6.5, although the enzyme activity was somewhat affected by the buffers used (MES buffer showed more than 1.4-fold the activity obtained with HEPES buffer at pH 6.5). The enzyme activity was stimulated by Mn^{2+} up to 2 mM and decreased at higher concentrations (Tone et al. 1999). In addition, the K_m value of the recombinant enzyme for UDP-GlcA was lower (29.3 µM) than the value (100 µM) for the chick cartilage enzyme, although Galβ1-3Gal was used as an acceptor in the previous study (Helting and Rodén 1969) instead of Galβ1-3Galβ1-4Xyl.

Recombinant human GlcAT-I utilizes only the linkage region trisaccharide derivatives such as Galβ1-3Galβ1-4Xyl and Galβ1-3Galβ1-4Xylβ1-O-Ser; and little incorporation is observed with other substrates containing a terminal GalNAc or Gal residue (i.e., polymer chondroitin), longer oligosaccharide-serines derived from the linkage region, N-acetyllactosamine, lactose, asialoorosomucoid, Galβ1-3GlcNAc, Galβ1-3GalNAc, and notably Galβ1-3Galβ1-O-benzyl (Table 1) (Kitagawa et al. 1998; Tone et al. 1999). Thus, the transfer of GlcA to the linkage trisaccharide primer is mediated by GlcAT-I, distinct from the enzyme termed glucuronyltransferase II, which is involved in formation of the repeating disaccharide units of chondroitin sulfate (Kitagawa et al. 1997) as described above. The above substrate specificity indicates that the minimum structural requirement for the acceptor substrate of the GlcAT-I is the trisaccharide sequence Galβ1-3Galβ1-4Xyl (K_m 80.4 µM), and the enzyme recognizes up to the third saccharide residue (Xyl) from the nonreducing end (Tone et al. 1999). The deduced trisaccharide recognition disagrees with the findings of Wei et al. (1999), who claimed that Galβ1-3Galβ1-O-naphthalenemethanol and Galβ1-3Galβ1-O-benzyl served as acceptors for the recombinant hamster GlcAT-I, which they obtained by expression cloning. Recently X-ray crystallographic analysis revealed the

Table 1. Acceptor specificity of GlcAT-I

Acceptor	Activity
Galβ1-3Galβ1-4Xyl	+
Galβ1-3Galβ1-4Xylβ1-*O*-Ser	+
GalNAcβ1-4GlcAβ1-3Galβ1-3Galβ1-4Xylβ1-*O*-Ser	−
GalNAcβ1-4GlcAβ1-3GalNAcβ1-4GlcAβ1-3Galβ1-3Galβ1-4Xylβ1-*O*-Ser	−
Galβ1-3Galβ1-*O*-benzyl	±[b]
Chondroitin (GalNAcβ1-4GlcA)$_n$	−
Lactose (Galβ1-4Glc)	−
N-Acetyllactosamine (Galβ1-4GlcNAc)	−
Asialoorosomucoid (Galβ1-4GlcNAc-R)[a]	−
Galβ1-4GlcNAcβ1-*O*-naphthalene methanol	−
Galβ1-3GlcNAc	−
Galβ1-3GalNAc	−

[a] R represents the remainder of the *N*-linked oligosaccharide chain
[b] The reported substrate specificities of human GlcAT-I (Tone et al. 1999) and hamster GlcAT-I (Wei et al. 1999) have been contradictory in terms of utilization of Galβ1-3Galβ1-*O*-benzyl

crystal structure of recombinant human GlcAT-I in the presence of both the donor substrate product UDP and the acceptor substrate analog Galβ1-3Galβ1-4Xyl, and identified the key residues involved in the catalysis and the binding to the two Gal residues (Pedersen et al. 2000).

Preparation

The enzyme is widely distributed in various vertebrate tissues and cells. The native enzymes have been partially purified from embryonic chick brains (Brandt et al. 1969) and mouse mastocytoma cells (Helting 1972) by DEAE-cellulose or gel filtration chromatography, respectively. With these purification procedures the enzyme was solubilized from cell membranes with detergents such as Triton X-100 and Tween 20. The recombinant enzymes have been expressed in COS cells, and the purified enzymes have been obtained from the expression systems as described above (Kitagawa et al. 1998; Wei et al. 1999).

Biological Aspects

Chinese hamster ovary (CHO) cell mutants defective in GlcAT-I have been reported (Bai et al. 1999). The mutants failed to synthesize heparan sulfate (HS) or chondroitin sulfate (CS); and transfection of the mutants with a GlcAT-I cDNA completely restored GlcAT-I activity and GAG synthesis (Bai et al. 1999). Thus, a single GlcAT-I gene appears to be responsible for biosynthesis of the common linkage region for both HS and CS in CHO cells.

Herman et al. (1999) have isolated mutations that perturb *Caenorhabditis elegans* vulval invagination without affecting the vulval cell lineage and the eight defined genes, *sqv-1* to *sqv-8* (squashed vulva). The predicted SQV-8 protein is similar in sequence to three mammalian glucuronyltransferases: GlcAT-I, GlcAT-P, GlcAT-D (Herman and Horvitz 1999). Recently, glucuronyltransferase activity of SQV-8 has been demonstrated (Bulik et al. 2000), the mutant phenotype suggests that glycosaminoglycan (GAG) might be required for normal vulval development, as the predicted SQV-3 protein has been described as β1,4-galactosyltransferase I (GalT-I), which is involved in biosynthesis of the GAG–protein linkage region Galβ1-4Xylβ1-O-Ser (Almeida et al. 1999).

GlcAT-I and HNK-1 glucuronyltransferases (GlcAT-P and GlcAT-D) exhibit distinct, with no overlapping, acceptor substrate specificities in vitro; and GlcAT-I shows strict specificity for Galβ1-3Galβ1-4Xyl, exhibiting negligible incorporation into other galactoside substrates (Tone et al. 1999; Shimoda et al. 1999; Y. Tone, H. Kitagawa, K. Sugahara, unpublished data). Nevertheless, transfection of the GlcAT-I cDNA into COS cells induces expression of the HNK-1 epitope, although the intensity is somewhat weaker than that of the GlcAT-P transfectant (Tone et al. 1999; Wei et al. 1999). Although the seeming discrepancy between the in vitro and in vivo substrate specificities remains to be clarified, similar phenomena were observed for other glycosyltransferases. When vectors are used under the control of a strong promoter to express an exogenous cDNA, a significant amount of the protein is occasionally expressed in the cells (Tsuji 1996). Hence, if a large amount of an enzyme is expressed in the Golgi apparatus, negligible in vitro activity may become meaningful. In the above case, considering the observations that the mock-transfected COS cells are not stained with the HNK-1 antibody, although the cells exhibit endogenous GlcAT-I gene expression and GlcAT-I activity (Tone et al. 1999), it is reasonable to suggest that GlcAT-I does not participate in formation of the HNK-1 epitope under physiological conditions. The GlcAT-I gene is ubiquitously expressed in virtually every human tissue examined (Kitagawa et al. 2001), which is in good agreement with the observations that proteoglycans are distributed on the surfaces of most cells and the extracellular matrices in virtually every tissue, but it is in sharp contrast to the findings that expression of the HNK-1 epitope is highly restricted and is temporally and spatially regulated during development of the nervous system (Schwarting et al. 1987). Therefore, GlcAT-I appears to be mainly involved in biosynthesis of the GAG–protein linkage region of proteoglycans.

It is possible, however, that tissues that strongly express the GlcAT-I gene synthesize the HNK-1 epitope GlcA(3-O-sulfate)β1-3Galβ1-4GlcNAc, as transfection of the GlcAT-I cDNA into COS-1 cells induces significant expression of the HNK-1 epitope, and the HNK-1 3-O-sulfotransferase is widely distributed in various tissues (Bakker et al. 1997; Ong et al. 1998). Hence, GlcAT-P and GlcAT-D are most likely the key enzymes that regulate expression of the HNK-1 epitope. Thus, at certain specific developmental stages or under pathological conditions including cancer, GlcAT-I may participate in synthesis of the HNK-1 epitope. In view of the finding that COS-1 cells transfected with the GlcAT-I cDNA expressed no GlcAT-P activity using either asialoorosomucoid or N-acetyllactosamine as an acceptor, which exhibited GlcAT-I activity more than 10-fold that of the mock-transfected cells (Tone et al. 1999), the possibility cannot be ruled out that COS-1 cells contain an as yet unidentified acceptor substrate for GlcAT-I to form the HNK-1 epitope. In future studies it is necessary to determine the carbohydrate structure of the HNK-1 antibody-reactive product in COS-1 cells transfected with GlcAT-I cDNA.

Future Perspectives

Although the GlcAT-I gene is ubiquitously expressed in virtually every human and mouse tissue examined, the gene exhibits dramatic differential expression among the tissues (Wei et al. 1999; Kitagawa et al. 2001). In addition, transfection of the CHO cell mutant (defective in GlcAT-I) with GlcAT-I cDNA augments GAG synthesis to levels approximately twofold higher than those observed in wild-type cells (Bai et al. 1999), suggesting that GlcAT-I is rate-limiting for GAG biosynthesis. Therefore, it is important to investigate the regulatory mechanism of the GlcAT-I gene at the transcriptional and translational levels in normal and cancerous tissues (Kitagawa et al. 2001). In addition, conditional knockout of the gene in mice would provide essential information regarding the biological functions of sulfated GAGs in each tissue.

References

Almeida R, Levery SB, Mandel U, Kresse H, Schwientek T, Bennett EP, Clausen H (1999) Cloning and expression of a proteoglycan UDP-galactose:β-xylose β1,4-galactosyltransferase I: a seventh member of the human β4-galactosyltransferase gene family. J Biol Chem 274:26165–26171

Bai X, Wei G, Sinha A, Esko JD (1999) Chinese hamster ovary cell mutants defective in glycosaminoglycan assembly and glucuronosyltransferase I. J Biol Chem 274:13017–13024

Bakker H, Friedmann I, Oka S, Kawasaki T, Nifant'ev N, Schachner M, Mantei N (1997) Expression cloning of a cDNA encoding a sulfotransferase involved in the biosynthesis of the HNK-1 carbohydrate epitope. J Biol Chem 272:29942–29946

Brandt AE, Distler J, Jourdian GW (1969) Biosynthesis of chondroitin sulfate-protein linkage region: purification and properties of a glucuronosyltransferase from embryonic chick brain. Proc Natl Acad Sci USA 64:374–380

Bulik DA, Wei G, Toyoda H, Kinoshita-Toyoda A, Waldrip WR, Esko JD, Robbins PW, Selleck SB (2000) sqv-3, -7 and -8, a set of genes affecting morphogenesis in Caenorhabditis elegans, encode enzymes required for glycosaminoglycan biosynthesis. Proc Natl Acad Sci USA 97:10838–10843

Curenton T, Ekborg G, Rodén L (1991) Glucuronosyl transfer to galactose residues in the biosynthesis of HNK-1 antigens and xylose-containing glycosaminoglycans: one or two transferases? Biochem Biophys Res Commun 179:416–422

Helting T (1972) Biosynthesis of heparin: solubilization and partial purification of uridine diphosphate glucuronic acid; acceptor glucuronosyltransferase from mouse mastocytoma. J Biol Chem 247:4327–4332

Helting T, Rodén L (1969) Biosynthesis of chondroitin sulfate. II. Glucuronosyl transfer in the formation of the carbohydrate-protein linkage region. J Biol Chem 244:2799–2805

Herman TH, Horvitz HR (1999) Three proteins involved in Caenorhabditis elegans vulval invagination are similar to components of a glycosylation pathway. Proc Natl Acad Sci USA 96:974–979

Herman TH, Hartwieg E, Horvitz HR (1999) sqv Mutants of Caenorhabditis elegans are defective in vulval epithelial invagination. Proc Natl Acad Sci USA 96:968–973

Kitagawa H, Tone Y, Tamura J, Neumann KW, Ogawa T, Oka S, Kawasaki T, Sugahara K (1998) Molecular cloning and expression of glucuronyltransferase I involved in the biosynthesis of the glycosaminoglycan-protein linkage region of proteoglycans. J Biol Chem 273:6615–6618

Kitagawa H, Ujikawa M, Tsutsumi K, Tamura J, Neumann KW, Ogawa T, Sugahara K (1997) Characterization of serum β-glucuronyltransferase involved in chondroitin sulfate biosynthesis. Glycobiology 7:905–911

Kitagawa H, Taoka M, Tone Y, Sugahara K (2001) Human glycosaminoglycan glucuronyltransferase I gene and a related processed pseudogene: genomic structure, chromosomal mapping and characterization. Biochem J 358:539–546

Ong E, Yeh J-C, Ding Y, Hindsgaul O, Fukuda M (1998) Expression cloning of a human sulfotransferase that directs the synthesis of the HNK-1 glycan on the neural cell adhesion molecule and glycolipids. J Biol Chem 273:5190–5195

Pedersen LC, Tsuchida K, Kitagawa H, Sugahara K, Darden TA, Negishi M (2000) Heparan/chondroitin sulfate biosynthesis. Structure and mechanism of human glucuronyltransferase I. J Biol Chem 275:34580–34585

Schwarting GA, Jungalwala FB, Chou DK, Boyer AM, Yamamoto M (1987) Sulfated glucuronic acid-containing glycoconjugates are temporally and spatially regulated antigens in the developing mammalian nervous system. Dev Biol 120:65–76

Seiki T, Oka S, Terayama K, Imiya K, Kawasaki T (1999) Molecular cloning and expression of a second glucuronyltransferase involved in the biosynthesis of the HNK-1 carbohydrate epitope. Biochem Biophys Res Commun 255:182–187

Shimoda Y, Tajima Y, Nagase T, Harii K, Osumi N, Sanai Y (1999) Cloning and expression of a novel galactoside β1,3-glucuronyltransferase involved in the biosynthesis of HNK-1 epitope. J Biol Chem 274:17115–17122

Terayama K, Oka S, Seiki T, Miki Y, Nakamura A, Kozutsumi Y, Takio K, Kawasaki T (1997) Proc Natl Acad Sci USA 94:6093–6098

Tone Y, Kitagawa H, Imiya K, Oka S, Kawasaki T, Sugahara K (1999) Characterization of recombinant human glucuronyltransferase I involved in the biosynthesis of the glycosaminoglycan-protein linkage region of proteoglycans. FEBS Lett 459:415–420

Tsuji S (1996) Molecular cloning and functional analysis of sialyltransferases. J Biochem 120:1–13

Wei G, Bai X, Sarkar AK, Esko JD (1999) Formation of HNK-1 determinants and the glycosaminoglycan tetrasaccharide linkage region by UDP-GlcUA:galactose β1,3-glucuronosyltransferases. J Biol Chem 274:7857–7864

UDP-Glucose Dehydrogenase

Introduction

Uridine diphosphate (UDP)-glucose dehydrogenase (UDPGDH) represents a key enzyme required for synthesis of glycosaminoglycans (GAGs) and for detoxification of toxins, drugs, and endogenous substances such as steroids, heme pigments, and thyroxine via glucuronidation. UDPGDH is a four-electron transferring NAD-oxidoreductase that oxidizes UDP-Glc to UDP-GlcA through two separate but linked reactions. The product of the first half-reaction remains covalently bound to the enzyme and is the substrate for the second half-reaction. The active UDPGDH (the form best characterized is from bovine liver) is an apparent homohexamer of 52-kilodalton (kDa) subunits functioning as a trimer of dimers. The primary product of UDPGDH action, UDP-GlcA, is critical to the synthesis of GAGs and to the function of proteoglycans. UDP-GlcA is converted to UDP-Xyl, which is required for initiation of the GAG chain on proteoglycan cores. Furthermore, UDP-GlcA is required for polymerization of the GAGs, hyaluronan (HA), heparan sulfate (HS), chondroitin sulfate (CS), and dermatan sulfate (DS).

Genes encoding UDPGDH have been identified in a viral genome, in Archaea, and in numerous prokaryotic and eukaryotic species, including gram-negative and gram-positive bacteria, plants, invertebrates, and vertebrates. Although the highest expression of UDPGDH is observed in the liver, where UDPGDH activity is essential to the glucuronidation of toxins, drugs, and endogenous substances, UDPGDH is expressed to varying degrees by most cell types. Indeed, UDPGDH is essential to the normal development of the higher metazoan body plan. Increases and decreases in UDPGDH activity have been correlated with certain disease states.

Andrew P. Spicer

Center for Extracellular Matrix Biology, Texas A&M University System Health Science Center, Institute of Biosciences and Technology, 2121 West Holcombe Boulevard, Houston, TX 77030, USA
Tel. +1-713-677-7575; Fax +1-713-677-7576
e-mail: aspicer@ibt.tamu.edu

Databanks

UDP-glucose dehydrogenase

NC-IUBMB enzyme classification: E.C.1.1.1.22

Species	Gene	Protein	mRNA	Genomic
Homo sapiens	*UGDH*	O60701	AF061016	–
		–	AJ007702	–
		–	AF049126	–
		CAB75891	–	AJ272273–AJ272281 (exons 1–12)
Mus musculus	*Ugdh*	O70475	AF061017	–
Rattus norvegicus	*ugdh*	O70199	AB013732	–
Bos taurus	*ugdh*	P12378	AF095792	–
		A17150	–	–
Drosophila melanogaster	*sgl, kiwi, ska*	O02373	AF007870	AF001311–AF001312 (exons 1–2)
		–	AF000570	–
		–	AF009013	–
		–	AF001310	–
Caenorhabditis elegans	F29F11.1	Q19905	Z73974	–
Acetobacter xylinum	*aceM*	O31279	Y11203	–
Acinetobacter lwoffii	*ugd*	CAB57210	AJ243431	–
Archaeoglobus fulgidus	AF0302	O29940	AE001084	–
		F69287	–	–
Archaeoglobus fulgidus	AF0596	O29659	AE001084	–
		D69324	–	–
Bacillus subtilis	*tuaD*	F69727	AF015609	–
Burkholderia pseudomallei	*udg*	AAD43344	AF159428	–
Escherichia coli K12	*udg*	P76373	AE000294	–
		–	D90840	–
		–	D90841	–
Escherichia coli K5	*kfiD*	Q47329	X77617	–
Glycine max	–	Q96558	U53418	–
Methanococcus jannischii	–	AAB99056	U67548	–
Paramecium bursaria Chlorella virus 1 (PBCV-1)	A609L	O41091	U42580	–
Pasteurella multocida	*hyaC*	O85458	AF067175	–
Populus tremula X	*ugdh*	AAF04455	AF053973	–
Pseudomonas aeruginosa	*udg*	O86422	AJ010734	–
Pyrococcus abyssi	*ugd*	CAB50069	AJ248286	–
Rhizobium meliloti	*rkpK*	O54068	AJ222661	–
Rickettsia prowazekii	udg (RP779)	O05973	Y11785	–
		–	AJ235273	–
Salmonella typhimurium	*udg*	Q04873	Z17278	–
		S31606	–	–
		S33671	–	–
Shigella flexeri	*udg*	P37791	X71970	–
Streptococcus pneumoniae	*cap3A*	Q57346	Z47210	–
	(*cps3D*)	–	Z12159	–
		–	U15171	–
Streptococcus pyogenes	*hasB*	Q07172	L08444	–
Sulfolobus solfataricus	C39_020	CAB57493	Y18930	–

(Continued)

Species	Gene	Protein	mRNA	Genomic
Synechocystis sp. (PCC 6803)		P72834	D90901	–
		S74698	–	–
Xanthomonas campestris		Q56812	X79772	–
Zymomonas mobilis	ugd	AAF23790	AF213822	–
Arabidopsis thaliana	ugd	–	–	AB010694

Name and History

UDP-glucose dehydrogenase is abbreviated as UDPGDH but can also be referred to as UDP-Glc dehydrogenase or UDP-glucose 6-dehydrogenase. UDPGDH was first detected in bovine liver (Strominger et al. 1954) and was subsequently purified to homogeneity 15 years later (Zalitis and Feingold 1969). The mechanism of action of UDPGDH was proposed by Ordman and Kirkwood (1977). The first half-reaction is dependent on the formation of a Schiff's base, between the C6 position of UDP-Glc and a catalytic lysine residue. The second half-reaction is catalyzed by a conserved cysteine, which attacks the Schiff's base, forming a thiohemiacetal intermediate; the latter is oxidized to a thiolester that is ultimately hydrolyzed to yield UDP-GlcA. It is apparent that the two half-reactions proceed on different subunits (Eccleston et al. 1979), as is the case for another four-electron transferring dehydrogenase, histidinol dehydrogenase (EC 1.1.1.23).

Based on cloning data and sequence analyses, UDPGDH is known to cluster with other four-electron-transferring nucleotide sugar dehydrogenases, such as GDP-mannose dehydrogenase and UDP-N-acetylmannosaminuronic acid dehydrogenase. All of these enzymes share an N-terminal coenzyme (NAD) binding fold, based on the consensus sequence GlyXGlyXXGly, plus additional invariant residues including the catalytic cysteine and two conserved lysine residues. UDP-glucose acts as an effector for NAD binding to UDPGDH (Franzen et al. 1983). The competitive inhibitor UDP-Xyl (Gainey and Phelps 1975) competes with UDP-Glc for the UDP-sugar binding site, leading to release of the NAD. The deduced amino acid sequence for the first mammalian UDPGDH was reported in 1994 by Hempel et al. Subsequently, genes encoding UDPGDH have been identified in numerous organisms, including a virus (Landstein et al. 1998), Archaea, and many prokaryotes and eukaryotes. Notably, the yeast Saccharomyces cerevisiae lacks the UDPGDH gene and consequently is unable to synthesize GAGs.

UDPGDH is functionally similar to the alcohol and aldehyde dehydrogenases. The first half-reaction, which produces UDP-6-aldehydo-D-glucose, is similar to that of alcohol dehydrogenases. UDPGDH is most like the short-chain alcohol dehydrogenases/reductases (SDR), which do not require a metal ion cofactor and are proposed to depend on a catalytic tyrosine and lysine residue (Jornvall et al. 1995). In addition, the short-chain alcohol dehydrogenases contain a conserved coenzyme binding fold (GlyXXXGlyXGly) close to their N-termini (Jornvall et al. 1995). The inducible UDPGDH from the French bean (Phaseolus vulgaris L.) has been shown to have alcohol dehydrogenase activity and may be bifunctional in vivo (Robertson et al. 1996).

The mechanism of action of the catalytic cysteine of UDPGDH, required for the second half-reaction, parallels that of the catalytic cysteine of aldehyde dehydrogenases (reviewed by Pietruszko et al. 1993). In addition, similarities in thiol reagent reactivity of UDPGDH and the aldehyde dehydrogenases suggest structural and functional similarities (Gainey et al. 1972). The ability to oxidize both alcohols and aldehydes has prompted speculation that the four-electron transferring dehydrogenases may have evolved as the consequence of an ancient gene fusion between an alcohol dehydrogenase and an aldehyde dehydrogenase.

Enzyme Activity Assay and Substrate Specificity

UDPGDH catalyzes the reaction

$$UDP\text{-}Glc + 2NAD^+ + H_2O \rightarrow UDP\text{-}GlcA + 2NADH + 2H^+$$

UDPGDH activity is most readily assayed spectrophotometrically through measuring reduction of NAD in the presence of UDP-Glc at 340 nm. For each 1 μmol of UDP-Glc converted to UDP-GlcA, 2 μmol of NAD are reduced. Thus, one international unit (IU) of UDPGDH activity is defined as the amount of enzyme required to reduce 2 μmol of NAD/min/ml at 30°C.

A standard assay for UDPGDH activity utilizes a given amount of recombinant protein, cell lysate, tissue extract, and so on (not to exceed 0.01 unit of UDPGDH activity) in a total volume of 1 ml (1 cm path length cuvet) containing 1 mM UDP-Glc, 100 mM sodium glycine pH 8.7, and 1 mM NAD. Reactions are performed at 30°C and are prewarmed to this temperature prior to initiation of the reaction by adding NAD or UDP-Glc to 1 mM. Absorbance is measured at 340 nm over a 5-min period, with particular emphasis on the first 30–60 s of the reaction, as the reaction is slowed by product inhibition. The change in absorbance at 340 nm (ΔA340/min) is calculated and converted to international units per milligram of protein.

The bovine liver UDPGDH had no detectable NAD reducing activity with the substrates UDP-galactose, UDP-mannose, or GDP-mannose (Zalitis and Feingold, 1969), whereas the Chlorella virus PBCV-1 UDPGDH had weak but significant activity with UDP-galactose (Landstein et al. 1998). UDPGDH activity is inhibited by UDP-Xyl (Gainey and Phelps 1975) (inhibition is complete when 1 mM UDP-Xyl is included in spectrophotometric assays) and by high concentrations of the reaction products NADH and UDP-GlcA. The typically cited pH optimum for vertebrate UDPGDH is pH 8.7, although the enzyme appears to be active over a comparatively wide pH range (for discussion of this area see Mehdizadeh et al. 1991).

A highly sensitive microassay for UDPGDH has also been described (Burrows and Cintron 1983). This assay determines the amount of UDP-GlcA produced by a given tissue extract by incubating the reaction products with 3-hydroxybenzo(a)pyrene and guinea pig microsomal membranes. Under these conditions the glucuronosyltransferase activity of the microsomes converts UDP-GlcA and 3-hydroxybenzo(a)pyrene to D-glucuronosylbenzo(a)pyrene, which can be measured fluorometrically. It is estimated that this reaction is 500 times more sensitive than the standard spectrophotometric assay and is thus applicable to small tissue or sample sizes.

UDPGDH activity can also be assessed in situ on tissue sections (Mehdizadeh et al. 1991). With this approach, precipitation of formazan from nitroblue tetrazolium (NBT) is used as a colorimetric and quantitative indicator of NAD reduction in the presence of UDP-glucose. Briefly, tissue sections are incubated in medium containing 5.3 mM UDP-Glc and 1.5 mM NAD in 30% (w/v) polyvinyl alcohol, in 50 mM sodium glycine buffer pH 7.8. NBT 3.7 mM is added to the reaction, which is allowed to proceed at 37°C for 40 min, after which sections are washed in water, dried, and mounted in an aqueous mounting medium. Relative levels of UDPGDH activity can be determined using a scanning microdensitometer at 560 nm.

Preparation

The enzyme is widely expressed by many embryonic and adult tissues in invertebrates and vertebrates and by most bacteria. The highest activity is found in the mammalian liver. The enzyme is most commonly purified from the liver (Zalitis and Feingold 1969), and bovine liver UDPGDH can be obtained from commercial sources (e.g., Sigma) in purified form. It is likely that recombinant UDPGDH will soon become the major source of commercially available enzyme. Bovine liver UDPGDH is generally purified through tissue homogenization, heat treatment, ion exchange, and molecular sieve chromatography. This process results, on average, in 500-fold purification. Enzyme assays can be performed on purified native or recombinant enzymes, relatively crude cell lysates, or tissue homogenates or on tissue sections (as described above). The *Escherichia coli* K5 UDPGDH, kfaC has been expressed and purified in recombinant form using chelation affinity chromatography (De Luca et al. 1996). For assays on prokaryotic and eukaryotic cell lysates, cells can be lysed in a detergent-based buffer (such as Tris-buffered saline or phosphate-buffered saline supplemented with 0.1% Triton X-100) supplemented with protease inhibitors or mechanically disrupted in a phosphate- or Tris-based buffer at pH 8.7 supplemented with protease inhibitors (e.g., Spicer et al. 1998). Lysates are centrifuged to remove nuclei and large membrane fragments or other debris, and UDPGDH activity is measured directly. Lysates can be dialyzed to remove endogenous nucleotide sugars and NAD^+ if necessary.

Biological Aspects

UDPGDH is of critical importance in GAG biosynthesis and the detoxification of various substances including toxins, drugs, steroid hormones, and heme pigments. In *Drosophila*, UDPGDH (*sugarless*, *sgl*) cDNAs were cloned independently by three groups attempting to identify novel genes involved in pattern formation and signaling pathways (Lander and Selleck 2000). Mutations in *sugarless* were identified based on their phenocopy of the classical *wingless* (*wg*) mutation. Loss of UDPGDH and consequently HS from the cell surface resulted in defects in developmental pathways dependent on growth factor signaling, including *wg*, fibroblast growth factor (FGF), and hedgehog (*hh*) dependent pathways.

UDPGDH is highly conserved in invertebrates and vertebrates. *Drosophila* UDPGDH shares 68% overall amino acid identity to its mammalian ortholog. Human and mouse UDPGDH share 98% amino acid identity. Based on the recent elucidation of the complete sequence for *Drosophila melanogaster, sugarless* is located on chromosome 3 and is encoded on 3 exons. The mouse *Ugdh* gene is located on mouse chromosome 5 approximately 39 centimorgans (cM) from the centromere (Spicer et al. 1998). The human *UGDH* gene is located at an equivalent position within the human genome 4p15.1 (Marcu et al. 1999). In the mouse, the *Ugdh* gene is encoded on at least 10 exons (Spicer et al. 1998), whereas the gene appears to be encoded by 12 exons in humans. One exon–intron boundary is conserved between *Drosophila sgl* and the human and mouse *UGDH* genes, confirming their orthologous relationship and evolution from a common ancestral gene. Gene cloning studies indicate that, despite earlier reports suggesting the existence of tissue-specific isoforms of UDPGDH in the rat and chicken (Darrow and Hendrickson 1971; Bardoni et al. 1989), UDPGDH is encoded by a single gene in most organisms. Tissue specificity, with respect to enzymic properties such as pH optima and isoelectric point, suggest that posttranslational modifications play a role in enzyme function. In this respect, it is interesting to note that the amino acid sequence determined for the purified bovine liver enzyme (Hempel et al. 1994) lacks the C-terminal 24 amino acids encoded by the cDNA (Spicer et al. 1998; Lind et al. 1999) but not necessary for enzyme function (Lind et al. 1999). It is possible that tissue-specific variability during the processing of the enzyme to remove this 24-amino-acid region, which carries a net positive charge, may lead to isoforms with differing isoelectric points and pH optima.

Expression analyses have indicated that UDPGDH acts in an immediate early manner in human fibroblast cultures (Spicer et al. 1998). Treatment of these cultures with the protein synthesis inhibitor cycloheximide resulted in rapid induction of UGDH mRNAs. Two transcripts were expressed in each instance, migrating at approximately 3.4 and 2.8 kb, respectively. It is likely that the size difference corresponds to alternate polyA-signal usage. A single transcript of approximately 2.4 kb was identified in the mouse. Treatment of fibroblast cultures with the proinflammatory cytokine interleukin-1β (IL-1β) also led to up-regulation of UGDH mRNAs.

Recently, UGDH was independently cloned through a differential display polymerase chain reaction (PCR) approach, which was attempting to identify novel androgen-responsive genes expressed by the human breast carcinoma line ZR-75-1 (Lapointe and Labrie 1999). In this instance it was suggested that up-regulation of UGDH expression may reflect the increased requirement for UDP-GlcA for inactivation of sex steroids via glucuronidation. It also suggested that these carcinoma cells have an increased capacity for GAG biosynthesis. This may be particularly important for the free GAG HA, which presumably draws off a cytosolic pool of UDP-sugars (Spicer et al. 1998). Indeed, high UDPGDH activity has been associated with cells in tissues that are synthesizing particularly high levels of HA.

Future Perspectives

Despite the purification of UDPGDH more than 30 years ago, it is only recently that renewed interest in this enzyme has been shown. This is primarily due to the identification of UDPGDH mutations that phenocopy developmental defects in *Drosophila*. It is evident that UDPGDH represents a key enzyme in the biosynthesis of GAGs.

Furthermore, it is becoming increasingly evident that the availability of substrate represents an important facet in the regulation of GAG biosynthesis. UGDH responds in an immediate early fashion and can be regulated by cytokines, steroids, and growth factors (Spicer et al. 1998; Lapointe and Labrie 1999; Smith and Spicer, unpublished data). This suggests that, rather than being encoded by a housekeeping gene, UDPGDH and consequently UDP-GlcA levels are dynamically regulated within the cell, reflecting the importance of GAGs in the regulation of various cell behaviors. The availability of cloned sequences for UDPGDH from many species suggests that recombinant enzymes will be available in the near future. In addition, through genetic manipulation of UGDH we may be able to identify new developmental pathways in which GAGs play a critical role.

Further Reading

The following references are recommended: For methodology on expression and purification of recombinant UDPGDH in *E. coli*, see De Luca et al. (1996). For data on cloning and characterization of mammalian UDPGDH cDNAs and in vitro enzyme assays, see Spicer et al. (1998) and Lind et al. (1999). For methodology on in situ detection of UDPGDH activity on tissue sections, see Mehdizadeh et al. (1991). For a review of proteoglycan synthesis and sugarless mutations see Lander and Selleck (2000).

References

Bardoni A, Pagliula MP, Pallavicini G, Rindi S, Castellani AA, De Luca G (1989) Biosynthesis of glycosaminoglycan precursors: evidences for different tissue specific forms of UDP-glucose dehydrogenase. Ital J Biochem 38:360–368

Burrows RB, Cintron C (1983) A microassay for UDP-glucose dehydrogenase. Anal Biochem 130:376–378

Darrow RA, Hendrickson WM (1971) UDP-glucose dehydrogenase from the chick embryo: tissue-specific forms of the enzyme. Biochem Biophys Res Commun 43:1125–1131

De Luca C, Lansing M, Crescenzi F, Martini I, Shen GJ, O'Regan M, Wong CH (1996) Overexpression, one-step purification and characterization of UDP-glucose dehydrogenase and UDP-*N*-acetylglucosamine pyrophosphorylase. Bioorg Med Chem 4:131–141

Eccleston ED, Thayer ML, Kirkwood S (1979) Mechanisms of action of histidinol dehydrogenase and UDP-Glc dehydrogenase: evidence that the half-reactions proceed on separate subunits. J Biol Chem 254:11399–11404

Franzen JS, Marchetti PS, Lockhart AH, Feingold DS (1983) Special effects of UDP-sugar binding to bovine liver uridine diphosphoglucose dehydrogenase. Biochim Biophys Acta 746:146–153

Gainey PA, Phelps CF (1975) Interactions of uridine diphosphate glucose dehydrogenase with the inhibitor uridine diphosphate xylose. Biochem J 145:129–134

Gainey PA, Pestell TC, Phelps CF (1972) A study of the subunit structure and the thiol reactivity of bovine liver UDP-glucose dehydrogenase. Biochem J 129:821–830

Hempel J, Perozich J, Romovacek H, Hinich A, Kuo I, Feingold DS (1994) UDP-glucose dehydrogenase from bovine liver: primary structure and relationship to other dehydrogenases. Protein Sci 3:2074–2080

Jornvall H, Persson B, Krook M, Atrian S, Gonzalez-Duarte R, Jeffrey J, Ghosh D (1995) Short-chain dehydrogenases/reductases (SDR). Biochemistry 34:6003–6013

Lander AD, Selleck SB (2000) The elusive functions of proteoglycans: in vivo veritas. J Cell Biol 148:227–232

Landstein D, Graves MV, Burbank DE, DeAngelis P, Van Etten JL (1998) Cholrella virus PBCV-1 encodes functional glutamine:fructose-6-phosphate amidotransferase and UDP-glucose dehydrogenase enzymes. Virology 250:388–396

Lapointe J, Labrie C (1999) Identification and cloning of a novel androgen-responsive gene, uridine diphosphoglucose dehydrogenase, in human breast cancer cells. Endocrinology 140:4486–4493

Lind T, Falk E, Hjertson E, Kusche-Gullberg M, Lidholt K (1999) cDNA cloning and expression of UDP-glucose dehydrogenase from bovine kidney. *Glycobiology* 9:595–600

Marcu O, Stathakis DG, Marsh JL (1999) Assignment of the UGDH locus encoding UDP-glucose dehydrogenase to human chromosome band 4p15.1 by radiation hybrid mapping. Cytogenet Cell Genet 86:244–245

Mehdizadeh S, Bitensky L, Chayen J (1991) The assay of uridine diphosphoglucose dehydrogenase activity: discrimination from xanthine dehydrogenase activity. Cell Biochem Function 9:103–110

Ordman AB, Kirkwood S (1977) Mechanism of action of UDP-glucose dehydrogenase: evidence for an essential lysine residue at the active site. J Biol Chem 252:1320–1326

Pietruszko R, Abriola DP, Blatter EE, Mukerjee N (1993) Aldehyde dehydrogenase: aldehyde dehydrogenation and ester hydrolysis. Adv Exp Med Biol 328:221–231

Robertson D, Smith C, Bolwell GP (1996) Inducible UDP-glucose dehydrogenase from the French bean (*Phaseolus vulgaris* L.) locates to vascular tissue and has alcohol dehydrogenase activity. Biochem J 313:311–317

Spicer AP, Kaback LA, Smith TJ, Seldin MF (1998) Molecular cloning and characterization of the human and mouse UDP-glucose dehydrogenase genes. J Biol Chem 273:25117–25124

Strominger JL, Kalckar HM, Axelrod J, Maxwell ES (1954) Enzymatic oxidation of uridine diphosphate glucose to uridine disphosphate glucuronic acid. J Am Chem Soc 76:6411–6412

Zalitis J, Feingold DS (1969) Purification and properties of UDPGDH from beef liver. Arch Biochem Biophys 132:457–465

GAG Synthesis

α4-N-Acetylhexosaminyltransferase (EXTL2)

Introduction

α1,4-N-Acetylhexosaminyltransferase (EXTL2) transfers GalNAc/GlcNAc from UDP-GalNAc/GlcNAc to the tetrasaccharide representing the common glycosaminoglycan (GAG)–protein linkage region (GlcAβ1-3Galβ1-3Galβ1-4Xyl), which is most likely the critical enzyme that determines and initiates heparin/heparan sulfate (HS) synthesis, distinguishing it from the chondroitin sulfate (CS)/dermatan sulfate (DS) synthesis (Kitagawa et al. 1999b). The enzyme is composed of 330 amino acids with one N-glycan and has a type II transmembrane protein topology characteristic of many other glycosyltransferases (Wuyts et al. 1997; Kitagawa et al. 1999b). The enzyme is encoded by the multiple exostoses-like gene *EXTL2*, a member of the hereditary multiple exostoses (EXT) gene family of tumor suppressors (Kitagawa et al. 1999b). The enzyme protein is approximately half the size of the other EXT family members that have 676–919 amino acids. The protein shows significant homology with the carboxy-terminus of the other members of the family (Wuyts et al. 1997).

The enzyme resides in the *cis*-Golgi apparatus (Miura and Freeze 1998). Because the NH$_2$-terminal amino acid sequence of the purified enzyme from the serum-free culture medium of a human sarcoma (malignant fibrous histiocytoma) cell line is found 54 amino acids from the putative start site for translation of the enzyme, the enzyme also exists as a soluble truncated form that has lost its transmembrane domain and has subsequently been released from the enzyme-producing cell, as observed for several other glycosyltransferases (Kitagawa et al. 1999b).

Hiroshi Kitagawa and Kazuyuki Sugahara

Department of Biochemistry, Kobe Pharmaceutical University, 4-19-1 Motoyamakita-machi, Higashinada-ku, Kobe 658-8558, Japan
Tel. +81-78-441-7570; Fax +81-78-441-7569
e-mail: k-sugar@kobepharma-u.ac.jp

Databanks

α4-N-Acetylhexosaminyltransferase (EXTL2)

No EC number has been allocated.

Species	Gene	Protein	mRNA	Genomic
Homo sapiens	*EXTL2*	–	AF000416	NT004308
Mus musculus	*EXTL2*	–	AB032170	–
			AB032171	–

Name and History

α1,4-N-Acetylhexosaminyltransferase was first discovered while searching for the key enzyme involved in the biosynthetic sorting of CS/DS from heparin/HS, as α-N-acetylgalactosaminyltransferase in fetal bovine sera and in mouse mastocytoma cells, which transferred an α-GalNAc residue from UDP-GalNAc to GlcAβ1-3Galβ1-3Galβ1-4Xylβ1-O-Ser derived from the GAG–protein linkage region (Kitagawa et al. 1995; Lidholt et al. 1997). The α-N-acetylgalactosaminyltransferase was also described as the enzyme that might account for the accumulation of GalNAcα1-4GlcAβ1-3Galβ1-3Galβ1-4Xylβ1-R in cultured cells fed with artificial GAG initiators, the β-D-xylosides (Manzi et al. 1995; Salimath et al. 1995). The α-N-acetylgalactosaminyltransferase activity was subsequently found in Golgi fractions prepared from several cultured cell types (Miura and Freeze 1998; Miura et al. 1999) and in the culture medium of a human sarcoma cell line (Kitagawa et al. 1999a). The structure of the reaction product has not been found in any natural GAG chain. The enzyme has been purified from the serum-free culture medium of a human sarcoma cell line, and peptide sequence analysis of the purified enzyme revealed 100% sequence identity to the protein encoded by the multiple exostoses-like gene *EXTL2* (Kitagawa et al. 1999b), a unique member of the *EXT* gene family (Wuyts et al. 1997). A recombinant enzyme had a dual catalytic activity of α1,4-N-acetylgalactosaminyltransferase and α1,4-N-acetylglucosaminyltransferase, that is, an α1,4-N-acetylhexosaminyltransferase that transferred GalNAc/GlcNAc to the core tetrasaccharide representing the GAG–protein linkage region or its synthetic analog GlcAβ1-3Galβ1-O-naphthalenemethanol that had been previously demonstrated as a good acceptor for N-acetylglucosaminyltransferase-I (GlcNAcT-I) (Fritz et al. 1994). Thus, the enzyme was revealed to be identical to the previously described GlcNAcT-I, which determines and initiates the biosynthesis of HS (Fritz et al. 1994), and most likely heparin as well.

Enzyme Assay and Substrate Specificity

α1,4-N-Acetylhexosaminyltransferase catalyzes the transfer of a GalNAc/GlcNAc residue from UDP-GalNAc/GlcNAc to the core oligosaccharide representing the GAG–protein linkage region, GlcAβ1-3Galβ1-3Galβ1-4Xyl, or its synthetic analogs through an α1,4-linkage (see below).

GlcAβ1-3Galβ1-3Galβ1-4Xylβ1-O-R

$\xrightarrow{\text{UDP-GalNAc}}$ GalNAcα1-4GlcAβ1-3Galβ1-3Galβ1-4Xylβ1-O-R

Despite the single enzyme with a dual catalytic activity of α1,4-N-acetylgalactosaminyltransferase and α1,4-N-acetylglucosaminyltransferase (GlcNAcT-I), the acceptor substrate specificities of the two enzymatic activities are not identical: α1,4-N-Acetylgalactosaminyltransferase activity was detected with a variety of acceptor substrates containing nonreducing terminal β-linked GlcA residues, such as N-acetylchondrosine (GlcAβ1-3GalNAc), GlcAβ1-3Galβ1-O-naphthalenemethanol, and the tetrasaccharide-serine GlcAβ1-3Galβ1-3Galβ1-4Xylβ1-O-Ser, representing the GAG–protein linkage region (Kitagawa et al. 1995, 1999a,b; Miura et al. 1999). In contrast, α1,4-N-acetylglucosaminyltransferase (GlcNAcT-I) activity was observed only with acceptor substrates with appropriate aglycones such as GlcAβ1-3Galβ1-O-naphthalenemethanol (Fritz et al. 1994), GlcAβ1-3Galβ1-3Galβ1-4Xylβ1-O-naphthalenemethanol, or GlcAβ1-3Galβ1-3Galβ1-4Xylβ1-O-benzyl (Fritz et al. 1997) but not with N-acetylheparosan (-4GlcAβ1-4GlcNAcα1-)$_n$ (Fritz et al. 1994) or the tetrasaccharide-serine (Lidholt et al. 1997; Kitagawa et al. 1999b). Therefore, it was suggested that transfer of the first GlcNAc residue to the linkage tetrasaccharide primer is mediated by α1,4-N-acetylglucosaminyltransferase (GlcNAcT-I), which is distinct from the enzyme that has been termed HS-polymerase and is involved in formation of the repeating disaccharide units of HS (Fritz et al. 1994). It was also suggested that the enzyme directly recognizes a specific sequence in the core protein or an aglycone structure attached to the tetrasaccharide linkage (reviewed by Esko and Zhang 1996; Fritz et al. 1997).

Among its acceptor substrates described above, only N-acetylchondrosine (GlcAβ1-3GalNAc) is commercially available from Seikagaku Corp. (Tokyo, Japan). The standard assay mixture for α1,4-N-acetylgalactosaminyltransferase activity contained 1–10 μl of an enzyme solution, 10 nmol N-acetylchondrosine GlcAβ1-3GalNAc, 20 μM UDP-[^{3}H]GalNAc (about 5.0×10^5 dpm), 20 mM MnCl$_2$, 171 μM ATP in a total volume of 35 μl of 50 mM MES buffer pH 6.5. ATP was included to prevent enzymatic degradation of UDP-GalNAc. The reaction mixtures were incubated at 37°C for 1–4h, then diluted with 1 ml of 5 mM sodium phosphate pH 6.8. ^{3}H-Labeled products were separated from UDP-[^{3}H]GalNAc by passing the reaction mixture through a Pasteur pipette column containing Dowex 1-X8 (PO$_4^{2-}$, 100–400 mesh; Bio-Rad) and quantified by liquid scintillation counting as described previously (Kitagawa et al. 1997a).

The enzyme exhibited a relatively broad optimum range from pH 6.5 to 7.5 depending on the buffers used, with the highest activity in a 50 mM MES buffer pH 6.5 (Kitagawa et al. 1999a; Miura et al. 1999). Divalent cations were essential for the enzymatic reaction, and 10 mM EDTA completely abolished the activity (Kitagawa et al. 1999a; Miura et al. 1999). Mn^{2+} exhibited the highest activity under standard assay conditions, and Co^{2+} was 50% as effective as Mn^{2+} (Kitagawa et al. 1999a; Miura et al. 1999). The optimal concentration of Mn^{2+} was approximately 20 mM (Kitagawa et al. 1999a). Addition of dithiothreitol at 20 mM had no effect on enzyme activity (Miura et al. 1999). The apparent K_m values for N-acetylchondrosine, the linkage tetrasaccharide-serine, and UDP-GalNAc were 1060, 188, and 27 μM, respectively.

The assay method for α1,4-*N*-acetylglucosaminyltransferase (GlcNAcT-I) activity has been described using chemically synthesized oligosaccharides with artificial aglycons, none of which is commercially available (Fritz et al. 1994). The enzyme activity also shows a relatively broad optimum range of pH 5–8. Divalent cations were essential for the enzymatic reaction, and Mn^{2+} (15–20 mM) exhibited the highest activity (Fritz et al. 1994). The apparent K_m value for UDP-GlcNAc was $36 \pm 4\,\mu M$ (Fritz et al. 1994).

Preparation

The enzyme is widely distributed in various mammalian tissues and cells and in serum. The enzyme has been purified from serum-free culture medium of a human sarcoma cell line to near-homogeneity mainly by affinity chromatography on heparin-Sepharose and UDP-hexanolamine-Sepharose (Kitagawa et al. 1999b). The recombinant enzyme has been expressed in COS-1 cells, and the purified enzyme has been obtained from the expression systems (Kitagawa et al. 1999b).

Biological Aspects

α1,4-*N*-Acetylhexosaminyltransferase (EXTL2) is most likely the key enzyme that determines and initiates the heparin/HS synthesis, distinguishing it from CS/DS synthesis. The *EXTL2* gene, like other *EXT* gene family members, is ubiquitously expressed in virtually every human tissue (Wuyts et al. 1997; Kitagawa et al. 1999b), which is in accordance with the observations that HS proteoglycans are distributed on the surfaces of most cells and the extracellular matrices in virtually every tissue. In view of the findings of the involvement of EXTL2 in HS biosynthesis together with the findings of Lind et al. (1998), who reported that *EXT1* and *EXT2* each encoded an HS-polymerase required for HS biosynthesis, the expression of HS appears to play an important role in tumor suppressor function, although the precise mechanism remains unclear.

Because *EXTL2* has been assigned to human chromosome 1p11-p12, and this region is involved in chromosomal rearrangements in a variety of tumors (Wuyts et al. 1997), *EXTL2* is a serious candidate gene for involvement in one of the tumors associated with this region. Considering the probability that deletion of the gene would cause complete elimination of HS and heparin unless functional redundancy with other genes exists, it is likely that germline mutations inactivating the enzymatic activity result in embryonic lethality and that somatic mutations cause much more serious defects than those caused by *EXT1* and *EXT2*, leading to the progression of various tumors or to lethal disorders. In this regard, it should be noted that another *EXT* gene family member EXTL3 protein was also shown to possess GlcNAcT-I activity (Kim et al. 2001). Congenital deficiency in HS, even in enterocytes, results in severe clinical problems and eventually death (Murch et al. 1996).

The possible role of α1,4-*N*-acetylgalactosaminyltransferase activity of the enzyme in GAG biosynthesis remains unclear, as no α-GalNAc-capped structure has been reported in naturally occurring GAG chains. The α-GalNAc-capped pentasaccharide serine GalNAcα1-4GlcAβ1-3Galβ1-3Galβ1-4Xylβ1-*O*-Ser, a reaction product of α1,4-

N-acetylgalactosaminyltransferase, is not utilized as an acceptor for a glucuronyl-transferase involved in CS biosynthesis (Kitagawa et al. 1997b). Therefore, it was once suggested that addition of an α-GalNAc unit to the tetrasaccharide core of the linkage region of proteoglycans might serve as a stop signal that precludes further chain elongation, creating a part-time proteoglycan (Fransson 1987). To search for the α-GalNAc-capped pentasaccharide linkage structure GalNAcα1-4GlcAβ1-3Galβ1-3Galβ1-4Xyl on part-time proteoglycans, the O-linked oligosaccharides on recombinant human α-thrombomodulin, a part-time proteoglycan, were isolated and characterized (Nadanaka et al. 1998). Structural analysis unexpectedly revealed the occurrence of an immature truncated sequence GlcAβ1-3Galβ1-3Galβ1-4Xyl on the α-thrombomodulin (Nadanaka et al. 1998). Although further study of the carbohydrate structure on other part-time proteoglycans is necessary to confirm the generality of the finding, it is unlikely that the α1,4-N-acetylgalactosaminyltransferase is involved in the formation of a part-time proteoglycan.

Because the apparent K_m values of the α1,4-N-acetylgalactosaminyltransferase for GlcA-containing oligosaccharides are low, as described above, and a substantial amount of α1,4-N-acetylgalactosaminyltransferase activity has been detected in Golgi fractions prepared from several cell types (Lidholt et al. 1997; Miura and Freeze 1998), it is possible that the α1,4-N-acetylgalactosaminyltransferase participates in the biosynthesis of other classes of GlcA-containing glycoconjugates (e.g., N-linked glycoproteins and glycolipids). Indeed, considering the similarity in the terminal structures of the linkage tetrasaccharide (GlcAβ1-3Galβ1-3Galβ1-4Xyl) and the precursor saccharide sequence (GlcAβ1-3Galβ1-4GlcNAc-R) for the HNK-1 carbohydrate epitope on glycoproteins or glycolipids, it is reasonable to assume that the α1,4-N-acetylgalactosaminyltransferase utilizes these classes of glycoconjugates as acceptor substrates. In this regard, when glucuronylneolactotetraosylceramide (GlcAβ1-3Galβ1-4GlcNAcβ1-3Galβ1-4Glcβ1-1Cer) and sulfoglucuronylneolacto-tetraosylceramide [GlcA(3-O-sulfate)β1-3Galβ1-4GlcNAcβ1-3Galβ1-4Glcβ1-1Cer] were tested, only the former compound served as an acceptor substrate for the α1,4-N-acetylgalactosaminyltransferase (Kitagawa et al. 1999b). Miura et al. (1999) reported that GlcAβ1-3Galβ1-4GlcNAcβ-octyl also served as an acceptor substrate for the α1,4-N-acetylgalactosaminyltransferase, and that the apparent K_m value for this substrate was 0.5 mM. Although these α-GalNAc-capped structures have not been reported in naturally occurring glycoproteins or glycolipids, α1,4-N-acetylgalactosaminyltransferase might play an important role in regulating expression of the HNK-1 carbohydrate epitope. Identification of α-GalNAc-capped structures in naturally occurring glycoconjugates would provide supportive evidence.

Future Perspectives

Glycoaminoglycans are synthesized as proteoglycans on specific Ser residues in the so-called GAG–protein linkage region GlcAβ1-3Galβ1-3Galβ1-4Xylβ1-O-Ser, which is common to GAGs including HS, heparin, CS, and DS. Heparin/HS is synthesized once GlcNAc is transferred to the common linkage region, and CS/DS is formed if GalNAc is first added. The two distinct transferases that catalyze the transfer of GlcNAc or GalNAc, respectively, to the common linkage region are the key enzymes that determine the type of GAGs to be synthesized. However, biosynthetic sorting mechanisms of the various GAG chains remain enigmatic. Amino acid sequences that regulate the HS assembly have been characterized (reviewed by Esko and Zhang 1996), so α1,4-N-acetylhexosaminyltransferase (EXTL2) should recognize the amino acid sequences flanking the attachment sites for HS. In the future it is necessary to determine how the enzyme binds to the common linkage region and regulates HS biosynthesis.

Further Reading

Esko and Zhang (1996) described the importance of the core protein sequence on the HS assembly and provide further information for understanding how cells regulate assembly of the GAG chains.

References

Esko JD, Zhang L (1996) Influence of core protein sequence on glycosaminoglycan assembly. Curr Opin Struct Biol 6:663–670

Fransson LÅ (1987) Structure and function of cell-associated proteoglycans. Trends Biochem Sci 12:406–411

Fritz TA, Agrawal PK, Esko JD, Krishna NR (1997) Partial purification and substrate specificity of heparan sulfate α-N-acetylglucosaminyltransferase. I. Synthesis, NMR spectroscopic characterization and in vitro assays of two aryl tetrasaccharides. Glycobiology 7:587–595

Fritz TA, Gabb MM, Wei G, Esko JD (1994) Two N-acetylglucosaminyltransferases catalyze the biosynthesis of heparan sulfate. J Biol Chem 269:28809–28814

Kim BT, Kitagawa H, Tamura J, Saito T, Kusche-Gullberg M, Lindahl U, Sugahara K (2001) Human tumor suppressor EXT gene family members $EXTL1$ and $EXTL3$ encode alpha 1,4-N-acetylglucosaminyltransferases that likely are involved in heparan sulfate/heparin biosynthesis. Proc Natl Acad Sci USA 98:7176–7181

Kitagawa H, Kano Y, Shimakawa H, Goto F, Ogawa T, Okabe H, Sugahara K (1999a) Identification and characterization of a novel UDP-GalNAc:GlcAβ-R α1,4-N-acetylgalactosaminyltransferase from a human sarcoma cell line. Glycobiology 9:697–703

Kitagawa H, Shimakawa H, Sugahara K (1999b) The tumor suppressor EXT-like gene EXTL2 encodes an α1,4-N-acetylhexosaminyltransferase that transfers N-acetylgalactosamine and N-acetylglucosamine to the common glycosaminoglycan-protein linkage region: the key enzyme for the chain initiation of heparan sulfate. J Biol Chem 274:13933–13937

Kitagawa H, Tanaka Y, Tsuchida K, Goto F, Ogawa T, Lidholt K, Lindahl U, Sugahara K (1995) N-Acetylgalactosamine (GalNAc) transfer to the common carbohydrate-protein linkage region of sulfated glycosaminoglycans: identification of UDP-GalNAc:chondro-oligosaccharide α-N-acetylgalactosaminyltransferase in fetal bovine serum. J Biol Chem 270:22190–22195

Kitagawa H, Tsutsumi K, Ujikawa M, Goto F, Tamura J, Neumann KW, Ogawa T, Sugahara K (1997a) Regulation of chondroitin sulfate biosynthesis by specific sulfation: acceptor specificity of serum β-GalNAc transferase revealed by structurally-defined oligosaccharides. Glycobiology 7:531–537

Kitagawa H, Ujikawa M, Tsutsumi K, Tamura J, Neumann KW, Ogawa T, Sugahara K (1997b) Characterization of serum β-glucuronyltransferase involved in chondroitin sulfate biosynthesis. Glycobiology 7:905–911

Lidholt K, Fjelstad M, Lindahl U, Goto F, Ogawa T, Kitagawa H, Sugahara K (1997) Assessment of glycosaminoglycan-protein linkage tetrasaccharides as acceptors for GalNAc- and GlcNAc-transferases from mouse mastocytoma. Glycoconj J 14:737–742

Lind T, Tufaro F, McCormick C, Lindahl U, Lidholt K (1998) The putative tumor suppressors EXT1 and EXT2 are glycosyltransferases required for the biosynthesis of heparan sulfate. J Biol Chem 273:26265–26268

Manzi A, Salimath PV, Spiro RC, Keifer PA, Freeze HH (1995) Identification of a novel glycosaminoglycan core-like molecule I: 500 MHz ^{1}H NMR analysis using a nano-NMR probe indicates the presence of a terminal α-GalNAc residue capping 4-methylumbelliferyl-β-D-xyloside. J Biol Chem 270:9154–9163

Miura Y, Freeze HH (1998) α-N-Acetylgalactosamine-capping of chondroitin sulfate core region oligosaccharides primed on xylosides. Glycobiology 8:813–819

Miura Y, Ding Y, Manzi A, Hindsgaul O, Freeze HH (1999) Characterization of mammalian UDP-GalNAc:glucuronide α1-4-N-acetylgalactosaminyltransferase. Glycobiology 9:1053–1060

Murch SH, Winyard PJD, Koletzko S, Wehner B, Cheema HA, Risdon RA, Philips AD, Meadows N, Klein NJ, Walker-Smith JA (1996) Congenital enterocyte heparan sulphate deficiency with massive albumin loss, secretory diarrhoea, and malnutrition. Lancet 347:1299–1301

Nadanaka S, Kitagawa H, Sugahara K (1998) Demonstration of the immature glycosaminoglycan tetrasaccharide sequence GlcAβ1-3Galβ1-3Galβ1-4Xyl on recombinant soluble human α-thrombomodulin: an oligosaccharide structure on a "part-time" proteoglycan. J Biol Chem 273:33728–33734

Salimath PV, Spiro RC, Freeze HH (1995) Identification of a novel glycosaminoglycan core-like molecule II: α-GalNAc-capped xylosides can be made by many cell types. J Biol Chem 270:9164–9168

Wuyts W, Van Hul W, Hendrickx J, Speleman F, Wauters J, De Boulle K, Van Roy N, Van Agtmael T, Bossuyt P, Willems PJ (1997) Identification and characterization of a novel member of the EXT gene family, EXTL2. Eur J Hum Genet 5:382–389

55

Hyaluronan Synthase-1, -2, and -3

Introduction

Hyaluronan (HA) is a high-molecular-weight linear polysaccharide composed of β-1,4-linked repeating disaccharides of glucuronic acid β-1,3-linked to *N*-acetylglucosamine. It is found in the extracellular matrices of most vertebrate tissues and in the capsules of certain bacterial pathogens. In vertebrates hyaluronan plays an important role not only in maintaining tissue architecture and function but also in modulating cell migration, cell adhesion, wound healing, and tumor invasion through its association with cell surface receptors. Despite considerable efforts, the HA biosynthesis mechanism had remained unclear owing to difficulty in the biochemical isolation of the active enzyme. Since the gene for bacterial HA synthase was first isolated from *Streptococcus pyogenes* several years ago (DeAngelis et al. 1993), rapid progress has been made to elucidate the enzyme properties and the reaction mechanisms. Three distinct yet highly conserved genes encoding mammalian HA synthases (HAS1, HAS2, HAS3) have been cloned (Itano and Kimata, 1996a,b; Shyjan et al. 1996; Spicer et al. 1996, 1997a,b); Watanabe and Yamaguchi 1996; Spicer and McDonald 1998), which has raised new questions about how they are different from each other.

Koji Kimata

The Institute for Molecular Science of Medicine, Aichi Medical University, Nagakute, Aichi 480-1195, Japan
Tel. +81-52-264-4811 (ext. 2088); Fax +81-561-63-3532
e-mail: Kimata@amugw.aichi-med-u.ac.jp

Databanks

Hyaluronan synthase -1, -2, and -3

No EC number has been allocated.

Species	Gene	Protein	mRNA	Genomic
HAS1				
Homo sapiens	*hsHAS1*	Q92839	D84424	–
		–	U59269	–
Mus musculus	*mmHAS1*	Q61647	D82964	AB005226, AB005227
Gallus gallus		P13563	–	–
Xenopus laevis		Q14470	M22249	–
HAS2				
Homo sapiens	*hsHAS2*	Q92819	U54804	–
Mus musculus	*mmHAS2*	Q62405	U52524	–
		P70312	U69695	–
		P70411	U53222	–
Rattus norvegicus	–	–	AF008201	–
Bos taurus	–	O97711	AJ004951	–
		–	AB017804	–
Gallus gallus	–	O57424	AF106940	–
		–	AF015776	–
Xenopus laevis	–	O57427	AF168465	–
		–	AF015779	–
HAS3				
Homo sapiens	*hsHAS3*	O00219	U86409	–
Mus musculus	*mmHAS3*	O08650	U86408	–
Gallus gallus	–	O57425	AF015777	–
Xenopus laevis	–	O57426	AF015779	–

Name and History

Hyaluronic acid was first used by Meyer and Palmer (1934) to name a uronic acid-containing polysaccharide from the vitreous (Greek: hyaloid) of the eye. Balazs et al. (1986) suggested that the name "hyaluronan" be used when the polysaccharide is mentioned in general terms because names of polysaccharides should end with -an according to accepted terminology. The enzyme for HA biosynthesis was designated HA synthase (Weigel et al. 1997). The first success in isolating the cDNA clone from mammalian cells created the abbreviation HAS for this enzyme (Itano and Kimata 1996a,b). The other isozymes of vertebrate HAS were subsequently found, and now the three vertebrate isozymes are numbered 1, 2, and 3 in order of their discovery. According to Weigel et al. (1997), to distinguish HASs from various sources the first letter of the genus and the species name should be put first, such as hsHAS1 for *Homo sapiens* hyaluronan synthase 1.

Enzyme Activity Assay and Substrate Specificity

Several lines of evidence have suggested that HAS resides at the plasma membrane. Although a single protein of recombinant HAS is capable of synthesizing HA from the sugar donor substrates UDP-GlcA and UDP-GlcNAc, some membranous components or hydrophobic environments markedly enhance the activity (Yoshida et al. 2000). Any isoform of HASs utilizes divalent metal ions (Mg^{2+} or Mn^{2+}), probably by forming the substrate complex through coordinating the ions with the phosphate groups of the UDP-sugars. HAS activity in vitro can be easily measured by incubating a preparation of crude membranous proteins in the reaction mixture containing buffer pH 7.1, 15 mM $MgCl_2$, 1 mM UDP-GlcNAc, and 0.05 mM radiolabeled UDP-GlcA. The radioactivity incorporated into the *Streptomyces hyaluronidase*-sensitive polysaccharide is then determined (Itano and Kimata, 1996a) by paper chromatography or gel filtration. There are no reports yet on the inhibitor or activator of the HAS enzyme activity, but molecular biological techniques such as expression of the anti-sense RNA have been successfully applied at the cellular level (Nishida et al. 1999). Various cell-growth factors such as transforming growth factor-β (TGFβ) and interleukin-6(IL-6) influence the mRNA levels dramatically. Endo's group in Japan found that incubation with 4-methylumbelliferone reduced HA synthesis in cultured cells, although the mechanism has not been elucidated (Nakamura et al. 1995).

Preparation

There were some attempts to purify vertebrate HAS to homogeneity, but they encountered the loss of enzyme activity (DeAngelis 1999). Transfection of isolated cDNAs into certain cells (e.g., COS-1 cells) revealed that expressed proteins could be HAS because the transfected cells and their membrane preparations showed marked increases in HA synthesis and synthetic activity, respectively (Itano and Kimata, 1996a). Clear evidence has been obtained by isolating a recombinant HAS1 protein as a FLAG-tagged fusion protein in COS-1 cells by solubilization with CHAPS detergent followed by anti-FLAG affinity chromatography, which is capable of synthesizing HA when incubated with both UDP-sugars as donor substrates with no further addition (Yoshida et al. 2000).

Biological Aspects

Northern analysis revealed a similar temporal expression pattern of HAS genes in developing mouse and frog embryos in a way that they are expressed in the order HAS1, HAS2, and HAS3 during development (Spicer and McDonald, 1998). Enzyme properties and products of three isoforms of vertbrate HAS have been compared and characterized (Itano et al. 1999b). The pericellular coats formed by HAS1 transfectants (COS-1 cells and rat 3Y1 fibroblasts) were significantly smaller than those formed by HAS2 or HAS3 transfectants. The recombinant HAS1 protein exhibited higher apparent K_m values for the substrates UDP-GlcNAc and UDP-GlcA than those of the recombinant HAS2 and HAS3 proteins, although among the three isoforms the K_m values for UDP-GlcNAc were always higher than those for UDP-GlcA. HAS1 and

HAS2 synthesized in vitro HA with relative molecular weights (Mr) of 2×10^5 to $>2 \times 10^6$ daltons, and HAS3 synthesized HA with Mr of 1×10^5 to 1×10^6 daltons, which was apparently shorter than those of the HA chains by HAS1 and HAS2. Although the results have not appeared in the literature, knockout (KO) mice have already been raised; interestingly, only the HAS2 KO mouse is lethal, which suggests dintinct differences in function among the three isoforms. So far there are no reports on disease involvement except for the implication of HAS1 expression in highly metastatic behaviors of some cancer cells (Itano et al. 1999a). The genomic gene structure and the consensus *cis*-elements-binding sites on the promoter region of the HAS1 gene were reported (Yamada et al. 1998).

Future Perspectives

As described above, the occurrence of three HAS isoforms with distinct enzymatic characteristics may provide the cells with flexibility in the control of HA biosynthesis and functions (Itano et al. 1999b). Therefore, regulation of each expression is an important subject. Techniques for regulation should reveal mechanisms for HA involvement in some basic cellular behaviors, such as cell migration and cell differentiation.

Further Reading

For a review of hyaluronan synthases, including fascinating glycosyltransferases from vertebrates, bacterial pathogens, and algal viruses, see DeAngelis (1999).

For understanding why the purification of HAS had long been unsuccessful, see Yoshida et al. (2000). These authors also discussed the identification of residues critical for HAS activity.

References

Balazs EA, Laurent C, Jeanloz RW (1986) Nomenclature of hyaluronic acid. Biochem J 235:903

DeAngelis PH (1999) Hyaluronan synthases: fascinating glycosyltransferases from vertebrates, bacterial pathogens, and algal viruses. Cell Mol Life Sci 56:670–682

DeAngelis PL, Papaconstantinou J, Weigel PH (1993) Iolation of a *Streptococcus pyogenes* gene locus that directs hyaluronan biosynthesis in acapsular mutants and in heterologous bacteria. J Biol Chem 268:14568–14571

Itano N, Kimata K (1996a) Expression cloning and molecular characterization of HAS protein, a eukaryotic hayluronan synthase. J Biol Chem 271:9875–9878

Itano N, Kimata K (1996b) Molecular cloning of human hyaluronan synthase. Biochem Biophys Res Commun 222:816–820

Itano N, Sawai T, Miyaishi O, Kimata K (1999a) Relationship between hyaluronan production and metastatic potential of mouse mammary carcinoma cells. Cancer Res 59:2499–2504

Itano N, Sawai T, Yoshida M, Lenas P, Yamada Y, Imagawa M, Shinomura T, Hamaguchi M, Yoshida Y, Onhuki Y, Miyauchi S, Spicer AP, McDonald JA, Kimata K (1999b) J Biol Chem 274:25085–2509

Meyer K, Palmer JW (1934) The polysaccharide of the vitreous humor. J Biol Chem 107:629–634

Nakamura T, Takagaki K, Shibata S, Tanaka K, Higuchi T, Endo M (1995) Hyaluronic-acid-deficient extracellular matrix induced by addition of 4-methylumbelliferone to the medium of cultured human skin fibroblasts. Biochem Biophys Res Commun 208:470–475

Nishida Y, Knudson CB, Nietfeld JJ, Margulis A, Knudson W (1999) Antisense inhibition of hyaluronan synthase-2 in human articular chondrocytes inhibits proteoglycan retention and matrix assembly. J Biol Chem 274:21893–21899

Shyjan AM, Heldin P, Butcher EC, Yoshino, Briskin MJ (1996) Functional cloning of the cDNA for a human hyaluronan synthase. J Biol Chem 271:23395–23399

Spicer AP, Augustine ML, McDonald JA (1996) Molecular cloning and characterization of a putative mouse hyaluronan synthase. J Biol Chem 271:23400–23406

Spicer AP, Olson JS, McDonald JA (1997a) Molecular cloning and characterization of a cDNA endoding the third putative mammalian hyaluronan synthase. J Biol Chem 272:8957–8961

Spicer AP, Seldin MF, Olsen AS, Brown N, Wells DE, Doggett NA, Itano N, Kimata K, Inazawa J, McDonald JA (1997b) Chromosomal localization of the human and mouse hyaluronan synthase genes. Genomics 41:493–497

Spicer AP, McDonald JA (1998) Characterization and molecular evolution of a vertebrate hyaluronan synthase gene family. J Biol Chem 273:1923–1932

Watanabe K, Yamaguchi Y (1996) Molecular identification of a putative human hyaluronan synthase. J Biol Chem 271:22945–22948

Weigel PH, Hascall VC, Tammi M (1997) Hyaluronan synthases. J Biol Chem 272: 13997–14000

Yamada Y, Itano N, Zako M, Yoshida M, Lenas P, Niimi A, Ueda M, Kimata K (1998) The gene structure and promoter sequence of mouse hyaluronan synthase 1 (*mHAS1*). 330:1223–1227

Yoshida M, Itano N, Yamada Y, Kimata K (2000) In vitro synthesis of hyaluronan by a single protein derived from mouse HAS1 gene and characterization of amino acid residues essential for the activity. J Biol Chem 275:497–506

Heparan Sulfate GlcA/GlcNAc Transferase

Introduction

Heparan sulfates are sulfated glycosaminoglycans distributed on the cell surfaces and in the extracellular matrices of most tissues. The structurally related polysaccharide heparin is a highly sulfated variant of heparan sulfate that is exclusively produced by connective tissue mast cells. Heparan sulfate and heparin are synthesized as proteoglycans, and their biosynthesis is initiated by formation of a tetrasaccharide linkage region attached to a serine residue in the core protein (GlcAβ1-3Galβ1-3Galβ1-4Xylβ1-O-Ser). After addition of a single α-GlcNAc residue, elongation proceeds by the action of glycosyltransferases, which add β1,4-GlcA and α1,4-GlcNAc units in alternating sequence to the nonreducing end of the growing polymer. Bifunctional glycosyltransferases, denoted EXT1 and EXT2, are believed to be involved in the sequential addition of GlcA and GlcNAc. Concomitant with chain elongation, further modifications occur through several enzymatic steps that generate a complex polysaccharide containing N-acetylated and N-sulfated GlcN residues, GlcA and IdoA acid units, and O-sulfate groups in various positions.

MARION KUSCHE-GULLBERG and ULF LINDAHL

Department of Medical Biochemistry and Microbiology, Uppsala University, The Biomedical Center, Box 582, SE-751 23 Uppsala, Sweden
Tel. +46-18-471-4196; Fax +46-18-471-4209
e-mail: Ulf.Lindahl@medkem.uu.se

Databanks

Heparan sulfate GlcA/GlcNAc transferase

No EC number has been allocated

Species	Gene	Protein	mRNA	Genomic
EXT1				
Homo sapiens	*EXT1*	–	Q16394	–
Mus musculus	*EXT1*	–	P97464	–
Drosophila melanogaster	*tout-velu (EXT1)*	–	AAC32397	–
Caenorhabditis elegans	*EXT1*	–	O01704	–
EXT2				
Homo sapiens	*EXT2*	–	Q93063	–
Mus musculus	*EXT2*	–	P70428	–
Bos taurus	*EXT2*	–	AF089748	–
Drosophila melanogaster	*DEXT2*	–	–	–
Caenorhabditis elegans	*EXT1*	–	O01705	–

Name and History

GlcA/GlcNAc transferase is also referred to as heparan sulfate-polymerase or GlcA/GlcNAc co-polymerase. In the earliest studies of heparin biosynthesis, during the 1960s, Silbert demonstrated formation of a nonsulfated heparin precursor composed of (GlcA-GlcNAc)$_n$ after incubation of a microsomal fraction from a heparin-producing mouse mastocytoma with UDP-GlcA and UDP-GlcNAc (Silbert 1963). Subsequent studies, particularly those by Lindahl and coworkers using well-defined oligosaccharides or carbohydrate-serine compounds as exogenous substrates for solubilized microsomal enzymes, demonstrated that assembly of the heparin chain occurred by the alternating transfer of GlcA and GlcNAc from their respective UDP derivatives to the nonreducing end of the growing polymer (Helting and Lindahl 1971, 1972; Forsee and Rodén 1981). According to a current model of heparin/heparan sulfate (HS) biosynthesis (Lidholt and Lindahl 1992), chain elongation occurs simultaneously with polymer modification. Indirect evidence of a bifunctional polymerizing enzyme was obtained from studies of a Chinese hamster ovary (CHO) cell mutant with a single mutation that was defective in both GlcNAc- and GlcA-transferase activities (Lidholt et al. 1992). Subsequently, a protein that catalyzed the in vitro transfer of both GlcA and GlcNAc to the appropriate substrates was purified from bovine serum (Lind et al. 1993). Cloning of the corresponding cDNA identified the GlcA/GlcNAc transferase as EXT2, a member of the exostosin family of tumor suppressors (Lind et al. 1998). The identified protein was referred to as HS-POL (*h*eparan sulfate *pol*ymerase). The same study also implicated EXT1, an additional member of the same family, with the GlcA/GlcNAc transferase reactions following the observation that EXT1 may restore HS biosynthesis in certain mutant cells deficient in this capacity (McCormick et al. 1998). Interestingly, mutational defects in either of the *EXT1* and *EXT2* genes cause multiple exostoses, an autosomal hereditary disorder characterized by excess bony outgrowth at the ends of the long bones (Stickens and Evans 1998).

Enzyme Activity Assay and Substrate Specificity

The mechanism of chain elongation in heparin/HS biosynthesis is illustrated in Fig. 1. The process involves inversion of the α-glucuronidic linkage in UDP-GlcA to form a β-linked product, whereas the α-linkage of GlcNAc is retained. Although solubilized microsomal enzymes and purified EXT proteins can catalyze the transfer of single GlcA or GlcNAc residues to exogenous substrates in vitro, polymerization seems to require an intact membrane-bound biosynthetic complex. Most of the information about the catalytic properties of the GlcA/GlcNAc transferase derives from in vitro studies of crude enzyme preparations from mouse mastocytoma (Lidholt and Lindahl 1992). Oligosaccharides derived from *Escherichia coli* K5 capsular polysaccharide were used as acceptor substrates in transferase assays. The K5 polysaccharide has the same structure ($[-4GlcA\beta1-4GlcNAc\alpha1]_n$) as the nonsulfated heparin/HS precursor molecule; the oligosaccharide derivatives thereof, with nonreducing terminal GlcA or GlcNAc residues, thus serve as acceptors in the GlcNAc- and GlcA-transferase reactions. Both reactions show absolute dependence on the appropriate acceptor structures, as oligosaccharides derived from hyaluronan ($[-4GlcA\beta1-3GlcNAc\beta1]_n$) or chondroitin ($[-4GlcA\beta1-3GalNAc\beta1]_n$) fail to serve as substrates.

Assays based on oligosaccharide substrates give an apparent K_m of about 5 µM for GlcNAc transfer and 500 µM for GlcA transfer (Lidholt and Lindahl 1992). Interestingly, the K_m for GlcNAc transfer is greatly influenced by the length of the acceptor molecule, such that the K_m increases with decreasing acceptor size. By contrast, GlcA transfer appeared relatively unaffected by acceptor size. The GlcNAc transferase reaction requires Mn^{2+} ions, whereas GlcA transfer is stimulated by Ca^{2+} and Mg^{2+} but is not affected by Mn^{2+} ions. Both glycosyltransferases are active within a broad pH interval with an optimum between pH 6.5 and 7.4. A standard assay mixture for both transferase activities contains all three cations at pH 7.2.

Microsomal heparin/HS chain elongation is stimulated by concomitant sulfation of the product (Lidholt et al. 1989). Although the exact coupling between polysaccharide chain elongation and modification has not yet been defined, it is noted that the GlcA transfer reaction shows a preference for substrates containing a penultimate nonreducing-terminal *N*-sulfated GlcN unit (in GlcNAc-GlcA-GlcNSO₃- sequences), presumably generated by *N*-deacetylation followed by *N*-sulfation of GlcNAc residues during biosynthetic polymer modification. By comparison, GlcNAc transfer appears unaffected by similar acceptor sulfation (Lidholt and Lindahl 1992). The effect of

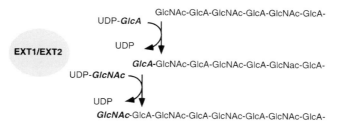

Fig. 1. Polymerization reactions of heparan sulfate biosynthesis

N-sulfation on the actual polymerization process may therefore be explained by a decrease in K_m for GlcA transfer due to the presence of sulfated acceptor sequences, which reduces the 100-fold difference in K_m between the GlcNAc and GlcA transfer reactions.

Although studies by Lind et al. (1998) tentatively suggested that EXT1 and EXT2 are co-polymerases involved in HS biosynthesis, it has been difficult to pinpoint the individual catalytic activities of the two proteins. No increase in glycosyltransferase activities could thus be detected after overexpression of either protein in mammalian cells. The problem is further complicated by the tendency of the EXT proteins to associate with each other and to remain associated after purification of recombinant proteins (Kobayashi et al. 2000; McCormick et al. 2000; Senay et al. 2000). However, expression of the EXT proteins in yeast, which does not synthesize HS, revealed that EXT1 and EXT2 are bifunctional enzymes that harbor both GlcA and GlcNAc transferase activities (Senay et al. 2000). Intriguingly, co-expression experiments showed that EXT1 and EXT2 form hetero-oligomers and that the EXT1/EXT2 complex thus formed has substantially higher glycosyltransferase activities than either EXT1 or EXT2 alone (McCormick et al. 2000; Senay et al. 2000).

Structural Chemistry

EXT1 and EXT2 are type II membrane-bound glycoproteins with an amino-terminal cytoplasmic tail, a transmembrane domain, a stem region, and a large luminal catalytic domain with two potential N-glycosylation sites. In addition to EXT1 and EXT2, three other members of the EXT family, designated EXT-like (EXTL) 1–3, have been cloned. The functions of EXTL1 and EXTL3 are not known, whereas, interestingly, EXTL2 has been identified as an α-GlcNAc transferase catalyzing the addition of the first GlcNAc unit onto the GlcA-Gal$_2$-Xyl-Ser linkage region (Kitagawa et al. 1999). The various members of the EXT family of proteins have highly conserved C-terminal regions with no sequence homology with other families of glycosyltransferases. Biochemical and immunocytochemical studies indicate that overexpressed EXT1 or EXT2 proteins alone are predominantly located in the endoplasmic reticulum (ER), whereas the hetero-oligomers formed upon co-expression are located in the Golgi apparatus (Kobayashi et al. 2000; McCormick et al. 2000; Senay et al. 2000). Furthermore, potentiation of the two catalytic activities observed after co-expression (see above) does not depend on the membrane-bound state of the EXT1 and EXT2 proteins, as a dramatic increase in glycosyltransferase activities was observed also after co-expression of soluble EXT1 and EXT2 (Senay et al. 2000).

Preparation

Both EXT1 and EXT2 are ubiquitously and abundantly expressed in mammalian tissues, as determined by Northern blotting. The EXT proteins have also been described in *Drosophila* (The et al. 1999). A truncated secreted active form of EXT2 (possibly in association with EXT1) has been purified from bovine serum (Lind et al. 1993, 1998). EXT1 and EXT2 have both been expressed in COS cells (Lind et al. 1998; Kobayashi et al. 2000), sog9 cells (McCormick et al. 2000), and the yeast strain *Pichia pastoris* (Senay et al. 2000); and purified recombinant enzymes have been obtained with the yeast system.

Biological Aspects

Mutations in *EXT1* and *EXT2* are associated with the human hereditary multiple exostoses syndrome, characterized by multiple cartilage-capped benign bone tumors around the growth plates in affected individuals (Stickens and Evans 1998). Hereditary multiple exostoses is one of the most common skeletal disorders, with an estimated frequency of 1–2/100 000. Malignant degeneration to chondro- or osteosarcomas occurs in about 2% of patients. In addition to *EXT1* and *EXT2*, mutations in a third gene not yet cloned (*EXT3*) appear to be connected with some cases of multiple exostoses. The *EXTL* genes have not been linked to hereditary multiple exostoses but may be associated with other tumor types. The tumor suppressor function of the *EXT* genes appears to be lost along with the loss of heterozygosity for the *EXT1* and *EXT2* loci.

Mutations affecting the glycosyltransferase EXT1 homolog in *Drosophila* (*tout-velu*) dramatically reduce the amount of HS in the fly embryo (The et al. 1999; Toyoda et al. 2000). The *tout-velu* mutations also impair diffusion of the transcription factor hedgehog (Bellaiche et al. 1998), suggesting, by analogy, that cell surface HS may play a critical role also for proper function of the vertebrate Indian hedgehog, involved in the regulation of normal bone development in humans. HS binds to a multitude of proteins and may thus affect a variety of important biological processes, including specific cell-signaling pathways. It is therefore not surprising that mutations in *EXT1* or *EXT2*, altering heparan sulfate formation, influence proper bone development.

Several related questions remain unresolved. How does the interaction between EXT1 and EXT2 promote enzyme translocation from the endoplasmic reticulum to the Golgi compartment? In what way do EXT1 and EXT2 interact with the other enzymes involved in HS biosynthesis? Finally, how does tumor formation due to defective EXT proteins relate to HS biosynthesis? These and similar questions can hopefully be approached after a more detailed analysis of the EXT proteins and their various interactions at molecular and cellular levels.

References

Bellaiche Y, The I, Perrimon N (1998) *Tout-velu* is a *Drosophila* homologue of the putative tumour suppressor *EXT-1* and is needed for Hh diffusion. Nature 394:85–88
Forsee WT, Rodén L (1981) Biosynthesis of heparin: transfer of *N*-acetylglucosamine to heparan sulfate oligosaccharides. J Biol Chem 256:7240–7247
Helting T, Lindahl U (1971) Occurrence and biosynthesis of β-glucuronidic linkages in heparin. J Biol Chem 246:5442–5447
Helting T, Lindahl U (1972) Biosynthesis of heparin: transfer of *N*-acetylglucosamine and glucuronic acid to low-molecular weight heparin fragments. Acta Chem Scand 26:3515–3523
Kitagawa H, Shimakawa H, Sugahara K (1999) The tumor suppressor EXT-like gene *EXTL2* encodes an α1,4-N-acetylhexosaminyltransferase that transfers *N*-acetylgalactosamine and *N*-acetylglucosamine to the common glycosaminoglycan-

protein linkage region: the key enzyme for the chain initiation of heparan sulfate. J Biol Chem 274:13933–13937

Kobayashi S-i, Morimoto K-i, Shimizu T, Takahashi M, Kurosawa H, Shirasawa T (2000) Association of EXT1 and EXT2, hereditary multiple exostoses gene products, in Golgi apparatus. Biochem Biophys Res Commun 268:860–867

Lidholt K, Lindahl U (1992) Biosynthesis of heparin: the D-glucuronosyl- and N-acetyl-D-glucosaminyltransferase reactions and their relation to polymer modification. Biochem J 289:21–29

Lidholt K, Kjellén L, Lindahl U (1989) Biosynthesis of heparin: relationship between the polymerization and sulphation processes. Biochem J 261:999–1007

Lidholt K, Weinke JL, Kiser CS, Lugemwa FN, Bame KJ, Cheifetz S, Massagué J, Lindahl U, Esko JD (1992) A single mutation affects both N-acetylglucosaminyltransferase and glucuronosyltransferase activities in a Chinese hamster ovary cell mutant defective in heparan sulfate biosynthesis. Proc Natl Acad Sci USA 89:2267–2271

Lind T, Lindahl U, Lidholt K (1993) Biosynthesis of heparin/heparan sulfate: identification of a 70-kDa protein catalyzing both the D-glucuronosyl- and the N-acetyl-D-glucosaminyltransferase reactions. J Biol Chem 268:20705–20708

Lind T, Tufaro F, McCormick C, Lindahl U, Lidholt K (1998) The putative tumor suppressors EXT1 and EXT2 are glycosyltransferases required for the biosynthesis of heparan sulfate. J Biol Chem 273:26265–26268

McCormick C, Duncan G, Goutsos KT, Tufaro F (2000) The putative tumor suppressors EXT1 and EXT2 form a stable complex that accumulates in the Golgi apparatus and catalyzes the synthesis of heparan sulfate. Proc Natl Acad Sci USA 97:668–673

McCormick C, Leduc Y, Martindale D, Mattison K, Esford L, Dyer A, Tufaro F (1998) The putative tumour suppressor EXT1 alters the expression of cell-surface heparan sulfate. Nat Genet 19:158–161

Senay C, Lind T, Muguruma K, Tone Y, Kitagawa H, Sugahara K, Lidholt K, Lindahl U, Kusche-Gullberg M (2000) The EXT1/EXT2 tumor suppressors; catalytic activies and role in heparan sulfate biosynthesis. EMBO Reports 1:282–286

Silbert JE (1963) Incorporation of ^{14}C and ^{3}H from nucleotide sugars into a polysaccharide in the presence of a cell-free preparation from mouse mast cell tumors. J Biol Chem 238:3542–3546

Stickens D, Evans G (1998) A sugar fix for bone tumours? Nat Genet 19:110–111

The I, Bellaiche Y, Perrimon N (1999) Hedgehog movement is regulated through *tout velu*-dependent synthesis of a heparan sulfate proteoglycan. Mol Cell 4:633–639

Toyoda H, Kinoshita-Toyoda A, Selleck SB (2000) Structural analysis of glycosaminoglycans in *Drosophila* and *Caenorhabditis elegans* and demonstration that *tout-velu*, a *Drosophila* gene related to EXT tumor suppressors, affects heparan sulfate in vivo. J Biol Chem 275:2269–2275

57

D-Glucuronyl C5-Epimerase in Heparin/Heparan Sulfate Biosynthesis

Introduction

Glucuronyl C5 epimerase is a key enzyme in the biosynthesis of heparin and heparan sulfate (HS), which are complex sulfated glycosaminoglycans (GAGs) composed of alternating hexuronic acid and glucosamine residues. Both species are synthesized as proteoglycans but on different core proteins (Esko 1991). Heparin is produced by connective tissue-type mast cells, whereas HS is generated by most other mammalian (and many nonmammalian) cells. The process is initiated by glycosylation reactions that generate saccharide sequences composed of alternating D-glucuronic acid (GlcA) and N-acetyl-D-glucosamine (GlcNAc) units, covalently bound to the respective core protein. The resulting polymer of $(GlcA\beta1,4\text{-}GlcNAc\alpha1,4)_n$ disaccharide repeats is modified through a series of reactions that includes N-deacetylation/N-sulfation of GlcNAc residues, C5-epimerization of the GlcA units to yield L-iduronic acid (IdoA) residues, and finally O-sulfation at various positions (Lindahl et al. 1998). The modification reactions occur in a stepwise manner, such that early steps provide the substrates for subsequent reactions. Heparin is extensively sulfated and has a high content of IdoA (typically 50%–90% of the total hexuronic acid), whereas HS has a more varied structure, generally less sulfated than heparin and with lower IdoA content (typically 30%–55% of the total hexuronic acid) (Taylor et al. 1973). The corresponding enzymes have all recently been cloned (see also Chapter 67–69, 70–72).

Jin-Ping Li and Ulf Lindahl

Department of Medical Biochemistry and Microbiology, Section of Medical Biochemistry, Uppsala University, Biomedical Center, Box 582, Husargatan 3, S-751 23 Uppsala, Sweden
Tel. +46-18-4714196 (U. L.); +46-18-4714241 (J-P. Li); Fax +46-18-4714209
e-mail: Ulf.Lindahl@imbim.uu.se; Jin-Ping.Li@imbim.uu.se

Databanks

D-glucuronyl C5-epimerase in heparin/heparan sulfate biosynthesis

No EC number has been allocated

Species	Gene	Protein	mRNA	Genomic
Homo sapiens	–	–	AB020643	–
Bos taurus	–	–	AF003927	–
Caenorhabditis elegans	–	P46555	–	B0285.5
Mus musculus			AF330049	

Name and History

Early investigators of GAG biosynthesis tacitly assumed that the incorporation of IdoA units into GAGs (heparin/HS and dermatan sulfate) would involve the corresponding sugar nucleotide (UDP-IdoA), from which the sugar moiety could be transferred to a growing polysaccharide chain. Although such a sugar nucleotide was reported, its existence has never been confirmed. Instead, experiments using microsomal enzymes from a heparin-producing mouse mastocytoma demonstrated that IdoA units are formed by C5-epimerization of GlcA residues at the polymer level (Höök et al. 1974). Specifically, a $(GlcA-GlcNAc/GlcNH_2)_n$ precursor polysaccharide was generated by incubating the microsomal preparation with labeled UDP-GlcA and unlabeled UDP-GlcNAc. Chase-incubation of the product in the presence of the sulfate donor 3′-phosphoadenosine 5′-phosphosulfate (PAPS) resulted in sulfation of the polysaccharide along with conversion of a fraction of the GlcA to IdoA units.

This conversion of GlcA to IdoA units is catalyzed by glucuronyl C5-epimerase, or heparosan-*N*-sulfate-D-glucopyranosyluronate 5-epimerase, the latter designation distinguishing the enzyme involved in heparin/HS biosynthesis from the corresponding enzyme in dermatan sulfate formation (Malmström and Åberg 1982). A third type of endo-hexuronyl C5-epimerase converts D-mannuronic acid to L-guluronic acid during biosynthesis of alginate in algae and bacteria (Ertesvåg et al. 1995).

Enzyme Activity Assay and Substrate Specificity

To elucidate the mechanism of the GlcA C5-epimerization reaction, microsomal polysaccharide modification was monitored using variously ³H-labeled UDP-GlcA precursor. In the absence of *N*-sulfation, and hence no IdoA formation, the label was quantitatively retained in the incorporated GlcA units irrespective of label position. Conversion of GlcA to IdoA residues, induced by the addition of PAPS, led to loss of ³H from the 5-position but not from the 2-, 3-, or 4-positions of the HexA units (Prihar et al. 1980). Moreover, incubation of the appropriate polysaccharide substrate with solubilized and purified C5-epimerase (from bovine liver) in the presence of ³H₂O resulted in ³H incorporation at C5 of both GlcA and IdoA residues. These results indicated that the inversion of configuration at C5 is freely reversible and occurs by abstraction and readdition of the C5 hydrogen atom, presumably through formation

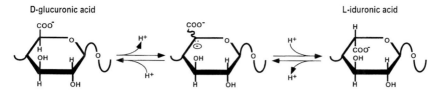

D-glucuronic acid   L-iduronic acid

Fig. 1. Basic mechanism for the hexuronic acid C5 epimerization reaction

of a carbanion intermediate (Fig. 1). Labeling kinetics during incubations in ^3H_2O suggested that the hydrogen exchange process is effected by two polyprotic bases, most likely lysine residues (Hagner-McWhirter et al. 2000b). Equilibrium (~65% GlcA/~35% IdoA) favors retention of D-gluco configuration. By contrast, IdoA formation in the cell-free microsomal system was not accompanied by any significant loss of ^3H from potentially susceptible, but nonepimerized GlcA residues (Jacobsson et al. 1979). Moreover, modification of an individual chain is completed within <30 s, apparently less time than that required for equilibrium to be established. These findings suggest that in the intact (mast) cell interaction between the epimerase and its GlcA target residue is restricted to a single encounter, which leads to the formation of IdoA. Such a mechanism is in accord with a processive mode of polymer modification, as proposed in more recent models of the biosynthetic process (Salmivirta et al. 1996).

The reversible C5 hydrogen exchange mechanism has been exploited in assays of the epimerase activity. Incubation of either $[4GlcA\beta1\text{-}4GlcNSO_3\alpha1\text{-}]_n$ (obtained by chemical N-deacetylation/N-sulfation of *Escherichia coli* K5 capsular polysaccharide) or $[4IdoA\alpha1\text{-}4GlcNSO_3\alpha1\text{-}]_n$ (predominant component of chemically O-desulfated heparin) with epimerase, in the presence of ^3H_2O, thus yields polysaccharide products with 5-^3H-labeled GlcA and IdoA residues (Hagner-McWhirter et al. 2000b). Moreover, an epimerase substrate with exclusively 5-^3H-labeled GlcA units can be prepared following metabolic labeling of the *E. coli* K5 capsular polysaccharide using [5-^3H]glucose (Hagner-McWhirter et al. 2000a). All of these products readily release their label as ^3H_2O upon incubation with the C5-epimerase. The ^3H_2O generated under assay conditions is quantified by scintillation counting after recovery through distillation (Jacobsson et al. 1979) or, more conveniently, after separation from the polysaccharide substrate by partition in a biphasic scintillation system (Campbell et al. 1983) or anion-exchange chromatography (Hagner-McWhirter et al. 2000a).

The C5-epimerase is active over a broad pH range, with the optimum at pH 7.0–7.5. Salt is required for maximal activity (50 mM NaCl or 100 mM KCl); higher concentrations are inhibitory. The enzyme shows no requirement for divalent cations (Malmström et al. 1980). A range of K_m values (4–70 μM expressed as substrate disaccharide units) was found for both the *E. coli* K5- and heparin-based substrates, the K_m unexpectedly increasing with increasing enzyme concentration (Hagner-McWhirter et al. 2000a). The enzyme is fairly stable in purified form when stored in 10% glycerol at 4°C.

Structural analysis of heparin and HS has revealed the occurrence of -GlcNSO$_3$-IdoA-GlcNSO$_3$- and -GlcNSO$_3$-IdoA-GlcNAc-, but not of -GlcNAc-IdoA-GlcNSO$_3$-

sequences (Lindahl et al. 1998). IdoA thus occurs in contiguous N-sulfated domains of the HS chain, as well as in regions composed of interspersed N-acetylated and N-sulfated disaccharide units (Maccarana et al. 1996). Accordingly, enzymatic radio-isotope labeling experiments indicated that substrate recognition requires a potential HexA target residue ($4GlcNSO_3\alpha1$-$4HexA\beta1$-$4GlcNAc/SO_3\alpha1$-) to be linked at C4 to an N-sulfated GlcN neighbor, whereas the unit linked at C1 may be either N-acetylated or N-sulfated (Jacobsson et al. 1984). O-Sulfation at C2 of a HexA unit or at C6 of one or both of the neighboring GlcN residues appears to preclude C5-epimerization. The C5-epimerase involved in heparin/HS formation does not attack chondroitin [$4GlcA\beta1$-$3GalNAc\beta1$-]$_n$, nor does the C5-epimerase involved in der-matan sulfate biosynthesis cross react with N-sulfo-heparosan (i.e., the heparin/HS precursor polysaccharide). A further distinction is that the latter enzyme shows an absolute functional requirement for divalent cations (Malmström and Åberg 1982).

Preparation

The GlcA C5-epimerase, apparently committed to heparin biosynthesis, was first demonstrated in mouse mastocytoma tissue (Höök et al. 1974). Owing to limited amounts of tissue, only partially purified enzyme was obtained from this source. More recent studies showed appreciable epimerase activity in a variety of mammalian tissues, and the enzyme was finally purified to homogeneity, about 1 million-fold, from bovine liver (Campbell et al. 1994). Analysis of the purified protein by sodium dodecyl sulfate-polyacrylamide gel electrophoresis (SDS-PAGE) showed a band with an apparent M_r of about 52 kDa, which was reduced to about 47 kDa following digestion with peptide N-glycosidase F. Peptide sequence information obtained from the purified protein was used to clone the enzyme, initially from a bovine lung cDNA library. A fragment of about 3-kb cDNA was identified with an open reading frame corresponding to a 444-amino-acid residue (M_r 50473) polypeptide. The deduced structure has three potential glycosylation sites, in accord with the size shift observed on SDS-PAGE following deglycosylation. The recombinant protein corresponding to the bovine cDNA clone was expressed in Sf9 insect cells (using a BacPAK8 plasmid vector) and showed weak but definite catalytic activity (Li et al. 1997). This species, apparently a truncated form, lacked a proper transmembrane domain, hence the N-terminal sequence of the full-length protein. The complete DNA sequence was obtained through analysis of murine genomic DNA using bovine cDNA as a probe (Li et al. 2001). Based on this information, new primers were designed and used for poly-merase chain reaction (PCR) cloning of the epimerase from a mouse liver cDNA (Clontech). A cDNA species was obtained encompassing the entire open reading frame for the epimerase, a type II transmembrane protein of 618 amino acid residues. Expression of this protein in the insect system yielded a product with much higher catalytic activity than that of the recombinant bovine protein. Aligning the deduced murine and bovine amino acid sequences indicated 97% identity.

The single gene found for the GlcA C5-epimerase in mouse contains 3 exons, spanning an approximately 14-kb nucleotide sequence. In contrast to several of the other enzymes involved in heparin/HS biosynthesis ["polymerase" (Chapter 54, 56); GlcNAc N-deacetylase/N-sulfotransferase (Chapter 70–72); GlcN 6-O- and 3-O-

sulfotransferases (Chapter 68, 69)], the C5-epimerase has been found to occur only as a single species without isoforms. Notably, semipurified epimerase preparations from bovine liver and mouse mastocytoma showed similar substrate specificity and kinetic properties, in accord with the notion that the same epimerase enzyme is involved in the formation of both HS and heparin (Hagner-McWhirter et al. 2000a).

Biological Aspects

A striking feature of, in particular, HS is its structural diversity, with discrete variants (as assessed by disaccharide composition) occurring in various tissues, with disease conditions, or as a result of aging (Lindahl et al. 1998). Distinct saccharide epitopes, recognized by anti-HS-antibodies, may be expressed by different cells in a composite tissue. This variability is believed to reflect the functional diversity of HS proteoglycans, which bind proteins to the HS chains and thus influence their functional properties. A variety of saccharide structures have been implicated in interactions with proteins, such as antithrombin, growth factors and their receptors, chemokines, enzymes, and viral proteins (Salmivirta et al. 1996; Rosenberg et al. 1997; Lindahl et al. 1998). These regions invariably contain one or more IdoA residues. The IdoA units are believed to promote the interactions due to their marked conformational flexibility (Casu et al. 1988). The reaction catalyzed by the GlcA C5-epimerase thus appears to be crucial for the biological functions of HS (and heparin).

Although the GlcA C5-epimerase has so far been demonstrated only in mammalian tissues, a gene that encodes a protein with homology to the epimerase occurs in *Caenorhabditis elegans* (GenBank accession number B0285.5). Indeed, HS-related polysaccharides appear to be widely spread through the animal kingdom, in vertebrates as well as invertebrates, and have obviously been preserved through major parts of animal evolution (Nader et al. 1999; Lander and Selleck 2000). Further work is needed to elucidate the structural and functional properties of IdoA units in such polysaccharides, as well as of the C5-epimerases involved in their formation.

Future Perspectives

Several intriguing aspects of the GlcA C5-epimerase remain to be clarified and developed. Crystallization and mutational analysis are required to define the active site of the enzyme and the precise catalytic mechanism. Further studies of the gene and its regulatory elements will provide more detailed information on the role of the enzyme in the complex enzymatic machinery involved in the assembly of HS chains. Efforts should be made to identify other proteins, enzymes, or auxiliary components in the ER/Golgi compartments with which the epimerase interacts.

A biotechnological aspect of potential importance should be considered. Heparin is a major antithrombotic drug obtained through large-scale processing of mammalian tissues. An alternative source of heparin manufacture, not involving animal tissues, would be desirable from several viewpoints. Furthermore, novel information regarding functional involvement of HS sequences in processes such as inflammatory reactions, growth factor action, neural development and plasticity, among other factors may potentially lead to the development of drugs based on specific saccharide sequences that contain IdoA residues. Previous work (see above) has identified the *E. coli* K5 capsular polysaccharide as an appropriate target for chemical/enzymatic modification aiming at generating heparin/HS-related products. Notably, whereas many of the *N*- and *O*-sulfation steps required in such processes can be achieved through chemical methods, IdoA residues can be produced only through enzymatic catalysis. The application of recombinant C5-epimerase in such an enterprise may therefore be anticipated.

References

Campbell P, Feingold DS, Jensen JW, Malmström A, Rodén L (1983) New assay for uronosyl 5-epimerases. Anal Biochem 131:146–152

Campbell P, Hannesson HH, Sandbäck D, Rodén L, Lindahl U, Li J-P (1994) Biosynthesis of heparin/heparan sulfate: purification of the D-glucuronyl C-5 epimerase from bovine liver. J Biol Chem 269:26953–26958

Casu B, Petitou M, Provasoli M, Sinaÿ P (1988) Conformational flexibility: a new concept for explaining binding and biological properties of iduronic acid-containing glycosaminoglycans. Trends Biochem Sci 13:221–225

Ertesvåg H, Høidal HK, Hals IK, Rian A, Doseth B, Valla S (1995) A family of modular type mannuronan C-5-epimerase genes controls alginate structure in *Azotobacter vinelandii*. Mol Microbiol 16:719–731

Esko JD (1991) Genetic analysis of proteoglycan structure, function and metabolism. Curr Opin Cell Biol 3:805–816

Hagner-McWhirter A, Hannesson HH, Campbell P, Westley J, Rodén L, Lindahl U, Li J-P (2000a) Biosynthesis of heparin/heparan sulfate: kinetic studies of the glucuronyl C5-epimerase with *N*-sulfated derivatives of the *Escherichia coli* K5 capsular polysaccharide as substrates. Glycobiology 10:159–171

Hagner-McWhirter Å, Lindahl U, Li J-P (2000b) Biosynthesis of heparin/heparan sulphate: mechanism of epimerization of glucuronyl C-5. Biochem J 347:69–75

Höök M, Lindahl U, Bäckström G, Malmström A, Fransson L (1974) Biosynthesis of heparin. 3. Formation of iduronic acid residues. J Biol Chem 249:3908–3915

Jacobsson I, Bäckström G, Höök M, Lindahl U, Feingold DS, Malmström A, Rodén, L (1979) Biosynthesis of heparin: assay and properties of the microsomal uronosyl C-5 epimerase. J Biol Chem 254:2975–2982

Jacobsson I, Lindahl U, Jensen JW, Rodén L, Prihar H, Feingold DS (1984) Biosynthesis of heparin: substrate specificity of heparosan *N*-sulfate D-glucuronosyl 5-epimerase. J Biol Chem 259:1056–1063

Lander AD, Selleck SB (2000) The elusive functions of proteoglycans: in vivo veritas. J Cell Biol 148:227–232

Li J-P, Hagner-McWhirter Å, Kjellén L, Palgi J, Jalkanen M, Lindahl U (1997) Biosynthesis of heparin/heparan sulfate: cDNA cloning and expressing of D-glucuronosyl C5-epimerase from bovine lung. J Biol Chem 272:28158–28163

Li J-P, Gong F, Darwish KE, Jalkanen M, Lindahl U (2001) Characterization of the D-glucuronyl C5-epimerase involved in the biosynthesis of heparin and heparan sulfate. J Biol Chem 276:20069–20077

Lindahl U, Kusche-Gullberg M, Kjellén L (1998) Regulated diversity of heparan sulfate. J Biol Chem 273:24979–24982

Maccarana M, Sakura Y, Tawada A, Yoshida K, Lindahl U (1996) Domain structure of heparan sulfates from bovine organs. J Biol Chem 271:17804–17810

Malmström A, Åberg, L (1982) Biosynthesis of dermatan sulfate: assay and properties of the uronyl C-5 epimerase. Biochem J 201:489–493

Malmström A, Rodén L, Feingold DS, Jacobsson J, Bäckström G, Lindahl U (1980) Biosynthesis of heparin: partial purification of the uronosyl C-5 epimerase. J Biol Chem 255:3878–3883

Nader HB, Chavante SF, dos-Santos EA, Oliveira TW, de Paiva JF, Jeronimo SM, Medeiros GF, de Abreu LR, Leite EL, de Sousa-Filho JF, Castro RA, Toma L, Tersariol IL, Porcionatto MA, Dietrich CP (1999) Heparan sulfates and heparins: similar compounds performing the same functions in vertebrates and invertebrates? Braz J Med Biol Res 32:529–538

Prihar HS, Campbell P, Feingold DS, Jacobsson I, Jensen JW, Lindahl U, Rodén L (1980) Biosynthesis of heparin: hydrogen exchange at carbon 5 of the glycuronosyl residues. Biochemistry 19:495–500

Rosenberg RD, Shworak NW, Liu J, Schwartz JJ, Zhang L (1997) Heparan sulfate proteoglycans of the cardiovascular system: specific structures emerge but how is synthesis regulated? J Clin Invest 99:2062–2070

Salmivirta M, Lidholt K, Lindahl U (1996) Heparan sulfate: a piece of information. FASEB J 10:1270–1279

Taylor RL, Shively JE, Cifonelli JA (1973) Uronic acid composition of heparins and heparan sulfates. Biochemistry 12:3633–3637

Sulfotransferases

Chondroitin 6-Sulfotransferase

Introduction

Chondroitin sulfate chains are composed of GlcAβ1-3GalNAcβ1-4 repeating unit with sulfate groups on GalNAc residues or on both GalNAc and GlcA residues. Sulfate groups attached to the backbone sugar chain produce various chondroitin sulfate isomers. Chondroitin 6-sulfate, which contains sulfate at position 6 of the GalNAc residue, is one of the major components of aggrecan and is thought to be involved in the formation and maintenance of cartilage. Chondroitin 6-sulfate is also contained in other various proteoglycans, such as phosphacan/PTPζ, and is proposed to be involved in neuronal cell interactions with growth factors pleiotrophin and midkine (Maeda et al. 1996, 1999). Chondroitin 6-sulfotransferase (C6ST) catalyzes the transfer of sulfate to position 6 of GalNAc residues of chondroitin. C6ST also sulfates the 6-position of Gal residue of keratan sulfate and sialyl N-acetyllactosamine oligosaccharides.

Databanks

Chondroitin 6-sulfotransferase

NC-IUBMB enzyme classification: E.C.2.8.2.17

Species	Gene	Protein	mRNA	Genomic
Homo sapiens	–	–	AB012192	AB017915
Mus musculus	C6st	–	AB008937	–
Gallus gallus	–	A57397	D49915	–

OSAMI HABUCHI

Department of Life Science, Aichi University of Education, Igaya-cho, Kariya, Aichi 448-8542, Japan
Tel. +81-566-26-2642; Fax +81-566-26-2649
e-mail: ohabuchi@auecc.aichi-edu.ac.jp

Name and History

Sulfotransferase activity that sulfates exogenous chondroitin was found in hen oviduct (Suzuki and Strominger 1960). Sulfotransferase specific to position 6 of the GalNAc residue of chondroitin was partially purified from mouse liver (Monberg et al. 1972), chick embryo cartilage (Habuchi and Miyata 1980), and human serum (Inoue et al. 1986). C6ST was purified to homogeneity from the serum-free culture medium of chick embryo chondrocytes (Habuchi et al. 1993) and chick serum (Sugumaran et al. 1995).

Enzyme Activity Assay and Substrate Specificity

C6ST catalyzed the transfer of sulfate to position 6 of the GalNAc residue of chondroitin (Habuchi et al. 1993) and to position 6 of the Gal residue of keratan sulfate (Habuchi et al. 1996) and sialyl N-acetyllactosamine oligosaccharides (Habuchi et al. 1997).

$$PAPS + (GlcA\beta1\text{-}3GalNAc\beta1\text{-}4)_n \rightarrow PAP + (GlcA\beta1\text{-}3GalNAc(6SO_4)\beta1\text{-}4)_n$$

$$PAPS + (GlcNAc\beta1\text{-}3Gal\beta1\text{-}4)_n \rightarrow PAP + (GlcNAc\beta1\text{-}3Gal(6SO_4)\beta1\text{-}4)_n$$

$$PAPS + NeuAc\alpha2\text{-}3Gal\beta1\text{-}4GlcNAc \rightarrow PAP + NeuAc\alpha2\text{-}3Gal(6SO_4)\beta1\text{-}4GlcNAc$$

where PAPS represents 3'-phosphoadenosine 5'-phosphosulfate, and PAP is 3'-phosphoadenosine 5'-phosphate.

For the C6ST assay, [^{35}S]PAPS is used as a donor and chondroitin as an acceptor (Habuchi et al. 1993). After incubation the sulfated glycosaminoglycan was isolated with ethanol precipitation and gel chromatography on Sephadex G-25 (fast desalting column, Amersham Pharmacia Biotech), and the radioactivity was counted. To determine the sulfation of positions 6 and 4 of the GalNAc residue ^{35}S-labeled glycosaminoglycan was digested with chondroitinase ACII, and unsaturated disaccharides were separated with paper chromatography or Partisil SAX HPLC. To determine the activity toward sialyl N-acetyllactosamine oligosaccharides, the sulfated products were isolated with Superdex 30 gel chromatography (Habuchi et al. 1997).

C6ST is stimulated with basic proteins and polyamines, such as protamine, histone, and spermine. In the presence of basic protein or polyamines, the K_m for PAPS was decreased (Habuchi and Miyata 1980; Habuchi and Miyashita 1982). Optimum protamine concentration varied with the acceptors used: 0.025 mg/ml for chondroitin and 0.075 mg/ml for keratan sulfate (Habuchi et al. 1993). Among divalent cations, Ca^{2+}, Sr^{2+}, Ba^{2+}, Mn^{2+}, and Co^{2+} showed a stimulatory effect. Stimulation with Co^{2+} was observed only in the presence of imidazole buffer. The optimal pH was 5.8 for chondroitin and 6.2 for keratan sulfate (Habuchi et al. 1996). Sulfhydryl compounds such as dithiothreitol showed no significant effect (Habuchi et al. 1991).

In addition to chondroitin and keratan sulfate, chondroitin sulfate A and chondroitin sulfate C served as acceptors. Chondroitin sulfate E from squid cartilage, dermatan sulfate, and heparan sulfate hardly served as acceptors of C6ST. When chondroitin sulfate A was used as an acceptor, sulfate was transferred to position 6 of the GalNAc residue but not to position 6 of the GalNAc 4-sulfate residue (Habuchi et al. 1993). Sulfation of the Gal residue in sialyl *N*-acetyllactosamine oligosaccharides was stimulated when the sulfate moiety was attached to the GlcNAc residue adjacent to the targeted Gal residue. Sialyl Lewis X tetrasaccharide did not serve as acceptor, indicating that Fuc attached to the GlcNAc residue may inhibit sulfation of the Gal residue (Habuchi et al. 1997).

Preparation

Chondroitin 6-sulfotransferase (C6ST) was purified from the serum-free culture medium of chick embryo chondrocytes with affinity chromatography on heparin-Sepharose CL-6B, wheat germ agglutinin-agarose, and 3′,5′-ADP-agarose (Habuchi et al. 1993). The purified C6ST showed a broad protein band of 75 kDa by sodium dodecyl sulfate polyacrylamide gel electrophoresis (SDS-PAGE). The enzyme has been expressed in COS-7 cells (Fukuta et al. 1995) and COS-1 cells (Tsutsumi et al. 1998).

Biological Aspects

A C6ST message of 7.8 kb was widely expressed in various human tissues including heart, placenta, skeletal muscle, and pancreas (Fukuta et al. 1998). Mouse C6ST was expressed strongly in the spleen and in decreasing amounts in lung, eye, and stomach (in that order). Chondroitin 6-sulfate, the product of C6ST, may play an important role in hematopoiesis, as C6ST mRNA was expressed in bone marrow and in stromal cells in the marginal zone and red pulp of the spleen (Uchimura et al. 1998a). During the development of chick embryo brain, expression of C6ST was found to be decreased (Kitagawa et al. 1997). C6ST null mice were generated (Uchimura et al. 1999). These mice were fertile and showed no obvious pathological phenotype, although corneal healing was delayed.

C6ST was secreted from cultured chondrocytes. Because the same size of C6ST was obtained from chicken serum, secretion of C6ST into the extracellular space seem to occur in the physiological state as well (Sugumaran et al. 1995). During the secretion C6ST may be cleaved at the transmembrane domain because the amino acid sequence of the NH_2-terminal of the purified C6ST was found in the transmembrane domain (Fukuta et al. 1995).

Nervous system-involved sulfotransferase (NSIST, accession number AF079875), which was cloned from the *Torpedo* electric organ by expression cloning (Nastuk et al. 1998), showed high sequence homology with C6ST and was thought to be an isoform of C6ST. Monoclonal antibody (mAb) 3B3, used for expression cloning, recognizes agrin-binding proteins (Bowe et al. 1994). One of the epitope-bearing proteins was found to be dystroglycan. The epitope for the mAb was deduced to be not glycosaminoglycans but oligosaccharides with sialic acid. The message of NSIST with 2.4 kb is expressed mainly in the electric organ. These observations indicate that NSIST participates in nervous system-specific posttranslational modification of membrane proteins (Nastuk et al. 1998).

C6ST is a prototype of the carbohydrate 6-sulfotransferase family. In this family, C6ST, keratan sulfate Gal-6-sulfotransferase (Fukuta et al. 1997), GlcNAc 6-*O*-sulfotransferase (Uchimura et al. 1998b), HEC-GlcNAc 6-*O*-sulfotransferase or LSST

(Bistrup et al. 1999; Hiraoka et al. 1999), and I-GlcNAc 6-O-sulfotransferase (Lee et al. 1999) have been cloned. These sulfotransferases showed high sequence homology in three regions—5'-PSB domain, 3'-PB domain, C-terminal region—and share substrate specificity such that sulfate is transferred to position 6 of the targeted sugar residues: GalNAc, GlcNAc, or Gal residues.

Future Perspectives

C6ST catalyzes transfer of sulfate to both chondroitin and keratan sulfate in vitro. It remains to be determined to what extent C6ST is involved in the biosynthesis of keratan sulfate. C6ST and C4ST are both secreted into the culture medium from chondrocytes or chondrosarcoma cells, but the physiological roles of the secreted sulfotransferases are not known because PAPS has not been detected in the extracellular space. During secretion, C6ST and C4ST are truncated at their amino-terminal region by proteolytic cleavage; characterization of the proteases involved in such cleavage during secretion remains to be determined. Further analysis of C6ST knockout mice may clarify the biological functions of C6ST.

Further Reading

See Habuchi (2000) for a review.

References

Bistrup A, Bhakta S, Lee JK, Belov YY, Gunn MD, Zuo F-R, Huang C-C, Kannagi R, Rosen SD, Hemmerich S (1999) Sulfotransferases of two specificities function in the reconstitution of high endothelial cell ligand for L-selectin. J Cell Biol 145:899–910

Bowe MA, Deyst KA, Leszyk JD, Fallon J (1994) Identification and purification of an agrin receptor from Torpedo postsynaptic membranes: a heteromeric complex related to the dystroglycans. Neuron 12:1173–1180

Fukuta M, Inazawa J, Torii T, Tsuzuki K, Shimada E, Habuchi O (1997) Molecular cloning and characterization of human keratan sulfate Gal-6-sulfotransferase. J Biol Chem 272:32321–32328

Fukuta M, Kobayashi Y, Uchimura K, Kimata K, Habuchi O (1998) Molecular cloning and expression of human chondroitin 6-sulfotransferase. Biochim Biophys Acta 1399: 57–61

Fukuta M, Uchimura K, Nakashima K, Kato M, Kimata K, Shinomura T, Habuchi O (1995) Molecular cloning and expression of chick chondrocyte chondroitin 6-sulfotransferase. J Biol Chem 270:18575–18580

Habuchi O (2000) Diversity and functions of glycosaminoglycan sulfotransferases. Biochim Biophys Acta 1474:115–127

Habuchi O, Miyashita N (1982) Separation and characterization of chondroitin 6-sulfotransferase and chondroitin 4-sulfotransferase from chick embryo cartilage. Biochim Biophys Acta 717:414–421

Habuchi O, Miyata K (1980) Stimulation of glycosaminoglycan sulfotransferase from chick embryo cartilage by basic proteins and polyamines. Biochim Biophys Acta 616:208–217

Habuchi O, Hirahara Y, Uchimura K, Fukuta M (1996) Enzymatic sulfation of galactose residue of keratan sulfate by chondroitin 6-sulfotransferase. Glycobiology 6:51–57

Habuchi O, Matsui Y, Kotoya Y, Aoyama Y, Yasuda Y, Noda M (1993) Purification of chondroitin 6-sulfotransferase secreted from cultured chick embryo chondrocytes. J Biol Chem 268:21968–21974

Habuchi O, Suzuki Y, Fukuta M (1997) Sulfation of sialyl lactosamine oligosaccharides by chondroitin 6-sulfotransferase. Glycobiology 7:405–412

Habuchi O, Tsuzuki M, Takeuchi I, Hara M, Matsui Y, Ashikari S (1991) Secretion of chondroitin 6-sulfotransferase and chondroitin 4-sulfotransferase from cultured chick embryo chondrocytes. Biochim Biophys Acta 1133:9–16

Hiraoka N, Petryniak B, Nakayama J, Tsuboi S, Suzuki M, Yeh JC, Izawa D, Tanaka T, Miyasaka M, Lowe JB, Fukuda M (1999) A novel, high endothelial venule-specific sulfotransferase expresses 6-sulfo sialyl Lewis(x), an L-selectin ligand displayed by CD34. Immunity 11:79–89

Inoue H, Otsu K, Yoneda M, Kimata K, Suzuki S, Nakanishi Y (1986) Glycosaminoglycan sulfotransferases in human and animal sera. J Biol Chem 261:4460–4469

Kitagawa H, Tsutsumi K, Tone Y, Sugahara K (1997) Developmental regulation of the sulfation profile of chondroitin sulfate chains in the chicken embryo brain. J Biol Chem 272:31377–31381

Lee JK, Bhakta S, Rosen SD, Hemmerich S (1999) Cloning and characterization of mammalian N-acetylglucosamine-6-sulfotransferase that is highly restricted to intestinal tissue. Biochem Biophys Res Commun 263:543–549

Maeda N, Ichihara-Tanaka K, Kimura T, Kadomatsu K, Muramatsu T, Noda M (1999) A receptor-like protein-tyrosine phosphatase PTPζ/RPTPβ binds a heparin-binding growth factor midkine. J Biol Chem 274:12474–12479

Maeda N, Nishiwaki T, Shintani T, Hamanaka H, Noda M (1996) 6B4 proteoglycan/phosphacan, an extracellular variant of receptor-like protein-tyrosine phosphatase ζ/RPTPβ, binds pleiotrophin/heparin-binding growth-associated molecule (HB-GAM). J Biol Chem 271:21446–21452

Monberg M, Stuhlsatz HW, Kisters R, Greiling H (1972) Isolierung und Substratspezifität einer 3'-Phpsphoadenylylsulfat: Chondroitin-6-Sulfotransferase aus der Mäuseleber. Hoppe Seylers Z Physiol Chem 353:1351–1361

Nastuk MA, Davis A, Yancopoulos GD, Fallon JR (1998) Expression cloning and characterization of NSIST, a novel sulfotransferase expressed by a subset of neurons and postsynaptic targets. J Neurosci 18:7167–7177

Sugumaran G, Katsman M, Drake RR (1995) Purification, photoaffinity labeling, and characterization of a single enzyme for 6-sulfation of both chondroitin sulfate and keratan sulfate. J Biol Chem 270:22483–22484

Suzuki S, Strominger JL (1960) Enzymatic sulfation of mucopolysaccharides in hen oviduct. I. Transfer of sulfate from 3'-phosphoadenosine 5'-phosphosulfate to mucopolysaccharides. J Biol Chem 235:257–266

Tsutsumi K, Shimakawa H, Kitagawa H, Sugahara K (1998) Functional expression and genomic structure of human chondroitin 6-sulfotransferase. FEBS Lett 441:235–241

Uchimura K, Kadomatsu K, Fan Q-W, Muramatsu H, Kurosawa N, Kaname T, Yamamura K, Fukuta M, Habuchi O, Muramatsu T (1998a) Mouse chondroitin 6-sulfotransferase: molecular cloning, characterization and chromosomal mapping. Glycobiology 8: 489–496

Uchimura K, Muramatsu H, Kadomatsu K, Fan Q-W, Kurosawa N, Mitsuoka C, Kannagi R, Habuchi O, Muramatsu T (1998b) Molecular cloning and characterization of an N-acetylglucosamine-6-O-sulfotransferase. J Biol Chem 273:22577–22583

Uchimura K, Kadomatsu K, Muramatsu H, Ishihama H, Nakamura E, Kurosawa N, Habuchi O, Muramatsu T (1999) Targeted disruption of the mouse chondroitin 6-sulfotransferase gene. Abstracts of International Conference on Molecular Interactions of Proteoglycans, Shonan Village Center, Kanagawa, Japan

59

Keratan Sulfate Gal-6-Sulfotransferase

Introduction

Keratan sulfate (KS) is composed of the repeating disaccharide unit of Galβ1-4GlcNAc (poly-*N*-acetyllactosamine) with sulfate groups at the 6-position of the Gal and GlcNAc residues. Most GlcNAc residues and half of the Gal residues of corneal KS are sulfated, suggesting that sulfation of the Gal residue occurs after sulfation of the GlcNAc residue and determines the extent of sulfation of KS. Keratan sulfate Gal-6-sulfotransferase (KSGal6ST) catalyzes the transfer of sulfate to position 6 of the Gal residue of KS. Therefore it has been thought that KSGal6ST is responsible for producing the highly sulfated type of KS. KSGal6ST is also able to sulfate sialyl *N*-acetyllactosamine oligosaccharides. KSGal6ST may be engaged in the biosynthesis of not only KS but also glycoproteins bearing sulfated sialyl *N*-acetyllactosamine or sulfated sialyl Lewis X oligosaccharides.

Databanks

Keratan sulfate Gal-6-sulfotransferase

NC-IUBMB enzyme classification: E.C.2.8.2.21

Species	Gene	Protein	mRNA	Genomic
Homo sapiens	–	–	AB003791	–
		–	U65637	–

MASAKAZU FUKUTA and OSAMI HABUCHI

Department of Life Science, Aichi University of Education, Igaya-cho, Kariya, Aichi 448-8542, Japan
Tel. +81-566-26-2642; Fax +81-566-26-2649
e-mail: ohabuchi@auecc.aichi-edu.ac.jp

Name and History

A partially purified sulfotransferase from bovine corneal cells was reported to catalyze the transfer of sulfate to desulfated KS but not to chondroitin (Rütter and Kresse 1984). Because C6ST is able to sulfate both chondroitin and KS (Habuchi et al. 1993), an unidentified sulfotransferase with distinct substrate specificity from C6ST was expected to be present in the cornea. A cDNA was cloned from the fetal human brain library by cross-hybridization with chick chondroitin 6-sulfotransferase (C6ST) cDNA. The amino acid sequence of the protein deduced from the human cDNA showed 37% homology with chick C6ST, and the protein expressed in COS-7 cells catalyzed the transfer of sulfate to position 6 of the Gal residues of KS but not to the GalNAc residue of chondroitin. Based on these substrate specificities, this protein was named KS Gal-6-sulfotransferase (KSGal6ST) (Fukuta et al. 1997).

Enzyme Activity Assay and Substrate Specificity

KSGal6ST catalyzed the transfer of sulfate not only to position 6 of the Gal residue of KS but also to position 6 of the Gal residue of sialyl N-acetyllactosamine oligosaccharides.

$$PAPS + [Gal\beta1\text{-}4GlcNAc(6\text{-}SO_4)\beta1\text{-}3]_n \rightarrow$$
$$PAP + [Gal(6\text{-}SO_4)\beta1\text{-}4GlcNAc(6\text{-}SO_4)\beta1\text{-}3]_n$$

$$PAPS + NeuAc\alpha2\text{-}3Gal\beta1\text{-}4GlcNAc \rightarrow PAP + NeuAc\alpha2\text{-}3Gal(6\text{-}SO_4)\beta1\text{-}4GlcNAc$$

where PAPS is 3'-phosphoadenosine 5'-phosphosulfate (PAPS), and PAP is 3'-phosphoadenosine 5'-phosphate.

In the KSGal6ST assay, [^{35}S]PAPS is used as a donor and KS as an acceptor. After incubation the sulfated glycosaminoglycan (GAG) was isolated with ethanol precipitation and gel chromatography on Sephadex G-25 (fast desalting column, Amersham Pharmacia Biotech), and the radioactivity was counted. To determine the activity toward sialyl N-acetyllactosamine oligosaccharides, sulfated products were isolated with Superdex 30 gel chromatography (Torii et al. 2000). KSGal6ST is stimulated with divalent cations such as Mn^{2+}, Ca^{2+}, Co^{2+}, Sr^{2+}, Ba^{2+}, and Mg^{2+}. Among these ions, Mn^{2+} and Ca^{2+} showed the highest stimulatory effects. The optimal Ca^{2+} concentration was 10 mM. Protamine, which was the best activator for C6ST, was less effective than Ca^{2+}. Optimal pH was around 6.5. Sulfhydryl compounds such as dithiothreitol showed no significant effect (Torii et al. 2000).

Keratan sulfate and partially desulfated KS served as acceptors. Chondroitin, chondroitin sulfate A, chondroitin sulfate C, dermatan sulfate, and completely desulfated N-resulfated heparin hardly served as acceptors of KSGal6ST. When KS and partially desulfated KS were used as acceptors, sulfate was transferred to position 6 of the Gal residue, and no sulfation of GlcNAc residue was observed (Fukuta et al. 1997). When partially desulfated KS [the ratio of Galβ1-4GlcNAc/Galβ1-4GlcNAc(6SO$_4$) is 0.73] was used as an acceptor, sulfate was transferred mainly to the Gal residue adjacent to GlcNAc(6SO$_4$), suggesting that the sulfate moiety attached to GlcNAc stimulates the rate of sulfation of the Gal residue (Fukuta et al. 1997). The expressed KSGal6ST

Table 1. Incorporation of $^{35}SO_4$ into sialyl N-acetyllactosamine oligosaccharides, fetuin and keratan sulfate

Acceptor	Relative incorporation of $^{35}SO_4$ (%)	K_m (mM)	Vmax (pmol/min/mg protein)
LN	ND	–	–
LNS	ND	–	–
3'SLN	4.0	–	–
SLe^x	ND	–	–
L1L1	7.9	–	–
SL1L1	6.7	–	–
L2L4	48.0	10.40	85.0
SL2L4	250.0	0.65	20.0
Fetuin	8.8	–	–
Keratan sulfate	100.0	0.38[a]	6.1

From Torii et al. 2000, LN, Galβ1-4GlcNAc; LNS, Galβ1-4GlcNAc(6SO₄); 3'SLN, NeuAcα2-3Galβ1-4GlcNAc; SLe^x, NeuAcα2-3Galβ1-4(Fucα1-3)GlcNAc; L1L1, Galβ1-4GlcNAcβ1-3Galβ1-4GlcNAc; SL1L1, NeuAcα2-3Galβ1-4GlcNAcβ1-3Galβ1-4GlcNAc; L2L4, Galβ1-4GlcNAc(6SO₄)β1-3Gal(6SO₄)β1-4GlcNAc(6SO₄); SL2L4, NeuAcα2-3Galβ1-4GlcNAc(6SO₄)β1-3Gal(6SO₄)β1-4GlcNAc(6SO₄)
ND, not detected
[a] Expressed as the concentration of repeating disaccharide units

catalyzed transfer of sulfate not only to position 6 of the Gal residues of KS but also to position 6 of the Gal residue of sialyl N-acetyllactosamine oligosaccharides and fetuin oligosaccharides (Table 1). When KSGal6ST was expressed as a fusion protein with FLAG peptide and purified with an anti-FLAG affinity column, the affinity purified-protein sulfated both KS and SL2L4 (abbreviations of oligosaccharides are indicated in the footnote of Table 1) but did not sulfate chondroitin at all, indicating that a single protein catalyzes sulfation of both KS and SL2L4. Moreover, unlike C6ST, KSGal6ST is unable to utilize chondroitin as acceptor. The relative rate of sulfation of SL2L4 was much higher than the rate of sulfation of KS. A comparison of the kinetic parameters for SL2L4 and L2L4 indicates that sialic acid attached to the nonreducing end of the sulfated oligosaccharide caused a marked decrease in K_m. On the other hand, a comparison of the kinetic parameters between SL2L4 and KS indicates that the higher incorporation of sulfate into SL2L4 was due mainly to the increase in Vmax. Sialyl Lewis X tetrasaccharide did not serve as an acceptor, indicating that Fuc attached to the GlcNAc residue may inhibit sulfation of the Gal residue (Torii et al. 2000).

Preparation

Recombinant KSGal6ST was expressed in COS-7 cells (Fukuta et al. 1997). FLAG-KSGal6ST was purified with anti-FLAG monoclonal antibody-conjugated affinity column (Torii et al. 2000).

Biological Aspects

The KSGal6ST message of 2.8 kb is expressed in human brain (Fukuta et al. 1997) and various immunologically relevant tissues such as spleen, lymph nodes, thymus, and appendix (Torii et al. 2000). KSGal6ST may participate in the synthesis of KS in the cornea because a message that was cross-hybridized with human KSGal6ST cDNA was expressed in chick embryo cornea (Fukuta et al. 1997).

Macular corneal dystrophy (MCD) is an inherited disorder characterized by corneal opacity. Corneas of MCD type I patients were reported to synthesize nearly normal amounts of poly-N-acetyllactosamine backbone structures but failed to sulfate the polysaccharide chain (Nakazawa et al. 1984; Midura et al. 1990). The sulfation of KS thus appears to be crucial for maintaining the proper spatial organization of type I collagen fibrils and thereby results in corneal transparency. This hypothesis has been partly supported by knockout mice that lack the core protein of lumican, one of the corneal KS proteoglycans (Chakravarti et al. 1998). In these mice the regularity of spaced collagen fibrils in the cornea was found to be partially missing. It has now been shown that GlcNAc 6-O-sulfotransferase activity was markedly decreased in the cornea of MCD patients, whereas Gal 6-O-sulfotransferase activity was not affected (Hasegawa et al. 1999). From the substrate specificity of KSGal6ST described above, 6-sulfation of Gal might be decreased when sulfation of GlcNAc is decreased or disappears.

Highly sulfated KS was reported to be reduced in the brains of Alzheimer patients compared to that of normal controls (Lindahl et al. 1996). As a brain KS proteoglycan, SV2 is known to exist in synaptic vesicle membranes and to extrude KS chains into the vesicle lumen (Bajjalieh et al. 1992; Feany et al. 1992; Scranton et al. 1993). The negatively charged KS of SV2 might form a charged gel matrix, conferring enough negative charge to store high concentrations of positively charged neurotransmitter, such as acetylcholine, in synaptic vesicles (Rahamimoff and Fernandez 1997). The substrate specificity of KSGal6ST and its strong expression in the brain suggest that KSGal6ST is involved in biosynthesis of the highly sulfated KS, which is reduced in the brains of Alzheimer patients and possibly involved in the storage of neurotransmitters in synaptic vesicles. The knockout mice deficient in SV2A (one of the SV2 isoforms) (Crowder et al. 1999) or both SV2 isoforms (SV2A and SV2B) (Janz et al. 1999) have been reported; these mice exhibit severe seizures and die postnatally. Although multiple functions are suggested for SV2, these results may indicate an essential role of KS in normal nervous system functioning.

KSGal6ST may function in the reconstitution of high endothelial cell ligand for L-selectin: When KSGal6ST was introduced into CHO cells previously transfected with fucosyl transferase VII and core 2 β1-6GlcNAc transferase, enhanced binding activity to L-selectin/immunoglobulin M (IgM) chimera was detected (Bistrup et al. 1999).

Future Perspectives

KSGal6ST transfers sulfate in vitro to the Gal residue of both KS and sialyl N-acetyllactosamine oligosaccharides at nearly the same rate. C6ST also transfers sulfate to both acceptors, but the rate of sulfation of sialyl N-acetyllactosamine oligosaccharides was much lower than the sulfation rate for KS. The physiological acceptor for KSGal6ST and the extent to which KSGal6ST and C6ST are involved in the sulfation of KS remain to be determined. KSGal6ST knockout mice, if generated, can address these issues and provide further evidence about the physiological roles of KS in the brain and cornea.

Further Reading

See Habuchi (2000) for a review.

References

Bajjalieh SM, Peterson K, Shinghal R, Sheller RH (1992) SV2, a brain synaptic vesicle protein homologous to bacterial transporters. Science 257:1271–1273

Bistrup A, Bhakta S, Lee JK, Belov YY, Gunn MD, Zuo F-R, Huang C-C, Kannagi R, Rosen SD, Hemmerich S (1999) Sulfotransferases of two specificities function in the reconstitution of high endothelial cell ligand for L-selectin. J Cell Biol 145:899–910

Chakravarti S, Magnuson T, Lass JH, Jepsen KJ, LaMantia C, Carroll H (1998) Lumican regulates collagen fibril assembly: skin fragility and corneal opacity in the absence of lumican. J Cell Biol 141:1277–1286

Crowder KM, Gunther JM, Jones TA, Hale BD, Zhang HZ, Peterson MR, Sheller RH, Chavkin C, Bajjalieh SM (1999) Abnormal neurotransmission in mice lacking synaptic vesicle protein 2A. Proc Natl Acad Sci USA 96:15268–15273

Feany MB, Lee S, Edwards RH, Buckley KM (1992) The synaptic vesicle protein SV2 is a novel type of transmembrane transporter. Cell 70:861–867

Fukuta M, Inazawa J, Torii T, Tsuzuki K, Shimada E, Habuchi O (1997) Molecular cloning and characterization of human keratan sulfate Gal-6-sulfotransferase. J Biol Chem 272:32321–32328

Habuchi O (2000) Diversity and functions of glycosaminoglycan sulfotransferases. Biochim Biophys Acta 1474:115–127

Habuchi O, Matsui Y, Kotoya Y, Aoyama Y, Yasuda Y, Noda M (1993) Purification of chondroitin 6-sulfotransferase secreted from cultured chick embryo chondrocytes. J Biol Chem 268:21968–21974

Hasegawa N, Torii T, Kato T, Miyajima H, Nakayasu K, Kanai A, Habuchi O (1999) Determination of activities of galactose 6-O-sulfotransferase and N-acetylglucosamine 6-O-sulfotransferase in the cornea of patients with macular corneal dystrophy. Glycoconj J 16:S155

Janz R, Goda Y, Geppert M, Missler M, Südhof TC (1999) SV2A and SV2B function as redundant Ca^{2+} regulators in neurotransmitter release. Neuron 24:1003–1016

Lindahl B, Erickson L, Spillmann D, Caterson B, Lindahl U (1996) Selective loss of cerebral keratan sulfate in Alzheimer's disease. J Biol Chem 271:16991–16994

Midura RJ, Hascall VC, MacCallum DK, Meyer RF, Thonar EJ-MA, Hassell JR, Smith CF, Klintworth GK (1990) Proteoglycan biosynthesis by human corneas from patients with type 1 and 2 macular corneal dystrophy. J Biol Chem 265:15947–15955

Nakazawa K, Hassel JR, Hascall VC, Lohmander S, Newsome DA, Krachmer J (1984) Defective processing of keratan sulfate in macular corneal dystrophy. J Biol Chem 259:13751–13757

Rahamimoff R, Fernandez JM (1997) Pre- and postfusion regulation of transmitter release. Neuron 18:17–27

Rütter ER, Kresse H (1984) Partial purification and characterization of 3'-phosphoadenylylsulfate:keratan sulfate sulfotransferases. J Biol Chem 259:11771–11776

Scranton TW, Iwata M, Carlson SS (1993) The SV2 protein of synaptic vesicles is a keratan sulfate proteoglycan. J Neurochem 61:29–44

Torii T, Fukuta M, Habuchi O (2000) Sulfation of sialyl N-acetyllactosamine oligosaccharides and fetuin oligosaccharides by keratan sulfate Gal-6-sulfotransferase. Glycobiology 10:203–211

Corneal *N*-Acetylglucosamine 6-*O*-Sulfotransferase

Introduction

Keratan sulfate glycosaminoglycan, which is found mainly in the cornea and in cartilage tissues, consists of a linear poly-*N*-acetyllactosamine chain that carries sulfate residues on its C-6 position of GlcNAc and Gal. So far two human sulfotransferases have been reported to transfer sulfate on poly-*N*-acetyllactosamine and to produce keratan sulfate. One of the sulfotransferases, corneal GlcNAc 6-*O*-sulfotransferase (C-GlcNAc6ST, GlcNAc6ST-5) is known to be involved in a hereditary eye disease, macular corneal dystrophy. This enzyme has an important role in processing keratan sulfate.

Databanks

Corneal *N*-acetylglucosamine 6-*O*-sulfotransferase

NC-IUBMB enzyme classification: EC 2.8.2.21

Species	Gene	Protein	mRNA	Genomic
Homo sapiens	CHST6	–	AF219990	AF219991

Name and History

The gene *CHST6* (*carbohydrate sulfotransferase 6*) has been identified as being responsible for macular corneal dystrophy (MCD) (Akama et al. 2000). MCD is an autosomal recessive hereditary disease that presents with progressive punctate opacities in the cornea. It is classified into two subtypes (I and II) defined by the respective absence and presence of sulfated keratan sulfate (KS) in the patient's serum

Tomoya O. Akama and Michiko N. Fukuda

Glycobiology Program, The Burnham Institute, 10901 North Torrey Pines Road, La Jolla, CA 92037, USA
Tel. +1-858-646-3100 (ext. 3682); Fax +1-858-646-3193
e-mail: takama@burnham-inst.org

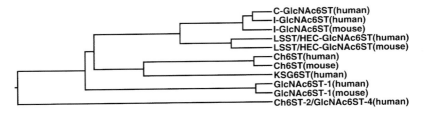

Fig. 1. Amino acid sequence relations among human and mouse carbohydrate sulfotransferases. Corneal GlcNAc 6-O-sulfotransferase (C-GlcNAc6ST), intestinal GlcNAc 6-O-sulfotransferase (I-GlcNAc6ST), L-selectin ligand sulfotransferase (LSST), or high endothelial-cell GlcNAc 6-O-sulfotransferase (HEC-GlcNAc6ST), chondroitin 6-O-sulfotransferase (Ch6ST), keratan sulfate Gal 6-O-sulfotransferase (KSG6ST), GlcNAc 6-O-sulfotransferase-1 (GlcNAc6ST-1), and chondroitin 6-O-sulfotransferase-2 (Ch6ST-2) or GlcNAc 6-O-sulfotransferase-4 (GlcNAc6ST-4) were aligned using the Clustal W program and are shown as a phylogenic tree

(Edward et al. 1988; Yang et al. 1988). The gene responsible for MCD has been mapped to chromosome 16q22 by linkage analyses (Vance et al. 1996; Liu et al. 1998). Biochemical analyses indicate that the cornea from MCD patients synthesizes normal levels of poly-N-acetyllactosamine but does not have keratan sulfate, suggesting that the sulfation step of keratan sulfate may be impaired in MCD (Hassel et al. 1980; Nakazawa et al. 1984).

The gene responsible for MCD has been cloned by the positional candidate approach (Akama et al. 2000). The gene (*CHST6*) was mapped to chromosome 16q22 by radiation hybrid analysis. *CHST6* encodes a type II membrane protein that is highly homologous to the carbohydrate sulfotransferases and is particularly homologous to human intestinal GlcNAc 6-O-sulfotransferase (I-GlcNAc6ST) (Fig. 1). Identities in the coding sequences of the gene and I-GlcNAc6ST are 90.6% at the nucleotide levels and 89.2% in the amino acid sequences, strongly suggesting that the gene encodes a carbohydrate sulfotransferase. Because this transcript is expressed in the cornea, the gene product was named corneal GlcNAc-6-sulfotransferase (C-GlcNAc6ST).

Enzyme Activity Assay and Substrate Specificity

There are no reports on the enzymatic activity of C-GlcNAc6ST. Based on the results that mutations of *CHST6* were found in the genomic DNAs from MCD patients who have no keratan sulfate in the their cornea and serum, C-GlcNAc6ST is suggested to have sulfotransferase activity that transfers sulfate on the C-6 position of GlcNAc and produces keratan sulfate (Akama et al. 2000). Nakazawa et al. (1998) reported that a protein fraction from chick cornea has GlcNAc 6-O sulfotransferase activity that transfers sulfate on GlcNAcβ1-3Galβ1-4Glc-PA, GlcNAcβ1-3(GlcNAcβ1-6)Galβ1-4Glc-PA, and GlcNAcβ1-3Galβ1-4GlcNAc-PA in vitro. Hasegawa et al. (2000) demonstrated that a protein extract from human cornea has GlcNAc 6-O-sulfotransferase activity that uses GlcNAcβ1-3Galβ1-4GlcNAc as a substrate. Because C-GlcNAc6ST is found in the human cornea, the enzyme in these reports is likely to be C-GlcNAc6ST. The reaction condition Hasegawa et al. (2000) used is follows: 2 µg protein from an enzyme source in 50 μl of 50 mM imidazole HCl (pH 6.8), 10 mM MnCl$_2$, 2 mM 5'-AMP, 20 mM NaF, 1 nM [^{35}S] 3'-phosphoadenosine 5'-phosphosulfate (PAPS), and 0.5 mM carbohydrate substrate. The reaction mixture was incubated at 20°C for up to 40 h. Reaction products were separated by gel chromatography, and incorporation of ^{35}S into the substrates was measured by scintillation counting.

Preparation

Because C-GlcNAc6ST mRNA is found mainly in the cornea, spinal cord, and trachea, the enzyme may be prepared from these tissues. Previous reports indicated that C-GlcNAc6ST can be prepared from the cornea (Nakazawa et al. 1998; Hasegawa et al. 2000). Transfected cells are also a good source of C-GlcNAc6ST (Hemmerich et al. 2001).

Biological Aspects

Mutation Analysis of CHST6 in MCD Patients

By mutation search study, frameshift, deletion, and missense mutations were found in the coding region of CHST6 in MCD type I patients (Akama et al. 2000). All of the missense mutations were located around conserved regions among carbohydrate sulfotransferases, suggesting that lack of sulfation on GlcNAc in keratan sulfate leads to an MCD type I phenotype.

On the other hand, deletion and substitutional DNA rearrangements were found on the upstream region of CHST6 in MCD type II patients. Because the upstream region of CHST6 is likely to have a gene regulatory element that affects the expression of C-GlcNAc6ST, MCD type II phenotypes may be caused by lack of CHST6 expression in corneal cells. It is possible that the mutations found in MCD type II patients do not affect the expression of CHST6 except in corneal tissues. The presence of keratan sulfate in the serum of MCD type II patients can be explained by this hypothesis (Fig. 2).

Evolutional Difference of C-GlcNAc6ST Between Humans and Mice

The CHST5 gene, which encodes I-GlcNAc6ST protein, is located about 30 kbp upstream of CHST6 in the same orientation (Akama et al. 2000). These two genes have regions that are highly homologous to each other not only in the coding region but also in the untranslated and upstream regions, suggesting that CHST5 and CHST6 were created by gene duplication.

A mouse C-GlcNAc6ST homolog has not yet been identified. The mouse sulfotransferase closest to human C-GlcNAc6ST is mouse I-GlcNAc6ST, which is thought to be a homolog of human I-GlcNAc6ST. It is possible that the mouse genome has only one sulfotransferase gene, which is involved in keratan sulfate production; and the gene was duplicated and generated as two sulfotransferase genes, CHST5 and CHST6, in the human genome during evolution.

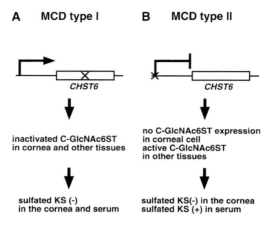

Fig. 2. Distinct mutations of *CHST6* in macular corneal dystrophy (MCD) type I and type II. **A** Mutations that affect the enzymatic activity of C-GlcNAc6ST (e.g., missense mutation and frameshift mutation) likely cause inactivation of C-GlcNAc6ST not only in the cornea but also in other tissues. Therefore an MCD type I patient lacks sulfated KS in his or her serum. **B** Mutations in the gene regulatory region of *CHST6* abolish expression of C-GlcNAc6ST in corneal cells but not in other tissues. Hence, an MCD type II patient shows the same clinical phenotype as type I patients in the cornea but has sulfated keratan sulfate (*KS*) in serum

Biological Features of Human C-GlcNAc6ST

All of the GlcNAc 6-O-sulfotransferases analyzed for their enzymatic activity transfer sulfate on the nonreducing terminal GlcNAc but not on the internal GlcNAc of a carbohydrate chain (Degroote et al. 1997; Uchimura et al. 1998, 2000). This suggests that C-GlcNAc6ST also transfers sulfate only on the nonreducing terminal GlcNAc and that this sulfation step may be coupled to its carbohydrate chain elongation (Fig. 3). Structure analyses of keratan sulfate suggested that sulfation of GlcNAc residue in keratan sulfate is coupled to the elongation step of the carbohydrate chain (Keller et al. 1983; Oeben et al. 1987). These results support the hypothesis presented in Fig. 3. Sulfation of Gal in keratan sulfate may follow GlcNAc sulfation because keratan sulfate Gal 6-O-sulfotransferase (KSG6ST), which transfers sulfate to C-6 of Gal in keratan sulfate, preferentially adds sulfate to a Gal residue adjacent to the sulfated GlcNAc (Fukuta et al. 1997) (Fig. 3).

Future Perspectives

Because C-GlcNAc6ST has been identified only recently, its biochemical characteristics are not well understood. It is important to analyze the enzymatic and biological characteristics, especially the substrate specificity, of C-GlcNAc6ST. Such information is necessary for revealing the sulfation step in keratan sulfate biosynthesis.

Fig. 3. Synthetic pathway of keratan sulfate. GlcNAc sulfation by C-GlcNAc6ST is coupled to the elongation step of poly-*N*-acetyllactosamine by GlcNAc- and Gal-transferases. Following sulfation of Gal residues by KSG6ST completes the processing of keratan sulfate production

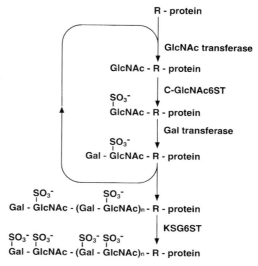

Further Reading

Akama et al. (2000) and Hemmerich et al. (2001) reported the cDNA and genomic structure of C-GlcNAc6ST (also called GST4b). Hasegawa et al. (2000) prepared a protein fraction, which may contain C-GlcNAc6ST, from human cornea and reported that its GlcNAc 6-*O*-sulfotransferase activity is decreased in the cornea of MCD patients.

References

Akama TO, Nishida K, Nakayama J, Watanabe H, Ozaki K, Nakamura T, Dota A, Kawasaki S, Inoue Y, Maeda N, Yamamoto S, Fujiwara T, Thonar EJMA, Shimomura Y, Kinoshita S, Tanigami A, Fukuda MN (2000) Macular corneal dystrophy type I and II are caused by distinct mutations in a new sulphotransferase gene. Nat Genet 26:237–241

Degroote S, Lo-Guidice JM, Strecker G, Ducourouble MP, Roussel P, Lamblin G (1997) Characterization of an *N*-acetylglucosamine-6-*O*-sulfotransferase from human respiratory mucosa active on mucin carbohydrate chains. J Biol Chem 272:29493–29501

Edward DP, Yue BYJT, Sugar J, Thonar EJMA, SunderRaj N, Stock EL, Tso MOM (1988) Heterogeneity in macular corneal dystrophy. Arch Ophthalmol 106:1579–1583

Fukuta M, Inazawa J, Torii T, Tsuzuki K, Shimada E, Habuchi O (1997) Molecular cloning and characterization of human keratan sulfate Gal-6-sulfotransferase. J Biol Chem 272:32321–32328

Hasegawa N, Torii T, Kato T, Miyajima H, Furuhata A, Nakayasu K, Kanai A, Habuchi O (2000) Decreased GlcNAc 6-*O*-sulfotransferase activity in the cornea with macular corneal dystrophy. Invest Ophthalmol Vis Sci 41:3670–3677

Hassell JR, Newsome DA, Krachmer JH, Rodrigues MM (1980) Macular corneal dystrophy: failure to synthesize a mature keratan sulfate proteoglycan. J Biol Chem 77:3705–3709

Hemmerich S, Lee JK, Bhakta S, Bistrup A, Ruddle NR, Rosen SD (2001) Glycobiology 11:75–87

Keller R, Driesch R, Stein T, Momburg M, Stuhlsatz HW, Greiling H, Franke H (1983) Biosynthesis of proteokeratan sulfate in the bovine cornea: isolation and characterization of a keratan sulfotransferase and the role of sulfation for the chain termination. Hoppe Seylers Z Physiol Chem 364:239–252

Liu NP, Baldwin J, Jonasson F, Dew-Knight S, Stajich JM, Lennon F, Pericak-Vance MA, Klintworth GK, Vance JM (1998) Haplotype analysis in Icelandic families defines a minimal interval for the macular corneal dystrophy type I gene. Am J Hum Genet 63:912–917

Nakazawa K, Hassell JR, Hascall VC, Lohmander S, Newsome DA, Krachmer J (1984) Defective processing of keratan sulfate in macular corneal dystrophy. J Biol Chem 259:13751–13757

Nakazawa K, Takahashi I, Yamamoto Y (1998) Glycosyltransferase and sulfotransferase activities in chick corneal stromal cells before and after in vitro culture. Arch Biochem Biophys 359:269–282

Oeben M, Keller R, Stuhlsatz HW, Greiling H (1987) Constant and variable domains of different disaccharide structure in corneal keratan sulphate chains. Biochem J 248:85–93

Uchimura K, Muramatsu H, Kadomatsu K, Fan QW, Kurosawa N, Mitsuoka C, Kannagi R, Habuchi O, Muramatsu T (1998) Molecular cloning and characterization of an N-acetylglucosamine-6-O-sulfotransferase. J Biol Chem 273:22577–22583

Uchimura K, Fasakhany F, Kadomatsu K, Matsukawa T, Yamakawa T, Kurosawa N, Muramatsu T (2000) Diversity of N-acetylglucosamine-6-O-sulfotransferases: molecular cloning of a novel enzyme with different distribution and specificities. Biochem Biophys Res Commun 274:291–296

Vance JM, Jonasson F, Lennon F, Sarrica J, Damji KF, Stauffer J, Pericak-Vance MA, Klintworth GK (1996) Linkage of gene for macular corneal dystrophy to chromosome 16. Am J Hum Genet 58:757–762

Yang CJ, SunderRaj N, Thonar EJMA, Klintworth GK (1988) Immunohistochemical evidence of heterogeneity in macular corneal dystrophy. Am J Ophthalmol 106:65–71

61

N-Acetylglucosamine 6-*O*-Sulfotransferase

Introduction

6-*O*-Sulfation of GlcNAc is found not only in keratan sulfate but also in asparagine-linked and mucin-type glycans, such as thyroglobulin (Spiro and Bhoyroo 1988), human immunodeficiency virus envelope glycoprotein, gp120 (Shilatifard et al. 1993), respiratory mucosa mucin (Lo-Guidice et al. 1994), and GlyCAM-1, an endothelial ligand of the L-selectin, which is a cell-adhesion molecule implicated in lymphocyte homing to lymph nodes (Rosen 1999). *N*-Acetylglucosamine-6-*O*-sulfotransferase (GlcNAc6ST) was cloned in the mouse and humans (Uchimura et al. 1998b,c). The expressed enzyme transferred sulfate to nonreducing GlcNAc but not to internally located GlcNAc. This specificity is consistent with that of the enzyme previously reported (Spiro et al. 1996; Degroote et al. 1997). It has been concluded that the cloned GlcNAc6ST participates in synthesis of 6-sulfo sialyl Lewis X structure, suggesting that the enzyme is involved in formation of L-selectin ligand (Uchimura et al. 1998b; Kimura et al. 1999). cDNAs encoding various molecular species of GlcNAc6STs was cloned subsequently (Bistrup et al. 1999; Hiraoka et al. 1999; Lee et al. 1999).

Databanks

N-Acetylglucosamine-6-*O*-sulfotransferase

No EC number has been allocated.

Species	Gene	Protein	mRNA	Genomic
GlcNAc6ST-1				
Homo sapiens	hGlcNAc6ST	JE0261	AB014679	–
	CHST2	–	AF083066	–
Mus musculus	mGlcNAc6ST	–	AB011451	–

Kᴇɴᴊɪ Uᴄʜɪᴍᴜʀᴀ and Tᴀᴋᴀsʜɪ Mᴜʀᴀᴍᴀᴛsᴜ

Department of Biochemistry, Nagoya University School of Medicine, 65 Tsurumai-cho, Showa-ku, Nagoya 466–8550, Japan
Tel. +81-52-744-2059; Fax +81-52-744-2065
E-mail tmurama@tsuru.med.nagoya-u.ac.jp

(Continued)

Species	Gene	Protein	mRNA	Genomic
GlcNAc6ST-2				
Homo sapiens	*hHEC-GlcNAc6ST*	–	AF131235	–
Mus musculus	*mHEC-GlcNAc6ST*	–	AF131236	–
	LSST	–	AF109155	–
GlcNAc6ST-3				
Homo sapiens	*hI-GlcNAc6ST*	–	AF176839	–
Mus musculus	*mI-GlcNAc6ST*	–	AF176841	–
Gallus Gallus				
GlcNAc6ST-4				
Homo sapiens	–	–	HS71L16	–

Name and History

The cDNA of GlcNAc6ST was first cloned from mouse embryo cDNA library (Uchimura et al. 1998b) based on sequence homology with chondroitin 6-sulfotransferase (Fukuta et al. 1995; Uchimura et al. 1998a). Cloned enzyme catalyzes the transfer of sulfate to nonreducing GlcNAc of oligosaccharides (Uchimura et al. 1999b). The human homolog was isolated and referred to human GlcNAc6ST (Uchimura et al. 1998c) or CHST2 (Li and Tedder 1999). cDNAs encoding GlcNAc6ST with related properties but with preferential expression in high endothelial venules of lymph nodes were subsequently isolated and named HEC-GlcNAc6ST (Bistrup et al. 1999) or LSST (Hiraoka et al. 1999). The third member of the GlcNAc6ST family, preferentially expressed in the small intestine, is called I-GlcNAc6ST (Lee et al. 1999). The fourth GlcNAc6ST has now been cloned in humans and mice (Uchimura et al. 2000). The enzyme is expressed in the heart, pancreas, spleen, ovary, and peripheral blood leukocytes. Herein the initially cloned GlcNAc6ST is called GlcNAc6ST-1; other GlcNAc6STs are called GlcNAc6ST-2 GlcNAc6ST-3, and GlcNAc6ST-4 according to their order of discovery.

Table 1. Substrate specificities of GlcNAc6ST-1 and GlcNAc6ST-4

Acceptor	Relative activity (%)	
	GlcNAc6ST-1*	GlcNAc6ST-4**
GlcNAcβ1-6Man-*O*-Me	100	100
GlcNAcβ1-2Man	128	91
GlcNAcβ1-6[Galβ1-3]GalNAc-pNP (core 2)	191	25
GlcNAcβ1-3GalNAc-pNP (core 3)	ND	N.D.
GlcNAcβ1-3Galβ1-4Glc	21	12

ND, not detected; Me, methyl; pNP, *p*-nitrophenyl

* Uchimura et al. unpublished

** Uchimura et al. 2000

Enzyme Activity Assay and Substrate Specificity

The reaction mixture contained 1 µmol Tris-HCl pH 7.2, 0.2 µmol $MnCl_2$, 0.04 µmol AMP, 2 µmol NaF, 280 pmol [^{35}S] 3'-phosphoadenosine 5'-phosphosulfate (PAPS) (about 3×10^6 cpm), 20 nmol oligosaccharides, 0.05% Triton X-100, and 2 µl microsome preparation in a final volume of 20 µl. The reaction mixture is incubated at 30°C for 1 h; and the ^{35}S-labeled products separated by thin-layer (TLC) or high-performance liquid (HPLC) chromatography are quantitated (Uchimura et al. 1998b). Other assay methods have been described as well (Spiro at al. 1996; Degroote et al. 1997).

The pH optimum of GlcNAc6ST-1 is about 6.5–7.0. Manganese (10 mM) allows more than two-fold more activity than the standard assay without adding any divalent cations. However, no effects of 10 mM EDTA for the activity indicates that divalent cations are not essential for the enzymatic reaction. The K_m values for GlcNAcβ1-6Man-O-methyl as an acceptor and PAPS as the sulfate donor are 1.25 mM and 11 µM, respectively.

GlcNAc6ST-1 acts on GlcNAcβ1-6[Galβ1-3]GalNAc-p-nitrophenyl, GlcNAcβ1-6Man-O-methyl, GlcNAcβ1-2Man, and GlcNAcβ1-3Galβ1-4Glc (Table 1), suggesting that GlcNAc6ST-1 plays a role in the sulfation of N-linked and O-linked glycans (Uchimura et al. unpublished). GlcNAcβ1-3GalNAc-p-nitrophenyl does not serve as an acceptor (Table 1). GlcNAc6ST-4 shows a similar specificity, and the relative enzymatic activities with different acceptors are different for GlcNAc6ST-1 and GlcNAc6ST-4 (Table 1) (Uchimura et al. 2000).

Preparation

High levels of GlcNAc6ST activity are obtained using microsomal preparations from CHO cells in which cloned GlcNAc6ST-1 is expressed.

Biological Aspects

Mouse and human GlcNAc6ST-1 mRNAs are expressed in a variety of adult tissues including brain, eye, pancreas, and lymph nodes. In situ hybridization localized the mRNA in specific regions of the mouse brain (Uchimura et al. 1998c). During mouse embryogenesis, GlcNAc6ST-1 mRNA is frequently expressed in one of two tissues undergoing epithelial–mesenchymal interactions and in specific regions of the central nervous system, indicating that GlcNAc6ST-1 plays a critical role in mammalian development (Fan et al. 1999).

GlcNAc6ST-1 and GlcNAc6ST-2 are expressed in high endothelial venules of lymph nodes and are thought to be involved in the sulfation of GlyCAM-1 to enhance the tethering and rolling of lymphocytes (Uchimura et al. 1998b; Bistrup et al. 1999; Hiraoka et al. 1999). Indeed, L-selectin ligand has been reconstituted by co-transfection of GlcNAc6ST-1 or GlcNAc6ST-2 and fucosyltransferase VII (Hiraoka et al. 1999; Kimura et al. 1999). Furthermore, Tangemann et al. (1999) reported that sulfation of GlyCAM-1 on the C-6 position of GlcNAc significantly enhances the efficiency of in vivo homing.

Future Perspectives

Targeting each cloned GlcNAc6ST gene will provide new insights about the physiological and developmental roles of 6-sulfo-*N*-acetylglucosamine-containing structures. The approach will also answer which GlcNAc6ST is important in the formation of L-selectin ligand and whether any of the GlcNAc6STs so far cloned are involved in the biosynthesis of keratan sulfate. Studies on GlcNAc6ST in *Drosophila* and *Caenorhabditis elegans*, if present, will also yield interesting results.

References

Bistrup A, Bhakta S, Lee JK, Belov YY, Gunn MD, Zuo FR, Huang CC, Kannagi R, Rosen SD, Hemmerich S (1999) Sulfotransferases of two specificities function in the reconstitution of high endothelial cell ligands for L-selectin. J Cell Biol 145:899–910

Degroote S, Lo-Guidice JM, Strecker G, Ducourouble MP, Roussel P, Lamblin G (1997) Characterization of an *N*-acetylglucosamine-6-*O*-sulfotransferase from human respiratory mucosa active on mucin carbohydrate chains. J Biol Chem 272:29493–29501

Fan QW, Uchimura K, Yuzawa Y, Matsuo S, Mitsuoka C, Kannagi R, Muramatsu H, Kadomatsu K, Muramatsu T (1999) Spatially and temporally regulated expression of *N*-acetylglucosamine-6-*O*-sulfotransferase during mouse embryogenesis. Glycobiology 9:947–955

Fukuta M, Uchimura K, Nakashima K, Kato M, Kimata K, Shinomura T, Habuchi O (1995) Molecular cloning and expression of chick chondrocyte chondroitin 6-sulfotransferase. J Biol Chem 270:18575–18580

Hiraoka N, Petryniak B, Nakayama J, Tsuboi S, Suzuki M, Yeh JC, Izawa D, Tanaka T, Miyasaka M, Lowe JB, Fukuda M (1999) A novel, high endothelial venule-specific sulfotransferase expresses 6-sulfo sialyl Lewis X, an L-selectin ligand displayed by CD34. Immunity 11:79–89

Kimura N, Mitsuoka C, Kanamori A, Hiraiwa N, Uchimura K, Muramatsu T, Tamatani T, Kansas GS, Kannagi R (1999) Reconstitution of functional L-selectin ligands on a cultured human endothelial cell line by cotransfection of α1-3 fucosyltransferase VII and newly cloned GlcNAcβ:6-sulfotransferase cDNA. Proc Natl Acad Sci USA 96:4530–4535

Lee JK, Bhakta S, Rosen SD, Hemmerich S (1999) Cloning and characterization of a mammalian *N*-acetylglucosamie-6-*O*-sulfotransferase that is highly restricted to intestinal tissue. Biochem Biophys Res Commun 263:543–549

Li X, Tedder TF (1999) CHST1 and CHST2 sulfotransferase expressed by human vascular endothelial cells: cDNA cloning, expression, and chromosomal localization. Genomics 55:345–347

Lo-Guidice JM, Wieruszeski JM, Lemoine J, Verbert A, Roussel P, Lamblin G (1994) Sialylation and sulfation of the carbohydrate chains in respiratory mucins from a patient with cystic fibrosis. J Biol Chem 269:18794–18813

Rosen SD (1999) Endothelial ligands for L-selectin. Am J Pathol 155:1013–1020

Shilatifard A, Merkle RK, Helland DE, Welles JL, Haseltine WA, Cummings RD (1993) Complex-type *N*-linked oligosaccharides of gp120 from human immunodeficiency virus type 1 contain sulfated *N*-acetylglucosamine. J Virol 67:943–952

Spiro RG, Bhoyroo VD (1988) Occurrence of sulfate in the asparagine-linked complex carbohydrate units of thyroglobulin. J Biol Chem 263:14351–14358

Spiro RG, Yasumoto Y, Bhoyroo V (1996) Characterization of a rat liver Golgi sulphotransferase responsible for the 6-*O*-sulphation of *N*-acetylglucosamine residues in β-linkage to mannose: role in assebly of sialyl-galactosyl-*N*-acetylglucosamine 6-sulphate sequence of *N*-linked oligosaccharides. Biochem J 319:209–216

Tangemann K, Bistrup A, Hemmerich S, Rosen SD (1999) Sulfation of a high endothelial venule-expressed ligand for L-selectin: effects on tethering and rolling of lymphocytes. J Exp Med 190:935–941

Uchimura K, Kadomatsu K, Fan QW, Muramatsu H, Kurosawa N, Kaname T, Yamamura K, Fukuta M, Habuchi O, Muramatsu T (1998a) Mouse chondroitin 6-sulfotransferase: molecular cloning, characterization and chromosomal mapping. Glycobiology 8: 489–496

Uchimura K, Muramatsu H, Kadomatsu K, Fan QW, Kurosawa N, Mitsuoka C, Kannagi R, Habuchi O, Muramatsu T (1998b) Molecular cloning and characterization of an *N*-acetylglucosamine-6-*O*-sulfotransferase. J Biol Chem 273:22577–22583

Uchimura K, Muramatsu H, Kaname T, Ogawa H, Yamakawa T, Fan QW, Mitsuoka C, Kannagi R, Habuchi O, Yokoyama I, Yamamura K, Ozaki T, Nakagawara A, Kadomatsu K, Muramatsu T (1998c) Human *N*-acetylglucosamine-6-*O*-sulfotransferase involved in the biosynthesis of 6-sulfo sialyl Lewis X: molecular cloning, chromosomal mapping, and expression in various organs and tumor cells. J Biochem 124:670–678

Uchimura K, Fasakhany F, Kadomatsu K, Matsukawa T, Yamakawa T, Kurosawa N, Muramatsu T (2000) Diversity of N-Acetylglucosamine-6-*O*-sulfotransferases: molecular cloning of a novel enzyme with different distribution and specificities. Biochem Biophys Res Commun 274:291–296

62

Intestinal *N*-Acetylglucosamine 6-*O*-Sulfotransferase

Introduction

Intestinal *N*-acetylglucosamine 6-*O*-sulfotransferase (I-GlcNAc6ST, GST-4, GlcNAc6ST-3) (Lee et al. 1999; Hemmerich and Rosen 2000) is the most recently discovered member of a novel family of carbohydrate sulfotransferases termed galactose, *N*-acetylgalactosamine, or *N*-acetylglucosamine 6-*O*-sulfotransferases (GST) (Rosen et al. 1999; Hemmerich and Rosen 2000). This is a class of five enzymes to date that have a type II membrane organization with a short cytoplasmic tail at the *N*-terminus, a transmembrane domain, and a large luminal *C*-terminal catalytic domain. I-GlcNAc6ST is one of the three *N*-acetylglucosamine 6-*O*-sulfotransferases contained in this family; the other two isozymes are the ubiquitous *N*-acetylglucosamine 6-*O*-sulfotransferase (GlcNAc6ST, GST-2, CHST-2) (Li and Tedder 1999; Uchimura et al. 1998b,c) and the high endothelial cell *N*-acetylglucosamine 6-*O*-sulfotransferase (HEC-GlcNAc6ST, GST-3) (Bistrup et al. 1999; Hiraoka et al. 1999). The other two enzymes of this family, chondroitin 6-*O*-sulfotransferase (C6ST, GST-0) (Fukuta et al. 1995, 1998; Uchimura et al. 1998a; Hemmerich and Rosen 1999) and keratan sulfate 6-*O*-sulfotransferase (KSGal6ST, GST-1, CHST-1) (Fukuta et al. 1997; Li and Tedder 1999; Hemmerich and Rosen 2000), facilitate sulfation at C-6 of galactose within the context of *N*-acetyllactosamine (Habuchi et al. 1996, 1997; Bistrup et al. 1999), with C6ST also catalyzing sulfation at C-6 of *N*-acetylgalactosamine within chondroitin (Fukuta et al. 1997). The three GlcNAc6ST isozymes exhibit more than 55% similarity to each other on the amino acid level, and protein sequence similarity across the entire enzyme family is more than 40% (Hemmerich and Rosen 2000). As indicated in the nomenclature, the three *N*-acetylglucosamine 6-*O*-sulfotransferases show distinct expression patterns. Thus GlcNAc6ST (GST-2) is abundantly expressed in most

STEFAN HEMMERICH

Department of Respiratory Diseases, Roche Bioscience, 3401 Hillview Avenue, Palo Alto, CA 94304, USA
Tel. +1-650-354-7169; Fax +1-650-354-7554
Present address: Thios Biotechnology Inc., 747 Fifty Second Street, Oakland, CA94609, USA
Tel. +1-510-601-5182; Fax +1-510-601-5201
e-mail: stefan@thiosbiotech.com

tissues and cell types (Uchimura et al. 1998; A. Bistrup, S. Hemmerich, and S.D. Rosen, unpublished), whereas HEC-GlcNAc6ST and I-GlcNAc6ST are remarkably restricted to high endothelial cells or intestinal tissue, respectively (Bistrup et al. 1999; Lee et al. 1999).

Sulfation has been shown to be an essential structural requirement for binding the lectin-like adhesion molecule L-selectin to glycoprotein ligands (vascular addressin) expressed on high endothelial venules (HEVs) in secondary lymphoid organs (Imai et al. 1993; Hemmerich et al. 1994b; Hemmerich and Rosen 2000). This adhesive interaction underlies the tethering and rolling of lymphocytes on HEVs, which is the prerequisite first step in the multistep process of lymphocyte homing (Butcher and Picker 1996; Vestweber and Blanks 1999). The necessary sulfations have been shown to occur at C-6 of galactose and N-acetylglucosamine within the context of sialyl lactosamine or sialyl Lewis X (Hemmerich and Rosen 1994; Hemmerich et al. 1994a, 1995). Therefore the GSTs have been postulated to play a prominent role in biosynthesis of these HEV ligands (Bistrup et al. 1999; Kimura et al. 1999). HEC-GlcNAc6ST may be the dominant sulfotransferase in the synthesis of the vascular addressin in peripheral lymph nodes and tonsils (Bistrup et al. 1999; Hiraoka et al. 1999). Gut-associated lymphoid organs also elaborate sulfated L-selectin ligands (Berg et al. 1993). Therefore, I-GlcNAc6ST, with its remarkable restriction to intestinal tissue, may play a role in the synthesis of mucosal HEV ligands for L-selectin (Lee et al. 1999). The presence of I-GlcNAcT in HEVs of gut-associated lymphoid organs has not been established conclusively.

Databanks

Intestinal N-acetylglucosamine 6-O-sulfotransferase

No EC number has been allocated.

Species	Gene	Protein	mRNA	Genomic
Homo sapiens	*I-GlcNAc6ST*	–	AF176838	AF219991
Mus musculus		–	AF176840	AF176841

Name and History

Intestinal N-acetylglucosamine 6-O-sulfotransferase (I-GlcNAc6ST, GST-4, GlcNAc6ST-3) is named for its particular expression pattern (restricted to intestinal tissue and various neoplasms) and acceptor specificity (Lee et al. 1999). It was discovered as the fourth novel enzyme of its class by screening (tBLASTn) (Altschul et al. 1997) public and proprietary EST databases (Genbank dbEST and LifeSeq; Incyte Pharmaceuticals, Palo Alto, CA, USA) using the protein sequence of HEC-GlcNAc6ST (GST-3) as the probe. The complete open reading frames of human and mouse I-GlcNAc6ST were determined on genomic DNA contained in bacterial artificial chromosomes (BACs), which had been identified by screening appropriate BAC libraries with the corresponding ESTs. The open reading frames encoding human and mouse I-GlcNAc6ST were found to be contained within a single exon in the human and mouse genome.

Enzyme Activity Assay and Substrate Specificity

The activity and acceptor specificity of I-GlcNAc6ST was determined by co-expression with a secreted sialomucin (GlyCAM-1), in COS-7 cells (Lee et al. 1999). Briefly, cells were transfected with identical amounts of two plasmids directing the expression of full-length I-GlcNAc6ST and a GlyCAM-1/immunoglobulin G (IgG) fusion protein. Following transfection the cells were grown in medium supplemented with [^{35}S]sulfate for 72 h; then recombinant fusion protein was isolated and subjected to partial acid hydrolysis (0.1 M H_2SO_4, 30 min). The charged mono- and disaccharides in the hydrolysis were resolved by high pH anion chromatography on a Dionex high-performance liquid chromatography (HPLC) system and assigned as N-acetyllactosamine-6-[^{35}S]sulfate (Galβ1-4[$^-O_3$S-6]GlcNAc) and N-acetylglucosamine-6-[^{35}S] sulfate (GlcNAc-6-SO_3^-). The acceptor specificity of I-GlcNAc6ST on defined oligosaccharide substrates in cell-free assays has not been determined to date.

Preparation

To date, I-GlcNAc6ST has been expressed in COS-7 cells only as complete open reading frame driven by the cytomegalovirus promoter in the expression vector pcDNA3.1 (Invitrogen). The recombinant enzyme was not isolated. Enzymatic activity corresponding to I-GlcNAc6ST so far has not been measured in intestinal tissue extracts, although GlcNAc-6-O-sulfotransferase activities have been reported in rat liver, human bronchial mucosa, and porcine lymph nodes (Spiro et al. 1996; Degroote et al. 1997; Bowman et al. 1998).

Biological Aspects

The human I-GlcNAc6ST cDNA (2215 bp) (Lee et al. 1999) encodes a 390-amino-acid type 2 transmembrane protein with three potential sites for N-linked glycosylation. Based on the coding cDNA, mouse I-GlcNAc6ST is predicted to have 395 amino acids and to be 76% identical to the human enzyme (Lee et al. 1999). I-GlcNAcT shares three regions of high homology with the other four enzymes of the GST family; in the human enzyme they are residues 48–78 (I), 191–212 (II), and 262–281 (III). Based on their homologies to the equivalent regions defined in other sulfotransferases by Kakuta et al. (1998), regions I and II are presumed to comprise the binding sites for 5'-phosphosulfate and 3'-phosphate, respectively. As implied in its name, I-GlcNAc6ST is strongly expressed in intestinal tissue and a number of neoplasms (Lee et al. 1999). Analysis of its expression pattern in gut-associated lymphoid tissue or inflamed gut is pending. Because I-GlcNAc6ST is able to sulfate C-6 of GlcNAc in mucin substrates (Lee et al. 1999), it may be involved in the biosynthesis of mucosal L-selectin ligands.

Future Perspectives

To date, the characterization of I-GlcNAc6ST is incomplete. Expression and purification of the enzyme as recombinant soluble protein must be pursued to generate sufficient amounts of well-characterized material for proper enzymological analysis.

Investigation of its acceptor specificity on oligosaccharide substrates in cell-free assays is pending (Bowman et al., in preparation). Its role in lymphocyte trafficking and possibly gut-inflammation needs to be probed by in situ hybridization and immunohistochemistry. A final definition of a biological role for I-GlcNAc6ST awaits the targeted deletion of this gene in mice.

Further Reading

See Lee et al. (1999) for discussion of the cloning, expression, and initial characterization of I-GlcNAc6ST.

A short review by Rosen and Bertozzi (1996) summarizes the role of sulfation in L- and P-selectin-mediated cell adhesion.

A review by Hemmerich and Rosen (2000) is the first comprehensive treatise on the role of carbohydrate sulfotransferases in L-selectin-mediated lymphocyte migration. Starting out with a short overview of sulfotransferase enzymes in general, the review summarizes current data on the structure of the sulfated L-selectin recognition epitopes displayed on sialomucins on high endothelial cells and then proceeds to the GST family of carbohydrate sulfotransferases and their function in biosynthesis of L-selectin ligands.

References

Altschul SF, Madden TL, Schäffer AA, Zhang J-H, Zhang Z, Miller W, Lipman DJ (1997) Gapped BLAST and PSI-BLAST: a new generation of protein database search programs. Nucleic Acids Res 25:3389–3402

Berg EL, McEvoy LM, Berlin C, Bargatze RF, Butcher EC (1993) L-Selectin-mediated lymphocyte rolling on MAdCAM-1. Nature 366:695–698

Bistrup A, Bhakta S, Lee JK, Belov YY, Gunn MD, Zuo FR, Huang CC, Kannagi R, Rosen SD, Hemmerich S (1999) Sulfotransferases of two specificities function in the reconstitution of high endothelial cell ligands for L-selectin. J Cell Biol 145:899–910

Bowman KG, Hemmerich S, Bhakta S, Singer MS, Bistrup A, Rosen SD, Bertozzi CR (1998) Identification of an N-acetylglucosamine-6-O-sulfotransferase activity specific to lymphoid tissue: an enzyme with a possible role in lymphocyte homing. Chem Biol 5:447–460

Butcher EC, Picker LJ (1996) Lymphocyte homing and homeostasis. Nature 272:60–66

Degroote S, Lo-Guidice J-M, Strecker G, Ducourouble, M-P, Roussel P, Lamblin G (1997) Characterization of an N-acetylglucosamine-6-O-sulfotransferase from human respiratory mucosa active on mucin carbohydrate chains. J Biol Chem 272:29493–29501

Fukuta M, Inazawa J, Torii T, Tsuzuki K, Shimada E, Habuchi O (1997) Molecular cloning and characterization of human keratan sulfate Gal-6-sulfotransferase. J Biol Chem 272:32321–32328

Fukuta M, Kobayashi Y, Uchimura K, Kimata K, Habuchi O (1998) Molecular cloning and expression of human chondroitin 6-sulfotransferase. Biochim Biophys Acta 1399:57–61

Fukuta M, Uchimura K, Nakashima K, Kato M, Kimata K, Shinomura T, Habuchi O (1995) Molecular cloning and expression of chick chondrocyte chondroitin 6-sulfotransferase. J Biol Chem 270:18575–18580

Habuchi O, Hirahara Y, Uchimura K, Fukuta M (1996) Enzymatic sulfation of galactose residue of keratan sulfate by chondroitin 6-sulfotransferase. Glycobiology 6:51–57

Habuchi O, Suzuki Y, Fukuta M (1997) Sulfation of sialyl lactosamine oligosaccharides by chondroitin 6-sulfotransferase. Glycobiology 7:405–412

Hemmerich S, Rosen SD (1994) 6'-Sulfated, sialyl Lewis X is a major capping group of GlyCAM-1. Biochemistry 33:4830–4835

Hemmerich S, Rosen SD (2000) Carbohydrate sulfotransferases in lymphocyte homing. Glycobiology 10:849–856

Hemmerich S, Bertozzi CR, Leffler H, Rosen SD (1994a) Identification of the sulfated monosaccharides of GlyCAM-1, an endothelial-derived ligand for L-selectin. Biochemistry 33:4820–4829

Hemmerich S, Butcher EC, Rosen SD (1994b) Sulfation-dependent recognition of HEV-ligands by L-selectin and MECA 79, an adhesion-blocking mAb. J Exp Med 180:2219–2226

Hemmerich S, Leffler H, Rosen SD (1995) Structure of the O-glycans in GlyCAM-1, an endothelial-derived ligand for L-selectin. 270:12035–12047

Hiraoka N, Petryniak B, Nakayama J, Tsuboi S, Suzuki M, Yeh J-C, Izawa D, Tanaka T, Miyasaka M, Lowe JB, Fukuda M (1999) A novel, high endothelial venule-specific sulfotransferase expresses 6-sulfo sialyl lewis x, an L-selectin ligand displayed by CD34. Immunity 11:79–89

Imai Y, Lasky LA, Rosen SD (1993) Sulphation requirement for GlyCAM-1, an endothelial ligand for L-selectin. Nature 361:555–557

Kakuta Y, Pedersen LG, Pedersen LC, Negishi M (1998) Conserved structural motifs in the sulfotransferase family. Trends Biochem Sci 23:129–130

Kimura N, Mitsuoka C, Kanamori A, Hiraiwa N, Uchimura K, Muramatsu T, Tamatani T, Kansas GS, Kannagi R (1999) Reconstitution of functional L-selectin ligands on a cultured human endothelial cell line by cotransfection of α(1,3) fucosyltransferase VII and newly cloned GlcNAcβ:6-sulfotransferase cDNA. Proc Natl Acad Sci USA 96:4530–4535

Lee JK, Bhakta S, Rosen SD, Hemmerich S (1999) Cloning and characterization of a mammalian N-acetylglucosamine-6-sulfotransferase that is highly restricted to intestinal tissue. Biochem Biophys Res Commun 263:543–549

Li X, Tedder TF (1999) CHST1 and CHST2 sulfotransferases expressed by human vascular endothelial cells: cDNA cloning, expression, and chromosomal localization. Genomics 55:345–347

Rosen SD, Bertozzi CB (1996) Leukocyte adhesion: two selectins converge on sulphate. Curr Biol 6:261–264

Rosen SD, Bistrup A, Hemmerich S (2000) Carbohydrate sulfotransferases. In: Ernst B, Sinaÿ P, Hart G (eds) Oligosaccharides in chemistry and biology. vol. 2. Weinheim, Wiley-VCH, pp 245–260

Spiro RG, Yasumoto Y, Bhoyroo V (1996) Characterization of a rat liver Golgi sulphotransferase responsible for the 6-O-sulphation of N-acetylglucosamine residues in beta-linkage to mannose: role in assembly of sialyl-galactosyl-N-acetylglucosamine 6-sulphate sequence of N-linked oligosaccharides. Biochem J 319:209–216

Uchimura K, Kadomatsu K, Fan QW, Muramatsu H, Kurosawa N, Kaname T, Yamamura K, Fukuta M, Habuchi O, Muramatsu T (1998a) Mouse chondroitin 6-sulfotransferase: molecular cloning, characterization and chromosomal mapping. Glycobiology 8:489–496

Uchimura K, Muramatsu H, Kadomatsu K, Fan QW, Kurosawa N, Mitsuoka C, Kannagi R, Habuchi O, Muramatsu T (1998b) Molecular cloning and characterization of an N-acetylglucosamine-6-O-sulfotransferase. J Biol Chem 273:22577–22583

Uchimura K, Muramatsu H, Kaname T, Ogawa H, Yamakawa T, Fan QW, Mitsuoka C, Kannagi R, Habuchi O, Yokoyama I, Yamamura K, Ozaki T, Nakagawara A, Kadomatsu K, Muramatsu T (1998c) Human N-acetylglucosamine-6-O-sulfotransferase involved in the biosynthesis of 6-sulfo sialyl Lewis X: molecular cloning, chromosomal mapping, and expression in various organs and tumor cells. J Biochem (Tokyo) 124:670–678

Vestweber D, Blanks JE (1999) Mechanisms that regulate the function of the selectins and their ligands. Physiol Rev 79:181–213

High Endothelial Cell *N*-Acetylglucosamine 6-*O*-Sulfotransferase

Introduction

High endothelial cell (HEC)-specific *N*-acetylglucosamine 6-sulfotransferase (HEC-GlcNAc6ST) (Bistrup et al. 1999), or L-selectin ligand sulfotransferase (LSST) (Hiraoka et al. 1999), is a transmembrane protein of 386 amino acids with a short, N-terminal cytoplasmic tail and a large luminal C-terminal catalytic domain. This enzyme belongs to a recently identified family of carbohydrate sulfotransferases that modify the 6-hydroxyl of Gal, GalNAc, or GlcNAc residues (GST family) (Rosen et al. 1999; Hemmerich and Rosen 2000). In addition to HEC-GlcNAc6ST, there are two other GlcNAc-6 sulfotransferases in this family: *N*-acetylglucosamine 6-sulfotransferase [GlcNAc6ST (Uchimura et al. 1998a, b) or CHST2 (Li and Tedder 1999)] and intestinal *N*-acetylglucosamine 6-sulfotransferase (I-GlcNAc6ST) (Lee et al. 1999). Within their sulfotransferase domains, these three enzymes are more than 27% identical at the amino acid level and more than 56% similar. HEC-GlcNAc6ST is expressed in a limited number of tissues and is highly enriched in HECs (Bistrup et al. 1999; Hiraoka et al. 1999). In contrast, GlcNAc6ST is expressed ubiquitously (Uchimura et al. 1998b), and I-GlcNAc6ST is expressed almost exclusively in intestinal tissues (Lee et al. 1999). A recently identified chondroitin 6-*O*-sulfotransferase, designated C6ST-2 (Kitagawa et al. 2000), exhibits GalNAc 6-sulfotransferase activity toward chondroitin (Kitagawa et al. 2000). The other members of this novel family, keratan sulfate 6-sulfotransferase [KSGal6ST (Fukuta et al. 1997); CHST1 (Li and Tedder 1999)] and chondroitin 6-sulfotransferase (C6ST) (Fukuta et al. 1998), have galactose 6-sulfotransferase activity; in addition, C6ST catalyzes sulfation at C-6 of GalNAc in chondroitin sulfate (Fukuta et al. 1997).

Specific carbohydrate sulfation has been shown to be required for binding of the leukocyte adhesion molecule L-selectin to its sialomucin ligands expressed on the high endothelial venules (HEVs) of secondary lymphoid organs (Imai et al. 1993;

Annette Bistrup and Steven D. Rosen

Department of Anatomy and Program in Immunology, University of California, 513 Parnassus Avenue, San Francisco, CA 94143-0452, USA
Tel. +1-415-476-1579; Fax +1-415-476-2526
e-mail: sdr@itsa.ucsf.edu

Hemmerich et al. 1994; Hemmerich and Rosen 2000). These adhesive interactions allow for the tethering and rolling of lymphocytes on HEVs, a requisite first step for lymphocyte homing to these organs (Vestweber and Blanks 1999). Structural analysis of several L-selectin ligands indicated that sulfation occurred exclusively at C-6 of Gal and GlcNAc, in the context of sialyl lactosamine or sialyl Lewis[x] (Hemmerich and Rosen 1994, unpublished observations; Hemmerich et al. 1994, 1995). Recent studies in mice genetically deficient in HEC-GlcNAc6ST indicate that this enzyme is involved in the biosynthesis of L-selectin ligands in vivo (Hemmerich et al. 2001).

Databanks

High Endothelial Cell *N*-acetylglucosamine 6-*O*-sulfotransferase

No EC number has been allocated.

Species	Gene	Protein	mRNA	Genomic
Homo sapiens	–	–	AF131235	–
Mus musculus	–	–	AF131236	–
			AF109155	

Name and History

HEC-GlcNAc6ST was discovered by screening public (GenBank dbEST) and proprietary EST databases (LifeSeq of Incyte Pharmaceuticals, Palo Alto, CA, USA) using probes based on the amino acid and nucleotide sequence encoding chicken chondroitin/keratan sulfate sulfotransferase (C6/KSST) (Fukuta et al. 1995) and human keratan sulfate sulfotransferase (KSST) (Fukuta et al. 1997). The cDNA corresponding to human HEC-GlcNAc6ST was isolated from a cDNA expression library derived from HECs by polymerase chain reaction (PCR)-mediated screening (Bistrup et al. 1999). The cDNA corresponding to mouse HEC-GlcNAc6ST was obtained from mouse embryo poly(A)[+] by a combination of reverse transcription (RT)-PCR and 5′ rapid amplification of cDNA ends (RACE) (Hiraoka et al. 1999). Subsequent screening of DNA contained in bacterial artificial chromosomes (BACs) has confirmed both the human and mouse cDNA sequences (Hemmerich and Rosen, unpublished observations). The open reading frames encoding human and mouse HEC-GlcNAc6ST are each contained in single exons (Hemmerich et al. submitted). HEC-GlcNAc6ST derives its name from its acceptor specificity (Bistrup et al. 1999; Hiraoka et al. 1999) and its markedly restricted expression pattern. HEC-GlcNAc6ST is expressed at low to undetectable levels in most adult tissues (mouse and human) but is highly enriched in the HECs of peripheral lymph nodes (Bistrup et al. 1999; Hiraoka et al. 1999), the site at which L-selectin ligands are synthesized.

Enzyme Activity Assay and Substrate Specificity

The activity and acceptor specificity of HEC-GlcNAc6ST were determined by co-expression of the enzyme with various glycoproteins in cell lines (Bistrup et al. 1999; Hiraoka et al. 1999). Briefly, a cDNA encoding HEC-GlcNAc6ST was co-transfected into COS or CHO cells with a cDNA encoding an immunoglobulin G (IgG) chimera of candidate glycoprotein acceptors, and the cells were cultured in the presence of [^{35}S]-sulfate. The secreted chimeric glycoprotein was then isolated from the conditioned medium, and sulfation incorporation was assayed by sodium dodecyl sulfate-polyacrylamide gel electrophoresis (SDS-PAGE). HEC-GlcNAc6ST was found to efficiently catalyze the sulfation of several mucin-like molecules known to serve as ligands for L-selectin, namely, CD34, glycosylation-dependent cell adhesion molecule-1 (GlyCAM-1), and mucosal addressin cell adhesion molecule-1 (MAdCAM-1) (Bistrup et al. 1999; Hiraoka et al. 1999). Additional substrates include intercellular adhesion molecule-1 (ICAM-1) and neural cell adhesion molecule-1 (NCAM-1) (Lee and Rosen, unpublished observations).

The regioselectivity of sulfation was established by chromatographic analysis of acid hydrolysates of the glycoprotein acceptors (GlyCAM-1/IgG and CD34/IgG) (Bistrup et al. 1999; Hiraoka et al. 1999, respectively). The only activity determined to be conferred by HEC-GlcNAc6ST was sulfation at the 6-hydroxyl of GlcNAc residues, thus establishing this enzyme as a GlcNAc-6 sulfotransferase.

The acceptor specificity of HEC-GlcNAc6ST was also determined in vitro by testing the ability of lysates from COS cells transfected with a cDNA encoding HEC-GlcNAc6ST to transfer [^{35}S]-sulfate from [^{35}S]-3′-phosphoadenosine 5′-phosphosulfate (PAPS) to various acceptors (Hiraoka et al. 1999). Keratan sulfate and chondroitin sulfate did not serve as acceptors for sulfation by HEC-GlcNAc6ST (Hiraoka et al. 1999). Furthermore, HEC-GlcNAc6ST did not transfer sulfate to complex glycan acceptor structures such as Galβ1,4GlcNAcβ1,6Manα1,6Manβ→octyl, GlcNAcβ1,3Galβ1,4GlcNAcβ1,6Manα1,6-Manβ1→octyl, and GlcNAcβ1,6(Galβ1,3)GalNAcα→p-nitrophenol (Hiraoka et al. 1999). Keratan sulfate and chondroitin sulfate did not serve as acceptors for sulfation by HEC-GlcNAc6ST (Hiraoka et al. 1999). It remains to be determined whether, in vivo, the enzyme has a preference or exclusive requirement for O-linked carbohydrate acceptor structures within mucin-like glycoproteins.

Preparation

To date, HEC-GlcNAc6ST has been expressed in COS cells as the complete open reading frame, driven by the cytomegalovirus (CMV) promoter in the vectors pcDNA 1.1 and 3.1 (Invitrogen) (Bistrup et al. 1999; Hiraoka et al. 1999). The enzyme has also been expressed from COS cells in soluble form from a variant of the pcDNA 3.1 vector, which directs the expressed product to be secreted. The secreted recombinant enzyme was capable of transferring sulfate from PAPS to the disaccharide acceptor GlcNAcβ1,6Gal→octyl (Bowman et al. in preparation). A related activity has been shown to be present in rat liver, human bronchial mucosa, and porcine lymph node (Spiro et al. 1996; Degroote et al. 1997; Bowman et al. 1998). The porcine lymph node GlcNAc-6 sulfotransferase activity was highly enriched in the HECs of the lymph node (Bowman et al. 1998), raising the intriguing and likely possibility that this activity corresponds to HEC-GlcNAc6ST.

Biological Aspects

The cDNA corresponding to human HEC-GlcNAc6ST encodes a predicted type II transmembrane protein of 386 amino acids, with three potential sites for N-linked glycosylation (Bistrup et al. 1999). The murine cDNA predicts a type II transmembrane protein of 388 amino acids with three potential sites of N-linked glycosylation (Bistrup et al. 1999; Hiraoka et al. 1999). Identity between human and murine HEC-GlcNAc6ST is 73% at the amino acid level. The HEC-GlcNAc6ST amino acid sequence contains three regions of sequence in which the homology with other members of the GST family of sulfotransferases (Rosen et al. 1999; Hemmerich and Rosen 2000) is high. In the human sequence these are residues 42–78 (region I), 192–216 (region II), and 264–283 (region III). Regions I and II are presumed to comprise residues that conform to the consensus binding motif for the high-energy sulfate donor PAPS (Kakuta et al. 1998).

The recombinant enzyme, when co-expressed with recombinant CD34 (Bistrup et al. 1999; Hiraoka et al. 1999) or GlyCAM-1/IgG (Tangemann et al. 1999), can contribute to L-selectin ligand activity on these mucin-like glycoproteins. This was demonstrated in an equilibrium binding assay and in two other assays that measure the kinetic parameters of L-selectin binding to ligands. Enhanced binding under equilibrium conditions was demonstrated by a flow cytometry assay in which CHO cells expressing CD34 sulfated by HEC-GlcNAc6ST bound an L-selectin/IgM chimera more avidly than CHO cells expressing CD34 but not HEC-GlcNAc6ST (Bistrup et al. 1999). The contribution of HEC-GlcNAc6ST to the generation of L-selectin ligand activity was also measured in a parallel plate flow chamber, which recapitulates the dynamics of physiological vascular flow (Lawrence et al. 1995). With two recombinant ligands for L-selectin, GlyCAM-1 (Tangemann et al. 1999) and CD34 (Hiraoka et al. 1999), as substrates for sulfation by HEC-GlcNAc6ST, enhanced binding by L-selectin to the sulfated ligands was demonstrated in terms of increased tethering rate, reduced velocity, and increased resistance to shear-dependent detachment (Hiraoka et al. 1999; Tangemann et al. 1999). Lymph nodes from HEC-GlcNAc6ST deficient mice did not support binding of recombinant L-selectin or adhesion of Lymphocytes in vitro. Further more, in vivo homing of lymphocytes was significantly reduced in the HEC-GlcNAc6ST −/− mice.

The expression pattern of HEC-GlcNAc6ST appears to be highly restricted. In Northern blots of human organs, expression was detected only in pancreas, fetal and adult liver, and lymph nodes (Bistrup et al. 1999). High levels of HEC-GlcNAc6ST mRNA were detected in isolated human HECs, whereas human umbilical vein endothelial cells did not express this mRNA (Bistrup et al. 1999). In the mouse, expression of HEC-GlcNAc6ST transcripts was detected in embryonic tissues (Hiraoka et al. 1999) and HECs (Hiraoka et al. 1999; Bistrup et al. 1999). Transcripts corresponding to HEC-GlcNAc6ST were also expressed in endothelial cells in HEV-like blood vessels in the hyperplastic thymus of the AKR/J mouse strain (Hiraoka et al. 1999). Concomitant expression of L-selectin ligand activity in these vessels indicates that this enzyme may be induced in vessels acquiring an HEV-like phenotype. Based on its expression pattern and the phenotype of the HEC-GlcNAc6ST −/− mice, this enzyme may be involved in the biosynthesis of L-selectin ligands at sites of chronic inflammation.

Future Perspectives

As a relatively recently cloned enzyme, HEC-GlcNAc6ST is not yet extensively characterized. A full investigation of its enzymatic activities and substrate specificities requires production of recombinant-soluble protein and the availability of various oligosaccharide acceptor structures. Substrate specificity at the level of macromolecules can be further investigated in cell lines by co-transfection of the HEC-GlcNAc6ST cDNA with various mucin-like and non-mucin-like proteins. The contribution of specific amino acid residues to the binding of both the sulfate acceptor (substrate) and donor (PAPS) is also of interest. Localization of the enzyme in the cell (Golgi apparatus versus trans-Golgi network, for example) will be informative with respect to its role in the biosynthesis of sulfated carbohydrate structures, as the ordering of this biosynthesis is of significance. For example, during biosynthesis of sulfated sialyl Lewis[x] on L-selectin ligands, a candidate GlcNAc6ST would have to act prior to several other glycosyl transferases that participate in the synthesis of this structure (e.g., fucosyltransferase VII) (Natsuka and Lowe 1994). The role of HEC-GlcNAc6ST in the trafficking of lymphocytes to peripheral lymph nodes and of leukocytes to extralymphoid sites of inflammation has not been established. Ultimately, the biological role for HEC-GlcNAc6ST will require inactivation of the gene in mice by targeted deletion at the genomic level.

References

Bistrup A, Bhakta S, Lee JK, Belov YY, Gunn MD, Zuo FR, Huang CC, Kannagi R, Rosen SD, Hemmerich S (1999) Sulfotransferases of two specificities function in the reconstitution of high endothelial cell ligands for L-selectin. J Cell Biol 145:899–910

Bowman KG, Hemmerich S, Bhakta S, Singer MS, Bistrup A, Rosen SD, Bertozzi CR (1998) Identification of an N-acetylglucosamine-6-O-sulfotransferase activity specific to lymphoid tissue: an enzyme with a possible role in lymphocyte homing. Chem Biol 5:447–460

Degroote S, Lo-Guidice JM, Strecker G, Ducourouble MP, Roussel P, Lamblin G (1997) Characterization of an N-acetylglucosamine-6-O-sulfotransferase from human respiratory mucosa active on mucin carbohydrate chains. J Biol Chem 272:29493–29501

Fukuta M, Inazawa J, Torii T, Tsuzuki K, Shimada E, Habuchi O (1997) Molecular cloning and characterization of human keratan sulfate Gal-6-sulfotransferase. J Biol Chem 272:32321–32328

Fukuta M, Kobayashi Y, Uchimura K, Kimata K, Habuchi O (1998) Molecular cloning and expression of human chondroitin 6-sulfotransferase. Biochim Biophys Acta 1399:57–61

Fukuta M, Uchimura K, Nakashima K, Kato M, Kimata K, Shinomura T, Habuchi O (1995) Molecular cloning and expression of chick chondrocyte chondroitin 6-sulfotransferase. J Biol Chem 270:18575–18580

Hemmerich S, Rosen SD (1994) 6'-Sulfated sialyl Lewis x is a major capping group of GlyCAM-1. Biochemistry 33:4830–4835

Hemmerich S, Rosen SD (2000) Carbohydrate sulfotransferases in lymphocyte homing. Glycobiology 10:849–856

Hemmerich S, Bertozzi CR, Leffler H, Rosen SD (1994) Identification of the sulfated monosaccharides of GlyCAM-1, an endothelial-derived ligand for L-selectin. Biochemistry 33:4820–4829

Hemmerich S, Bhakta S, Lee J-K, Bistrup A, Ruddle N, Rosen SD (2001) Chromosomal localization and genomic organization for the galactose/N-acetylgalactosamine/N-acetylglucosamine 6-O-sulfotransferase gene family. Glycobiology 11:75–87

Hemmerich S, Bistrup A, Singer MS, van Zante A, Lee JK, Tsay D, Peters M, Carminati JL, Brennan TJ, Carver-Moore K, Leviten M, Fuentes ME, Ruddle NH, Rosen SD (2001) Sulfation of L-selectin ligands by an HEV-restricted sulfotransferase regulates lymphocyte homing to lymoph nodes. Immunity 15:237–247

Hemmerich S, Butcher EC, Rosen SD (1994) Sulfation-dependent recognition of high endothelial venules (HEV)-ligands by L-selectin and MECA 79, and adhesion-blocking monoclonal antibody. J Exp Med 180:2219–2226

Hemmerich S, Leffler H, Rosen SD (1995) Structure of the O-glycans in GlyCAM-1, an endothelial-derived ligand for L-selectin. J Biol Chem 270:12035–12047

Hiraoka N, Petryniak B, Nakayama J, Tsuboi S, Suzuki M, Yeh JC, Izawa D, Tanaka T, Miyasaka M, Lowe JB, Fukuda M (1999) A novel, high endothelial venule-specific sulfotransferase expresses 6-sulfo sialyl Lewis(x), an L-selectin ligand displayed by CD34. Immunity 11:79–89

Imai Y, Lasky LA, Rosen SD (1993) Sulphation requirement for GlyCAM-1, an endothelial ligand for L-selectin. Nature 361:555–557

Kakuta Y, Pedersen LG, Pedersen LC, Negishi M (1998) Conserved structural motifs in the sulfotransferase family. Trends Biochem Sci 23:129–130

Kitagawa H, Fujita M, Ito N, Sugahara K (2000) Molecular cloning and expression of a novel chondroitin 6-O-sulfotransferase. J Biol Chem 275:21075–21080

Lawrence MB, Berg EL, Butcher EC, Springer TA (1995) Rolling of lymphocytes and neutrophils on peripheral node addressin and subsequent arrest on ICAM-1 in shear flow. Eur J Immunol 25:1025–1031

Lee JK, Bhakta S, Rosen SD, Hemmerich S (1999) Cloning and characterization of a mammalian N-acetylglucosamine-6-sulfotransferase that is highly restricted to intestinal tissue. Biochem Biophys Res Commun 263:543–549

Li X, Tedder TF (1999) CHST1 and CHST2 sulfotransferases expressed by human vascular endothelial cells: cDNA cloning, expression, and chromosomal localization. Genomics 55:345–347

Natsuka S, Lowe JB (1994) Enzymes involved in mammalian oligosaccharide biosynthesis. Curr Opin Struct Biol 4:683–691

Rosen SD, Bistrup A, Hemmerich S (2000) Carbohydrate sulfotransferases. In: Ernst B, Sinaÿ P, Hart G (eds) Carbohydrates in chemistry and biology. Wiley-VCH, Weinheim 3(part II) pp 245–246

Spiro RG, Yasumoto Y, Bhoyroo V (1996) Characterization of a rat liver Golgi sulphotransferase responsible for the 6-O-sulphation of N-acetylglucosamine residues in beta-linkage to mannose: role in assembly of sialyl-galactosyl-N-acetylglucosamine 6-sulphate sequence of N-linked oligosaccharides. Biochem J 319:209–216

Tangemann K, Bistrup A, Hemmerich S, Rosen SD (1999) Sulfation of a high endothelial venule-expressed ligand for L-selectin: effects on tethering and rolling of lymphocytes. J Exp Med 190:935–942

Uchimura K, Muramatsu H, Kadomatsu K, Fan QW, Kurosawa N, Mitsuoka C, Kannagi R, Habuchi O, Muramatsu T (1998a) Molecular cloning and characterization of an N-acetylglucosamine-6-O-sulfotransferase. J Biol Chem 273:22577–22583

Uchimura K, Muramatsu H, Kaname T, Ogawa H, Yamakawa T, Fan QW, Mitsuoka C, Kannagi R, Habuchi O, Yokoyama I, Yamamura K, Ozaki T, Nakagawara A, Kadomatsu K, Muramatsu T (1998b) Human N-acetylglucosamine-6-O-sulfotransferase involved in the biosynthesis of 6-sulfo sialyl Lewis X: molecular cloning, chromosomal mapping, and expression in various organs and tumor cells. J Biochem (Tokyo) 124:670–678

Vestweber D, Blanks JE (1999) Mechanisms that regulate the function of the selectins and their ligands. Physiol Rev 79:181–213

Chondroitin 4-Sulfotransferase

Introduction

Chondroitin 4-sulfate, which contains a sulfate group on position 4 of GalNAc residues, is a major component of cartilage proteoglycan (aggrecan). Chondroitin 4-sulfate is also implicated in the celluar interaction with *Plasmodium falciparum*-infected erythrocytes (Rogerson et al. 1995; Fried and Duffy 1996). Chondroitin 4-sulfotransferase (C4ST) catalyzes the transfer of sulfate to position 4 of GalNAc residues of chondroitin and is involved in the biosynthesis of chondroitin 4-sulfate. cDNAs encoding mouse C4ST and human C4ST-1 and C4ST-2 have been cloned (Hiraoka et al. 2000; Yamauchi et al. 2000) mouse C4ST-1 is considered to be the mouse counterpart of human C4ST-1. All of the enzymes encoded by these cDNAs add 4-O-sulfate on dermatan sulfate using an in vitro assay. It remains to be studied whether C4ST is also involved in in vivo biosynthesis of dermatan sulfate.

Databanks

Chondroitin 4-sulfotransferase

NC-IUBMB enzyme classification: E.C.2.8.2.5

Species	Gene	Protein	mRNA	Genomic
Homo sapiens	C4ST-1	–	AB042326	–
			AF239820	
			AJ269537	
	C4ST-2	–	AF239822	–
Mus musculus	C4ST	–	AB030378	–

Osami Habuchi[1] and Nobuyoshi Hiraoka[2]

[1] Department of Life Science, Aichi University of Education, Igaya, Kariya, Aichi 448-8542, Japan
Tel. +81-566-26-2642; Fax +81-566-26-2649
e-mail: ohabuchi@auecc.aichi-edu.ac.jp
[2] Glycobiology Program, Cancer Research Center, The Burnham Institute, La Jolla, CA 92037, USA

Name and History

This enzyme was found in chick embryo cartilage (Robinson 1969; Kimata et al. 1973; Habuchi and Miyashita 1982), mouse mastocytoma (Sugumaran and Silbert 1988), and chick chondrocytes (Habuchi et al. 1991). C4ST was purified to homogeneity from the serum-free culture medium of rat chondrosarcoma cells (Yamauchi et al. 1999).

Enzyme Activity Assay and Substrate Specificity

C4ST catalyzed transfer of sulfate to position 4 of GalNAc residue of chondroitin and desulfated dermatan sulfate:

$$PAPS + (GlcA\beta1\text{-}3GalNAc\beta1\text{-}4)_n \rightarrow PAP + (GlcA\beta1\text{-}3GalNAc(4\text{-}SO_4)\beta1\text{-}4)_n$$

where PAPS is 3'-phosphoadenosine 5'-phosphosulfate, and PAP is 3'-phosphoadenosine 5'-phosphate. In the chondroitin 4-sulfotransferase assay [^{35}S]PAPS is used as the donor and chondroitin as the acceptor. The assay system included 2.5 µmol imidazole HCl (pH 6.8), 1.25 µg protamine chloride, 0.1 µmol dithiothreitol, 25 nmol (as GalNAc) squid skin chondroitin, 50 pmol [^{35}S]PAPS (5×10^5 cpm), and enzyme in a final volume of 50 µl (Habuchi et al. 1993). After incubation the sulfated glycosaminoglycan was isolated via precipitation with 66% ethanol containing 1.3% potassium acetate and gel chromatography on Sephadex G-25 (Fast Desalting column, Amersham Pharmacia Biotech); the radioactivity was then counted. To determine the sulfation of positions 6 and 4 of GalNAc residue, ^{35}S-labeled glycosaminoglycan was digested with chondroitinase ACII, and unsaturated disaccharides were separated with paper chromatography or Partisil SAX high-performance liquid chromatography (HPLC).

C4ST activity is stimulated with basic proteins such as protamine and histone. Spermidine and spermine also stimulate C4ST activity. The K_m for PAPS is decreased in the presence of basic protein or polyamines (Habuchi and Miyata 1980; Habuchi and Miyashita 1982). The optimum protamine concentration varies with the acceptor used: 0.025 mg/ml for chondroitin and 0.2 mg/ml for desulfated dermatan sulfate. Among divalent cations, Ca^{2+}, Fe^{2+}, Mn^{2+}, Ba^{2+}, and Sr^{2+} showed a stimulatory effect. Co^{2+}, which activated C6ST, is inhibitory. The optimum pH is around 7.2 (Yamauchi et al. 1999). C4ST requires sulfhydryl compounds such as dithiothreitol, 2-mercaptoethanol, and glutathione (Habuchi and Miyashita 1982; Habuchi et al. 1991; Yamauchi et al. 1999). The presence of Cys residue in the putative 5'-phosphosulfate binding (5'-PSB) domain may be relevant to the requirement for sulfhydryl compounds (Yamauchi et al. 2000).

Chondroitin and desulfated dermatan sulfate served as acceptors. Chondroitin sulfate A and chondroitin sulfate C are poor acceptors. Analysis of products indicate that only 4-sulfate GalNAc is synthesized by the enzymes that have been cloned (Hiraoka et al. 2000; Yamauchi et al. 2000). The enzymes do not add a sulfate to chondroitin 6-sulfate and thus do not play a role in the synthesis of chondroitin 4,6-disulfate. Chondroitin sulfate E from squid cartilage, dermatan sulfate, heparan sulfate, and completely desulfated N-resulfated heparin hardly served as acceptors of C4ST. When ^{35}S-labeled desulfated dermatan sulfate was digested with chondroitinase ACII, about 50% of the ^{35}S radioactivity was recovered in ΔDi-4S; the remaining radioactivity was bound to tetrasaccharide, hexasaccharide, and larger oligosaccharides. No radioactivity was observed in polysaccharide fractions. This observation indicates that C4ST preferentially transfers sulfate to GalNAc residues in the GlcA-rich portion of the desulfated dermatan sulfate (Yamauchi et al. 1999; Hiraoka et al. 2000). Analysis of ^{35}S-labeled tetrasaccharide formed from ^{35}S-labeled desulfated dermatan sulfate by chondroitinase ACII digestion showed that ^{35}SO$_4$ was located exclusively at the nonreducing side of the GalNAc residue (Habuchi et al., unpublished data). Because the sequence of the tetrasaccharide could be presumed to be D-GlcA-GalNAc-IdoA-GalNAc from the specificity of chondroitinase ACII, C4ST should absolutely require GlcA residue on the nonreducing side of the acceptor GalNAc residue.

Preparation

C4ST was purified from the serum-free culture medium of rat chondrosarcoma cells with affinity chromatography on heparin-Sepharose CL-6B, RedA-agarose, and 3′,5′-ADP-agarose (Yamauchi et al. 1999). The purified C4ST showed broad protein bands of 50 and 54 kDa on sodium dodecyl sulfote-polyacrylamide gel electrophoresis (SDS-PAGE). The recombinant enzyme has been expressed in COS-7 cells and purified with anti FLAG peptide monoclonal antibody (FLAG) affinity chromatography (Yamauchi et al. 2000).

Two human C4STs have been cloned based on the search of the EST database with human HNK-1ST sequences as probes (Hiraoka et al. 2000). The amino acid sequences of one of them, human C4ST-1, is 96% identical to that of the mouse C4ST mentioned above, so it is thought to be a human homolog of mouse C4ST. The other, C4ST-2, is 42% identical to human C4ST-1 on the amino acid level, but certain domains are highly homologous, as shown in Fig. 1. These homologous clusters, such as 5′-PSB and 3′-PB (A and B in Fig. 1, are also conserved in HNK-1ST; thus C4STs and HNK-1ST form a subfamily of sulfotransferase (see also Hiraoka et al. 2000). The amino acid sequences of C4ST-1 and C4ST-2 are 32% and 31% identical, respectively, to that of HNK-1ST. It is tempting to speculate that the active sites of both HNK-1ST and C4ST may approach the acceptor from above the plane of GlcAβ1→3Galβ1→4GlcNAcβ1→R (for HNK-1ST) and GlcAβ1→3GalNAcβ1→4GlcAβ1→R (for C4ST). The enzymatic activity of human C4ST-1 and C4ST-2 are almost the same as that of mouse C4ST, and the activity of C4ST-1 is stronger than that of C4ST-2 when expressed in CHO cells.

Biological Aspects

Chondroitin 4-sulfate, the product of the reaction with C4ST, is one of the major components of aggrecan and is involved in the formation and maintenance of cartilage tissue. Chondroitin 4-sulfate has also been implicated in various cellular and molecular interactions, such as receptor for malaria-infected erythrocytes (Rogerson et al. 1995; Fried and Duffy 1996), receptor for CD44 (Naujokas et al. 1993; Toyama-Sorimachi et al. 1995), cell adhesion to fibronectin via integrin (Yamagata et al. 1989; Sugiura et al. 1993), and protection of high-density lipoprotein against copper-dependent oxidation (Albertini et al. 1999). The mouse C4ST message of 5.7 kb was mainly expressed in the brain and kidney among mouse adult tissues. C4ST purified

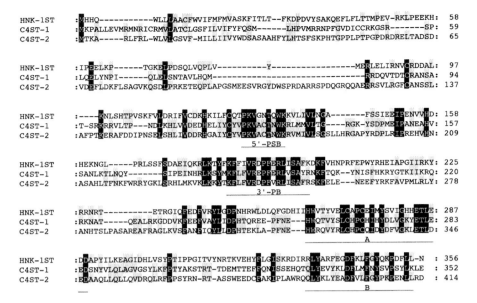

Fig. 1. Comparison of amino acid sequences of human HNK-1ST, C4ST-1, and C4ST-2. Introduced gaps are shown as *hyphens*, and aligned identical residues are *boxed* (*black* for all sequences, *dark gray* for two sequences). Putative binding sites for the 5′-phosphosulfate group (5′-PSB) and 3′-phosphate group (3′-PB) and two highly conserved sequences (A and B) are denoted

from the culture medium of chondrosarcoma cells seems to be a truncated form because the amino acid sequence of the NH_2-terminal of the purified C4ST was found in the stem region. Proteolytic cleavage at the stem region may occur during secretion. C4ST showed no significant homology in the amino acid sequence with C6ST, although both C4ST and C6ST utilize chondroitin as a common acceptor. In contrast, C4ST showed significant homology in the amino acid sequence with HNK-1 sulfotransferase (Hiraoka et al. 2000; Yamauchi et al. 2000).

The tissue distribution of human C4ST-1 transcript is mostly in peripheral blood leukocytes and hematopoietic tissues such as spleen, thymus, lymph node and bone marrow; it is also found in lung and placenta. Although human C4ST-2 transcript is more widely expressed in various tissues, it is expressed significantly more in heart, thyroid, pituitary gland, adrenal gland, small intestine, spleen, peripheral blood leukocytes, fetal kidney and fetal lung (Hiraoka et al. 2000).

Future Perspectives

When desulfated dermatan sulfate was used as an acceptor, the purified and cloned C4ST preferentially transferred sulfate to GalNAc residue located in the GlcA-rich region of the acceptor polysaccharide. Such specificity of C4ST suggests that 4-sulfation of GalNAc residue may precede epimerization of adjacent uronic acid residue; however, it has not been shown whether C4ST is involved in the biosynthesis of dermatan sulfate in vivo.

Overexpression or suppression of C4ST in dermatan sulfate-producing cells can provide information about the involvement of C4ST in dermatan sulfate synthesis. Among the various glycosaminoglycan sulfotransferases so far cloned, only C4ST is stimulated with sulfhydryl compounds such as dithiothreitol, suggesting possible participation of sulfhydryl compounds in the regulation of 4-sulfation. The amino acid sequence deduced from the cDNA showed putative PAPS binding motif in C4ST: 5'-PSB and 3'-PB. Cys is found in 5'-PSB of C4ST but not in 5'-PSB of other glycosaminoglycan sulfotransferases so far cloned. These observation may be relevant to the fact that C4ST requires sulfhydryl compounds for the activity. Biosynthesis of chondroitin 4-sulfate is regulated during the development of animals. If C4ST knockout mice are generated, the physiological function of chondroitin 4-sulfate and C4ST can be clarified.

Further Reading

See Habuchi (2000) and Hiraoka et al. (2000).

References

Albertini R, De Luca G, Passi A, Moratti R, Abuja PM (1999) Chondroitin-4-sulfate protects high-density lipoprotein against copper-dependent oxidation. Arch Biochem Biophys 365:143–149

Fried M, Duffy PE (1996) Adherence of *Plasmodium falciparum* to chondroitin sulfate A in the human placenta. Science 272:1416–1417

Habuchi O (2000) Diversity and functions of glycosaminoglycan sulfotransferases. Biochim Biophys Acta 1474:115–127

Habuchi O, Miyashita N (1982) Separation and characterization of chondroitin 6-sulfotransferase and chondroitin 4-sulfotransferase from chick embryo cartilage. Biochim Biophys Acta 717:414–421

Habuchi O, Miyata K (1980) Stimulation of glycosaminoglycan sulfotransferase from chick embryo cartilage by basic proteins and polyamines. Biochim Biophys Acta 616:208–217

Habuchi O, Matsui Y, Kotoya Y, Aoyama Y, Yasuda Y, Noda M (1993) Purification of chondroitin 6-sulfotransferase secreted from cultured chick embryo chondrocytes. J Biol Chem 268:21968–21974

Habuchi O, Tsuzuki M, Takeuchi I, Hara M, Matsui Y, Ashikari S (1991) Secretion of chondroitin 6-sulfotransferase and chondroitin 4-sulfotransferase from cultured chick embryo chondrocytes. Biochim Biophys Acta 1133:9–16

Hiraoka N, Nakagawa H, Ong E, Akama OT, Fukuda MN, Fukuda M (2000) Molecular cloning and expression of two distinct human chondroitin 4-O-sulfotransferases that belong to the HNK-1 sulfotransferase gene family. J Biol Chem 275:20188–20196

Kimata K, Okayama M, Oohira A, Suzuki S (1973) Cytodifferentiation and proteoglycan biosynthesis. Mol Cell Biochem 1:211–228

Naujokas MF, Morin M, Anderson MS, Peterson M, Miller J (1993) The chondroitin sulfate form of invariant chain can enhance stimulation of T cell responses through interaction with CD44. Cell 74:257–268

Robinson HC (1969) The sulphation of chondroitin sulphate in embryonic chicken cartilage. Biochem J 113:543–549

Rogerson SJ, Chaiyaroj SC, Ng K, Reeder JC, Brown GV (1995) Chondroitin sulfate A is a cell surface receptor for *Plasmodium falciparum*-infected erythrocytes. J Exp Med 182:15–20

Sugiura N, Sakurai K, Hori Y, Karasawa K, Suzuki S, Kimata K (1993) Preparation of lipid-derivatized glycosaminoglycans to probe a regulatory function of the carbohydrate moieties of proteoglycans in cell-matrix interaction. J Biol Chem 268:15779–15787

Sugumaran G, Silbert JE (1988) Sulfation of chondroitin: specificity, degree of sulfation, and detergent effects with 4-sulfating and 6-sulfating microsomal systems. J Biol Chem 263:4673–4678

Toyama-Sorimachi N, Sorimachi H, Tobita Y, Kitamura F, Yagita H, Suzuki K, Miyasaka M (1995) A novel ligand for CD44 is serglycin, a hematopoietic cell lineage-specific proteoglycan. Possible involvement in lymphoid cell adherence and activation. J Biol Chem 270:7437–7444

Yamagata M, Suzuki S, Akiyama SK, Yamada KM, Kimata K (1989) Regulation of cell-substrate adhesion by proteoglycans immobilized on extracellular substrates. J Biol Chem 264:8012–8018

Yamauchi S, Hirahara Y, Usui H, Takeda Y, Hoshino M, Fukuta M, Kimura JH, Habuchi O (1999) Purification and characterization of chondroitin 4-sulfotransferase from the culture medium of a rat chondrosarcoma cell line. J Biol Chem 274:2456–2463

Yamauchi S, Mita S, Matsubara T, Fukuta M, Habuchi H, Kimata K, Habuchi O (2000) Molecular cloning and expression of chondroitin 4-sulfotransferase. J Biol Chem 275: 8975–8981

65

HNK-1 Sulfotransferase

Introduction

The HNK-1 epitope was originally described as a marker of a subfraction of human natural killer cells (Abo and Balch 1981). Despite its name, the HNK-1 carbohydrate epitope is found in many tissues but is predominantly expressed in the mammalian nervous system. The expression pattern of the HNK-1 carbohydrate in both the central and peripheral nervous system is spatially and developmentally regulated. In the nervous system the HNK-1 carbohydrate epitope is carried by many neural recognition glycoproteins and is involved in homophilic and heterophilic binding of these proteins (Schachner and Martini 1995). The epitope is characterized by the structure SO_4-3GlcAβ1-3Galβ(1-4GlcNAc) at the nonreducing end of glycans found on glycoproteins (N- and O-linked) (Voshol et al. 1996; Wakabayashi et al. 1999) and glycolipids (Chou et al. 1986). The key enzymes in the biosynthesis of HNK-1 carbohydrates are a glucuronyltransferase (GlcA-transferase) (Terayama et al. 1997) (see Chapter 51) and a sulfotransferase, as the underlying lactosamine structure is commonly found in glycoconjugates. The sulfotransferase enzyme catalyzing the transfer of sulfate from the donor substrate 3'-phosphoadenosine 5'-phosphosulfate (PAPS) to the glucuronic acid containing glycolipid or glycoprotein acceptor was first characterized by Chou and Jungalwala (1993), and the cDNA was first cloned by Bakker et al. (1997).

To date, rat DNAs (Bakker et al. 1997) and human cDNAs (Ong et al. 1998) are the only cloned sequences that have been proven to encode an HNK-1 sulfotransferase, but at least four other groups of homologous ESTs (see unigene sequences below) have been reported. The mouse homolog of one of them was recently determined to encode

HANS BAKKER

Plant Research International, Wageningen University and Research Centre, PO Box 16, 6700 AA Wageningen, The Netherlands
Tel. +31-317-477000; Fax +31-317-418094
Present address: Institut für Physiologische Chemie, Proteinstruktur-OE4315, Medizinische Hochschule Hannover, Carl-Neuberg-Strasse 1, 30625 Hannover, Germany
Tel. +49-511-532-9802; Fax +49-511-532-3956
e-mail: Hans@mh-hannover.de

a chondroitin 4-sulfotransferase (Yamauchi et al. 2000 and Chapter 64). Three homologs are also present in the recently sequenced *Drosophila* genome. Although homologs of HNK-1 GlcA-transferase are present in *Caenorhabditis elegans* and *Arabidopsis thaliana*, no close homologs are found for the HNK-1 sulfotransferase in these organisms. There is no overall homology of the HNK-1 sulfotransferase to other known carbohydrate sulfotransferases, but two shared sequence motifs, also with non-carbohydrate sulfotransferases, can be found (Ong et al. 1998, 1999). Mutations in these motifs indeed influence the sulfotransferase activity and substrate binding.

Databanks

HNK-1 sulfotransferase

No EC number has been allocated.

Species	Gene	Protein	mRNA	Genomic
Homo sapiens	–	AAC04707	AF033827	–
		–	Hs.155553 (Unigene)	–
Rattus noevegicus	–	AAB88123	AF022729	–
Drosophila melanogaster	–	AAF53584	–	AE003653
		AAF52238	–	AE003609
		AAF47588	–	AE003473

Name and History

The name HNK-1 originates from a monoclonal antibody that recognizes an epitope on human natural killer cells (Abo and Balch 1981). Later, this antibody was shown to recognize a carbohydrate epitope that is also carried by neural recognition glyco-proteins, such as the neural cell adhesion molecule (NCAM) and myelin-associated glycoprotein (MAG) (Kruse et al. 1984), as well as glycolipids in the nervous system (Chou et al. 1986). The structure of the glycolipid was first described by these authors as SO_4-3GlcAβ1-3Galβ1-4GlcNAcβ1-3Galβ1-4Glcβ1-1ceramide (sulfated glucuronyl glyco-lipids). On the *N*-linked glycans of bovine peripheral myelin glycoprotein P0, the SO_4-3GlcAβ1-3Galβ1-4GlcNAc structure was found on the common core structure (Voshol et al. 1996). HNK-1-like structures can also be found in *O*-linked glycans (Wakabayashi et al. 1999), but there the terminal sulfated disaccharide is β1,3-linked to Galβ1-4Xyl. Based on these studies, the HNK-1 epitope can be described as SO_4-3GlcAβ1-3Gal present at the nonreducing end of glycoconjugates. The key enzymes involved in its synthesis are HNK-1 GlcA-transferase and HNK-1 sulfotransferase. The sulfotransferase was first characterized by Chou and Jungalwala (1993). Later, rat (Bakker et al. 1997) and human (Ong et al. 1998) cDNAs were cloned by expression cloning using a cell line that is negative for the HNK-1 epitope. A cDNA library made from tissue expressing the sulfotransferase was expressed together with the GlcA-transferase clone. The HNK-1 antibody was used to select cells expressing the HNK-1 epitope. It also showed that these two enzymes are both required and are sufficient for expression of the epitope.

Enzyme Activity Assay and Substrate Specificity

The HNK-1 or glucuronyl-glycolipid sulfotransferase activity was first described and characterized by Chou and Jungalwala (1993). They used a natural glycolipid acceptor (GlcAnLcOse$_4$Cer) and an enzyme preparation from rat brain to determine the optimal reaction conditions. The enzyme requires Mn^{2+} (10 mM optimum); and as the enzyme is membrane-bound, a detergent such as triton X-100 (0.1%) must be added for activity. ATP is added in the reaction mixture at 2.5 mM. Different buffers can be used to measure the sulfotransferase activity. The authors reported a pH optimum of 7.2 using 100 mM Tris. However, being at the border of the buffering range of Tris, no lower pH was tested. Using 100 mM bis-Tris Bakker et al. (1997) obtained a pH optimum of 6.6.

The enzyme uses 3'-phosphoadenosine 5'-phosphosulfate (PAPS) as a donor substrate and various acceptor molecules having a glucuronic acid at the nonreducing end: PAPS + GlcA-R → PAP + SO$_4$-3-GlcA-R. The enzymatic assay is usually performed using radiolabeled [^{35}S]PAPS (New England Nuclear) and an oligosaccharide acceptor substrate conjugated to a hydrophobic group. This can be either the natural glycolipid acceptor (Chou and Jungalwala 1993) or synthetic substrates (Bakker et al. 1997; Ong et al. 1998). The transfer of sulfate can be measured by taking advantage of the difference in hydrophobicity between PAPS and the reaction product. The product is retained on a C-18 column (e.g., Sep-pak C-18 cartridges from Waters) whereas PAPS is not. The eluate can then be directly measured by liquid scintillation counting or, alternatively, run on silica gel thin layer chromatography (TLC) to quantify and characterize the product (Chou and Jungalwala 1993). After separation by TLC in chloroform/methanol/0.25% CaCl$_2$ (5:4:1), the plates can be exposed to X-ray film or analyzed by phosphor imaging. The reaction product can also be directly analyzed by TLC, without prior C-18 purification, as PAPS does not migrate from the origin (Bakker et al. 1997).

The presence of terminal glucuronic acid in the acceptor substrate is an absolute requirement for the sulfotransferase activity. When GlcA is replaced by Glc or 6-methyl GlcA in the natural glycolipid acceptor, molecules are no longer a substrate (Chou and Jungalwala 1993). On the other hand, the underlying structure can be largely omitted. The enzyme is even able to use the monosaccharide acceptor GlcAβ1-pNP. However, under the specified assay conditions, the disaccharide acceptor GlcAβ1-3Galβ1-R, containing a different hydrophobic aglycon, was five times more reactive (Bakker et al. 1997). The use of larger structures as substrate did not increase the sulfotransferase activity (personal observation). The trisaccharide GlcAβ1-3Galβ1-4GlcNAcβ1-octyl used by Ong et al. (1998, 1999) is also a suitable acceptor. However, except for GlcAβ1-pNP (Sigma, Fluka), none of these acceptor molecules is commercially available.

Preparation

The natural source of the enzyme mainly used is rat brain or, more specifically, cerebral cortex of young rats. The sulfotransferase activity can be measured using either a microsomal fraction (Chou and Jungalwala 1993) or crude homogenate (Bakker et al. 1997). Coming from a natural source, this enzyme preparation contains other sulfotransferases, especially sulfatide sulfotransferase. Higher purity enzyme preparation can be obtained from recombinant enzyme. As for most glycosyltransferases, an easy way to obtain purified enzyme is to express it as a fusion protein with protein A, as described in Bakker et al. (1997) and Ong et al. (1998). The enzyme is secreted in the medium of a mammalian cell culture and can easily be purified by immunoglobulin affinity. A disadvantage of this technique is that the recombinant enzyme is no longer membrane bound, unlike the natural enzyme, and might therefore have different characteristics.

Biological Aspects

Although several HNK-1 GlcA-transferases have been cloned and various enzymes are expected to be responsible for biosynthesis of the glycolipid and glycoprotein form of the epitope, there seems to be only one HNK-1 sulfotransferase. The acceptor specificity indicates that the enzyme can act on acceptors that mimic the natural glycolipid epitope; and the cloned rat and human enzymes are also active on glycoproteins (Bakker et al. 1997; Ong et al. 1998). The sulfotransferase seems therefore capable of synthesizing all known HNK-1 carbohydrate structures. Expression of the sulfotransferase mRNA does not seem to regulate the occurrence of the HNK-1 epitope. Its expression pattern is indeed much less restricted than that of the HNK-1 GlcA-transferase (Ong et al. 1998), which is therefore thought to regulate the occurrence of HNK-1 carbohydrates. An interesting novel structure has been described: SO_4-3GlcAβ1-3Galβ1-3Galβ1-4Xyl (Wakabayashi et al. 2000). Without sulfate, this structure is normally the proteoglycan linkage tetrasaccharide. In this case the sulfate has been suggested to form a stop signal for proteoglycan synthesis. The GlcA-transferase is almost certainly different from the one involved in HNK-1 biosynthesis, but at least according to the in vitro activity the same sulfotransferase could be involved.

Future Perspectives

There might be therapeutic applications for HNK-1, which plays important roles in the nervous system (e.g., in selective neurite outgrowth). The cloned sulfotransferase could be employed for biosynthesis of HNK-1 carbohydrates by in vitro enzyme-assisted synthesis and by expressing the enzyme together with the GlcA-transferase in cell lines in which HNK-1 carrying glycoproteins can be produced.

Obviously, now that the cDNAs encoding the enzymes involved in biosynthesis of the HNK-1 carbohydrate are cloned, knocking-out the genes becomes feasible. Although the expression pattern of the epitope predicts a function in development and maintenance of the nervous system, it will hopefully give more conclusive evidence for the exact role.

There are still three human genes with unknown function that show homology to the HNK-1 sulfotransferase. It is difficult to predict the enzymatic activity of these enzymes. The two known members of the homologous family transfer sulfate to the C-3 position of GlcA (HNK-1 sulfotransferase) and C-4 of *N*-acetylgalactosamine (chondroitin 4-sulfotransferase).

Further Reading

A review of Schachner and Martini (1995) is a good starting point for understanding the involvement of HNK-1 glycans and glycoconjugates in neural recognition. Two papers about cloning HNK-1 sulfotransferase (Bakker et al. 1997; Ong et al. 1998) describe the activity of the recombinant enzyme and outline elegant expression cloning strategies. A nice sequence comparison, showing conserved amino acids in the PAPS binding domains of various sulfotransferases, can be found in Ong et al. (1999).

Acknowledgments

The author's work on the HNK-1 sulfotransferase has been carried out in the laboratory of Prof. Melitta Schachner at the Swiss Federal Institute of Technology, Zürich, Switzerland. Her group is now at the Universität Hamburg, Germany.

References

Abo T, Balch CM (1981) A differentiation antigen of human NK and K cells identified by a monoclonal antibody (HNK-1). J Immunol 127:1024–1029

Bakker H, Friedmann I, Oka S, Kawasaki T, Nifant'ev N, Schachner M, Mantei N (1997) Expression cloning of a cDNA encoding a sulfotransferase involved in the biosynthesis of the HNK-1 carbohydrate epitope. J Biol Chem 272:29942–29946

Chou DKH, Jungalwala FB (1993) Characterization and developmental expression of a novel sulfotransferase for the biosynthesis of sulfoglucuronyl glycolipids in the nervous system. J Biol Chem 268:330–336

Chou DKH, Ilyas AA, Evans JE, Costello C, Quarles RH, Jungalwala FB (1986) Structure of sulfated glucuronyl glycolipids in the nervous system reacting with HNK-1 antibody and some IgM paraproteins in neuropathy. J Biol Chem 261:11717–11725

Kruse J, Mailhammer R, Wernecke H, Faissner A, Sommer I, Goridis C, Schachner M (1984) Neural cell adhesion molecules and myelin-associated glycoprotein share a common carbohydrate moiety recognized by monoclonal antibodies L2 and HNK-1. Nature 311:153–155

Ong E, Yeh J-C, Ding Y, Hindsgaul O, Fukuda M (1998) Expression cloning of a human sulfotransferase that directs the synthesis of the HNK-1 glycan on the neural cell adhesion molecule and glycolipids. J Biol Chem 273:5190–5195

Ong E, Yeh J-C, Ding Y, Hindsgaul O, Pedersen LC, Negishi M, Fukuda M (1999) Structure and function of HNK-1 sulfotransferase: identification of donor and acceptor binding sites by site-directed mutagenesis. J Biol Chem 274:25608–25612

Schachner M, Martini R (1995) Glycans and the modulation of neural-recognition molecule function. Trends Neurosci 18:183–191

Terayama K, Oka S, Seiki T, Miki Y, Nakamura A, Kozutsumi Y, Takio K, Kawasaki T (1997) Cloning and functional expression of novel glucuronyltransferase involved in the biosynthesis of the carbohydrate epitope, HNK-1. Proc Natl Acad Sci USA 94:6093–6098

Voshol H, van Zuylen CWEM, Orberger G, Vliegenthart JFG, Schachner M (1996) Structure of the HNK-1 carbohydrate epitope on bovine peripheral myelin glycoprotein P0. J Biol Chem 271:22957–22960

Wakabayashi H, Natsuka S, Mega T, Otsuki N, Isaji M, Naotsuka M, Koyama S, Kanamori T, Sakai K, Hase S (1999) Novel proteoglycan linkage tetrasaccharides of human urinary soluble thrombomodulin, SO_4-3GlcAβ1-3Galβ1-3($\pm$Siaα2-6)Galβ1-4Xyl. J Biol Chem 274:5436–5442

Yamauchi S, Mita S, Matsubara T, Fukuta M, Habuchi H, Kimata K, Habuchi O (2000) Molecular cloning and expression of chondroitin 4-sulfotransferase. J Biol Chem 275:8975–8981

Galactosaminoglycan Uronyl 2-Sulfotransferase

Introduction

The glycosaminoglycan (GAG) side chains of proteoglycans exhibit structural diversity that allows participation in numerous biologic functions. Structural diversity is in part dependent on the abundance and placement of 2-sulfated uronyl residues. These components are highly abundant in the GAGs heparan sulfate and heparin but are infrequent in the galactosaminoglycans. For example, dermatan sulfate is predominantly composed of IdoA→GalNAc-4S and to a lesser extent GlcA→GalNAc-4. However, the 2-sulfated unit IdoA-2S→GalNAc-4S can also occur in minor amounts (5%–10% of total disaccharides) (Maimone and Tollefsen 1991). Similarly, chondroitin sulfate principally contains GlcA→GalNAc-4S, GlcA→GalNAc-6S, or both; low amounts of GlcA-2S→GalNAc-6S may also be present (Cheng et al. 1994). Such paucity makes 2-sulfated uronyl residues ideal candidates for regulating selected biologic activities of dermatan sulfate/chondroitin sulfate. This chapter describes the recent isolation and characterization of the enzyme galactosaminoglycan uronyl 2-sulfotransferase, which is capable of adding this rare substituent to dermatan sulfate and chondroitin sulfate (Kobayashi et al. 1999).

Nicholas W. Shworak[1,2]* and Robert D. Rosenberg[1,2]

[1] Department of Medicine, Harvard Medical School, Angiogenesis Research Center, SL-418, Beth Israel Deaconess Medical Center, 330 Brookline Avenue, Boston, MA 02215, USA
[2] Department of Biology, Massachusetts Institute of Technology, 77 Massachusetts Avenue, Cambridge MA 02139, USA
* Present address: Department of Medicine, Dartmouth Medical School, Angiogenesis Research Center, HB7504 Dartmouth Mitchcock Medical Center, One Medical Center Drive, Lebanon, NH 03756, USA
Tel. +1-603-650-6401; Fax +1-603-653-0510

Databanks

Galactosaminoglycan uronyl 2-sulfotransferase (DS/CS2ST)

No EC number has been allocated.

Species	Gene	Protein	mRNA	Genomic
Homo sapiens	*UST*	NP_005706	AB020316	–
		–	NM_005715	–
		–	Hs.134015 (Unigene)	–

Name and History

Galactosaminoglycan uronyl 2-sulfotransferase (2ST) has also been referred to as dermatan/chondroitin sulfate 2-sulfotransferase (DS/CS2ST) and uronyl 2-sulfotransferase. Galactosaminoglycan 2ST cDNA was initially identified by a homology search designed to identify potential uronyl 2-sulfotransferases (Kobayashi et al. 1999). The deduced amino acid sequence of the heparan sulfate uronyl 2-sulfotransferase (see Chapter 67) was used to probe a databank of cDNA clones (the National Center for Biotechnology Information databank of I.M.A.G.E. Consortium expressed sequence tagged clones). The search identified a partial-length insert of a novel clone (clone ID HE9MJ06) isolated from a cDNA library generated from a 9-week-old human embryo. The insert was then used as a probe in Northern blot experiments to evaluate transcript expression levels in multiple tissues and cell types. This analysis showed maximal expression in Ragi cells, so a Ragi cDNA library was then screened to isolate a full-length cDNA clone (Kobayashi et al. 1999).

Enzyme Activity Assay and Substrate Specificity

Enzymatic properties have only been analyzed using crude preparations of recombinantly expressed enzyme (Kobayashi et al. 1999). Insect cells were generated to express either the galactosaminoglycan 2ST cDNA or the appropriate control vector. The corresponding cell extracts were incubated with [^{35}S] 3′-phosphoadenosine 5′-phosphosulfate (PAPS) (20 µM) plus various GAG acceptors (0.5 mg/ml) in imidazole HCl buffer (50 mM, pH 6.8) with protamine sulfate (75.2 µg/ml). It is important to note that the composition of the cell lysis buffer introduced to the reaction 30 mM NaCl, 0.1% Triton-X 100, 2 mM MgCl$_2$, 0.4 mM CaCl$_2$, and 4% glycerol. The resulting profile of ^{35}S incorporation (control-corrected) revealed that the sulfotransferase activity was substrate-specific. The acceptors tested were the naturally occurring GAGs dermatan sulfate (from porcine skin) and chondroitin sulfate C (from shark cartilage), their corresponding partially desulfated forms (chemically generated), and chemically desulfated N-resulfated heparin. Dermatan sulfate showed maximal incorporation, which was twofold greater than chemically desulfated dermatan sulfate and 16-fold greater than chondroitin sulfate. The remaining GAGs lacked significant sulfotransferase activity above background. However, heparan sulfate and heparin cannot be completely excluded as potential substrates, as the control extracts exhibited extremely high incorporation toward the chemically modified heparin acceptor, which

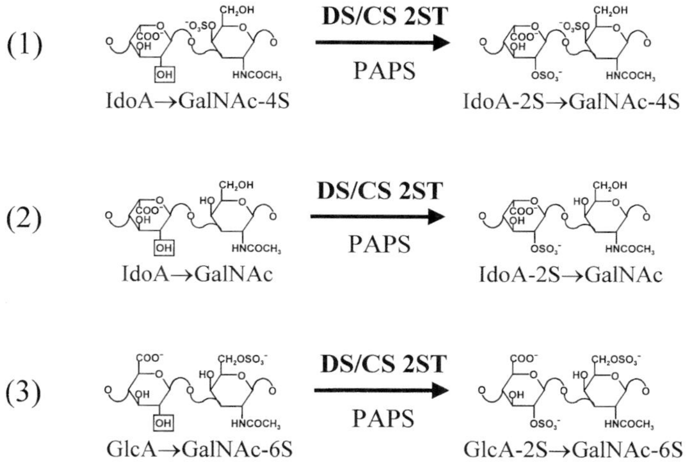

(1)

IdoA→GalNAc-4S   **DS/CS 2ST** / PAPS   IdoA-2S→GalNAc-4S

(2)

IdoA→GalNAc   **DS/CS 2ST** / PAPS   IdoA-2S→GalNAc

(3)

GlcA→GalNAc-6S   **DS/CS 2ST** / PAPS   GlcA-2S→GalNAc-6S

Fig. 1. Galactosaminoglycan uronyl 2-sulfotransferase (2ST) is known to perform three major reactions. Reaction (1) and to a lesser extent reaction (2) occur during the biosynthesis of dermatan sulfate. Reaction (3) can occur during production of chondroitin sulfate C

precluded sensitive detection of minor sulfotransferase activities (Kobayashi et al. 1999).

The structures of the major sulfated products were indirectly deduced at the disaccharide level. The uronic acid position was established based on susceptibility to cleavage with chondroitinase B versus chondroitinase AC (the former cleaves prior to iduronyl residues, whereas the later cleaves prior to glucuronyl residues). The positions of ^{35}S-labeled sulfate (2ST generated) and unlabeled sulfate (preexisting) groups were determined by incubating the lyase-liberated unsaturated disaccharides with disaccharide 2-sulfatase, 4-sulfatase, 6-sulfatase, or no enzyme. The reaction products were identified by co-resolution with standards on a high-performance liquid chromatography (HPLC) polyamine column. The data show that this uronyl 2ST modifies galactosaminoglycans by three major reactions (Fig. 1).

The optimal acceptor, dermatan sulfate, predominantly provides substrate for reaction (1) (Fig. 1), with more than 90% of the sulfate being incorporated into IdoA-2S→GalNAc-4S residues. However, partially desulfated dermatan is an efficient substrate for reactions (1) and (2), with about equal generation of IdoA-2S→GalNAc and IdoA-2S→GalNAc-4S. The unmodified substrate did not allow synthesis of IdoA-2S→GalNAc because only meager quantities of the appropriate precursor disaccharide (IdoA→GalNAc) occur in porcine skin dermatan sulfate (less than 1% of total disaccharides) (Maimone and Tollefsen 1990). Chondroitin sulfate C almost exclusively underwent reaction (3) with more than 98% of sulfate incorporating into GlcA-2S→GalNAc-6S products. Similar analyses of chondroitin sulfate A suggested little to no production of GlcA-2S→GalNAc-4S. Thus, the uronyl 2ST favors modification of selected IdoA residues with adjacent reducing side GalNAc ± 4S and favors modification of certain GlcA residues with adjacent reducing side GalNAc-6S. A more in-depth understanding of relative specificities would require kinetic evaluation of purified enzyme with structurally defined substrates. Nevertheless, the data are consistent with the 2-sulfation pattern of uronyl residues found in connective tissue from many species. Moreover, the enzyme specificity suggests an order of galactosaminoglycan biosynthesis where 2-sulfation of IdoA follows GalNAc 4-sulfation and that 2-sulfation of GlcA follows GalNAc 6-sulfation (Kobayashi et al. 1999).

Preparation

The enzyme has never been purified, and activity has only been demonstrated by cDNA expression in insect cells. Specifically, the coding region of the galactosaminoglycan 2ST cDNA was initially subcloned into the pFASTBAC HTa donor plasmid. This places a viral polyhedrin promoter upstream of the cDNA. Furthermore, the entire transcriptional unit is flanked with mini-Tn7 transposable elements. The donor plasmid was then transformed to DH10BAC. This *Escherichia coli* strain harbors a recipient bacmid with a mini-*att*Tn7 target site. In vivo transposition transfers the transcriptional unit from the donor plasmid to the target bacmid DNA from the resulting expression bacmid was then isolated and transfected into Sf9 insect cells. These cells replicate the bacmid DNA, express the encoded viral proteins, and secrete infectious recombinant baculovirus particles. Culture medium that contains the viral stock was collected and used to infect additional Sf9 cells for expression of the galactosaminoglycan 2ST cDNA. After 3 days the infected cell monolayers were washed, harvested by scraping, and lysed with a buffer containing Triton X-100. The lysate was centrifuged, and the supernatant was used as a crude source of enzyme (Kobayashi et al. 1999).

Biological Aspects

Dermatan sulfate proteoglycans and chondroitin sulfate proteoglycans are present in most tissues, and the GAG fine structure is thought to regulate various biologic activities. Consistent with this notion, Northern blot analysis shows that all major human tissues express galactosaminoglycan 2ST mRNAs (Kobayashi et al. 1999). Expression is particularly high in heart, pancreas, and brain; and levels are lowest in liver. Ubiquitous tissue expression, however, does not mandate ubiquitous cell-type expression.

Examination of poly(A)$^+$ RNA from several cell lines showed undetectable expression in the promyelocytic leukemia line HL-60 and the chronic myelogenous leukemia cell K-652. All tissues and expressive cell types show a major transcript of 5.1 kb and a minor species of 2.0 kb. Multiple transcripts are observed for many GAG biosynthetic enzymes and usually reflect differences in the untranslated regions (Shworak et al. 1999). Distinct untranslated region sequences might engender enhanced regulatory control through alteration of translational efficiency or mRNA accumulation.

Numerous roles for dermatan sulfate (DS) are suggested because of its ability to interact with several heparan-binding factors, including basic fibroblast growth factor (bFGF), histidine-rich glycoprotein, platelet factor 4, fibronectin, interleukin-7, protein C inhibitor, and interferon-γ (Brooks et al. 2000). This subject was reviewed by Penc et al. (1999). Heparan–protein interactions frequently involve 2-sulfated iduronyl residues (Turnbull et al. 1992; Spillmann et al. 1998), which might account for the affinity of the above proteins for DS. Although such interactions are considered weak (micromolar dissociation constants), it does not preclude biologic activity. Indeed, for fluid from human wounds DS is responsible for most of the FGF-2 dependent effects, and heparan sulfate plays only a minor cofactor role (Penc et al. 1998). DS also shows strong interactions with heparin cofactor II and with hepatocyte growth factor/scatter factor. The former protein requires multiple 2-sulfated iduronyl residues for binding, whereas binding by the latter factor is independent of such residues (Maimone and Tollefsen 1990; Lyon et al. 1998). Thus, biosynthetic regulation of galactosaminoglycan 2-ST activity may serve to regulate distinct biologic activities of DS.

Chondroitins with 2-sulfated glucuronyl residues can be detected with the monoclonal antibodies MO-225 and 473HD (Yamagata et al. 1987; Faissner et al. 1994). These rare moieties have been detected in diverse tissues, including developing limb bud, tooth germ basement membrane, mast cells, and neural tissue (Avnur and Geiger 1984; Davidson et al. 1990; Mark et al. 1990). The disaccharide GlcA-2S→GalNAc-6S serves as a phenotypic marker of mast cell subsets, specifically identifying cells derived from immune lymph nodes (Davidson et al. 1990). Biologic activity has been demonstrated in neural development. Immature glial cells of the central nervous system express GlcA-2S, which activates neurite outgrowth. Outgrowth can be neutralized with the 473HD antibody and then recovered by supplementation with chondroitin-containing GlcA-2S (Faissner et al. 1994; Nadanaka et al. 1998). This result is tantalizing, given that GlcA-2S occurs in adult brain and that galactosaminoglycan 2ST mRNA has high brain expression (Avnur and Geiger 1984; Kobayashi et al. 1999). Thus, galactosaminoglycan 2ST may participate in regulating synaptic plasticity.

Future Perspectives

The rarity of 2-sulfated uronyl residues within galactosaminoglycans make this modification ideal for regulating selected biologic activities. An analogous situation occurs during biosynthesis of GAGs, where placement of the rare 3-sulfate group in one sequence context creates antithrombin binding sites (Shworak et al. 1996) and in a distinct sequence creates binding sites for the herpes virus glycoprotein D (Shukla et al. 1999). Although study of the galactosaminoglycan 2ST is in its infancy, we expect future investigations to show that this enzyme regulates many critical biologic functions.

Further Reading

See Kobayashi et al. (1999) for isolation and expression of the cDNA; Faissner et al. (1994) for neurite outgrowth activity of 2-sulfated chondroitin sulfate; and Maimone and Tollefsen (1990) for elucidation of the dermatan structure that binds heparin cofactor II.

References

Avnur Z, Geiger B (1984) Immunocytochemical localization of native chondroitin-sulfate in tissues and cultured cells using specific monoclonal antibody. Cell 38:811–822

Brooks B, Briggs DM, Eastmond NC, Fernig DG, Coleman JW (2000) Presentation of IFN-gamma to nitric oxide-producing cells: a novel function for mast cells. J Immunol 164:573–579

Cheng F, Heinegard D, Malmstrom A, Schmidtchen A, Yoshida K, Fransson LA (1994) Patterns of uronyl epimerization and 4-/6-O-sulphation in chondroitin/dermatan sulphate from decorin and biglycan of various bovine tissues. Glycobiology 4:685–696

Davidson S, Gilead L, Amira M, Ginsburg H, Razin E (1990) Synthesis of chondroitin sulfate D and heparin proteoglycans in murine lymph node-derived mast cells: the dependence on fibroblasts. J Biol Chem 265:12324–12330

Faissner A, Clement A, Lochter A, Streit A, Mandl C, Schachner M (1994) Isolation of a neural chondroitin sulfate proteoglycan with neurite outgrowth promoting properties. J Cell Biol 126:783–799

Kobayashi M, Sugumaran G, Liu J, Shworak NW, Silbert JE, Rosenberg RD (1999) Molecular cloning and characterization of a human uronyl 2-sulfotransferase that sulfates iduronyl and glucuronyl residues in dermatan/chondroitin sulfate. J Biol Chem 274:10474–10480

Lyon M, Deakin JA, Rahmoune H, Fernig DG, Nakamura T, Gallagher JT (1998) Hepatocyte growth factor/scatter factor binds with high affinity to dermatan sulfate. J Biol Chem 273:271–278

Maimone MM, Tollefsen DM (1990) Structure of a dermatan sulfate hexasaccharide that binds to heparin cofactor II with high affinity. J Biol Chem 265:18263–18271

Maimone MM, Tollefsen DM (1991) Structure of a dermatan sulfate hexasaccharide that binds to heparin cofactor II with high affinity. J Biol Chem 266:14830

Mark MP, Baker JR, Kimata K, Ruch JV (1990) Regulated changes in chondroitin sulfation during embryogenesis: an immunohistochemical approach. Int J Dev Biol 34:191–204

Nadanaka S, Clement A, Masayama K, Faissner A, Sugahara K (1998) Characteristic hexasaccharide sequences in octasaccharides derived from shark cartilage chondroitin sulfate D with a neurite outgrowth promoting activity. J Biol Chem 273:3296–3307

Penc SF, Pomahac B, Eriksson E, Detmar M, Gallo RL (1999) Dermatan sulfate activates nuclear factor-κb and induces endothelial and circulating intercellular adhesion molecule-1. J Clin Invest 103:1329–1335

Penc SF, Pomahac B, Winkler T, Dorschner RA, Eriksson E, Herndon M, Gallo RL (1998) Dermatan sulfate released after injury is a potent promoter of fibroblast growth factor-2 function. J Biol Chem 273:28116–28121

Shukla D, Liu J, Blaiklock P, Shworak NW, Bai X, Esko JD, Cohen GH, Eisenberg RJ, Rosenberg RD, Spear PG (1999) A novel role for 3-O-sulfated heparan sulfate in herpes simplex virus 1 entry. Cell 99:13–22

Shworak NW, Fritze LMS, Liu J, Butler LD, Rosenberg RD (1996) Cell-free synthesis of anticoagulant heparan sulfate reveals a limiting activity which modifies a nonlimiting precursor pool. J Biol Chem 271:27063–27071

Shworak NW, Liu J, Petros LM, Zhang L, Kobayashi M, Copeland NG, Jenkins NA, Rosenberg RD (1999) Multiple isoforms of heparan sulfate D-glucosaminyl 3-O-sulfotransferase: isolation, characterization, and expression of human cDNAs and identification of distinct genomic loci. J Biol Chem 274:5170–5184

Spillmann D, Witt D, Lindahl U (1998) Defining the interleukin-8-binding domain of heparan sulfate. J Biol Chem 273:15487–15493

Turnbull JE, Fernig DG, Ke Y, Wilkinson MC, Gallagher JT (1992) Identification of the basic fibroblast growth factor binding sequence in fibroblast heparan sulfate. J Biol Chem 267:10337–10341

Yamagata M, Kimata K, Oike Y, Tani K, Maeda N, Yoshida K, Shimomura Y, Yoneda M, Suzuki S (1987) A monoclonal antibody that specifically recognizes a glucuronic acid 2-sulfate-containing determinant in intact chondroitin sulfate chain. J Biol Chem 262:4146–4152

Heparan Sulfate 2-Sulfotransferase

Introduction

Heparan sulfate proteoglycans (HSPGs) are ubiquitously present on cell surfaces and in extracellular matrices including basement membranes; they have divergent structures and functions. The heparan sulfate chains in HSPGs are known to interact with a variety of proteins, such as heparin-binding growth factors, extracellular matrix components, protease inhibitors, proteases, and lipoprotein lipase. These interactions are implicated not only in various dynamic cellular behaviors including cell proliferation, differentiation, adhesion, migration, and morphogenesis during development but also in physiological phenomena such as inflammation, blood coagulation, and tumor cell invasion and malignancy. Moreover, some pathogens (e.g., bacteria, parasites, viruses) are known to interact with the cell surface HSPGs on host cells; and the heparan sulfate chains in most cases are ligands for the coat proteins or cell surface proteins of pathogens.

The interactions of heparan sulfate with various ligands take place on the specific regions of heparan sulfate. Heparan sulfate 2-O-sulfotransferase (HS2ST) catalyzes the transfer reaction of sulfate from 3'-phosphoadenosine 5'-phosphosulfate (PAPS) to position 2 of the iduronic acid residue and appear to be essential for synthesis of the fibroblast growth factor-2 (FGF-2) binding structure.

HIROKO HABUCHI and KOJI KIMATA

The Institute for Molecular Science of Medicine, Aichi Medical University, Nagakute, Aichi 480-1195, Japan
Tel. +81-52-264-4811 (ext. 2088); Fax +81-561-63-3532
e-mail: Kimata@amugw.aichi-med-u.ac.jp

Databanks

Heparan sulfate 2-sulfotransferase

No EC number has been allocated.

Species	Gene	Protein	mRNA	Genomic
Homo sapiens	HS2ST	–	AB024568	–
Mus musculus	HS2ST	–	AF060178	–
Cricetulus griseus	HS2ST	–	E17300	–
Xenopus laevis	HS2ST	–	AF060179	–
Drosophila melanogaster	Sd	–	AF143860	–
	pipe	–	AF102136	–

Name and History

The activity of the *O*-sulfotransferase of heparan sulfate (heparan sulfate *O*-sulfotransferase) was detected in heparin-producing mouse mastocytoma using completely desulfated *N*-resulfated heparin (CDSNS-heparin) as acceptor substrate. The partially purified enzyme from the tissues exhibited both *N*-sulfoglucosaminyl 6-*O*-sulfotransferase and iduronyl 2-*O*-sulfotransferase activities (Wlad et al. 1994). However, the enzyme purified to apparent homogeneity from cultured Chinese hamster ovary (CHO) cells showed only heparan sulfate 2-*O*-sulfotransferase activity, not *N*-sulfoglucosaminyl 6-*O*-sulfotransferase activity. Therefore, it was designated HS2ST (Kobayashi et al. 1996). The peptide sequences from the purified HS2ST made it possible to perform cDNA cloning (Kobayashi et al. 1997).

Enzyme Activity Assay and Enzyme Properties

HS2ST transfers sulfate to position 2 of L-iduronic acid residue and with less activity to position 2 of the D-glucuronic acid residue in heparan sulfate as follows.

$$PAPS + (IdoA\alpha 1\text{-}4GlcNSO_3\alpha 1\text{-}4)_n \rightarrow PAP + [IdoA(2SO_4)\alpha 1\text{-}4GlcNSO_3\alpha 1\text{-}4]_n$$

Where PAP is 3′-phosphoadenosine 5′-phosphate.

For the HS2ST activity assay the standard reaction mixture contained 2.5 μmol imidazole-HCl pH 6.8, 3.75 μg protamine chloride, 25 nmol (as hexosamine) completely desulfated and *N*-resulfated (CDSNS)-heparin, 50 pmol [^{35}S]PAPS (about 5 × 10^5 cpm), and enzyme, in a final volume of 50 μl. After incubation at 37°C, ^{35}S-labeled polysaccharides were precipitated in cold 70% (v/v) ethanol containing 1% potassium acetate and 0.2 mM EDTA. The precipitates were subjected to gel chromatography using a Fast desalting column (Pharmacia). The ^{35}S radioactivity in the polysaccharide fraction was counted. To determine the sulfation of position 2 of the hexuronic acid residue, the ^{35}S-labeled polysaccharide was digested with a mixture of heparitinases I, II, and III. The digested products were analyzed together with the standard unsaturated disaccharides (derived from heparin and heparan sulfate; Seikagaku Corp.) by high-performance liquid chromatography (HPLC) on a column of PAMN (YMC, Kyoto, Japan). To determine the isomeric hexuronic acid, iduronic acid or glucuronic acid, the ^{35}S-labeled glycosaminoglycans were treated with nitrous acid at pH 1.5 and then were reduced with NaBH$_4$ (Shively and Conrad 1976). The ^{35}S-labeled products were analyzed by HPLC on a Partisil-10 SAX column.

The purified HS2ST showed the following enzyme properties: The optimal pH was around 5.5. The enzyme activity was little affected by dithiothreitol (DTT) up to 10 mM. NaCl stimulated the enzyme activity, and the maximal activity was around 100 mM NaCl. Protamine also stimulated the enzyme activity to the maximum level at a concentration of 0.05 mg/ml. The K_m value for PAPS was 0.2 μM. When cultured CHO cells were subjected to detection of HS2ST activity, most HS2ST was retained in the cell layer, while more than 90% of the HS6ST activity was secreted into the culture medium (Kobayashi et al. 1996). HS2ST activity was rarely detected in mouse serum or human serum (Habuchi et al. 1998a,b).

Substrate Specificity

CDSNS-heparin and mouse EHS tumor heparan sulfate were good acceptor substrates, whereas heparin, pig arterial heparan sulfate, and bovine liver heparan sulfate were poor substrates. Chondroitin, chondroitin sulfate, dermatan sulfate, and keratan sulfate did not serve as acceptors at all. Structural analyses of those products showed that HS2ST transfers sulfate to position 2 of the iduronic acid residue of the IdoAα1-4GlcNSO$_3$ unit but not to that of the IdoAα1-4GlcNSO3(6SO$_4$) unit (Kobayashi et al. 1996). Low but significant activity was detected when N-sulfated heparosan that had (GlcAβ1-4GlcNSO$_3$α1-4) repeating units was used as acceptor; and analysis of the products showed that it could transfer sulfate to position 2 of the glucuronic acid residue of GlcAβ1-4GlcNSO$_3$ (Habuchi et al. 1999). A CHO cell mutant, pgsF-17, which was shown to be defective in 2-O-sulfation of iduronic acid of heparan sulfate (HS), produced HS lacking not only IdoA(2SO$_4$)-GlcNSO$_3$ and IdoA(2SO$_4$)-GlcNSO3(6SO$_4$) but also GlcA(2SO$_4$)-GlcNSO$_3$ (Bai and Esko 1996). When the HS2ST cDNA was transfected into the mutant cell, the syntheses of these structures were recovered (Bai et al. 1997). The observation suggests that a single HS2ST catalyzes sulfation of both IdoA and GlcA. Considering the specificity of HS2ST together with that of HS6ST described in Chapter 68, the trisulfated structure of IdoA(2SO$_4$)-GlcNSO$_3$(6SO$_4$), the richest unit in heparin, may be synthesized by the following sequence: First, 2-O-sulfation at the IdoA residues takes place; and then the GlcNSO$_3$ residue of the same unit is 6-O sulfated.

Preparation

HS2ST was purified from the detergent-solubilized extracts of cultured CHO cells with heparin- and 3′,5′-ADP-affinity chromatography and Superose 12 gel chromatography. When the purified enzyme preparation was separated with sodium dodecyl polyacrylamide gel electrophoresis (SDS-PAGE), two protein bands of 47 and 44 kDa, respectively, were observed. After N-glycanase digestion these bands disappeared, but new protein bands of 38 and 34 kDa appeared (Kobayashi et al. 1996). The recombinant protein of CHO-HS2ST was expressed in COS-7 cells using FLAG-CMV-2 vector; and it was purified by anti-FLAG antibody affinity chromatography (Kobayashi et al.

1997). By cross-hybridization with the cDNA of CHO-HS2ST, human, chicken and mouse counterparts were obtained from the human fetal brain cDNA library, the chick embryo limb bud cDNA library, and mouse brain (Habuchi et al. 1999) and mastocytoma cDNA libraries (Rong et al. 2000).

Biological Aspects

Heparan sulfate (HS) has been shown to be essential for fibroblast growth factor (FGF) signaling (Yayon et al. 1991; Rapraeger and Olwin 1991). HS forms a ternary complex with FGF and FGF receptor (FGFR) (Kan et al. 1993; Plotnikov et al. 1999). FGF-2 interacts with a cluster for $IdoA(2SO_4)$-$GlcNSO_3$ in HS (Habuchi et al. 1992; Turnbull et al. 1992; Maccarana et al. 1993; Faham et al. 1996) and its high-affinity receptor appears to interact with some specific regions containing the $IdoA(2SO_4)$-$GlcNSO_3(6SO_4)$ unit in HS (Guimond, et al. 1993; Ishihara et al. 1995; DiGabriele et al. 1998; Pye et al. 1998).

Genetic screening and analyses are providing the evidence that HS chains play a pivotal role not only in mammals but also in *Drosophila* (Perrimon and Bernfield 2000; Selleck 2000). Mutations of enzymes involved in biosynthesis of the HS chain in *Drosophila* resulted in the phenotype of the *Wingless/Wnt* mutations, *FGF* family mutations, and *Hedgehog* mutations. The *pipe* mutation in *Drosophila* affected the formation of embryonic dorsoventral polarity; and because of the high identity the *pipe* gene may encode an enzyme similar to heparan sulfate 2-O-sulfotransferase (HS2ST) in *Drosophila* (Sen et al. 1998). It is interesting that destruction of a mouse gene encoding HS2ST by the gene-trapping method caused bilateral renal agenesis (resulting from a failure of uteric bud branching and mesenchymal condensation) and defects of the eye and skeleton (Bullock et al. 1998). The relation between these defects and the abnormal HS structures remains to be studied.

The expression of HS2ST in various mouse tissues was examined by Northern blot analysis. Variously sized HS2ST transcripts were widely expressed, and a high level of expression was observed in lung (Rong et al. 2000). HS2ST transcripts in various human tissues were also widely expressed, but high expression was observed in pituitary gland, adrenal gland, and testis (Habuchi et al. 1999).

Future Perspectives

HS2ST exhibited high activity for the $IdoA$-$GlcNSO_3$ unit and very low activity for the $GlcA$-$GlcNSO_3$ unit. HSs obtained from some tissues contained fairly large amounts of $GlcA(2SO_4)$ residues. For example, heparan sulfate from adult human cerebral cortex contained more $GlcA(2SO_4)$ than $IdoA(2SO_4)$ (Lindahl et al. 1995). The factors or mechanisms that regulate the proportion of 2-O-sulfation of GlcA and IdoA units are interesting and important subjects. Various concerted organizations of the HS modification enzymes (including GlcA C5-epimerase, NDSTs, HS2ST, HS6STs, and HS3STs) may account for the structural diversity of HS. To clarify the biosynthetic routes and mechanisms of the HS fine structures, it may be essential to study how the expressions of these enzymes are regulated and how these enzymes are organized in the Golgi compartments. Furthermore, disruption of genes related to HS biosynthesis could reveal the critical roles of HS during development in the mouse and *Drosophila*. As described above, mice homozygous for HS2ST mutations died during the neonatal period. A detailed analysis could provide further information on HS2ST function.

Further Reading

See Habuchi et al. (1998b) and Habuchi (2000) for review.

References

Bai X, Esko JD (1996) An animal cell mutant defective in heparan sulfate hexuronic acid 2-O-sulfation. J Biol Chem 271:17711–17717

Bai X, Bame KJ, Habuchi H, Kimata K, Esko JD (1997) Turnover of heparan sulfate depend on 2-O-sulfation of uronic acid. J Biol Chem 272:23172–23179

Bullock SL, Fletcher JM, Beddington RS, Wilson VA (1998) Renal agenesis in mice homozygous for a gene trap mutation in the gene encoding heparan sulfate 2-sulfotransferase. Genes Dev 12:1894–1906

DiGabriele AD, Lax I, Chen DI, Svahn CM, Jaye M, Schlessinger J, Hendrickson WA (1998) Structure of a heparin-linked biologically active dimer of fibroblast growth factor. Nature 393:812–817

Faham S, Hileman RE, Fromm JR, Lidhardt RJ, Rees DC (1996) Heparin structure and interactions with basic fibroblast growth factor. Science 271:1116–1120

Guimond S, Maccarana M, Olwin BB, Lindahl U, Rapraeger AC (1993) Activating and inhibitory heparin sequences for FGF-2. *J Biol* Chem 268:23906–23914

Habuchi H, Habuchi O, Kimata K (1998b) Biosynthesis of heparan sulfate and heparin: how are the multifunctional glycosaminoglycans built up? Trends Glycosci Glycotechnol 10:65–80

Habuchi H, Kobayashi M, Kimata K (1998a) Molecular characterization and expression of heparin-sulfate 6-sulfotransferase: complete cDNA cloning in human and partial cloning in Chinese hamster ovary cells. J Biol Chem 273:9208–9213

Habuchi H, Suzuki S, Saito T, Tamura T, Harada T, Yoshida K, Kimata K (1992) Structure of a heparan sulphate oligosaccharide that binds to basic fibroblast growth factor Biochem J 285:805–813

Habuchi H, Tanaka M, Matsuda Y, Habuchi O, Yoshida K, Suzuki H, Kimata K (1999) Three isoforms of heparan sulfate 6-O-sulfotransferase (HS6ST) having different specificities for hexuronic acid adjacent to the targeted N-sulfoglucosamine. Glycoconj J 16:S40

Habuchi O (2000) Diversity and functions of glycosaminoglycan sulfotransferases Biochim Biophys Acta 1474:115–127

Ishihara M, Takano R, Kanda T, Hayashi K, Hara S, Kikuchi H, Yoshida K (1995) Importance of 6-O-sulfate group of glucosamine residues in heparin for activation of FGF-1 and FGF-2. J Biochem 118:1255–1260

Kan M, Wang F, Xu J, Crabb JW, Hou J, McKeehan WL (1993) An essential heparin-binding domain in the fibroblast growth factor receptor kinase. Science 259:1918–1921

Kobayashi M, Habuchi H, Habuchi O, Saito M, Kimata K (1996) Purification and characterization of heparan sulfate 2-sulfotransferase from cultured Chinese hamster ovary cells. J Biol Chem 271:7645–7653

Kobayashi M, Habuchi H, Yoneda M, Habuchi O, Kimata K (1997) Molecular cloning and expression of Chinese hamster ovary cell heparan-sulfate 2-sulfotransferase. J Biol Chem 272:13980–13985

Lindahl B, Eriksson L, Lindahl U (1995) Structure of heparan sulfate from human brain, with special regard to Alzheimer's disease. Biochem J 306:177–184

Maccarana M, Casu B, Lindahl U (1993) Minimal sequence in heparin/heparan sulfate required for binding of basic fibroblast growth factor. J Biol Chem 268:23898–23905

Perrimon N, Bernfield M (2000) Specificities of heparan sulphate proteoglycans in developmental processes. Nature 404:725–728

Plotnikov AN, Schlessinger J, Hubbard SR, Mohammadi M (1999) Structural basis for FGF receptor dimerization and activation. Cell 98:641–650

Pye DA, Vives RR, Turnbull JE, Hyde P, Gallagher JT (1998) Heparan sulfate oligosaccharides require 6-O-sulfation for promotion of basic fibroblast growth factor mitogenic activity. J Biol Chem 273:22936–22942

Rapraeger AC, Olwin BB (1991) Requirement of heparan sulfate for bFGF-mediated fibroblast growth and myoblast differentiation. Science 252:1705–1708

Rong J, Habuchi H, Kimata K, Lindahl U, Kusche-Gullberg, M (2000) Expression of heparan sulphate L-iduronyl 2-O-sulphotransferase in human kidney 293 cells results in increased D-glucuronyl 2-O-sulphation. Biochem J 346:463–468

Selleck SB (2000) Proteoglycans and pattern formation: sugar biochemistry meets developmental genetics. Trends Genet 16:206–212

Sen J, Goltz JS, Stevens L, Stein D (1998) Spatially restricted expression of *pipe* in the *Drosophila* egg chamber defines embryonic dorsal-ventral polarity. Cell 95:471–481

Shively JE, Conrad HE (1976) Formation of anhydrosugars in the chemical depolymerization of heparin. Biochemistry 15:3932–3942

Turnbull JE, Fernig DG, Ke Y, Wilkinson MG, Gallagher JT (1992) Identification of the basic fibroblast growth factor binding sequence in fibroblast heparan sulfate. J Biol Chem 267:10337–10341

Wlad H, Maccarana M, Eriksson I, Lindahl U (1994) Biosynthesis of heparin: different molecular forms of O-sulfotransferases. J Biol Chem 269:24538–24541

Yayon A, Klagsbrun M, Esko JD, Ledeer P, Ornitz DM (1991) Cell surface heparin-like molecules are required for binding of basic fibroblast growth factor to its high affinity receptor. Cell 64:841–848

Heparan Sulfate 6-Sulfotransferase

Introduction

The carbohydrate backbone in heparan sulfate (HS) and heparin is composed of HexA-GlcNAc/GlcNSO$_3$ repeating unit in which HexA is either D-GlcA or L-IdoA. Sulfate groups are present on the various positions of the repeating units: position 2 of HexA, position 6 of GlcNAc/GlcNSO$_3$, and position 3 of GlcNAc/GlcNSO$_3$. Heparan sulfate 6-O-sulfotransferase (HS6ST) transfers sulfate to position 6 of the N-sulfated glucosamine residue in HS and heparin. HS6STs serve as enzymes catalyzing the final modification step for the formation of the IdoA/IdoA(2SO$_4$)-GlcNSO$_3$(6SO$_4$) unit and the GlcA-GlcNSO$_3$(6SO$_4$) unit in HS and heparin. HS chains in heparan sulfate proteoglycans (HSPGs) are known to interact with a variety of proteins, such as heparin-binding growth factors, extracellular matrix components, protease inhibitors, proteases, and lipoprotein lipase. As described in Chapter 67, these interactions are implicated not only in various dynamic cellular behaviors such as cell proliferation, cell differentiation, cell adhesion, cell migration, and cell morphology but also in various physiological phenomena such as inflammation, blood coagulation, and tumor cell invasion and malignancy. Moreover, infections of host cells with some living pathogens such as viruses and parasites have been shown to occur through the interactions with cell surface HSs on host cells. The interactions of HS with such various ligands seem to be mediated by particular regions of HS chains where the 6-O-sulfated sugar residues play important roles as factors to determine their specific structures (Ashikari et al. 1995).

HIROKO HABUCHI and KOJI KIMATA

The Institute for Molecular Science of Medicine, Aichi Medical University, Nagakute, Aichi 480-1195, Japan
Tel. +81-52-264-4811 (ext. 2088); Fax +81-561-63-3532
e-mail: kimata@amugw.aichi-med-u.ac.jp

Databanks

Heparan sulfate 6-sulfotransferase

No EC number has been allocated.

Species	Gene	Protein	mRNA	Genomic
Homo sapiens	*HS6ST-1*	–	AB006179	–
Cricetulus griseus	*HS6ST-1*	–	AB006180	–
Mus musculus	*HS6ST-1*	–	AB024566	–
	HS6ST-2	–	AB024565	–
	HS6ST-3	–	AB024567	–

Name and History

Heparan sulfate 6-sulfotransferase (HS6ST) is an enzyme that catalyzes the transfer of sulfate to position 6 of *N*-sulfoglucosamine in HS. HS *O*-sulfotransferase activity, which transfers sulfate to completely desulfated *N*-resulfated heparin, was found in heparin-producing mouse mastocytoma. The partially purified enzyme contained both *N*-sulfoglucosaminyl 6-*O*-sulfotransferase and iduronyl 2-*O*-sulfotransferase activities (Wlad et al. 1994). However, HS6ST purified to apparent homogeneity from the serum-free culture medium of Chinese hamster ovary (CHO) cells showed only *N*-sulfoglucosaminyl 6-*O*-sulfotransferase activity (Habuchi et al. 1995). The peptide sequence information obtained from the purified HS6ST yielded success with cDNA cloning (Habuchi et al. 1998). It has been shown that HS6ST consists of at least three isoforms. The cDNAs for the novel two forms were first obtained from the mouse brain cDNA library by cross-hybridization with human HS6ST cDNA (Habuchi et al. 2000). The original form corresponding to the HS6ST originally cloned from a human fetal brain cDNA library was designated HS6ST-1 and the novel two forms HS6ST-2 and HS6ST-3 (Habuchi et al. 2000). They showed different specificities and different expression patterns; and they were derived from different genes. The amino acid sequence of HS6ST-1 was 51% and 57% identical to those of HS6ST-2 and HS6ST-3, respectively. HS6ST-2 and HS6ST-3 had 50% amino acid sequence identity. The recombinant proteins of the three isoforms have been expressed in COS-7 cells using FLAG-CMV-2 vector and have been purified by anti-FLAG antibody affinity chromatography.

Enzyme Activity Assay and Substrate Specificity

HS6ST transfers sulfate to position 6 of the *N*-sulfoglucosamine residue in heparan sulfate but not to the *N*-acetylglucosamine residues.

$$PAPS + (IdoA\alpha1\text{-}4GlcNSO_3\alpha1\text{-}4)_n \rightarrow PAP + [IdoA\alpha1\text{-}4GlcNSO_3(6SO_4)\alpha1\text{-}4]_n$$

where PAPS is 3′-phosphoadenosine 5′-phosphosulfate, and PAP is 3′-phosphoadenosine 5′-phosphate.

To assay the heparan sulfate 6-*O*-sulfotransferase activity, the standard reaction mixture (50 μl) contained 2.5 μmol imidazole HCl pH 6.8, 3.75 μg protamine chloride,

25 nmol (as hexosamine) glycosaminoglycans (GAGs), 50 pmol [^{35}S]PAPS (about 5×10^5 cpm), and enzyme. After incubation at 37°C the ^{35}S-labeled polysaccharides were isolated by precipitation with 70% (v/v) ethanol containing 1% (w/v) potassium acetate and 0.2 mM EDTA; they were then subjected to gel chromatography on a Fast desalting column. The radioactivity in the high-molecular-weight fractions was then counted. To determine the sulfation of position 6 of the N-sulfoglucosamine residue, ^{35}S-labeled polysaccharides were digested with a mixture of 10 mU of heparitinase I, 1 mU heparitinase II, and 10 mU heparitinase III in 40 μl 50 mM Tris HCl pH 7.2, 1 mM CaCl$_2$, and 4 μg bovine serum albumin at 37°C for 2 h. The digests were subjected to gel chromatography of Superdex pg 30 columns equilibrated with 0.2 M ammonium acetate. The ^{35}S-labeled fractions corresponding to disaccharides were collected and then injected together with standard unsaturated disaccharides into a column of PAMN. The radioactivity of each fraction (0.6 ml) was measured, and the fractions corresponding to standard disaccharide units were determined.

The properties of HS6ST-1 were as follows. The optimal pH for enzyme activity was around 6.3. Protamine markably activated the activity. NaCl stimulated enzyme activity. Maximum activity was observed at around 175 mM NaCl. Dithiothreitol (DTT), which had no effect on HS2ST, inhibited the enzyme activity to 19% of the control value at a concentration of 10 mM. The K_m for PAPS was 0.44 μM (Habuchi et al. 1995). In contrast to HS2ST (more than 95% of which was retained in the cell layer), more than 90% of the HS6ST activity was secreted into the culture medium of CHO cells (Kobayashi et al. 1996). The HS6ST activity was also detected in mouse serum and human serum (Habuchi et al. 1998).

The purified three isoforms could transfer sulfate to heparan sulfate and heparin but not to chondroitin and keratan sulfate (Table 1) (Habuchi et al. 1995). Each

Table 1. Acceptor substrate specificities of the recombinant 6-O-sulfotransferases purified by anti-FLAG antibody affinity column chromatography

Substrate	Relative activity of sulfotransferases[a] (%)		
	mHS6ST-1	mHS6ST-2	mHS6ST-3
CDSNS-heparin	100	100	100
CDSNAc-heparin	1.5	4.7	5.3
Heparin	2.9	8.6	8.5
NS-heparosan	16.0	185.0	80.0
Heparan sulfate (moue EHS tumor)	50.0	97.0	108.0
Heparan sulfate (bovine liver)	7.3	29.0	27.0
Heparan sulfate (pig aorta)	12.0	65.0	52.0
Keratan sulfate	0.2	1.1	0.4
Chondroitin	0.5	1.6	0.4

Sulfotransferase activities were assayed using various glycosaminoglycans as acceptors. Recombinant sulfotransferases were expressed in COS-7 cells transfected with pFLAG-CMV2-mHS6ST-1, pFLAG-CMV2-mHS6ST-2, and pFLAG-CMV2-mHS6ST-3, respectively, and purified on the anti-FLAG antibody column

[a] The values indicate the relative rates of incorporation into various substrates to those into CDSNS-heparin. The amount of each isoform used was designed to give the same rate of ^{35}S incorporation into CDSNS-heparin (0.35 pmol/min)

isoform showed a different specificity toward the isomeric hexuronic acid adjacent to the targeted N-sulfoglucosamine; HS6ST-1 appeared to prefer iduronosyl N-sulfoglucosamine, whereas HS6ST-2 had a different preference, depending on the substrate concentrations (it preferred the glucuronosyl residue at high concentrations but the iduronosyl residue at low concentrations). HS6ST-3 acted on either substrate equally. These three isoforms preferred CDSNS-heparin to completely desulfated N-acetylated heparin as an acceptor substrate. The analysis of heparitinase digests of their ^{35}S-labeled products showed that these three isoforms could transfer sulfate to position 6 of N-sulfoglucosamine residue adjacent to the reducing side of iduronic acid 2-sulfate. None of these isoforms could transfer sulfate to the N-acetylglucosamine residues (Habuchi et al. 2000), suggesting that a sulfotransferase other than HS6ST might be involved in the synthesis of the GlcAβ1-4GlcNAc(6SO$_4$) unit present in the HS chain.

Preparation

HS6ST-1 was purified from the serum-free cultured medium of CHO cells. The isolation procedure included affinity chromatography of the first heparin-Sepharose CL-6B column (stepwise elution), 3′,5′-ADP-agarose, and the second heparin-Sepharose CL-6B column (gradient elution). The purified enzyme showed two protein bands with molecular masses of 52 and 45 kDa, respectively (Habuchi et al. 1995). The human HS6ST-1 has been expressed as a fusion protein in COS-7 cells transfected with FLAG-CMV-2 vector and purified by anti-FLAG antibody affinity chromatography (Habuchi et al. 1998). The recombinant three isoforms of mouse HS6ST (HS6ST-1, HS6ST-2, HS6ST-3) have also been expressed as fusion proteins in COS-7 cells using FLAG-CMV-2 vector; and they have been successfully purified by anti-FLAG antibody affinity chromatography (Habuchi et al. 2000). The amino acid sequence of HS6ST-1 was 51% and 57% identical to those of HS6ST-2 and HS6ST-3, respectively. HS6ST-2 and HS6ST-3 had 50% amino acid sequence identity.

Biological Aspects

Most but not all of the interactions between HS and various functional proteins occur on some regions of the HS with specific sulfated monosaccharide sequences (Habuchi et al. 1998; Rosenberg et al. 1998). Such regional structures are also required for the activation and subsequent signaling of functional proteins such as basic fibroblast growth factor (bFGF) dimerization and signaling (Lyon and Gallagher 1998; Ornitz 2000). Formation of the ternary complex of bFGF/FGF receptor (FGFR)/HS and its signal transduction also need the IdoA(2SO$_4$)-GlcNSO$_3$(6SO$_4$) unit (Guimond et al. 1993; Pye et al. 1998). In addition, the structure for hepatocyte growth factor (HGF) binding needs a cluster of IdoA(2SO$_4$)-GlcNSO$_3$(6SO$_4$) (Guimond et al. 1993; Lyon et al. 1994; Ashikari et al. 1995). Therefore, HS6STs are key enzymes for generating these structures.

Several reports have described changes in the HS sulfation pattern in association with development and certain pathological processes. The HS accumulated in internal organs in amyloid A protein (AA) amyloidosis indicated a switch in O-sulfation

(Lindahl and Lindahl 1997). Colon carcinoma cells change the HS structure during cell differentiation (Jayson et al. 1998), and the differentiated cell synthesizes HS composed of a larger proportion of the IdoA(2SO$_4$) unit and a smaller amount of the GlcNSO$_3$(6SO$_4$) unit (Salmivirta et al. 1998). These structural changes in HS might be reflections of changes in HS6ST expression. It is likely that the interactions of HS with growth factors and extracellular matrix components are affected by such structural changes. In fact, a switch from FGF-2 to FGF-1 during development of the embryonic brain appears to be mediated by changes in the HS structure (Nurcombe et al. 1993), and neural precursor cells differentiate to neurons with alterations in the 6-O-sulfation pattern in HS (Brickman et al. 1998).

The expression patterns of mouse HS6ST-1, HS6ST-2, and HS6ST-3 were quite different from each other (Habuchi et al. 2000). A single transcript of HS6ST-1 with 3.9 kb was expressed mainly in liver, whereas a single transcript of HS6ST-2 with 4.6 kb was expressed mainly in the brain and spleen. In contrast to these two examples, transcripts of HS6ST-3 with multiple sizes (1.6, 2.0, 2.8, 4.0, and 6.5 kb) were expressed rather ubiquitously in most tissues examined, and their sizes tended to vary tissue-dependently. These tissue-specific expression patterns of the three isoforms might regulate the production of HSs with tissue-specific structures. For example, the abundant presence of the GlcA-GlcNSO$_3$(6SO$_4$) unit in brain HS may be related to the observation that HS6ST-2 is mainly expressed in brain (Lindahl et al. 1995). On the other hand, the highest expression of HS6ST-1 in liver may result in increased synthesis of the IdoA(2SO$_4$)-GlcNSO$_3$(6SO$_4$) unit, which is abundant in liver HS (Ashikari et al. 1995).

Future Perspectives

To clarify the roles of these isoforms in the biosynthesis of HS it is important to characterize their substrate specificity in further detail and to investigate their assembly within the cytoplasmic organelles involved in the processes and regulation mechanisms of their expression. The latter may require promoter analyses of these three isoforms. Needless to say, generation of isoform-specific knockout mice and targeted gene mutations in *Drosophila* and *Canorhabditis elegans* would provide important information about their biological roles.

Further Reading

For detailed information on the three isoforms of HS6ST see Habuchi et al. (2000), and for a review see Habuchi et al. (1998).

References

Ashikari S, Habuchi H, Kimata K (1995) Characterization of heparan sulfate oligosaccharides that bind to hepatocyte growth factor. J Biol Chem 270:29586–29593
Brickman YG, Ford MD, Gallagher JT, Nurcombe V, Bartlett PF, Turnbull JE (1998) Structural modification of fibroblast growth factor-binding heparan sulfate at a determinative stage of neural development. J Biol Chem 273:4350–4359

Guimond S, Maccarana M, Olwin BB, Lindahl U, Rapraeger AC (1993) Activating and inhibitory heparin sequences for FGF-2 (basic FGF). J Biol Chem 268:23906–23914

Habuchi H, Habuchi O, Kimata K (1995) Purification and characterization of heparan sulfate 6-sulfotransferase from the culture medium of Chinese hamster ovary cells. J Biol Chem 270:4172–4179

Habuchi H, Habuchi O, Kimata K (1998) Biosynthesis of heparan sulfate and heparin: how are the multifunctional glycosaminoglycans built up? Trends Glycosci Glycotechnol 10:65–80

Habuchi H, Kobayashi M, Kimata K (1998) Molecular characterization and expression of heparan-sulfate 6-sulfotransferase: complete cDNA cloning in human and partial cloning in Chinese hamster ovary cells. J Biol Chem 273:9208–9213

Habuchi H, Tanaka M, Habuchi O, Yoshida K, Suzuki H, Ban K, Kimata K (2000) The occurrence of three isoforms of heparan sulfate 6-O-sulfotransferase having different specificities for hexuronic acid adjacent to the targeted N-sulfoglucosamine. J Biol Chem 275:2859–2868

Jayson GC, Lyon M, Paraskeva C, Turnbull JE, Deakin JA, Gallagher JT (1998) Heparan sulfate undergoes specific structural changes during the progression from human colon adenoma to carcinoma in vitro. J Biol Chem 273:51–57

Kobayashi M, Habuchi H, Habuchi O, Saito M, Kimata K (1996) Purification and characterization of heparan sulfate 2-sulfotransferase from cultured Chinese hamster ovary cells. J Biol Chem 271:7645–7653

Lindahl B, Lindahl U (1997) Amyloid-specific heparan sulfate from human liver and spleen. J Biol Chem 272:26091–26094

Lindahl B, Eriksson L, Lindahl U (1995) Structure of heparan sulfate from human brain, with special regard to Alzheimer's disease. Biochem J 306:177–184

Lyon M, Gallagher JT (1998) Bio-specific sequences and domains in heparan sulfate and the regulation of cell and adhesion. Matrix Biol 17:485–493

Lyon M, Deakin JA, Mizuno K, Nakamura T, Gallagher JT (1994) Interaction of hepatocyte growth factor with heparan sulfate: elucidation of the major heparan sulfate structural determinants. J Biol Chem 269:11216–11223

Nurcombe V, Ford MD, Wildschut JA, Bartlett PF (1993) Developmental regulation of neural response to FGF-1 and FGF-2 by heparan sulfate proteoglycan. Science 260:103–106

Ornitz DM (2000) FGFs, heparan sulfate and FGFRs: complex interactions essential for development. Bioessays 22:108–112

Pye DA, Vives RR, Turnbull JE, Hyde P, Gallagher JT (1998) Heparan sulfate oligosaccharides require 6-O-sulfation for promotion of basic fibroblast growth factor mitogenic activity. J Biol Chem 273:22936–22942

Rosenberg RD, Shworak NW, Liu J, Schwartz JJ, Zhang L (1998) Heparan sulfate proteoglycans of the cardiovascular system: specific structures emerge but how is synthesis regulated? J Clin Invest 99:2062–2072

Salmivirta M, Safaiyan F, Prydz K, Andresen MS, Ayan M, Kolset SO (1998) Differentiation-associated modulation of heparan sulfate structure and function in CaCo-2 colon carcinoma cells. Glycobiology 8:1029–1036

Wlad H, Maccarana M, Eriksson I, Lindahl U (1994) Biosynthesis of heparin: different molecular forms of O-sulfotransferases. J Biol Chem 269:24538–24541

Heparan Sulfate D-Glucosaminyl 3-O-Sulfotransferase-1, -2, -3, and -4

Introduction

Heparan sulfate D-glucosaminyl 3-O-sulfotransferase (HS3ST) is a class of sulfo-transferase that transfers the sulfate from 3'-phosphoadenosine 5'-phosphosulfate (PAPS) to the 3-OH of the glucosamine residue. HS3ST was originally exclusively designated the key sulfotransferase to synthesize the antithrombin-binding site in heparan sulfate (HS) to confer its anticoagulant activity (Kusche et al. 1990; Shworak et al. 1996). However, recent studies suggest that HS3ST is present in at least five isoforms (Shworak et al. 1999). Among these isoforms HS3ST-1, HS3ST-2, HS3ST-3A, and HS3ST-3B have been characterized, and the results allow us to distinguish these enzymes by their DNA/amino acid sequence, biochemical, and biological aspects. Although HS3ST isoforms have similar amino acid sequences at the C-terminus, they differ mainly in the following characteristics: (1) gene locations in the human genome; (2) mRNA expression levels in various human tissues; (3) substrate specificity; and (4) biological functions of enzyme-modified HS (Rosenberg et al. 1997).

Databanks

Heparan sulfate D-glucosaminyl 3-O-sulfotransferase-1, -2, -3, and -4

NC-IUBMB enzyme classification: EC 2.8.2.23

Species	Gene	Protein	mRNA	Genomic
HS3ST-1				
Homo sapiens	*HS3ST1*	–	NM_005114	–
Mus musculus	*Hs3st1*	–	NM_010474	–

JIAN LIU[1] and ROBERT D. ROSENBERG[2]

[1] Division of Medicinal Chemistry and Natural Products, School of Pharmacy, University of North Carolina, Chapel Hill, NC 27599, USA
Tel. +1-919-843-6511; Fax +1-919-843-5432
e-mail: jian_liu@unc.edu
[2] Biology Department, Massachusetts Institute of Technology, 31 Ames Street, Cambridge, MA 02139, USA

(Continued)

Species	Gene	Protein	mRNA	Genomic
HS3ST-2				
Homo sapiens	*HS3ST2*	–	NM_006043	–
HS3ST-3A				
Homo sapiens	*HS3ST3A1*	–	NM_006042	–
HS3ST-3B				
Homo sapiens	*HS3ST3B1*	–	NM_006041	–
Mus musculus		–	AAF04505	–
HS3ST-4				
Homo sapiens	*HS3ST4*	–	AAD30210	–

Name and History

The product of HS3ST-modified HS contains 3-*O*-sulfated glucosamine residue, which is a rare constituent in heparin and HS. 3-*O*-Sulfated glucosamine was discovered to be part of heparin in 1969 by Danishefsky et al. (Danishefsky et al. 1969) Subsequently, the presence of this particular residue was linked to the anticoagulant activities of heparin and HS (Lindahl et al. 1980; Atha et al. 1985), and it was reported that 3-*O*-sulfated glucosamine contributes to the antiproliferative activity of heparin (Garg et al. 1996). Anticoagulant HS also binds to fibroblast growth factor-2 receptor (McKeehan et al. 1999), but it is still unknown whether antithrombin and fibroblast growth factor-2 receptor interact with the same saccharide sequence in the HS. Given the fact that 3-*O*-sulfated glucosamine residue is always present in anticoagulant HS, McKeehan and colleagues' results implied that this residue is involved in the binding to fibroblast growth factor-2 receptor.

Among the HS3ST isoforms, HS3ST-1 was discovered earliest because of its role in synthesizing anticoagulant HS. The activity was first reported in mouse mastocytoma cells and was speculated to be an essential activity to generate the antithrombin-binding site (Kusche et al. 1990). Furthermore, the results from studying the biosynthesis of anticoagulant HS in various genetically modified L-cell mutants suggested that there is a specific biosynthetic pathway for generating anticoagulant HS that also requires a "limiting factor" (Colliec-Jouault et al. 1994; Shworak et al. 1994). The critical role of HS3ST-1 in making anticoagulant HS was not confirmed until the enzyme was purified from a large amount of L-cell condition medium (Liu et al. 1996) followed by cloning of the cDNA (Shworak et al. 1997). Using purified HS3ST-1 enzyme to prepare the 3-*O*-[^{35}S]sulfated HS, we obtained both antithrombin-binding 3-*O*-sulfated HS and the 3-*O*-sulfated HS, which does not bind to antithrombin. The latter results were confirmed by an in vivo experiment suggesting that the biosynthesis of antithrombin-binding HS also requires an appropriate HS precursor structure (Zhang et al. 1998, 1999).

A recent study demonstrated that the 3-*O*-sulfated glucosamine residue is a part of the HS that serves as an entry receptor for herpes simplex virus-1 (HSV-1) (Shukla et al. 1999). This particular type of 3-*O*-sulfated HS is generated by HS3ST-3, including HS3ST-3A and HS3ST-3B. HS3ST-3-modified HS specifically binds to glycoprotein D (gD) of herpes simplex virus 1 (HSV-1) to result in otherwise nonpermissive CHO cells susceptible to HSV-1 infection. The complete saccharide sequence for the

Table 1. Specificities of HS3STs and the biological functions of enzyme-modified heparan sulfate

Enzyme	Sulfation site	Biological functions of enzyme-modified heparan sulfate
HS3ST-1	-GlcA-GlcNS ± 6S**3S**[a,b]	Anticoagulant activity
HS3ST-2	-GlcA2S-GlcNS (or NH$_2$?)**3S**- and -IdoA2S-GlcNS (or NH$_2$?)**3S**-	Unknown
HS3ST-3A and HS3ST-3B	-IdoA2S-GlcNH$_2$ ± 6S**3S**-	HSV-1 entry receptor
HS3ST-4	Unknown	Unknown

HSV-1, herpes simplex virus type 1

[a] GlcA, glucuronic acid; GlcNS ± 6S, N-sulfoglucosamine 6-O-sulfate or 6-OH; GlcA2S, glucuronic acid 2-O-sulfate; GlcNS (or NH$_2$?), N-sulfogluosamine (but it is not known if HS3ST-2 sulfates N-unsubstituted glucosamine); IdoA2S, iduronic acid 2-O-sulfate; GlcNH$_2$, N-unsubstituted glucosamine

[b] The 3-O-sulfation is in boldface and underlined to emphasize the modification

gD-binding site is still unknown, but the binding site must contain a disaccharide with a structure of 2-O-sulfated iduronic acid (IdoA2S) linked to 3-O-sulfated glucosamine (GlcNH$_2$ ± 6S3S). It is important to note that this type of disaccharide is rarely found in HS from a natural source, and it is not a part of the antithrombin-binding site (Liu et al. 1999b). A similar disaccharide structure has been reported in the HS only from bovine glomerular basement membrane (Edge and Spiro, 1990). The study provides strong evidence to demonstrate that HSV-1 utilizes a specific saccharide sequence to infect target cells. Furthermore, it becomes the first evidence that HS polysaccharide can serve as an entry receptor for HSV-1, provided previously characterized entry receptors for HSV-1 are all protein-based receptors (Shukla et al. 1999).

Enzyme Activity Assay and Substrate Specificity

The method for determining the enzymatic activities of HS3ST-1 and other HS3ST isoforms differs in how to analyze the products generated by these enzymes. The product of HS3ST-1-modified HS is analyzed by an antithrombin affinity column, and the products of HS3ST-2- and HS3ST-3-modified HS are determined by the disaccharide compositional analysis of nitrous acid-degraded 3-O-[^{35}S]sulfated HS on reversed-phase ion-pairing high-performance liquid chromatography (HPLC) (Liu et al. 1999b). It is generally thought that the method for assaying HS3ST-1 activity is much more sensitive than the one used for assaying HS3ST-2 and HS3ST-3 because it was specifically designed to detect the enzyme activity that converts nonanticoagulant [^{35}S]HS to anticoagulant [^{35}S]HS (Shworak et al. 1996). In the assay for HS3ST-1, we utilized the high specific radioactivity of nonanticoagulant [^{35}S]HS (3.8 × 10^{19} cpm/mol) from 33-cells, an L-cell variant, as the substrate and unlabeled PAPS as the sulfate donor. After transferring the unlabeled sulfate from PAPS to the 3-OH position of glucosamine in HS by HS3ST-1 activity, we observed an increased amount of [^{35}S]HS bound to the antithrombin affinity column. Furthermore, the increase in the amount of antithrombin binding [^{35}S]HS after enzymatic modification related proportionally to the amount of enzyme employed in the reaction mixture (Shworak et al. 1996; Liu et al. 1996). The detection limit of this assay is about 4×10^{-17} mol. HS3ST-1 activity is sensitive to sodium chloride, and we estimated that the IC_{50} was 30 mM. We also found that dextran sulfate is another potent inhibitor of this assay (Liu et al. 1996).

To determine the activities of HS3ST-2 and HS3ST-3, the cellular extracts containing the enzyme activities were incubated with unlabeled HS from 33-cells as the substrate and [^{35}S]PAPS as the [^{35}S]sulfate donor. The resultant [^{35}S]HS was degraded with nitrous acid at pH 1.5, and the [^{35}S]disaccharides were analyzed on reversed-phase ion-pairing HPLC. The chromatograms of the disaccharide compositional analysis are shown in Fig. 1. HS3ST-2-modified HS contains two [^{35}S]disaccharides with structures of IdoA2S-[^{35}S]AnMan3S and GlcA2S-[^{35}S]AnMan3S. (IdoA2S represents 2-O-sulfated iduronic acid residue; AnMan3S represents 3-O-sulfated 2,5-anhydromanitol; and GlcA2S represents 2-O-sulfated glucuronic acid residue.) HS3ST-3-modified HS contains one [^{35}S]disaccharide with a structure of IdoA2S-[^{35}S]AnMan3S. HS3ST-3A and HS3ST-3B sulfate are identical disaccharides. Further detailed structural analysis of HS3ST-3-modified HS found that it also contains a [^{35}S]disaccharide with a structure of IdoA2S-[^{35}S]AnMan3S6S (AnMan3S6S represents 3,6-disulfated 2,5-anhydromannitol) by using a higher concentration of ammonium phosphate monobasic and a higher concentration of acetonitrile in the eluent (Liu et al. 1999a; Shukla et al. 1999). We have found that 150 mM sodium chloride in the assay buffer helps maximize the activity of HS3ST-3A (Liu et al. 1999a). It is also interesting to note that HS3ST-3A enzyme sulfates N-unsubstituted glucosamine residue (Liu et al. 1999a). The N-unsubstituted glucosamine residue represents only 1%–7% of the glucosamine the heparan sulfate isolated from various tissues (Toida et al. 1997), suggesting that HS3ST-3A generates a unique saccharide sequence.

Preparation

The HS3ST-1 enzyme was purified from conditioned medium of 33 cells, and the activity was also found in mouse mastocytoma cells. The natural sources of HS3ST-2, HS3ST-3A, and HS3ST-3B enzymes are not known. One can speculate on the potential locations of these enzymes based on the mRNA level in various human tissues as described above. The HS3STs were largely obtained by expressing recombinant proteins in various systems. All of the known HS3STs were successfully expressed in COS-7 cells using transient transfection (Shworak et al. 1997; Liu et al. 1999b). HS3ST-1 was also expressed by using an in vitro translation system in the presence of canine microsomes (Shworak et al. 1997). HS3ST-1 and truncated HS3ST-3A were also expressed in SF9 insect cells using a baculovirus expression approach with high expression level (Liu et al. 1999a; Hernaiz et al. 2000). Both recombinant HS3ST-1 and HS3ST-3A, expressed in a baculovirus/insect cell system, were successfully purified using heparin-Toyopearl Gel and 3',5'-ADP-agarose chromatography (Liu et al. 1999a). Anti-HS3ST-3A (PB1437), a chicken immunoglobulin Y (IgY), is available against a linear peptide sequence, residue 93-111, of HS3ST-3A. This antibody does not cross-react with purified HS3ST-1. A specific antibody against HS3ST-1 or other HS3ST isoforms is still not available.

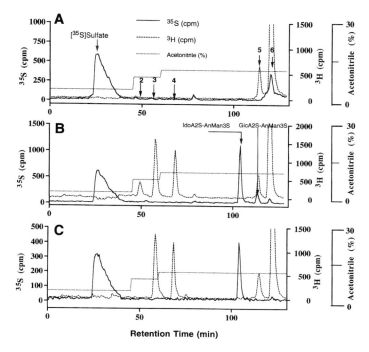

Fig. 1A–C. Chromatograms of reversed-phase ion-pairing high-performance liquid chromatography (HPLC) of the disaccharide analysis of nitrous acid-degraded HS3ST-2- and HS3ST-3A-modified heparan sulfate ([^{35}S]HS). The [^{35}S]HS was prepared by incubating 3′-phosphoadenosine 5′-phosphosulfate ([^{35}S]PAPS) and unlabeled HS from 33 cells with the cellular extracts from COS-7 cells transfected with pcDNA3 alone, HS3ST-2, or HS3ST-3A. The [^{35}S]HS samples were degraded with nitrous acid at pH 1.5 followed by reduction with sodium borohydride. The resultant disaccharides were analyzed on reversed-phase ion-pairing HPLC. The three panels represent the chromatograms of ^{35}S-labeled disaccharides derived from [^{35}S]HS prepared with the cell extracts of COS-7 cells transfected with pcDNA3 (**A**), HS3ST-2 (**B**), or HS3ST-3A (**C**). The *arrows* indicate the elution positions of ^{3}H-labeled disaccharide standards. *2*, IdoA2S-[^{3}H]AnMan; *3*, GlcA-[^{3}H]AnMan6S; *4*, IdoA-[^{3}H]AnMan6S; *5*, GlcA-[^{3}H]AnMan3S6S; *6*, IdoA2S-[^{3}H]AnMan6S. (From Liu et al. 1999a, with permission)

Biological Aspects

Molecular Characteristics of HS3STs

The HS3STs, except for HS3ST-1, are type II membrane-bound proteins. They have more than 60% homology in *C*-terminal amino acid sequences in about 260 residues. In particular, HS3ST-3A and HS3ST-3B have 99.2% homology in this region. The highly homologous *C*-terminal region is also found to be present in other HS sulfotransferases, and it was proposed to be the sulfotransferase domain with enzyme activity (Shworak et al. 1997). This hypothesis has been proved by expressing a carboxyl-terminal fragment of HS *N*-deacetylase/*N*-sulfotransferase, showing that the

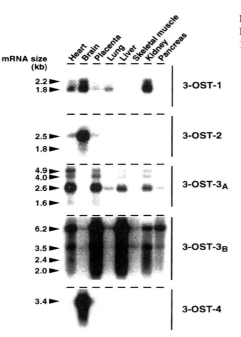

Fig. 2. Tissue-specific expression of human HS3STs (3-OSTs). (From Shworak et al. 1999, with permission)

truncated enzyme contains *N*-sulfotransferase but lacks *N*-deacetylase activity (Berninsone and Hirschberg, 1998). An SPLAG region, representing an enriched region of Serine, Proline, Leucine, Alanine, and Glycine, was found to be highly variable among HS3STs (Shworak et al. 1999). The SPLAG domain is between the sulfotransferase domain and the *N*-terminus, which might act as a flexible arm to link the catalytic domain and the membrane anchor (Shworak et al. 1999).

The HS3ST isoforms have distinct expression levels in various human tissues determined by Northern analysis, as shown in Fig. 2. HS3ST-1 is expressed at high levels in brain and kidney, intermediate levels in heart and lung, and detectable levels in various other tissues. HS3ST-2 is predominantly expressed in human brain, and HS3ST-4 is almost exclusively expressed in human brain. HS3ST-3A and HS3ST-3B are present in variously sized transcripts and are expressed ubiquitously in human tissues. It is interesting to note that both HS3ST-3A and HS3ST-3B are expressed at a very low level in human brain, whereas HS3ST-2 and HS3ST-4 are predominantly expressed.

Genomic Locations of HS3STs

Southern analyses of human and mouse genomic DNA with isoform-specific probes reveal that HS3ST-1, HS3ST-2, and HS3ST-4 are single-copy genes. Both HS3ST-3A and HS3ST-3B are present in two copies in the human genome; thus, the human genome consists of four HS3ST3 genes. Probing the mouse-expressed sequence tag database with human HS3ST3A and HS3ST3B sequences suggests that there is at least one copy of HS3ST3a and HS3ST3b in the murine genome.

The mouse chromosomal location of each HS3ST locus was determined by inter-specific back-cross analysis using progeny derived from matings of [(C57BL/6 × *Mus spretus*)F1 × C57BL/6J] mice. HS3ST1 mapped to the proximal region of mouse chromosome 5; HS3ST2 and HS3ST4 mapped to the distal region of chromosome 7; and HS3ST3a and HS3ST3b mapped to the central region of mouse chromosome 11. The human chromosomal locations of HS3STs are as follows: HS3ST1 is located in human chromosome 4p; HS3ST2 and HS3ST4 map to human chromosome 16p; and HS3ST3 localizes in human chromosome 5q or chromosome 17p (Shworak et al. 1999).

Biosynthesis of Biologically Active HS

Biosynthesis of the anticoagulant HS and an HSV-1 entry receptor HS (simply called gD-binding HS) is regulated by two factors, including the HS3ST-1 and HS3ST-3 enzyme levels and the availability of the appropriate HS precursors for 3-O-sulfation. Because 3-O-sulfation occurs at low frequency in any given HS, it is likely that HS3ST isoforms recognize specific saccharide sequences as the sites to transfer the sulfate group. Given the critical roles of 3-O-sulfation in making anticoagulant HS and gD-binding HS, only those saccharide regions that can be recognized and modified by HS3ST-1 and HS3ST-3 have the possibility to become the HS with desired biological activities. We designate the HS containing these regions the biologically active HS precursor. Indeed, based on our recent studies, we found that the anticoagulant active precursor requires the presence of 6-O-sulfated glucosamine residue near the HS3ST-1 sulfation site. Structural analyses of the HS3ST-1-modified anticoagulant HS and HS3ST-1-modified nonanticoagulant HS suggest that loss of the 6-O-sulfation site results in the inability to bind HS to antithrombin (Zhang et al. 1999).

In addition, the presence of 2-O-sulfated iduronic acid residue may also determine the biosynthesis of gD-binding HS through the effect on the level of the precursor for HS3ST-3 modification. This conclusion was based on studying the synthesis of HS3ST-3-modified HS from CHO F17 cells, which is a mutant known to be defective in making 2-O-sulfated iduronic acid residue. We have found a 20- to 25-fold decrease in the levels of two characteristic 3-O-sulfated disaccharides in the HS3ST-3A-modified HS from CHO F17 cells from an in vitro preparation. Furthermore, HS3ST-3A-modified HS from CHO F17 cells does not bind to gD. As expected, the CHO F17 cells carrying a plasmid expressing HS3ST-3 enzyme are resistant to HSV-1 infection (Shukla et al. 1999). The simplest explanation for these observations is that the biosynthesis of gD-binding HS is defective in CHO F17 cells via significantly perturbing the synthesis of HS precursor structures for HS3ST-3 sulfation, and a 2-O-sulfated iduronic acid residue is likely involved in the HS precursor. Taken together, the level of the precursor/key enzyme could play an important role in regulating the synthesis of gD-binding HS. It remains to be investigated whether such a "dual regulating" mechanism also controls biosynthesis of the specific saccharide sequences without 3-O-sulfated glucosamine residue.

Future Perspectives

Our recent results from the characterization of HS3STs demonstrate that 3-O-sulfated glucosamine is present in various saccharide sequences and is generated by different isoforms of HS3STs. The various 3-O-sulfated glucosamines containing saccharide sequences exhibit distinct biological functions: HS3ST-1-modified HS contains anti-coagulant activity; HS3ST-3 (including HS3ST-3A and HS3ST-3B)-modified HS is an entry receptor for HSV-1. The biological functions of HS3ST-2- and HS3ST-4-modified HS remain to be investigated. Given their unique substrate specificities and tissue-specific distribution, HS3ST isoforms give us an excellent tool with which to study the fine saccharide structures of HS in a specific tissue and related biological functions.

Further Reading

See Rosenberg et al. (1997) for a review. Their article describes the biological activities of HS and a possible mechanism for the biosynthesis of HS with specific saccharide sequences.

References

Atha DH, Lormeau JC, Petitou M, Rosenberg RD, Choay J (1985) Contribution of monosac-charide residues in heparin binding to antithrombin III. Biochemistry 24:6723–6729

Berninsone P, Hirschberg CB (1998) Heparan sulfate/heparin N-deacetylase/N-sulfotransferase: the N-sulfotransferase activity domain is at the carboxyl half of the holoenzyme. J Biol Chem 273:25556–25559

Colliec-Jouault S, Shworak NW, Liu J, De Agostini AI, Rosenberg RD (1994) Characteriza-tion of a cell mutant specifically defective in the synthesis of anticoagulantly active heparan sulfate. J Biol Chem 271:24953–24958

Danishefsky I, Steiner H, Bella AJ, Friedlander A (1969) Investigation on the chemistry of heparin. J Biol Chem 244:1741–1745

Edge ASB, Spiro RG (1990) Characterization of novel sequences containing 3-O-sulfated glucosamine in glomerular basement membrane heparan sulfate and localization of sulfated disaccharides to a peripheral domain. J Biol Chem 265:15874–15881

Garg HG, Joseph PAM, Yoshida K, Thompson BT, Hales CA (1996) Antiproliferative role of 3-O-sulfate glucosamine in heparin on cultured pulmonary artery smooth muscle cells. Biochem Biophys Res Commun 224:468–473

Hernaiz M, Liu J, Rosenberg RD, Linhardt RJ (2000) Enzymatic modification of heparan sulfate on a biochip promotes its interaction with antithrombin III. Biochem Biophys Res Commun 276:292–297

Kusche M, Torri G, Casu B, Lindahl U (1990) Biosynthesis of heparin availability of glucosaminyl 3-O-sulfation sites. J Biol Chem 265:7292–7300

Lindahl U, Backstrom G, Thunberg L, Leder IG (1980) Evidence for a 3-O-sulfated D-glucosamine residue in the antithrombin-binding sequence of heparin. Proc Natl Acad Sci USA 77:6551–6555

Liu J, Shriver Z, Blaiklock P, Yoshida K, Sasisekharan R, Rosenberg RD (1999b) Heparan sulfate D-glucosaminyl 3-O-sulfotransferase-3A sulfates N-unsubstituted glucosamine residues. J Biol Chem 274:38155–38162

Liu J, Shworak NW, Fritze LMS, Edelberg JM, Rosenberg RD (1996) Purification of heparan sulfate D-glucosaminyl 3-O-sulfotransferase. J Biol Chem 271:27072–27082

Liu J, Shworak NW, Sinaÿ P, Schwartz JJ, Zhang L, Fritze LMS, Rosenberg RD (1999b) Expression of heparan sulfate D-glucosaminyl 3-O-sulfotransferase isoforms reveals novel substrate specificities. J Biol Chem 274:5185–5192

McKeehan WL, Wu X, Kan M (1999) Requirement for anticoagulant heparan sulfate in the fibroblast growth factor receptor complex. J Biol Chem 274:21511–21514

Rosenberg RD, Showrak NW, Liu J, Schwartz JJ, Zhang L (1997) Heparan sulfate proteoglycans of the cardiovascular system: specific structures emerge but how is synthesis regulated? J Clin Invest 99:2062–2070

Shukla D, Liu J, Blaiklock P, Shworak NW, Bai X, Esko JD, Cohen GH, Eisenberg RJ, Rosenberg RD, Spear PG (1999) A novel role for 3-O-sulfated heparan sulfate in herpes simplex virus 1 entry. Cell 99:13–22

Shworak NW, Fritze LMS, Liu J, Butler LD, Rosenberg RD (1996) Cell-free synthesis of anticoagulant heparan sulfate reveals a limiting converting activity that modifies an excess precursor pool. J Biol Chem 271:27063–27071

Shworak NW, Liu J, Fritze LMS, Schwartz JJ, Zhang L, Logear D, Rosenberg RD (1997) Molecular cloning and expression of mouse and human cDNAs encoding heparan sulfate D-glucosaminyl 3-O-sulfotransferase. J Biol Chem 272:28008–28019

Shworak NW, Liu J, Petros LM, Zhang L, Kobayashi M, Copeland NG, Jenkins NA, Rosenberg RD (1999) Diversity of the extensive heparan sulfate D-glucosaminyl 3-O-sulfotransferase (HS3ST) multigene family. J Biol Chem 274:5170–5184

Shworak NW, Shirakawa M, Colliec-Jouault S, Liu J, Mulligan RC, Birinyi LK, Rosenberg RD (1994) Pathway-specific regulation of the biosynthesis of anticoagulantly active heparan sulfate. J Biol Chem 269:24941–24952

Toida T, Yoshida H, Toyoda H, Koshiishi T, Imanari T, Hileman RE, Fromm JR, Linhardt RJ (1997) Structural differences and the presence of unsubstituted amino groups in heparan sulphates from different tissues and species. Biochem J 322:499–506

Zhang L, Schwartz JJ, Miller J, Liu J, Fritze LMS, Shworak NW, Rosenberg RD (1998) The retinoic acid and cAMP-dependent up-regulation of 3-O-sulfotransferase-1 leads to a dramatic augmentation of anticoagulantly active heparan sulfate biosynthesis in F9 embryonal carcinoma cells. J Biol Chem 273:27998–28003

Zhang L, Yoshida K, Liu J, Rosenberg RD (1999) Anticoagulant heparan sulfate precursor structures in F9 embryonic carcinoma cells. J Biol Chem 274:5681–5691

70

Heparan Sulfate/Heparin N-Deacetylase/N-Sulfotransferase-1

Introduction

Heparan sulfate N-sulfotransferase-1 (NST-1) was first identified as a Golgi enzyme that catalyzes N-sulfation of the glucosamine moiety (GlcN) of the glycosaminoglycan chain of heparan sulfate (HS). This enzyme was later demonstrated to catalyze another reaction, the N-deacetylation of N-acetylglucosamine (GlcNAc), the step immediately preceding N-sulfation (Wei et al. 1993) (Fig. 1). This enzyme, known now as heparan sulfate N-deacetylase/N-sulfotransferase-1 (NDST-1), thus catalyzes two sequential reactions that are obligatory initiation steps in the modification of the HS glycosaminoglycan (Lindahl et al. 1998; Bernfield et al. 1999).

The glycosaminoglycan is initially biosynthesized as a polysaccharide chain composed of alternating D-glucuronic acid and N-acetylglucosamine units, which is then deacetylated and sulfated by NDST-1. The resulting polysaccharide chain is further modified by C-5 epimerization of D-glucuronic acid and O-sulfation at various positions of the disaccharide unit (Fig. 1), generating a mature HS glycosaminoglycan. The mature product has been demonstrated to bind and modulate the activity of various proteins, including growth factors and morphogens important for cellular differentiation and embryogenesis (Bernfield et al. 1999; Perrimon and Bernfield 2000). Loss of NDST-1 function results in deficient production of mature HS, resulting in abnormal differentiation and embryogenesis. For instance, mutation of a putative NDST-1 homolog in *Drosophila melanogaster* causes abnormal morphogenesis (Lin and Perrimon 1999; Tsuda et al. 1999); and targeted disruption of the mouse NDST-1 gene results in pulmonary hypoplasia (Fan et al. 2000).

Yasuhiro Hashimoto[1] and Carlos B. Hirschberg[2]

[1] RIKEN Frontier Research System, Supra-biomolecular System Group, Glyco-chain Functions Laboratory, 2-1 Hirosawa, Wako-shi, Saitama 351-0198, Japan
Tel. +81-48-467-9613; Fax +81-48-462-4690
e-mail: yasua@postman.riken.go.jp
[2] Department of Molecular and Cell Biology, Boston University Goldman School of Dental Medicine, Center for Advanced Biomedical Research, 700 Albany Street W201, Boston, MA 02118-2526, USA

Fig. 1. Modification of alternating GlcA and GlcNAc units. The GlcNAc residue is deacetylated to GlcN and subsequently N-acetylated by N-deacetylase/N-sulfotransferase. The disaccharide units are further modified by C5-epimerization and O-sulfation of various positions

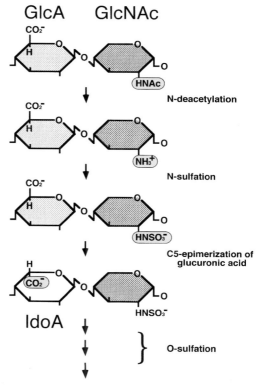

Databanks

Heparan sulfate/heparin N-deacetylase/N-sulfotransferase-1

NC-IUBMB enzyme classification: EC 2.8.2.8 for N-sulfotransferase.
No EC number has been allocated for N-deacetylase.

Species	Gene	Protein	mRNA	Genomic
Homo sapiens	*hsst1*	P52848	U18918	–
	hsst1	A57169	–	–
	hsst	G02129	U36600	–
	ndst1	G01581	U17970	–
Mus musculus		A49733	U02304	–
		–	AF074926	–
Rattus norvegicus	*hsst1, hsst*	Q02353	M92042	–
Drosophila melanogaster	*sfl*	–	AF175689	–

Name and History

Heparan sulfate N-sulfotransferase (NST, N-HSST) was discovered to be one of several enzymes that catalyze transfer of a sulfate group from 3′-phosphoadenosine 5′-phosphosulfate (PAPS) to heparan sulfate. It was first purified from rat liver (Brandan and Hirschberg 1988), and 4 years later its cDNA was cloned (Hashimoto et al. 1992). When the enzyme was expressed in COS cells by cDNA transfection, it had N-deacetylase, in addition to N-sulfotransferase, activity (Wei et al. 1993). The enzyme was then redesignated N-acetylglucosaminyl N-deacetylase/N-sulfotransferase (NDANST, NDST, HSNdAc/NST). Because the N-deacetylation/N-sulfation reaction occurs during both HS and heparin biosynthesis, identification of a comparable NDST in the heparin biosynthetic pathway was undertaken. A putative NDST responsible for heparin biosynthesis was purified from mastocytoma cells (Petterson et al. 1991) in which heparin was actively biosynthesized. Cloning of mastocytoma NDST cDNA revealed significant sequence similarity to liver NDST, but it had different enzymatic properties (Orellana and Hirschberg 1994). These two enzymes are isozymes; they are encoded by different genes and exhibit tissue-specific expression (Kusche-Gullberg et al. 1998; Aikawa and Esko 1999a). Thus, the liver isozyme was designated NDST-1, and the species isolated from mastocytoma cells was designated NDST-2. NDST-2 is thought to be responsible for heparin biosynthesis and is highly expressed in mast cells specific to connective tissues (Eriksson et al. 1994; Orellana and Hirschberg 1994). cDNA coding for another isozyme, NDST-3, has now been isolated from human brain (Aikawa and Esko 1999a), and the presence of NDST-4 has been suggested in a preliminary report (Aikawa and Esko 1999b).

Enzyme Activity Assay and Substrate Specificity

Biosynthesis of the glycosaminoglycan chain of HS begins with glycosylation reactions that form a large polysaccharide chain composed of alternating GlcA and GlcNAc units. The resulting disaccharide repeats are modified by a series of reactions initiated by N-deacetylation and N-sulfation of GlcNAc residues (Fig. 1) (Lindahl et al. 1998; Bernfield et al. 1999). NDST catalyzes both reactions, and the activities can be assessed independently using various substrates as described below.

For the N-deacetylase assay, an N-[^{3}H]acetylated polysaccharide is generated by reacetylation of a chemically deacetylated capsular polysaccharide from *Escherichia coli* K5 with [^{3}H]acetic anhydride (Navia et al. 1983). The labeled K5 polysaccharide is incubated with NDST, and the release of [^{3}H]acetic acid from the substrate is used as a measure of N-deacetylase activity. The N-deacetylase activity is maximal at pH 6.5; it is activated by dithiothreitol and inhibited by N-ethylmaleimide, suggesting that a sulfhydryl group(s) is required for activity. Inhibition is also achieved with sodium chloride, with more than 90% activity being lost at physiological concentrations (Wei et al. 1993; Wei and Swiedler 1999).

For the N-sulfotransferase assay, a chemically N-deacetylated K5 polysaccharide, heparin, or HS is typically used as a substrate. Even intact HS can be used, as it contains an unsubstituted GlcN nitrogen. Incorporation of [^{35}S]sulfate from [^{35}S]PAPS into the polysaccharide is used as a measure of N-sulfotransferase activity. The optimal pH for the reaction is around 7.5. In contrast to N-deacetylase activity, N-sulfotransferase activity is not inhibited by sodium chloride at physiological concentrations or by N-ethylmaleimide, but it is sensitive to dithiothreitol (Wei et al. 1993; Wei and Swiedler 1999). The differential sensitivity to these inhibitors suggests that the two catalytic sites behave independently. The two sites are thought to occupy different regions of the polypeptide chain of the enzyme; the deacetylase activity has been localized to the amino-terminal half, whereas the sulfotransferase activity has been isolated to the carboxyl-terminal half (Wei and Swiedler 1999).

The structurally related isozymes NDST-2 and NDST-3, which also catalyze the deacetylation and sulfotransferase reactions, exhibit similar domain organization. When the C-terminal half of NDST-2 is expressed in COS-7 cells, the expressed polypeptide catalyzes the N-sulfation reaction, indicating that this half is sufficient for N-sulfotransferase activity (Berninsone and Hirschberg 1998). There are, however, obvious differences in enzymatic properties among these isozymes. NDST-2 exhibits a 4- to 10-fold higher ratio of N-deacetylation /N-sulfation activity than do NDST-1 and NDST-3 (Orellana and Hirschberg 1994; Toma et al. 1998; Aikawa and Esko 1999a). Kinetic analysis has revealed that NDST-2 has a 25-fold higher K_m value for the N-deacetylated K5 polysaccharide than NDST-1. In addition to these kinetic differences, the three NDST isozymes are expressed in a tissue-specific manner (Kusche-Gullberg et al. 1998; Toma et al. 1998; Aikawa and Esko 1999a). The relative contribution of each isozyme to HS glycosaminoglycan biosynthesis in specific tissues has yet to be fully explored.

Preparation

NDST-1 was purified from the membrane fraction of rat liver. The enzyme was solubilized with Triton X-100 and then purified by successive chromatography on DEAE-Sephacel, heparin-agarose, 3′,5′-ADP-agarose, and wheat germ-Sepharose columns and finally by a glycerol gradient (Brandan and Hirschberg 1988). The purified enzyme was subjected to peptide sequence analysis, and the resulting sequence information was utilized for molecular cloning of NDST-1 cDNA (Hashimoto et al. 1992). The cloned cDNA was inserted in a mammalian expression vector, and it was expressed in COS cells and in mutant cells lacking NDST-1; the resulting increase in enzyme activity was confirmed (Ishihara et al. 1993). Because NDST-1 is a type II transmembrane protein, a soluble fusion protein comprised of protein A and NDST-1 lacking putative cytosolic and transmembrane domain was also expressed in COS cells. The fusion protein was purified from cell culture medium using immunoglobulin G (IgG)-agarose, and its enzymatic properties were characterized (Wei et al. 1993; Toma et al. 1998).

Biological Aspects

A key enzyme for HS biosynthesis, NDST-1 is critical in HS-dependent biological processes such as morphogenesis, wound healing, the immune response, and tumor metastasis (Bernfield et al. 1999). These processes are mediated via interactions with morphogens, growth factors, cytokines, and components of the extracellular matrix.

The interaction of HS with fibroblast growth factors (FGFs) has been characterized extensively at the molecular level. FGF-2 has distinct binding regions for HS and the FGF-2 receptor. FGF-2 forms a dimer with HS glycosaminoglycan on the cell surface, after which it can bind with high affinity to the receptor, transducing a signal inside the cell. FGF-2 interacts with HS glycosaminoglycan through a binding motif that includes several sulfate residues in a defined spatial relation. *N*-Sulfation, catalyzed by NDST-1, is a prerequisite for all consequent *O*-sulfation reactions. Therefore, NDST is essential for FGF-2 signaling; mutant cells lacking NDST activity produce only poorly sulfated glycosaminoglycans lacking the FGF-2 binding motif (Ishihara et al. 1993).

In addition, HS glycosaminoglycan interacts with a morphogen, Wingless (Wg), implicating NDST-1 in morphogen signaling (Perrimon and Bernfield 2000). Wg, a member of the Wnt family of proteins, regulates many developmental processes in *Drosophila melanogaster* by binding to HS glycosaminoglycan constructed on a core protein, Dally. The complex of Wg and HS glycosaminoglycan transduces a signal to its receptor, *Drosophila* fizzled 2 (Dfz2). This signaling cascade is essential for developmental processes such as wing formation and naked cuticle formation in epithelial cells. Mutants of either Wg, Dally, or Dfz2 exhibit similar developmental abnormalities (Lin and Perrimon 1999; Tsuda et al. 1999). A putative NDST-1 mutant, sulfateless (sfl), exhibits a similar phenotype, suggesting that NDST-1 is essential for Wg signaling through the catalysis of HS glycosaminoglycan biosynthesis on the Dally protein.

Mice homozygous for a targeted disruption of the NDST-1 gene do not develop fully matured type II pneumocytes, resulting in neonatal death due to respiratory distress (Fan et al. 2000). The development of other organs did not appear to be impaired, suggesting that other isozymes (e.g., NDST-2 and NDST-3) compensate for the lack of NDST-1 activity. NDST-1 and NDST-2 in both mice and humans are widely distributed in various tissues (Kusche-Gullberg et al. 1998; Aikawa and Esko 1999a). Targeted disruption of NDST-2 resulted in abnormalities only in mast cells specific to connective tissues, where the isozyme is expressed at a high level (Forsberg et al. 1999; Humphries et al. 1999). This result suggests a redundancy in the functions of these isozymes. Double (or triple) disruptions of these enzymes can shed further light on their biological relevance in vivo.

Future Perspectives

Molecular cloning of NDST-1 and other isozymes has revealed that several NDSTs catalyze the biosynthesis of HS glycosaminoglycans, but little is known about how their expression is controlled and about their relative contribution to its biosynthetic activity in each tissue. Analysis of promoter regions of these genes may provide important information for understanding mechanisms by which the expression of these enzymes is regulated. NDSTs identified so far are structurally diverse in the *N*-terminal portion, which is assumed to define the topology of these enzymes in the Golgi apparatus. The exact subcellular localization of these enzymes and the functional consequences of the distribution must be determined. Manipulation of HS biosynthesis could represent a new therapeutic target for the treatment of various pathological processes such as atherosclerosis, angiostenosis, cirrhosis of various organs, tumor invasion, and metastasis. NDSTs, including NDST-1, may represent a particularly attractive pharmacological target.

Further Reading

For reviews see Bernfield et al. (1999) and Perrimon and Bernfield (2000).

References

Aikawa J, Esko JD (1999a) Molecular cloning and expression of a third member of the heparan sulfate/heparin GlcNAc N-deacetylase/N-sulfotransferase family. J Biol Chem 274:2690–2695

Aikawa J, Esko JD (1999b) Molecular cloning and expression of the third and fourth members of the heparan sulfate/heparin N-deacetylase/N-sulfotransferase family. Glycoconj J 16:S40

Bernfield M, Gotte M, Park PW, Reizes O, Fitzgerald ML, Lincecum J, Zako M (1999) Functions of cell surface heparan sulfate proteoglycans. Annu Rev Biochem 68:729–777

Berninsone P, Hirschberg CB (1998) Heparan sulfate/heparin N-deacetylase/N-sulfotransferase. J Biol Chem 273:25556–25559

Brandan E, Hirschberg CB (1988) Purification of rat liver N-heparan-sulfate sulfotransferase. J Biol Chem 263:2417–2422

Eriksson I, Sandback D, Ek B, Lindahl U, Kjellen L (1994) cDNA cloning and sequencing of mouse mastocytoma glucosaminyl N-deacetylase/N-sulfotransferase, and enzyme involved in the biosynthesis of heparin. J Biol Chem 269:10438–10443

Fan G, Xiao L, Cheng L, Wang X, Sun B, Hu G (2000) Targeted disruption of NDST-1 gene leads to pulmonary hypoplasia and neonatal respiratory distress in mice. FEBS Lett 467:7–11

Forsberg E, Pejler G, Ringvall M, Lunderius C, Tomasini-Johansson B, Kusche-Gullberg M, Eriksson I, Ledin J, Hellman L, Kjellen L (1999) Abnormal mast cells in mice deficient in a heparin-synthesizing enzyme. Nature 400:773–776

Hashimoto Y, Orellana A, Gil G, Hirschberg CB (1992) Molecular cloning and expression of rat liver N-heparan sulfate sulfotransferase. J Biol Chem 267:15744–15750

Humphries DE, Wong GW, Friend DS, Gurish MF, Qiu W-T, Huang C, Sharpe AH, Stevens RL (1999) Heparin is essential for the storage of specific granule proteases in mast cells. Nature 400:769–772

Ishihara M, Guo Y, Wei Z, Yang Z, Swiedler SJ (1993) Regulation of biosynthesis of the basic fibroblast growth factor binding domains of heparan sulfate by heparan sulfate-N-deacetylase/N-sulfotransferase expression. J Biol Chem 268:20091–20095

Kusche-Gullberg M, Eriksson I, Pikas DS, Kjellen L (1998) Identification and expression in mouse of two heparan sulfate glucosaminyl N-deacetylase/N-sulfotransferase genes. J Biol Chem 273:11902–11907

Lin X, Perrimon N (1999) Dally cooperates with Drosophila frizzled 2 to transduce wingless signaling. Nature 400:281–284

Lindahl U, Kusche-Gullberg M, Kjellen L (1998) Regulated diversity of heparan sulfate. J Biol Chem 273:24979–24982

Navia JL, Reisenfeld J, Vann WF, Lindahl U, Roden L (1983) Assay of N-acetylheparosan deacetylase with a capsular polysaccharide from Escherichia coli K5 as substrate. Ann Biochem 135:134–140

Orellana A, Hirschberg CB (1994) Molecular cloning and expression of a glycosaminoglycan N-acetylglucosaminyl N-deacetylase/N-sulfotransferase from a heparin-producing cell line. J Biol Chem 269:2270–2276

Perrimon N, Bernfield M (2000) Specificities of heparan sulfate proteoglycans in developmental processes. Nature 404:725–728

Pettersson I, Kusche M, Unger E, Wlad H, Nylund L, Lindahl U, Kjellen L (1991) Biosynthesis of heparin. J Biol Chem 266:8044–8049

Tsuda M, Kamimura K, Nakato H, Archer M, Staatz W, Fox B, Humphrey M, Olson S, Futch T, Kaluza V, Siegfried E, Stam L, Selleck SB (1999) The cell-surface proteoglycan Dally regulates Wingless signaling in *Drosophila*. Nature 400:276–280

Wei Z, Swiedler SJ (1999) Functional analysis of conserved cysteines in heparan sulfate *N*-deacetylase-*N*-sulfotransferases. J Biol Chem 274:1966–1970

Wei Z, Swiedler SJ, Ishihara M, Orellana A, Hirschberg CB (1993) A single protein catalyzes both *N*-deacetylation and *N*-sulfation during the biosynthesis of heparan sulfate. Proc Natl Acad Sci USA 90:3885–3888

Heparan Sulfate/Heparin *N*-Deacetylase/*N*-Sulfotransferase-2

Introduction

Glucosaminyl *N*-deacetylase/*N*-sulfotransferase-2 (NDST-2) is a modification enzyme involved in heparin/heparan sulfate (HS) biosynthesis. This bifunctional enzyme catalyzes replacement of the acetyl group in GlcNAc with a sulfate group, which is the first modification step following polymerization of the $(GlcNAc-GlcA)_n$ precursor of heparin/HS. It is a key step during biosynthesis, as subsequent modifications, including GlcA epimerization and *O*-sulfation at various positions occur only in the vicinity of *N*-sulfate groups. Four mammalian NDST isoforms have been identified. At present, the specific roles of the various isoforms involved in heparin/HS biosynthesis are not known. Single orthologs are present in Drosophila melanogaster and Caenorhabditis elogans.

Databanks

Heparan sulfate/heparin *N*-deacetylase/*N*-sulfotransferase-2

NC-IUBMB enzyme classification: EC 2.8.2.8 for N-sulfotransferase.
No EC number has been allocated for N-deacetylase.

Species	Gene	Protein	mRNA	Genomic
Homo sapiens	NDST2	–	U36601	–
			XM005875	NT024037
			AF042084	
Mus musculus	–	–	U02304	–
		–	X75885	–
		–	AF074925	–
Bos taurus			AF064825	
Drosophila melanogaster	–	–	AF175689	–
Caenorhabditis elogans			AB038044	

Lena Kjellén

Department of Medical Biochemistry and Microbiology, University of Uppsala, The Biomedical Center, Husargatan 3, Box 582, SE-751 23 Uppsala, Sweden
Tel. +46-18-471-4217; Fax +46-18-471-4244
e-mail: lena.kjellen@vmk.slu.se

Name and History

N-Deacetylase and N-sulfotransferase activities were previously believed to reside in different protein molecules. During purification of N-deacetylase from a heparin-synthesizing mouse mastocytoma it was discovered that N-deacetylase and N-sulfotransferase activities were associated with the same purified 110-kDa protein (Pettersson et al. 1991). However, the N-deacetylase activity was dependent on the auxiliary action of an additional unidentified macromolecule (Pettersson et al. 1991). Later it was shown that this unidentified component could be replaced by synthetic polycations (Kjellén et al. 1992; Eriksson et al. 1994) demonstrating that both enzyme activities were present in the purified protein. The bifunctional nature of the protein and the presence of several NDST isoforms were discovered during the 1990s. Therefore, older names referred to one of the two enzyme activities (e.g., N-acetyl-D-glucosaminyl N-deacetylase, N-acetylheparosan deacetylase, heparan sulfate N-sulfotransferase, N-heparan sulfate sulfotransferase). Now the four isoforms are named N-deacetylase/N-sulfotransferase 1, 2, 3, and 4, abbreviated NDST-1 and so on. Often "heparan sulfate/heparin" and/or "glucosaminyl-" are used as prefixes.

Enzyme Activity Assay and Substrate Specificity

A convenient assay for N-deacetylase activity was developed by Navia et al. (1983). With this method the release of ^{3}H-labeled acetyl groups from chemically N-[^{3}H]-acetyl-labeled *Escherichia coli* K5 capsular polysaccharide substrate is quantitated. N-Sulfotransferase activity is assayed by measuring incorporation of [^{35}S]sulfate groups from [^{35}S]3'-phosphoadenosine 5'-phosphosulfate (PAPS) into N-deacetylated K5 polysaccharide (Sandbäck Pikas et al. 2000). N-Desulfated and N,O-desulfated heparin were previously used as substrates. Because the presence of O-sulfate groups and iduronic acid may influence the N-sulfotransferase activity, the N-deacetylated K5 polysaccharide is a better substrate.

The two enzyme activities have different pH optimums when measured in vitro. A pH of about 6.2 is optimal for the N-deacetylase reaction (Riesenfeld et al. 1980), whereas the pH optimum for N-sulfotransferase activity is about 7.5 (Jansson et al. 1975). When the concerted N-deacetylation/N-sulfation reaction was studied using [^{35}S]PAPS and unlabeled K5 polysaccharide as substrate, the pH optimum was close to pH 6 (i.e., similar to the preferred pH for the N-deacetylation reaction (D. Sandbäck Pikas, L. Kjellén, U. Lindahl, unpublished). The N-deacetylase activity of NDST-2 is enhanced by polycations and is inhibited by N-ethylmaleimide (Pettersson et al. 1991; Eriksson et al. 1994). As shown for NDST-1 (Wei and Swiedler 1999), the inhibitory action of N-ethylmaleimide is probably due to alkylation of the free SH group of Cys486 (corresponding to Cys484 in NDST-2).

Preparation

Native NDST-2 has been purified from mouse mastocytoma (Pettersson et al. 1991). The purification procedure involved solubilization of microsomal membranes in Triton X-100 containing buffer followed by affinity chromatography on wheat germ agglutinin-Sepharose, blue Sepharose, and 3′,5′-ADP-agarose. The protein has also been overexpressed with retained enzymatic activities in COS cells in its full-length transmembrane form and as a soluble secreted protein A chimera (Orellana et al. 1994).

Biological Aspects

Because NDST-2 was first identified in mastocytoma cells (Pettersson et al. 1991), it was initially believed that this isoform was responsible for heparin biosynthesis in mast cells only, and that other isoforms participated in HS biosynthesis (Eriksson et al. 1994; Orellana et al. 1994). However, the wide distribution of the NDST-2 transcript in tissues and its presence in nonmast cells (Kusche-Gullberg et al. 1998; Toma et al. 1998) indicated a role for NDST-2 in HS biosynthesis as well. It was therefore surprising that targeted disruption of the murine NDST-2 gene did not affect the structure of HS in liver, where large amounts of NDST-2 normally are present (Forsberg et al. 1999). However, NDST-2-deficient mice, which have a normal life-span and are fertile, displayed a severe mast cell defect caused by complete lack of sulfated heparin in the secretory granules (Forsberg et al. 1999; Humphries et al. 1999). The absolute requirement for NDST-2 in mast cell heparin biosynthesis was thus established. How is NDST-2 involved in HS biosynthesis? Mice that lack NDST-1 die neonatally with a condition resembling respiratory distress syndrome (Fan et al. 2000; Ringvall et al. 2000). In addition, the HS structure in large parts of the body seems to be affected by the lack of this NDST isoform (Ringvall et al. 2000). Ongoing studies (K. Holmborn, E. Forsberg, unpublished) demonstrate that lack of both NDST-1 and NDST-2 cause early embryonic death, indicating that NDST-2 indeed is involved also in HS biosynthesis.

Transfection of the human embryonic kidney 293 cell line with murine NDST-2 and NDST-1, respectively, results in different HS N-sulfation patterns (Cheung et al. 1996; Sandbäck Pikas et al. 2000). Although an increased level of sulfation is achieved in both cases, HS synthesized in the presence of NDST-2 has a higher N-sulfate content, and the N-sulfated domains are more extended. It is concluded from these studies that the level of enzyme expression and the NDST isoform are important for determining the N-sulfation pattern of HS. Interestingly, the increase in N-sulfation in the NDST-transfected cells was accompanied by an increased polysaccharide chain length (Sandbäck Pikas et al. 2000), indicating coupling between the polymerization and sulfation processes, as suggested by previous studies in a mastocytoma microsomal system (Lidholt et al. 1989).

Future Perspectives

Results are now accumulating that describe the structural features and catalytic mechanisms for individual enzymes in the biosynthesis machinery responsible for HS biosynthesis, but almost nothing is known about the organization of the enzymes within the Golgi compartment. Studies on localization of the enzymes and how they cooperate to synthesize the polysaccharide is an area ripe for investigators in the field. Regulation of the process is another important issue.

Further Reading

See Lindahl et al. (1998) for a review on heparan sulfate and its biosynthesis.

References

Aikawa J, Esko JD (1999) Molecular cloning and expression of a third member of the heparan sulfate/heparin GlcNAc N-deacetylase/N-sulfotransferase family. J Biol Chem 274:2690–2695

Cheung W-F, Eriksson I, Kusche-Gullberg M, Lindahl U, Kjellén L (1996) Expression of mouse mastocytoma glucosaminyl N-deacetylase/N-sulfotransferase in human kidney 293 cells results in increased N-sulfation of heparan sulfate. Biochemistry 35:5250–5256

Eriksson I, Sandbäck D, Ek B, Lindahl U, Kjellén L (1994) cDNA cloning and sequencing of mouse mastocytoma N-deacetylase/N-sulfotransferase. J Biol Chem 269:10438–10443

Fan G, Xiao L, Cheng L, Wang X, Sun B, Hu G (2000) Targeted disruption of NDST-1 gene leads to pulmonary hypoplasia and neonatal respiratory distress in mice. FEBS Lett 467:7–11

Forsberg E, Pejler G, Ringvall M, Lunderius C, Tomasini-Johansson B, Kusche-Gullberg M, Eriksson I, Ledin J, Hellman L, Kjellén L (1999) Abnormal mast cells in mice deficient in a heparin-synthesizing enzyme. Nature 400:773–776

Humphries DE, Lanciotti J, Karlinsky JB (1998) cDNA cloning, genomic organization and chromosomal localization of human heparan glucosaminyl N-deacetylase/N-sulpho-transferase-2. Biochem J 332:303–307

Humphries DE, Wong GW, Friend DS, Gurish MF, Qiu WT, Huang C, Sharpe AH, Stevens RL (1999) Heparin is essential for the storage of specific granule proteases in mast cells. Nature 400:769–772

Jansson L, Höök M, Wasteson A, Lindahl U (1975) Biosynthesis of heparin: solubilization and partial characterization of N- and O-sulphotransferases. Biochem J 149:49–55

Kjellén L, Pettersson I, Unger E, Lindahl U (1992) Two enzymes in one: N-deacetylation and N-sulfation in heparin biosynthesis are catalyzed by the same protein. Adv Exp Med Biol 313:107–111

Kusche-Gullberg M, Eriksson I, Sandbäck Pikas D, Kjellén L (1998) Identification and expression in mouse of two heparan sulfate glucosaminyl N-deacetylase/N-sulfo-transferase genes. J Biol Chem 273:11902–11907

Lidholt K, Kjellén L, Lindahl U (1989) Biosynthesis of heparin: relationship between the polymerization and sulphation processes. Biochem J 261:999–1007

Lin X, Perrimon N (1999) Dally cooperates with Drosophila frizzled 2 to transduce wingless signalling. Nature 400:281–284

Lindahl U, Kusche-Gullberg M, Kjellén L (1998) Regulated diversity of heparan sulfate. J Biol Chem 273:24979–24982

Navia JL, Riesenfeld J, Vann WF, Lindahl U, Rodén L (1983) Assay of N-acetylheparosan deacetylase with a capsular polysaccharide from Escherichia coli K5 as substrate. Anal Biochem 135:134–140

Orellana A, Hirschberg CB, Wei Z, Swiedler SJ, Ishihara M (1994) Molecular cloning and expression of a glycosaminoglycan N-acetylglucosaminyl N-deacetylase/N-sulfo-transferase from a heparin-producing cell line. J Biol Chem 269:2270–2276

Pettersson I, Kusche M, Unger E, Wlad H, Nylund L, Lindahl U, Kjellén L (1991) Biosynthesis of heparin: purification of a 110 kDa mouse mastocytoma protein required for both glucosaminyl N-deacetylation and N-sulfation. J Biol Chem 266:8044–8049

Riesenfeld J, Höök M, Lindahl U (1980) Biosynthesis of heparin: assay and properties of the microsomal N-acetyl-D-glucosaminyl N-deacetylase. J Biol Chem 255:922–928

Ringvall M, Ledin J, Holmborn K, Kuppevelt T, Ellin F, Eriksson I, Olofsson A-M, Kjellén L, Forsberg E (2000) Defective heparan sulfate biosynthesis and neonatal lethality in mice lacking N-deacetylase/N-sulfotransferase-1. J Biol Chem 275:25926–25930

Sandbäck Pikas D, Eriksson I, Kjellén L (2000) Overexpression of different isoforms of glucosaminyl N-deacetylase/N-sulfotransferase results in distinct heparan sulfate N-sulfation patterns. Biochemistry 39:4552–4558

Toma L, Berninsone P, Hirschberg CB (1998) The putative heparin-specific N-acetylglucosaminyl N-deacetylase/N-sulfotransferase also occurs in non-heparin-producing cells. J Biol Chem 273:22458–22465

Wei Z, Swiedler SJ (1999) Functional analysis of conserved cysteines in heparan sulfate N-deacetylase-N-sulfotransferases. J Biol Chem 274:1966–1970

72

Heparan Sulfate/Heparin *N*-Deacetylase/ *N*-Sulfotransferase-3 and -4

Introduction

This chapter describes the third and fourth members of HS/heparin *N*-deacetylase/ *N*-sulfotransferase (NDST). Refer to Chapters 70 and 71 for information about NDST-1 and NDST-2, respectively. Searching databases, NDST-3 and NDST-4 are discovered in the EST and genomic databases, respectively, as two new homologs to well known NDST-1 and NDST-2. Preparation of the cDNA clone to cover the complete open reading frame revealed that both NDST-3 and NDST-4 encode 872–873 amino acid residues (Aikawa and Esko 1999a,b) and exhibit a putative domain structure common with that of NDST-1 and NDST-2. It is composed of five domains as follows: cytoplasm, transmembrane, stem, *N*-deacetylase, and *N*-sulfotransferase domain in that order from the *N*-terminus (Berninsone and Hirschberg 1998; Sueyoshi et al. 1998). Enzyme activities for GlcNAc *N*-deacetylation and GlcN *N*-sulfation were detected in a recombinant form of human NDST-3 (Aikawa and Esko 1999a), and studies of those for NDST-4 are currently in progress (Aikawa and Esko 1999b). Because NDST-3 and NDST-4 can be prepared from both human and mouse, the presence of two new isozymes seems common in mammals.

JUN-ICHI AIKAWA

Cellular Biochemistry Laboratory, RIKEN, 2-1 Hirosawa, Wako-shi, Saitama 351-0198, Japan
Tel. +81-48-467-9372; Fax +81-48-462-4670
e-mail: aikawa@postman.riken.go.jp.

Databanks

Heparan sulfate/heparin *N*-deacetylase/*N*-sulfotransferase-3, and -4

NC-IUBMB enzyme classification: EC 2.8.2.8 for N-sulfotransferase.
No EC number has been allocated for N-deacetylase.

Species	Gene	Protein	mRNA	Genomic
Homo sapiens	NDST3	–	AF074924	–
	NDST4	–	AB036429	–
Mus musculus	NDST3	–	AF221095	–
	NDST4	–	AB036838	–

Name and History

The definition of HS/heparin NDST has appeared in Chapters 70 and 71. Following two isozymes of the NDST gene family, NDST-1 (Hashimoto et al. 1992) and NDST-2 (Eriksson et al. 1994; Orellana et al. 1994), NDST-3 was identified as the third member of this group. Using the sequence of human NDST-1 as a probe, some cDNA sequences of human NDST-3 were first found in the EST database, the longest clone of which still lacks the initiation codon. The remained portion was then prepared and analyzed by 5′ rapid amplification of complementary DNA ends (RACE). Human NDST-3 exhibited 70% and 65% amino acid identity with those of NDST-1 and NDST-2, respectively. Its mouse homolog was prepared from brain tissue and showed 93% identity to that of human NDST-3.

NDST-4, the fourth member of the NDST gene family, was first discovered in the genomic database (Aikawa and Esko 1999b). A possible exon sequence in the fragment showed strong similarity to that of other NDSTs. A full-length open reading frame was prepared by both 5′ and 3′ RACE (Aikawa et al. 2001). The amino acid sequence of human NDST-4 showed the closest similarity to that of NDST-3 among three isoforms in this group. The murine counterpart was detected in some mouse tissues.

Enzyme Activity Assay and Substrate Specificity

Because neither NDST-3 nor NDST-4 has been prepared from any tissues, utilization of a recombinant technology is the only way to obtain enzymes to evaluate their catalytic function. Enzyme activities of protein A chimera of human NDST-3 were detected. When *N*-deacetylase activity was measured using release of ^{3}H-acetate from a substrate (^{3}H-acetyl heparosan) (Bame et al. 1991b), human NDST-3 showed activity comparable to that of rat NDST-1 at 37°C. The enzyme also exhibited *N*-sulfation activity similar to that of the control enzyme, measuring transfer of radioactivity from [^{35}S]3′-phosphoadenosine 5′-phosphosulfate (PAPS) to *N*-desulfo-heparin (Bame and Esko 1989). Further analysis is in progress. Assays for NDST-4 have not yet been examined. The substrate specificity of NDST-3 along with NDST-4 will be studied.

Preparation

Preparation and purification of native NDST-3 or NDST-4 polypeptide have not been attempted from any mammalian tissue. A strategy to construct a soluble chimera of NDST-3 is essentially the same as that for rat NDST-1 (Wei et al. 1993). Briefly, recombinant DNA was constructed to fuse a soluble form of NDST that lacks N-terminal cytoplasm and a transmembrane domain to signal-peptide and bacterial protein A portion. That DNA unit was placed under human cytomegarovirus enhancer/promoter and then introduced into simian COS7 cells. After 3 days of culture the fused protein was secreted into the culture medium, recovered from the supernatant, and purified using immunoglobulin G (IgG)–agarose. Enzymes conjugated with beads are used as an enzyme source directly for assay of catalytic activities for N-deacetylation and N-sulfation.

Biological Aspects

The expression of NDST in human tissues was tested using the reverse transcriptase-polymerase chain reaction (RT-PCR) technique. Specific primers to each NDST were carefully designed to avoid nonspecific amplification of isozymes other than the target. Expression of both NDST-1 and NDST-2 was detected in most human tissues tested. Expression of NDST-3 was detected in greatest abundance in brain, kidney, liver, and lung of both adult and fetal tissues. NDST-4 was expressed in an organ-specific manner but differently from that of NDST-3 (Aikawa et al. 2001).

There is no information on whether deficient expression of either NDST-3 or NDST-4 is related to human disease. Their knockout mice have not yet been established. Cell transfectants for which expression of either NDST-3 or NDST-4 is enhanced are being studied, and the results will be reported in the future.

Cultured CHO cells, which have been used to generate deficient mutants in one of the glycosaminoglycan biosynthetic enzymes (Esko 1991; Bai and Esko 1996; Bai et al. 1999), did not express NDST-3, although a transcript of both NDST-1 and NDST-2 was detected in this cell line (Aikawa and Esko 1999a). No information is available about expression of NDST-4 there.

Future Perspectives

Because NDST-3 and NDST-4 were identified only recently, molecular and biochemical studies are still in progress. At least two major points should be solved in future. First, do NDST-3 and NDST-4 exhibit any differences in substrate recognition and formation of N-sulfation block (Lindahl et al. 1998) from NDST-1 and NDST-2 (Orellana et al. 1994; Pikas et al. 2000)? As N-sulfation regulates polymerization (Lidholt et al. 1989) and other modification steps (Bame et al. 1991a) in the biosynthetic pathway, these characteristics are of biochemical significance from the point of view of the formation of binding sequences to fibroblast growth factor-2 (FGF-2) (Yayon et al. 1991; Habuchi et al. 1992) and antithrombin III (Grootenhuis and van Boeckel 1991), among others. Second, what is the biological significance of NDST-3 and NDST-4? Based on the RT-PCR analysis (see above), expression of the NDSTs overlaps in many tissues, which suggests that the function of each NDST would be compensated by that of other NDSTs. However, their expression in individual cells have not been examined, so, we have no idea about the contribution of two new isozymes to the development, maintenance, and death of each cell. When each tissue is investigated in the knockout mice of these NDSTs, that question will be solved.

Further Reading

Aikawa J and Esko JD (1999a) described cloning of human NDST3, characterization of the enzyme activity, and the RT-PCR analysis of three NDSTs in human tissues

References

Aikawa J, Esko JD (1999a) Molecular cloning and expression of a third member of the heparan sulfate/heparin GlcNAc N-deacetylase/N-sulfotransferase family. J Biol Chem 274:2690–2695

Aikawa J, Esko JD (1999b) Molecular cloning and expression of the third and fourth members of the heparan sulfate/heparin N-deacetylase/N-sulfotransferase family. Glycoconj J 16:S40

Aikawa J, Grobe K, Tsujimoto M, Esko JD (2001) Multiple isozymes of heparan sulfate/heparin GlcNAc N-deacetylase/GlcN N-sulfotransferase. Structure and activity of the fourth member, NDST4. J Biol Chem 276:5876–5882

Bai X, Esko JD (1996) An animal cell mutant defective in heparan sulfate hexuronic acid 2-O-sulfation. J Biol Chem 271:17711–17717

Bai X, Wei G, Sinha A, Esko JD (1999) Chinese hamster ovary cell mutants defective in glycosaminoglycan assembly and glucuronosyltransferase I. J Biol Chem 274:13017–13024

Bame KJ, Esko JD (1989) Undersulfated heparan sulfate in a Chinese hamster ovary cell mutant defective in heparan sulfate N-sulfotransferase. J Biol Chem 264:8059–8065

Bame KJ, Lidholt K, Lindahl U, Esko JD (1991a) Biosynthesis of heparan sulfate: coordination of polymer-modification reactions in a Chinese hamster ovary cell mutant defective in N-sulfotransferase. J Biol Chem 266:10287–10293

Bame KJ, Reddy RV, Esko JD (1991b) Coupling of N-deacetylation and N-sulfation in a Chinese hamster ovary cell mutant defective in heparan sulfate N-sulfotransferase. J Biol Chem 266:12461–12468

Berninsone P, Hirschberg CB (1998) Heparan sulfate/heparin N-deacetylase/N-sulfotransferase: the N-sulfotransferase activity domain is at the carboxyl half of the holoenzyme. J Biol Chem 273:22556–22559

Eriksson I, Sandback D, Ek B, Lindahl U, Kjellen L (1994) cDNA cloning and sequencing of mouse mastocytoma glucosaminyl N-deacetylase/N-sulfotransferase, an enzyme involved in the biosynthesis of heparin. J Biol Chem 269:10438–10443

Esko JD (1991) Genetic analysis of proteoglycan structure, function and metabolism. Curr Opin Cell Biol 3:805–816

Grootenhuis PDJ, van Boeckel CAA (1991) Constructing a molecular model of the interaction between antithrombin III and a potent heparin analogue. J Am Chem Soc 113:2743–2747

Habuchi H, Suzuki S, Saito T, Tamura T, Harada T, Yoshida K, Kimata K (1992) Structure of a heparan sulphate oligosaccharide that binds to basic fibroblast growth factor. Biochem J 285:805–813

Hashimoto Y, Orellana A, Gil G, Hirschberg CB (1992) Molecular cloning and expression of rat liver N-heparan sulfate sulfotransferase. J Biol Chem 267:15744–15750

Lidholt K, Kjellen L, Lindahl U (1989) Biosynthesis of heparin: relationship between the polymerization and sulphation processes. Biochem J 261:999–1007

Lindahl U, Kusche-Gullberg M, Kjellen L (1998) Regulated diversity of heparan sulfate. J Biol Chem 273:24979–24982

Orellana A, Hirschberg CB, Wei Z, Swiedler SJ, Ishihara M (1994) Molecular cloning and expression of a glycosaminoglycan N-acetylglucosaminyl N-deacetylase/N-sulfotransferase from a heparin-producing cell line. J Biol Chem 269:2270–2276

Pikas DS, Eriksson I, Kjellen L (2000) Overexpression of different isoforms of glucosaminyl N-deacetylase/N-sulfotransferase results in distinct heparan sulfate N-sulfation patterns. Biochemistry 39:4552–4558

Sueyoshi T, Kakuta Y, Pedersen LC, Wall FE, Pedersen LG, Negishi M (1998) A role of Lys614 in the sulfotransferase activity of human heparan sulfate N-deacetylase/N-sulfotransferase. FEBS Lett 433:211–214

Wei Z, Swiedler SJ, Ishihara M, Orellana A, Hirschberg CB (1993) A single protein catalyzes both N-deacetylation and N-sulfation during the biosynthesis of heparan sulfate. Proc Natl Acad Sci USA 90:3885–3888

Yayon A, Klagsbrun M, Esko JD, Leder P, Ornitz DM (1991) Cell surface, heparin-like molecules are required for binding of basic fibroblast growth factor to its high affinity receptor. Cell 64:841–848

βGal 3-O-Sulfotransferase-1, -2, -3, and -4

Introduction

βGal 3-O-sulfotransferases (Gal3STs) are Golgi-membrane sulfotransferases catalyzing the transfer of a sulfate group from the donor substrate PAPS to the C3 position of the nonreducing terminal Gal residue of carbohydrate chains. Four Gal3STs have been identified so far and their primary structures are similar to each other (Honke et al. 1997; Honke et al. 2001; Suzuki et al. 2001; El-Fasakhany et al. 2001; Seko et al. 2001). The original Gal3ST gene was cloned as a gene that encodes cerebroside sulfotransferase (CST) (Honke et al. 1997) based on the amino acid sequence of the purified enzyme (Honke et al. 1996). The second through fourth Gal3ST genes were very recently cloned by making use of homology to the *CST* gene (Honke et al. 2001; Suzuki et al. 2001; El-Fasakhany et al. 2001; Seko et al. 2001). Although the characteristics and biological function of CST have been extensively studied (Honke et al. 1997; Ishizuka 1997), reports on the other Gal3STs are limited. Therefore, I mainly focus on the original Gal3ST, CST, in this chapter.

Databanks

βGal 3-O-sulfotransferase-1, -2, -3, and -4

NC-IUBMB enzyme classification: E.C.2.8.2.11

Species	Gene	Protein	mRNA	Genomic
Gal3ST-1				
Homo sapiens	*CST*	–	D88667	AB029900–AB029901
Mus musculus	*Cst*	–	AB032939	AB032940

Koichi Honke

Department of Biochemistry, Osaka University Medical School, 2-2 Yamadaoka, Suita, Osaka 565-0871, Japan
Tel. +81-6-6879-3421; Fax +81-6-6879-3429
e-mail: khonke@biochem.med.osaka-u.ac.jp

(Continued)

Species	Gene	Protein	mRNA	Genomic
Gal3ST-2				
Homo sapiens	*GP3ST*	–	AB040610	AF048727
Gal3ST-3				
Homo sapiens	–	–	AY026481	AC008102
Gal3ST-4				
Homo sapiens	–	–	AK022178	AC073842

Name and History

Because glycolipid sulfotransferase was identified as an enzyme to catalyze the transfer of a sulfate group from PAPS to the C3 position of the galactose residue of GalCer (Farooqui et al. 1977), it was originally called cerebroside sulfotransferase (cerebroside is the old name of GalCer) and therefore abbreviated CST. Later, it was demonstrated that the sulfotransferase acts not only on GalCer but also on galactosylglycerolipids, LacCer, and other glycolipids (Lingwood 1985; Tennekoon et al. 1985; Honke et al. 1996). Recently, CST was purified to apparent homogeneity from human renal cancer cells (Honke et al. 1996), and its cDNA was cloned therefrom (Honke et al. 1997). Then the human *CST* gene was assigned on chromosome 22q12 (Tsuda et al. 2000).

When the *CST* gene was cloned, it showed no homologous genes, suggesting that CST has a unique evolutionary origin (Honke et al. 1997). Very recently, the second Gal3ST (termed GP3ST) gene that acts on Galβ1-3/4GlcNAc-R oligosaccharides was identified based on its similarity to the *CST* gene (Honke et al. 2001). This finding indicated the existence of the Gal3ST gene family. Subsequently, two other members of this family have been identified (Suzuki et al. 2001; El-Fasakhany et al. 2001; Seko et al. 2001) and the terminology of Gal3ST-1 through 4 was proposed (Suzuki et al. 2001). Table 1 summarizes the Gal3ST family, and Fig. 1 shows its phylogenetic tree.

Table 1. βGal 3-*O*-Sulfotransferase family

Name	Substrate	Distribution	Molecular cloning	Human chromosome
CST (Gal3ST-1)	GalCer, LacCer, GalEAG (glycolipid specific)	Brain, kidney, testis, GI tract	Honke et al. 1997	22q12
GP3ST (Gal3ST-2)	Galβ1-3GlcNac-R, Galβ1-4GlcNAc-R, Galβ1-3GalNAcα-Bzl	Various tissues	Honke et al. 2001	2q27.3
Gal3ST-3	Galβ1-4GlcNAc-R	Thyroid, brain, kidney	Suzuki et al. 2001 El-Fasakhany et al. 2001	11q13
Gal3ST-4	Galβ1-3GalNAcα-R, Galβ1-3(GlcNAcβ1-6)GalNAcα-R (*O*-glycan specific)	Various tissues	Seko et al. 2001	7q22

Fig. 1. The phylogenetic tree of the βGal 3-O-sulfotransferase (Gal3ST) gene family. The alignment of amino acid sequences of Gal3ST was carried out using *Clustal W (ver. 1.7)*. The branch length indicates the mean number of differences per site in an alignment and reflects evolutionary distances between different genes

βGal 3-*O*-Sulfotransferase Family

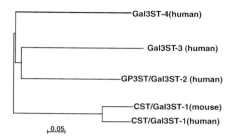

Enzyme Activity Assay and Substrate Specificity

A convenient assay method for CST (Gal3ST-1) activity was developed using anion-exchange chromatography (Kawano et al. 1989). The reaction mixture contained 5 nmol of GalCer, 0.5 μmol of $MnCl_2$, 1 nmol of [^{35}S]PAPS (100 cpm/pmol), 0.5 mg of Lubrol PX, 12.5 nmol of dithiothreitol, 0.25 μmol of NaF, 0.1 μmol of ATP, 20 mg of bovine serum albumin (BSA), and enzyme protein in 25 mM Na cacodylate HCl, pH 6.5, in a total volume of 50 μl. After incubation at 37°C for 1 h, the reaction was terminated with 1 ml of chloroform/methanol/water (30:60:8). The reaction product was isolated on a DEAE-Sephadex A-25 column and assayed for radioactivity using a liquid scintillation counter. The values were corrected for a blank value, which was obtained using a reaction mixture devoid of the acceptor.

The substrate specificity of CST purified from human renal cancer cells was investigated (Honke et al. 1996). GalCer is the best acceptor, although LacCer, galactosyl 1-alkyl-2acyl-*sn*-glycerol (the precursor for seminolipid), and galactosyl diacylglycerol are also good acceptors. GlcCer, Gg3Cer, Gg4Cer, Gb4Cer, and nLc4Cer serve as acceptors, although the relative activities are low. On the other hand, the enzyme cannot act on Gb3Cer, which has α-galactoside at the nonreducing terminus. Neither galactose nor lactose serves as an acceptor. These observations suggest that CST prefers β-galactoside at the nonreducing termini of sugar chains attached to a lipid moiety.

A novel sulfotransferase assay system for GP3ST (Gal3ST-2) using pyridylaminated oligosaccharides as acceptors was developed (Honke et al. 2001). This method does not require a radioisotope-labeled donor but is very sensitive. Furthermore, the reaction products can be isolated by a combination HPLC system, subjected to further analysis such as two-dimensional ^{1}H-NMR (Honke et al. 2001), and used as acceptors for other glycosyltransferases (Ikeda et al. 2001). GP3ST acts on type 1 (Galβ1-3GlcNAc-R), type 2 (Galβ1-4GlcNAc-R), and O-glycan core1 (Galβ1-3GalNAcα-R) oligosaccharides (Honke et al. 2001).

Gal3ST-3 activity was assayed using [^{35}S]PAPS as a donor (Suzuki et al. 2001; El-Fasakhan et al. 2001). The reaction products were separated from nonreacted donor substrate and its degraded by-products by reverse-phase column chromatography (Suzuki et al. 2001) or gel filtration (El-Fasakhany et al. 2001). Gal3ST-3 exclusively acts on the type 2 chain in N- and O-glycans (Suzuki et al. 2001; El-Fasakhany et al. 2001) but also acts on Galβ1-4(sulfo-6)GlcNAc-R oligosaccharide (Suzuki et al. 2001).

Gal3ST-4 activity was also assayed using [^{35}S]PAPS and the reaction products were isolated by paper electrophoresis (Seko et al. 2001). Gal3ST-4 recognizes Galβ1-3GalNAc-R and Galβ1-3(GlcNAcβ1-6)GalNAc-R as good acceptors but not Galβ1-3/4GlcNAc-R (Seko et al. 2001). When asialofetuin was used as a substrate, sulfation was exclusively found in O-linked glycans containing the Galβ1-3GalNAc structure (Seko et al. 2001). Neither GP3ST (Gal3ST-2), Gal3ST-3, nor Gal3ST-4 acted on GalCer (Honke et al. 2001; Suzuki et al. 2001; El-Fasakhany et al. 2001; Seko et al. 2001) in good agreement with the finding obtained in CST-knockout mice that a single enzyme is responsible for the biosynthesis of sulfatide (Honke et al. unpublished results).

Preparation

CST (Gal3ST-1) activity has been demonstrated in the Golgi-rich fraction from brain, kidney, testis, gastric mucosa, lung, endometrium, and cultured cells such as MDCK (Ishizuka 1977).

Human renal cell carcinoma tissue (Sakakibara et al. 1989) and a cell line (SMKT-R3) derived therefrom (Kobayashi et al. 1993d) show remarkably high CST activity. Furthermore, the sulfotransferase level in the cell line is significantly enhanced by the action of EGF (Kobayashi et al. 1993b). Therefore, CST was purified from EGF-treated SMKT-R3 cells by a combination of affinity chromatographies on heparin-Sepharose, galactosylsphingosine-Sepharose, and HiTrap-3′,5′-bisphosphoadenosine (Honke et al. 1996). The purified sulfotransferase showed a specific activity of 1.2 μmol/min/mg. Homogeneity of the purified sulfotransferase was supported by the facts that the enzyme preparation showed a single protein band with an apparent molecular mass of 54 kDa on reducing SDS-PAGE and that protein bands coincided with the enzyme activity on both native PAGE and nonreducing SDS-PAGE.

To obtain recombinant enzymes, human and mouse CST (Gal3ST-1) cDNAs were inserted into the expression vector pSVK3 and transfected into COS-1 cells (Honke et al. 1997; Hirahara et al. 2000). A GP3ST (Gal3ST-2) open reading frame fragment was also inserted into the pSVK3 vector and introduced into COS-1 cells (Honke et al. 2001). CHO cells were transfected with pcDNA3.1/HSH-Gal3ST-3, and the conditioned medium was used as an enzyme source (Suzuki et al. 2001). Triton extract of microsomal fractions from CHO cells that had been transfected with pCXNGalST was also utilized as a Gal3ST-3 source (El-Fasakhany et al. 2001). The crude membrane fraction of COS-7 cells transfected with pCMV-SPORT-Gal3ST-4 was used as a Gal3ST-4 source (Seko et al. 2001).

Biological Aspects

Sulfoglycolipids comprise a class of acidic glycolipids containing sulfate esters on their oligosaccharide chains. They have been implicated in a variety of physiological functions through their interactions with extracellular matrix proteins, cellular adhesion receptors, blood coagulation systems, complement activation systems, cation transport systems, and microorganisms (Vos et al. 1994; Ishizuka 1997). The distribution of sulfoglycolipids is tissue specific, and they are abundant in myelin sheath, spermatozoa, renal tubular cells, and epithelial cells of the gastrointestinal tract (Vos et al. 1994; Ishizuka 1997).

CST (Gal3ST-1) activity has been demonstrated in brain, testis, kidney, gastric mucosa, lung, and endometrium of various mammals (Ishizuka 1997) where sulfoglycolipids are expressed. Northern blot analysis and subquantitative RT-PCR analysis of mouse tissues showed that the *CST* gene is preferentially transcribed in stomach, small intestine, brain, kidney, lung, and testis, in that order (Hirahara et al. 2000). Moreover, tissue-specific expression of the *CST* gene is explained by alternative usage of multiple 5′-UTR exons flanked with tissue-specific promoters (Hirahara et al. 2000).

To elucidate the biological function of sulfoglycolipids, CST-null mice were generated by gene targeting (Honke et al., unpublished results). The CST-deficient mice manifested some neurological disorders due to myelin dysfunction and spermatogenesis arrest. These findings indicate that sulfation of glycolipids is essential for myelin function and spermatogenesis (Honke et al., unpublished results). So far, however, no human neuropathological diseases have been mapped in the vicinity of the *CST* locus.

Qualitative and quantitative alterations of cell-surface glycolipids occur during neoplastic transformation (Hakomori 1985). Sulfoglycolipids have been found to increase in many human cancer tissues. In particular, human renal cell carcinoma tissue (Sakakibara et al. 1989) and a cell line derived therefrom (SMKT-R3) (Kobayashi et al. 1993d) accumulate sulfoglycolipids secondary to marked elevation of CST activity. CST activity in SMKT-R3 cells is enhanced by the actions of growth factors such as EGF (Kobayashi et al. 1993b), transforming growth factor-α (TGFα) (Kobayashi et al. 1993a), and hepatocyte growth factor (HGF) (Kobayashi et al. 1994). On the other hand, tyrosine kinase inhibitors, protein kinase C inhibitors, and chronic treatment with 12-O-tetradecanoylphorbol-13-acetate reduce CST activity (Kobayashi et al. 1993c; Balbaa et al. 1996), and Ras is involved in the regulatory pathway of EGF-induced CST activity (Yabunaka et al. 1997) in SMKT-R3 cells. Furthermore, it has been shown that aberrant usage of transcription initiation sites flanked with promoters/enhancers (Tsuda et al. 2000) is involved in cancer-associated overexpression of the *CST* gene (Honke et al. 1998).

Carbohydrate structures with 3′-sulfo-β-Gal linkage have been found in both N-linked and O-linked glycoproteins. Feizi and Glaustian (1999) have demonstrated that chemically synthesized sulfo-3Galβ1-3(Fucα1-4)GlcNAc-R (3′-sulfo Le^a) and sulfo-3Galβ1-4(Fucα1-3)GlcNAc-R (3′-sulfo Le^x) oligosaccharides are more potent ligands for both L- and E-selectin than 3′-sialylated-Le^a and -Le^x determinants. Expression of the 3′-sulfo Le^a epitope decreases with increasing depth of invasion of human colon carcinomas (Yamachika et al. 1997), and human colon carcinoma cells expressing the 3′-sulfo Le^a epitope show lower tumorigenicity in nude mice (Vavasseur et al. 1994). On the other hand, 3′-sulfo-Le^a and/or -Le^x determinants are detected in cancer cells as well as in surrounding nonmalignant epithelia in human colon cancer tissues (Izawa et al. 2000), and the 3′-sulfo Le^x epitope is a major carbohydrate motif in a human colon carcinoma cell line with a high metastatic tendency (Capon et al. 1997). These findings imply that 3′-sulfated Lewis epitopes serve as relevant ligands for selectins *in vivo* and that their expression modulates tumor progression in human colon cancer. Now that the genes of the relevant Gal3STs are available, we can approach to these issues at the molecular level (Ikeda et al. 2001).

Future Perspectives

The biological functions and gene expression mechanisms of Gal3STs are the most important questions at the present time. Gene targeting may be useful to elucidate their physiological roles, as shown in CST. Their relationships to human diseases such as hereditary deficiency, cancer, inflammation, and infection are also important from the point of view of medicine. Investigation of these diseases will enable us to understand the biological functions of sulfated carbohydrates more fully and may provide a hint for curing these disastrous diseases.

The evolutionary aspects of the genes are interesting. Sulfatides have been reported only from Deuterostomia (Ishizuka 1997). Therefore, the expression and function of CST in echinoderms (the most primitive deuterostomia) must be clarifed. Studies on the expression and biological functions of the other Gal3ST family members in invertebrates may be required to understand the true meaning of sulfation. The evolutionary relations between the *CST* and *CGT* genes and among the Gal3ST family members are intriguing questions as well.

Further Reading

For a comprehensive review of sulfoglycolipids and their metabolism see Ishizuka (1997).

References

Balbaa M, Honke K, Makita A (1996) Regulation of glycolipid sulfotransferase by tyrosine kinases in human renal cancer cells. Biochim Biophys Acta 1299:141–145

Brockhausen I, William K (1997) Trends Glycosci Glycotechnol 9:379–398

Capon C, Wieruszeski JM, Lemoine J, Byrd JC, Leffler H, Kim YS (1997) Sulfated Lewis x determinants as a major structural motif in glycans from LS174T-HM7 human colon carcinoma mucin. J Biol Chem 272:31957–31968

El-Fasakhany FM, Uchimura K, Kannagi R, Muramatsu T (2001) A novel human Gal-3-O-sulfotransferase: molecular cloning, characterization and its implications in biosynthesis of 3′-sulfo Lewis^x. J Biol Chem 276:26988–26994

Farooqui AA, Rebel G, Mandel P (1977) Sulphatide metabolism in brain. Life Sci 20:569–583

Feizi, T, Glaustian C (1999) Novel oligosaccharide ligands and ligand-processing pathways for the selectins. Trends Biochem Sci 24:369–372

Hakomori S (1985) Aberrant glycosylation in cancer cell membranes as focused on glycolipids: overview and prospectives. Cancer Res 45:2405–2414

Hirahara Y, Tsuda M, Wada Y, Honke K (2000) cDNA cloning, genomic cloning, and tissue-specific regulation of mouse cerebroside sulfotransferase. Eur J Biochem 267: 1909–1916

Honke K, Tsuda M, Hirahara, Y, Ishii A, Makita A, Wada Y (1997) Molecular cloning and expression of cDNA encoding human 3′- phosphoadenylylsulfate:galactosylceramide 3′-sulfotransferase. J Biol Chem 272:4864–4868

Honke K, Tsuda M, Hirahara Y, Miyao N, Tsukamoto T, Satoh M, Wada Y (1998) Cancer-associated expression of glycolipid sulfotransferase gene in human renal cell carcinoma cells. Cancer Res 58:3800–3805

Honke K, Tsuda M, Koyota S, Wada Y, Tanaka N, Ishizuka I, Nakayama J, Taniguchi N (2001) Molecular cloning and characterization of a human β-Gal 3'-sulfotransferase which acts on both type 1 and type 2 (Galβ1,3/1,4GlcNAc-R) oligosaccharides. J Biol Chem 276:267–274

Honke K, Yamane M, Ishii A, Kobayashi T, Makita A (1996) Purification and characterization of 3'-phosphoadenosine-5'-phosphosulfate:GalCer sulfotransferase from human renal cancer cells. J Biochem 119:421–427

Ikeda N, Eguchi H, Nishihara S, Narimatsu H, Kannagi R, Irimura T, Ohta M, Matsuda H, Taniguchi N, Honke K (2001) A remodeling system of 3'-sulfo Lewis a and 3'-sulfo Lewis x epitopes. J Biol Chem (in press)

Ishizuka I (1997) Chemistry and functional distribution of sulfoglycolipids. Prog Lipid Res 36:245–319

Izawa M, Kumamoto K, Mitsuoka C, Kanamori A, Ohmori K, Ishida H, Nakamura S, Kurata-Miura K, Sasaki K, Nishi T, Kannagi R (2000) Expression of sialyl 6-sulfo Lewis x is inversely correlated with conventional sialyl Lewis x expression in human colorectal cancer. Cancer Res 60:1410–1416

Kawano M, Honke K, Tachi M, Gasa S, Makita A (1989) An assay method for ganglioside synthase using anion-exchange chromatography. Anal Biochem 182:9–15

Kobayashi T, Honke K, Gasa S, Imai S, Tanaka J, Miyazaki T, Makita A (1993a) Regulation of activity levels of glycolipid sulfotransferases by transforming growth factor α in renal cell carcinoma cells. Cancer Res. 53:5638–5642

Kobayashi T, Honke K, Gasa S, Kato N, Miyazaki T, Makita A (1993b) Epidermal growth factor elevates the activity levels of glycolipid sulfotransferases in renal-cell-carcinoma cells. Int J Cancer 55:448–452

Kobayashi T, Honke K, Gasa S, Sugiura M, Miyazaki T, Ishizuka I, Makita A (1993c) Involvement of protein kinase C in the regulation of glycolipid sulfotransferase activity levels in renal cell carcinoma cells. Cancer Res. 53:2484–2489

Kobayashi T, Honke K, Kamio K, Sakakibara N, Gasa S, Miyao N, Tsukamoto T, Ishizuka I, Miyazaki T, Makita A (1993d) Sulfolipids and glycolipid sulfotransferase activities in human renal cell carcinoma cells. Br J Cancer 67:76–80

Kobayashi T, Honke K, Gasa S, Miyazaki T, Tajima H, Matsumoto K, Nakamura T, Makita A (1994) Hepatocyte growth factor elevates the activity levels of glycolipid sulfotransferases in renal cell carcinoma cells. Eur J Biochem. 219:407–413

Lingwood CA (1985) Developmental regulation of galactoglycerolipid and galactosphingolipid sulphation during mammalian spermatogenesis. Biochem J 231:393–400

Sakakibara N, Gasa S, Kamio K, Makita A, Koyanagi T (1989) Association of elevated sulfatides and sulfotranasferase activities with human renal cell carcinoma. Cancer Res 49:335–339

Seko A, Hara-Kuge S, Yamashita K (2001) Molecular cloning and characterization of a novel human galactose 3-O-sulfotransferase that transfers sulfate to Galβ1-3GalNAc residue in O-glycans. J Biol Chem 276:25697–25704

Suzuki A, Hiraoka N, Suzuki M, Angata K, Misra AK, McAuliffe J, Hindsgaul O, Fukuda M (2001) Molecular cloning and expression of a novel human β-Gal-3-O-sulfotransferase that acts preferentially on N-acetyllactosamine in N- and O-glycans. J Biol Chem 276:24388–24395

Tennekoon G, Aitchison S, Zaruba M (1985) Purification and characterization of galactocerebroside sulfotransferase from rat kidney. Arch Biochem Biophys 240:932–944

Tsuda M, Egashira M, Niikawa N, Wada Y, Honke K (2000) Cancer-associated alternative usage of multiple promoters of human GalCer sulfotransferase gene. Eur J Biochem 267:2672–2679

Vavasseur F, Dole K, Yang J, Matta KL, Myerscough N, Cornfield A, Paraskeva C, Brock-hausen I (1994) O-Glycan biosynthesis in human colorectal adenoma cells during progression to cancer. Eur J Biochem 222:415–424

Vos JP, Lopes-Cardozo M, Gadella BM (1994) Metabolic and functional aspects of sulfogalactolipids. Biochim Biophys Acta 1211:125–149

Yabunaka N, Honke K, Ishii A, Ogiso Y, Kuzumaki N, Agishi Y, Makita A (1997) Involvement of Ras in the expression of glycolipid sulfotransferase in human renal cancer cells. Int J Cancer 71:620–623

Yamachika T, Nakanishi H, Inada K, Kitoh K, Kato T, Irimura T, Tatematsu M (1997) Reciprocal control of colon-specific sulfomucin and sialosyl-Tn antigen expression in human colorectal neoplasia. Virchows Arch 431:25–30

Nucleotide Sugar Transporters

UDP-Gal Transporter-1 and -2

Introduction

UDP-Gal transporter is a multiple-segment transmembrane protein of the Golgi apparatus that delivers UDP-Gal, synthesized in the cytosol, into the Golgi lumen to provide various Gal transferases with their substrate for the elongation of carbohydrate chains (Kawakita et al. 1998). Gal transferases, whose catalytic sites face the Golgi lumen, cannot add Gal residues to growing carbohydrate chains unless UDP-Gal is supplied through UDP-Gal transporter. Thus, mutant cultured cell lines defective in UDP-Gal transporter were shown to accumulate Glc-Cer instead of lactosylceramide (Taki et al. 1991), and truncated N-linked oligosaccharide chains terminated at GlcNAc (Hara et al. 1989).

The structure of UDP-Gal transporter, deduced from the nucleotide sequence of its cDNA, is similar to the structures of other nucleotide sugar transporters such as CMP-sialic acid transporter and UDP-GlcNAc transporter. The human UDP-Gal transporter occurs in two isoforms: hUGT1 and hUGT2 (Ishida et al. 1996; Miura et al. 1996). They are 393 and 396 amino acid residues long, respectively, and similar in structure in that only the C-terminal five amino acid residues in hUGT1 are replaced in hUGT2 by a different stretch of eight amino acid residues. The two isoforms are produced from a single gene, which has been mapped to band Xp11.22–p11.23 on the X chromosome (Hara et al. 1993) through alternative splicing (Ishida et al. 1996). Both hUGT1 and hUGT2 cDNAs were able to correct the UDP-Gal transporter-defective phenotype of mouse Had-1 cells (Yoshioka et al. 1997).

MASAO KAWAKITA* and NOBUHIRO ISHIDA

Department of Physiological Chemistry, Tokyo Metropolitan Institute of Medical Science, 3-18-22 Hon-Komagome, Bunkyo-ku, Tokyo 113-8613, Japan
Tel. +81-3-3823-2101 (ext. 5285); Fax +81-3-3823-2965
e-mail: kawakita@rinshoken.or.jp
* Present address: Department of Applied Chemistry, Kogakuin University, 1-24-2 Nishi-shinjuku, Shinjuku-ku, Tokyo 163-8677, Japan
Tel. +81-3-3340-2731; Fax +81-3-3340-0147
e-mail: bt13004@ns.kogakuin.ac.jp

Databanks

UDP-Gal transporter -1, and -2

No EC number has been allocated.

Species	Gene	Protein	mRNA	Genomic
UGT1				
Homo sapiens	*hUGT1*	P78381	D84454	AF207550
		JC4903	–	AB042425
Mus musculus	*mUGT1*	–	AB027147	–
			AF229634	
Schizosaccharomyces pombe	*gms1*	T41140	AB023425	D89616
		–	–	AL022598
UGT2				
Homo sapiens	*hUGT2*	JC5022	D88146	AB042425

Name and History

UDP-Gal transporter is also referred to as UDP-Gal translocator. The occurrence of an activity to transport UDP-Gal specifically from the cytosol to the Golgi lumen was first inferred during the course of studies on lactose synthesis in the microsomes from mammary glands of lactating rats (Kuhn and White 1976). UDP-Gal transporting activity of Golgi membranes was directly demonstrated several years later in membrane vesicles isolated from rat liver (Brandan and Fleischer 1982). Evidence indicating the presence of a transporter highly specific for UDP-Gal and the indispensability of the transporter in glycoconjugate biosynthesis was provided through analyses of mutant cell lines devoid of UDP-Gal transporter. Such cells were found to have pleiotropic defects in the addition of Gal residues to glycoproteins and glycolipids. Microsomal vesicles isolated from these cells were unable to transport UDP-Gal but were able to transport other nucleotide sugars (Deutscher and Hirschberg 1986; Yoshioka et al. 1997). Cloning of a cDNA and expression of the cDNA in mammalian and yeast cells finally confirmed the transporter's specificity for UDP-Gal and its localization in the Golgi membranes (Yoshioka et al. 1997; Sun-Wada et al. 1998).

Enzyme Activity Assay and Substrate Specificity

UDP-Gal is transported into Golgi vesicles in a one-to-one exchange for luminal UMP (Brandan and Fleischer 1982). The UDP-Gal transporter is specific for UDP-Gal. Thus, microsomal vesicles obtained from UDP-Gal transporter-deficient mutant cells (murine FM3A-derived Had-1 and CHO-derived Lec8) were specifically incapable of transporting UDP-Gal but were competent in UDP-GlcNAc and CMP-sialic acid transport (Deutscher and Hirschberg 1986; Yoshioka et al. 1997). Expression of either hUGT1 or hUGT2 in Had-1 cells restored the UDP-Gal transporting activity of microsomal vesicles (Yoshioka et al. 1997). The microsomal vesicles prepared from hUGT1- or hUGT2-expressing *Saccharomyces cerevisiae* cells exhibited UDP-Gal but not CMP-sialic acid transporting activity (Sun-Wada et al. 1998). The apparent K_m values for UDP-Gal (1.2 and 2.0 µM with hUGT1 and hUGT2, respectively) (Sun-Wada et al. 1998) were similar to the value (2.4 µM) obtained with the UDP-Gal transport system of rat liver Golgi membranes (Milla et al. 1992). UDP-Gal transport into rat liver Golgi vesicles was more than 90% inhibited by 200 µM 5'-iodo-dUMP (Milla et al. 1992). UMP (20 µM) inhibited the hUGT expressed in *S. cerevisiae* and Had-1 cells by 50% and 70%–80%, respectively (Sun-Wada et al. 1998; Yoshioka, unpublished observations). The mechanism of the UDP-Gal transport reaction is similar to the mechanisms of other nucleotide-sugar transporters. Further information concerning the effects of substrate analogs and other inhibitors may be obtained by referring to the results obtained with other nucleotide sugar transporters (summarized in Hirschberg and Snider 1987). It should be kept in mind, however, that, for instance, 4,4'-diisothiocyano-2,2'-disulfonic acid stilbene (DIDS), which significantly inhibits some nucleotide sugar transporters at 20 µM, only weakly inhibited UDP-Gal transporter (Milla et al. 1992).

Direct measurement of the transport activity is carried out in three steps: (1) incubation of membrane vesicles with radiolabeled UDP-Gal; (2) separation and recovery of the vesicles from the incubation mixture; and (3) determination of the radioactivity trapped in the vesicles, where step (2) may be carried out in any of three ways. During the filtration procedure (Sun-Wada et al. 1998) the reaction is stopped by 10-fold dilution with an ice-cold solution containing nonradioactive UDP-Gal, and the vesicles are recovered onto a nitrocellulose membrane filter. The procedure is rapid, simple, and accurate. During the centrifugation procedure (Perez and Hirschberg 1987) the reaction is stopped by threefold dilution with an ice-cold buffer, and the reaction mixture is centrifuged at 100000 g for 25 min at 4°C to recover membrane vesicles in a pellet. The amount of UDP-Gal remaining free in the lumen of the vesicles and the amount of Gal moiety that has been transferred to macromolecules can be separately estimated by this procedure by determining the acid-soluble and acid-insoluble radioactivity, respectively, recovered in the pellet of membrane vesicles. Finally, during the chromatographic separation procedure the reaction mixture after incubation with UDP-Gal is applied to a column of Sephadex G-50 or Dowex-1, and the amount of UDP-Gal trapped in the membrane vesicles, which is eluted earlier than free UDP-Gal, is determined (Milla and Hirschberg 1989; Mayinger and Meyer 1993).

Preparation

Most earlier studies were carried out using Golgi-enriched microsomal membrane vesicles from rat liver (Brandan and Fleischer 1982) and rat mammary gland (Kuhn and White 1976). The development of cDNA expression systems, as described below, made it possible to study UDP-Gal transport using the microsomal membrane vesicles isolated from UDP-Gal transporter cDNA-transformants of cultured mammalian cells and the yeast *S. cerevisiae* (Yoshioka et al. 1997; Sun-Wada et al. 1998).

Expression systems for hUGT1 and hUGT2 cDNAs in UDP-Gal transporter-deficient mutant cells, such as Had-1 and Lec8, were developed using an expression vector, pMKIT-neo, a derivative of the SRα vector (Miura et al. 1996). Transformants that express UDP-Gal transporter are selected based on their lectin sensitivity. UDP-

Gal transporter-deficient cells that expose GlcNAc residues at the termini of their *N*-linked oligosaccharides are sensitive to the cytotoxic effects of *Griffonia simplicifolia* lectin GS-II, whereas the transformants in which galactosylation of oligosaccharide chains takes place normally are rendered GS-II-resistant and are able to grow in the presence of the lectin. hUGT1 and hUGT2 cDNAs were also expressed in *S. cerevisiae* (Sun-Wada et al. 1998). DNA fragments containing the coding regions of hUGT1 or hUGT2 were inserted into the expression vector pKTΔATG, a modified version of pKT10, under control of the GAPDH promoter. The expression plasmids were transfected into *S. cerevisiae* strain YPH501, and transformants were selected for growth in uracil-free medium. UDP-Gal transporter of *Schizosaccharomyces pombe* was also expressed heterologously in Lec8 cells, and it complemented the genetic defect of the mutant cells (Segawa et al. 1999).

Solubilization of the transporter from the Golgi membranes and reconstitution into proteoliposomes were reported, but a purified preparation of the transporter has not yet been obtained (Milla et al. 1992).

Biological Aspects

Gal transferases, whose catalytic sites face the Golgi lumen, require UDP-Gal, which is supplied through UDP-Gal transporter, for the addition of Gal residues to the growing ends of various glycoconjugates. UDP-Gal transporter-deficient cells therefore show a pleiotropic aberration in the structure of both cell surface glycoproteins and glycolipids. A drastic reduction in galactosylation as a result of a severe shortage of the substrate leads to a decrease in the amounts of sialylated glycoconjugates and lactosylceramide, with a concomitant increase in the amount of GlcNAc-terminated sugar chains and Glc-Cer (Hara et al. 1989; Taki et al. 1991). The galactosylation at different linkages is not uniformly affected by the defect in UDP-Gal transporter. Analysis of *N*-linked oligosaccharides of Had-1 and parental FM3A cells revealed that Galβ1-4GlcNAc linkages amounting to 10%–15% of those in the parental cells persisted in the mutant cells, whereas Galα1-3Gal linkages were totally absent in the mutant (Fig. 1) (Hara et al. 1989). The incorporation of galactose residues into the link tetrasaccharide of proteoglycans seems to be surprisingly resistant to the deficiency of UDP-Gal, as the amounts of chondroitin sulfate and heparan sulfate were not reduced significantly in UDP-Gal transporter-deficient cells derived from MDCK cells (Toma et al. 1996). Keratan sulfate, which contains Gal residues in its glycosaminoglycan moiety, is sensitive to UDP-Gal deficiency (Toma et al. 1996). UDP-Gal transporter-deficient mutant cell lines derived from cancer cell lines exhibit decreased tumorigenicity or metastatic potential in vivo (Dennis et al. 1984; Taki et al. 1991). One such cell line, murine Had-1, is not tumorigenic in C3H/He mice, in contrast to parental FM3A carcinoma cells. Interestingly, prior inoculation of Had-1 cells into mice suppressed the tumorigenicity of FM3A cells inoculated later into the same mice (Taki et al. 1991).

Immunofluorescence microscopy using antibodies against a C-terminal peptide of hUGT1 detected the hUGT1 protein in the Golgi region of hUGT1 transformants of Had-1 cells (Yoshioka et al. 1997). Endogenous hUGT1 protein present in the Golgi apparatus of HeLa cells is also detected with this antibody (Yoshioka et al. 1997).

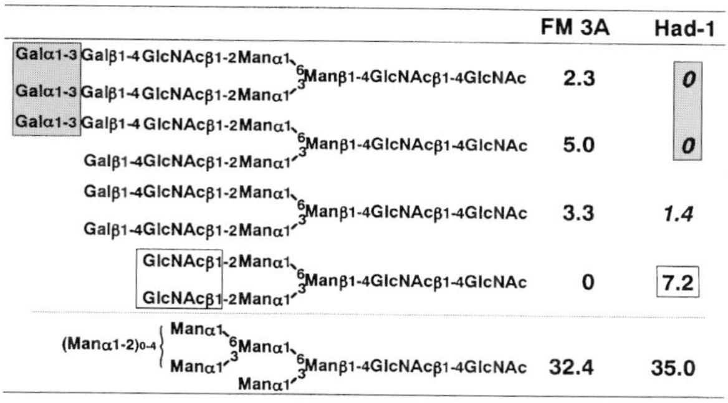

	FM 3A	Had-1
Galα1-3Galβ1-4GlcNAcβ1-2Manα1↘ 　　　　　　　　　　　　　　　６Manβ1-4GlcNAcβ1-4GlcNAc Galα1-3Galβ1-4GlcNAcβ1-2Manα1↗³	2.3	0
Galα1-3Galβ1-4GlcNAcβ1-2Manα1↘ 　　　　　　　　　　　　　　　６Manβ1-4GlcNAcβ1-4GlcNAc Galβ1-4GlcNAcβ1-2Manα1↗³	5.0	0
Galβ1-4GlcNAcβ1-2Manα1↘ 　　　　　　　　　　　　６Manβ1-4GlcNAcβ1-4GlcNAc Galβ1-4GlcNAcβ1-2Manα1↗³	3.3	1.4
GlcNAcβ1-2Manα1↘ 　　　　　　　６Manβ1-4GlcNAcβ1-4GlcNAc GlcNAcβ1-2Manα1↗³	0	7.2
Manα1↘ (Manα1-2)₀₋₄ {　　　　　６Manα1↘ 　　　　　 Manα1↗³　　　６Manβ1-4GlcNAcβ1-4GlcNAc 　　　　　　 Manα1↗³	32.4	35.0

Fig. 1. Asparagine-linked oligosaccharides of murine FM3A and Had-1 cells. Oligosaccharides released by hydrazinolysis and treated with sialidase were analyzed. The results with biantennary complex-type and high mannose-type oligosaccharides are shown. The values represent percent molar ratio relative to the total amount of N-linked oligosaccharide chains. (Modified from Hara et al. 1989)

Northern blot analysis revealed that both hUGT1 and hUGT2 mRNAs have been expressed ubiquitously in every human tissue so far examined (Ishida, unpublished observations). Two species of transcripts (2.6 and 1.7 kb), which can be detected using an open reading frame (ORF) portion of the mUGT cDNA, whose expression corrected the genetic defects of Lec 8 cells, as a probe, were expressed ubiquitously in mouse tissues. They may correspond to the two isozymes identified in humans. The 2.6 and 1.7 kb species were expressed from an early stage of development, in 7- to 17-day embryos (Ishida et al. 1999).

Future Perspectives

The significance of the occurrence of the hUGT1 and hUGT2 isoforms remains obscure. They are expressed ubiquitously in the body, as noted above; but the half-lives of hUGT1 and hUGT2 mRNAs have not been compared. It has been noted that the hUGT1 mRNA contains three mRNA-destabilizing AUUUA sequences in its 3'-UTR (Miura et al. 1996).

Controlling the UDP-Gal transport activity may have practical implications. The Galα1-3Gal linkage, a constituent of the so-called α-galactosyl epitope that has recently been a subject of considerable discussion in relation to xenotransplantation and retroviral-mediated gene therapy, is highly sensitive to a reduction in the UDP-Gal concentration in mouse cells (Hara et al. 1989). This implies that effective control over UDP-Gal transport activity may alter expression of the epitope by limiting the availability of the substrate for its synthesis.

Recently, an α1,3-mannosyltransferase-deficient Δmnn1 mutant of S. cerevisiae was transfected with hUGT2 and S. pombe α1,2-galactosyltransferase cDNAs, and it was demonstrated that the transformed mutant yeast cells expressed N- and O-linked oligosaccharides on their cell surface. Expression of UDP-Gal transporter was indispensable for the efficient occurrence of galactosylation (Kainuma et al. 1999). This may represent a significant step toward the production of human-type oligosaccharides in yeast.

Further Reading

For further information regarding nucleotide sugar transporters, see Kawakita et al. (1998).

References

Brandan E, Fleischer B (1982) Orientation and role of nucleoside diphosphatase and 5'-nucleotidase in Golgi vesicles from rat liver. Biochemistry 21:4640–4645

Dennis JW, Carver JP, Schachter H (1984) Asparagine-linked oligosaccharides in murine tumor cells: comparison of a WGA-resistant (WGA^r) nonmetastatic mutant and a related WGA-sensitive (WGA^s) metastatic line. J Cell Biol 99:1034–1044

Deutscher SL, Hirschberg CB (1986) Mechanism of galactosylation in the Golgi apparatus: a Chinese hamster ovary cell mutant deficient in translocation of UDP-galactose across Golgi membranes. J Biol Chem 261:96–100

Hara T, Endo T, Furukawa K, Kawakita M, Kobata A (1989) Elucidation of the phenotypic change on the surface of Had-1 cell, a mutant cell line of mouse FM3A carcinoma cells selected by resistance to Newcastle disease virus infection. J Biochem 106:236–247

Hara T, Yamauchi M, Takahashi E, Hoshino M, Aoki K, Ayusawa D, Kawakita M (1993) The UDP-galactose translocator gene is mapped to band Xp11.23–p11.22 containing the Wiskott-Aldrich syndrome locus. Somat Cell Mol Genet 19:571–575

Hirschberg CB, Snider MD (1987) Topography of glycosylation in the rough endoplasmic reticulum and Golgi apparatus. Annu Rev Biochem 56:63–87

Ishida N, Miura N, Yoshioka S, Kawakita M (1996) Molecular cloning and characterization of a novel isoform of the human UDP-galactose transporter, and of related complementary DNAs belonging to the nucleotide-sugar transporter gene family. J Biochem 120:1074–1078

Ishida N, Yoshioka S, Iida M, Sudo K, Miura N, Aoki K, Kawakita M (1999) Indispensability of transmembrane domains of Golgi UDP-galactose transporter as revealed by analysis of genetic defects in UDP-galactose transporter-deficient murine Had-1 mutant cell lines and construction of deletion mutants. J Biochem 126:1107–1117

Kainuma M, Ishida N, Yoko-o T, Yoshioka S, Takeuchi M, Kawakita M, Jigami Y (1999) Coexpression of α1,2 galactosyltransferase and UDP-galactose transporter efficiently galactosylates N- and O-glycans in Saccharomyces cerevisiae. Glycobiology 9:133–141

Kawakita M, Ishida N, Miura N, Sun-Wada G-H, Yoshioka S (1998) Nucleotide sugar transporters: elucidation of their molecular identity and its implication for future studies. J Biochem 123:777–785

Kuhn NJ, White A (1976) Evidence for specific transport of uridine diphosphate galactose across the Golgi membrane of rat mammary gland. Biochem J 154:243–244

Mayinger P, Meyer DI (1993) An ATP transporter is required for protein translocation into the yeast endoplasmic reticulum. EMBO J 12:659–666

Milla ME, Hirschberg CB (1989) Reconstitution of Golgi vesicle CMP-sialic acid and adenosine 3'-phosphate 5'-phosphosulfate transport into proteoliposomes. Proc Natl Acad Sci USA 86:1786–1790

Milla ME, Clairmont CA, Hirschberg CB (1992) Reconstitution into proteoliposomes and partial purification of the Golgi apparatus membrane UDP-galactose, UDP-xylose, and UDP-glucuronic acid transport activities. J Biol Chem 267:103–107

Miura N, Ishida N, Hoshino M, Yamauchi M, Hara T, Ayusawa D, Kawakita M (1996) Human UDP-galactose translocator: molecular cloning of a complementary DNA that complements the genetic defect of a mutant cell line deficient in UDP-galactose translocator. J Biochem 120:236–241

Perez M, Hirschberg CB (1987) Transport of sugar nucleotides into the lumen of vesicles derived from rat liver rough endoplasmic reticulum and Golgi apparatus. Methods Enzymol 138:709–715

Segawa H, Ishida N, Takegawa K, Kawakita M (1999) *Schizosaccharomyces pombe* UDP-galactose transporter: identification of its functional form through cDNA cloning and expression in mammalian cells. FEBS Lett 451:295–298

Sun-Wada G-H, Yoshioka S, Ishida N, Kawakita M (1998) Functional expression of the human UDP-galactose transporters in the yeast *Saccharomyces cerevisiae*. J Biochem 123:912–917

Taki T, Ogura K, Rokukawa C, Hara T, Kawakita M, Endo T, Kobata A, Handa S (1991) Had-1, a uridine 5′-diphosphogalactose transport-defective mutant of mouse mammary tumor cell FM3A: composition of glycolipids, cell growth inhibition by lactosylceramide, and loss of tumorigenicity. Cancer Res 51:1701–1707

Toma L, Pinhal MAS, Dietrich CP, Nader HB, Hirschberg CB (1996) Transport of UDP-galactose into the Golgi lumen regulates the biosynthesis of proteoglycans. J Biol Chem 271:3897–3901

Yoshioka S, Wada G-H, Ishida N, Kawakita M (1997) Expression of the human UDP-galactose transporter in the Golgi membranes of murine Had-1 cells that lack the endogenous transporter. J Biochem 122:691–695

75

UDP-GlcNAc Transporter

Introduction

The Golgi and rough endoplasmic reticulum (ER) membranes have an activity that transports UDP-GlcNAc from the cytosol into their lumens (Perez and Hirschberg 1985; Bossuyt and Blanckaert 1994). The nucleotide sugar transported to the Golgi lumen serves as a substrate of GlcNAc transferases for the elongation of carbohydrate chains of glycoproteins, glycolipids, and proteoglycans. On the other hand, the reactions in which UDP-GlcNAc is utilized in the ER have not yet been definitively identified.

The UDP-GlcNAc transporter genes of the yeasts *Klyuveromyces lactis* (Abeijon et al. 1996b) and *Saccharomyces cerevisiae* (Roy et al. 2000) have been identified, as has the cDNA coding for dog (Guillen et al. 1998) and human (Ishida et al. 1999) UDP-GlcNAc transporter. Human, canine, *K. lactis*, and *S. cerevisiae* UDP-GlcNAc transporters are multiple-segment transmembrane proteins of 325, 326, 328, and 342 amino acid residues, respectively. The amino acid sequences of human and canine UDP-GlcNAc transporters are similar, being 98% identical, whereas canine UDP-GlcNAc transporter has only 22% amino acid sequence identity with its *K. lactis* counterpart. Interestingly, the structures of mammalian UDP-GlcNAc transporters much more closely resemble those of mammalian UDP-Gal and CMP-sialic acid transporters, which are designed for transporting substances with distinct structures, than that of *K. lactis* UDP-GlcNAc transporter.

UDP-GlcNAc transporter-encoding DNAs (*K. lactis* gene and canine cDNA) complement the genetic defect of UDP-GlcNAc transporter-deficient *K. lactis* mutant mnn2-2 (Abeijon et al. 1996b; Guillen et al. 1998). Canine and human UDP-GlcNAc

Masao Kawakita* and Nobuhiro Ishida

Department of Physiological Chemistry, Tokyo Metropolitan Institute of Medical Science, 3-18-22 Hon-Komagome, Bunkyo-ku, Tokyo 113-8613, Japan
Tel. +81-3-3823-2101 (ext. 5285); Fax +81-3-3823-2965
e-mail: kawakita@rinshoken.or.jp
* Present address: Department of Applied Chemistry, Kogakuin University, 1-24-2 Nishi-shinjuku, Shinjuku-ku, Tokyo 163-8677, Japan
Tel. +81-3-3340-2731; Fax +81-3-3340-0147
e-mail: bt13004@ns.kogakuin.ac.jp

transporters were expressed in yeast cells, and their transport activity was confirmed directly using microsomal membrane vesicles prepared from the transporter cDNA transformants *of K. lactis* mnn2-2 or *S. cerevisiae* (Guillen et al. 1998; Ishida et al. 1999).

Databanks

UDP-GlcNAc transporter

No EC number has been allocated.

Species	Gene	Protein	mRNA	Genomic
Homo sapiens	hUGTrel2	–	AB021981	–
Canis familiaris		O77592	AF057365	–
Kluyveromyces lactis	MNN2-2	Q00974	–	AF106080
Saccharomyces cerevisiae	YEL004W/YEA4	P40004	–	U18530

Name and History

UDP-GlcNAc transporter is also called UDP-GlcNAc translocator. UDP-GlcNAc synthesized in the cytosol is transported across the Golgi and ER membranes into their luminal spaces. This was demonstrated directly using membrane vesicles prepared from rat liver (Perez and Hirschberg 1985). Although the significance of the transporter in the Golgi membrane in providing the substrate for glycoconjugate synthesis is clear, the physiological role of the one in the ER membranes remains obscure.

Mammalian cells defective in UDP-GlcNAc transporter have not been obtained, but a yeast mutant deficient in the transporter *K. lactis* mnn2-2 was identified (Abeijon et al. 1996a). This led to isolation of the *K. lactis* UDP-GlcNAc transporter gene based on complementation of the genetic defect of the mutant (Abeijon et al. 1996b). Canine and human UDP-GlcNAc transporter cDNAs were subsequently identified by heterologous expression of these transporters in *K. lactis* and other yeast cells that do not have endogenous UDP-GlcNAc transporter (Guillen et al. 1998; Ishida et al. 1999). A heterologous expression system in *S. cerevisiae* also allowed us to demonstrate that the transporter is specific for UDP-GlcNAc (Ishida et al. 1999). Subsequent expression of human UDP-GlcNAc transporter in CHO cells revealed its localization in the Golgi membranes (Ishida et al. 1999).

Enzyme Activity Assay and Substrate Specificity

UDP-GlcNAc transporter is an antiporter similar to other nucleotide sugar transporters. It transports UDP-GlcNAc across the Golgi and ER membranes in a one-to-one exchange for UMP (Waldman and Rudnick 1990). Initial overshoot of UDP-GlcNAc transport was noted using rat liver microsomal membrane vesicles. This overshoot phenomenon reflects *trans*-stimulation by the countersubstrate UMP. The amount of UDP-GlcNAc taken up in the vesicles at equilibrium was linearly related to the initial solute concentration in the extravesicular medium and to the volume of the intravesicular space (Bossuyt and Blanckaert 1994). The transporter does not need

ATP hydrolysis or a pH gradient as the driving force of the reaction, but the reaction is sensitive to acidification of UMP-containing compartments. Thus, the influx of UDP-GlcNAc into UMP-preloaded membrane vesicles was diminished when the luminal pH was lowered from 7.50 to 5.45, and UDP-GlcNAc efflux in exchange for extravesicular UMP was suppressed when the extravesicular pH was lowered from 7.50 to 5.65 (Waldman and Rudnick 1990). Lowering the pH alters primarily the relative concentration of UMP^{2-} and UMP^-, and UMP^- becomes the predominant species at pH 5.45 and 5.65. This strongly suggests that the antiport reaction is an electroneutral exchange of UDP-GlcNAc with UMP^{2-}.

The human UDP-GlcNAc transporter cDNA product specifically transported UDP-GlcNAc. The microsomal membrane vesicles prepared from human UDP-GlcNAc transporter-expressing *S. cerevisiae* cells exhibited UDP-GlcNAc but not UDP-Gal or CMP-sialic acid transporting activity (Ishida et al. 1999), substantiating the strict substrate specificity of this transporter. The apparent K_m values for UDP-GlcNAc of the Golgi and the rough ER UDP-GlcNAc transport systems of rat liver were reported to range from 4.5 to 8.0 µM and from 2.5 to 3.0 µM, respectively (Perez and Hirschberg 1985; Bossuyt and Blanckaert 1994). Golgi membrane vesicles prepared from wild-type *K. lactis* cells transported UDP-GlcNAc with an apparent K_m of 5.5 µM (Abeijon et al. 1996a).

3′-Azido-3′-deoxythymidine monophosphate (AZTMP) inhibited the synthesis of glycoconjugates by inhibiting UDP-GlcNAc and other pyrimidine nucleotide-sugar transport into the Golgi-enriched microsomal membranes (Hall et al. 1994). The inhibition was competitive with respect to UDP-GlcNAc (K_i 4 µM). Preincubation with UDP-2′,3′-dialdehyde inactivated the UDP-GlcNAc transporting activity without affecting the GlcNAc transferase activity in a permeabilized thymocyte preparation (Cecchelli et al. 1985). The mechanism of the UDP-GlcNAc transport reaction is similar to the mechanisims of other nucleotide-sugar transporters. Further information concerning the effects of substrate analogs and inhibitors may therefore be obtained by referring to the results obtained with other nucleotide sugar transporters (summarized in Hirschberg and Snider 1987).

Direct measurement of the transport activity is carried out in three steps: (1) incubation of membrane vesicles with radiolabeled UDP-GlcNAc; (2) separation and recovery of the vesicles from the incubation mixture; and (3) determination of the radioactivity trapped in the vesicles, where step (2) may be carried out in any of three ways: filtration (Waldman and Rudnick 1990; Sun-Wada et al. 1998), centrifugation (Perez and Hirschberg 1987), or chromatographic separation (Milla and Hirschberg 1989; Mayinger and Meyer 1993). These procedures are outlined in Chapter 74.

Preparation

Most earlier studies were carried out using Golgi- or rough ER-enriched microsomal membrane vesicles from rat liver (Perez and Hirschberg 1985; Waldman and Rudnick 1990; Bossuyt and Blanckaert 1994). Thymocytes whose plasma membranes were rendered permeable to nucleotide sugars by isotonic ammonium choloride treatment were also used in early studies (Cecchelli et al. 1985). Development of heterologous cDNA expression systems (described below) made it possible to study mammalian UDP-GlcNAc transport using membrane vesicles isolated from UDP-GlcNAc transporter cDNA-transformed yeast cells (Guillen et al. 1998; Ishida et al. 1999). In a typical example, a DNA fragment containing the coding region of human UDP-GlcNAc transporter with three repeated hemagglutinin (HA)-epitope sequences attached to its C-terminal portion was inserted into the expression vector pYEX-BX under the control of a copper-inducible promoter. The expression plasmid was transfected into S. cerevisiae strain YPH501, and transformants were selected for growth in uracil-free medium (Ishida et al. 1999). No purification to any significant extent of solubilized transporter molecules in a reconstitutable active form has yet been reported.

Biological Aspects

As the catalytic sites of GlcNAc transferases face the Golgi lumen, UDP-GlcNAc, which is synthesized in the cytosol, must be transported across the Golgi membranes by UDP-GlcNAc transporter before its GlcNAc moiety can be added to the growing ends of various glycoconjugates. A UDP-GlcNAc transporter-deficient mutant yeast strain, K. lactis mnn2-2, thus lacked terminal GlcNAc residues in its mannoprotein, though the mutant had normal levels of GlcNAc transferase activity (Abeijon et al. 1996a). A UDP-GlcNAc transporter gene-deleted S. cerevisiae mutant contained less chitin than the wild-type yeast. The effects of UDP-GlcNAc transporter deficiency on the structure of glycoconjugates in mammalian cells have not yet been determined, as mutant mammalian cells defective in UDP-GlcNAc transporter are not available.

Human UDP-GlcNAc transporter protein and its C-terminally HA-tagged version were detected by immunofluorescence microscopy using anti-C-terminal peptide polyclonal antibody and anti-HA antibody, respectively. Both proteins were located in the Golgi region, but not in the ER region, of a transporter cDNA-transformant of CHO cells (Fig. 1) (Ishida et al. 1999). This is not consistent with previous biochemical data indicating that the Golgi and rough ER membranes are capable of transporting UDP-GlcNAc (Perez and Hirschberg 1985; Bossuyt and Blanckaert 1994). The discrepancy between the immunocytochemical and biochemical observations cannot be reconciled at present.

Northern blot analysis revealed two human UDP-GlcNAc transporter mRNA species, 2.4 and 6.2 kb in length, suggesting that there are at least two isoforms for

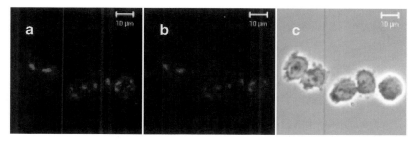

Fig. 1. Subcellular localization of hemagglutinin epitope (HA)-tagged human UDP-GlcNAc transporter. CHO-K1 cells expressing human UDP-GlcNAc transporter protein were fixed and permeabilized with cold methanol. Analysis by indirect immunofluorescence was performed using anti-HA rat monoclonal antibody 3F10 (a) and rabbit anti-α-mannosidase II antiserum (b) together with appropriate secondary antibodies. (c) Phase-contrast view. Bars 10 μm

human UDP-GlcNAc transporter. The transporter cDNA described above likely corresponds to the 2.4-kb species. The 6.2-kb species has not been characterized any further. Both mRNA species were expressed ubiquitously in every tissue examined (Ishida et al. 1999).

The genes for UDP-GlcNAc transporter and GlcNAc transferase are contiguous in the *K. lactis* genome (Guillen et al. 1999). The human UDP-GlcNAc transporter gene has been mapped to band p21 on chromosome 1 (Ishida et al. 1999).

Future Perspectives

Many basic issues remain to be settled concerning the identity of UDP-GlcNAc transporter in mammalian cells. The human UDP-GlcNAc transporter cDNA product was shown to be localized exclusively in the Golgi region, as described above. This raises the question of whether UDP-GlcNAc transporters in the Golgi and rough ER membranes represent identical or distinct molecular entities, as a UDP-GlcNAc transport system is certainly operating in the ER membranes. So far, only one UDP-GlcNAc transporter from each species has been identified, but the possibility remains that yet-unidentified molecular species of UDP-GlcNAc transporter has thus far escaped detection, and that these species are responsible for the UDP-GlcNAc transporting activity of the rough ER membranes. A search for additional proteins capable of transporting UDP-GlcNAc would be an interesting task in this respect. On the other hand, it is also possible that the already-isolated UDP-GlcNAc transporter cDNA of a given species represents the sole UDP-GlcNAc transporter of that species. If this is the case, identical protein products of these cDNAs must be distributed in appropriate proportions to both the Golgi and ER membranes. It is therefore necessary to answer the questions of how a single protein molecule can be sorted to different target organelles and how such sorting processes are regulated. When these issues are settled, we will have a clearer understanding of the mechanisms underlying the regulation of glycoconjugate synthesis at the level of GlcNAc addition.

Further Reading

For further information regarding nucleotide sugar transporters, see Kawakita et al. (1998).

References

Abeijon C, Mandon EC, Robbins PW, Hirschberg CB (1996a) A mutant yeast deficient in Golgi transport of uridine diphosphate *N*-acetylglucosamine. J Biol Chem 271: 8851–8854

Abeijon C, Robbins PW, Hirschberg CB (1996b) Molecular cloning of the Golgi apparatus uridine diphosphate-*N*-acetylglucosamine transporter from *Kluyveromyces lactis*. Proc Natl Acad Sci USA 93:5963–5968

Bossuyt X, Blanckaert N (1994) Functional characterization of carrier-mediated transport of uridine diphosphate *N*-acetylglucosamine across the endoplasmic reticulum membrane. Eur J Biochem 223:981–988

Cecchelli R, Cacan R, Verbert A (1985) Accumulation of UDP-GlcNAc into intracellular vesicles and occurrence of a carrier-mediated transport: study with plasma-membrane-permeabilized mouse thymocytes. Eur J Biochem 153:111–116

Guillen E, Abeijon C, Hirschberg CB (1998) Mammalian Golgi apparatus UDP-*N*-acetylglucosamine transporter: molecular cloning by phenotypic correction of a yeast mutant. Proc Natl Acad Sci USA 95:7888–7892

Guillen E, Abeijon C, Hirschberg CB (1999) The genes for the Golgi apparatus *N*-acetylglucosaminyltransferase and the UDP-*N*-acetylglucosamine transporter are contiguous in *Kluyveromyces lactis*. J Biol Chem 274:6641–6646

Hall ET, Yan J-P, Melançon P, Kuchta RD (1994) 3′-Azido-3′-deoxy thymidine potently inhibits protein glycosylation. J Biol Chem 269:14355–14358

Hirschberg CB, Snider MD (1987) Topography of glycosylation in the rough endoplasmic reticulum and Golgi apparatus. Annu Rev Biochem 56:63–87

Ishida N, Yoshioka S, Chiba Y, Takeuchi M, Kawakita M (1999) Molecular cloning and functional expression of the human Golgi UDP-*N*-acetylglucosamine transporter. J Biochem 126:68–77

Kawakita M, Ishida N, Miura N, Sun-Wada G-H, Yoshioka S (1998) Nucleotide sugar transporters: elucidation of their molecular identity and its implication for future studies. J Biochem 123:777–785

Mayinger P, Meyer DI (1993) An ATP transporter is required for protein translocation into the yeast endoplasmic reticulum. EMBO J 12:659–666

Milla ME, Hirschberg CB (1989) Reconstitution of Golgi vesicle CMP-sialic acid and adenosine 3′-phosphate 5′-phosphosulfate transport into proteoliposomes. Proc Natl Acad Sci USA 86:1786–1790

Perez M, Hirschberg CB (1985) Translocation of UDP-*N*-acetylglucosamine into vesicles derived from rat liver rough endoplasmic reticulum and Golgi apparatus. J Biol Chem 260:4671–4678

Perez M, Hirschberg CB (1987) Transport of sugar nucleotides into the lumen of vesicles derived from rat liver rough endoplasmic reticulum and Golgi apparatus. Methods Enzymol 138:709–715

Roy SK, Chiba Y, Takeuchi M, Jigami Y (2000) Characterization of yeast Yea4p, a uridine diphosphate-*N*-acetylglucosamine transporter localized in the endoplasmic reticulum and required for chitin synthesis. J Biol Chem 275:13580–13587

Sun-Wada G-H, Yoshioka S, Ishida N, Kawakita M (1998) Functional expression of the human UDP-galactose transporters in the yeast *Saccharomyces cerevisiae*. J Biochem 123:912–917

Waldman BC, Rudnick G (1990) UDP-GlcNAc tansport across the Golgi membrane: electroneutral exchange for dianionic UMP. Biochemistry 29:44–52

76

CMP-Sialic Acid Transporter

Introduction

The compartmentalization of cytidine 5′-monophosphate (CMP)-sialic acid synthesis in the cell nucleus (for review Kean 1991; Münster et al. 1998) and its metabolism in the lumen of the Golgi apparatus demonstrates that a system that transports the nucleotide sugar across the Golgi membrane is required. This function is accomplished by the CMP-sialic acid transporter (CST), a highly hydrophobic type III membrane protein (Hirschberg and Snider 1987; Hirschberg et al. 1998). The membrane topology has been identified for murine CST (Eckhardt et al. 1999), but so far no information is available on the tertiary and potential quaternary structure of the protein. The CST provides a key element in the cellular sialylation pathway, and defects in the CST gene lead to drastically reduced levels (Stanley and Siminovitch 1976; Briles et al. 1977) or complete loss (Eckhardt et al. 1996) of sialylated glycoconjugates. The lack of clinical manifestations caused by defects in the CST gene suggests that mutations in this important structure are lethal.

Databanks

CMP-sialic acid transporter

NO EC number has been allocated.

Species	Gene	Protein	mRNA	Genomic
Homo sapiens	*SLC35A1*	P78382	D87969	AL049697.9
		JC5023	–	–
Mus musculus	*SLC35A1*	Q61420	Z71268	–
Cricetulus griseus	*SLC35A1*	O08520	Y12074	–

RITA GERARDY-SCHAHN

Institut für Physiologische Chemie, Proteinstruktur Medizinische Hochschule Hannover, Carl-Neuberg-Strasse 1, 30625 Hannover, Germany
Tel. +49-511-532-9801; Fax +49-511-532-3956
e-mail: rgs1@gmx.de

Name and History

The names given to nucleotide sugar transporters reflect their substrate specificity. The systematic name cytidine 5′-monophosphate-sialic acid transporter is mostly abbreviated to CMP-sialic acid transporter to CMP-Neu5Ac transporter because *N*-acetylneuraminic acid (Neu5Ac) is the most abundant sialic acid derivative in the animal kingdom. In this review CST is used as an abbreviation for the transport protein. First evidence for the existence of the CMP-sialic acid transporter was provided in 1981 by the demonstration that Golgi vesicles, after incubation in CMP-[^{14}C]sialic acid-containing buffers, incorporated radioactivity into luminal structures (Hirschberg and Snider 1987; Hirschberg et al. 1998). In a consecutive series of experiments carried out during the early 1980s CST was further characterized biochemically and was shown to localize to the Golgi membranes exclusively (Hirschberg and Snider 1987). Its key role in the cellular sialylation pathway was unambiguously demonstrated when the asialo phenotype of Lec2 (Stanley and Siminovitch 1976) and 1021 cells (Briles et al. 1977) could be explained by the inability of the cells to translocate CMP-sialic acid into the Golgi lumen (Deutscher et al. 1984).

By complementation of the defect in Lec2 cells, the murine (Eckhardt et al. 1996) and hamster CST genes (Eckhardt and Gerardy-Schahn 1997) were identified in expression cloning approaches. The human homolog was readily isolated by a polymerase chain reaction (PCR)-based approach (Ishida et al. 1998). The availability of the cDNA sequences facilitated investigations on the structure–function relations in the CMP-sialic acid transport protein (Eckhardt et al. 1998, 1999; Aoki et al. 1999).

Enzyme Activity Assay and Substrate Specificity

The CMP-sialic acid transporter supplies the Golgi sialyltransferases with CMP-sialic acid. The solute can be actively transported against its concentration gradient. The driving force for this intraorganelle concentration is not completely understood but can be explained in part by the countertransport of the nucleoside monophosphate CMP, which follows its concentration gradient. Hence, CST (as well as all other nucleotide sugar transporters) is an antiporter (Hirschberg and Snider 1987; Hirschberg et al. 1998).

Assay systems used to study the kinetic and functional properties of the CST directly measure the uptake of the radioactively labeled nucleotide sugar CMP-[^{14}C]sialic acid into the Golgi lumen of either intact and highly purified Golgi vesicles (Hirschberg et al. 1998) or proteoliposomes, artificial membrane vesicles in which the solubilized and partially purified transport protein has been reconstituted (Milla and Hirschberg 1989). The assay systems described differ with respect to (1) reaction volumes, (2) procedures used to separate free radioactivity from radioactivity transported into the vesicular lumen, and (3) control methods used to determine nonspecifically the adsorbed radioactivity from radioactivity that has been transported into the organelle lumen.

A centrifugation assay developed by Hirschberg and coworkers (Perez and Hirschberg 1987) has been used most frequently. In this system vesicles are incubated separately with penetrating and nonpenetrating radioactive standard solutes. After incubation, vesicles are separated by centrifugation and the amount of each solute is measured in the pellets. The volumes accessible for each standard and for CMP-[^{14}C]sialic acid are calculated via the specific activity. Values obtained for CMP-[^{14}C]sialic acid are compared to those for the standards. A filtration assay has been described by Brandan and Fleischer (1982). After incubation with the radioactive nucleotide sugar, the vesicles are filtered through Millipore membranes, and total radioactivity associated with the vesicle fraction is calculated. Nonspecific adsorption is determined after disrupting the vesicles with low concentrations of detergent. Most recently an assay system has been developed that allows high throughput screening of substrates and inhibitors of the CST (Tiralongo et al. 2000). With this system the free radioactivity is separated by gel filtration using mini-spin columns prepared in 96-well filter plates (Millipore Multiscreen). To correct for nonspecific binding, 4,4′-diisothiocyanostilbene-2,2-disulfonate (DIDS), a known inhibitor of anion transport (Capasso and Hirschberg 1984), is added to control samples.

Kinetic and functional characteristics of the CST have been determined based on these techniques. The transporter has been shown to be highly substrate-specific and to localize to Golgi membranes exclusively (Hirschberg et al. 1998). CMP-[^{14}C]sialic acid transport has been shown to be temperature-dependent and saturable with a K_m in the low micromolar range (Milla and Hirschberg 1989; Tiralongo et al. 2000). Addition of mononucleotide phosphates (e.g., CMP) or CMP-sialic acid substrate analogs to the external (*cis*) compartment competitively inhibited CMP-[^{14}C]sialic acid transport (Capasso and Hirschberg 1984; Tiralongo et al. 2000), whereas addition of CMP in the *trans* (internal) compartment drastically increased the rate of CMP-[^{14}C]sialic acid uptake into proteoliposomes (Milla and Hirschberg 1989). These findings provide strong evidence that CMP-sialic acid transport by CST occurs via an antiport mechanism, with CMP being the antiport ligand in vivo (Capasso and Hirschberg 1984).

To describe the kinetics of the murine CST further, Mark von Itzstein and coworkers (Griffith University, Gold Cost, Australia) have commenced an investigation on the effect of *trans* concentrations of substrates and inhibitors on the CMP-[^{14}C]sialic acid uptake into proteoliposomes containing rat liver Golgi CST (Tiralongo et al. 2000). Additional analyses were carried out with the recombinant murine CST expressed in *Escherichia coli* (J. Tiralongo, S. Abo, M. Eckhardt, R. Gerardy-Schahn, and M. von Itzstein, in preparation). Results from these studies demonstrate that the binding of both CMP and CMP-sialic acid at the *trans* face accelerates the uptake of *cis*-added CMP-[^{14}C]sialic acid into the vesicular lumen. This effect, called *trans* acceleration, provides support for the notion that CST functions as a simple mobile carrier (Tiralongo et al. 2000). According to this model the substrate binding site of the CST alternates between the *cis* and *trans* faces. The mobility of the substrate–transporter complex across the membrane is thereby greater than transition of the free transport protein (Tiralongo et al. 2000; Tiralongo et al., in preparation).

Furthermore, pyrimidine and purine-based nucleosides and nucleotides have been synthesized and analyzed for their ability to inhibit the transport of CMP-sialic acid into Golgi vesicles and proteoliposomes containing rat liver Golgi CST (Tiralongo et al. 2000) and bacterial expressed murine CST (Tiralongo et al., in preparation). These results, along with those reported in previous studies (Capasso and Hirschberg 1984), demonstrate that pyrimidine nucleotides are much more efficient in blocking CST activity than are purine-based nucleotides. This finding indicates that the major structural element recognized by the transport protein is the nucleotide/nucleoside base. Modifications in the sugar moiety had no influence on the binding. The α-sialosyl nucleotides Neu4,5,7,8,9Ac$_5$1Me-α-O-5-fluorouridine (KI-8110) (Harvey and Thomas 1993) and its S-linked analogs Neu4,5,7,8,9Ac$_5$1Me-α-S-5-fluorouridine and Neu4,5,7,8,9Ac$_5$1Me-α-S-5-fluorouridine (isoprop) (Tiralongo et al. 2000) were found to be the most efficient inhibitors. The latter two compounds were two to three times more effective than CMP itself. This observation is particularly curious because the natural substrate of the CMP-sialic acid transporter is the β-sialosyl nucleotide. Moreover, in the case of the bacterially expressed murine CST the change from β- to α-linkage was found to reduce the inhibitory activity in all other uridine-based compounds tested (Tiralongo et al. 2000; Tiralongo et al., in preparation). The finding that pyrimidine-based nucleotides and pyrimidine-based nucleoside sugars interfere with CST transport activity substantiates recent observations suggesting that glyco-sylation processes are metabolically controlled at the cellular level (Hirschberg et al. 1998).

Preparation

Purification of a CST to homogeneity has so far not been described. cDNAs encoding the murine (*Mus musculus*) (Eckhardt et al. 1996) and hamster (*Cricetulus griseus*) (Eckhardt et al. 1997) CMP-sialic acid transport proteins have been isolated by expression cloning in CHO mutants of the complementation group Lec2. The human (*Homo sapiens*) homolog has been isolated by a PCR-based approach (Ishida et al. 1998). Functionally active CMP-sialic acid transport proteins have been expressed in mammalian cells (Eckhardt et al. 1996, 1997; Ishida et al. 1998) and yeast (Berninsone et al. 1997; Tiralongo et al. 2000). Tiralongo et al. (in preparation) demonstrated that the bacterially expressed murine CST can be functionally reconstituted in artificial membranes.

Biological Aspects

Terminal additions of sialic acids to cell surface glycoconjugates affect the chemical and functional properties of animal cell surfaces. More than 30 natural sialic acid derivatives exist that can be involved in the formation of cellular contact and com-munication structures. Modulations in the sialylation patterns are used to realize rapid changes in cellular recognition events, such as those required during the course of development, inflammation, regeneration, and neural plasticity (Kelm and Schauer 1997). Furthermore, a correlation exists between increased cell surface sialylation and the metastatic potential of cancer cells (Fukuda 1996). In line with this, one study has shown that hypersialylated cells are less susceptible to apoptotic factors, resulting in conversion of tumor cells to a more malignant phenotype (Keppler et al. 1999).

Cells exhibiting genetic defects that interfere with expression of an active CST express asialo cell surfaces (Stanley and Siminovitch 1976; Briles et al. 1977; Eckhardt et al. 1996) and confirm the key role played by the CST in the sialylation pathway. Based on the knowledge that sialylated structures occur early in development and no clinical manifestations have been observed so far that can be explained by CST defects, it is tempting to speculate that mutations in this gene cause a lethal phenotype.

Analyzing the molecular defects responsible for the loss of CST function in CHO cells of the genetic complementation group Lec2, we identified extended deletions in most of the clones. The resulting proteins were found to be unstable or mistargeted in the cells (or both). In one subclone, inactivity of the CST could be attributed to a point mutation in the coding region, leading to the exchange of glycine-189 against glutamic acid at the primary sequence level (Eckhardt et al. 1998). The mutant protein was found to be stable and correctly localized to the Golgi apparatus. Glycine-189 is part of an highly conserved stretch of amino acids present in all mammalian nucleotide sugar transporters irrespective of their substrate specificity. An equal mutation carried out in the hamster UDP-galactose transporter gene inactivated the translation product but in this case had no influence on the fundamental architecture of the protein. Thus our data suggest that glycine-189 is part of a primary sequence element essential for formation of the transport-active unit.

Future Perspectives

The CST has been well characterized at both biochemical and primary sequence levels, and a study has demonstrated that the protein consists of 10 transmembrane domains. The N- and C-termini are on the cytosolic side of the Golgi membrane (Hirschberg et al. 1998). Data on the mode of function and on primary sequence elements that dictate the tertiary and quaternary organization of the functional transporter are still missing. Moreover, protein domains involved in specific substrate recognition, mediating the translocation process, and regulating the transport activity are yet to be determined. The development in the expression of a recombinant murine CST in bacteria (Tiralongo et al., manuscript in preparation) and a high-throughput assay system (Tiralongo et al. 2000) may provide the opportunity to generate and analyze large numbers of randomly or site-specifically mutated transport proteins to identify amino acid residues involved in the functions mentioned above. Moreover, with the potential of generating large quantities of recombinant transport proteins in bacteria, the possibility of a crystal structure of the CST may be close at hand.

Increasing evidence suggests that glycoconjugate synthesis is controlled also at the metabolic level (for review see Hirschberg et al. 1998). Considering the many functional epitopes that are built in conjunction with sialic acids (Kelm and Schauer 1997), regulation of sialylation patterns via modulation of the CST activity provides an attractive perspective in terms of drug design. Last but not least, functional expression of mammalian nucleotide sugar transporters in homologous and heterologous (e.g., yeast, bacteria) cell systems enables the establishment of improved cellular expression systems to be used in biotechnological approaches.

Further Reading

For further reading see the book *Transporter of nucleotide sugars*;
Hirschberg CB (2001) Golgi nucleotide sugar transport and leukocyte adhesion deficiency II. J Clin Invest 108(1):3–6

Gerardy-Schahn R, Oelmann S, Bakker H (2001) Nucleotide sugar transporters: Biological and functional aspects. Biochimie 83(8):775–82

Goto S, Taniguchi M, Muraoka M, Toyoda H, Sado Y, Kawakita M, Hayashi S (2001) UDP-sugar transporter implicated in glycosylation and processing of Notch. Nat Cell Biol 3(9):816–822

References

Aoki K, Sun-Wada GH, Segawa H, Yoshioka S, Ishida N, Kawakita M (1999) Expression and activity of chimeric molecules between human UDP-galactose transporter and CMP-sialic acid transporter. J Biochem (Tokyo) 126:940–950

Berninsone P, Eckhardt M, Gerardy-Schahn R, Hirschberg CB (1997) Functional expression of the murine Golgi CMP-sialic acid transporter in *Saccharomyces cerevisiae*. J Biol Chem 272:12616–12619

Brandan E, Fleischer B (1982) Orientation and role of nucleoside diphosphatase and 5'-nucleotidase in Golgi vesicles from rat liver. Biochemistry 21:4640–4645

Briles EB, Li E, Kornfeld S (1977) Isolation of wheat germ agglutinin-resistant clones of Chinese hamster ovary cells deficient in membrane sialic acid and galactose. J Biol Chem 252:1107–1116

Capasso JM, Hirschberg CB (1984) Effect of nucleotides on translocation of sugar nucleotides and adenosine 3'-phosphate 5'-phosphosulfate into Golgi apparatus vesicles. Biochim Biophys Acta 777:133–139

Deutscher SL, Nuwayhid N, Stanley P, Briles EI, Hirschberg CB (1984) Translocation across Golgi vesicle membranes: a CHO glycosylation mutant deficient in CMP-sialic acid transport. Cell 39:295–299

Eckhardt M, Gerardy-Schahn R (1997) Molecular cloning of the hamster CMP-sialic acid transporter. Eur J Biochem 248:187–192

Eckhardt M, Gotza B, Gerardy-Schahn R (1998) Mutants of the CMP-sialic acid transporter causing the Lec2 phenotype. J Biol Chem 273:20189–20195

Eckhardt M, Gotza B, Gerardy-Schahn R (1999) Membrane topology of the mammalian CMP-sialic acid transporter. J Biol Chem 274:8779–8787

Eckhardt M, Mühlenhoff M, Bethe A, Gerardy-Schahn R (1996) Expression cloning of the Golgi CMP-sialic acid transporter. Proc Natl Acad Sci USA 93:7572–7576

Fukuda M (1996) Possible roles of tumor-associated carbohydrate antigens. Cancer Res 56:2237–2244

Harvey BE, Thomas P (1993) Inhibition of CMP-sialic acid transport in human liver and colorectal cancer cell lines by a sialic acid nucleoside conjugate (KI-8110). Biochem Biophys Res Commun 190:571–575

Hirschberg CB, Snider MD (1987) Topography of glycosylation in the rough endoplasmic reticulum and Golgi apparatus. Annu Rev Biochem 56:63–87

Hirschberg CB, Robbins PW, Abeijon C (1998) Transporters of nucleotide sugars, ATP, and nucleotide sulfate in the endoplasmic reticulum and Golgi apparatus. Annu Rev Biochem 67:49–69

Ishida N, Ito M, Yoshioka S, Sun-Wada GH, Kawakita M (1998) Functional expression of human Golgi CMP-sialic acid transporter in the Golgi complex of a transporter-deficient Chinese hamster ovary cell mutant. J Biochem (Tokyo) 124:171–178

Kean EL (1991) Sialic acid activation. Glycobiology 1:441–447

Kelm S, Schauer R (1997) Sialic acids in molecular and cellular interactions. Int Rev Cytol 175:137–240

Keppler OT, Peter ME, Hinderlich S, Moldenhauer G, Stehling P, Schmitz I, Schwartz-Albiez R, Reutter W, Pawlita M (1999) Differential sialylation of cell surface glycoconjugates in a human B lymphoma cell line regulates susceptibility for CD95 (APO-1/Fas)-

mediated apoptosis and for infection by a lymphotropic virus. Glycobiology 9: 557–569

Milla ME, Hirschberg CB (1989) Reconstitution of Golgi vesicle CMP-sialic acid and adenosine 3'-phosphate 5'-phosphosulfate transport into proteoliposomes. Proc Natl Acad Sci USA 86:1786–1790

Münster AK, Eckhardt M, Potvin B, Mühlenhoff M, Stanley P, Gerardy-Schahn R (1998) Mammalian cytidine 5'-monophosphate N-acetylneuraminic acid synthetase: a nuclear protein with evolutionarily conserved structural motifs. Proc Natl Acad Sci USA 95:9140–9145

Perez M, Hirschberg CB (1987) Transport of sugar nucleotides into the lumen of vesicles derived from rat liver rough endoplasmic reticulum and Golig apparatus. Methods Enzymol 138:709–715

Stanley P, Siminovitch L (1976) Selection and characterization of Chinese hamster ovary cells resistant to the cytotoxicity of lectins. In Vitro 12:208–215

Tiralongo J, Abo S, Danylec B, Gerardy-Schahn R, von Itzstein M (2000) A high-throughput assay for rat liver Golgi and Saccharomyces cerevisiae-expressed murine CMP-N-acetylneuraminic acid transport proteins. Anal Biochem 285:21–32

Dolichol Pathway/GPI-Anchor

GPI-GlcNAc Transferase: Complex of PIG-A, PIG-C, PIG-H, hGPI1, and PIG-P

Introduction

Glycosylphosphatidylinositols (GPIs) anchor proteins to the cellular plasma membrane in a wide range of eukaryotic organisms from yeasts to mammals (Kinoshita et al. 1997). The GPI-anchor is transferred en bloc to the carboxyl-terminus of protein precursors in the endoplasmic reticulum (ER). Attachment of the GPI-anchor to the proteins is essential for their expression on the cell surface. Synthesis of the GPI-anchor is initiated by transferring GlcNAc to phosphatidylinositol (PI) and proceeds by de-*N*-acetylation of GlcNAc, acylation of *myo*inositol, and addition of three mannoses and three ethanolamine phosphates in mammalian cells (see Appendix Map-5). The basic structure and biosynthetic pathway are conserved in various organisms. The UDP-GlcNAc:PI-α1-4GlcNAc transferase (GPI-GlcNAc transferase, or GPI-GnT) catalyzes the first reaction in the biosynthesis of the GPI-anchor.

Some trypanosomatid parasites such as *Leishmania* have GPI-related molecules that are not linked to proteins (Ferguson 1999). There are three families of GPI-related molecules (type 1, type 2, hybrid GPIs) characterized by Manα1-6Manα1-4GlcNAc-PI, Manα1-3Manα1-4GlcNAc-PI, and Manα1-6(Manα1-3)Manα1-4GlcNAc-PI motifs, respectively. Glycoinositol phospholipids (GIPLs) bearing type 1, type 2, or hybrid GPIs are abundantly expressed in a number of trypanosomatids, forming dense layers on the cell surface. The *Leishmania* cells also express a type 2 GPI-related molecule called lipophosphoglycan (LPG), which has a long phosphosaccharide-repeat domain with species-specific side chains. The type 1 motif is shared by the GPI-anchored protein. The syntheses of these GPI-related molecules are also initiated by transferring GlcNAc to PI. The GPI-GlcNAc transferase should be involved in the common first step in synthesis of the GPI-related molecules and the protein GPI-anchor.

Norimitsu Inoue and Taroh Kinoshita

Department of Immunoregulation, Research Institute for Microbial Diseases, Osaka University, 3-1 Yamadaoka, Suita, Osaka 565-0871, Japan
Tel: +81-6-6879-8329; Fax +81-6-6875-5233
e-mail: inoue@biken.osaka-u.ac.jp

Databanks

GPI-GlcNAc transferase: Complex of PIG-A, PIG-C, PIG-H, hGPI1, and PIG-P

No EC number has been allocated.

Species	Gene	Protein	mRNA	Genomic
PIG-A				
Homo sapiens	PIG-A	BAA02019	D11466	D28787–D28791
				(exons 1–6)
		P37287	–	–
		BAA05966	–	–
Mus musculus	Pig-a	BAA05047	D26047	D31858–D31863
				(exon 1–6)
		A55731	–	–
		BAA06663	–	–
Saccharomyces	SPT14/GPI3/CWH6	P32363	–	NC_001148
cerevisiae	/YPL175w	NP_015150	–	–
PIG-H/GPI-H				
Homo sapiens	PIG-H	AAA03545	L19783	–
		NP_004560	NM_004569	–
PIG-C				
Homo sapiens	PIG-C	BAA12812	D85418	AB000360
		NP_002633	NM_002642	–
		BAA22866	–	–
Saccharomyces	GPI2/YPL076w	AAA79518	U23788	NC_001148
cerevisiae		NP_015249	–	–
GPI1				
Homo sapiens	GPI1	BAA24948	AB003723	–
		NP_004195	NM_004204	–
		–	NM_011822	–
Mus musculus	Gpi1	BAA84658	AF030178	AB008915–AB008920
				(exons 1–10)
		NP_035952	NM_011822	–
		NP_034467	NM_010337	–
			AB008921	–
Saccharomyces	GPI1/YGR216c	NP_011732	–	NC_001139
cerevisiae				
Schizosaccharomyces	GPI1	AAC49650	U77355	–
pombe				
		O14357	–	–
PIG-P				
Homo sapiens	PIG-P	–	AB039659	–
Mus musculus	Pig-p/DCRC-1	AAF32294	AF16306	–
Saccharomyces	YDR437w	NP_010725	–	NC_001136
cerevisiae				

Name and History

Mammalian GPI-GlcNAc transferase consists of at least five components. Mutants of three complementation groups (class A, C, and H) are defective in the first step of the reaction; and three genes (*PIG-A*, *PIG-C*, *PIG-H/GPI-H*) that complement their mutations, respectively, have been cloned (Kamitani et al. 1993; Miyata et al. 1993; Inoue et al. 1996). Three temperature-sensitive mutants of *Saccharomyces cerevisiae* (*gpi1*, *gpi2*, *gpi3/spt14/cwh6*) are also defective in the first step (Leidich et al. 1994; Vossen et al. 1995). *GPI3* and *GPI2* are homologous to *PIG-A* and *PIG-C*, respectively (Leidich et al. 1995; Schonbachler et al. 1995; Inoue et al. 1996). Human *GPI1* was cloned as a homolog of yeast GPI1 (Leidich and Orlean 1996; Watanabe et al. 1998). There is no gene highly homologous to human PIG-H in *S. cerevisiae* DNA or protein databases, but part of YNL038wp has weak homology to human PIG-H, suggesting that YNL038wp may be a candidate PIG-H homolog. Recently the fifth gene, *PIG-P*, was cloned through identification of a new protein in the GPI-GnT complex (Watanabe et al. 2000). *PIG-P* complemented a new T cell mutant established by Mina D. Marmor and Michael Julius.

Enzyme Activity Assay and Substrate Specificity

GPI-GnT catalyzes the transfer of *N*-acetylglucosamine from UDP-GlcNAc to PI to form GlcNAc-PI. The partially purified mammalian GPI-GnT shows specificity to some types of PI in that it used bovine PI bearing mainly stearic and arachidonic acids 100 times more efficiently than soybean PI bearing mainly palmitic and linoleic acids (Watanabe et al. 1998). Bovine lysoPI was an inefficient substrate, and a large amount of lysoPI inhibited the utilization of endogenous PI, whereas polar analogs of PI (i.e., inositol, glycerophosphate, glycerophosphoinositol) did not inhibit the reaction (Watanabe et al. 1998). These results indicate that fatty acyl chains especially at the sn-2 position of PI are important for an efficient reaction. Phosphatidylcholine, sphingomyelin, and phosphatidylserine inhibited the activity of partially purified GPI-GnT, whereas phosphatidylethanolamine, ceramide and phosphatidic acid had no effect (Watanabe et al. 1998).

This enzyme reaction should occur on the cytoplasmic side of the ER because the catalytic domain of PIG-A faces the cytoplasmic side (Watanabe et al. 1996). Moreover, the product of this reaction, GlcNAc-PI, is present in the cytoplasmic leaflet.

Structural Chemistry

The mammalian GPI-GnT components (PIG-A, PIG-C, PIG-H, GPI1, PIG-P) have been well characterized. PIG-A, PIG-C, PIG-H, and PIG-P, but not GPI1, are essential for this enzyme reaction.

PIG-A consists of 484 amino acids and has a transmembrane region with an amino-terminal large cytoplasmic portion and a carboxyl-terminal small luminal portion (Watanabe et al. 1996). PIG-A has a catalytic region homologous to bacterial LPS GlcNAc transferases, RfaK, in the middle of the cytoplasmic portion and is a member of nucleotide-diphospho-sugar glycosyltransferase family 4 (Ullman and Perkins 1997). The luminal juxtamembrane 23-amino-acid residues are essential for the enzyme reaction and are involved in targeting PIG-A to the ER membrane (Watanabe et al. 1996).

PIG-H consists of 188 amino acids and is a cytoplasmic protein associated with the ER (Watanabe et al. 1996). PIG-H has no homology to proteins of known function in the DNA and protein databases. PIG-H binds to PIG-A and GPI1 (Watanabe et al. 1998).

PIG-C consists of 297 amino acids and has multiple transmembrane domains (Inoue et al. 1996). PIG-C tightly binds to GPI1 but not to PIG-A, PIG-H, or PIG-P. PIG-C weakly binds to a complex of PIG-A and PIG-H (Watanabe et al. 1998).

GPI1 consists of 581 amino acids and has several transmembrane regions (Watanabe et al. 1998). A mouse *GPI1*-deficient cell line was established with a homologous recombination technique (Hong et al. 1999). Although the activity of GPI-GnT was not detected in the mutant cell line in vitro, the GPI-anchored proteins were still slightly expressed on the surface membrane, indicating that GPI1 is not essential for the enzyme activity of GPI-GnT (Hong et al. 1999). GPI1 directly binds to all other components of the enzyme, and the expression levels of PIG-H and PIG-C decreased in GPI1-deficient cell lines (Hong et al. 1999). These results indicate that GPI1 acts as a stabilizer of the enzyme complex. Overexpression of GPI2, but not GPI3, suppressed mutation of *gpi1* in a yeast system (Leidich et al. 1995). GPI1 may act as a bridge between PIG-C and the complex of PIG-A, PIG-H, and PIG-P.

PIG-P consists of 134 amino acids and has two transmembrane regions. PIG-P binds to PIG-A strongly and to GPI1 weakly (Watanabe et al. 2000). There is no significant homology between PIG-P and other proteins in the database.

Preparation

GPI-GnT complexes were prepared from cell lines transfected with an expression vector of epitope-tagged PIG-A. The complexes from the ER membrane were stable in 1% digitonin solution but not in 1% NP40 (Watanabe et al. 1998). Preparation of the enzyme complex from yeast or protozoa has not been reported.

Biological Aspects

Many membrane surface proteins of yeast and parasites have the GPI-anchors that are essential for their expression on the surface membrane. In yeast, GPI is essential for viability and is involved in formation of the cell wall (Lipke and Ovalle 1998). A number of GPI-anchored proteins are expressed on the plasma membrane, where the GPI anchor is cleaved in the glycan portion and the proteins are transferred to the branched β1,6-glucan, which is linked to the β1,3-glucan and chitin complex of the cell wall (Lipke and Ovalle 1998). Incorporation of the GPI-anchored cell wall proteins, which are highly mannosylated, is affected in the gpi3/spt14/cwh6 mutant (Vossen et al. 1995). *GPI1*-disrupted yeast mutant grows at a permissive temperature but forms large, round, multiply budded cells with dividing defects at a semipermissive temperature (Leidich and Orlean 1996), indicating that GPIs play a role in the morphogenesis. The *gpi3* mutant shows down-regulation of inositolphosphoceramide synthesis and inhibition of serine-palmitoyltransferase activity, whereas addition of an inhibitor for serine-palmitoyltransferase (myriocin) does not inhibit the activity of GPI-GnT (Schonbachler et al. 1995).

Mutants of the GPI-GnT components have not been established in parasites, although mutants of *Trypanosoma* and *Leishmania* defective in the later step of the GPI-anchor biosynthesis have been analyzed. Biosynthesis of the GPI-anchor is not essential for growth of mammalian cells in vitro. However, GPI-anchor-deficient mice are embryonic lethal. *Pig-a* knockout mice generated using the Cre/loxp system have been analyzed (Nozaki et al. 1999). Embryos bearing mostly *Pig-a*-disrupted cells had a more transparent yolk sac and were smaller than wild-type embryos by one-quarter to one-half at 9.5 dpc (Nozaki et al. 1999). Developmental defects occur after differentiation of the three germ layers. Partially *Pig-a*-disrupted embryos exhibit abnormalities in the central nervous system including exencephaly and facial primordia (Nozaki et al. 1999).

Functions of Pig-a in skin development have been studied using skin-specific conditional knockout mice (Tarutani et al. 1997). The skin of these mice looked more wrinkled and scaly, and the epidermal horny layer was packed and thickened (Tarutani et al. 1997). The mice died within a few days after birth. These results indicate that GPI or GPI-anchored proteins are important for the development of skin.

In humans the deficiency of GPI-anchored complement regulators—decay-accelerating factor (DAF) and CD59—causes paroxysmal nocturnal hemoglobinuria (PNH), an acquired hemolytic disease. All patients with PNH have somatic mutations in the PIG-A gene that result in a loss or decrease of GPI-GnT activity. The reason mutations in other GPI-anchor biosynthesis genes have not been found in any PNH patients is that PIG-A is X-linked (Xp22.1) and functionally haploid, whereas all other genes are autosomal and require two mutations to lose their function (Takeda et al. 1993). In patients with PNH, PIG-A-deficient hematopoietic cells predominate over normal cells. The analysis of conditional knockout mice indicates that mutation of *Pig-a* is not sufficient for dominance of hematopoietic stem cells, and other abnormalities are involved (Murakami et al. 1999; Tremml et al. 1999).

Future Perspectives

Because the purified complex consists of at least five components, it is necessary to clarify the roles of components other than PIG-A and GPI1. The enzyme has a different specificity for fatty acyl chains of mammalian and plant PIs. The enzymes for parasites and yeast may also have specificities different from those of the mammalian enzyme. The molecular basis of the different specificity should be determined.

Further Reading

See Kinoshita et al. (1997) and Ferguson (1999).

References

Ferguson MA (1999) The structure, biosynthesis and functions of glycosylphosphatidylinositol anchors, and the contributions of trypanosome research. J Cell Sci 112: 2799–2809

Hong Y, Ohishi K, Watanabe R, Endo Y, Maeda Y, Kinoshita T (1999) GPI1 stabilizes an enzyme essential in the first step of glycosylphosphatidylinositol biosynthesis. J Biol Chem 274:18582–18588

Inoue N, Watanabe R, Takeda J, Kinoshita T (1996) PIG-C, one of the three human genes involved in the first step of glycosylphosphatidylinositol biosynthesis is a homologue of *Saccharomyces cerevisiae* GPI2. Biochem Biophys Res Commun 226: 193–199

Kamitani T, Chang HM, Rollins C, Waneck GL, Yeh ETH (1993) Correction of the class H defect in glycosylphosphatidylinositol anchor biosynthesis in Ltk-cells by a human cDNA clone. J Biol Chem 268:20733–20736

Kinoshita T, Ohishi K, Takeda J (1997) GPI-anchor synthesis in mammalian cells: genes, their products, and a deficiency. J Biochem 122:251–257

Leidich SD, Orlean P (1996) Gpi1, a *Saccharomyces cerevisiae* protein that participates in the first step in glycosylphosphatidylinositol anchor synthesis. J Biol Chem 271: 27829–27837

Leidich SD, Drapp DA, Orlean P (1994) A conditionally lethal yeast mutant blocked at the first step in glycosyl phosphatidylinositol anchor synthesis. J Biol Chem 269: 10193–10196

Leidich SD, Kostova Z, Latek RR, Costello LC, Drapp DA, Gray W, Fassler JS, Orlean P (1995) Temperature-sensitive yeast GPI anchoring mutants gpi2 and gpi3 are defective in the synthesis of *N*-acetylglucosaminyl phosphatidylinositol: cloning of the GPI2 gene. J Biol Chem 270:13029–13035

Lipke PN, Ovalle R (1998) Cell wall architecture in yeast: new structure and new challenges. J Bacteriol 180:3735–3740

Miyata T, Takeda J, Iida Y, Yamada N, Inoue N, Takahashi M, Maeda K, Kitani T, Kinoshita T (1993) Cloning of PIG-A, a component in the early step of GPI-anchor biosynthesis. Science 259:1318–1320

Murakami Y, Kinoshita T, Maeda Y, Nakano T, Kosaka H, Takeda J (1999) Different roles of glycosylphosphatidylinositol in various hematopoietic cells as revealed by model mice of paroxysmal nocturnal hemoglobinuria. Blood 94:2963–2970

Nozaki M, Ohishi K, Yamada N, Kinoshita T, Nagy A, Takeda J (1999) Developmental abnormalities of glycosylphosphatidylinositol-anchor-deficient embryos revealed by Cre/loxP system. Lab Invest 79:293–299

Schonbachler M, Horvath A, Fassler J, Riezman H (1995) The yeast spt14 gene is homologous to the human PIG-A gene and is required for GPI anchor synthesis. EMBO J 14:1637–1645

Takeda J, Miyata T, Kawagoe K, Iida Y, Endo Y, Fujita T, Takahashi M, Kitani T, Kinoshita T (1993) Deficiency of the GPI anchor caused by a somatic mutation of the PIG-A gene in paroxysmal nocturnal hemoglobinuria. Cell 73:703–711

Tarutani M, Itami S, Okabe M, Ikawa M, Tezuka T, Yoshikawa K, Kinoshita T, Takeda J (1997) Tissue specific knock-out of the mouse Pig-a gene reveals important roles for GPI-anchored proteins in skin development. Proc Natl Acad Sci USA 94:7400–7405

Tremml G, Dominguez C, Rosti V, Zhang Z, Pandolfi PP, Keller P, Bessler M (1999) Increased sensitivity to complement and a decreased red cell life span in mice mosaic for a nonfunctional Piga gene. Blood 94:2945–2962

Ullman CG, Perkins SJ (1997) A classification of nucleotide-diphospho-sugar glycosyltransferases based on amino acid sequence similarities. Biochem J 326:929–942

Vossen JH, Ram AF, Klis FM (1995) Identification of SPT14/CWH6 as the yeast homologue of hPIG-A, a gene involved in the biosynthesis of GPI anchors. Biochim Biophys Acta 1243:549–551

Watanabe R, Inoue N, Westfall B, Taron CH, Orlean P, Takeda J, Kinoshita T (1998) The first step of glycosylphosphatidylinositol biosynthesis is mediated by a complex of PIG-A, PIG-H, PIG-C and GPI1. EMBO J 17:877–885

Watanabe R, Murakami Y, Marmor MD, Inoue N, Maeda Y, Hino J, Kangawa K, Julius M, Kinoshita T (2000) Initial enzyme for glycosylphosphatidylinositol biosynthesis requires PIG-P and is regulated by DPM2. EMBO J 19:4402–4411

Watanabe R, Kinoshita T, Masaki R, Yamamoto A, Takeda J, Inoue N (1996) PIG-A and PIG-H, which participate in glycosylphosphatidylinositol anchor biosynthesis, form a protein complex in the endoplasmic reticulum. J Biol Chem 271:26868–26875

78

Dolichol Phosphate-Mannose Synthase (DPM1 and DPM2)

Introduction

Dolichol phosphate-mannose synthase (Dol-P-Man synthase), which generates dolichol phosphate-mannose (Dol-P-Man) from GDP-mannose (GDP-Man) and dolichol phosphate (Dol-P), is an essential enzyme for all eukaryotic cells. The Dol-P-Man synthase is located in the endoplasmic reticulum (ER), through two kinds of binding to the membrane depending on the species (Colussi et al. 1997; Maeda et al. 1998). One group comprises *Saccharomyces cerevisiae*, *Ustilago maydis*, and *Trypanosoma brucei* enzymes, which have a C-terminal hydrophobic domain; and the other group includes the human, mouse, *Schizosaccharomyces pombe*, and *Caenorhabditis briggsiae* synthases, which lack a C-terminal hydrophobic domain. In the latter group, at least two polypeptides form the Dol-P-Man synthase complex, where the catalytic subunit DPM1 binds to the membrane protein DPM2, which has two membrane spanning regions (Maeda et al. 1998). The Dol-P-Man synthesized by Dol-P-Man synthase is used for the five mannose moieties of *N*-linked oligosaccharide, the first mannose moiety of fungal *O*-linked oligosaccharide, three mannose moieties of glycosyl phosphatidylinositol (GPI) anchoring, and C-mannosylation (Orlean 1990; Doucey et al. 1998). Recently, a deficiency of the *DPM1* gene was found to cause a congenital disorder of glycosylation (CDG) type Ie, giving rise to mental and psychomotor retardation, dysmorphism, and blood coagulation defects.

YOH-ICHI SHIMMA and YOSHIFUMI JIGAMI

Institute of Molecular and Cell Biology (IMCB), National Institute of Advanced Industrial Science and Technology (AIST), 1-1 Higashi, Tsukuba-city, Ibaraki 305-8566, Japan
Tel. +81-298-61-6212; Fax +81-298-61-6220
e-mail: shimma@nibh.go.jp

Databanks

Dolichol phosphate-mannose synthase (DPM1 and DPM2)

NC-IUBMB enzyme classification: E.C.2.4.1.83

Species	Gene	Protein	mRNA	Genomic
DPM1				
Homo sapiens	*DPM1*	–	D86198	–
Mus musculus	*DPM1*	–	AV312164	–
Rattus norvegicus	*DPM1*	–	AA900880	–
Cricetulus griseus	*DPM1*	–	AF121895	–
Caenorhabditis briggsae	*DPM1*	–	AF007874	–
		–	R02847	–
Saccharomyces cerevisiae	*DPM1*	P14020	J04184	–
		A32122	–	–
Schizosaccharomyces pombe	*DPM1*	–	AF007873	–
Ustilago maydis	*DPM1*	P54856	U54797	–
		S71642	–	–
Trypanosoma brucei	*DPM1*	S70643	Z54162	–
DPM2				
Homo sapiens	*DPM2*	Q94777	AB013361	–
Mus musculus	*DPM2*	Q9Z324	AB013360	–
Rattus norvegicus	*DPM2*	Q9Z325	AB013359	–
Cricetulus griseus	*DPM2*	Q9Z1P1	AF115410	–

Name and History

During the 1960s it was found that a lipophilic mannosyl intermediate is the immediate precursor for mannan biosynthesis (Caccam et al. 1969; Tanner 1969). The lipid was identified as dolichol phosphate, and characterization of the responsible enzyme, dolichol-phosphate-mannose synthase (Dol-P-Man synthase), or GDP-mannose:dolichol-phosphate O-β-D-mannosyltransferase, was carried out (Behrens and Leloir 1970; Babczinski and Tanner 1973; Jung and Tanner 1973). Several mutants were isolated that showed a deficiency in Dol-P-Man synthase activity, and the corresponding genes were cloned. First, the gene with overexpressed Dol-P-Man synthase activity was cloned from the yeast *S. cerevisiae* genomic DNA library, designated *DPM1* (Orlean et al. 1988). Yeast *DPM1* is a membrane protein with a transmembrane (TM) domain at the carboxyl-terminus. *DPM1* is the structural gene for Dol-P-Man synthase, as *Escherichia coli* transformants harboring this yeast gene express Dol-P-Man synthase activity. In mammalian cells, mutants with decreased Dol-P-Man synthase activity are reported as Thy-1-negative class E and Lec15 cells (Chapman et al. 1980; Stoll et al. 1982). Human *DPM1* was cloned based on the homology with yeast *DPM1* (Colussi et al. 1997), and it lacks a TM region at the C-terminus. The human *DPM1* gene complements Thy-1 but not Lec15 mutant phenotype (Tomita et al. 1998), whereas yeast *DPM1* complements both Thy-1 and Lec15/B4-2-1 mutant phenotype (Beck et al. 1990). The *DPM2* gene was cloned by complementation of the Lec15 mutant phenotype (Maeda et al. 1998). DPM2 protein has two TM regions and an ER retention signal, and it forms a complex with DPM1 protein. The known Dol-P-Man synthases can be divided into two classes. One contains *S. cerevisiae*, *Ustilago maydis*, and *Trypanosoma brucei* enzymes, which have a COOH-terminal hydrophobic domain; the other contains the human, *S. pombe*, and *Caenorhabditis briggsiae* synthases, which lack a hydrophobic COOH-terminal domain. The two classes of synthase are functionally equivalent because *S. cerevesiae* *DPM1* and its human counterpart complement the lethal null mutation in *S. pombe* *dpm1+* (Colussi et al. 1997). More recently the *DPM3* gene was cloned, and it consists of 92 amino acids and contains two TM regions at the N-terminus (Maeda et al. 2000). DPM3 protein binds to DPM1 and DPM2 proteins. These membrane proteins are required for proper localization and stable expression of Dol-P-Man synthase activity.

Enzyme Activity Assay and Substrate Specificity

Dol-P-Man synthase catalyzes the following reaction.

$$\text{GDP-Man} + \text{Dol-P} \rightleftarrows \text{Dol-P-Man} + \text{GDP}$$

where Mn^{2+} and Mg^{2+} are required (Sharma et al. 1974). Yeast synthase has a K_m value (for GDP-Man) of $7\,\mu M$, an optimum pH of 7.3, and a maximum activity of $140\,\mu M$ (Babczinski et al. 1980; Haselbeck and Tanner 1982). Rat enzyme has K_m values of $0.69\,\mu M$ for GDP-Man and $0.3\,\mu M$ for Dol-P and an optimum pH of 7.5 (Jensen and Schutzbach 1985).

The transfer of radioactivity from GDP-[14C]Man to exogenous Dol-P was measured for the detergent-solubilized yeast enzyme (Haselbeck and Tanner 1982). The standard incubation in a total volume of $70\,\mu l$ contained $7\,mM$ $MgCl_2$, $5\,mM$ $MnCl_2$, $0.05\,\mu Ci$ GDP-[14C]Man, $5\,mM$ Tris-HCl (pH 7.4), $5\,\mu g$ of Dol-P, 0.35% Triton X-100, and 0.1–$0.3\,mg$ of membrane protein. When the solubilized enzyme preparations were used, the protein content was 25–$50\,\mu g$ in the case of Triton X-100 extract and 0.3–$10.0\,\mu g$ in the case of partially purified enzyme after column chromatography. Dol-P, dissolved in acetone, was dried together with $10\,\mu l$ of Mg-EDTA ($0.1\,M$) under a stream of nitrogen before the other ingredients were added. Incubation time was $50\,s$ at room temperature.

Enzyme activity integrated into liposomes was assayed in a mixture of $25\,\mu l$ of liposomes ($215\,\mu g$ of phopholipid) in $10\,mM$ potassium phosphate, pH $6.8/100\,mM$ KCl plus $45\,\mu l$ of a solution containing $12.5\,mM$ Tris-HCl (pH 7.4), $7\,mM$ $MgCl_2$, and $0.05\,\mu Ci$ GDP-[14C]Man. The liposomes contained $1.12\,\mu g$ of protein or $0.12\,\mu g$ of protein plus $3\,\mu g$ of Dol-P.

The activity of mammalian enzyme is measured as follows (Maeda et al. 1998): Cells were destroyed hypotonically by a Teflon homogenizer in a buffer ($20\,mM$ Tris-HCl pH 7.4, $10\,mM$ NaCl, leupeptin $2\,\mu g/ml$, and $1\,mM$ p-APMSF) on ice. After removal of cell debris and nuclei by centrifugation at $1500\,g$ for $10\,min$, membranes were collected by centrifugation at $100\,000\,g$ for $1\,h$ and suspended in a reaction buffer consisting of $50\,mM$ HEPES-KOH (pH 7.4), $25\,mM$ KCl, $5\,mM$ $MgCl_2$, and $5\,mM$ $MnCl_2$. Dol-P ($10\,\mu g$) was first added to a tube in chloroform/methanol (2:1) solution and dried under a nitrogen stream; the membranes ($100\,\mu g$ of protein) and GDP-[3H]Man

(0.16 μM, 0.8 μCi) were then mixed vigorously in a final volume of 100 μl of reaction buffer. The mixture was incubated for 10 min at 37°C and then 0.5 ml chloroform/methanol (2:1) was added; it was washed once with 0.5 ml of chloroform/methanol (2:1)-saturated water and then evaporated. The dried materials were extracted with 30 μl of chloroform/methanol (2:1), and the extracts were separated by thin-layer chromatography (TLC) on Kieselgel 60 with a solvent system of chloroform/methanol/H$_2$O (10:10:3). The radiolabeled lipids were analyzed by an Image Analyzer BAS 1500 after 2–4 days of exposure.

The purified enzyme preparation was unstable in the presence of detergent (Jensen and Schutzbach 1985). Detergent-free enzyme was active in the presence of phosphatidylethanolamine (PE) and in the presence of phospholipid mixtures of PE and phosphatidylcholine (PC). The enzyme was inactive in the presence of PC alone. These results suggest that Dol-P-Man synthase is optimally active in a phospholipid matrix containing phospholipid components that prefer nonbilayer structural organization.

Phytanyl (3,7,11,15-tetramethylhexadecanyl) phosphate was utilized at 60%–70% of the efficiency of the natural dolichyl lipid during transfer of mannose from GDP-Man, whereas addition of S-3-methyloctadecanyl phosphate, which is of a length similar to that of the phytanyl analog but with only one branch, resulted in approximately 25% of the incorporation of the natural substarate (Wilson et al. 1993). Based on their findings and other data, Wilson et al. concluded that branching of lipid phosphates is essential for substrate recognition by Dol-P-Man synthase.

When microsomal membranes were phosphorylated in vitro by a cAMP-dependent protein kinase, an increase in Dol-P-Man synthase activity was reported with a moderate change in K_m for GDP-Man and a twofold higher Vmax (Banerjee et al. 1987).

Amphomycin inhibited mannose transfer from GDP-Man to Dol-P (Kang et al. 1978). Amphomycin interacted with Dol-P by forming a complex, thereby preventing its participation at the enzymatic reaction (Banerjee 1989).

Preparation

The yeast Dol-P-Man synthase was solubilized with Triton X-100 (0.5%) (Babczinski et al. 1980). Clear supernatants were obtained after 45 min of centrifugation at 48 000 g. The extract was fractionated on hydoxyapatite, a stepwise gradient of potassium phosphate (pH 6.8) containing 0.5% Triton X-100. The protein peak eluted with 150 mM salt contained the Dol-P-Man synthase. The enzyme was recovered in the void volume on a DEAE-cellulose column equilibrated with 10 mM potassium phosphate (pH 6.8) containing 0.5% Triton X-100.

Rat liver enzyme was solubilized by 2.5% Nonidet P-40 (5 ml) to 8 ml of the microsomal suspension (Jensen and Schutzbach 1985). The mixture was diluted to 51 ml by adding a solution containing water (28 ml), glycerol (10 ml), and mercaptoethanol (0.1 ml); and it was centrifuged at 100 000 g for 60 min. The supernatant was applied to a column of DEAE-cellulose. The enzyme was eluted with a linear gradient of 0.01–0.10 M Tris/acetate (pH 7.0) containing 10% glycerol, 0.1% Nonidet P-40, and 0.5 mM dithiothreitol (DTT).

Biological Aspects

Defects in the attachment of carbohydrate to protein give rise to mental and psychomoter retardation, dysmorphism, and blood coagulation defects. These symptoms are referred to as congenital disorders of glycosylation, or carbohydrate-deficient glycoprotein syndrome (CDG). All patients exhibit altered isoelectric focusing paterns of multiple serum glycoproteins that result from under-sialylation of the N-linked oligosaccharides. CDG-type Ie patients have a point mutation in $DPM1$, and the K_m for GDP-Man of the mutant enzyme is sixfold higher than that of normal enzyme (Imbach et al. 2000; Kim et al. 2000).

Future Perspectives

Human $DPM1$ mutations that cause CDG have been found in only two cases. Further single nucleatide polymorphisms (SNPs) of $hDPM1$ will be found. $DPM1$ mutations cause a wide range of phenotypes. Symptoms of CDG may depend on SNPs. $DPM2$ and $DPM3$ mutations will certainly be found in the near future.

Further Reading

Kim et al. (2000) and Imbach et al. (2000) first reported that $DPM1$ mutations cause CDG type Ie. The symptoms appeared mainly in the nervous system. The role of oligosaccharides in the development of the nervous system will be interesting.

References

Babczinski P, Tanner W (1973) Involvement of dolicholmonophosphate in the formation of specific mannosyl-linkages in yeast glycoproteins. Biochem Biophys Res Commun 54:1119–1124

Babczinski P, Haselbeck A, Tanner W (1980) Yeast mannosyl transferases requiring dolichyl phosphate and dolichyl phosphate mannose as substrate. Eur J Biochem 105: 509–515

Banerjee DK (1989) Amphomycin inhibits mannosylphosphoryldolichol synthesis by forming a complex with dolichylmonophosphate. J Biol Chem 264:2024–2028

Banerjee DK, Kousvelari EE, Baum BJ (1987) cAMP-mediated protein phosphorylation of microsomal membranes increases mannosylphosphodolichol synthase activity. Proc Natl Acad Sci USA 84:6389–6393

Beck PJ, Orlean P, Albright C, Robbins PW, Gething MJ, Sambrook JF (1990) The *Saccharomyces cerevisiae* DPM1 gene encoding dolichol-phosphate-mannose synthase is able to complement a glycosylation-defective mammalian cell line. Mol Cell Biol 10:4612–4622

Behrens NH, Leloir JF (1970) Dolichol monophosphate glucose: an intermediate in glucose transfer in liver. Proc Natl Acad Sci USA 66:153–159

Caccam JF, Jackson JJ, Eylar EH (1969) The biosynthesis of mannose-containing glycoproteins: a possible lipid intermediate. Biochem Biophys Res Commun 35:505–511

Chapman A, Fujimoto K, Kornfeld S (1980) The primary glycosylation defect in class E Thy-1-negative mutant mouse lymphoma cells is an inability to synthesize dolichol-P-mannose. J Biol Chem 255:4441–4446

Colussi PA, Taron CH, Mack JC, Orlean P (1997) Human and *Saccharomyces cerevisiae* dolichol phosphate mannose synthases represent two classes of the enzyme, but both function in *Schizosaccharomyces pombe*. Proc Natl Acad Sci USA 94:7873–7878

Doucey M-A, Hess D, Cacan R, Hofsteenge J (1998) Protein C-mannosylation is enzyme-catalysed and uses dolichyl-phosphate-mannose as a precursor. Mol Biol Cell 9: 291–300

Haselbeck A, Tanner W (1982) Dolichyl phosphate-mediated mannosyl transfer through liposomal membranes. Proc Natl Acad Sci USA 79:1520–1524

Imbach T, Schenk B, Schollen E, Burda P, Stutz A, Grunewald S, Bailie NM, King MD, Jaeken J, Matthijs G, Berger EG, Aebi M, Hennet T (2000) Deficiency of dolichol-phosphate-mannose synthase-1 causes congenital disorder of glycosylation type Ie. J Clin Invest 105:233–239

Jensen JW, Schutzbach JS (1985) Activation of dolichyl-phospho-mannose synthase by phospholipids. Eur J Biochem 153:41–48

Jung P, Tanner W (1973) Identification of the lipid intermediate in yeast mannan biosynthesis. Eur J Biochem 37:1–6

Kang MS, Spencer JP, Elbein AD (1978) Amphomycin inhibition of mannose and GlcNAc incorporation into lipid-linked saccharides. J Biol Chem 253:8860–8866

Kim S, Westphal V, Srikrishna G, Mehta DP, Peterson S, Filiano J, Karnes PS, Patterson MC, Freeze HH (2000) Dolichol phosphate mannose synthase (*DPM1*) mutations define congenital disorder of glycosylation Ie (CDG-Ie). J Clin Invest 105:191–198

Maeda Y, Tomita S, Watanabe R, Ohishi K, Kinoshita T (1998) DPM2 regulates biosynthesis of dolichol phosphate-mannose in mammalian cells: correct subcellular localization and stabilization of DPM1, and binding of dolichol phosphate. EMBO J 17:4920–4929

Maeda Y, Tanaka S, Hino J, Kangawa K, Kinoshita T (2000) Human dolichol-phosphate-mannose synthase consists of three suhunits, DPM1, DPM2 and DPM3. EMBO J 19:2475–2482

Orlean P (1990) Dolichol phosphate mannose synthase is required in vivo for glycosyl phosphatidylinositol membrane anchoring, *O* mannosylation, and *N* glycosylation of protein in *Saccharomyces cerevisiae*. Mol Cell Biol 10:5796–5805

Orlean P, Albright C, Robbins PW (1988) Cloning and sequencing of the yeast gene for dolichol phosphate mannose synthase, an essential protein. J Biol Chem 263: 17499–17507

Sharma CB, Babczinski P, Lehle L, Tanner W (1974) The role of dolicholmonophosphate in glycoprotein biosynthesis in *Saccharomyces cerevisiae*. Eur J Biochem 46:35–41

Stoll J, Robbins AR, Krag SS (1982) Mutant of Chinese hamster overy cells with altered mannose 6-phosphate receptor activity is unable to synthesize mannosylphosphoryl-dolichol. Proc Natl Acad Sci USA 79:2296–2300

Tanner W (1969) A lipid intermediate in mannan biosynthesis in yeast. Biochem Biophys Res Commun 35:144–150

Tomita S, Inoue N, Maeda Y, Ohishi K, Takeda J, Kinoshita T (1998) A homologue of *Saccharomyces cerevisiae* Dpm1p is not sufficient for synthesis of dolichol-phosphate-mannose in mammalian cells. J Biol Chem 273:9249–9254

Wilson IBH, Taylor JP, Webberley MC, Turner NJ, Flitsch SL (1993) A novel mono-branched lipid phosphate acts as a substrate for dolichyl phosphate mannose synthetase. Biochem J 295:195–201

79

PIG-B, GPI-Man Transferase III, Man-(Ethanolaminephosphate)Man-GlcN-(Acyl)PI Mannosyltransferase

Introduction

The core backbone of the glycosylphosphatidylinositol (GPI) anchors contains three mannoses, each linked in different bonds. All three mannoses are donated by dolichol-phosphate-mannose. Three dolichol-phosphate-mannose-dependent mannosyltransferases are therefore involved in biosynthesis of the GPI-anchors. PIG-B is necessary for transfer of the third mannose and is most likely α1,2-mannosyltransferase itself. *PIG-B* cDNA was expression-cloned using a mouse mutant cell line that is defective in transfer of the third mannose to GPI. *PIG-B* encodes a 554-amino-acid protein expressed in the endoplasmic reticulum (ER), where GPI is assembled. It has multiple hydrophobic regions, some of which may be transmembrane domains.

Gpi10p of *Saccharomyces cerevisiae*, an ortholog of PIG-B, consists of 616 amino acids with 33% identity to human PIG-B. Gpi10p is essential for growth of *S. cerevisiae*, and its deletion can be complemented by human *PIG-B* cDNA.

Three open reading frames in the genome of *S. cerevisiae* have homology to Gpi10p: Alg9p (555 amino acids with 18% identity); Smp3p (517 amino acids with 24% identity); and YNR030W (552 amino acids with 18% identity). Alg9p is necessary for transfer of the seventh mannose to *N*-glycan precursor that is linked in α1,2 bond and is donated by dolichol-phosphate-mannose (Burda et al. 1996). Functions of Smp3p and YNR030W have not been reported. It is likely that this group of proteins represent a family of α-mannosyltransferases that utilize dolichol-phosphate-mannose.

Taroh Kinoshita and Norimitsu Inoue

Department of Immunoregulation, Research Institute for Microbial Diseases, Osaka University, 3-1 Yamadaoka, Suita, Osaka 565-0871, Japan
Tel. +81-6-6879-8328; Fax +81-6-6875-5233
e-mail: tkinoshi@biken.osaka-u.ac.jp

Databanks

PIG-B, GPI-Man transferase III, Man-(ethanolaminephosphate)Man-GlcN-(acyl)PI mannosyltransferase

No EC number has been allocated.

Species	Gene	Protein	mRNA	Genomic
Homo sapiens	*PIG-B*	BAA07709	D42138	–
		NP_004846	NM_004855	–
Mus musculus	*Pig-b*	–	D84436	–
Saccharomyces cerevisiae	*GPI10/YGL142C*	NP_011373	–	NC_001139

Name and History

PIG-B cDNA was obtained by means of expression-cloning using a GPI-anchor-deficient mouse T lymphoma cell line, S1Ab (Takahashi et al. 1996). S1Ab was classified as complementation class B of GPI-anchor-deficient mutants (Hyman 1988) and was shown to be defective in transfer of the third mannose (Sugiyama et al. 1991; Puoti and Conzelmann 1993). PIG-B was named for *p*hosphatidyl*i*nositol *g*lycan-class B (Takahashi et al. 1996).

Gpi10p was identified based on sequence homology to PIG-B (Sutterlin et al. 1998). The role of *GPI10* during transfer of the third mannose was verified by depleting its expression and confirming accumulation of intermediates bearing two mannoses (Sutterlin et al. 1998).

Enzyme Activity Assay and Substrate Specificity

In mammalian cells and *S. cerevisiae*, a donor substrate is dolichol-phosphate-mannose and an acceptor substrate is mannoseα1,6-(ethanolaminephosphate-2)mannoseα1,4-glucosamineα1,6-myoinositolphospholipid (Menon et al. 1990; Canivenc-Gansel et al. 1998; Sutterlin et al. 1998). When the ethanolaminephosphate side chain is not added in mammalian cells because of the absence of the responsible enzyme (Hong et al. 1999) or there is an inhibitor of the enzyme (Sutterlin et al. 1997), mannoseα1,6-mannoseα1,4-glucosamineα1,6-myoinositol-phospholipid acts as an acceptor substrate for PIG-B (Hong et al. 1999).

There is no report on in vitro enzyme assay. Enzyme activity in mammalian cells is measured by metabolically labeling the cells with [^{3}H]mannose and detecting the radiolabeled product by thin-layer chromatography (Sugiyama et al. 1991; Takahashi et al. 1996). Human *PIG-B* cDNA complemented *S. cerevisiae gpi10* disruptant, indicating that PIG-B and Gpi10p are interchangeable (Sutterlin et al. 1998).

Preparation

No procedures for preparation of the active enzyme have been reported. Because GPI-anchors are ubiquitous in eukaryotic cells and the third mannose is conserved (Kinoshita et al. 1997), various tissues and cells from various organisms should express PIG-B or its orthologous protein.

Biological Aspects

Biosynthesis of GPI-anchor is essential for growth in yeast. Consistent with this, *gpi10*-disrupted *S. cerevisiae* did not grow (Sutterlin et al. 1998). This finding also indicates that there is no functionally redundant gene.

PIG-B is a membrane protein expressed in the endoplasmic reticulum. Of 554 amino acids of human PIG-B, the amino-terminal 60 residues are on the cytoplasmic side, followed by a transmembrane domain. The amino-terminal cytoplasmic portion was deleted without loss of function. The carboxyl-terminus is present in the ER lumen. The carboxyl-terminal portion of 470 amino acids was protected from proteinase K acted from the cytoplasmic side of the microsomes, being concluded that this portion is on the luminal side and that transfer of the third mannose occurs in the ER lumen (Takahashi et al. 1996).

Future Perspectives

Several lines of evidence indicate that PIG-B is the mannosyltransferase, but direct demonstration of enzyme activity using pure protein and substrates is necessary.

It is possible that PIG-B/Gpi10p has multiple transmembrane domains, as predicted from the hydropathy profiles. This should be experimentally determined to confirm reported luminal orientation of transfer of the third mannose.

Further Reading

Kinoshita et al., 1997

References

Burda P, te Heesen S, Brachat A, Wach A, Dusterhoft A, Aebi M (1996) Stepwise assembly of the lipid-linked oligosaccharide in the endoplasmic reticulum of *Saccharomyces cerevisiae*: identification of the ALG9 gene encoding a putative mannosyl transferase. Proc Natl Acad Sci USA 93:7160–7165

Canivenc-Gansel E, Imhof I, Reggiori F, Burda P, Conzelmann A, Benachour A (1998) GPI anchor biosynthesis in yeast: phosphoethanolamine is attached to the α1,4-linked mannose of the complete precursor glycophospholipid. Glycobiology 8: 761–770

Hong Y, Maeda Y, Watanabe R, Ohishi K, Mishkind M, Riezman H, Kinoshita T (1999) Pig-n, a mammalian homologue of yeast Mcd4p, is involved in transferring phosphoethanolamine to the first mannose of the glycosylphosphatidylinositol. J Biol Chem 274:35099–35106

Hyman R (1988) Somatic genetic analysis of the expression of cell surface molecules. Trends Genet 4:5–8

Kinoshita T, Ohishi K, Takeda J (1997) GPI-anchor synthesis in mammalian cells: genes, their products, and a deficiency. J Biochem 122:251–257

Menon AK, Mayor S, Schwarz RT (1990) Biosynthesis of glycosyl-phosphatidylinositol lipids in *Trypanosoma brucei*: involvement of mannosyl-phosphoryldolichol as the mannose donor. EMBO J 9:4249–4258

Puoti A, Conzelmann A (1993) Characterization of abnormal free glycophosphatidylinos-itols accumulating in mutant lymphoma cells of classes B, E, F, and H. J Biol Chem 268:7215–7224

Sugiyama E, DeGasperi R, Urakaze M, Chang HM, Thomas LJ, Hyman R, Warren CD, Yeh ETH (1991) Identification of defects in glycosylphosphatidylinositol anchor biosyn-thesis in the Thy-1 expression mutants. J Biol Chem 266:12119–12122

Sutterlin C, Escribano MV, Gerold P, Maeda Y, Mazon MJ, Kinoshita T, Schwarz RT, Riezman H (1998) *Saccharomyces cerevisiae* GPI10, the functional homologue of human PIG-B, is required for glycosylphosphatidylinositol-anchor synthesis. Biochem J 332:153–159

Sutterlin C, Horvath A, Gerold P, Schwarz RT, Wang Y, Dreyfuss M, Riezman H (1997) Iden-tification of a species-specific inhibitor of glycosylphosphatidylinositol synthesis. EMBO J 16:6374–6383

Takahashi M, Inoue N, Ohishi K, Maeda Y, Nakamura N, Endo Y, Fujita T, Takeda J, Kinoshita T (1996) PIG-B, a membrane protein of the endoplasmic reticulum with a large lumenal domain, is involved in transferring the third mannose of the GPI anchor. EMBO J 15:4254–4261

80

Dolichol Phosphate GlcNAc-1-P Transferase

Introduction

Asparagine-linked glycosylation is a covalent modification of secretory and integral membrane proteins in eukaryotes that modulates the structure and function of these proteins. Dolichol-linked oligosaccharides are transferred en bloc to selected asparagine residues of nascent polypeptides. Biosynthesis of the dolichol-bound oligosaccharides is initiated by formation of GlcNAc-P-P-dolichol from UDP-GlcNAc and dolichol phosphate. This reaction is catalyzed by the endoplasmic reticulum (ER) residential enzyme UDP-GlcNAc:dolichol phosphate *N*-acetylglucosamine-1-phosphate transferase (GPT). The two substrates, dolichol phosphate and UDP-GlcNAc, can serve other pathways, but the product of GPT activity can only participate in lipid-bound oligosaccharide biosynthesis. Thus, GPT catalyzes the committed step of the dolichol cycle involved in asparagine-linked glycosylation.

Databanks

Dolichol phosphate GlcNAc-1-P transferase

NC-IUBMB enzyme classification: E.C.2.7.8.15

Species	Gene	Protein	mRNA	Genomic
Homo Sapiens	–	–	Z82022	–
Mus musculus	–	P42867	U03603	–
			X65603	
Cricetulus griseus	–	P24140	U09453	–
			M36899	
Cricetulus longicaudatus	–	P23338	J05590	–
			M22755	

Andreas Hübel and Ralph T. Schwarz

Center for Hygiene and Medical Microbiology, Philipps-University Marburg, Robert-Koch-Strasse 17, 35037 Marburg, Germany
Tel. +49-6421-2865149; Fax +49-6421-2868976
e-mail: schwarz@mailer.uni-marburg.de

(Continued)

Species	Gene	Protein	mRNA	Genomic
Caenorhabditis briggsae	–	–	–	Y60A3.I
				Y60A3.H
Saccharomyces cerevisiae	–	P07286	–	Y00126
				Z36112
Schizosaccharomyces pombe	–	P42881	U09454	AL03139
Coprinus cinereus	–	–	–	X98860
Leishmania mexicana	–	P42864	M96635	–
Plasmodium falciparum	–	–	–	AL034559
Toxoplasma gondii	–	–	N59933	–
Arabidopsis thaliana	–	–	–	D88036–D88037

Name and History

The structure of dolichol-linked oligosaccharides and the order of their biosynthesis are highly conserved among eukaryotes. The latter is initialized by GlcNAc-1-phosphate transferase, which is also abbreviated GPT. GPT activity was found in a number of unicellular organisms and a variety of cells and tissues of multicellular species.

GPT can be specifically inhibited by tunicamycin (Tn), and GPT candidate genes were obtained by screening for resistance to tunicamycin. The first tunicamycin-resistance gene (TRG) was cloned from a *Saccharomyces cerevisiae* genomic library whose overexpresion conferred increased GPT activity in yeast as well as 75-fold resistance to Tn (Rine et al. 1983). The identified protein coding region encoded a protein of 448 amino acids (Hartog and Bishop 1987), which corresponds to the yeast ALG7 gene, to which dominant mutations that lead to Tn resistance have been mapped (Barnes et al. 1984). The *S. cerevisiae* ALG7 gene is also referred to as TUR1, and its protein coding region is designated YBR243C in the yeast genome project.

To identify TRGs in mammalian cells, CHO cells were adapted to high levels of Tn resistance, which also correlated with 10- to 20-fold higher GPT activity and 40- to 50-fold higher copy number of genes that cross-hybridize to the yeast ALG7 gene (Lehrman et al. 1988; Scocca and Krag 1990). Full-length cDNAs were isolated, and their overexpression in COS cells led to increased GPT activity in microsomal membranes (Zhu and Lehrmann 1990). Biochemical and immunochemical studies using anti-TRG antisera and CHO cells that overexpress TRG ruled out alternative explanations for the identity of TRG and further supported the conclusion that TRG encodes GPT (Zhu et al. 1992). Furthermore, the predicted amino acid sequence of hamster TRG and yeast ALG7p are similar.

Later, human GPT was isolated from a human lung fibroblast cDNA library by heterologous complementation of a conditional-lethal yeast strain (YPH-A7-GAL) in which the endogenous ALG7 promoter was replaced by the regulatable GAL1 promoter (Eckert et al. 1998). Despite a moderate identity of only 42% between yeast and human GPT functional expression of the latter gene in *S. cerevisiae* YPH-A7-GAL restored Tn-sensitive GlcNAc-P-P-dolichol biosynthesis at nonpermisssive growth conditions.

Analogous to isolation of the GPT gene from CHO cells, the GlcNAc-1-P transferase gene of the human pathogenic protozoan *Leishmania mexicana amazonensis* was identified owing to the extrachromosomal amplification of a single 63-kbp chromosomal region in Tn-resistant parasite lines (Detke et al. 1988). Tn resistance correlated not only with elevated GPT activity but also with virulence in these pathogens (Kink and Chang 1987).

For GPT genes cloned from other species, refer to the Database section. The increasing number of genome projects will undoubtedly donate additional GPT homolog genes to the databases.

Enzyme Activity Assay and Substrate Specificity

GPT catalyzes the reaction:

$$UDP\text{-}GlcNAc + dolichol\ phosphate \rightarrow GlcNAc\text{-}P\text{-}P\text{-}dolichol + UMP$$

Although it was shown in vitro that this reaction is reversible in principle (Heifetz et al. 1979), it is unlikely to occur in vivo because the UMP levels are not expected to reach the required level.

GPT activity was measured from intact membranes as well as solubilized enzyme preparations with a K_m of about 1×10^{-6} M (Keller et al. 1979). The enzyme requires divalent cations, preferable Mg^{2+} or Mn^{2+} (Kean 1983; Kaushal and Elbein 1985). α-Saturated polyprenyl phosphates were much better acceptors than the corresponding allylic polyprenyl phosphates of either yeast or rat liver GPT (Palamarczyk et al. 1980). Dolichol phosphates of 20 or 35 carbon atoms were not accepted by GPT, whereas dolichol phosphates with increasing chain lengths between C55 and C100 had similar Vmax values but decreasing K_m values.

Phospholipids are required for mammalian GPT activity, as inferred from enzyme purified from rat liver cells or mammary gland (Kaushal and Elbein 1985; Shailubhai et al. 1988). In particular phosphatidylglycerol, glycerol, UDP-GlcNAc, and dolichol phosphate are needed to maintain enzyme stability. A 1:1 mixture of phosphatidylglycerol and phosphatidylcholine exhibited even more activity than phosphatidylglycerol alone. In yeast only phosphatidylinositol (no other phospholipid) was capable of restoring GPT activity after inositol starvation (Hanson and Lester 1982).

Activation of GPT in vitro can be achived by exogenous dolichol phosphate mannose or GDP-mannose but not by other nucleotide sugars or dolichol phosphate glucose (Kean 1985). Furthermore, porcine aorta cells contain a factor that activates GPT four- to five-fold. This unknown factor is a heat-stable, pronase- and RNAase-resistant, trichloroacetic acid (TCA)-precipitable compound that is insoluble in organic solvent. It could not be replaced by phospholipids or dolichol phosphate mannose or by inhibitors of UDP-GlcNAc degradation (Kaushal and Elbein 1985).

Tunicamycin is a potent, specific inhibitor of GPT in vitro and in vivo and has been applied for analyzing the biological activities of glycoproteins (Tamura 1982). It has structural similarities to dolichol phosphate and UDP-GlcNAc (Elbein 1987). Tn binds to GPT with an 50% inhibitory concentration (IC_{50}) of 7×10^{-9} M. Its replaceability by

UDP-GlcNAc but not by dolichol phosphate indicates that Tn is a transition state analog that binds to the catalytic site of GPT.

Although not specifically, formation of GlcNAc-P-P-dolichol can be inhibited by amphomycin. Amphomycin forms a complex with dolichol monophosphate and thus inhibits any dolichol phosphate-specific glycosyltransferases (Banerjee 1989).

Preparation

GPT has been detected in many eukaryotic organisms, ranging from the ameba to humans. It is widely distributed in various mammalian tissues such as retina, aorta, liver, lymphoma, and mammary gland. It is a protein of the ER with presumably 10 membrane-spanning domains (Dan et al. 1996). Thus, most biochemical data were obtained from microsomal or membrane preparations. Nonionic and ionic detergents were used successfully to solubilize the GPT activity (Kaushal and Elbein 1985). The enzyme was partially purified from porcine aorta (Kaushal and Elbein 1985) and to homogeneity from bovine mammary gland, where GPT activity is increased 9.5-fold at mid-lactation (Shailubhai et al. 1988).

Biological Aspects

Expression of N-linked glycoproteins and GPT is developmentally and hormonally regulated in mammary glands. Lactogenic hormones, insulin, glucocorticoids, and prolactin induce transcription of mouse mammary GPT (Rajput et al. 1994). A negative regulatory element was identified in transient transfection experiments with primary mouse mammary epithelial cells using GPT promoter/luciferase chimera (Ma et al. 1996). Deletion of this distal region enhanced hormonal induction of reporter gene expression. Two pentameric direct repeat motifs (AGGAA and GAAAC) were identified in this negative regulatory element and were proven to repress GPT gene expression in vivo.

In addition to the transcriptional activation of GPT expression in mammals, post-transcriptional mechanisms were proposed to regulate GPT expression during mouse mammary development (Rajput et al. 1994). Accordingly, it was shown that two short sequences (3 amino acids close to the carboxyl-terminus and the last 11 amino acids) of hamster GPT are sufficient for expression, but neither is necessary (Zara and Lehrman 1994). Removal of the last 11 amino acids led to an enzymatically active but thermolabile protein. Removal or scrambling of an additional 3 amino acids eliminated enzyme expression. However, reattachment of the last 11 amino acids to the scrambled GPT restored expression. Thus, the 3-amino-acid motif Phe^{395}-Ser^{396}-Ile^{397} is thought to stabilize GPT.

Chemical cross-linking experiments indicate that GPT forms functional homo-dimers, and they influence each other, as inferred from dominant-negative effects caused by co-expression of an enzymatically inactive GPT mutant (Dan and Lehrman 1997). This oligomerization was favored by nonionic detergent rather than promotion of nonspecific aggregation. The functional relevance of oligomerization in vivo remains to be elucidated.

Transcript heterogeneity of GPT genes has been observed in CHO cells and in yeast (Kukuruzinska and Robbins 1987; Huang et al. 1998). Different mRNAs (1.5 and 1.9 kb) in mammalian cells might lead to related protein isoforms with different amino-terminal portions. In yeast the two major transcripts (1.4 and 1.6 kb) seem to be involved in the regulation of ALG7 gene expression, as mRNA ratios change depending on Alg7 expression levels and in mutants defective in lipid-linked saccharide-donor synthesis.

Comparison of amino acid sequences of ALG7, dolichol phosphate mannose transferase, and ALG1 revealed a consensus sequence of a 13-amino-acid residue in the presumed transmembrane domains. This consensus sequence is belived to be involved in dolichol recognition, as these proteins are known to bind the isoprenoid region of dolichol phosphates (Albright et al. 1989) and are found in yeast and mamalian TRGs.

In *Saccharomyces cerevisiae* ALG7 expression is highly regulated during the cell cycle (Kukuruzinska and Lennon-Hopkins 1999). ALG7 mRNA accumulates if cells are stimulated to resume proliferation after being arrested in G_0, but decreases in a time-dependent manner in response to antimitotic drugs that lead to cell cycle arrest in G_0. Deregulation of ALG7 gene expression interferes with the cell cycle by maintaining elevated cyclin mRNA levels.

Gene disruption experiments and promoter replacement studies demonstrated that TRG expression is essential for the viability of yeast (Kukuruzinska and Robbins 1987; Eckert et al. 1998).

Future Perspectives

Inhibiton of dolichol-linked oligosaccharide biosynthesis by cycloheximide or puromycin (known inhibitors of protein biosynthesis), actinomycin D (an inhibitor of RNA polymerase), or certain amino acid analogs indicate a regulatory role of dolichol-linked oligosaccharides in the biosynthesis of *N*-linked glycoproteins (Lehrman 1991, and references therein). In addition, cell cycle-dependent regulation of ALG7 expression in yeast, which resembles properties of mammalian early growth response genes, indicates a role for ALG7 in cell proliferation and differentiation. The mechanisms underlying this regulation are important aspects to study in the near future.

Further Reading

For comprehensive reviews of the biochemical properties of GlcNac-1-phosphate transferases and biological importance, see Lehrman (1991), Burda and Aebi (1999), and Kukuruzinska and Lennon (1998).

Acknowledgments

The work of the authors was supported by Deutsche Forschungsgemeinschaft, Fonds der Chemischen Industrie, and Stiftung P.E. Kempkes.

References

Albright CF, Orlean P, Robbins PW (1989) A 13-amino acid peptide in three yeast glycosyltransferases may be invoved in dolichol recognition. Proc Natl Acad Sci USA 86:7366–7369

Banerjee DK (1989) Amphomycin inhibits mannosylphosphoryldolichol synthesis by forming a complex with dolichylmonophosphate. J Biol Chem 264:2024–2028

Barnes G, Hansen WJ, Holcomb CL, Rine J (1984) Asparagine-linked glycosylation in *Saccharomyces cerevisiae*: genetic analysis of an early step. Mol Cell Biol 4:2381–2388

Burda P, Aebi M (1999) The dolichol pathway of *N*-linked glycosylation. Biochim Biophys Acta 1426:239–257

Dan N, Lehrman MA (1997) Oligomerization of hamster UDP-GlcNAc:dolichol-P GlcNAc-1-P transferase, an enzyme with multiple transmembrane spans. J Biol Chem 272:14214–14219

Dan N, Middleton RB, Lehrman MA (1996) Hamster UDP-*N*-acetylglucosamine:dolichol-P *N*-acetylglucosmaine-1-P transferase has multiple transmembrane spans and a critical cytosolic loop. J Biol Chem 271:30717–30724

Detke S, Chaudhuri G, Kink JA, Chang KP (1988) DNA amplification in tunicamycin-resistant *Leishmania mexicana*: multicopies of a single 63-kilobase supercoiled molecule and their expression. J Biol Chem 263:3418–3424

Eckert V, Blank M, Mazhari-Tabrizi R, Mumberg D, Funk M, Schwarz RT (1998) Cloning and functional expression of the human GlcNAc-1-P transferase, the enzyme for the committed step of the dolichol cycle, by heterologous complementation in *Saccharomyces cerevisiae*. Glycobiology 8:77–85

Elbein AD (1987) Inhibitors of the biosynthesis and processing of *N*-linked oligosaccharide chains. Annu Rev Biochem 56:497–534

Hanson BA, Lester RL (1982) Effect of inositol starvation on the in vitro synthesis of mannan and *N*-acetylglucosaminylpyrophosphoryldolichol in *Saccharomyces cerevisiae*. J Bacteriol 151:334–342

Hartog KO, Bishop B (1987) Genomic sequences coding for tunicamycin resistance in yeast. Nucleic Acids Res 15:3627

Heifetz A, Keenan RW, Elbein AD (1979) Mechanism of action of tunicamycin on the UDP-GlcNAc:dolichol-phosphate GlcNAc-1-phosphate transferase. Biochemistry 18:2186–2192

Huang GT, Lennon K, Kukuruzinska MA (1998) Characterization of multiple transcripts of the hamster dolichol-P-dependent *N*-acetylglucosmanie-1-P transferase suggests functionally complex expression. Mol Cell Biol 18:97–106

Kaushal GP, Elbein AD (1985) Purification and properties of UDP-GlcNAc:dolichol-phosphate GlcNAc-1-phosphate transferase. J Biol Chem 260:16303–16309

Kean EL (1983) Influence of metal ions on the biosynthesis of *N*-acetylglucosaminyl polyprenols by the retina. Biochim Biophys Acta 750:268–273

Kean EL (1985) Stimulation by dolichol phosphate-mannose and phospholipids of the biosynthesis of *N*-acetylglucosaminylpyrophosphoryl dolichol. J Biol Chem 260:12561–12571

Keller RK, Boon DY, Crum FC (1979) *N*-Acetylglucosamine-1-phosphate transferase from hen oviduct: solubilization, characterization, and inhibition by tunicamycin. Biochemistry 18:3946–3952

Kink JA, Chang KP (1987) Tunicamycin-resistant *Leishmania mexicana amazonensis*: expression of virulence associated with an increased activity of *N*-acetylglucosaminyl transferase and amplification of its presumptive gene. Proc Natl Acad Sci USA 84:1253–1257

Kukuruzinska MA, Lennon K (1998) Protein *N*-glycosylation: molecular genetics and functional significance. Crit Rev Oral Biol Med 9:415–448

Kukuruzinska MA, Lennon-Hopkins K (1999) ALG gene expression and cell progression. Biochim Biophys Acta 1426:359–372

Kukuruzinska MA, Robbins PW (1987) Protein glycosylation in yeast: transcript heterogeneity of the ALG7 gene. Proc Natl Acad Sci USA 84:2145–2149

Lehrman MA (1991) Biosynthesis of N-acetylglucosamine-P-P-dolichol, the committed step of asparagine-linked oligosaccharide assembly. Glycobiology 1:553–562

Lehrman MA, Zhu X, Khounlo S (1988) Amplification and molecular cloning of the hamster tunicamycin-sensitive N-acetylglucosamine-1-phosphate transferase gene: the hamster and yeast enzyme share a common peptide sequence. J Biol Chem 263:19796–19803

Ma J, Saito H, Oka T, Vijay IK (1996) Negative regulatory element involved in the hormonal regulation of GlcNAc-1-P transferase in mouse mammary gland. J Biol Chem 271:11197–11203

Palamarczyk G, Lehle L, Mankowski T, Chonacki T, Tanner W (1980) Specificity of solubilized yeast glycosyltransferases for polyprenyl derivatives. Eur J Biochem 263: 17499–17507

Rajput B, Muniappa N, Vijay IK (1994) Developmental and hormonal regulation of UDP-GlcNAc:dolichol phosphate GlcNAc-1-P transferase in mouse mammary gland. J Biol Chem 269:16054–16061

Rine J, Hansen W, Hardman E, Davis RW (1983) Targeted selesction of recombinant clones through gene dosage effects. Proc Natl Acad Sci USA 80:6750–6754

Scocca JR, Krag SS (1990) Sequence of a cDNA that specifies the uridine diphosphate N-acetyl-D-glucosamine:dolichol phosphate N-acetylglucosamine-1-phosphate transferase from Chinese hamster ovary cells. J Biol Chem 265:20621–26026

Shailubhai K, Dong-Yu B, Saxena ES, Vijay IK (1988) Purification and characterization of UDP-N-acetylglucosamine:dolichol phosphate N-acetyl-D-glucosamine-1-phosphate transferase involved in the biosynthesis of asparagine-linked glycoprotein in the mammary gland. J Biol Chem 263:15964–15972

Tamura G (ed) (1982) Tunicamycin. Japan Scientific Societies Press, Tokyo

Zar J, Lehrman MA (1994) Role of the carboxyl terminus in stable expression of hamster UDP-GlcNAc:dolichol phosphate GlcNAc-1-P transferase. J Biol Chem 269: 19108–19115

Zhu X, Lehrmann MA (1990) Cloning, sequence and expression of a cDNA encoding hamster UDP-GlcNAc:dolichol phosphate N-acatylglucosamine-1-phosphate transferase. J Biol Chem 265:14250–14255

Zhu X, Zeng Y, Lehrman MA (1992) Evidence that the hamster tunicamycin resistance gene encodes UDP-GlcNAc:dolichol phosphate N-acetylglucosamine-1-phosphate transferase. J Biol Chem 267:8895–8902

81

ALG3 Mannosyltransferase

Introduction

N-Linked glycosylation is an essential eukaryotic protein modification. An oligosaccharide consisting of 14 sugar residues is synthesized at the endoplasmic reticulum (ER) membrane before being transferred to a protein (Kornfeld and Kornfeld 1985; Varki 1996; Burda and Aebi 1999). Dolichol phosphate is used as a lipid carrier for the synthesis, which is initiated by the addition of phospho-GlcNAc forming GlcNAc-pyrophosphate-dolichol (GlcNAc-PP-Dol). This is followed, in order, by additions of another GlcNAc, nine mannose, and three glucose residues. On the cytoplasmic side of the ER, activated sugars (UDP-GlcNAc and GDP-Man) are the donors for the two GlcNAc and first five mannose additions, respectively. The $Man_5GlcNAc_2$-PP-Dol is then translocated across the membrane into the ER lumen, where dolichol-phosphate-mannose-dependent mannosyltransferases sequentially add four mannose residues. (For further information on Dol-P-Man, see Chapter 78). The synthesis is completed by the transfer of three glucose residues from Dol-P-Glc to the oligosaccharide. (For a description of Dol-P-Glc-dependent transferases, see Chapter 82). The $Glc_3Man_9GlcNAc_2$ oligosaccharide is transferred to nascent glycoproteins by the oligosaccharyltransferase complex (see Chapter 83).

Several *Saccharomyces cerevisiae* mutants accumulating biosynthetic intermediates of the N-glycosylation pathway have been generated and characterized (see below). Deficiencies in these proteins cause accumulation of characteristic lipid-linked oligosaccharide intermediates. To date, 11 yeast genes have been determined to be required for 11 of the 14 monosaccharide transfer reactions. One gene has been found for each transfer reaction, prompting the "one linkage, one glycosyltranferase" hypothesis (Schachter 1995). At least three genes (*ALG3, ALG9, ALG12*) are required for the ER-luminal mannosyltransferase reactions (Fig. 1).

JONNE H. HELENIUS and CLAUDE A. JAKOB

Insitute for Microbiology, ETH Zürich, Schmelzbergstrasse 7, CH-8092 Zürich, Switzerland
Tel. +41-1-632-3327; Fax +41-1-632-1148
e-mail: jakob@micro.biol.ethz.ch

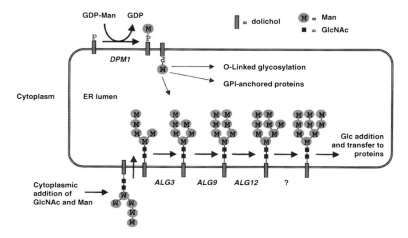

Fig. 1. Mannosyltransferases of the endoplasmic reticulum (ER) lumen. *ALG3*, *ALG9*, and *ALG12* are required for the addition of the first three of four mannose residues to Man$_5$GlcNAc$_2$-PP-Dol inside the lumen. Substrates Man$_5$GlcNAc$_2$-PP-Dol and Dol-P-Man are synthesized on the cytoplasmic face of the ER and translocated into the lumen. Fully mannosylated Man$_9$GlcNAc$_2$-PP-Dol is subsequently glucosylated, and the oligosaccharide is transferred to secretory proteins

The *ALG3* gene is essential for the first of the four dolichol-phosphate-mannose-dependent mannosyltransfer reactions (Aebi et al. 1996). The transferase adds a mannose residue via an α1,3 linkage to the α1,6-linked mannose of the B-arm of the Man$_5$GlcNAc$_2$-PP-Dol oligosaccharide. Alg3p likely represents the mannosyltransferase, but the final in vitro experiment with purified reconstituted components remains to be performed. Homologs of the yeast *ALG3* locus have been found in several eukaryotes including humans. The genes involved in oligosaccharide biosynthesis have gained attention because deficiency can cause congenital disorder of glycosylation (CDG) type I. A mutation in the human homolog of yeast *ALG3* has been identified as the cause of CDG type Id (Korner et al. 1999).

Databanks

ALG3 mannosyltransferase

EC number: 4.1.130

Species	Gene	Protein	mRNA	Genomic
Homo sapiens	Not56-like	Q92685	Y09022	–
Mus musculus	–	–	AA215144	–
Drosophila melanogaster	Not56	CAA64532	X95243	–
Caenorhabditis elegans	–	CAB70171	–	Z83234
Saccharomyces cerevisiae	ALG3	S45424	Y13134	Z35844
Schizosaccharomyces pombe	–	CAB16723	–	Z99532
Arabidopsis thaliana	Not56-like	AAC63631	–	AC005309
				AE002093

Name and History

Various yeast screening techniques have led to the isolation of loci involved in N-linked oligosaccharide biosynthesis. Robbins and coworkers performed a mannose-suicide selection screen to isolate yeast strains with deficiencies in N-linked glycosylation (Huffaker and Robbins 1981). Mutants deficient in mannose incorporation were selected. Chemically mutated yeast cells were labeled with [^{3}H]-mannose and frozen for 35 days. Survivors displaying intact translation but reduced glycosylation of secretory proteins and increased levels of truncated lipid-linked oligosaccharides were named *alg* (*a*sparagine-*l*inked *g*lycosylation) mutants (Huffaker and Robbins 1981). The screen resulted in mutation in six *ALG* loci including *alg3-1* (Huffaker and Robbins 1983).

Other screens have identified mutations in *ALG* genes by various methods. Screens making use of increased sensitivity to aminoglycosides (Dean 1994; Shimma et al. 1997) and resistance to *Hansenula mrakii* killer toxin (Kasahara et al. 1994) or calcofluor white (Ram et al. 1994) due to defects in glycosylation have been performed. Further loci were identified by purification of protein activity with subsequent cloning of the gene (Orlean et al. 1988) and by comparing genome sequences with genes involved in N-glycosylation (Burda and Aebi 1999).

The *alg3-1* mutant strain accumulates the Man$_5$GlcNAc$_2$ form of lipid-linked oligosaccharide and underglycosylates invertase, but it lacks a growth phenotype (Huffaker and Robbins 1983). By combining the *alg3-1* mutation with a mutation in

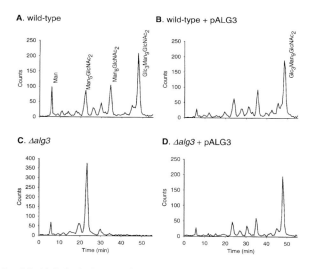

Fig. 2. Analysis of lipid-linked oligosaccharides of *ALG3*-proficient and *ALG3*-deficient yeast cells. In vivo determination of lipid-linked oligosaccharides was performed by labeling yeast strains with [^{3}H]mannose followed by organic phase extraction and high-performance liquid chromatography (HPLC) separation (Aebi et al. 1996). Glc$_3$Man$_9$GlcNAc$_2$ is the predominant form in wild-type yeast (**A**), and deletion of *ALG3* results in accumulation of the Man$_5$GlcNAc$_2$ form (**C**). Addition of a plasmid bearing *ALG3* restores the pathway (**D**) in the Δ*alg3* strain, whereas it does not affect the wild-type strain (**B**)

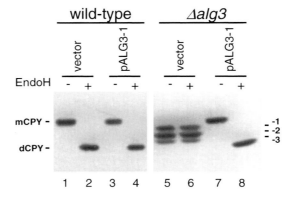

Fig. 3. Hypoglycosylation of carboxypeptidase Y (CPY) in an *ALG3*-deficent yeast strain. CPY glycosylation was visualized by Western analysis using a specific antiserum. The mature CPY (mCPY) is converted to its deglycosylated form (dCPY) by EndoH treatment (lanes 1–4). Deletion of *ALG3* causes underglycosylation of CPY (lane 5), and oligosaccharides (Man$_5$GlcNAc$_2$) are resistant to EndoH digestion (lane 6). Both effects are restored by introduction of a plasmid containing the *ALG3* locus (lanes 7 and 8)

the oligosaccharyltransferase subunit, *STT3*, a synthetic phenotype results. The *ALG3* gene was identified by complementing the temperature-sensitive growth defect (Aebi et al. 1996). The *ALG3* gene restores the biochemical defect of the *alg3-1* mutant. Moreover, deletion of the nonessential *ALG3* gene results in accumulation of lipid-linked Man$_5$GlcNAc$_2$ (Fig. 2), underglycosylation of secretory proteins, and protein-bound carbohydrates that were resistant to endoglycosidase H (Fig. 3). Three potential defects lead to the accumulation of the Man$_5$GlcNAc$_2$ form of lipid-linked oligosaccharide: (1) a defect of the Alg3 transferase reaction; (2) a lack of Dol-P-Man; and (3) retention of Man$_5$GlcNAc$_2$-PP-Dol on the cytoplasmic side of the ER membrane (Fig. 1). Dol-P-Man is not limited by a deficiency in *ALG3*, as O-linked glycosylation and glycosylphosphatidylinositol (GPI)-anchored proteins are unaffected (Conzelmann et al. 1991; Orlean 1994) and Man$_5$GlcNAc$_2$-PP-Dol must be present inside the ER as it is transferred to proteins in *alg3-1* cells (Verostek et al. 1991). Alg3p is assumed to be the transferase (Aebi et al. 1996).

Concurrent with cloning of *S. cerevisiae ALG3*, the *Drosophila melanogaster* homologs were sequenced and named l(2)not+ [*lethal* (2) *n*eighbor to *t*id] or NOT56 (Kurzik-Dumke et al. 1997). Thereafter, the same investigators sequenced a similar transcript in human and termed the protein NOT56-like. As the Databases section demonstrates, homologs of *ALG3* have be found in several eukaryotes.

Enzyme Activity Assay and Substrate Specificity

Alg3p has been shown to be essential for addition of the sixth mannose residue of the lipid-linked oligosaccharide in the reaction below (Aebi et al. 1996).

$$Dol\text{-}P\text{-}Man + Man_5GlcNAc_2\text{-}PP\text{-}Dol \rightarrow Man_6GlcNAc_2\text{-}PP\text{-}Dol + Dol\text{-}P$$

The mannose residue is transferred from the Dol-P-Man carrier, forming an α1,3 linkage with $Man_5GlcNAc_2$-PP-Dol (Verostek et al. 1991). The transfer likely occurs on the luminal side of the ER membrane, as demonstrated for the Dol-P-Glc-dependent transferases of the pathway (Rush and Waechter 1998).

Sharma et al. (1990) partially purified and characterized the Alg3 transferase activity of pig aorta. The activity was assayed in the presence of detergent using purified $Man_5GlcNAc_2$-PP-Dol. To determine the transfer rates, the lipid-linked oligosaccharide product was separated from the radiolabeled monosaccharide donor by organic phase extraction. Among the tested donors, only Dol-P-Man showed activity. The Dol-P-Man-dependent activity was shown to have a pH optimum of 6.5 and to be Ca^{2+}-dependent. Dol-P-[^{14}C]Man and $Man_5GlcNAc_2$-PP-Dol substrates required enzymatic synthesis using pig aorta extracts.

To determine the efficiency of mannose transfer onto $Man_5GlcNAc_2$-PP-Dol in human fibroblasts, a microsome assay was developed (Korner et al. 1999). Microsomes isolated from cultured fibroblasts are incubated with purified [^{3}H]$Man_5GlcNAc_2$ lipid-linked oligosaccharides (isolated from labeled $\Delta alg3$ yeast strain) and Dol-P-[^{14}C]Man (isolated from labeled yeast). After incubation the lipid-linked oligosaccharides are extracted and analyzed by high-performance liquid chromatography (HPLC), allowing better characterization of the products. Similar assays using yeast microsomes have been performed.

Specific inhibitors of the Dol-P-Man-dependent transferase reaction are not known. The activity can be indirectly inhibited by amphomycin, which inhibits Dol-P-Man synthesis by forming a complex with Dol-P (Banerjee 1989). As amphomycin blocks formation of Dol-P-Man and Dol-P-Glc, it inhibits several N-linked glycosylation reactions (Elbein 1991). Furthermore, as GPI anchoring and O-linked glycosylation require Dol-P-Man (Orlean 1990), they are also affected.

Biological Aspects

Alg3 protein is a constituent of the essential N-glycosylation pathway. Once transferred to protein, the oligosaccharide has been shown to be involved in a wide variety of vital cellular processes, including protein folding, protein sorting, cell wall integrity, and cell signaling (Ellgaard et al. 1999). CDGs comprise a group of severe genetic diseases caused by defects in the synthesis and processing of N-linked carbohydrates. (For further information on CDGs see Chapter 82). Patients suffer from a diverse set of pleiotropic symptoms, including developmental problems, convulsive disease, liver dysfunction, and mental retardation (Orlean 2000). CDG type Id is caused by a mutation in the NOT56-like gene (see http://www.ncbi.nlm.nih.gov/omim/, accession no. 601110) (Korner et al. 1999).

Transcription data in yeast indicate that synthesis of lipid-linked oligosaccharides is regulated at the initial committed step—transfer of the first GlcNAc residue to dolichol phosphate—encoded by the *ALG7* locus (Kukuruzinska and Lennon-Hopkins 1999).

Absence of the mannose added by the *ALG3*-dependent transferase affects many subsequent glycosyl transfer reactions. The named mannose residue not only allows extension of the B-branch, it is required for initiation of the C-branch (Huffaker and

Robbins 1983). With the presence of this mannose residue, the efficiencies of ER glucosyltransferases (Alg6p, Alg8p, Alg10p) and trimming glucosidases (Gls1p and Gls2p) reach near wild-type levels, even in the absence of further mannose additions. Furthermore, in the Golgi Och1p, activity is significantly and specifically increased by the presence of this mannose residue (Cipollo and Trimble 2000).

Whereas N-linked glycosylation is essential in yeast, *ALG3* is not (Aebi et al. 1996). Deficiencies in monosaccharide transfer reactions occurring on the cytoplasmic side of the ER are lethal or growth inhibiting, whereas the lack of luminal glycosyltransferases does not affect yeast growth (Burda and Aebi 1999). An essential step is the translocation ("flipping") of the oligosaccharide into the lumen of the ER. In the absence of fully assembled lipid-linked oligosaccharides, truncated oligosaccharides are transferred to proteins, though less efficiently. For the transfer of oligosaccharides to occur, however, they must be present inside the ER where the oligosaccharyltransferase is. Hence, the absence of Alg3p activity results in underglycosylation but not growth arrest.

Higher eukaryotes are more sensitive to defects in protein glycosylation. The *ALG3* homolog in *Drosophila melanogaster*, NOT56, was found in a lethality screen and was shown to be translated ubiquitously throughout embryonic life (Kurzik-Dumke et al. 1997). The defect of a CDG type Id patient suffering from debilitating symptoms was shown to be caused by a mutation in the NOT56-like gene. By heterologous expression in a Δ*alg3* yeast strain it was demonstrated that the point mutation in NOT56-like homolog reduced but did not abolish the transfer reaction (Korner et al. 1999).

Related Mannosyltransferases

Yeast genes *ALG9* and *ALG12* have been shown to be required for transfer of the seventh and eighth mannoses to lipid-linked oligosaccharides (Fig. 1) (Burda et al. 1996, 1999). The mannose transferase reactions shown are similar to the reaction requiring *ALG3*, and all three share the same NC-IUBMB enzyme classification (EC2.4.1.130) despite catalyzing the addition of mannose residues via different linkages.

$$ALG9\text{: Dol-P-Man} + Man_6GlcNAc_2\text{-PP-Dol} \rightarrow Man_7GlcNAc_2\text{-PP-Dol} + Dol\text{-P}$$

$$ALG12\text{: Dol-P-Man} + Man_7GlcNAc_2\text{-PP-Dol} \rightarrow Man_8GlcNAc_2\text{-PP-Dol} + Dol\text{-P}$$

Both transfer reactions likely occur at the luminal side of the ER and use Dol-P-Man as the substrate. *ALG 9* and *ALG12* show slight homology to each other but not to *ALG3*. In vitro assays have not been developed for either reaction, but microsome assays resembling the *ALG3* assay are possible. A gene required for addition of the final mannose remains to be identified. No yeast mutants accumulating the $Man_8GlcNAc_2$ form have been found.

Note Added in Proof

Recent studies using purified protein in an in vitro assay prove that the *Saccharomyces cerevisiae ALG3* gene alone encodes the Dol-P-Man:$Man_5GlcNAc_2$-PP-Dol mannosyltransferase [Sharma CB, Knauer R, Lehle L (2001) Biosynthesis of lipid-linked oligosaccharides in yeast: the ALG3 gene encodes the Dol-P-Man:Man$_5$GlcNAc$_2$-PP-Dol mannosyltransferase. J Biol Chem 382:321–328].

Further Reading

See Burda and Aebi (1999) and Orlean (2000).

References

Aebi M, Gassenhuber J, Domdey H, te Heesen S (1996) Cloning and Characterization of the *ALG3* gene of *Saccharomyces cerevisiae*. Glycobiology 6:439–444

Banerjee DK (1989) Amphomycin inhibits mannosylphosphoryldolichol synthesis by forming a complex with dolichylmonophosphate. J Biol Chem 264:2024–2028

Burda P, Aebi M (1999) The dolichol pathway of N-glycosylation. Biochim Biophys Acta 1426:239–257

Burda P, Jakob CA, Beinhauer J, Hegemann JH, Aebi M (1999) Ordered assembly of the asymmetrically branched lipid-linked oligosaccharide in the endoplasmic reticulum is ensured by the substrate specificity of the individual glycosyltransferases. Glycobiology 9:617–625

Burda P, te Heesen S, Aebi M (1996) Stepwise assembly of the lipid-linked oligosaccharide in the endoplasmic reticulum of *Saccharomyces cerevisiae*: identification of the *ALG9* gene encoding a putative mannosyl transferase. Proc Natl Acad Sci USA 93:7160–7165

Cipollo JF, Trimble RB (2000) The accumulation of Man(6)GlcNAc(2)-PP-dolichol in the *Saccharomyces cerevisiae* Δalg9 mutant reveals a regulatory role for the Alg3p α1,3-Man middle-arm addition in downstream oligosaccharide-lipid and glycoprotein glycan processing. J Biol Chem 275:4267–4277

Conzelmann A, Fankhauser C, Puoti A, Desponds C (1991) Biosynthesis of glycophosphoinositol anchors in *Saccharomyces cerevisiae*. Cell Biol Int Rep 15:863–873

Dean N (1994) Yeast glycosylation mutants are sensitive to aminoglycosides. Proc Natl Acad Sci USA 92:1287–1291

Elbein AD (1991) Glycosidase inhibitors: inhibitors of *N*-linked oligosaccharide processing. FASEB J 5:3055–3063

Ellgaard L, Molinari M, Helenius A (1999) Setting the standards: quality control in the secretory pathway. Science 286:1882–1888

Huffaker TC, Robbins PW (1981) Temperature-sensitive yeast mutants deficient in asparagine-linked glycosylation. J Biol Chem 257:3203–3210

Huffaker TC, Robbins PW (1983) Yeast mutants deficient in protein glycosylation. Proc Natl Acad Sci USA 80:7466–7470

Kasahara S, Yamada H, Mio T, Shiratori Y, Miyamoto C, Yabe T, Nakajima T, Ichishima E, Furuichi Y (1994) Cloning of the *Saccharomyces cerevisiae* gene whose overexpression overcomes the effects of HM-1 killer toxin, which inhibits beta-glucan synthesis. J Bacteriol 176:1488–1499

Korner C, Knauer R, Stephani U, Marquardt T, Lehle L, von Figura K (1999) Carbohydrate deficient glycoprotein syndrome type IV: deficiency of dolichyl-P-Man:Man(5)GlcNAc(2)-PP-dolichyl mannosyltransferase. EMBO J 18:6816–6822

Kornfeld R, Kornfeld S (1985) Assembly of asparagine-linked oligosaccharides. Annu Rev Biochem 54:631–664

Kukuruzinska MA, Lennon-Hopkins K (1999) ALG gene expression and cell cycle progression. Biochim Biophys Acta 1426:359–372

Kurzik-Dumke U, Kaymer M, Gundacker D (1997) Gene within a gene configuration and expression of the *Drosophila melanogaster* genes lethal(2) neighbour of tid [l(2)not] and lethal (2) relative of tid. Genetics 200:45–58

Orlean P (1990) Dolichol phosphate mannose synthase is required in vivo for glycosyl phosphatidylinositol membrane anchoring, *O* mannosylation, and *N* glycosylation of protein in *Saccharomyces cerevisiae*. Mol Cell Biol 10:5796–5805

Orlean P (1994) *DPM1*. In: Book Rothblatt J, Novick P, Stevens T (eds) Guidebook to the secretory pathway. Oxford University Press, Oxford, pp 54–56

Orlean P (2000) Congenital disorders of glycosylation caused by defects in mannose addition during *N*-linked oligosaccharide assembly. J Clin Invest 105:131–132

Orlean P, Albright C, Pobbins PW (1988) Cloning and sequencing of the yeast gene for dolichol phosphate mannose synthase, an essential protein. J Biol Chem 263: 17499–17507

Ram AF, Wolters A, Ten HR, Klis FM (1994) A new approach for isolating cell wall mutants in *Saccharomyces cerevisiae* by screening for hypersensitivity to calcofluor white. Yeast 10:1019–1030

Rush J, Waechter CJ (1998) Topological studies on the enzymes catalyzing the biosynthesis of Glc-P-dolichol and the triglucosyl cap of Glc3Man9GlcNAc2-P-P-dolichol in microsomal vesicles from pig brain: use of the processing glucosidases I/II as latency markers. Glycobiology 8:1207–1213

Schachter H (1995) Biosynthesis. In: Book Montreuil J, Schachter H, Vliegenhart JFG (eds) Glycoproteins. Elsevier Science, Amsterdam, pp 123–126

Sharma CB, Kausthal GP, Pan YT, Elbein AD (1990) Purification and characterization of dolichyl-P-mannose:Man5(GlcNAc)2-PP-dolichol mannosyltransferase. Biochemistry 29:8901–8907

Shimma Y, Nishikawa A, bin Kassim B, Eto A, Jigami Y (1997) A defect in GTP synthesis affects mannose outer chain elongation in *Saccharomyces cerevisiae*. Mol Gen Genet 256:469–480

Varki A (1996) "Unusual" modifications and variations of vertebrate oligosaccharides: are we missing the flowers for the trees? Glycobiology 6:707–710

Verostek MF, Atkinson PH, Trimble RB (1991) Structure of *Saccharomyces cerevisiae alg3*, *sec18* mutant oligosaccharides. J Biol Chem 266:5547–5551

ALG6 Glucosyltransferase

Introduction

Lipid-linked oligosaccharides are assembled in the endoplasmic reticulum (ER) by a series of glycosyltransferases. In the lumen of the ER, four mannose residues are successively added to the $Man_5GlcNAc_2$ oligosaccharide by Alg3p, Alg9p, Alg12p and an unknown mannosyltransferase (for details, see Chapter 81). The assembly is completed via transfer of three glucose residues by the glucosyltransferases Alg6p, Alg8p, and Alg10p (Fig. 1). In contrast to the glycosyltransferases acting on the cytoplasmic face of the ER using nucleotide-activated sugars, the ER luminal glycosyltransferases require dolichol-phosphate-activated sugars. As such, the ER luminal glucosyltransferases Alg6p, Alg8p, and Alg10p require dolichol-phosphate-glucose (Dol-P-Glc) as a sugar donor (Kornfeld and Kornfeld 1985; Herscovics and Orlean 1993; Burda and Aebi 1999).

The substrate Dol-P-Glc is provided by the Alg5 protein Dol-P-Glc synthase (Palamarczyk et al. 1990; te Heesen et al. 1994). This 38-kDa transmembrane protein transfers a glucose residue from UDP-Glc to dolichol phosphate. The synthesis is thought to occur on the cytoplasmic side of the ER membrane (Rush and Waechter 1998) (Fig. 1). Activated glucose is thus available in close proximity to the ER membrane where transfer reactions by Alg6p, Alg8p, and Alg10p occur.

The Alg6 protein initiates the transfer of glucose residues to the high-mannose lipid-linked oligosaccharide. The glucose residue is attached by an $\alpha1,3$-linkage to the terminal mannose of the A-branch.

Cells with reduced activity of Alg5 or Alg6 protein have the same phenotype. Reduced activity in the Alg5 protein depletes the ER of Dol-P-Glc, the substrate of the subsequent Alg6 protein. Glycoproteins lacking one or multiple N-linked oligosaccharides are observed in cells with reduced activity of either Alg5 or Alg6 protein. In both cases, oligosaccharides of the $Man_9GlcNAc_2$ structure devoid of all glucose

Barbara Schenk and Claude A. Jakob

Institute for Microbiology, ETH Zürich, Schmelzbergstrasse 7, CH-8092 Zürich, Switzerland
Tel. +41-1-632-3327; Fax +41-1-632-1148
e-mail: jakob@micro.biol.ethz.ch

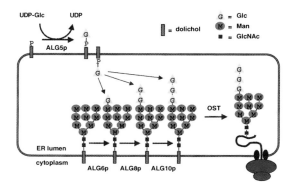

Fig. 1. Glucosyltransferases in the endoplasmic reticulum (ER) lumen. Alg6, Alg8, and Alg10 proteins represent the three glucosyltransferases acting in the lumen of the ER and add the triglucosyl cap to the Man₉GlcNAc₂ oligosaccharide. The substrate dolichol-phosphate-glucose (Dol-P-Glc) used by these glucosyltransferases is provided by Dol-P-Glc synthase (Alg5 protein), which catalyzes the formation of Dol-P-Glc from dolichylphosphate and UDP-glucose. Finally, the completed oligosaccharide is transferred to protein by the enzyme complex oligosaccharyl-transferase (OST)

residues are detected. Therefore, mutations in the *ALG5* and *ALG6* loci can be distinguished only by measuring enzymatic activity and by complementation analysis (see below).

Alg8 protein is highly homologous to Alg6 protein. Alg8 glucosyltransferase attaches the second glucose residue also in an α1,3-linked manner.

Alg10 glucosyltransferase transfers the third and last glucose residue to the A-branch of the high-mannose lipid-linked oligosaccharide. This glucose residue is attached by α1,2-linkage to the preceding glucose residue.

Databanks

ALG6 glucosyltransferase

No EC number has been allocated

Species	Gene	Protein	mRNA	Genomic
Homo sapiens	*hALG6*	–	AF102851	–
		–	NM_013339	–
Caenorhabditis elegans	C08B11.8	T19071	–	Z46676
		Q09226	–	–
Saccharomyces cerevisiae	*ALG6*	S61985	–	U43491
		Q12001	–	Z74910
Schizosaccharomyces pombe	SPBC3F6.06c	–	–	AL022019

The Alg6 protein shows sequence similarity to the ER glucosyltransferase Alg8p catalyzing the addition of an α1,3-linked glucose. A search of the available databases with the *S. cerevisiae* Alg6 amino acid sequence (using TBLASTN) identified genes in *Arabidopsis thaliana* on chromosome V (AB005248) and chromosome II (AAC00362). Whether these represent the *ALG6* locus or the *ALG8* locus remains to be determined. Similarly, in *Drosophila melanogaster* (http://www.flybase.org/cgi-bin/EST/community_query/blastReport.pl; GenBank accession nos. AA952053, AI512288, AI512289) and in *Mus musculus* (http://www.ncbi.nlm.nih.gov/blast/blast.cgi?Jform=0; GenBank accession nos. AI316687, AI132109), several DNA fragments with weak homology to the yeast Alg6 protein were found using the TBLASTN algorithm.

Name and History

An *alg6* mutant strain of *S. cerevisiae* was isolated among other *alg* mutants in the [^{3}H]mannose suicide selection screen performed by Huffaker and Robbins (1983). This screen was designed to find mutants defective in the biosynthesis of lipid-linked oligosaccharides. The *alg* mutants (mutant for *a*sparagine-*l*inked *g*lycosylation) accumulate characteristic lipid-linked oligosaccharide intermediates. For example, *alg5-1* and *alg6-1* mutant yeast cells accumulate Man$_9$GlcNAc$_2$-PP-Dol. In these mutant cells, incompletely assembled Man$_9$GlcNAc$_2$ oligosaccharide is transferred to protein, albeit with reduced efficiency (Huffaker and Robbins 1983; Runge et al. 1984). Therefore, glycoproteins such as carboxypeptidase Y (CPY) and invertase are hypoglycosylated. Apart from these biochemical phenotypes, cells with mutations in the *ALG5* or *ALG6* locus grow normally under various conditions.

The yeast *ALG6* locus was isolated by functional complementation of the synthetic growth phenotype of *alg6-1 wpb-1* double mutations (Reiss et al. 1996). A yeast strain deleted for the *ALG6* locus shows a hypoglycosylation phenotype similar to that of *alg6-1* cells (Fig. 2, lane 3). When the isolated *ALG6* locus is expressed in this mutated strain, the hypoglycosylation phenotype is complemented (Fig. 2, lane 2).

Enzyme Activity Assay and Substrate Specificity

The Alg6 protein is proposed to represent a Dol-P-Glc-dependent α1,3-glucosyltransferase. To date, enzyme activity assays have been performed only with partially purified enzyme or microsomes and not in a reconstituted system with the purified components; hence there remains the possibility that the Alg6 protein represents only an essential component of the glucosyltransferase.

The putative Alg6 glucosyltransferase catalyzes the transfer of a glucose residue from Dol-P-Glc to the acceptor Man$_9$GlcNAc$_2$-PP-Dol (Fig. 1) according to the following equation.

$$\text{Dol-P-Glc} + \text{Man}_9\text{GlcNAc}_2\text{-PP-Dol} \rightarrow \text{Glc}_1\text{Man}_9\text{GlcNAc}_2\text{-PP-Dol} + \text{Dol-P}$$

For proper function of the Alg6 protein, a supply of Dol-P-Glc as substrate is crucial. Dol-P-Glc is synthesized by the Alg5 protein (Alg5 Dol-P-Glc synthase), which transfers a glucose residue from UDP-Glc to the lipid carrier dolichol phosphate. It has long been accepted, but not proven, that the Alg6 glucosyltransferase reaction takes place

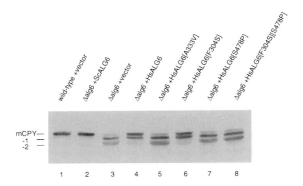

Fig. 2. Complementation of the yeast *alg6* defect by the human ALG6 cDNA. Wild-type yeast strains fully glycosylate mature, vacuolar carboxypeptidase Y (CPY) (*lane 1*). In a strain deleted for the *ALG6* locus, CPY is formed lacking one or two *N*-glycan chains (*lane 3*). Single copy expression of *ScALG6* under the control of its endogenous promoter fully complements the *ALG6* deletion (*lane 2*). Overexpression of the normal human *ALG6* cDNA (*hALG6*) is able partially to complement the Δ*alg6* defect (*lane 4*). In contrast, cDNA isolated from a patient with congenital disorder of glycosylation (CDG) type Ic containing a A333V mutation does not complement the *alg6* defect (*lane 5*). A second CDG type Ic patient carried two point mutations, F304S and S478P, in one allele (*lane 8*). Separating the mutations shows F304S to be silent, and the S478P mutation was responsible for the CDG phenotype (*lanes 6 and 7*)

in the lumen of the ER. Performing topological studies on the Dol-P-Glc synthase and the ER glucosyltransferases, Rush and Waechter (1998) were able to provide evidence that the activity of the ER glucosyltransferases Alg6p, Alg8p, and Alg10p is located in the ER lumen and that Dol-P-Glc is synthesized at the cytoplasmic side of the ER. This implies that the Dol-P-Glc must be translocated into the ER lumen by a mechanism that remains to be elucidated.

By overexpression studies in yeast, it has been shown that $Man_9GlcNAc_2$-PP-Dol is the optimal substrate for Alg6 protein activity. The Alg8 protein is highly homologous to Alg6p, and both proteins are Dol-P-Glc-dependent $\alpha1,3$-glycosyltransferases. However, Alg8p is required for the addition of a glucose residue to $Glc_1Man_9GlcNAc_2$-PP-Dol. Overexpression of the *ALG8* gene in an *alg6*-deficient strain does not suppress the accumulation of incomplete oligosaccharide intermediates and vice versa. On the other hand, overexpression of Alg6 protein in an *alg3*-deficient yeast strain background yields glucosylation of incompletely mannosylated oligosaccharide. Similarly, in *ALG12*-deficient cells a considerable amount of fully glucosylated $Glc_3Man_7GlcNAc_2$ oligosaccharide is detected, demonstrating that the $\alpha1,3$-$\alpha1,2$-dimannose antenna is involved in substrate recognition by the Alg6 enzyme. Nevertheless, $Man_9GlcNAc_2$-PP-Dol is the preferred subtrate of the Alg6 protein. It has been postulated that this substrate specificity of the Alg6 protein makes sure that only completely mannosylated lipid-linked oligosaccharides are glucosylated (Burda et al. 1999). This point is important because oligosaccharyltransferase does not monitor for the presence of all nine mannose residues but requires three glucose residues for efficient recognition of the complete oligosaccharide and the subsequent transfer to protein (Burda and Aebi 1999). However, the stringency of substrate recognition may differ in other strain backgrounds and in higher eukaryotes.

An assay for the Alg6 glucosyltransferase was reported by D'Souza-Schorey and Elbein (1993). In brief, protein purified from porcine aorta was assayed in a buffer containing 0.06% NP-40, $Man_9GlcNAc_2$-PP-Dol as an acceptor, and Dol-P-[^{14}C]Glc as a substrate. The reaction mixture was incubated for 15 min at 37°C, and the reaction was terminated by adding chloroform/methanol (1:1). The products were analyzed by scintillation counting and by passing through a concanavalin A-Sepharose column.

Korner et al. (1998) reported a similar procedure using a microsomal membrane fraction obtained from human primary fibroblasts as the enzyme source. Dol-P-[^{14}C]Glc and [^{3}H]$Man_9GlcNAc_2$-PP-Dol were mixed, dried under nitrogen, and solubilized in a buffer containing 0.25% Nikkol. After adding microsomal protein the reaction was incubated at 25°C for 30 min and stopped with HCl. The lipid-linked oligosaccharides were then extracted and analyzed by high-performance liquid chromatography (HPLC).

Preparation

The Alg6 enzyme activity was partially purified from porcine aorta; it involved ammonium sulfate precipitation, ion exchange, gel filtration, and hydroxylapatite column chromatography (D'Souza-Schorey and Elbein 1993). The partially purified protein has a pH optimum of 6.4 and requires divalent cations. This activity is specific for Dol-P-Glc as a donor, and $Man_9GlcNAc_2$-PP-Dol is the optimal acceptor substrate. $Man_9GlcNAc_2$ and $Man_9GlcNAc_2$-glycoprotein are not used as acceptors.

The yeast and human Alg6 proteins share 32% identity and 51% similarity. Both proteins are highly hydrophobic and contain several potential membrane spanning domains. As shown by Imbach et al. (1999), the human *ALG6* ortholog can be heterologously expressed in *S. cerevisiae*. Expression of the human *ALG6* gene under control of the *GAL1* promotor considerably improved the hypoglycosylation of CPY in an *ALG6*-deficient yeast strain (Fig. 2). In humans the *ALG6* gene is differentially expressed, as shown by Northern blot analysis. The highest expression levels were found in pancreas, placenta, and heart; moderate expression levels were found in brain, liver, kidney, and skeletal muscle; and weak expression levels were detected in lung tissue (Imbach et al. 1999). Expression of *ALG6* and *ALG5* mRNA levels were found to be similar, implying that the expression of these genes is coordinately regulated.

Biological Aspects

Glycosyltransferases involved in the biosynthesis of lipid-linked oligosaccharide are highly specific for their substrate. This implies that they depend on the action of the previous glycosyltransferase. The result is a highly ordered assembly of lipid-linked oligosaccharide, which ensures that only complete oligosaccharide is transferred to protein. As already mentioned, only $Glc_3Man_9GlcNAc_2$-PP-Dol is an optimal substrate for the oligosaccharyltransferase.

In *S. cerevisiae*, the transfer of oligosaccharide to protein is immediately followed by trimming reactions. ER glucosidase I and II cut the terminal α1,2- and α1,3-linked glucoses, whereas ER mannosidase I removes an α1,2-linked mannose. This trimming results in a $Man_8GlcNAc_2$-glycoprotein (Jakob et al. 1998b).

In higher eukaryotes, a quality control system ensures that incorrectly folded glycoproteins are retained in the ER and are either degraded or refolded. Proteins involved in this quality control are glucosidase II, calnexin, calreticulin, and UDP-glucose:glycoprotein glucosyltransferase. The trimming intermediate $Glc_1Man_9GlcNAc_2$-glycoprotein plays an important role, as this monoglucosylated structure binds to calnexin. The UDP-glucose:glycoprotein glucosyltransferase reglucosylates unglucosylated, misfolded glycoproteins to enable calnexin binding and, thereby, exposure to a folding environment. Correctly folded proteins exit the ER to the Golgi apparatus and to their final destinations (Ellgaard et al. 1999).

In yeast the monoglucosylated oligosaccharide, at least under mild ER stress conditions, is necessary for the folding process (Jakob et al. 1998b). Because yeast does not contain UDP-glucose:glycoprotein glucosyltransferase activity, the addition of the first glucose by the Alg6 glucosyltransferase has extra significance. With in vivo studies of the trimming reactions, it was shown by Jakob et al. (1998a) that the first two glucose removal steps occur rapidly, and that removal of the mannose is the rate-limiting step. The degradation of misfolded proteins was dependent on specific trimming steps. Fully trimmed but misfolded $Man_8GlcNAc_2$-glycoprotein was degraded with highest efficiency, indicating a probable role for the $Man_8GlcNAc_2$-oligosaccharide in protein degradation.

Taken together, these results show the importance of the first $\alpha1,3$-linked glucose on the oligosaccharide in quality control and degradation of misfolded proteins. Consistent with this idea, Travers et al. (2000) reported elevated expression of the *ALG6* gene upon inducing the unfolded protein response.

Related Glucosyltransferases and Dol-P-Glc Synthase

Alg5 Dol-P-Glc Synthase

The Alg5 protein is classified as a hexose transferase with the NC-IUBMB number EC 2.4.1.117. Homologs of the *Saccharomyces cerevisiae ALG5* gene are found in *Mus musculus* (BAA25759), *Homo sapiems* (NP_003850), and *Caenorhabditis elegans* (CAA95792). The *ALG5* locus codes for Dol-P-Glc synthase, which synthesizes the substrate for the Alg6 protein (Fig. 1). With heterologous expression of Alg5 protein in *Escherichia coli*, Dol-P-Glc synthase activity is detected in these cells (te Heesen et al. 1994). The enzyme reaction is described by the following equation.

$$UDP\text{-}Glc + Dol \rightarrow Dol\text{-}P\text{-}Glc + UDP$$

The activity assay for the Alg5 protein is performed using exogenous Dol-P as an acceptor and radiolabeled UDP-Glc as a donor. According to Stoll and Krag (1988), microsomal protein is incubated for 4 min at 37°C with 0.9 μM UDP-[^{3}H]Glc and 21 mM Dol-P in a buffer 0.025% T-X-100. The reaction is stopped by adding chloroform/methanol (2:1), and the labeled lipid is extracted and analyzed.

Alg8 α1,3-Glucosyltransferase

The Alg8 protein is classified as a hexose transferase with the NC-IUBMB number EC 2.4.1.-. Homologs of the *S. cerevisiae ALG8* locus are found in *C. elegans* (CAA91145,

CAA86666) and *H. sapiens* (CAA12176, AAD41466). The Alg8 protein is an ER glucosyltransferase acting directly downstream of the Alg6 protein, adding a second glucose residue to the A-branch (Fig. 1). The Alg8 enzyme reaction is described by the following equation.

$$\text{Dol-P-Glc} + \text{Glc}_1\text{Man}_9\text{GlcNAc}_2\text{-PP-Dol} \rightarrow \text{Glc}_2\text{Man}_9\text{GlcNAc}_2\text{-PP-Dol} + \text{Dol-P}$$

The Alg8 protein activity was measured in the same assay for glucosyltransferase activities that was used for determining Alg6 activity (D'Souza-Schorey and Elbein 1993). Characterization of the reaction products at different steps of the protein purification suggested that there are three glucosyltransferase activities, identified as Alg6p, Alg8p, and Alg10p.

Alg10 α1,2-Glucosyltransferase

No NC-IUBMB number has been assigned to the Alg10 protein. Homologs of the *S. cerevisiae ALG10* locus are found in *C. elegans* (CAB03424) and *Rattus norvegicus* (AAC34249). The Alg10 protein, which catalyzes the addition of the third α1,2-linked glucose residue to the oligosaccharide (Fig. 1) is highly specific for the substrate $\text{Glc}_2\text{Man}_9\text{GlcNAc}_2\text{-PP-Dol}$. $\text{Glc}_1\text{Man}_9\text{GlcNAc}_2\text{-PP-Dol}$ is not a substrate for Alg10 protein. Alg10 protein shares no sequence similarity with the Alg6 or Alg8 protein. Although all catalyze a glucosyltransferase reaction, the linkage formed by Alg10 protein is different. The terminal α1,2-linked glucose is an important determinant for recognition of the completely assembled oligosaccharide by oligosaccharyltransferase (OST) and is essential for the efficient transfer of oligosaccharide to protein (see Chapter 83 for details). The enzyme reaction is described by the following equation.

$$\text{Dol-P-Glc} + \text{Glc}_2\text{Man}_9\text{GlcNAc}_2\text{-PP-Dol} \rightarrow \text{Glc}_3\text{Man}_9\text{GlcNAc}_2\text{-PP-Dol} + \text{Dol-P}$$

An enzyme assay has been reported for the Alg10 protein (Burda and Aebi 1998). In brief, microsomal protein was incubated with [^{3}H]mannose-labeled $\text{Glc}_2\text{Man}_9\text{GlcNAc}_2\text{-PP-Dol}$ for 10 min at 25°C in a buffer of 0.4% Nikkol. The reaction was stopped by adding chloroform/methanol/water (10:10:3). The lipid-linked oligosaccharides were extracted and analyzed by HPLC.

Future Perspectives

Recently, Alg6 glucosyltransferase has attracted attention as patients suffering from congenital disorders of glycosylation (CDG) type Ic (see OMIM database, http://www.ncbi.nem.gov/OMIM, nos. 603147 and 604565) were identified who accumulate $\text{Man}_9\text{GlcNAc}_2$ oligosaccharide (Fig. 3). The primary defect in these patients was shown to be a mutation in the *ALG6* locus resulting in reduced glucosyltransferase activity (Burda et al. 1998; Korner et al. 1998; Imbach et al. 1999). CDG is a class of severe genetic diseases characterized by defects in *N*-linked glycosylation. In CDG patients a large number of proteins are hypoglycosylated, resulting in various abnormalities such as dysmorphism, psychomotor retardation, and organ failure. The primary defect in such patients lies in the biosynthetic pathway of *N*-glycosylation, either in the synthesis of lipid-linked oligosaccharide precursor (type I) or the pro-

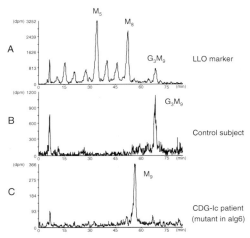

Fig. 3. Analysis of oligosaccharides isolated from fibroblasts obtained from a CDG type Ic patient. Lipid-linked oligosaccharides were labeled with [^{3}H]mannose. The oligosaccharides were isolated and analyzed by high-performance liquid chromatography (HPLC). **A** Oligosaccharides isolated from a wild-type yeast, serving as oligosaccharide standards. **B** Healthy human fibroblasts assemble full-length Glc$_3$Man$_9$GlcNAc$_2$ oligosaccharide. **C** Fibroblasts obtained from a CDG type Ic patient accumulate Man$_9$GlcNAc$_2$ oligosaccharide

cessing of *N*-glycan chains after transfer to protein (type II). It is important to note that CDG patients with *alg5* or *alg6* mutations cannot be distinguished by their oligosaccharide accumulation. Enzymatic assays are required for final identification and classification of CDG subtypes. Yeast has proven to be a useful model system for identifying the primary defects of untyped CDG patients. Yeast cells can be used for complementation studies with human cDNA. Moreover, oligosaccharides accumulating in a yeast strain with a known deficiency can be compared to the oligosaccharides accumulating in human fibroblasts.

Symptoms associated with a defect in the biosynthesis of lipid-linked oligosaccharide or in the processing of *N*-glycan chains are pleiotropic, as a wide variety of proteins are affected. This makes the diagnosis and typing of new patients difficult. With characterization of the molecular basis of the disease, it will become possible to screen patients for defects. Mutations in the *ALG8* and *ALG10* genes have not yet been reported. Based on our knowledge of growth and biochemical phenotypes in yeast and the biochemical manifestations of CDG patients, we believe that further CDG type I defects will be discovered originating from defects in *ALG8* or *ALG10* genes.

Further Reading

For reviews see Burda and Aebi (1999) and Freeze and Aebi (1999a,b).

References

Burda P, Aebi M (1998) The *ALG10* locus of *Saccharomyces cerevisiae* encodes the α-1,2 glucosyltransferase of the endoplasmic reticulum: the terminal glucose of the lipid-linked oligosaccharide is required for efficient *N*-linked glycosylation. Glycobiology 8:455–462

Burda P, Aebi M (1999) The dolichol pathway of N-linked glycosylation. Biochim Biophys Acta 1426:239–257

Burda P, Borsig L, de Rijk-van Andel J, Wevers R, Jaeken J, Carchon H, Berger EG, Aebi M (1998) A novel carbohydrate-deficient glycoprotein syndrome characterized by a deficiency in glucosylation of the dolichol-linked oligosaccharide. J Clin Invest 102: 647–652

Burda P, Jakob CA, Beinhauer J, Hegemann JH, Aebi M (1999) Ordered assembly of the asymmetrically branched lipid-linked oligosaccharide in the endoplasmic reticulum is ensured by the substrate specificity of the individual glycosyltransferases. Glycobiology 9:617–625

D'Souza-Schorey, Elbein AD (1993) Partial purification and properties of a glucosyltransferase that synthesizes $Glc_1Man_9(GlcNAc)_2$-pyrophosphoryldolichol. J Biol Chem 268:4720–4727

Ellgaard L, Molinari M, Helenius A (1999) Setting the standards: quality control in the secretory pathway. Science 286:1882–1888

Freeze HH, Aebi M (1999a) Molecular basis of carbohydrate-deficient glycoprotein syndromes type I with normal phosphomannomutase activity. Biochim Biophys Acta 1455:167–178

Freeze HH, Aebi M (1999b) Molecular basis of carbohydrate-deficient glycoprotein syndromes type I with normal phosphomannomutase activity. Biochim Biophys Acta 1455:167–178. Addendum. Biochim Biophys Acta 1500:349, 2000

Herscovics A, Orlean P (1993) Glycoprotein biosynthesis in yeast. FASEB J 7:540–550

Huffaker TC, Robbins PW (1983) Yeast mutants deficient in protein glycosylation. Proc Natl Acad Sci USA 80:7466–7470

Imbach T, Burda P, Kuhnert P, Wevers RA, Aebi M, Berger EG, Hennet T (1999) A mutation in the human ortholog of the *Saccharomyces cerevisiae ALG6* gene causes carbohydrate-deficient glycoprotein syndrome type-Ic. Proc Natl Acad Sci USA 96: 6982–6987

Jakob CA, Burda P, Roth J, Aebi M (1998a) Degradation of misfolded endoplasmic reticulum glycoproteins in *Saccharomyces cerevisiae* is determined by a specific oligosaccharide structure. J Cell Biol 142:1223–1233

Jakob CA, Burda P, te Heesen S, Aebi M, Roth J (1998b) Genetic tailoring of N-linked oligosaccharides: the role of glucose residues in glycoprotein processing of *Saccharomyces cerevisiae* in vivo. Glycobiology 8:155–164

Korner C, Knauer R, Holzbach U, Hanefeld F, Lehle L, von Figura K (1998) Carbohydrate-deficient glycoprotein syndrome type V: deficiency of dolichyl-P-Glc:$Man_9GlcNAc_2$-PP-dolichyl glucosyltransferase. Proc Natl Acad Sci USA 95:13200–13205

Kornfeld R, Kornfeld S (1985) Assembly of asparagine-linked oligosaccharides. Annu Rev Biochem 54:631–664

Palamarczyk G, Drake R, Haley B, Lennarz WJ (1990) Evidence that the synthesis of glucosylphosphodolichol in yeast involves a 35-kDa membrane protein. Proc Natl Acad Sci USA 87:2666–2670

Reiss G, te Heesen S, Zimmerman J, Robbins PW, Aebi M (1996) Isolation of the *ALG6* locus of *Saccharomyces cerevisiae* required for glucosylation in the N-linked glycosylation pathway. Glycobiology 6:493–498

Runge KW, Huffaker TC, Robbins PW (1984) Two yeast mutations in glucosylation steps of the asparagine glycosylation pathway. J Biol Chem 259:412–417

Rush JS, Waechter CJ (1998) Topological studies on the enzymes catalyzing the biosynthesis of Glc-P-dolichol and the triglucosyl cap of $Glc_3Man_9GlcNAc2$-P-P-dolichol in microsomal vesicles from pig brain: use of the processing glucosidases I/II as latency markers. Glycobiology 8:1207–1213

Stoll J, Krag SS (1988) A mutant of Chinese hamster ovary cells with a reduction in levels of dolichyl phosphate available for glycosylation. J Biol Chem 263:10766–10773

Te Heesen S, Lehle L, Weissmann A, Aebi M (1994) Isolation of the *ALG5* locus encoding the UDP-glucose:dolichyl-phosphate glucosyltransferase from *Saccharomyces cerevisiae*. Eur J Biochem 224:71–79

Travers KJ, Patil CK, Wodicka L, Lockhart DJ, Weissman JS, Walter P (2000) Functional and genomic analyses reveal an essential coordination between the unfolded protein response and ER-associated degradation. Cell 101:249–258

N-Glycan Processing Enzymes

Oligosaccharyltransferase Complex, Ribophorin-I, Ribophorin-II, OST48, and DAD1

Introduction

Mammalian *N*-glycoproteins have been implicated as affecting a variety of biological processes such as cell growth, cell development, and cell communication, as well as protein stability and the control of protein folding (Varki et al. 1999). *N*-Glycan diversity arises from a common $GlcNAc_2$-Man_9-Glc_3 precursor that is preassembled with dolichol-PP (Dol-PP), transferred en bloc to specific asparagine residues of the nascent polypeptide chain, and then remodeled by endoplasmic reticulum (ER)- and Golgi-resident α-glycosidases and glycosyltransferases (Kornfeld and Kornfeld 1985). Oligosaccharyltransferase (OST), a hetero-oligomeric protein complex associated with the ER membrane, occupies a central role in this pathway linking the Dol-PP-dependent reaction sequence of oligosaccharide precursor formation with the lipid-independent route of *N*-glycan processing and maturation.

Databanks

Oligosaccharyltransferase complex, Ribophorin I, Ribophorin II, OST48, and DAD1

NC-IUBMB enzyme classification: E.C.2.4.1.119

Species	Gene	Protein	mRNA	Genomic
OST48				
Homo sapiens		P39656	D29643	NT004610
Canis familiaris		Q05052	M98392	–
		A45139	–	–
Gallus gallus		P48440	–	–
Drosophila melanogaster		Q24319	X81999	–
			X81207	–
Mus musculus		BAA23671	D89063	–
Sus scrofa		CAC10570	AJ293581	–

ERNST BAUSE and BIRGIT HARDT

Institut für Physiologische Chemie, Universität Bonn, Nussallee 11, 53115 Bonn, Germany
Tel. +49 228 737081; Fax +49 228 732416
e-mail: bause@institut.physiochem.uni-bonn.de

(Continued)

Species	Gene	Protein	mRNA	Genomic
Ribophorin I				
Homo sapiens		P04843	Y00281	NC000987
		A26168	–	–
Rattus norvegicus		P07153	X05300	–
		A27274	M33508	–
Sus scrofa		CAC04096	AJ293582	–
Ribophorin II				
Homo sapiens		P04844	Y00282	NC001106
		B26168	–	–
Rattus norvegicus		P25235	X55298	–
		JN0065	–	–
Sus Scrofa		CAC10571	AJ293583	–
DAD1				
Homo sapiens		P46966	D15057	NC019583
			U84212	–
			U84213	–
Mus musculus		P46966	U22107	NC001208
			U81050–U81051	–
			U83628	–
			U84209–U84210	–
			Y13335	–
			AF051310	–
Rattus norvegicus		P46966	Y13336	–
Mesocricetus auratus		P46966	D15058	–
Sus scrofa		Q29036	D86562	–
Gallus gallus		O13113	U83627	–
Xenopas laevis		P46967	D15059	–
Caenorhabditis elegans		P52872	X89080	–
			AF039713	–
Arabidopsis thaliana		Q39080	X95585	–
Malus domestica		O24060	U68560	–
Oryza sativa		O50070	D89726–D89727	–
Picea mariana		O65085	AF05217	–
Zea mays		O81214	AF055909	–

Name and History

The detection of Dol-PP-GlcNAc$_2$-Man$_9$-Glc$_3$ during the 1960s and 1970s as the common glycosyl donor in OST-catalyzed *N*-glycosylation was a milestone in the elucidation of *N*-glycoprotein biosynthesis (Waechter and Lennarz 1976). At about this time it was predicted that the triplet sequence Asn-Xaa-Thr/Ser was necessary, based on sequence homologies at sugar attachment sites in *N*-glycoproteins, although not in itself a sufficient prerequisite for *N*-glycosylation (Marshall 1972). This hypothesis was verified experimentally some 10 years later with the demonstration that synthetic peptides containing the Asn-triplet could act as acceptors for OST-catalyzed reactions (Ronin et al. 1978; Bause 1979; Hart et al. 1979). This finding prompted a number of investigations that substantially increased our current knowledge of *N*-glycosylation and the properties of OST. In the literature OST is also referred to as asparagine-

N-glycosyltransferase or dolichyl-diphospho-oligosaccharide:protein-L-asparagine-oligosaccharyltransferase (EC 2.4.1.119).

Enzyme Activity Assay and Substrate Specificity

The reaction catalyzed by OST is shown in Fig. 1. Because of the specificity of the enzyme, its activity can be measured reliably in crude microsomal extracts or cell lysates using Dol-PP-linked oligosaccharides as glycosyl donors and Asn-Xaa-Thr/Ser-containing peptides as acceptors. The purified enzyme from pig liver displays a pH optimum between 6.9 and 7.1, with half-maximal activity observed at pH 6.4 and 8.2 (Breuer and Bause 1995). The steep increase in activity at acidic pH suggests that a histidine residue may be involved in catalysis. OST shows an absolute require-ment for divalent cations. Although inactive in the presence of EDTA, enzyme activ-ity can be restored by adding metal ions with octahedral coordination geometry such as Mn^{2+}, Fe^{2+}, Ca^{2+}, and Mg^{2+}, with Mn^{2+} being the most effective. No reactivation of the metal-depleted apoenzyme is seen with Zn^{2+}, Ni^{2+}, Cd^{2+}, or Co^{2+} (Hendrickson and Imperiali 1995).

Under cell-free conditions, OST glycosylates Asn-Xaa-Thr/Ser-containing peptides of different chain lengths and amino acid sequences efficiently so long as the Asn-triplet is not located at the N- or C-terminus (Bause and Hettkamp 1979). The minimum sequence requirement for a peptide to function as a glycosyl acceptor is the presence of the Asn-Xaa-Thr/Ser triplet, provided both the N- and C-termini are blocked (Hart et al. 1979). Incorporation of the Asn-Xaa-Thr/Ser triplet in a cyclic derivative, thereby promoting a constrained Asx-turn conformation at the glycosyla-tion site, improves the binding constant significantly, compared to linear peptide analogs, indicating that this structural motif may be important for substrate recogni-tion (Imperiali et al. 1994). OST does not glycosylate or bind peptides that contain a proline at position Xaa or on the C-terminal side of the consensus sequence; accep-tor properties are not impaired by a proline preceding the Asn triplet (Bause 1983). Asn-Pro-Thr/Ser and Asn-Xaa-Thr/Ser-Pro sites in proteins are generally not glyco-sylated in vivo, mimicking the conclusions from in vitro model studies (Gavel and von Heijne 1990).

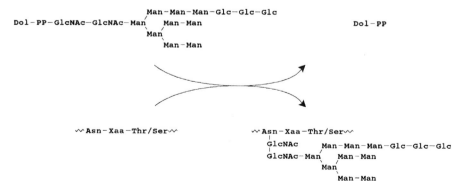

Fig. 1. Transglycosylation catalyzed by oligosaccharyltransferase

The functional significance of the amino acids in the Asn-Xaa-Thr/Ser consensus sequence has been investigated systematically using peptides as potential OST substrates in which asparagine and the hydroxyamino acid have been selectively replaced by natural or nonphysiological amino acids. These studies have shown that asparagine in the first position of the consensus sequence is invariably required for the peptide to function as a glycosyl acceptor. Peptide derivatives in which asparagine is replaced by glutamine, aspartic acid, aspartic acid methylester, or cysteinsulfonamide are neither inhibitory nor glycosylated (Ronin et al. 1978; Bause 1979; Bause et al. 1995). An approximately 10-fold reduction of V_{max} compared to the standard acceptor tripeptide, with no obvious effect on K_m, is observed on replacing asparagine by thioasparagine (Imperiali et al. 1992). Hydroxylation of asparagine at the β-C atom (*threo/erythro*-β-hydroxyasparagine) is accompanied, on the other hand, by a 50-fold increase in K_m and a threefold decrease in V_{max} (Bause et al. 1995). A similar effect on both binding and catalysis is also seen for a tripeptide containing *threo/erythro*-β-fluoroasparagine; V_{max}/K_m is approximately 100-fold lower than with the asparagine analog (Rathod et al. 1986). The poor acceptor properties of β-fluoroasparagine tripeptides are consistent with the inhibition of polypeptide *N*-glycosylation seen in cell-free translation/glycosylation studies when asparagine is replaced by *threo*-β-fluoroasparagine (Hortin et al. 1983; Tillmann et al. 1987).

In vitro the glycosylation rates for Asn-Xaa-Thr peptides are higher than for those containing serine instead of threonine. This is consistent with the in vivo observation that Thr-containing triplets in *N*-glycoproteins are twice as likely to be glycosylated as those containing Ser, even though both triplets are equally abundant (Gavel and von Heijne 1990). Peptides in which the hydroxyamino acid in the Asn triplet is replaced by either cysteine or by *threo*-α,β-diaminobutyric acid (DABA) bind to, and are glycosylated by, OST (Bause and Legler 1981; Bause et al. 1995). Glycosylation rates are lower, however, compared to the corresponding Asn-Xaa-Thr/Ser analogs. With the α,β-diaminobutyric acid derivative the pH optimum for glycosylation is shifted by +1 pH unit, indicating that the β-amino group must be deprotonated. Thus the β-thiol and the nonprotonated β-amino group can mimic the function of the β-hydroxy group for OST binding, catalysis, or both.

$GlcNAc_2$-Man_9-Glc_3, preassembled on Dol-PP by a reaction sequence referred to as the Dol-P-pathway, is the physiological glycosyl donor in the OST-catalyzed glycosylation reaction (Kornfeld and Kornfeld 1985). In vitro, however, the enzyme also accepts truncated oligosaccharide chains as short as the chitobiosyl unit (Dol-PP-$GlcNAc_2$), although less efficiently. Dol-PP-bound *N*-acetylglucosamine, on the other hand, fails to bind to OST (Lehle and Bause 1984). Thus the distal GlcNAc residue in the chitobiose core and the Glc_3 unit in Glc_3-Man_9-GlcNAc contribute to binding. The K_m values for acceptor peptides when determined with Dol-PP-$GlcNAc_2$ as donor, are approximately 10-fold higher than those using lipid-linked $GlcNAc_2$-Man_9-Glc_3, suggesting some form of cooperativity between the glycosyl donor and acceptor peptide (Breuer and Bause 1995). The possibility of using shorter Dol-PP-oligosaccharides as glycosyl donors, has greatly facilitated the characterization of the OST enzyme because these truncated substrates, particularly Dol-PP-$GlcNAc_2$, are readily accessible using enzymatic or chemical methods (Lehle and Tanner 1978; Imperiali and Zimmerman 1990).

Two assay variants have been reported in the literature for determination of OST activity. The method first described by Welply et al. (1983) used a crude microsomal extract as the enzyme source that was incubated in the absence of detergent with radioactive tripeptides such as N-[^{3}H]acetyl-Asn-Leu-Thr-NHCH$_3$. In this assay the pool of endogenous Dol-PP-oligosaccharides serves as the glycosyl donor. Radiolabeled reaction products and nonglycosylated [^{3}H]peptide are separated by Bio-Gel P4 or paper chromatography, followed by quantification. The assay does not work with highly polar or charged peptide substrates for which the microsomal membranes are impermeable. An alternative OST assay, which appears to be of more general applicability, utilizes detergent extracts of microsomes or cell lysates as the enzyme source. These are supplemented with radiolabeled Dol-PP-GlcNAc$_2$ (or larger lipid-linked oligosaccharides) and Asn-Xaa-Thr/Ser peptides. Here any peptide containing an Asn triplet can be used independent of its sequence, polarity, and chain length with the exceptions mentioned above. Radioactive glycopeptides are separated by ultrafiltration (Kumar et al. 1994a) or using chloroform/methanol/water (3:2:1, by volume) (Bause et al. 1982). The extraction procedure is easier to perform, allowing one to handle large numbers of samples in a reasonable time, a particular advantage for monitoring purification efficiency.

The OST activity is strongly inhibited by tripeptide derivatives in which the acceptor asparagine in the Asn-triplet is replaced by α,γ-diaminobutyric acid (Imperiali et al. 1992; Bause et al. 1995). The inhibition is concentration-dependent, with 50% inhibition being observed in the low micromolar range. K_i values in the nanomolar range are obtained when the tripeptide sequence is incorporated into a cyclic derivative, assumed to favor adoption of an Asx-turn conformation, as well as when this derivative is elongated by additional amino acids at the C-terminus (Hendrickson et al. 1996). A tripeptide in which α,γ-diaminobutyric acid in the asparagine position is replaced by homoserine, is not recognized by the enzyme (Bause et al. 1995).

A hexapeptide derivative that contains epoxyethylglycine in the hydroxyamino acid position of the Asn-triplet inactivates OST in a time- and concentration-dependent manner. Incubation of the pig liver enzyme in the presence of Dol-PP-[^{14}C]oligosaccharides and the N-3,5-dinitrophenylated epoxy inhibitor led to specific "double-labeling" of two protein species identical in size to the two OST oligomer subunits (48-kDa protein and ribophorin I) assumed to be responsible for catalytic activity (Bause et al. 1997).

N-Glycosylation of asparagine is thought to occur through nucleophilic attack by the β-amide electron pair on C-1 of the phosphoacetal-activated Dol-PP-oligosaccharide (Fig. 1). Several models dealing with possible mechanisms and involvement of the Asn-Xaa-Thr/Ser triplet have been proposed. One model postulates that the asparagine β-amide nitrogen acts as hydrogen bond donor and the threonine/serine hydroxy group as a hydrogen bond acceptor. In addition to enhancing β-amide nucleophilicity, the hydroxyamino acid is thought to function as a "proton vehicle" during catalysis, accepting a hydrogen from the asparagine β-amide and delivering its own β-hydroxy hydrogen to a conjugate base at the active site (Bause and Legler 1981; Bause et al. 1995). A second model assumes that an Asx-turn at the glycosylation site is essential as recognition motif (Imperiali et al. 1992, 1994). This Asx-turn is stabilized by hydrogen bonds between the β-carbonyl group of asparagine as acceptor and the α-NH and β-OH groups of the hydroxyamino acid as donors. The increased β-amide acidity resulting from these interactions facilitates deprotonation and formation of an imidate intermediate that is able to act as a nucleophile in the glycosylation reaction. Based on the experimental evidence in favor of an Asx-turn as an essential structural element for OST recognition, this mechanism appears at least possible. Failure of the β-fluoroasparagine-containing peptide derivatives to act as glycosyl acceptors, however, is difficult to explain with this model. A third catalytic model has been proposed that involves release of NH_3 from the asparagine β-amide group and formation of an oligosaccharylamine intermediate (Xu and Coward 1997). This mechanism seems highly unlikely, as cleavage of the β-C-N bond has never been demonstrated.

Preparation

Oligosaccharyltransferase is an integral component of the ER membrane. Detergent solubilization causes loss of activity, making purification difficult. Catalytic activity in the presence of detergent can, however, be stabilized by addition of phospholipids (Chalifour and Spiro 1988) and by sucrose or glycerol (Breuer and Bause 1995; Pathak et al. 1995). OST was first purified from hen oviduct by affinity chromatography using an Asn-Xaa-Thr/Ser-containing tetradecapeptide ligand (Aubert et al. 1982). The OST preparation contained three prominent proteins (69, 61, and 54 kDa) that were not characterized further. Using conventional techniques, OST has been purified from dog pancreas (Kelleher et al. 1992), hen oviduct and liver (Kumar et al. 1994b, 1995a), human liver (Kumar et al. 1995b), and pig liver (Breuer and Bause 1995). Except for the pig liver preparation, which contained an additional, ~40-kDa component, OST activity has always been found to be associated with a hetero-oligomeric protein complex consisting of three subunits (~48, ~63, and ~66 kDa), the two larger proteins being identical with ribophorin I (~66 kDa) and ribophorin II (~63 kDa). The tetrameric complex purified from pig liver was separated further by concanavalin A (ConA)-Sepharose chromatography, yielding an active enzyme consisting only of OST48 and ribophorin I. Thus catalytic activity must be associated with one or both of these two subunits. Crosslinking studies demonstrated that a ~12 kDa polypeptide (DAD1: "defender against death") is a further component of the mammalian OST complex (Kelleher and Gilmore 1997). This protein may have been present in previous OST preparations but was undetected owing to its small size. Catalytically active OST complexes have also been purified from yeast consisting of four subunits (Ost1p, Swp1p, Wbp1p, Ost3p) or six subunits (Ost1p, Swp1p, Wbp1p, Ost3p, Ost2p, Ost5p), with Wbp1p, Ost1p, Swp1p, and Ost2p displaying marked sequence similarities to the mammalian subunits OST48, ribophorin I, ribophorin II, and DAD1, respectively (Kelleher and Gilmore 1994; Knauer and Lehle 1994).

Biological Aspects

The components of the OST complex have been cloned from various mammalian tissues and species. Individal subunits display a high degree of sequence homology at the nucleotide level (>82%) and the amino acid level (>90%). The genes for the OST subunits were localized by fluorescence in situ hybridization, demonstrating ribophorin I on chromosome 3q21 and 6, ribophorin II on chromosome 20q12–q13 and 2, OST48 on chromosome 1p36.1, and DAD1 on chromosome 14q11–q12 and 14 in humans and mice, respectively (Behal et al. 1990; Loffler et al. 1991; Pirozzi et al. 1991; Apte et al. 1995; Yamagata et al. 1997). OST48, ribophorin I, and ribophorin II are type I membrane (glyco)proteins, with the bulk of their polypeptide chain directed toward the ER lumen, whereas DAD1 is a type IV protein and traverses the membrane twice with its N- and C-termini in the cytosol. Studies using the yeast two hybrid technique revealed no direct interaction between ribophorin I and II but, rather, an association of their luminal domain with the luminal domain of OST48. The cytosolic domain of OST48, on the other hand, binds to DAD1 (Fu et al. 1997). An interaction between ribophorin I and OST48 was also deduced from the observation that an OST48 protein mutant, transported to the plasma membrane after disruption of its cytosolic ER-retrieval signal, is retained in the ER on co-expression with ribophorin I (Fu and Kreibich 2000). Although DAD1 appears not to be directly involved in OST catalysis, its significance for the structural and functional stability of the membrane-associated enzyme complex has been highlighted by studies using a temperature-sensitive BHK cell line (tsBN7) expressing a defective DAD1 mutant protein (Sanjay et al. 1998) and by the observation that DAD1 knockout mice are not viable (Nishii et al. 1999). Thus, it appears that OST is necessary for life.

Future Perspectives

The specific function of each subunit in the hetero-oligomeric OST complex remains to be established in detail. Available data suggest that the OST subunits form a larger protein complex in the ER membrane, which may also involve components of the translocation machinery (Wang and Dobberstein 1999) and membrane-bound enzymes of the Dol-P-pathway. A close spatial and temporal relation between polypeptide translocation and OST activity must exist given the short space of 10–15 amino acid residues between the Asn-sequon and the ER membrane for N-glycosylation of the nascent polypeptide chain to take place (Nilsson and von Heijne 1993). The precise mechanism by which N-glycosylation of asparagine residues takes place also remains to be explained in detail.

Further Reading

Concise reviews covering oligosaccharyltransferase have been offered by Imperiali (1997), Knauer and Lehle (1999), and Silberstein and Gilmore (1996).

References

Apte SS, Mattei MG, Seldin MF, Olsen BR (1995) The highly conserved defender against death 1 (DAD1) gene maps to human chromosome 14q11–q12 and mouse chromosome 14 and has plant and nematode homologs. FEBS Lett 363:304–306

Aubert JP, Chiroutre M, Kerckaert JP, Helbecque N, Loucheux-Lefebvre MH (1982) Purification by affinity chromatography of the solubilized oligosaccharyltransferase from hen oviducts using a privileged secondary structure adopting peptide. Biochem Biophys Res Commun 104:1550–1559

Bause E (1979) Studies on the acceptor specifity of asparagine-*N*-glycosyltransferase from rat liver. FEBS Lett 103:296–299

Bause E (1983) Structural requirements of *N*-glycosylation of proteins: studies with proline peptides as conformational probes. Biochem J 209:331–336

Bause E, Hettkamp H (1979) Primary structural requirements for *N*-glycosylation of peptides in rat liver. FEBS Lett 108:341–344

Bause E, Legler G (1981) The role of the hydroxy amino acid in the triplet sequence Asn-Xaa-Thr (Ser) for the *N*-glycosylation step during glycoprotein biosynthesis. Biochem J 195:639–644

Bause E, Breuer W, Peters S (1995) Investigation of the active site of oligosaccharyltransferase from pig liver using synthetic tripeptides as tools. Biochem J 312:979–985

Bause E, Hettkamp H, Legler G (1982) Conformational aspects of *N*-glycosylation of proteins: studies with linear and cyclic peptides as probes. Biochem J 203:761–768

Bause E, Wesemann M, Bartoschek A, Breuer W (1997) Epoxyethylglycyl peptides as inhibitors of oligosaccharyltransferase: double-labelling of the active site. Biochem J 322:95–102

Behal A, Prakash K, D´Eustachio P, Adesnik M, Sabatini DD, Kreibich G (1990) Structure and chromosomal location of the rat ribophorin I gene. J Biol Chem 265:8252–8258

Breuer W, Bause E (1995) Oligosaccharyl transferase is a constitutive component of an oligomeric protein complex from pig liver endoplasmatic reticulum. Eur J Biochem 228:689–696

Chalifour RJ, Spiro RG (1988) Effect of phospholipids on thyroid oligosaccharyltransferase activity and orientation: evaluation of structural determinants for stimulation of *N*-glycosylation. J Biol Chem 263:15673–15680

Fu J, Kreibich G (2000) Retention of subunits of the oligosaccharyltransferase complex in the endoplasmic reticulum. J Biol Chem 275:3984–3990

Fu J, Ren M, Kreibich G (1997) Interactions among the subunits of the oligosaccharyltransferase complex. J Biol Chem 272:29687–29692

Gavel Y, von Heijne G (1990) Sequence differences between glycosylated and non-glycosylated Asn-X-Thr/Ser acceptor sites: implications for protein engineering. Protein Eng 3:433–442

Hart GW, Brew K, Grant GA, Bradshaw RA, Lennarz WJ (1979) Primary structural requirements for the enzymatic formation of the *N*-glycosidic bond in glycoproteins: studies with natural and synthetic peptides. J Biol Chem 254:9747–9753

Hendrickson T, Imperiali B (1995) Metal ion dependance of oligosaccharyl transferase: implications for catalysis. Biochem 34:9444–9450

Hendrickson T, Spencer JR, Kato M, Imperiali B (1996) Design and evaluation of potent inhibitors of asparagine-linked protein glycosylation. J Am Chem Soc 118:7636–7637

Hortin G, Stern AM, Miller B, Abeles RH, Boime I (1983) DL-*threo*-β-Fluoroasparagine inhibits asparagine-linked glycosylation in cell-free lysates. J Biol Chem 258:4047–4050

Imperiali B (1997) Protein glycosylation: the clash of the titans. Acc Chem Res 30:452–459

Imperiali B, Zimmerman JW (1990) Synthesis of dolichylpyrophospate-linked oligosaccharides. Tetrahedron Lett 31:6485–6488

Imperiali B, Shannon KL, Unno M, Rickett KW (1992) A mechanistic proposal for asparagine-linked glycosylation. J Am Chem Soc 114:7944–7945

Imperiali B, Spencer JR, Struthers MD (1994) Structural and functional characterization of a constrained Asx-turn motif. J Am Chem Soc 116:8424–8425

Kelleher DJ, Gilmore R (1997) DAD1, the defender against apoptotic cell death, is a subunit of the mammalian oligosaccharyltransferase. Proc Natl Acad Sci USA 94:4994–4999

Kelleher DJ, Gilmore R (1994) The *Saccharomyces cerevisiae* oligosaccharyltransferase is a protein complex composed of Wbp1p, Swp1p, and four additional polypeptides. J Biol Chem 269:12908–12917

Kelleher DJ, Kreibich G, Gilmore R (1992) Oligosaccharyltransferase activity is associated with a protein complex composed of ribophorin I and II and a 48 kd protein. Cell 69:55–65

Knauer R, Lehle L (1994) The N-oligosaccharyltransferase complex from yeast. FEBS Lett 344:83–86

Knauer R, Lehle L (1999) The oligosaccharyltransferase complex from yeast. Biochim Biophys Acta 1426:259–273

Kornfeld R, Kornfeld S (1985) Assembly of asparagine-linked oligosaccharides. Annu Rev Biochem 54:631–664

Kumar V, Heinemann FS, Ozols J (1994a) Microassay for oligosaccharyltransferase: separation of reaction components by partitioning in detergent solution followed by ultrafiltration. Anal Biochem 219:305–308

Kumar V, Heinemann FS, Ozols J (1994b) Purification and characterization of avian oligosaccharyltransferase: complete amino acid sequence of the 50-kDa subunit. J Biol Chem 269:13451–13457

Kumar V, Heinemann FS, Ozols J (1995a) Purification and characterization of hepatic oligosaccharyltransferase. Biochem Mol Biol Int 36:817–826

Kumar V, Korza G, Heinemann FS, Ozols J (1995b) Human oligosaccharyltransferase: isolation, characterization, and the complete amino acid sequence of 50-kDa subunit. Arch Biochem Biophys 320:217–223

Lehle L, Bause E (1984) Primary structural requirements for N- and O-glycosylation of yeast mannoproteins. Biochim Biophys Acta 799:246–251

Lehle L, Tanner W (1978) Glycosyl transfer from dolichyl phosphate sugars to endogenous and exogenous glycoprotein acceptors in yeast. Eur J Biochem 83:563–570

Loffler C, Rao VV, Hansmann I (1991) Mapping of the ribophorin II (RPNII) gene to human chromosome 20q12–q13.1 by in-situ hybridization. Hum Genet 87:221–222

Marshall RD (1972) Glycoproteins. Annu Rev Biochem 41:673–702

Nilsson I, von Heijne G (1993) Determination of the distance between the oligosaccharyltransferase active site and the endoplasmic reticulum membrane. J Biol Chem 268:5798–5801

Nishii K, Tsuzuki T, Kumai M, Takeda N, Koga H, Aizawa S, Nishimoto T, Shibata Y (1999) Abnormalities of developmental cell death in DAD1-deficient mice. Genes Cells 4:243–252

Pathak R, Hendrickson TL, Imperiali B (1995) Sulhydryl modification of the yeast Wbp1p inhibits oligosaccharyl transferase activity. Biochemistry 34:4179–4185

Pirozzi G, Zhou ZM, D´Eustachio P, Sabatini DD, Kreibich G (1991) Rat ribophorin II: molecular cloning and chromosomal localization of a highly conserved transmembrane glycoprotein of the rough endoplasmic reticulum. Biochem Biophys Res Commun 176:1482–1486

Rathod PK, Tashjian AH, Abeles RH (1986) Incorporation of β-fluoroasparagine into peptides prevents N-linked glycosylation: in vitro studies with synthetic fluoropeptides. J Biol Chem 261:6461–6469

Ronin C, Bouchilloux S, Granier C, van Rietschoten J (1978) Enzymatic N-glycosylation of synthetic Asn-X-Thr containing peptides. FEBS Lett 96:179–182

Sanjay A, Fu J, Kreibich G (1998) DAD1 is required for the function and structural integrity of the oligosaccharyltransferase complex. J Biol Chem 273:26094–26099

Silberstein S, Gilmore R (1996) Biochemistry, molecular biology, and genetics of the oligosaccharyltransferase. FASEB J 10:849–858

Tillmann U, Günther R, Schweden J, Bause E (1987) Subcellular location of enzymes involved in the N-glycosylation and processing of asparagine-linked oligosaccharides in Saccharomyces cerevisiae. Eur J Biochem 162:635–642

Varki A, Cummings R, Esko J, Freeze H, Hart G, Marth J (1999) Essential of Glycobiology. Cold Spring Harbor Laboratory Press, Cold Spring Harbor, New York

Waechter CJ, Lennarz WJ (1976) The role of polyprenol-linked sugars in glycoproteins. Annu Rev Biochem 45:95–112

Wang L, Dobberstein B (1999) Oligomeric complexes involved in translocation of proteins across the membrane of the endoplasmic reticulum. FEBS Lett 457:316–322

Welply J, Shenbagamurthi P, Lennarz WJ, Naider F (1983) Substrate recognition by oligosaccharyltransferase: studies on glycosylation of modified Asn-X-Thr/Ser tripeptides. J Biol Chem 258:11856–11863

Xu T, Coward K (1997) ^{13}C- and ^{15}N-labeled peptide substrates as mechanistic probes of oligosaccharyltransferase. Biochemistry 36:14683–14689

Yamagata T, Tsuru T, Momoi MY, Suwa K, Nozaki Y, Mukasa T, Ohashi H, Fukushima Y, Momoi T (1997) Genome organization of human 48-kDa oligosaccharyltransferase. Genomics 45:535–540

Phosphomannomutase

Introduction

The mannose donor during synthesis of the core oligosaccharide in the endoplasmic reticulum (ER) is GDP-α-mannose (GDP-Man). This activated mannose is generated in the cytosol from α-D-mannose-1-P (Man-1-P) and GTP by a specific pyrophosphorylase. Man-1-P originates from Glc-6-P through the action of three enzymes. Phosphoglucose isomerase (PGI) catalyzes the reversible conversion of Glc-6-P to Fru-6-P, and phosphomannose isomerase (PMI or MPI) catalyzes that of Fru-6-P to Man-6-P. Phosphomannomutase (PMM) is required for the formation of Man-1-P from Man-6-P.

Phosphomannomutases in yeast and higher organisms probably act as homodimers. They are members of a novel class of phosphotransferases that form a phosphoenzyme intermediate on an aspartate residue in a DXDX(T/V) motif (Collet et al. 1998). Phosphomannomutase deficiency is the cause of the most frequent of the congenital disorders of glycosylation: type IA (CDG-Ia) (Van Schaftingen and Jaeken 1995). Two PMM genes, *PMM1* and *PMM2*, have been identified in humans. They are the homologs of the yeast SEC53 and *Candida albicans* PMM. Only PMM2 is involved in the disease (Matthijs et al. 1997a).

With yeast SEC53, the identity is 54% and 58% at the amino acid level for human PMM1 and PMM2, respectively. The degree of identity with yeast SEC53 is thus higher for PMM2. An alignment of PMM1, PMM2, SEC53, and *Candida albicans* PMM was shown by Matthijs et al. (1997a). The identity between PMM1 and PMM2 is 65% at the nucleotide level and 66% at the amino acid level. PMM2 is shorter than PMM1

Gert Matthijs[1] and Emile Van Schaftingen[2]

[1] Center for Human Genetics, University of Leuven, UZGasthuisberg, Herestraat 49, B-3000 Leuven, Belgium
Tel. +32-16-346070; Fax +32-16-346060
e-mail: gert.matthijs@med.kuleuven.ac.be
[2] Laboratory of Physiological Chemistry, International Institute of Cellular and Molecular Pathology and University of Louvain, Avenue Hippocrate 75, B-1200 Brussels, Belgium
Tel. +32-2-7647564; Fax +32-2-7647598
e-mail: vanschaftingen@bchm.ucl.ac.be

and than yeast and *Candida albicans* PMM. The bacterial PMM genes show no sequence similarity.

Databanks

Phosphomannomutase

NC-IUBMB enzyme classification: E.C.5.4.2.8
PROSITE: PDOC00589, PS00710

Species	Gene	Protein	mRNA	Genomic
Homo sapiens	*PMM1*	Q92871	U62526	HS347H13
	PMM2	O15305	U85773	AH008020
Mus musculus	*Pmm1*	–	AAB62943	–
	Pmm2	–	AF043514	–
Caenorhabditis elegans	F52B11.2	T22485	CAB05198	Z82268
			T22485	
Drosophila melanogaster	CG10688	–	AAF49899	AE003541
Saccharomyces cerevisiae	*sec53*	BVBY53	BAA09196	NC_001138
		P07283		
Schizosaccharomyces pombe	*pmm1*	–	CAB61218	–
Candida albicans	*PMM1*	P31353	M96770	–
Babesia bovis	*pmm*	–	AAC27385	–
Azospirillum brasilense	*exoC*	P45632	U20583	–
Aeromonas hydrophila	*PMM*	–	AF148126	–
Escherichia coli	*rfbK1*	D40630	L27646	–
		P37755		
	rfbK2	D40630	L27632	
		P37755		
Pseudomonas aeruginosa	*algC*	A40013	M60873	–
		P26276		
Salmonella enterica	*cpsG*	P26341	X59886	–
Sfingomonas paucimobilis	*pmm*	–	AF167367	–
Shigella sonnei	*manB*	–	AF031957	–
Xanthomonas campestris	*xanA*	A43304	M83231	–
		P29955		
Yersinia pseudotuberculosis	*manB*	–	AJ270441–AJ270450	–
Prochlorothrix hollandica	*pmmA*	S78440	U23551	–
Arabidopsis thaliana	At2g45790	–	AAC28545	–
			T02468	

Name and History

The name of the enzyme describes the catalyzed reaction. The temperature-sensitive yeast mutant *sec53* (allelic to *alg4*) represents an early block in the synthesis of the oligosaccharide. The *sec53* cells, at their restrictive temperature of 37°C, accumulate incompletely glycosylated proteins in the lumen of the ER. The glycosylation defect can be corrected in cell extracts by adding GDP-mannose or a combination of GTP and Man-1-P, but not by a combination of GTP and Man-6-P. This, together with the results of the enzymatic assays, led to the conclusion that *SEC53* encodes the yeast PMM (Bernstein et al. 1985; Kepes and Schekman 1988). The yeast protein Sec53p is a soluble protein found in the cytosol that acts as a homodimer. The molecular mass of the Sec53p subunit is 29 kDa.

In prokaryotes the enzyme is a 50-kDa polypeptide that has the consensus phosphorylation site (TASHNP) of phosphoglucomutases (PGMs). In contrast, the fungal and mammalian PMMs are dimers of approximately 30-kDa subunits that have no sequence similarity with PGMs. It had long been supposed that the PGMs assumed the task in mammalian systems. However, also in mammalian systems, a specific PMM is required for the assembly of *N*-glycans. The first evidence for this came from experiments showing separation of PMM from PGM when rabbit brain extracts were chromatographed on an anion exchanger (Guha and Rose 1985). In 1995 Van Schaftingen and Jaeken showed that a specific PMM was deficient in tissues from patients with CDG-Ia (see below). Previously, Yamashita and coworkers had shown that the glycosylation defect in CDG-Ia patients consisted in a decrease in the number of oligosaccharides per molecule of transferrin, but that these oligosaccharides had a normal structure (Yamashita et al. 1993). Thus, the absence of PMM activity leads to reduced availability of the complete core oligosaccharide necessary for transfer by oligosaccharyltransferase (OST).

Subsequently, two human genes that encode active PMMs were identified on the basis of the sequence similarity of a number of expressed sequence tagged (EST) sequences in the public databank with the sequence of yeast PMM or Sec53 (Matthijs et al. 1997a,b). The first isozyme, PMM1, encoded by the *PMM1* gene on chromosome 22q13, has a low substrate specificity and low affinity for hexose-1,6-bisphosphates (Hansen et al. 1997; Matthijs et al. 1997b; Pirard et al. 1997). The second isozyme, PMM2, is a genuine PMM; and the *PMM2* gene is located on chromosome 16p13 (Matthijs et al. 1997a; Pirard et al. 1997). The genes have arisen by duplication, before mammalian irradiation (Schollen et al. 1998). More than 50 mutations in PMM2 have been described in patients with CDG-Ia to date (Matthijs et al. 2000).

Enzyme Activity Assay and Substrate Specificity

Phosphomannomutase catalyzes the reversible conversion of Man-6-P to Man-1-P. The equilibrium constant for PMM favors conversion of Man-1-P to Man-6-P, a reversal of the normal metabolic reaction. Like PGM, PMM requires a hexose-bisphosphate to be active, which can be either Man-1,6-P_2 or Glc-1,6-P_2. This cofactor serves to phosphorylate the enzyme on the first aspartate (Asp-19 in PMM1, Asp-12 in PMM2) of the conserved DVDXT motif, which is present close to the N-terminus of the enzyme (Collet et al. 1998). Conversion of Man-6-P to Man-1-P involves the two following (reversible) reactions.

$$\text{Man-6-P} + \text{phosphoenzyme} \rightarrow \text{Man-1,6-}P_2 + \text{dephosphoenzyme}$$

$$\text{Man-1,6-}P_2 + \text{dephosphoenzyme} \rightarrow \text{Man-1-P} + \text{phosphoenzyme}$$

Like other osyl-phosphates, Man-1-P is acid-labile at high temperatures, whereas Man-6-P is not. Kepes and Schekman (1988) took advantage of this property to measure PMM activity by the decrease in acid-labile phosphate when cell extracts are incubated with Man-1-P, Mg^{2+}, and Glc-1,6-P_2.

However, PMM is more conveniently measured with a spectrophotometric assay using Man-1-P as a substrate (Van Schaftingen and Jaeken 1995; Jaeken et al. 1997; Pirard et al. 1997, 1999a). The product of the reaction, Man-6-P, is successively converted to Fru-6-P, Glc-6-P, and 6-phosphogluconate in the presence of the appropriate auxiliary enzymes (phosphomannose isomerase, phosphoglucose isomerase, and Glc-6-P dehydrogenase). The last reaction forms stoichiometric amounts of NADPH, which is readily measured at 340 nm. The reaction requires the presence of Mg^{2+} and of Man-1,6-P_2, which can be prepared as in Van Schaftingen and Jaeken (1995), or Glc-1,6-P_2.

Human recombinant PMM1 catalyzes the conversion of Man-1-P and Glc-1-P to the corresponding hexose-6-phosphates with a K_m of, respectively, 3.2 and 6.0 μM, and a comparable V_{max} of approximately 46 μmol/min/mg protein (Pirard et al. 1997). For its activity the enzyme depends on the presence of a hexose-bisphosphate, with a K_a of 5.0 ± 0.2 μM for Man-1,6-P_2 and Glc-1,6-P_2. The lack of specificity for Man-1-P and Man-6-P contrasts with the narrow specificity of mammalian PGMs, which are at least 500-fold more active with Glc-1-P than with Man-1-P (Pirard et al. 1997). On the other hand, PMM2 is a specific PMM that converts Man-1-P to Man-6-P at the same rate as PMM1 (at approximately 46 μmol/min/mg protein) but about 20 times more rapidly than Glc-1-P into Glc-6-P. The K_m values are, respectively, 18 and 12 μM. The K_a values for Man-1,6-P_2 and Glc-1,6-P_2 of PMM2 are significantly lower (<1 μM) than those of PMM1 (Pirard et al. 1999a). Comparable values were obtained with rat liver PMM, suggesting that the major enzyme in liver is PMM2 (Pirard et al. 1997). In addition to the mutase reactions, PMM1 and PMM2 slowly catalyze the hydrolysis of various hexose bisphosphates at maximal rates of approximately 3.5% and 0.3% of their PMM activity, respectively (Pirard et al. 1999a). PMM1 is also activated by Fru-1,6-P_2; PMM2 is not.

The pH dependence for PMM1 and PMM2 is the same, with an optimum at pH 6.5 (Pirard et al. 1997). This is about 1 unit more acidic than rat liver PGM. All PMMs showed the same sensitivity to vanadate (K_I = 100 μM), a compound known to inhibit phosphotransferases that form a phosphoenzyme intermediate. This property again distinguishes PMMs for PGMs, which are much more sensitive to this inhibitor.

Preparation

Partial purification of Sec53p on a Sephacryl 200 column has been described by Kepes and Schekman (1988). The protein was eluted as an active dimer. Partial purification from rat liver was achieved by applying a 6%–22% poly(ethylene glycol) fraction, resuspended in HEPES buffer at pH 7.1, onto a DEAE-Sepharose column. The column was developed with a NaCl gradient, and the active fractions were further purified on a Sephacryl S-200 column (Pirard et al. 1997). Pirard et al. (1997, 1999a) used a comparable method to purify human recombinant PMM1 and PMM2 from *Escherichia coli* close to homogeneity. The recombinant PMM2 was retained on Q-Sepharose at pH 8.8. Several recombinant PMM1 and PMM2 mutants have been produced in *E. coli* BL21(DE3)pLys (Pirard et al. 1999b).

Biological Aspects

Expression Patterns

Enzymatic assays show widespread distribution of PMM activity in rat tissues, with particularly high levels in the intestinal mucosa. PMM2 appears to contribute more than 95% of the activity in most tissues, except for brain where it represents only about one-third of the activity and in lung where PMM1 contributes <13% (Pirard et al. 1999a). Northern blot analysis of human tissues showed the highest expression of the PMM2 gene in pancreas and liver (the intestinal mucosa was not tested), two organs with significant production of secreted proteins. The gene is also expressed in kidney and placenta, to a lesser extent in skeletal muscle and heart, and weakly in brain. On Northern blot analyses no expression of PMM2 is detected in lung (Matthijs et al. 1997a,b). It is not clear at present how these levels of expression correlate with the clinical picture, in which brain is one of the most severely affected organs. Immunohistological data are not yet available.

Regulation

An analysis of the sequence of the 5′ region of the gene suggests that PMM2 has a housekeeping promoter: The region upstream of the first ATG is characterized by a high GC content and contains several potential binding sites for the transcription factor Sp1. This is expected for an enzyme with a fundamental role in posttranslational processing. However, the variable expression suggests that the basal expression can be modulated in a tissue-specific manner. Thus, specific elements must be present in the promoter or in enhancers to regulate more precisely the expression in tissues with significant production of glycosylated (secreted) proteins and correspondingly high expression of the gene.

CDG-Ia is Caused by a PMM Deficiency and by Mutations in PMM2

The clinical picture of PMM deficiency includes three main features: a moderate to severe neurological disease; more or less typical dysmorphic signs; and variable involvement of organ systems (Jaeken et al. 2000). The neonatal period is characterized by severe failure to thrive and acute, life-threatening problems caused by liver failure, cardiac insufficiency, or nephrotic syndrome in some infants. Later in infancy involvement of the central nervous system becomes more obvious. All patients are moderately to severely mentally retarded, and most suffer from epilepsy and some from stroke-like episodes. The disease is largely nonprogressive, be it that skeletal deformities present at a later age and overt premature aging is apparent in some adults. Few adult patients have been described. Patients with a milder presentation of the disease have now been detected. The disease is definitely underrecognized despite the availability of an easy screening test [isoelectric focusing (IEF) of serum transferrin].

The pathogenesis of CDG-Ia is largely unknown. A large number of serum glycoproteins are abnormal, including transport proteins, glycoprotein hormones, lysosomal and other enzymes, and other glycoproteins. GDP-mannose and dolichol-P-mannose are also essential for biosynthesis of GPI anchors; thus, membrane-bound enzymatic and receptor functions, cell-to-cell signaling, and cell adhesion might also be affected by the deficit of Man-1-P. The fact that there is no sign of hemolysis in patients with CDG-Ia indicates that a PMM deficiency does not grossly perturb GPI-anchor synthesis. Mannose corrects the altered *N*-glycosylation in fibroblasts from CDG-Ia patients in culture (Panneerselvam and Freeze 1996) but does not alter the biochemical parameters in vivo or the clinical outcome.

Molecular data from 249 CDG-Ia patients from 23 countries have been collated from six laboratories involved in PMM2 screening (Matthijs et al. 2000). Two important conclusions have arisen. First, there is a plethora of missense mutations causing CDG-Ia. The two most common mutations are R141H and F119L, accounting for, respectively, 37% and 16% of all mutant chromosomes in the Caucasian population. The other mutations are scattered over the gene. The high prevalence of F119L results from a founder effect in the Scandinavian population. The R141H must be an extremely old mutation because it is also associated with a specific haplotype. Enzyme studies of some of the mutant proteins have shown that different mutants had a V_{max} of 0.2%–50% of the wild-type enzyme and were unstable (Pirard et al. 1999b). Second, the lack of PMM2 activity is incompatible with life. This conclusion stems from the observation that there is a total absence of patients homozygous for the frequent R141H mutation. The lack of homozygotes for R141H is easily explained by the severity of the mutation: R141H encodes the least active of the protein mutants that have been tested (Pirard et al. 1999b). Still, the carrier frequency of the R141H mutation is on the order of 1/50 to 1/130 in the western European population, indicating a heterozygote advantage for R141H.

Future Perspectives

Based on the structural data obtained with P-type ATPases, Aravind et al. (1998) delineated three motifs in the PMMs that may be directly involved in catalytic activity. The phosphorylation sites at Asp-19 in PMM1 and at Asp-12 in PMM2 (Collet et al. 1998; Pirard et al. 1999a) are indeed located in motif 1. Mutations at this position would probably result in zero residual activity and have not (yet) been identified in CDG-Ia patients. On the other hand, mutations in motif 3 (e.g., Asp to Glu at position 217 in PMM2) do not fully impair the enzyme. A more detailed structural analysis of the PMMs is required to understand the effect of this and other mutations on the enzymatic activity and the stability of PMM2.

In parallel, mouse models are needed to study further the pathogenesis of PMM deficiency. They will probably be available soon.

Further Reading

See Kepes and Schekman (1988) for the initial identification of Sec53p as PMM and references to the early literature, including citations on plant PMM.
See Van Schaftingen and Jaeken (1995) for the primary description of PMM deficiency in CDG-Ia.
See Matthijs et al. (1997a) for identification of the PMM2 gene as the gene mutated in CDG Ia and Matthijs et al. (2000) for a mutation update.

See Aravind et al. (1998) for identification of the large superfamily of proteins to which PMM belongs.

See Pirard et al. (1999a) for a comparative study of PMM1 and PMM2.

See Collet et al. (1998) for identification of the phosphorylated aspartate in PMM and identification of a family of phosphotransferases with a DXDX(T/V) motif.

See Jaeken et al. (2001) for a comprehensive review of the congenital disorders of glycosylation, including CDG-Ia.

References

Aravind L, Galperin MY, Koonin EV (1998) The catalytic domain of the P-type ATPase has the haloacid dehalogenase fold. Trends Biochem Sci 23:127–129

Bernstein M, Hoffmann W, Ammerer G, Schekman R (1985) Characterization of a gene product (Sec53p) required for protein assembly in the yeast endoplasmic reticulum. J Cell Biol 101:2374–2382

Collet JF, Stroobant V, Pirard M, Delpierre G, Van Schaftingen E (1998) A new class of phosphotransferases phosphorylated on an aspartate residue in a DXDXT/V motif. J Biol Chem 273:14107–14112

Guha SK, Rose ZB (1985) The synthesis of mannose 1-phosphate in brain. Arch Biochem Biophys 243:168–173

Hansen SH, Frank SR, Casanova JE (1997) Cloning and characterization of human phosphomannomutase, a mammalian homologue of yeast SEC53. Glycobiology 7:829–834

Jaeken J, Artigas J, Barone R, Fiumara A, de Koning TJ, Poll-The BT, de Rijk-van Andel JF, Hoffmann GF, Assmann B, Mayatepek E, Pineda M, Vilaseca MA, Saudubray JM, Schlüter B, Wevers R, Van Schaftingen E (1997) Phosphomannomutase deficiency is the main cause of carbohydrate-deficient glycoprotein syndrome with type I isoelectrofocusing pattern of serum sialotransferrins. J Inherit Metab Dis 20:447–449

Jaeken J, Matthijs G, Carchon H, Van Schaftingen E (2001) Defects of N-glycan synthesis. In: Scriver CR, Beaudet AL, Sly WS, Valle D (eds) The metabolic and molecular bases of inherited disease. McGraw-Hill New York 1601–1622

Kepes F, Schekman R (1988) The yeast SEC53 gene encodes phosphomannomutase. J Biol Chem 263:9155–9161

Matthijs G, Schollen E, Bjursell C, Erlandson A, Freeze H, Imtiaz F, Kjaergaard S, Martinsson T, Schwartz M, Seta N, Vuillaumier-Barrot S, Westphal V, Winchester B (2000) Mutation update: mutations in PMM2 cause congenital disorders of glycosylation, type Ia (CDG-Ia). Hum Mutat 16:386–394

Matthijs G, Schollen E, Pardon E, Veiga-Da-Cunha M, Jaeken J, Cassiman J-J, Van Schaftingen E (1997a) Mutations in PMM2, a phosphomannomutase gene on chromosome 16p13, in carbohydrate-deficient glycoprotein type I syndrome (Jaeken syndrome). Nat Genet 16:88–92

Matthijs G, Schollen EM, Budarf ML, Van Schaftingen E, Cassiman J-J (1997b) PMM (PMM1), the human homologue of SEC53 or yeast phosphomannomutase, is localized on chromosome 22q13. Genomics 40:41–47

Panneerselvam K, Freeze HH (1996) Mannose corrects altered N-glycosylation in carbohydrate-deficient glycoprotein syndrome fibroblasts. J Clin Invest 97:1478–1487

Pirard M, Collet JF, Matthijs G, Van Schaftingen E (1997) Comparison of PMM1 with the phosphomannomutases expressed in rat liver and in human cells. FEBS Lett 411:251–254

Pirard M, Achouri Y, Collet JF, Schollen E, Matthijs G, Van Schaftingen E (1999a) Kinetic properties and tissular distribution of mammalian phosphomannomutases isozymes. Biochem J 339:201–207

Pirard M, Matthijs G, Heykants L, Schollen E, Grünewald S, Jaeken J, Van Schaftingen E (1999b) Effects of mutations found in carbohydrate-deficient glycoprotein syndrome type IA on the activity of phosphomannomutase 2. FEBS Lett 452:319–322

Schollen E, Pardon E, Heykants L, Renard J, Doggett NA, Callen DF, Cassiman JJ, Matthijs G (1998) Comparative analysis of the phosphomannomutase genes PMM1, PMM2 and PMM2psi: the sequence variation in the processed pseudogene is a reflection of the mutations found in the functional gene. Hum Mol Genet 7:157–164

Van Schaftingen E, Jaeken J (1995) Phosphomannomutase deficiency is a cause of carbohydrate-deficient glycoprotein syndrome type I. FEBS Lett 377:318–320

Yamashita K, Ideo H, Ohkura T, Fukushima K, Yuasa I, Ohno K, Takeshita K (1993) Sugar chains of serum transferrin from patients with carbohydrate deficient glycoprotein syndrome: evidence of asparagine-*N*-linked oligosaccharide transfer deficiency. J Biol Chem 268:5783–5789

Phosphomannose Isomerase

Introduction

Phosphomannose isomerase (PMI) is a monomeric enzyme that converts fructose-6-P (Fru-6-P) and mannose-6-P (Man-6-P). It is the only known link between glucose catabolism and mannose activation for glycosylation. Man-6-P can also be produced directly via hexokinase from mannose. The enzyme has been purified, crystallized, and its structure determined (Gracy and Noltmann 1968; Proudfoot et al. 1994a,b; Cleasby et al. 1996). The genes have been cloned from several organisms. Loss of PMI is lethal in yeast (Smith et al. 1992) and causes a congenital disorder of glycosylation (CDG-Ib) in humans (Niehues et al. 1998; Freeze and Aebi 1999), but both can be rescued by providing exogenous mannose. Because loss of PMI is lethal in yeast, highly specific inhibitors of *Candida albicans* PMI were sought, as patients with acquired immunodeficiency disease (AIDS) frequently have serious *C. albicans* infections. No specific inhibitors have been found.

The activity is widely distributed and levels vary over a 10-fold range in various tissues in the mouse. Mannose is lethal to honeybees when provided as the only sugar (Saunders et al. 1969), and it can be teratogenic to fetal rats. In the first case, the low PMI/hexokinase ratio leads to formation of mannose-6-P, but only a small amount can enter glycolysis. The remainder is reconverted to mannose by a phosphatase, only to be phosphorylated again with an ever-diminishing supply of ATP. The same bottleneck occurs when 9- to 10-day rat embryos in vitro are given normal glucose and an equal amount of mannose. At this stage of development, most metabolic energy is derived from glycolysis, and the diminishing ATP leads to anoxia and teratogenesis (Freinkel et al. 1984).

HUDSON H. FREEZE

Glyciobiology Program, The Burnham Institute, 10901 North Torrey Pines Road, La Jolla, CA 92037, USA
Tel. +1-858-646-3142; Fax +1-858-646-3193
e-mail: hudson@burnham.org

Databanks

Phosphomannose isomerase

NC-IUBMB emzyme classification: E.C.5.3.1.8
MMDB Id: 4973 PDB Id: 1PMI
OMIM: 154550

Species	Gene	Protein	mRNA	Genomic
Homo sapiens	–	–	AF227216–AF227218	–
Mus musculus	–	–	AF244360	–
Saccharomyces cerevisiae	–	–	M85238	–
Candida albicans	–	–	X82024	–
Rhodospirillum rubrum	–	–	D12652	–
Salmonella typhimurium	–	–	X57117	–
Streptococcus mutans	–	–	D16594	–
Xanthomonas campestris	–	–	M83231	–

The mouse gene is on chromosome 9 (MGI accession 97075). The mammalian gene is composed of 8 exons spanning 5 kb. The intron–exon boundaries are the same in the mouse and humans. The human gene is located on chromosome 15q22-qter. The Online Mendelian Inheritance in Man (OMIM) (http://www3.ncbi.nlm.nih.gov:80/Omim/) designation is 154550.

Name and History

Phosphomannose isomerase or mannose phosphate isomerase (MPI) was first discovered in 1950 and first purified from brewer's yeast in 1968 (Gracy and Noltmann 1968). In yeast and higher organisms PMI is a monofunctional enzyme carrying out a single reaction: Fru-6-P→Man-6-P. In some prokaryotes such as *Pseudomonas aerugninosa*, *Rhodospirilium rubrum*, and *Xanthomonas campestris*, the gene codes for a bifunctional enzyme having both PMI and GDP-mannose pyrophosphorylase activities (Jensen and Reeves 1998). PMI provides the only known metabolic link between glycolysis and GDP-mannose needed for glycoconjugate synthesis.

Enzyme Activity Assay and Substrate Specificity

Reaction and Specificity

The PMI reaction is specific for Man-6-P and Fru-6-P with a K_{eq} near 1.0. No other natural substrates are known.

Assay Methods

The Zn^{+2}-requiring enzyme has a broad pH optimum near neutral. Because the equilibrium constant is near unity, it is important to drive the reaction in one direction by consuming the product formed. This is usually done by the Man-6-P→Fru-6-P→Glc-6-P→6-P-gluconate route with reduction of NADP to NADPH$^+$. Formation of the reduced cofactor is measured spectrophotometrically by an increase in absorbance at 340 nm. The coupling enzymes (phosphoglucomutase and glucose-6-P dehydrogenase), substrate (Man-6-P), and cofactor (NADP) are present in excess; and the change in absorbance is rate-limited by the activity of PMI.

Properties

The pH optimum is 6.0–8.0, and no activators are known. Because the enzyme requires Zn^{+2}, chelators such as EDTA, o-phenanthroline, and α,α'dipyridyl irreversibly inhibit activity. Reducing agents such as dithiothreitol (DTT) can inhibit over time owing to metal chelation. In 100 μM amounts erythrose-4-P and arabinose-4-P inhibit activity in vitro, but no other natural inhibitors are known for specific use in biological systems under physiological conditions. Mercuric ions and silver sulfadiazine irreversibly inhibit C. albicans PMI by reacting with a specific cysteine residue.

Preparation

Source (Biological/Commercial)

Phosphomannose isomerase is abundant in the yeast S. cerevisiae, the usual commercial source. It accounts for <0.1% of the cell protein. In mammalian cells the specific activity is 50–100 times less than in yeast (Gracy and Noltmann 1968; Proudfoot et al. 1994a,b).

Expression Systems

Human PMI has been expressed in Escherichia coli and purified to homogeneity. C. albicans PMI was also expressed in E. coli in sufficient amounts for crystallization and total structural analysis at 1.7 Å (Cleasby et al. 1996).

Isolation, Purification (Native/Recombinant)

Conventional purification of the yeast enzyme combines ammonium sulfate precipitation, ion exchange, and gel filtration and hydrophobic chromatography to give a 29 000-fold purified, homogeneous enzyme from human placenta. Similar methods are used for native enzyme from S. cerevisiae, C. albicans, and recombinant forms from E. coli. Electrospray mass spectrometric (MS) analysis of expressed human PMI indicates the mass predicted from the gene sequence 46 524 with no posttranslational modifications (Proudfoot et al. 1994a,b).

Biological Aspects

No gene promoters (transacting elements) have been described. No trafficking is known to occur. There are no known biological activators.

Distribution in tissues: In mammalian cells, PMI is present at low levels in liver, spleen, and lung. Rats and mice have high PMI levels in skeletal muscle, brain, and heart. PMI activity is 10–20 times higher in mouse tests than in liver. Northern blot analyses of human tissue give very weak signals owing to low abundance (Proudfoot et al. 1994b; Alton et al. 1998).

Normal function/substrates: In mammals PMI is thought to be responsible for the formation of most of the Man-6-P needed for glycosylation. However, recent studies in human fibroblasts and hepatoma cells show that about 80% of the mannose found in glycoproteins is derived directly from mannose via hexokinase, rather than from fructose-6-P through PMI. A mannose-specific transporter delivers mannose to the cells even in the presence of 100-fold excess glucose in the blood. In fact, little labeled glucose can be incorporated into mannose (Panneerselvam et al. 1997). In higher animals the proportion of mannose derived from glucose and from mannose is not known.

Disease involvement: A deficiency in PMI is responsible for a human genetic disorder called congenital disorder of glycosylation (CDG) type Ib (OMIM 602579). Specific point and frame shift mutations in the coding region and aberrant splicing account for the known defects that reduce residual activity to <10%. There are about 12 known patients, several of whom died of the disorder before diagnosis. Several other patients have been given daily oral doses of mannose supplements that reverse most of their clinical symptoms with only minimal side effects. Mannose substantially or completely corrects protein-losing enteropathy, hypoglycemia, coagulopathy, and accumulation of misfolded, underglycosylated proteins in the endoplasmic reticulum. A dose of 0.4 g mannose/kg/day is sufficient to alleviate most symptoms.

Knockout and transgenic mice: There appears to be only one PMI gene in mammals, and no transgenic lines have been reported. A PMI-knockout (KO) mouse strain is being developed. In contrast to the patients who have a small amount of residual activity, these mice have none. It is predicted that the PMI-KO mice will survive if supplied with sufficient exogenous mannose, but this has not been demonstrated. High levels of mannose in plasma and milk can be achieved by supplying mannose in the drinking water.

Future Perspectives

Two years after PMI was identified as the basis of CDG-Ib, there are about 12 known patients. Several died before the diagnosis was available. As awareness increases, it is likely that additional patients will be found. Mannose will probably continue to be a useful therapy with minimal side effects.

Further Reading

See Alton et al. (1998) for a description of mannose metabolism in the rat and the tissue distribution of PMI.
See Cleasby et al. (1996) for the crystallization and structure of *C. albicans* PMI.
Freeze (1998) discusses the basic science and clinical aspects of congenital disorders of glycosylation.
Gracy and Noltmann (1968) reported the first purification of PMI from yeast.
Niehues et al. (1998) described the first patients with CDG-Ib and the effective use of mannose therapy.

Panneerselvam et al. (1997) compared PMI-derived mannose with mannose for glycosylation.

Proudfoot et al. (1994a) compared PMI purification from various organisms.

Proudfoot et al. (1994b) discussed the cloning of human PMI and heterologous expression in yeast.

Smith et al. (1992) discussed the identification of PMI-40 in yeast and rescue with mannose.

References

Alton G, Hasilik M, Niehues R, Panneerselvam K, Etchison JR, Fana F, Freeze HH (1998) Direct utilization of mannose for mammalian glycoprotein biosynthesis. Glycobiology 8:286–295

Cleasby AA, Wonacott, et al (1996) The X-ray crystal structure of phosphomannose isomerase from *Candida albicans* at 1.7 Å resolution. Nat Struct Biol 3:470–479

Freeze HH (1998) Disorders in protein glycosylation and potential therapy: tip of an iceberg? J Pediatr 133:593–600

Freeze HH, Aebi MM (1999) Molecular basis of carbohydrate-deficient glycoprotein syndromes type I with normal phosphomannomutase activity. Biochim Biophys Acta 1455:167–178

Freinkel N, Lewis NJ, et al (1984) The honeybee syndrome: implications of the teratogenicity of mannose in rat-embryo culture. N Engl J Med 310:223–230

Gracy RW, Noltmann EA (1968) Studies on phosphomannose isomerase. I. Isolation, homogeneity measurements, and determination of some physical properties. J Biol Chem 243:3161–3168

Jensen SO, Reeves PR (1998) Domain organisation in phosphomannose isomerases (type I and type II). Biochim Biophys Acta 1382:5–7

Niehues R, Hasilik M, Alton G, Korner C, Schiebe-Sukumar M, Koch HG, Zimmer KP, Wu R, Harms E, Reiter K, von Figura K, Freeze HH, Harms HK, Marquardt T (1998) Carbohydrate-deficient glycoprotein syndrome type Ib: phosphomannose isomerase deficiency and mannose therapy. J Clin Invest 101:1414–1420

Panneerselvam K, Etchison JR, et al (1997) Human fibroblasts prefer mannose over glucose as a source of mannose for *N*-glycosylation: evidence for the functional importance of transported mannose. J Biol Chem 272:23123–23129

Proudfoot AE, Payton MA, Wells TN (1994a) Purification and characterization of fungal and mammalian phosphomannose isomerases. J Protein Chem 13:619–627

Proudfoot AE, Turcatti G, Wells TN, Payton MA, Smith DJ (1994b) Purification, cDNA cloning and heterologous expression of human phosphomannose isomerase. Eur J Biochem 219:415–423

Saunders SA, Gracy RW, Schnackerz KD, Noltmann EA (1969) Are honeybees deficient in phosphomannose isomerase? Science 164:858–859

Smith DJ, Proudfoot A, Friedli L, Klig LS, Paravicini G, Payton MA (1992) PMI40, an intron-containing gene required for early steps in yeast mannosylation. Mol Cell Biol 12:2924–2930

86

α-Mannosidase-II

Introduction

α-Mannosidase II, also known as Golgi mannosidase II, mannosyl-oligosaccharide 1,3-1,6-α-mannosidase, 1,3-(1,6)mannosyl-oligosaccharide α-D-mannohydrolase, and GlcNAc transferase I-dependent α1,3(α1,6)-mannosidase, is a glycosylhydrolase that catalyzes cleavage of the final two mannose residues during mammalian Asn-linked oligosaccharide processing and the removal of α1,3 and α1,6 mannose residues from $GlcNAcMan_5GlcNAc_2$-Asn to form $GlcNAcMan_3GlcNAc_2$-Asn. The enzyme is a member of glycosylhydrolase family 38.

Databanks

α-Mannosidase II

NC-IUBMB enzyme classification: E.C.3.2.1.114

Species	Gene	Protein	mRNA	Genomic
Homo sapiens	*Man2A1/Man II*	Q16706	U31520	–
		Q16767	D63998	
Mus musculus	*Man2A1/Man II*	P27046	X61172	–
Rattus norvegicus	*Man2A1/Man II*	P28494	M24353	–
Spodoptera frugiperda	–	O18497	AF005034	–
Drosophila melanogaster	–	Q24451	X77652	AJ132715
		O97043	AB018079	AE003682
		BAA75817		AE003710
		AAF55228		AC006414
Caenorhabditis elegans	–	O01574	–	U97015
		CAB00104		Z75954

KELLEY W. MOREMEN

University of Georgia, Athens, GA 30602, USA
Tel. +1-706-542-1705; Fax +1-706-542-1759
e-mail: moremen@arches.uga.edu

Name and History

α-Mannosidase II is also referred to by the systematic name 1,3-(1,6)mannosyl-oligosaccharide α-D-mannohydrolase and the NC-IUBMB recommended name mannosyl-oligosaccharide 1,3-1,6α-mannosidase, as well as the more common name Golgi mannosidase II (Tulsiani et al. 1982b), to contrast it with a distinct but related enzyme in the endoplasmic reticulum (ER) referred to as ER α-mannosidase II (Weng and Spiro 1996). The activity of Golgi mannosidase II was first discovered in rat liver Golgi membrane extracts (Dewald and Touster 1973), where it was initially shown to be able to cleave the synthetic substrate p-nitrophenyl-α-D-mannoside (pNP-Man). It was subsequently demonstrated to cleave two mannose residues from the oligosaccharide substrate GlcNAcMan$_5$GlcNAc$_2$ (Tabas and Kornfeld 1978; Tulsiani et al. 1982b).

Early studies on the processing of Asn-linked glycans indicated that mannose trimming from Man$_9$GlcNAc$_2$ to the common Man$_3$GlcNAc$_2$ core structure on complex-type oligosaccharides occurred through the action of at least two distinct types of enzyme (Tabas and Kornfeld 1978). The first series of enzyme reactions cleave Man$_9$GlcNAc$_2$ structures to Man$_5$GlcNAc$_2$. It is now known that this first phase of oligosaccharide processing occurs through the concerted action of at least five enzymes capable of cleaving α1,2-mannose linkages (Moremen et al. 1994; Moremen 2000). Several of these enzymes belong to glycosylhydrolase family 47. An indication that the second phase of oligosaccharide trimming requires the prior action of GlcNAc transferase I was first observed in the ricin-resistant CHO cell line, clone 15B (Tabas and Kornfeld 1978). In these cells, which are deficient in GlcNAc transferase I, trimming beyond Man$_5$GlcNAc$_2$ is aborted. In addition, the ricin-resistant BHK cell lines (Ric[R]15 and Ric[R]19) have been shown to contain reduced Golgi mannosidase II activity and to accumulate glycoproteins containing hybrid-type oligosaccharides (Hughes and Feeney 1986). These data indicated that Man$_5$GlcNAc$_2$ is not an effective substrate for the Golgi processing enzyme that carries out these final processing steps and that Golgi mannosidase II is the predominant processing hydrolase catalyzing these cleavage reactions. Substrate specificity studies have subsequently demonstrated that Golgi mannosidase II has a selective preference (>10-fold) for cleavage of GlcNAc-Man$_5$GlcNAc$_2$ over Man$_5$GlcNAc$_2$ (Tulsiani et al. 1982b).

Early assays employing the pNP-Man synthetic substrate resulted in the realization that there were at least three distinct pNP-Man-hydrolyzing enzyme activities in mammalian cells that differed in pH optimum, subcellular localization, and response in enzymatic activity to cations and EDTA (Dewald and Touster 1973). These activities included lysosomal α-mannosidase, with its characteristic low pH optimum (pH ~4.5) and stimulation by Zn^{+2}; a cytosolic/ER α-mannosidase characterized by a neutral pH optimum (pH ~6.5) and stimulation by Co^{+2}; and Golgi α-mannosidase, with its intermediate pH optimum (pH ~5.5) and its lack of response to cations. Following the subsequent isolation of cDNA clones encoding each of these enzymes, sequence comparisons of the enzymes and related enzymes from nonmammalian systems resulted in their common classification as family 38 glycosylhydrolases (Moremen 2000). This inclusion of three seemingly disparate enzyme activities into a single glycosylhydrolase family results from a conserved core of sequence, presumed structure, and presumed enzymatic mechanism among all of the family members despite their distinct subcellular localizations, substrate specificities, and pH optima.

In addition to the mammalian mannosidase sequences derived from cDNA cloning, additional homologs have been identified recently based on sequence queries of genome sequencing data from *Caenorhabditis elegans*, fungi, eubacteria, and archea as well as directed cDNA cloning efforts in insects, cellular slime molds, and fungi. These sequence similarity searches have indicated that the proteins within glycosyl-hydrolase family 38 are relatively large (about 90–135 kDa) and contain at least one completely conserved active-site acidic amino acid in a highly conserved approximately 180-amino-acid subregion (Moremen 2000). In addition, this 180-amino-acid region has a high degree of sequence similarity to an equivalent subregion of a collection of α-amylases/glucanotransferases from hyperthermophilic bacteria and archea (Laderman et al. 1993). Structure determination has not yet been accomplished for any of the family 38 members, so the correlation of sequence similarity with a conserved three-dimensional structure and enzyme mechanism has yet to be proven.

Enzyme Activity Assay and Substrate Specificity

A comparison of the rates of cleavage of various oligosaccharide substrates by rat liver Golgi mannosidase II has resulted in identification of GlcNAcMan$_5$GlcNAc$_2$ as the preferred substrate, rather than Man$_{9-5}$GlcNAc$_2$, GlcNAc$_2$Man$_5$GlcNAc$_2$ (with one bisecting GlcNAc), or GalGlcNAcMan$_5$GlcNAc$_2$ (Tulsiani et al. 1982b). Processing of the latter two structures in the absence of Golgi mannosidase II action would be predicted to lead to the production of bisected hybrid or hybrid-type oligosaccharides. Thus the fate of oligosaccharide processing is influenced by the relative activities of Golgi mannosidase II, GlcNAc transferase III, and β1,4-galactosyltransferase in the Golgi complex and their relative localization in the Golgi stacks, as each of these enzymes exclusively competes for the same processing intermediates. The preference for processing oligosaccharides to complex-type structures can be accounted for by the spatial separation of the respective enzymes in the Golgi stacks. Golgi mannosidase II has been localized to a broad distribution in the Golgi stacks, with a greatest abundance in the medial Golgi stacks, whereas β1,4-galactosyltransferase was found predominantly in the *trans*-Golgi cisternae. GlcNAc transferase III has been localized to the Golgi complex by immunofluorescence, but the sub-Golgi localization has not been determined (Moremen 2000).

Despite the data indicating a requirement for the sequential action of GlcNAc transferase I prior to the action of Golgi mannosidase II for processing Asn-linked glycans to complex-type structures, several lines of evidence indicate that alternate enzyme systems in mammalian cells provide a bypass for the Golgi mannosidase II step. Mice defective for the expression of Golgi mannosidase II through gene disruption are capable of synthesizing complex-type oligosaccharides in many adult tissues, although at reduced levels (Chui et al. 1997). Enzyme assays of tissues from these animals indicated the presence of an enzyme activity capable of cleaving Man$_{6-5}$GlcNAc$_2$ to Man$_3$GlcNAc$_2$ without the prior addition of a GlcNAc by GlcNAc transferase I. In fact, no hydrolytic activity toward GlcNAcMan$_5$GlcNAc$_2$ could be detected in tissues of the Golgi mannosidase II-deficient mice. An activity capable of cleaving Man$_{9-4}$GlcNAc$_2$ to Man$_3$GlcNAc$_2$ has been detected in rat brain (Tulsiani and Touster 1985) and liver (Bonay and Hughes 1991) microsomes. These activities cleave *p*NP-Man poorly and are poorly inhibited by swainsonine. The enzyme from rat brain and liver apparently differ in their ability to cleave GlcNAc Man$_5$GlcNAc$_2$. The rat brain enzyme is capable of cleaving this substrate (Tulsiani and Touster 1985), whereas the enzyme from rat liver does not (Bonay and Hughes 1991), consistent with the enzyme detected in liver tissues from the Golgi mannosidase II-deficient mice.

A cDNA and gene have been identified in both mouse and human tissues that have a high degree of sequence similarity to Golgi mannosidase II (Misago et al. 1995). These proteins have been designated Golgi mannosidase IIx based on their sequence similarity and their lack of fully characterized substrate specificity (see Chapter 87 for a discussion of α-mannosidase IIx). It is possible that the α-mannosidase IIx gene encodes the enzyme activities detected in rat liver and brain that are capable of bypassing the Golgi mannosidase II step in the Golgi mannosidase II-deficient mice (Moremen 2000). In the future the role of this alternate processing enzyme must be addressed relative to the classical Golgi mannosidase II to determine if it plays a major role in the processing of Asn-linked oligosaccharides.

Preparation

Golgi mannosidase II was first purified from rat liver Golgi membranes (Tulsiani et al. 1982b) and has subsequently been shown to be a dimer of about 135 kDa subunits (Moremen and Robbins 1991; Moremen et al. 1991). The enzyme has also been purified from mung bean seedlings and was shown to have catalytic characteristics similar to those of the mammalian enzyme (Kaushal et al. 1990). Several lines of evidence indicate that the enzyme is a type II transmembrane protein with a short (5-amino-acid) cytoplasmic tail, a single transmembrane domain, and a luminally oriented catalytic domain (Moremen and Robbins 1991). Part of the luminally oriented polypeptide is not essential for catalytic activity, as cleavage with a variety of proteases results in excision of the cytoplasmic tail, transmembrane domain, and part of the luminally oriented peptide, resulting in the formation of a proteolytically resistant catalytic domain (Moremen et al. 1991). The high proline content of the luminally oriented sequence that is removed by proteolysis has led to the hypothesis that this region represents a "stem domain" that allows flexibility for the catalytic domain to gain access to soluble and membrane-bound substrates in the lumen of the Golgi complex. The rat liver enzyme has been purified in its intact form and following the chymotrypsin cleavage to generate a soluble dimer of 110-kDa subunits comprising the catalytic domain (Moremen et al. 1991). The proteolytically cleaved form has been shown to retain all of the catalytic character of the intact enzyme.

Several inhibitors have been identified that are selective for family 38 glycosylhydrolases. The first of the inhibitors to be identified was swainsonine (Tulsiani et al. 1982a; Elbein 1991). This compound was originally isolated from toxic plants of the genuses *Astragalus* and *Oxytropis* (also known as locoweed), where it has been shown to cause a phenocopy of the lysosomal storage disease, α-mannosidosis, including proliferation of storage vacuoles containing oligosaccharides and neurologic symptoms in livestock that have ingested the plants. In addition to causing an accumulation of undegraded oligosaccharides as a result of inhibition of the lysosomal α-mannosidase, oligosaccharide processing is also altered as a result of inhibition of Golgi mannosidase II (Tulsiani et al. 1982a). Swainsonine has been shown to cause the synthesis of hybrid-type oligosaccharides in vitro and in vivo, and it is a potent [50% inhibitory concentration (IC_{50}) < 200 nM] inhibitor of both lysosomal mannosidase and Golgi mannosidase II. Surprisingly, swainsonine is a poor inhibitor of the other mammalian family 38 glycosylhydrolase, the cytosolic/ER mannosidase II (Weng and Spiro 1996). The structural basis for this difference in inhibitor response is not clear.

Mannostatin (Elbein 1991) and 1,4-dideoxy-1,4-imino-D-mannitol (DIM) (Weng and Spiro 1996) are furanose analogs of mannose that have been found to be potent inhibitors of family 38 mannosidases (Daniel et al. 1994). These inhibitors are to be contrasted with the pyranose mannose derivatives that are generally ineffective inhibitors of family 38 mannosidases but are potent inhibitors of family 47 mannosidases (Daniel et al. 1994). The distinctions between the two mannosidase families extend to the mechanisms of glycoside cleavage. Members of the glycosylhydrolase family 38 cleave mannoside linkages with retention of the anomeric configuration, whereas family 47 enzymes have an inverting mechanism (the cleaved sugar is released as β-anomer) (Moremen 2000).

Biological Aspects

A human genetic autosomal recessive deficiency in Golgi mannosidase II, termed congenital dyserythropoietic anemia type II (HEMPAS), has been shown to cause incomplete processing of Asn-linked glycoproteins and is characterized by ineffective erythropoiesis, bone marrow erythroid multinuclearity, and secondary tissue siderosis (Fukuda 1999). The two major erythrocyte cell surface glycoproteins, bands 3 and 4.5, are normally processed to complex-type oligosaccharides with extended polylactosamine structures, whereas in HEMPAS the oligosaccharides on these glycoproteins are processed to hybrid-type structures. In contrast to the protein-linked polylactosamine structures normally found on bands 3 and 4.5, extended polylactosamine structures are found on glycolipids instead. The genetic basis of the disease is complex. Although two patients were found to have defects in the Golgi mannosidase II locus, biochemical analyses of other patient samples have indicated the possibility of defects in GlcNAc transferase II. Linkage analysis of several Italian patients indicated that the defective gene was on chromosome 20q11.2 rather than linkage to the loci for GlcNAc transferase II, Golgi mannosidase II, or Golgi mannosidase IIx (Iolascon et al. 1997).

A gene disruption in the Golgi mannosidase II locus was generated in mice as a model for HEMPAS, and these animals reproduced many of the characteristics of the human disease (Chui et al. 1997). The mice showed anemia, splenomegaly, and immature erythrocytes or reticulocytes in peripheral blood; and lectin blots and cell sorting indicated an absence of complex-type structures on erythrocytes. In contrast, splenocytes and fibroblasts were shown to contain complex-type oligosaccharides, although at lower levels than in wild-type animals. Enzyme assays from a variety of mouse tissues indicated the presence the previously mentioned mannosidase activity (Tulsiani and Touster 1985; Bonay and Hughes 1991) that could bypass the Golgi mannosidase II step, but direct demonstration that this in vitro enzyme activity is responsible for the in vivo bypass of the Golgi mannosidase II deficiency remains to be demonstrated. Future studies on the gene disruption of Golgi mannosidase IIx and the proposed novel α-mannosidase activity that acts directly on $Man_5GlcNAc_2$ should help resolve the contributions of each of these enzymes during glycoprotein processing in mammalian tissues.

Future Perspectives

The family 38 mannosidases have surprisingly diverse members, including mammalian enzymes that are involved in glycoprotein maturation in the Golgi complex and enzymes involved in glycoprotein catabolism in the cytosol and lysosomes (Moremen 2000). Investigation of the structural features of this broad class of enzymes that allow their discrimination in substrate specificity and localization should provide an ongoing basis for future study.

The trimming phase of Asn-linked oligosaccharide maturation occurs through the action of a surprisingly large number of enzymes. The latter stage of oligosaccharide maturation was originally thought to occur by the action of a single enzyme, Golgi mannosidase II, as a result of studies on lectin-resistant cell lines. In contrast, mouse models for disruption in the Golgi mannosidase II locus and HEMPAS patients have shown the appearance of complex-type oligosaccharides in a variety of tissues, indicating relatively efficient bypass of the Golgi mannosidase II step. The nature of the enzyme that accomplishes this bypass step and the role of this alternate processing pathway in normal cells is a major subject for future study.

Further Reading

For review of the role of mannosidases on glycoprotein biosynthesis and catabolism the reader should see several reviews on the biochemistry and molecular biology of these enzymes (Moremen et al. 1994; Herscovics 1999; Moremen 2000).

References

Bonay P, Hughes RC (1991) Purification and characterization of a novel broad-specificity (α1-2, α1-3 and α1-6) mannosidase from rat liver. Eur J Biochem 197:229–238

Chui D, Oh-Eda M, Liao YF, Panneerselvam K, Lal A, Marek KW, Freeze HH, Moremen KW, Fukuda MN, Marth JD (1997) α-Mannosidase-II deficiency results in dyserythropoiesis and unveils an alternate pathway in oligosaccharide biosynthesis. Cell 90:157–167

Daniel PF, Winchester B, Warren CD (1994) Mammalian α-mannosidases—multiple forms but a common purpose? Glycobiology 4:551–566

Dewald B, Touster O (1973) A new α-D-mannosidase occurring in Golgi membranes. J Biol Chem 248:7223–7233

Elbein AD (1991) Glycosidase inhibitors: inhibitors of N-linked oligosaccharide processing. FASEB J 5:3055–3063

Fukuda MN (1999) HEMPAS: hereditary erythroblastic multinuclearity with positive acidified serum lysis test. Biochim Biophys Acta 1455:231–239

Herscovics A (1999) Glycosidases of the asparagine-linked oligosaccharide processing pathway. In: Pinto BM (ed) Comprehensive natural products chemistry. Elsevier, New York, pp 13–35

Hughes RC, Feeney J (1986) Ricin-resistant mutants of baby-hamster-kidney cells deficient in alpha-mannosidase-II-catalyzed processing of asparagine-linked oligosaccharides. Eur J Biochem 158:227–237

Iolascon A, Miraglia del Giudice E, Perrotta S, Granatiero M, Zelante L, Gasparini P (1997) Exclusion of three candidate genes as determinants of congenital dyserythropoietic anemia type II (CDA-II). Blood 90:4197–4200

Kaushal GP, Szumilo T, Pastuszak I, Elbein AD (1990) Purification to homogeneity and properties of mannosidase II from mung bean seedlings. Biochemistry 29:2168–2176

Laderman KA, Davis BR, Krutzsch HC, Lewis MS, Griko YV, Privalov PL, Anfinsen CB (1993) The purification and characterization of an extremely thermostable α-amylase from the hyperthermophilic archaebacterium *Pyrococcus furiosus*. J Biol Chem 268:24394–24401

Misago M, Liao Y, Kudo S, Eto S, Mattei M, Moremen K, Fukuda M (1995) Molecular cloning and expression of cDNAs encoding human α-mannosidase II and a previously unrecognized α-mannosidase IIx isozyme. Proc Natl Acad Sci USA 92:11766–11770

Moremen KW (2000) α-Mannosidases in asparagine-linked oligosaccharide processing and catabolism. In: Ernst B, Hart G, Sinay P (eds) Oligosaccharides in chemistry and biology: a comprehensive handbook, vol II: Biology of saccharides (part 1: Biosynthesis of glycoconjugates). Wiley, New York, pp 81–117

Moremen KW, Robbins PW (1991) Isolation, characterization, and expression of cDNAs encoding murine α-mannosidase II, a Golgi enzyme that controls conversion of high mannose to complex *N*-glycans. J Cell Biol 115:1521–1534

Moremen KW, Touster O, Robbins PW (1991) Novel purification of the catalytic domain of Golgi alpha-mannosidase II: characterization and comparison with the intact enzyme. J Biol Chem 266:16876–16885

Moremen KW, Trimble RB, Herscovics A (1994) Glycosidases of the asparagine-linked oligosaccharide processing pathway. Glycobiology 4:113–125

Tabas I, Kornfeld S (1978) The synthesis of complex-type oligosaccharides. III. Identification of an α-D-mannosidase activity involved in a late stage of processing of complex-type oligosaccharides. J Biol Chem 253:7779–7786

Tulsiani DR, Touster O (1985) Characterization of a novel α-D-mannosidase from rat brain microsomes. J Biol Chem 260:13081–13087

Tulsiani DR, Harris TM, Touster O (1982a) Swainsonine inhibits the biosynthesis of complex glycoproteins by inhibition of Golgi mannosidase II. J Biol Chem 257:7936–7939

Tulsiani DR, Hubbard SC, Robbins PW, Touster O (1982b) α-D-Mannosidases of rat liver Golgi membranes: mannosidase II is the GlcNAcMan$_5$-cleaving enzyme in glycoprotein biosynthesis and mannosidases IA and IB are the enzymes converting Man$_9$ precursors to Man$_5$ intermediates. J Biol Chem 257:3660–3668

Weng S, Spiro R (1996) Endoplasmic reticulum kifunensine-resistant α-mannosidase is enzymatically and immunologically related to the cytosolic α-mannosidase. Arch Biochem Biophys 325:113–123

α-Mannosidase-IIx

Introduction

Golgi α-mannosidase II (MII) is a processing enzyme involved in *N*-glycan biosynthesis (Tulsiani et al. 1982b; Moremen and Touster 1985; Moremen et al. 1994). MII-related enzyme, α-mannosidase IIx (MX), was identified in the human genome; and subsequently, MX cDNA was cloned (Misago et al. 1995).

Databanks

α-Mannosidase IIx

No EC number has been allocated.

Species	Gene	Protein	mRNA	Genomic
Homo sapiens	*Man2A2/Man IIX*	P49641	D55649	AC003004
		Q13754	L28821	Y08833
Mus musculus	*Man2A2/Man IIX*	Q9WU23	–	AF107018

Name and History

Golgi MII was found using *p*-nitrophenyl-α-D-mannoside as a substrate in rat liver Golgi membrane fractions (Tulsiani et al. 1982b). The activities of this enzyme are distinct from those of lysosomal α-mannosidase. MII has an extremely restricted specificity toward natural oligosaccharide substrates, removing only the terminal α1,6- and α1,3-linked mannose residues from $GlcNAc_1Man_5GlcNAc_2$ to yield $GlcNAc_1Man_3GlcNAc_2$ (Tulsiani et al. 1982b). The prior addition of a single GlcNAc residue to $Man_5GlcNAc_2$ by GlcNAc transferase-I is required for MII to act. MII has been purified to homogeneity from rat liver (Moremen et al. 1991). A partial cDNA

TOMOYA O. AKAMA, MASAYOSHI OH-EDA, and MICHIKO N. FUKUDA

The Burnham Institute, 10901 North Torrey Pines Road, La Jolla, CA 92037, USA
Tel. +1-858-646-3100 (ext. 3680) or +1-858-646-3143; Fax +1-858-646-3193
e-mail: michiko@burnham-inst.org

sequence of rat liver MII was obtained by a mixed oligonucleotide-based polymerase chain reaction (PCR) (Moremen 1989). Subsequently, a full-length amino acid sequence was determined (Moremen and Robbins 1991). Murine MII is a type II membrane protein made up of 1150 amino acid residues. The intact protein is a dimer of identical subunits. Proteolysis with chymotrypsin releases a soluble form of the 110-kDa subunits that retains all the catalytic character of the intact enzyme. Human MII is made up of 1144 amino acid residues. The MII gene has been mapped to human chromosome 5q21-22 (Misago et al. 1995).

The MII-related enzyme gene was found in the human genomic library by cross-hybridization using human MII cDNA as a probe. MII-related enzyme cDNA was then cloned. This MII-related enzyme was designated α-mannosidase IIx (MX). Human MX is a type II membrane protein consisting of 1139 amino acid residues. MII and MX are highly homologous to each other: 66% identical at the amino acid level. Whereas MX showed α-mannosidase activity with *p*-nitrophenyl-α-mannoside as a substrate (Misago et al. 1995), the substrate specificity of MX to natural oligosaccharides is unknown. The human MX gene is mapped to 15q25 (Misago et al. 1995).

Tulsiani and his colleagues found an MII-like enzyme in rat epididymis (Skudlark et al. 1991). They also detected novel α-mannosidase activity on the mouse sperm surface (Tulsiani et al. 1989). However, the peptide sequences of these enzymes are not known. No additional MII-related enzymes other than MX have been identified to date.

Enzyme Activity Assay and Substrate Specificity

Activity assays for MX are difficult, as this enzyme exhibits extremely weak activity in vitro. The cell lysate prepared from transfected COS cells with an MX expression vector showed elevated α-mannosidase activity (compared to that of the control) (Misago et al. 1995). The specificity of MX to oligosaccharide substrates is unknown.

When human MX is overexpressed in CHO cells, the asparagine-linked oligosaccharide patterns were shifted (Oh-eda et al. 2001). The MX overexpressing cells accumulated an unusual $Man_4GlcNAc_2$ oligosaccharide. The results suggested that MX hydrolyzes two α-mannosyl residues from $Man_6GlcNAc_2$ oligosaccharide to produce $Man_4GlcNAc_2$ oligosaccharide (Fig. 1).

Preparation

No purification has been carried out for human MX. A recombinant human MX protein (proteinA-MX) prepared in the same manner as MII shows weak activity to the synthetic substrate but does not hydrolyze oligosaccharides.

Biological Aspects

α-Mannosidase IIx is ubiquitously expressed in humans (Misago et al. 1995). Thus it may be a housekeeping gene, presumably playing a subsidiary role to MII in *N*-glycan biosynthesis. Examination of the active promoter region for the human MX gene

Fig. 1. Effect of overexpression of α-mannosidase IIx (MX) in N-glycan synthesis in CHO cells (Oh-eda et al. 2001). With CHO overexpression of MX, M_6Gn_2 was reduced and unusually, M_4Gn_2 levels were elevated, suggesting that MX hydrolyzes Manα1→6 and Manα1→3 residues of M_6Gn_2. In CHO cells, M_4Gn_2 was further converted to M_3Gn_2, suggesting the possibility that MX provides an alternate pathway independent from MII. This possibility remains to be verified by further studies

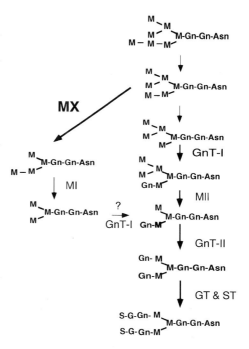

revealed that this region contains cytokine-responsive sequences. It is possible that MX expression is stimulated under physiological conditions such as inflammation (Ogawa et al. 1996).

The significance of MX in vivo is revealed by gene knockout mice in which the *Man2A2* gene is disrupted. MX-deficient mice were born and grew normally, but male mice were found to be infertile. Testes from MX-deficient mice are smaller than those from MX-heterozygous or wild-type litter mates. Many signs of failure of spermatogenesis were detected in MX-deficient mouse testes. These results suggest that MX plays an important role in synthesizing N-glycans essential for spermatogenesis (Akama et al. unpublished results).

References

Misago M, Liao Y-F, Kudo S, Eto S, Mattei M-G, Moremen KW, Fukuda MN (1995) Molecular cloning and expression of cDNAs encoding human a-mannosidase II and a novel a-mannosidase IIx isozyme. Proc Natl Acad Sci USA 92:11766–11770

Moremen KW (1989) Isolation of a rat liver Golgi mannosidase II clone by mixed oligonucleotide-primed amplification of cDNA. Proc Natl Acad Sci USA 86:5276–5280

Moremen KW, Robbins PW (1991) Isolation, characterization, and expression of cDNAs encoding murine α-mannosidase II, a Golgi enzyme that controls conversion of high mannose to complex N-glycans. J Cell Biol 115:1521–1534

Moremen KW, Touster O (1985) Biosynthesis and modification of Golgi mannosidase II in HeLa and 3T3 cells. J Biol Chem 260:6654–6662

Moremen KW, Touster O, Robbins PW (1991) Novel purification of the catalytic domain of Golgi α-mannosidase II. characterization and comparison with the intact enzyme. J Biol Chem 266:16876–16885

Moremen KW, Trimble RB, Herscovics A (1994) Glycosidases of asparagine-linked oligosaccharide processing pathway. Glycobiology 4:113–125

Ogawa R, Misago M, Fukuda MN, Kudo S, Tsukada J, Morimoto I, Eto S (1996) Structure and transcriptional regulation of human alpha-mannosidase IIx (alpha-mannosidase II isotype) gene. Eur J Biochem 242:446–453

Oh-eda M, Nakagawa H, Akama TO, Misago M, Moremen KW, Fukuda MN (2001) Over-expression of the Golgi-localized enzyme α-mannosidase IIx in Chinese hamster ovary cells results in the conversion of hexamannosyl-*N*-acetylchitobiose to tetramannosyl-*N*-acetylchitobiose in the *N*-glycan-processing pathway. Eur J Biochem 268:1280–1288

Skudlark MD, Orgebin-Crist M-C, Tulsiani RP (1991) Asparagine-linked glycoprotein biosynthesis in rat epididymis; presence of a mannosidase II-like enzyme. Biochem J 277:213–221

Tropea JE, Kaushal GP, Pastuszak I, Mitchell M, Aoyagi T, Molyneux RJ, Elbein AD (1990) Mannostatin A, a new glycoprotein-processing inhibitor. Biochemistry 29:10062–10069

Tulsiani DRP, Harris TM, Touster O (1982a) Swainsonine inhibits the biosynthesis of complex glycoproteins by inhibition of Golgi mannosidase II. J Biol Chem 257:7936–7939

Tulsiani DR, Hubbard SC, Robbins PW, Touster O (1982b) Alpha-D-mannosidases of rat liver Golgi membranes: mannosidase II is the GlcNAcMAN5-cleaving enzyme in glycoprotein biosynthesis and mannosidases Ia and Ib are the enzymes converting Man9 precursors to Man5 intermediates. J Biol Chem 257:3660–3668

Tulsiani DRP, Skudlarek MD, Orgebin-Crist M-C (1989) Novel alpha-D-mannosidase of rat sperm plasma membranes; characterization and potential role in sperm-egg interactions. J Cell Biol 109:1257–1267

Lysosomal Enzyme GlcNAc-1-Phosphotransferase

Introduction

GlcNAc-1-Phosphotransferase (UDP-GlcNAc; lysosomal enzyme *N*-acetylglu-cosamine-1-phosphotransferase; EC number 2.7.8.17) catalyzes the transfer of GlcNAc-1-phosphate from UDP-GlcNAc to the C-6 oxygen of selected mannose residues in high-mannose-type *N*-glycan. The product, mannose 6-phosphate (Man-6-P), which is modified consecutively by *N*-acetylglucosamine-1-phosphodiester α-*N*-acetylglucosaminidase (see Chapter 89) is the specific tag for the sorting system of the lysosomal enzymes.

GlcNAc-1-Phosphotransferase was purified about 490 000-fold from bovine mammary glands using an immunoaffinity purification procedure (Bao et al. 1996a). The purified enzyme was found to be a complex of six subunits and was composed of homodimers of 166-, 56-, and 51-kDa subunits. The *Km* for UDP-GlcNAc was 30 µM and that for the lysosomal enzyme cathepsin D was 18 µM (Bao et al. 1996b).

The enzyme activity is deficient in human I-cell disease.

Databanks

Lysosomal Enzyme GlcNAc-1-phosphotransferase

NC-IUBMB enzyme classification: E.C.2.7.8.17

Species	Gene	Protein	mRNA	Genomic
γ-subunit				
Homo sapiens	–	AAK61277	–	–

Atsushi Nishikawa

Department of Applied Biological Science, Tokyo University of Agriculture and Technology, 3-5-8 Saiwai-cho, Fuchu, Tokyo 183-8509, Japan
Tel. +81-42-367-5905; Fax: +81-42-367-5705
e-mail: nishikaw@cc.tuat.ac.jp

Name and History

N-Acetylglucosamine-1-phosphotransferase was first demonstrated in the membrane fraction of rat liver, skin fibroblasts, and CHO (Reitman and Kornfeld 1981a, b; Hasilik et al. 1981).

GlcNAc-1-Phosphotransferase phosphorylates acid hydrolases at least 100-fold more than nonlysosomal glycoproteins that contain identical oligosaccharide units (Reitman and Kornfeld 1981b; Lang et al. 1984; Ketcham and Kornfeld 1992). Heat-denatured forms of acid hydrolases as well as high-mannose-type oligosaccharides and glycopeptides were extremely poor substrates. Furthermore, deglycosylated lysosomal enzymes were potent inhibitors of the phosphorylation of intact lysosomal enzymes (Lang et al. 1984). These data suggest that GlcNAc-1-phosphotransferase recognizes a conformation-dependent protein domain that is common to all acid hydrolases but absent in nonlysosomal glycoproteins (Lang et al. 1984). Subsequently many laboratories have attempted to identify the recognition domain on lysosomal proteins (Baranski et al. 1990; Cuozzo and Sahagian 1994; Tikkanen et al. 1997).

Numerous researchers tried to purify and clone the enzyme, and in 1996 the group of Canfield succeeded in purifying the enzyme from bovine mammary glands (Bao et al. 1996a). It is a 540-kDa complex composed of three different subunits ($\alpha2\beta2\gamma2$). Recently one of these subunits of human GlcNAc-1-phosphotransferase has been cloned and its gene mapped on chromosome 16p (Raas-Rothschild et al. 2000).

N-acetylglucosamine-1-phosphotransferase is abbreviated as *N*-acetylglucosamine phosphotransferase, GlcNAc-1-phosphotransferase, GlcNAc-phosphotransferase, GlcNAc-P-T, and GPT.

Enzyme Activity Assay and Substrate Specificity

GlcNAc-1-Phosphotransferase catalyzes the transfer of GlcNAc-1-phosphate from UDP-GlcNAc to the C-6 oxygen of selected mannose residues on high-mannose-type *N*-glycan of lysosomal enzymes (Fig. 1). The assays of GlcNAc-1-phosphotransferase activity were carried out using $[\beta\text{-}^{32}\text{P}]$UDP-GlcNAc. While this sugar donor is not available commercially, it is the most suitable substrate for determination of GlcNAc-1-phosphotransferase activity and can be prepared from $[\gamma\text{-}^{32}\text{P}]$ATP using commercially available enzymes (Lang and Kornfeld 1984) and methyl α-mannoside. The phosphodiester product was separated from residual substrate using QAE-Sephadex (Reitman et al. 1984). When whole glycoproteins such as arylsulfatase (Waheed et al. 1982) and cathepsin L (Cuozzo and Sahagian 1994) were used as the acceptor substrate, GlcNAc-1-phosphate transferred products were separated by SDS-PAGE. Other assay methods, for example, using UDP-$[^3\text{H}]$GlcNAc as the sugar donor and oligosaccharides as the acceptor substrate, were reported in the literature (Reitman et al. 1984; Ketcham and Kornfeld 1992).

The enzyme activity is dependent on a divalent cation, Mg^{2+} or Mn^{2+} (Reitman and Kornfeld 1981a). The *Km* of bovine GlcNAc-1-phosphotransferase for UDP-GlcNAc is $30\,\mu M$ and that for the lysosomal enzyme cathepsin D is $18\,\mu M$. On the other hand, the *Km* for the nonlysosomal enzyme RNase B is $1.2\,mM$ and that for methyl α-mannoside is $64\,mM$ (Bao et al. 1996b).

A

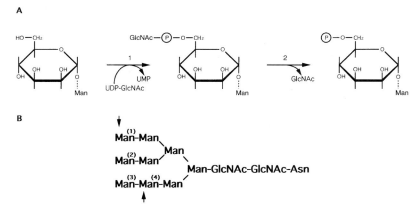

B

Fig. 1. Synthesis of mannose 6-phosphate (Man-6-P) on a lysosomal hydrolase. **A** The biosynthesis of Man-6-P on the oligosaccharides of lysosomal acid hydrolases is catalyzed by the sequential action of two enzymes. *1*, GlcNAc-1-Phosphotransferase catalyzes the transfer of GlcNAc-1-phosphate from UDP-GlcNAc to the C-6 oxygen of selected mannose residues. *2*, N-acetylglucosamine-1-phosphodiester α-N-acetylglucosaminidase catalyzes the cleavage of GlcNAc and creates Man-6-P. **B** Location of phosphorylated mannose residues on high-mannose-type oligosaccharide. *Arrows*, preferentially phosphorylated site, but mannose (*4*) is phosphorylated on the condition that mannose (*3*) is cleaved previously (A. Nishikawa et al. unpublished study)

The GlcNAc-1-phosphotransferase that recognizes the lysosomal hydrolases has separate catalytic and recognition sites (Reitman and Kornfeld 1981b; Waheed et al. 1982). The recognition site binds to a signal patch that is present in lysosomal hydrolases but absent in most secretory glycoproteins. Kornfeld's group demonstrated that lysine 203 and amino acids 265–292 of cathepsin D were the minimum number of amino acids on the signal patch for recognition of GlcNAc-1-phosphotransferase using chimeric proteins of cathepsin D and pepsinogen (Baranski et al. 1990; Baranski et al. 1991; Cantor et al. 1992). Furthermore, in experiments using mutants of DNase I, arginine 27 to lysine and asparagine 74 to lysine mutations caused a dramatic increase in oligosaccharide phosphorylation (Nishikawa et al. 1997, 1999). Other groups reported that the critical two lysine residues and oligosaccharide were positioned at a specific distance apart on the GlcNAc-1-phosphotransferase recognition domain of lysosomal hydrolases (Tikkanen et al. 1997; Cuozzo et al. 1998).

GlcNAc-1-Phosphate is transferred to selected mannose residues on high-mannose-type N-glycan at the catalytic site of GlcNAc-1-phosphotransferase. The first mannose to be phosphorylated is generally the terminal mannose on the α1,6 branch linked to the core mannose, while phosphorylated oligosaccharides formed later contain one or two Man-6-P residues located at different positions on the oligosaccharide (Goldberg and Kornfeld 1981; Natowicz et al. 1982; Varki and Kornfeld 1983).

Preparation

GlcNAc-1-Phosphotransferase has been demonstrated in membranes of numerous tissues and cells, including rat liver, skin fibroblasts, placenta, bovine pancreas, CHO cells, and COS-1 cells (Waheed et al. 1981; Hasilik et al. 1981; Reitman and Kornfeld 1981a, b; Nishikawa et al. 1997). Rat GlcNAc-1-phosphotransferase was partially purified using diethylaminoethanol-cellulose chromatography (Reitman and Kornfeld 1981b) and the bovine form was purified approximately 490 000-fold from mammary glands using monoclonal antibody immunoaffinity chromatography (Bao et al. 1996a). The purified bovine enzyme is a complex of six subunits ($\alpha2\beta2\gamma2$). The cDNA of the γ-subunit of human GlcNAc-1-phosphotransferase has been cloned by the Canfield group (Raas-Rothschild et al. 2000).

Biological Aspects

The phosphomannosyl targeting system for the delivery of acid hydrolases to lysosomes was first demonstrated by the presence of Man-6-P on soluble lysosomal enzymes (Kaplan et al. 1977). The interaction of GlcNAc-1-phosphotransferase with acid hydrolase results in the transfer of GlcNAc-1-phosphate to mannose residues on the asparagine-linked high-mannose oligosaccharides of the lysosomal hydrolases. The GlcNAc residues are then removed by *N*-acetylglucosamine-1-phosphodiester α-*N*-acetylglucosaminidase to generate phosphomannosyl residues that mediate the binding of the hydrolase to Man-6-P receptors present in the Golgi apparatus. These complexes are subsequently translocated via clathrin-coated vesicles to endosomes where the hydrolases are discharged for packaging into lysosomes (von Figura and Hasilik 1986; Kornfeld and Mellman 1989). The specificity of this pathway is determined by GlcNAc-1-phosphotransferase, which recognizes a conformation-dependent protein determinant that is present in lysosomal hydrolases but absent in most secretory glycoproteins (Reitman and Kornfeld 1981; Lang et al. 1984; Ketcham and Kornfeld 1992). Therefore, GlcNAc-1-phosphotransferase is selective for lysosomal enzymes and glycosyltransferase.

Lysosomal storage diseases are caused by genetic defects that affect one or more of the lysosomal acid hydrolases. In I-cell disease (mucolipidosis type II, ML-II) almost all of the lysosomal enzymes are missing from the fibroblasts and GlcNAc-1-phosphotransferase activity is defective (Hasilik et al. 1981; Reitman et al. 1981). Pseudo-Hurler polydystrophy (mucolipidosis type III, ML-III) is also known to be a disorder of lysosomal enzyme phosphorylation and localization (Varki et al. 1981; Reitman et al. 1981). Canfield's group has cloned the cDNA of the human GlcNAc-1-phosphotransferase γ-subunit and identified a frame-shift mutation in the γ-subunit gene (Raas-Rothschild et al. 2000).

Further Reading

The review of Kornfeld and Mellman (1989) and the review on P-type lectins (Varki 1999) are noteworthy to understand not only the characteristics of GlcNAc-1-phosphotransferase but also the entire delivery system of lysosomal enzymes.

References

Bao M, Booth JL, Elmendorf BJ, Canfield WM (1996a) Bovine UDP-*N*-acetylglucosamine: Lysosomal-enzyme *N*-acetylglucosamine-1-phosphotransferase. I. Purification and subunit structure. J Biol Chem 271:31437–31445

Bao M, Elmendorf BJ, Booth JL, Drake RR, Canfield WM (1996b) Bovine UDP-*N*-acetylglucosamine: Lysosomal-enzyme *N*-acetylglucosamine 1-phosphotransferase. II. Enzymatic characterization and identification of the catalytic subunit. J Biol Chem 271:31446–31451

Baranski TJ, Faust PL, Kornfeld S (1990) Generation of a lysosomal enzyme targeting signal in the secretory protein pepsinogen. Cell 63:281–291

Baranski TJ, Koelsch G, Hartsuck JA, Kornfeld S (1991) Mapping and molecular modeling of a recognition domain for lysosomal enzyme targeting. J Biol Chem 266: 23365–23372

Cantor AB, Baranski TJ, Kornfeld S (1992) Lysosomal enzyme phosphorylation. II. Protein recognition determinants in either lobe of procathepsin D are sufficient for phosphorylation of both the amino acid and carboxyl lobe oligosaccharides. J Biol Chem 267:23349–23356

Cuozzo JW, Sahagian GG (1994) Lysine is a common determinant for mannose phosphorylation of lysosomal proteins. J Biol Chem 269:14490–14496

Cuozzo JW, Tao K, Cygler M, Mort JS, Sahagian GG (1998) Lysine based structure responsible for selective mannose phosphorylation of cathepsin D and cathepsin L defines a common structural motif for lysosomal enzyme targeting. J Biol Chem 273: 21067–21076

Goldberg DE, Kornfeld S (1981) The phosphorylation of β-glucuronidase oligosaccharides in mouse P388D1 cells. J Biol Chem 256:13060–13067

Hasilik A, Waheed A, von Figura K (1981) Enzymatic phosphorylation of lysosomal enzymes in the presence of UDP-*N*-acetylglucosamine: Absence of the activity in I-cell fibroblasts. Biochem Biophys Res Commun 98:761–767

Kaplan A, Achord DT, Sly WS (1977) Phosphohexosyl components of a lysosomal enzyme are recognized by pinocytosis receptors on human fibroblasts. Proc Natl Acad Sci USA 74:2026–2030

Ketcham CM, Kornfeld S (1992) Characterization of UDP-*N*-acetylglucosamine: Glycoprotein *N*-acetylglucosamine-1-phosphotransferase from *Acanthamoeba castellanii*. J Biol Chem 1992:11654–11659

Kornfeld S, Mellman I (1989) The biogenesis of lysosomes. Annu Rev Cell Biol 5:483–525

Lang L, Kornfeld S (1984) A simplified procedure for synthesizing large quantities of highly purified uridine [β-^{32}P]diphospho-*N*-acetylglucosamine. Anal Biochem 140:264–269

Lang L, Reitman ML, Tang J, Roberts M, Kornfeld S (1984) Lysosomal enzyme phosphorylation: Recognition of a protein-dependent determinant allows specific phosphorylation of oligosaccharides present on lysosomal enzymes. J Biol Chem 259:14663–14671

Natowicz M, Baenziger JU, Sly WS (1982) Structural studies of the phosphorylated high mannose-type oligosaccharides on human β-glucuronidase. J Biol Chem 257:4412–4420

Nishikawa A, Gregory W, Frenz J, Cacia J, Kornfeld S (1997) The phosphorylation of bovine DNase I Asn-linked oligosaccharides is dependent on specific lysine and arginine residues. J Biol Chem 272:19408–19412

Nishikawa A, Nanda A, Gregory W, Frenz J, Kornfeld S (1999) Identification of amino acids that modulate mannose phosphorylation of mouse DNase I, a secretory glycoprotein. J Biol Chem 274:19309–19315

Raas-Rothschild A, Cormier-Daire V, Bao M, Genin E, Salomon R, Brewer K, Zeigler M, Mandel H, Toth S, Roe B, Munnich A, Canfield WM (2000) Molecular basis of variant pseudo-Hurler polydystrophy (mucolipidosis IIIC). J Clin Invest 105:673–681

Reitman ML, Kornfeld S (1981a) UDP-N-acetylglycosamine: glycoprotein N-acetylglucosamine-1-phosphotransferase: Proposed enyme for the phosphorylation of the high mannose oligosaccharide units of lysosomal enzymes. J Biol Chem 256:4275–4281

Reitman ML, Kornfeld S (1981b) Lysosomal enzyme targetting: N-Acetylglucosaminylphosphotransferase selectively phosphorylates native lysosomal enzymes. J Biol Chem 256:11977–11980

Reitman ML, Varki A, Kornfeld S (1981) Fibroblasts from patients with I-cell disease and pseudo-Hurler polydystrophy are deficient in uridine 5'-diphosphate-N-acetylglucosamine: glycoprotein N-acetylglucosaminyltransferase activity. J Clin Invest 67: 1574–1579

Reitman ML, Lang L, Kornfeld S (1984) UDP-N-acetylglucosamine: Lysosomal enzyme N-acetylglucosamine-1-phosphotransferase. Methods Enzymol 107:163–172

Tikkanen R, Peltola M, Oinonen C, Rouvinen J, Peltonen L (1997) Several cooperating binding sites mediate the interaction of a lysosomal enzyme with phosphotransferase. EMBO J 16:6684–6693

Varki A, Reitman ML, Kornfeld S (1981) Identification of a variant of mucolipidosis III (pseudo-Hurler polydystrophy): a catalytically active N-acetylglucosaminylphosphotransferase that fails to phosphorylate lysosomal enzymes. Proc Natl Acad Sci USA 78:7773–7777

Varki A, Kornfeld S (1983) The spectrum of anionic oligosaccharides released by endo-β-N-acetylglucosaminidase H from glycoproteins. J Biol Chem 258:2808–2818

Varki A (1999) P-type lectins. In: Varki A, Cummings R, Esko J, Freeze H, Hart G, Marth J (eds) Essentials of glycobiology. Cold Spring Harbor Laboratory, New York, pp 345–361

von Figura K, Hashilik A (1986) Lysosomal enzymes and their receptors. Annu Rev Biochem 55:167–193

Waheed A, Pohlmann R, Hasilik A, von Figura K (1981) Subcellular location of two enzymes involved in the synthesis of phosphorylated recognition markers in lysosomal enzymes. J Biol Chem 256:4150–4152

Waheed A, Hasilik A, von Figura K (1982) UDP-N-acetylglucosamine: Lysosomal enzyme precursor N-acetylglucosamine-1-phosphotransferase: Partial purification and characterization of the rat liver Golgi enzyme. J Biol Chem 257:12322–12331

GlcNAc-1-Phosphodiester α-*N*-Acetylglucosaminidase

Introduction

N-Acetylglucosamine-1-phosphodiester α-*N*-acetylglucosaminidase (phosphodiester α-GlcNAcase) is a type I membrane-spanning glycoprotein enzyme of the Golgi apparatus. It exists as a tetramer (272 kDa) composed of two dimers, each containing a pair of disulfide-linked monomers of 68 kDa. The enzyme catalyzes the second step in the formation of the mannose 6-phosphate recognition marker on lysosomal enzyme oligosaccharides. It removes the "covering" GlcNAc residue from GlcNAc-P added during the first step to C-6-hydroxyl groups of selected mannose residues by the enzyme UDP-*N*-acetylglucosamine:lysosomal enzyme *N*-acetylglucosamine-1-phosphotransferase (phosphotransferase). The Man-6-P moiety exposed by phosphodiester α-GlcNAcase action is responsible for the specific, high-affinity binding of lysosomal enzymes to one of the two mannose-6-P receptors in the trans-Golgi network (TGN) that transport the lysosomal enzymes to endosomes and subsequently to lysosomes. Phosphodiester α-GlcNAcase activity is present in all tissues of higher eukaryotes examined but is absent in the slime mold *Dictyostelium discoideum* and in *Acanthamoeba castellani*, each of which contains phosphotransferase activity (Couso et al. 1986).

Databanks

GlcNAc-1-phosphodiester α-*N*-acetylglucosaminidase

NC-IUBMB enzyme classification: E.C.3.1.4.45

Species	Gene	Protein	mRNA	Genomic
Homo sapiens	–	–	AF187072	AC007011
Mus musculus	–	–	–	AF187073

Rosalind Kornfeld

Division of Hematology, Washington University School of Medicine, 660 S. Euclid Avenue, Campus Box 8125, St. Louis, MO 63110, USA
Tel. +1-314-362-8835; Fax +1-314-362-8826
e-mail: rkornfeld@im.wustl.edu

Name and History

Tabas and Kornfeld (1980) first reported that the [2-³H] mannose-labeled biosynthetic intermediates of the lysosomal enzyme β-glucuronidase synthesized by mouse lymphoma cells contained high mannose oligosaccharides modified by phosphate groups covered by a moiety labile to mild acid treatment. Removal of the "covering" group from the oligosaccharides with mild acid or pig liver α-*N*-acetylglucosaminidase liberated *N*-acetylglucosamine and increased the net negative charge of the residual phosphorylated oligosaccharide, indicating that the phosphate was present in the form of a phosphodiester linkage. On the basis of their results they proposed a two-step process for biosynthesis of the mannose phosphate groups on lysosomal enzyme oligosaccharides, a later refined version of which is depicted in Fig. 1. Hasilik et al. (1980) reported similar observations on the lysosomal enzyme oligosaccharides secreted from radiolabeled human skin fibroblasts. Varki and Kornfeld (1980) first identified "uncovering enzyme" from rat liver Golgi membranes and called the enzyme α-*N*-acetylglucosaminyl phosphodiesterase. They differentiated it from lysosomal α-*N*-acetylglucosaminidase on the basis of subcellular localization, substrate and inhibitor specificities, and activity in mutant fibroblasts lacking α-*N*-acetylglucosaminidase. Varki and Kornfeld (1981) purified the enzyme 1800-fold from rat liver and studied its properties (Table 1). Varki et al. (1983) used this preparation

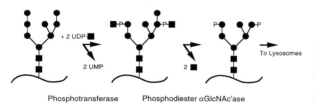

Fig. 1. Phosphorylation of mannose residues of lysosomal enzyme oligosaccharides. *Circles*, mannose; *Squares*, GlcNAc

Table 1. Affinity of substrates and inhibitors for phosphodiester α-GlcNAcase from various sources

Substance	Bovine liver[a]	Rat liver[b]	Human serum[c]	Human lymphoblasts[d]
Substrates, K_m (µM)				
GlcNAc α-P-Man α Me	740	190	450	761
UDPGlcNAc	734	1300	940	778
GlcNAc α-P-Man α 1,2 ManαMe	19	–	–	–
GlcNAc α-P-uteroferrin	20	–	–	42
GlcNAc α-P[Man$_{6-8}$] GlcNAc from uteroferrin	17	–	–	–
Competitive inhibitors, K_i (mM)				
GlcNAc	3.9	3.6	–	5.1
GlcNAc-1-P	1.7	1.3	–	1.58
UDPGlcNAc	0.39	2.4	–	–

Data are from [a] Mullis et al. (1994); [b] Varki and Kornfeld (1981); [c] Lee and Pierce (1995); [d] Page et al. (1996)

to demonstrate, by the ^{18}O enrichment method, that the enzyme cleaves the C—O bond rather than the O—P bond in its phosphodiester substrates, thus acting by a glycosidase-type mechanism. They renamed the enzyme α-*N*-acetyl D-glucosamine-1-phosphodiester *N*-acetylglucosaminidase.

Enzyme Activity Assay and Substrate Specificity

Phosphodiester α-GlcNAcase catalyzes the second step in the two-step reaction sequence shown in Fig. 1 by removing the "covering" GlcNAc residues from high mannose oligosaccharides bearing GlcNAc-P-Man. One may use metabolically labeled oligosaccharides isolated from the lysosomal enzymes of cultured cells as substrate for the enzyme activity assay, as was done in the early studies, or more conveniently the artificial substrate [^{3}H]GlcNAc-P-ManαMe. The latter substrate is synthesized in vitro as described by Mullis and Ketchum (1992) using UDP-[^{3}H]GlcNAc, α-methylmannoside, and phosphotransferase. UDP-[^{3}H]GlcNAc is also a substrate but is somewhat less specific when assaying crude extracts owing to the presence of other activities that split the nucleotide sugar. In all cases the neutral [^{3}H]-GlcNAc released in the assay can be separated from the negatively charged phosphodiester substrate and monoester product using a small column of anion-exchange resin, typically QAE-Sephadex equilibrated in 2 mM Tris pH 8. Table 1 shows a compilation of data from several sources of the kinetic constants (K_m) of a number of substrates for phosphodiester α-GlcNAcase as well as the K_i values for three competitive inhibitors. Although GlcNAcα-P-ManαMe and UDP-GlcNAc have relatively low affinities for the enzyme (K_m 190–1300 μM), the former has an especially high Vmax and is convenient for synthesizing in large quantities, making it the substrate of choice for routine assay during purification of the enzyme. Addition of the underlying α1,2-Man residue in GlcNAcα-P-Manα-1,2-ManαMe increases its affinity some 40-fold (K_m 19 μM), and this is apparently a key determinant for phosphodiester α-GlcNAcase tight binding because neither intact uteroferrin nor isolated uteroferrin oligosaccharides displays a lower K_m. This is in contrast to the specificity of phosphotransferase, which recognizes protein determinants in lysosomal enzymes to achieve high-affinity binding and reacts poorly with released oligosaccharides. The compound GlcNAc-1-P is not a substrate for phosphodiester α-GlcNAcase, indicating the importance of the diester linkage in substrate specificity, but it is a competitive inhibitor (K_i 1.3–1.7 mM). Other competitive inhibitors are GlcNAc and UDP-GlcNAc; and inorganic phosphate is a noncompetitive inhibitor, giving approximately 50% inhibition at 10 mM. The *N*-acetylglucosamine analog 6-acetamido-6-dideoxy castanospermine was shown by Page et al. (1996) to be a potent inhibitor of phosphodiester α-GlcNAcase (K_i 0.35 μM). Lee and Pierce (1995) reported that the enzyme could not split UDP-GalNAc or UDP-Glc, demonstrating the requirement for *N*-acetylglucosamine diester. A number of workers have shown that *p*-nitrophenyl α-GlcNAc, the standard substrate for assay of lysosomal α-*N*-acetylglucosaminidase, cannot be split by phosphodiester α-GlcNAcase; nor does it act as an inhibitor of the enzyme.

The pH optimum of phosphodiester α-GlcNAcase activity from a variety of sources has been determined and generally is reported to range broadly from pH 6 to pH 8. The assays have commonly been carried out at about pH 6.8. In contrast, Waheed et al. (1981) reported that the human placental enzyme had a pH optimum at 4.5–6.5, and they performed their assays at pH 5.5 in acetate buffer. Most other workers used citrate buffer for the low pH end of their pH activity measurements; and only fairly recently has this author discovered that citrate, especially in the 50 mM concentration used in buffers, is inhibitory to phosphodiester α-GlcNAcase. Hence the true pH optimum for the enzyme may be closer to pH 5.5–6.0, which may make better sense as it now appears that its site of action is in the more acidic TGN rather than the *cis*-Golgi.

The precise intracellular localization of the enzyme has been a subject of some controversy over the years, although all agree that it is a Golgi membrane protein. Various subcellular fractionation and density gradient methods have given conflicting results about which subregion of the Golgi contains the phosphodiester α-GlcNAcase. For example, Pohlmann et al. (1982) reported that sucrose density gradient centrifugation of rat liver membranes showed co-localization of phosphodiester α-GlcNAcase with the *cis*-Golgi marker enzyme α1,2-mannosidase in a denser fraction than the *trans*-Golgi marker galactosyl transferase. Goldberg and Kornfeld (1983) found a similar separation of activities upon sucrose gradient fractionation of lymphoma cell membranes. In contrast, when Deutscher et al. (1983) subfractionated a highly purified Golgi preparation of rat liver on a Percoll gradient, they found that phosphodiester α-GlcNAcase migrated to a denser fraction well separated from most of the α1,2-mannosidase activity but overlapping a significant peak of the sialyltransferase activity generally associated with the TGN. Recently we obtained evidence from immunolocalization in whole cells that phosphodiester α-GlcNAcase resides in the TGN (Rohrer and Kornfeld 2001).

Preparation

Phosphodiester α-GlcNAcase has been partially purified 1800-fold from rat liver (Varki and Kornfeld 1981), 820-fold from human placenta (Waheed et al. 1981), 3000-fold from bovine liver (Mullis et al. 1994), more than 6000-fold in a soluble form from human serum (Lee and Pierce 1995), and 6900-fold from human lymphoblasts by Page et al. (1996). Most recently the enzyme was purified 670 000-fold to apparent homogeneity from bovine liver using immunoaffinity chromatography and gel filtration (Kornfeld et al. 1998). The partial purifications were achieved using standard protein fractionation methods: detergent extraction of cell membranes; anion-exchange chromatography; lectin affinity chromatography utilizing ConA-Sepharose, lentil lectin-Sepharose or WGA-Sepharose; metal chelating Sepharose (Cu^{2+} or Zn^{2+}) chromatography; gel filtration fractionation and preparative sodium dodecyl sulfate-polyacrylamide gel electrophoresis (SDS-PAGE) on unboiled samples in the absence of reducing agents. The immunoaffinity purification was achieved using such a partially purified bovine phosphodiester α-GlcNAcase to generate a panel of murine monoclonal antibodies, which were screened for their ability to capture the enzyme with high affinity to immobilized Protein A Plus coated with rabbit anti-mouse immunoglobulin G (IgG). One of the antibodies (UC-1) had high affinity; and when coupled to Ultra Link beads, it provided a high-affinity matrix that could bind virtually all of the phosphodiester α-GlcNAcase solubilized from bovine liver membranes with 2% Triton X-100/0.5% deoxycholate. The enzyme could be eluted with 0.5 M NaHCO 0.3% Lubrol at pH 10 to give about 20000-fold purification in that one step alone. Further contaminants were removed by gel filtration on Superose 6 to yield a single band on SDS-PAGE in the presence of reducing agent that had a mobility of 68 kDa. This pure enzyme provided amino acid sequence data that allowed cloning of the cDNA, as described in the next section.

The availability of pure bovine liver phosphodiester α-GlcNAcase permitted Kornfeld et al. (1998) to discover that the native enzyme existed as a tetramer of about 242 kDa on gel filtration, was dissociated into S-S bridged dimers of 136 kDa by SDS-PAGE in the absence of reducing agents, and resolved into 68-kDa monomers by SDS-PAGE in the presence of reducing agents. The mobility of the monomers on SDS-PAGE was shifted to about 50 kDa following treatment with PNGase F but was not changed after treatment with endoglycosidase H, showing that each enzyme monomer contained five to seven complex-type oligosaccharides but no high mannose-type oligosaccharides. The pure enzyme was shown to contain 3.8 moles of sialic acid per mole of monomer, suggesting that it traveled to the TGN, site of sialyltransferase, during its intracellular itinerary.

Biological Aspects

The amino acid sequences of peptides derived from pure bovine liver phosphodiester α-GlcNAcase allowed Kornfeld et al. (1999) to isolate human cDNA and mouse genomic DNA clones encoding the enzyme. Subsequently, a databank search revealed that recently deposited 160 kb of human genomic sequence from chromosome 16 contained a 10-kb segment corresponding to the genomic sequence of human phosphodiester α-GlcNAcase. Kornfeld et al. (1999) found that the gene is organized into 10 exons, and that the protein sequence encoded by the clones shows 80% identity between human and mouse phosphodiester α-GlcNAcase but no homology to other known proteins. The deduced protein sequence defines a type I membrane-spanning glycoprotein of 515 amino acids with an amino-terminal hydrophobic signal sequence of 25 amino acids and a likely signal sequence cleavage site between Gly-25 and Leu 26. A luminal domain of 422 amino acids containing six potential N-linked glycosylation sites is followed by a hydrophobic 26-residue transmembrane region and a 41-residue cytoplasmic tail that contains both a tyrosine-based and an NPF internalization motif. The presence of [487]YHPL suggests that phosphodiester α-GlcNAcase travels to the plasma membrane and is returned to the TGN via coated vesicles; and the presence of NPFKD at the C-terminus of the enzyme suggests that the enzyme may undergo endocytosis from an endosomal compartment aided perhaps by a cytosolic protein with an EH domain. The data now available indicate that phosphodiester α-GlcNAcase probably acts in the TGN compartment to "uncover" lysosomal enzymes just prior to their binding to the mannose 6-phosphate receptors in the TGN, which target them to endosomes. Perhaps a certain amount of phosphodiester α-GlcNAcase gets carried along to both endosomes and plasma membrane and must be retrieved. Future studies will clarify the trafficking routes of phosphodiester α-GlcNAcase.

References

Couso R, Lang L, Roberts RM, Kornfeld S (1986) Phosphorylation of the oligosaccharides of uteroferrin by UDP-GlcNAc:glycoprotein *N*-acetylglucosamine-1-phosphotransferases from rat liver, *Acanthamoeba castellani*, and *Dictyostelium discoideum* requires α1,2 linked mannose residues. J Biol Chem 261:6326–6331

Deutscher SL, Creek KE, Merion M, Hirschberg CB (1983) Subfractionation of rat liver Golgi apparatus: separation of enzyme activities involved in the biosynthesis of the phosphomannosyl recognition marker in lysosomal enzymes. Proc Natl Acad Sci USA 80:3938–3942

Goldberg DE, Kornfeld S (1983) Evidence for extensive subcellular organization of asparagine-linked oligosaccharide processing and lysosomal enzyme phosphorylation. J Biol Chem 258:3159–3165

Hasilik A, Klein V, Waheed A, Strecker G, von Figura K (1980) Phosphorylated oligosaccharides in lysosomal enzymes: identification of α-*N*-acetylglucosamine(1) phospho(6) mannose diester groups. Proc Natl Acad Sci USA 77:7074–7078

Kornfeld R, Bao M, Brewer K, Noll C, Canfield WM (1998) Purification and multimeric structure of bovine *N*-acetylglucosamine-1-phosphodiester α-*N*-acetylglucosaminidase. J Biol Chem 273:23203–23210

Kornfeld R, Bao M, Brewer K, Noll C, Canfield W (1999) Molecular cloning and functional expression of two splice forms of human *N*-acetylglucosamine-1-phosphodiester α-*N*-acetylglucosaminidase. J Biol Chem 274:32778–32785

Lee JK, Pierce M (1995) Purification and characterization of human serum *N*-acetylglucosamine-1-phosphodiester α-*N*-acetylglucosaminidase. Arch Biochem Biophys 319: 413–425

Mullis KG, Ketcham CM (1992) The synthesis of substrates and two assays for the detection of *N*-acetylglucosamine-1-phosphodiester α-*N*-acetylglucosaminidase (uncovering enzyme). Anal Biochem 205:200–207

Mullis KG, Huynh M, Kornfeld RH (1994) Purification and kinetic parameters of bovine liver *N*-acetylglucosamine-1-phosphodiester α-*N*-acetylglucosaminidase. J Biol Chem 269:1718–1726

Page T, Zhao KW, Tao L, Miller AL (1996) Purification and characterization of human lymphoblast *N*-acetylglucosamine-1-phosphodiester α-*N*-acetylglucosaminidase. Glycobiology 6:619–626

Pohlmann R, Waheed A, Hasilik A, von Figura K (1982) Synthesis of phosphorylated recognition marker in lysosomal enzymes is located in the *cis* part of Golgi apparatus. J Biol Chem 257:5323–5325

Rohrer J, Kornfeld R (2001) Lysosomal Hydrolase Mannose 6-Phosphate Uncovering Enzyme Resides in the trans-Golgi Network. Mol Biol Cell 12:1623–1631

Tabas I, Kornfeld S (1980) Biosynthetic intermediates of β-glucuronidase contain high mannose oligosaccharides with blocked phosphate residues. J Biol Chem 255:6633–6639

Varki A, Kornfeld S (1980) Identification of a rat liver α-*N*-acetylglucosaminyl phosphodiesterase capable of removing "blocking" α-*N*-acetylglucosamine residues from phosphorylated high mannose oligosaccharides of lysosomal enzymes. J Biol Chem 255:8398–8401

Varki A, Kornfeld S (1981) Purification and characterization of rat liver α-*N*-acetylglucosaminyl phosphodiesterase. J Biol Chem 256:9937–9943

Varki A, Sherman W, Kornfeld S (1983) Demonstration of the enzymatic mechanisms of α-*N*-acetyl-D-glucosamine-1-phosphodiester *N*-acetylglucosaminidase (formerly called α-*N*-acetylglucosaminylphosphodiesterase) and lysosomal α-*N*-acetylglucosaminidase. Arch Biochem Biophys 222:145–149

Waheed A, Hasilik A, von Figura K (1981) Processing of the phosphorylated recognition marker in lysosomal enzymes: characterization and partial purification of a microsomal α-*N*-acetylglucosaminyl phosphodiesterase. J Biol Chem 256:5717–5721

Appendix

Map 1

Biosynthetic Pathways of *N*-Glycans

N-Glycans in mammals are initially synthesized as lipid-linked oligosaccharides (LLO) on the rough endoplasmic reticulum (ER) in the cells as shown in Map 1-1. The structure of this precursor can be written as $(Glc)_3(Man)_9(GlcNAc)_2$. This precursor oligosaccharide is linked by a pyrophosphoryl residue to dolichol, a long-chain (75–105 carbon atoms) polyisoprenoid lipid that acts as a carrier for the oligosaccharide. The dolichol pyrophosphoryl oligosaccharide is formed in the ER in a complex set of reactions utilizing membrane-attached enzymes of the rough ER, in which *N*-acetylglucosamine, mannose, and glucose residues are added one at a time to dolichol phosphate. The monosaccharide donors include UDP-GlcNAc and GDP-Man for the first seven linkage reactions that occur on the cytosolic side of the ER membrane. In the first two steps, GlcNAc residues are added to the Dol-P by GlcNAc-1-phosphotransferase and then by a GlcNAc-transferase. Next, five mannose residues are added using GDP-Man, and the chain flips across the membrane bilayer to become oriented in the lumen of the ER. Sequentially four mannose and three glucose residues are added using Dol-P-Man and Dol-P-Glc. The final dolichol pyrophosphoryl oligosaccharide is oriented so the dolichol portion is firmly embedded in the ER membrane and the oligosaccharide portion faces the ER lumen. The oligosaccharide is transferred en bloc by an ER enzyme, oligosaccharyltransferase, from the dolichol carrier to an asparagine residue on the nascent polypeptide. The asparagine residue must be in the tripeptide recognition sequence Asn-X-Ser or Asn-X-Thr, where X is any amino acid except proline. In some rare cases Asn-X-Cys is also utilized for *N*-glycosylation.

As summarized in Map 1-2, as soon as the oligosaccharide is transferred to the asparagine residue in the rough ER, all three glucose residues and one particular mannose residue of the oligosaccharide are removed by three enzymes, α-glucosidase I, α-glucosidase II, and ER α-mannosidase IA. Upon removal of the glucose residues by glucosidases I and II, the glycoprotein either is properly folded and ready for

Κατsuko Yamashita

Department of Biochemistry, Sasaki Institute, 2-2 Kanda-Surugadai, Chiyoda-ku, Tokyo 101-0062, Japan
Tel. +81-3-3294-3286; Fax +81-3-3294-2656
e-mail: yamashita@sasaki.or.jp

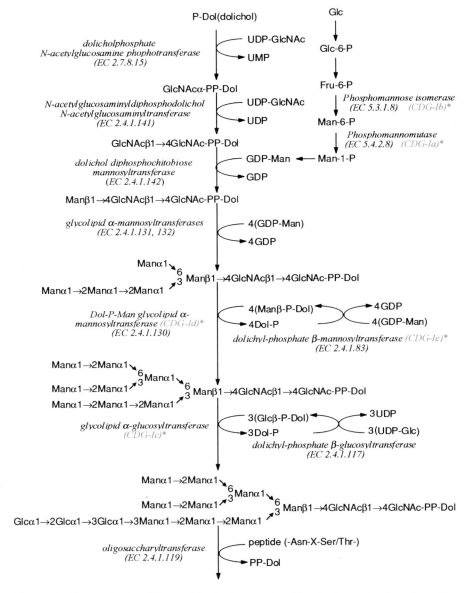

Map 1-1. Biosynthesis of lipid-linked oligosaccharides in the endoplasmic reticulum (ER)

* Enzyme deficiencies shown in blue are the subtypes of congenital disorders of glycosylation type I (CDG-I) which result in under-*N*-glycosylation

Map 1 Biosynthetic Pathways of *N*-Glycans   627

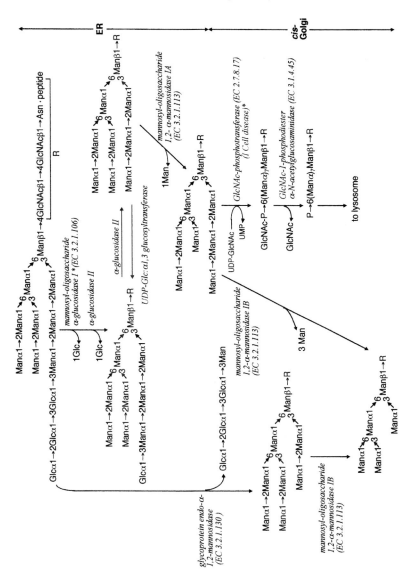

Map 1-2. Processing of *N*-glycans in the ER and *cis*-Golgi

* Studies of α-glucosidase I deficiency confirmed the existence of the alternative pathway via endo-α-mannosidase; and studies of GlcNAc-phosphotransferase deficiency (I-cell disease) suggested the existence of a trafficking signal directing lysosomal enzymes to the lysosome via the Man-6-P receptor

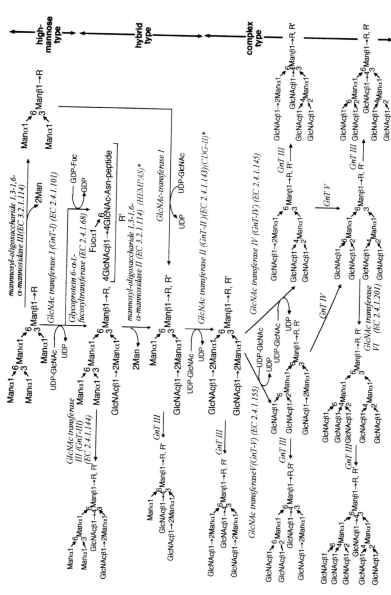

Map 1-3. Diversification of *N*-glycan biosynthesis in *medial* Golgi

* Glycoproteins from HEMPAS (α-mannosidase II deficiency) and CDG-II (GnT-II deficiency) patients cannot be fully processed to the complex type of glycan

Map 1 Biosynthetic Pathways of N-Glycans   629

further processing or is reglucosylated by a glucosyltransferase that acts on the Man$_9$GlcNAc$_2$-Asn of improperly folded proteins. Calnexin, which is a kind of molecular chaperone in the ER, binds preferentially to the glucose α1,3 linked to underlying mannose and retains the glycoprotein in the ER for folding. A few glucose-containing N-glycan precursors, escaping regulation by this pathway, have been observed occasionally in the Golgi. An endo-α-mannosidase acts on these precursors, generating the Man$_8$GlcNAc$_2$-Asn-peptide and Glc$_3$Man. On lysosomal glycoenzymes, the Man$_8$GlcNAc$_2$-Asn-peptide is modified by a GlcNAc-phosphotransferase and the GlcNAc residues are subsequently removed, yielding Man-6-P at the nonreducing terminus. In most cases distinct α-mannosidase enzymes IA and IB located in the ER and Golgi sequentially process the high-mannose type N-glycans to Man$_5$GlcNAc$_2$. This Man$_5$GlcNAc$_2$ glycan becomes a substrate for the diversification of N-glycans consisting of complex, high-mannose, and hybrid types.

As shown in Map 1–3, β-N-acetylglucosaminyltransferase-I (GnT-I) adds a GlcNAc in β1 $\rightarrow$ 2 linkage to produce a hybrid-type N-glycan structure. α-Mannosidase II acts specifically on the GlcNAc$_1$Man$_5$GlcNAc$_2$-Asn, hybrid-type N-glycan in the medial Golgi; and the processed N-glycan product is the specific substrate for GnT-II, which catalyzes the conversion of hybrid to complex-type N-glycans. As an alternate pathway, Man$_5$GlcNAc$_2$-Asn-peptide is directly converted to Man$_3$GlcNAc$_2$-Asn-peptide by α-mannosidase III, and sequentially N-acetylglucosamine is transferred to

Table 1. Various distal structural modifications of N-glycans

Structure	Enzyme	EC number
Galβ1$\rightarrow$4GlcNAc	GlcNAc:β1,4-galactosyltransferase	2.4.1.38, 90
Galβ1$\rightarrow$3GlcNAc	GlcNAc:β1,3-galactosyltransferase	
Galβ1$\rightarrow$4(HSO$_3$$\rightarrow$6)GlcNAc	HSO$_3$$\rightarrow$6GlcNAc:$\beta$1,4-galactosyltransferase	
Galα1$\rightarrow$3Galβ1$\rightarrow$4GlcNAc	Gal:α1,3-galactosyltransferase	2.4.1. 151
GalNAcβ1$\rightarrow$4GlcNAc	GlcNAc:β1,4-N-acetylgalactosaminyltransferase	
GalNAcβ1$\rightarrow$4(Siaα2$\rightarrow$3)Galβ1 $\rightarrow$4GlcNAc	Gal:β1,4-N-acetylgalactosaminyltransferase	
Fucα1$\rightarrow$2Galβ1$\rightarrow$4GlcNAc	Gal:α1,2-fucosyltransferase	2.4.1.69
Galβ1$\rightarrow$4(Fucα1$\rightarrow$3)GlcNAc	GlcNAc:α1,3-fucosyltransferase	2.4.1.65, 152
GlcNAcβ1$\rightarrow$3(Galβ1$\rightarrow$4GlcNAc)$_n$	Gal:β1,3-N-acetylglucosaminyltransferase	2.4.1.149
Galβ1$\rightarrow$4GlcNAcβ1$\rightarrow$ 3(GlcNAcβ1$\rightarrow$6)Galβ1 $\rightarrow$4GlcNAc	Gal:β1,6-N-acetylglucosaminyltransferase	2.4.1.150
GlcAβ1$\rightarrow$3Galβ1$\rightarrow$4GlcNAc	HNK-I β1$\rightarrow$3-glucuronyltransferase	
Siaα2$\rightarrow$3Galβ1$\rightarrow$4GlcNAc	Gal:α2,3-sialyltransferase	2.4.99.6
Siaα2$\rightarrow$6Galβ1$\rightarrow$4GlcNAc	Gal:α2,6-sialyltransferase	2.4.99.1
Galβ1$\rightarrow$3(Siaα2$\rightarrow$6)GlcNAc	GlcNAc:α2,6-sialyltransferase	
(Siaα2$\rightarrow$8)$_{5-100}$Siaα2$\rightarrow$3Galβ1 $\rightarrow$4GlcNAc	α2,8-Sialyltransferase	2.4.99.8
HSO$_3$$\rightarrow$4GalNAc$\beta$1$\rightarrow$4GlcNAc	GalNAc: $\rightarrow$4-sulfotransferase	2.8.2.7
HSO$_3$$\rightarrow$6GlcNAc$\beta$1$\rightarrow$Man	GlcNAc: $\rightarrow$6-sulfotransferase	
HSO$_3$$\rightarrow$3GlcA$\beta$1$\rightarrow$3Gal$\beta$1$\rightarrow$4GlcNAc	HNK-I: 3-sulfotransferase	

Underlined residues are transferred by the respective transferases it by GnT-I. When forming multiantennary N-glycan structures, GlcNAc residues are added to the trimannosyl core by six GlcNAc transferases (I–VI). GnT-I and α-mannosidase II are key enzymes required to produce a complex N-glycan.

GlcNAcMan$_3$GlcNAc$_2$-Asn is the competitive substrate for GnT-II, GnT-III, and GnT-IV; and GlcNAc$_2$Man$_3$GlcNAc$_2$-Asn is the competitive substrate for GnT-III, GnT-IV, and GnT-V, as indicated by the relative enzymatic activities. However, as soon as the bisecting N-acetylglucosamine residue is added to the respective branched N-glycans by GnT-III, the branching stops. GnT-VI action is uncommon, and it may require prior branching by GnT-II and GnT-V. Many N-glycans are modified by a fucosyltransferase that adds a fucosyl residue in an α1-6 linkage to the GlcNAc residue that is linked to asparagine. This modification is found on hybrid and complex-type N-glycans following GnT-I action.

Furthermore, multiple N-glycosylation sites on the same protein contain different glycan structures as site-specific oligosaccharide heterogeneity or microheterogeneity. Protein sequence or conformation may influence N-glycan diversification by affecting substrate accessibility for glycosyltransferases. Other factors include sugar nucleotide metabolism, transport rates in the ER and Golgi, and the localization and expression levels of glycosyltransferases in the assembly line of the Golgi. The various distal structural modifications of N-glycans in a given cell are affected by varied levels of expression of glycosyltransferases in different types of cells as summarized in Table 1. Because glycosyltransferase genes often contain promoter regions bearing transcription-factor elements that function in growth regulatory and oncogene transformation pathways, their activities are altered by cell development, cell transformation, cell differentiation, aging, and the cell cycle.

Map 2

Biosynthetic Pathways of *O*-Glycans

O-Linked mucin-type oligosaccharides are not preassembled on a dolichol derivative, but every sugar is transferred individually from a specific nucleotide sugar. *O*-Glycans are not processed by glycosidases.

Synthesis of Core Structures

The first step in the biosynthesis of mucin-type *O*-glycans is the enzymatic transfer of GalNAc from UDP-GalNAc to a Ser/Thr residue. The reaction is catalyzed by a family of UDP-GalNAc:polypeptide *N*-acetylgalactosaminyltransferases (GalNAc transferase), which occurs mainly in the *cis*-Golgi. This enzyme is expressed in all mammalian cells and is common to the synthesis of all mucin-type *O*-glycans. It appears that each member of the multiple GalNAc transferases has a slightly different, but overlapping, substrate specificity; and the repertoire of GalNAc transferases expressed in a cell controls the *O*-glycosylation pattern. No general consensus sequence of the peptide backbone for *O*-glycosylation has been found.

Subsequently, stepwise elongation by specific transferases yields eight core structures containing GlcNAc, Gal, or GalNAc substitutions of GalNAc 1α-Ser/Thr. If this elongation does not occur, GalNAc 1α-Ser/Thr, which is often referred to as the Tn antigen, is generated. The Tn antigen may be modified to sialyl Tn antigen by α2,6-SA transferase.

Core 1 is a common core structure in mucins and other secreted and cell surface glycoproteins. The enzyme synthesizing core1, core1β1,3 Gal transferase is a ubiquitous enzyme present in most mammalian cells. Core 1 as a terminal structure, often referred to as the T antigen, is prevalent in cancer cells and their secretions. Core 1 is not usually exposed in glycoproteins but is monosialylated (SAα2→3Galβ1→

Hiroshi Nakada

Department of Biotechnology, Faculty of Engineering, Kyoto Sangyo University, Kamigamo-Motoyama, Kita-ku, Kyoto 603-8555, Japan
Tel. +81-75-705-1897; Fax +81-75-705-1888
e-mail: hnakada@cc.kyoto-su.ac.jp

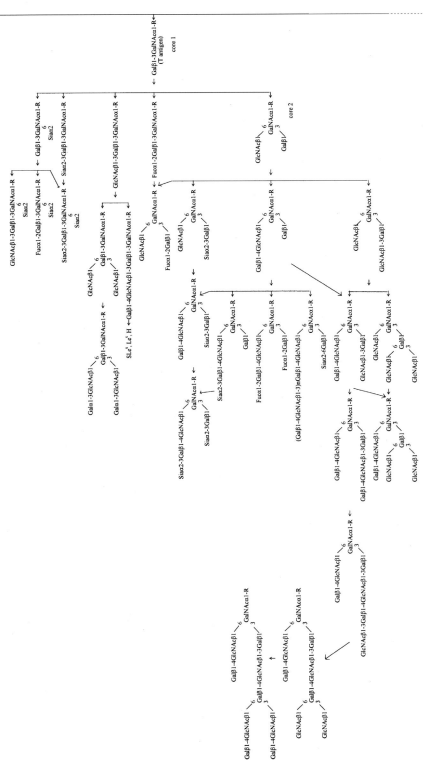

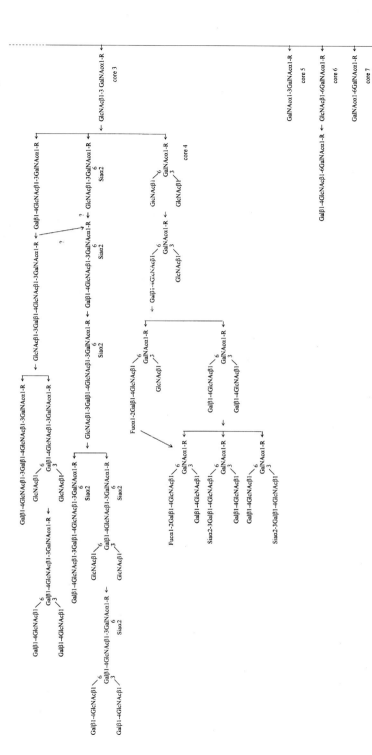

Map 2. Biosynthetic pathways of mucin-type *O*-glycans in mammals

3GalNAcα1→Ser/Thr and SAα2→6[Galβ1→3]GalNAcα1→Ser/Thr) or disialylated (SAα2→6[SAα2→3Galβ1→3]GalNAcα1→Ser/Thr).

Core 2 β6-GlcNAc transferase converts an unsubstituted core 1 to core 2. A number of related β6-GlcNAc transferases appear to exist with distinct substrate specificities. The core 2 β6-GlcNAc transferase does not act when core 1 is elongated, sialylated, sulfated, or fucosylated. The L-type core 2 β6-GlcNAc transferase occurs in leukocytes, leukemia cells, and many cancer cells and catalyzes only the conversion of core 1 to core 2. It is postulated that this enzyme is rate-limiting for biosynthesis of the poly-lactosamine-type structure on O-glycans. The enzyme activity is high in leukemia and activated T lymphocytes, and it is often altered in cancer cells. The M-type enzyme may act on three substrates—Gal β1→3GalNAcα1→Ser/Thr, GlcNAc β1→3GalNAcα1→Ser/Thr, and GlcNAc β1→3Galβ→, to form core 2, core 4, and I antigen, respectively. Other types of β1,6 GlcNAc transferase synthesize terminal or internal GlcNAc β1→6 branches as well as linear GlcNAc β1→6Gal/GalNAc structures of cores 2, 4, 6, and I determinants.

Core 3 is synthesized by core 3 β3-GlcNAc transferase. The enzyme activity has been found only in mucin-secreting tissues and is low in colon cancer tissue. Because synthesis of core 3 precedes that of core 4, synthesis of core 4 is limited by low activity and limited tissue distribution of core 3 β3-GlcNAc transferase.

The core 4 β6-GlcNAc transferase activity is found in colonic tissues and a number of cancer cell lines. Occurrence of the core 4 structure is restricted owing to the limiting and tissue-specific activity of core 3 β3-GlcNAc transferase as described above.

The core 5 structure has been found in mucins from several mammalian and human meconia. Core 5 α3-GalNAc transferase activity has been demonstrated in the colonic mucosa of a patient with adenocarcinoma.

Core 6 β6-GlcNAc transferase activity has been found in human ovarian tissues. The enzyme acts on GalNAc without prior substitution of GalNAc by a Gal β1→3 or GlcNAc β1→3 residue. Core 7 and 8 structures have been reported in bovine sub-maxillary mucin and in human bronchial mucin, respectively.

Synthesis of the core structures regulates the expression of functional terminal carbohydrate structures. Thus control of the biosynthetic pathways at early steps may have a great effect on the structures, properties, and functions of O-glycans. When terminal sialic acid or sulfate residues are added to the core structures, further elongation is inhibited, resulting in short acidic O-glycans.

Elongation of O-Glycans

Elongation β3-GlcNAc transferase catalyzes the elongation of core 1 or core 2 structure with an unsubstituted Gal β1→3 residue. β4-Gal transferase and i β3-GlcNAc transferase act sequentially to synthesize a common backbone structure, repeating Gal β1→4 GlcNAc β1→3 units (type 2 chain), leading to the synthesis of i antigen and poly-N-acetyllactosamine. These two enzymes are ubiquitous, and i β3-GlcNAc transferase differs from the elongation β3-GlcNAc transferase and from the core 3 β3-GlcNAc transferase by its tissue distribution and specificity. β4-Gal transferase is ubiquitous, and a number of related β4-Gal transferases exist.

Map 2 Biosynthetic Pathways of O-Glycans   635

Another backbone structure (i.e., repeating Gal β1→3 GlcNAc β1→3 units, or type 1 chain) may be introduced by a β3-Gal transferase. Thus at least two types of Gal transferase are involved in the synthesis of poly-*N*-acetyllactosamine.

GlcNAc β1→6 branches of the elongated backbone structure, which is referred to as the I antigen, are synthesized by a number of β6-GlcNAc transferases.

Terminal Glycosylation

Terminal sugars include SA, Fuc, Gal, GalNAc, and GlcNAc. *O*-Glycans may carry terminal epitopes such as Lewis-type or ABO-type antigens, which are synthesized probably by the same enzymes as those assembling these antigens on glycolipids and *N*-glycans.

A number of α3-SA transferases with various specificities toward Gal terminal acceptors act on core 1, core 2, and other *O*-glycan structures. Core 1 α3-SA transferase I (ST3GalI) acts on core 1. A family of α6-SA transferases have been cloned. α6-SA transferase I (ST6GalNAc-I) acts on substituted or unsubstituted GalNAc-Ser/Thr. α6-SA transferase II (ST6GalNAc-II) acts on the GalNAc residue of the core 1 structure. SAα2→6GalNAcα1→Ser/Thr is referred to as the sialyl Tn antigen. SAα2→3Galβ1→3GalNAc→R is a substrate for α6-SA transferases III and IV (ST6GalNAc-III and ST6GalNAc-IV).

There are two types of α2-Fuc transferases [i.e., hematopoietic type (FucT-I) and secretory type (FucT-II)] that synthesize the blood group H (O) determinant. A family of α3-Fuc transferases (FucT-III–VII and IX) act on internal GlcNAc residues with slightly different substrate specificities. These enzymes synthesize Lewis determinants in various cell types. α3-Fuc transferase III (FucT-III) has an unusual dual enzyme activity and is capable of synthesizing both Fucα1→3GlcNAc and Fucα1→4GlcNAc linkages. α3-Fuc transferases IV and VII (FucT-IV and -VII) appear to be responsible for the synthesis of sialyl- and sulfo-Le^x ligands for selectins.

Map 3

Biosynthetic Pathways of Glycosphingolipids

The biosynthetic pathways of glycosphingolipids are classified into eight series on the basis of the core structures: Gala-, Globo- (Gb), Isoglobo- (iGb), Ganglio- (Gg), Lacto- (Lc), Neolacto- (nLc), Arthro- (Ar), and Mollu- (Ml). The Arthro- and Mollu-series are coined on the basis of the animal species, arthropods and molluscs, respectively, that carry corresponding glycosphingolipids as major constituents. This is summarized as a species-specific feature of glycosphingolipids and suggests that animal species had a chance for the selection of monosaccharides as elements for glyco-chains of glycosphingolipids during the process of evolution. The selection itself is the subject of future research of glyco-science. The Isoglobo-series has been found only in rats. Another good example of species-specific glycosphingolipids is the Forssman antigen. The Forssman antigen is quite immunogenic to Forssman antigen-negative animals, the rabbit and rat; this is based on the fact that the animal species cannot express $Gb_4Cer:\alpha3GalNAc$ transferase or glycosyltransferases involved in the production of Gb_4Cer.

Other series are found in various mammals. Most of the glycosphingolipids are biosynthesized from GlcCer, with only limited members from GalCer (NeuAc-GalCer, sulfated GalCer, Galα4GalCer). Glycosphingolipids containing sialic acids are called gangliosides, and the maps in this chapter describe only NeuAc as the sialic acid species. Gangliosides obtained from visceral organs (but not the central nervous system) of mammals except humans contain NeuGc as well.

Glyceroglycolipids are not described in the maps because they are minor constituents in animal species. One exception is the glyceroglycolipids in the mammalian testis and spermatozoa: monoalkyl-monoacyl-galactosyl-glycerol and monoalkyl-monoacyl-3'-sulfogalactosyl-glycerol (seminolipid). This is a unique situation, and recently it was suggested that these two glyceroglycolipids play important but still unknown roles in the meiosis of spermatogenesis, as shown by gene targeting experiments.

Akemi Suzuki

RIKEN Frontier Research System, 2-1 Hirosawa, Wako-shi, Saitama 351-0198, Japan
Tel. +81-48-467-9615; Fax +81-48-462-4692
e-mail: aksuzuki@postman.riken.go.jp

Map 3 Biosynthetic Pathways of Glycosphingolipids   637

Terminal structures of glycosphingolipids are often functionally important because they are recognized by specific antibodies, such as anti-A, B, H, Le^a, Le^b, Le^x, sialyl Le^a, and sialyl Le^x antibodies or carbohydrate chain recognition molecules such as Selectins or Siglecs. This feature is associated not only with glycosphingolipids but with glycoproteins as well.

Glyco-chains of glycosphingolipids are mainly synthesized on the inner leaflet of the Golgi membrane. The exception is GlcCer, which is biosynthesized on the cytoplasmic leaflet of endoplasmic reticulum (ER) membranes by GlcCer synthetase. Most of the glycosphingolipids are biosynthesized from GlcCer, and further elongation of glyco-chains is carried out on the inner leaflet of the Golgi membrane. To complete the elongation, Glc-Cer must be translocated from the cytosolic leaflet to the inner leaflet of the ER membrane by a specific translocase, which has not been identified. This topological orientation of catalytic sites of glycosyltransferases is supported by the structures of glycosyltransferases. GlcCer synthase is a type III membrane protein with an N-terminal siganal-anchor sequence, and other glycosyltransferases are type II membrane proteins with an N-terminal membrane-spanning domain and catalytic domains at the C-terminal region.

As described elsewhere in this handbook, which glycosyltransferase among the various glycosyltransferases in one family (e.g., β4GalT-I of the β4Gal transferase family) is actually responsible for synthesis of a particular glycosphingolipid is an important issue. In vitro assay of glycosyltransferase activities using recombinant proteins as enzymes and glycosphingolipids as substrates gives basic information; and demonstration of biosynthesis at the cellular level is required. Glycosyltransferases involved in the biosynthesis of glycosphingolipids are classified into two groups. One group includes those specific for glycosphingolipids, and the other includes those taking both glycosphingolipids and glycoproteins as substrates. An example of the former group is GM3:β4GlcNAc transferase, which is required to make GM2; and examples of the latter group are α3GalNAc and α3Gal transferases of the ABO blood group antigens. In general, glycosyltransferases for outer or terminal glyco-chains take both substrates, glycosphinogolipids and glycoproteins; and glycosyltransferases for the inner and unique structures of glycosphingolipids are glycolipid-specific.

The composition of glycosphingolipids is tissue-specific, and this feature is produced by regulation of glycosyltransferase activities and partly by sugar hydrolase activities. Every tissue or cell has a unique set of glycosyltransferases, which can be regulated by tissue- or cell-specific transcriptional control or posttranslational modifications. The molecular mechanisms and physiological meanings of tissue- or cell-specific regulations are an important subject for future research.

Creation of a new paradigm in glycoscience will be brought about by the introduction of new research subjects, including *Caenorhabditis elegans* and *Drosophila* research. We have now accumulated data for glycosphingolipids in insects, as shown in Map 3-8.

These maps are revised versions of those published in 1992 in Protein, Nucleic Acid and Enzyme 37:2088–2106.

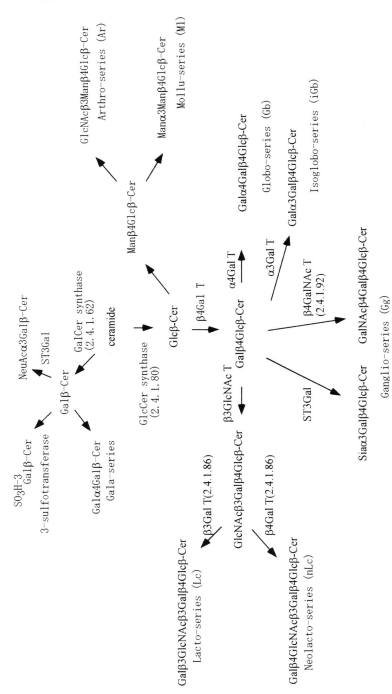

Map 3-1. Glycosphingolipid biosynthesis

Map 3 Biosynthetic Pathways of Glycosphingolipids   639

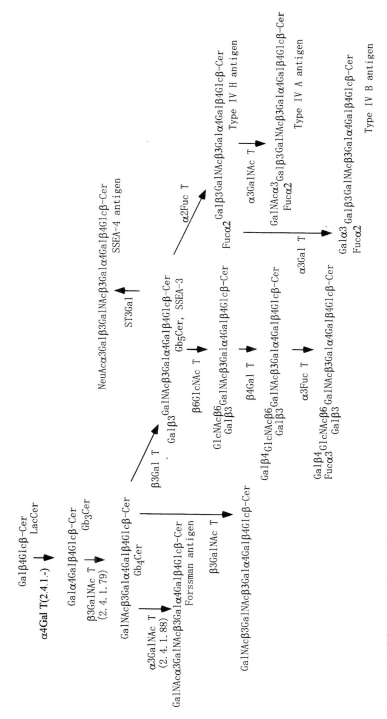

Map 3-2. Globo-series glycosphingolipids

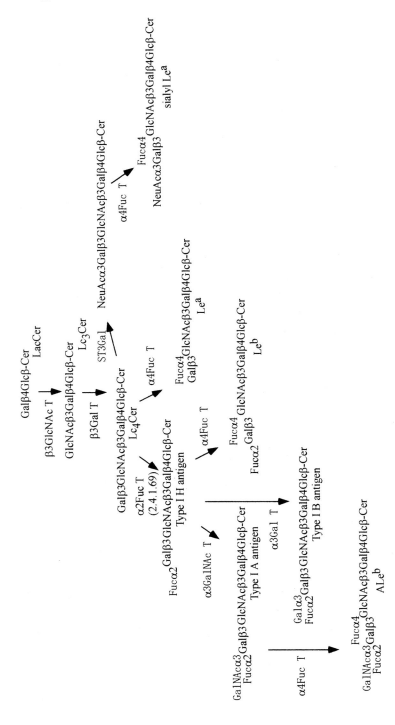

Map 3-3. Lacto-series glycosphingolipids

Map 3 Biosynthetic Pathways of Glycosphingolipids   641

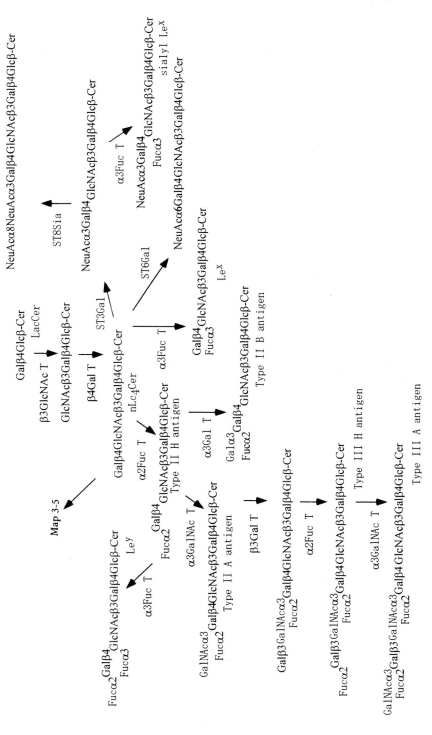

Map 3-4. Neolacto-series glycosphingolipids 1

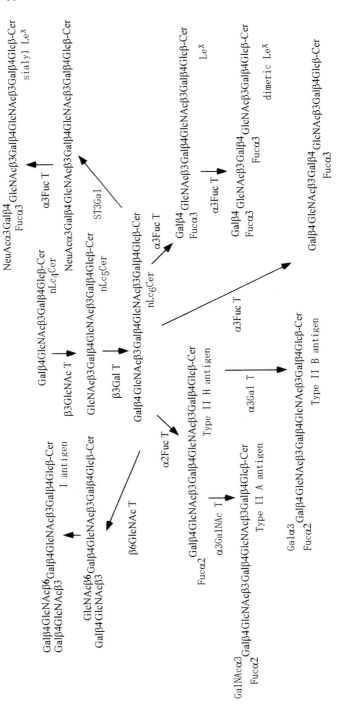

Map 3-5. Neolacto-series glycosphingolipids 2

Map 3 Biosynthetic Pathways of Glycosphingolipids   643

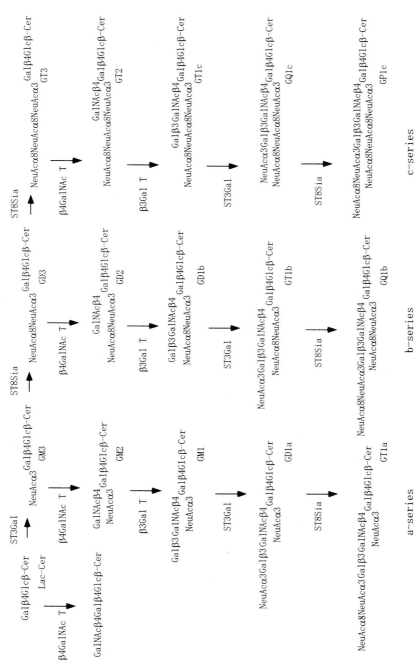

Map 3-6. Gangliosides 1

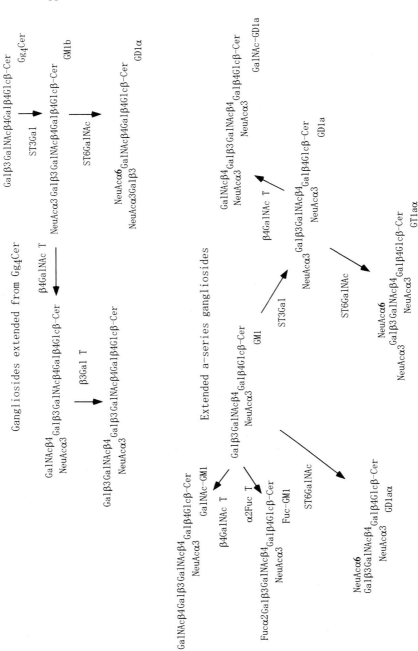

Map 3-7. Gangliosides 2

Map 3 Biosynthetic Pathways of Glycosphingolipids   645

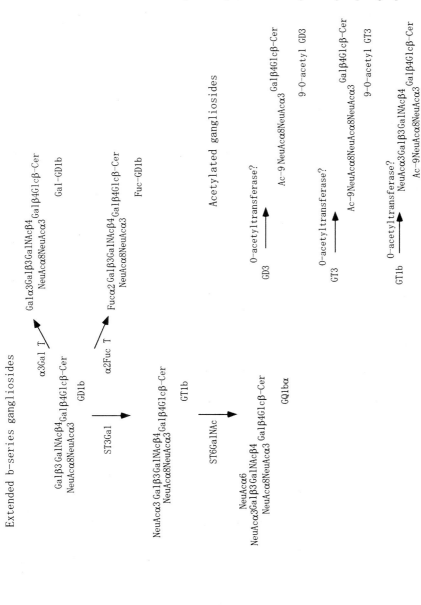

Map 3-8. Gangliosides 3

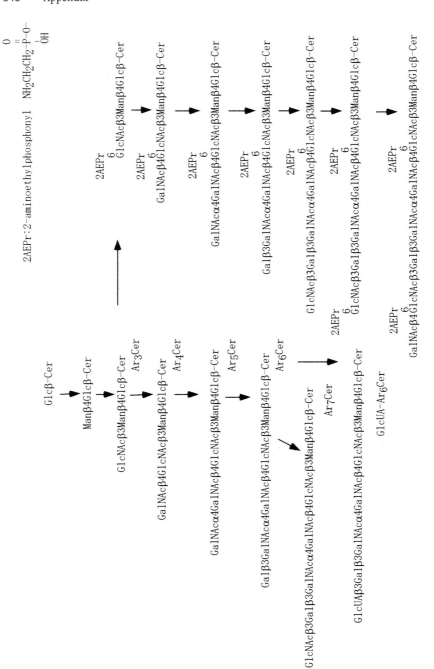

Map 3-9. Arthro-series

Map 4

Biosynthetic Pathways of Proteoglycans

Proteoglycans comprise a class of glycosylated proteins that have glycosaminoglycans as a carbohydrate component. Biosynthesis of glycosaminoglycans, consisting of repeating disaccharides, is built on specific oligosaccharide acceptors synthesized on core proteins. In the case of chondroitin sulfate (dermatan sulfate) and heparan sulfate (heparin), the acceptor oligosaccharide is a common tetrasaccharide (GlcA-Gal-Gal-Xyl) attached to Ser residues in the core protein. In the case of keratan sulfate (KS), the glycosaminoglycan is synthesized as a branch of complex N-linked oligosaccharides (KS I) or as a branch of O-linked oligosaccharides (KS II).

The carbohydrate structure of each glycosaminoglycan is depicted in the following biosynthetic maps, followed by a brief annotation on the nature of the carbohydrate transfer reactions and the enzymes involved in the reaction. The sites of sulfation, a key feature of the glycosaminoglycan modifications, are also indicated by adding symbols (2S, 3S, 6S, and NS—each indicating ester sulfate on the C-2, C-3, or C-6 position and N-sulfate, respectively) underneath sugar residues. Another map details the sequence of carbohydrate modification during heparan sulfate/heparin biosynthesis.

In addition to the glycosaminoglycans, the proteoglycans usually have both N-linked and O-linked oligosaccharides, as are found in ordinary glycoproteins. These oligosaccharides are expected to follow the same biosynthetic pathways as other complex carbohydrates (and are not discussed here).

Masaki Yanagishita

Department of Hard Tissue Engineering, Biochemistry, Division of Bio-Matrix, Graduate School, Tokyo Medical and Dental University, 1-5-45 Yushima, Bunkyo-ku, Tokyo 113-8549, Japan
Tel. +81-3-5803-5447; Fax +81-3-5803-0187
e-mail: m.yanagishita.bch@tmd.ac.jp

Map 4-1. Chondroitin sulfate (dermatan sulfate*)

⑦   ⑥   ⑤   ④   ③   ②   ①
$(\rightarrow \beta GlcA\text{-}1,3\rightarrow \beta GalNAc\text{-}1,4)_n \rightarrow \beta GlcA\text{-}1,3 \rightarrow \beta Gal\text{-}1,3 \rightarrow \beta Gal\text{-}1,4 \rightarrow \beta Xyl\text{-}1 \rightarrow Ser$
              4S (⑨),6S (⑩)

↓⑧*

αIdoA-1,3
  2S (⑪)

①–④: Synthesis of linkage tetrasaccharide (indicated by a single rule)
⑤–⑦: Synthesis of chondroitin sulfate backbone (indicated by a double rule)
⑧*: Conversion of glucuronic acid to iduronic acid to form dermatan sulfate
⑨–⑪: Formation of ester sulfate on GalNAc or IdoA
n: Typically 50–100 repeats

①: Xylosyltransferase
②: Galactosyltransferase I
③: Galactosyltransferase II
④: Glucuronyltransferase I
⑤: *N*-Acetylgalactosaminyltransferase I
⑥: Glucuronyltransferase I
⑦: *N*-Acetylgalactosaminyltransferase II
⑧*: Glucuronic acid C5-epimerase
⑨: GalN 4-*O* sulfotransferase
⑩: GalN 6-*O* sulfotransferase
⑪: IdoA 2-*O* sulfotransferase

Map 4-2. Heparin and heparan sulfate
See also Fig. 1 below.

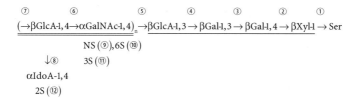

⑦   ⑥   ⑤   ④   ③   ②   ①
$(\rightarrow \beta GlcA\text{-}1,4\rightarrow \alpha GalNAc\text{-}1,4)_n \rightarrow \beta GlcA\text{-}1,3 \rightarrow \beta Gal\text{-}1,3 \rightarrow \beta Gal\text{-}1,4 \rightarrow \beta Xyl\text{-}1 \rightarrow Ser$
              NS (⑨),6S (⑩)
↓⑧   3S (⑪)
αIdoA-1,4
  2S (⑫)

①–④: Synthesis of linkage tetrasaccharide (indicated by a single rule)
⑤–⑦: Synthesis of heparan sulfate backbone (indicated by a double rule)
⑧: Conversion of glucuronic acid to iduronic acid
⑨: *N*-Deacetylation and *N*-sulfotransferation of GlcNAc
⑩⑪: Formation of ester sulfate on GlcNAc or GlcNS
⑫: Formation of ester sulfate on IdoA
n: Typically 50–100 repeats

* Chondroitin sulfate and dermatan sulfate share the common glycosaminoglycan backbone structure. In dermatan sulfate chains, some glucuronic acid residues are epimerized by enzyme reaction ⑧ to form iduronic acid. Biosynthetic reactions that occur only in dermatan sulfate synthesis are marked by asterisks

Map 4 Biosynthetic Pathways of Proteoglycans   649

①: Xylosyltransferase
②: Galactosyltransferase I
③: Galactosyltransferase II
④: Glucuronyltransferase I
⑤: N-Acetylglucosaminyltransferase I
⑥⑦: Heparan sulfate copolymerase (a single enzyme with both GlcNAc transferase and GlcA transferase activities)
⑧: Glucuronic acid C-5 epimerase
⑨: N-Deacetylase/N-sulfotransferase (a single enzyme with both N-deacetylase and N-sulfotransferase activities)
⑩: GlcN 6-O sulfotransferase
⑪: GlcN 3-O sulfotransferase
⑫: IdoA 2-O sulfotransferase

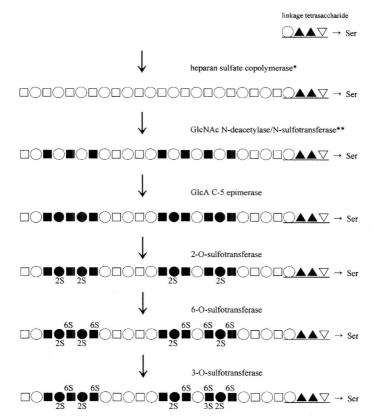

Fig. 1. Sequence of biosynthetic modifications of heparin and heparan sulfate

* A single enzyme has both GlcNAc transferase and GlcA transferase activities
** A single enzyme has both GlcNAc N-deacetylase and N-sulfotransferase activities
▽, Xyl; ▲, Gal; ○, GlcA; □, GlcNAc; ■, GlcNSO₃; ●, IdoA. The *2S*, *3S*, and *6S* below or above symbols indicate ester sulfate on C-2, C-3, and C-6 positions, respectively

Map 4-3. Keratan sulfate

A. KS I

        ②   ①

$(\beta Gal\text{-}1,4 \rightarrow \beta GlcNAc\text{-}1,3)_n \rightarrow \beta Gal\text{-}1,4 \rightarrow \beta GlcNAc\text{-}1,2 \rightarrow \alpha Man\text{-}1,6$ _

  6S (③)   6S (④) $\beta Man\text{-}1,4 \rightarrow \beta GlcNAc\text{-}1,4$

                                                              $\rightarrow \beta GlcNAc\text{-}1,4 \rightarrow Asn$

              $\alpha Sia\text{-}1,3 \rightarrow \beta Gal\text{-}1,4 \rightarrow \beta GlcNAc\text{-}1,2 \rightarrow \alpha Man\text{-}1,3$ _

B. KS II

          ②   ①

$(\beta Gal\text{-}1,4 \rightarrow \beta GlcNAc\text{-}1,3)_n \rightarrow \beta Gal\text{-}1,4 \rightarrow \beta GlcNAc\text{-}1,6$ _

  6S (③)   6S (④) $\beta GlcNAc\text{-}1 \rightarrow Ser/Thr$

                  $\alpha Sia\text{-}1,3 \rightarrow \beta Gal\text{-}1,3$ _

①②: Synthesis of keratan sulfate (indicated by a double rule) as a branch of a complex-type *N*-linked oligosaccharide (KS I) or an *O*-linked oligosaccharide (KS II)

③④: Formation of ester sulfate on Gal or GlcNAc

n: Typically 8–20 repeats

①: β1-3 *N*-Acetylglucosaminyltransferase

②: β1-4 Galactosyltransferase

③: Gal 6-*O* sulfotransferase

④: GlcNAc 6-*O* sulfotransferase (acts only on nonreducing terminal GlcNAc; therefore, it has to work simultaneously with chain elongation.)

Map 5

Biosynthetic Pathways of GPI-Anchor

The biosynthetic pathway of GPI-anchor and its transfer to protein comprises at least ten steps: transfer of GlcNAc to PI, de-N-acetylation, palmitoylation of the inositol ring, transfer of three mannoses and three ethanolamine phosphates, and transamidation to protein. Nineteen genes involved in this pathway already have been identified and characterized in mammals and yeast. The intermediates of GPI-anchor in the first three steps are synthesized on the cytoplasmic leaflet of the endoplasmic reticulum (ER) membrane. GlcN-(palmitoyl) PI probably is flip-flopped to the lumenal leaflet of the ER by a flippase that has not been identified. The later steps of the GPI-anchor synthesis and the transamidation occur on the lumenal side of the ER. Whereas the biosynthetic pathway of GPI-anchor is basically conserved in all eukaryotes, the yeast pathway is different from the mammalian pathway in that an addition of a fourth mannose is essential for proceeding to the next reaction in yeast. The transamidation is mediated by a transamidase enzyme complex that consists of at least four components (GPI8, GAA1, PIG-T, and PIG-S). Pre-assembled GPI is transferred to the carboxyl terminus of protein by replacing the GPI-attachment signal peptide. In the mammalian system, the palmitoyl chain on the inositol ring is removed in the ER after transfer of the GPI to protein. GPI is further modified by various side chains linked to the three mannoses dependent on proteins. In some yeast proteins, the lipid portion of GPI is remodeled to inositolphosphoceramide. These modifications occur in the ER or during transport to the cell surface. Enzymes involved in these modifications should be characterized in the future.

NORIMITSU INOUE and TAROH KINOSHITA

Department of Immunoregulation, Research Institute for Microbial Diseases, Osaka University, 3-1 Yamadaoka, Suita, Osaka 565-0871, Japan
Tel. +81-6-6879-8329; Fax +81-6-6875-5233
e-mail: inoue@biken.osaka-u.ac.jp

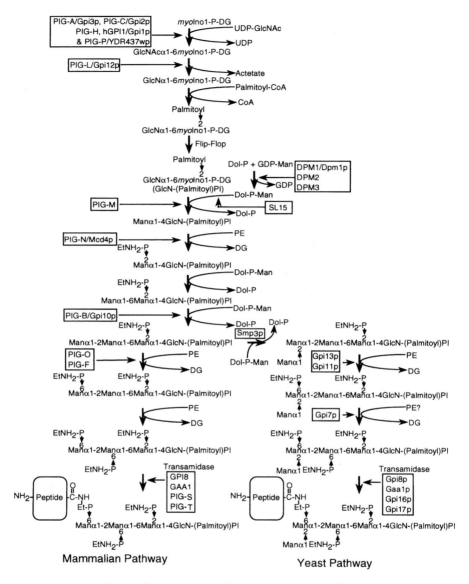

Map 5. Mammalian and yeast GPI anchor biosynthetic pathways

Boxes indicate enzyme components at each step. *DG*, diacylglycerol; *PI*, phosphatidylinositol (DG and PI are used as global abbreviations for structures including fatty alkyl chains); *Ino*, inositol; *EtNH₂*, ethanolamine; *P*, phosphate

Subject Index